AF879287

MARINES IN THE REVOLUTION

A History of the Continental Marines in the American Revolution 1775-1783

By
Charles R. Smith

Illustrated
By
Major Charles H. Waterhouse, USMCR

History and Museums Division, Headquarters, U.S. Marine Corps
Washington, D.C. 20380

1975

Library of Congress Catalog Card No. 75-1718

For Sale by the Superintendent of Documents, U.S. Government Printing Office,
Washington, D.C. 20402—Price $20.30
Stock Number 008-055-00083

FOREWORD

On 10 November 1775 the Second Continental Congress authorized the raising of two battalions of Marines. From this small beginning we have seen the United States Marine Corps grow into a powerful force for the nation's security.

In this volume, through the actions and words of the participants, we read of a small Marine force which promptly challenged Great Britain's control on both land and sea. Our first amphibious raid landed in the Bahamas on 3 March 1776, capturing gravely needed munitions, and proving to the foe that the infant American Marine Corps was a threat to be reckoned with. Likewise we read of small bands of Marines who dared to stand in the way of British troops at Princeton, New Jersey, and Charleston, South Carolina. And we also learn of the little-known Marine expedition down the Mississippi River under naval Captain James Willing in 1778.

Historians, past and present, all too often neglect completely the maritime history of the American Revolution or they pass over it with superficial attention. Mr. Charles R. Smith has corrected this omission with respect to Continental Marines. Based on extensive and careful research, the author has rescued from oblivion those actions from which our modern concepts of amphibious warfare have grown.

Mr. Smith has been a member of the staff of the History and Museums Division since July 1971. He holds Bachelor of Arts degrees in History and Political Science from the University of California, a Master of Arts degree in History from San Diego State University, and is working toward a Doctorate in History at The American University. He also served in the Republic of Vietnam as an Army artilleryman and field historian. Since joining the division, Mr. Smith has authored a historical pamphlet entitled, *A Brief History of the 12th Marines,* and articles for the division's newsletter, *Fortitudine.*

Original artwork used to illustrate this volume was prepared by Major Charles H. Waterhouse, USMCR. A free-lance illustrator, Major Waterhouse's work has appeared in a wide variety of publications, ranging from childrens' books to his forthright interpretations of the war in Vietnam, published under the titles, *Vietnam Sketchbook—Drawings from Delta to DMZ,* and *Vietnam War Sketches—From the Air, Land and Sea.*

The author and illustrator frequently consulted research materials in the hands of state historical societies, museums, libraries, and private collections. Everywhere they encountered full cooperation, unstinted assistance, and genuine interest. To those individuals and institutions who permitted use of their materials in this volume, we extend our sincerest appreciation.

E. H. SIMMONS
Brigadier General, U.S. Marine Corps (Ret.)
Director of Marine Corps History and Museums

PREFACE

The documents connected with the early history of the navy of the country were never kept with sufficient method, and the few that did exist have become much scattered and lost.

JAMES FENIMORE COOPER

In its various aspects our struggle to achieve independence has from the very beginning fired the imagination of popular writers, and has received the critical study of historical scholars. From social, economic, political, diplomatic, and military points of view this exciting period has been covered most thoroughly. Yet Marine activities during the war, except for a few brilliant episodes, have been neglected, although their importance in many instances is apparent.

The intention of this volume is not to present a complete study of all Marine activities during the period, rather it is the object of this work to trace the activities of one special group of Marines; those who served in Continental employ. Even here no attempt has been made to present a complete biographical study of any one individual or group of individuals. Instead the author means to present an objective analysis of individual and collective contributions, the successes and failures of the group as a whole, and the fundamental aspects of modern Marine amphibious doctrine which grew out of Continental Marine experience during the eight-year fight for American independence.

Surviving correspondence, ship logs, letter books, journals, and similar primary source materials of both an official and personal nature on Marine activities are fragmentary. More than one hundred years ago James Fenimore Cooper, in words which precede this preface, succinctly stated the obstacles confronting a researcher in early American naval history. However, for an investigator dealing with the history of Continental Marines, Cooper understates the obstacles. The few documents which have survived the ravages of time are still "much scattered"; the majority have been lost. On the other hand, new materials have been brought to light, often from the most unexpected of sources, and local historical societies, state and federal archives, and libraries have gathered, and in some instances published, valuable collections. One outstanding example is the indispensable series, *Naval Documents of the American Revolution*, presently being published by the Naval Historical Center in Washington. Conceived by the late William Bell Clark, and now under the expert editorship of Dr. William J. Morgan, the published volumes of this series have drawn together naval and maritime documents from all major foreign and domestic depositories.

Acknowledgements are also in order for a number of people who have extended willing and useful aid. Among them special appreciation goes to those members of the History and Museums Division, United States Marine Corps, who assisted in the preparation of this volume: Mr. Henry I. Shaw, Jr., Chief Historian, who reviewed both the draft and final manuscript; Mr. Richard A. Long, whose genealogical expertise was indispensable and whose biographical sketches of all known Continental Marine officers are appended; Mr. Ralph Donnelly, who critically reviewed all draft chapters; Miss Carolyn A. Tyson, whose aid on Continental uniforms was valuable; Mr. Charles L. Updegraph, Jr., who drafted a portion of Chapter X before leaving the division; Mrs. Barbara Rhenish Smith, who reviewed and assisted in processing the final manuscript; and Mr. Rowland P. Gill, whose preliminary research was invaluable.

Special thanks are due to the individuals of other governmental agencies for their research assistance. Outstanding among them were: Dr. William J. Morgan, Chief, Historical Research Section, Naval Historical Center, and his colleagues, Mr. Robert L.

Scheina, Mr. E. Gordon Bowen-Hassell, Mr. Robert I. Campbell, and Yeoman First Class Lenzie D. Crosby; Mr. W. Bart Greenwood, The Librarian, Department of the Navy Library; Dr. John Sellars, Library of Congress; Mr. James Walker, National Archives; Mr. Kenneth E. Harris, Center for the Documentary Study of the American Revolution, National Archives; Mr. Harry Schwartz, Old Military Records Branch, National Archives; Dr. Philip K. Lundeberg, Curator, Division of Naval History, Smithsonian Institution; and Mr. Donald E. Kloster, Assistant Curator, Division of Military History, Smithsonian Institution.

For additional aid and advice in his research, the author is greatly indebted to the following individuals and institutions: Director, Archivo General de Indias, Sevilla, Spain; Mr. Eugène Lelièpvre, Montrouge, France; The Secretary, Public Records Office, London, England; Mr. John L. Garland, Essex, England; Mrs. Gail Saunders, Archivist, Ministry of Education and Culture, Nassau, Bahamas; Dr. E. Paul Albury, Nassau, Bahamas; Mr. René Chartrand, Ottawa, Canada; Mr. Thomas Gaffney, Maine Historical Society; Mr. Ronald J. Kley, Head, Research and Collections, Maine State Museum Commission; Dr. David C. Smith, University of Maine; Miss Linda C. Webb, Bangor Public Library; Mr. Barry A. Macy and Mrs. N. B. Lacy, New Hampshire Historical Society; Miss Winifred Collins, Massachusetts Historical Society; Mr. and Mrs. Leo Flaherty, Massachusetts State Archives; Mrs. Nina Regis, New England Historic Genealogical Society; Mr. Frederick C. Hirst and Mr. Paul Sanford, Boston Public Library; Mrs. Phyllis Peloquin, Rhode Island Archives; Director, Rhode Island Historical Society; Captain Theodore Waterbury, Newport Historical Society; Mrs. John Nicholas Brown, Providence, Rhode Island; Mrs. J. Davidson Rider, Archivist, Brown University; Director, Connecticut Historical Society; Director, Marine Historical Association, Mystic Seaport; Miss Judith H. Brown, Ledyard Libraries; Mr. James Gregory, New York Historical Society; Mr. Kenneth A. Lohf, Columbia University; Mr. Paul A. Rugen, New York Public Library; Mrs. Marie T. Capps, United States Military Academy, West Point; Mr. Kenneth W. Richards, New Jersey State Library; Mr. Samuel S. Smith, Monmouth Beach, New Jersey; Mr. Nicholas B. Wainwright and Mr. Peter Parker, Historical Society of Pennsylvania; Dr. Whitfield J. Bell and Dr. Murphy D. Smith, American Philosophical Society; Mr. Thomas Waldman, University of Pennsylvania; Mr. Henry E. Bown, Pennsylvania Historical and Museum Commission; Mrs. Julia M. Cunningham, Carnegie Library of Pittsburgh; Miss Ruth Salisbury, Director, Historical Society of Western Pennsylvania; Mr. John W. Jackson, Flourtown, Pennsylvania; Dr. John J. McCusker, University of Maryland; Mr. Norman N. Rubin, Silver Springs, Maryland; Lieutenant Colonel Joseph K. Griffis, Jr., USMC, Quantico, Virginia; Mr. Frank H. Rathbun, Alexandria, Virginia; Dr. Louis H. Manarin, State Archivist, Virginia State Library; Miss Mattie Russell, Curator of Manuscripts, Duke University; Mr. J. Isaac Copeland, Director, Southern Historical Collection, University of North Carolina; Mrs. Granville T. Prior, Director, South Carolina Historical Society; Mr. E. L. Inabinett, Librarian, University of South Carolina; Mr. H. G. Jones and Miss Wylma Wates, Department of Archives and History, State of South Carolina; Mrs. Lilla M. Hewes, Director, Georgia Historical Society; Mr. John C. Dann, Curator of Manuscripts, William L. Clements Library, University of Michigan.

Chief Warrant Officer Joseph R. Fitzgerald and his predecessor as History and Museums Division Administrative Officer, Chief Warrant Officer Dennis Egan, ably handled the many exacting duties involved in processing the volume from first drafts through final printed form. The bulk of the early typescripts were prepared by Private First Class Carl W. Rice, who, with the assistance of Miss Kay P. Sue, also expertly handled the painstaking task of typing portions of the final manuscript for the printer. Miss Cathy Stoll, assisted by Private Denise Alexander, performed the meticulous task of abstracting and typing the index. Mrs. Joyce E. Bonnett, assisted by Miss Sue, performed the often-times hectic duties involved in rapidly corresponding with those individuals and institutions who assisted the author.

Charles R. Smith

CHARLES R. SMITH

CONTENTS

APPENDIXES

DESCRIPTIVE LIST OF ILLUSTRATIONS

INTRODUCTION

What is a Marine? He is a soldier who serves at sea on a vessel of war either as part of its crew or as part of a military expedition under naval supervision. At various times he has been called a "maritime soldier," a "sea soldier," and a "soldier of the ocean." But a study of the past reveals that whether a soldier is a Marine depends, not upon the name given him, but upon the character of duty he performs coupled with a familiarity of the sea and his subordination to naval jurisdiction. There have been soldiers performing the duties of Marines since the first fighting men served on board ships or in expeditions of a naval character. But these men were best prepared to carry out their mission as Marines when they were trained to the ways of the sea and were an integral part of the naval establishment.

Marines are as old as naval warfare itself. When Themistocles mobilized Athenian sea power against the invading Persians in 480 B.C., one of his first decrees was to order the enlistment of Marines for the fleet. These men, called *Epibatae*, or "heavy-armed sea soldiers," fought in the Greek triremes at Salamis which turned back Xerxes and saved Athens. Later, Rome had what Polybius described as *milites classiarii* (soldiers of the fleet), a category of Roman soldier organized and specially armed for duty on board warships, usually quinqueremes of the line. During the middle ages, ordinary soldiers were frequently embarked on board ship to provide a fighting backbone, but it was not until the naval wars of the 17th century that the distinct and organized role of Marines was almost simultaneously rediscovered by the British and Dutch, who raised the first two modern corps of Marines in 1664 and 1665, respectively.

Americans of the 17th and 18th centuries were notably a maritime people. The British colonies were close to the sea, but scattered along a coast line of more than a thousand miles, so that, in the absence of good roads, intercommunication was almost completely by water. Ocean trade also, chiefly with England and the West Indies, was extensive. Fishing was one of the most important industries, especially of the northeastern colonies, and the handling of small vessels on the Newfoundland Banks during all seasons of the year trained large numbers of men in seamanship. The whale-fisheries likewise furnished an unsurpassed school for mariners.

A considerable number of colonists, therefore, were at home upon the sea and, more than this, they were to some extent practiced in maritime warfare. England, during the 17th and 18th centuries, was at war with various European powers a great part of the time, and almost from the beginning of the colonial period American privateers and letters of marque scoured the ocean in search of French and Spanish prizes. Large fleets were also fitted out and manned by provincials for various expeditions against foreign-held territories. The first of these expeditions to employ Americans in a capacity as Marines was that launched by Admiral Edward Vernon against the Spaniards in the War of the Austrian Succession.

British Marines, after the Peace of Utrecht in 1713, were practically disbanded; only four invalid companies remained. However, with the outbreak of hostilities with Spain in 1739, King George II took measures to re-establish the Marines. On 15 November, in his address to the House of Commons which opened Parliament, the King stated that the prosecution of the war would require "a number of soldiers to serve on board the fleet," and he "judged it proper, that a body of Marines should be raised." The following month, an Order in Council decreed the formation of six regiments of Marines, each with an authorized strength of 1,100. Increases soon followed—among them were three regiments organized in the colonies and placed under the command of Colonel Alexander Spotswood of Virginia.[1]

Men for the war were scarce in England, and the

Crown seized upon the idea of using men from the colonies. In addition, it was supposed that the Americans, being acclimated, were better suited for the service upon which they were destined than Europeans. Therefore an order went out to the governors of 10 colonies to raise a number of American Marines to be commanded by Colonel Spotswood. These Americans were to be organized into 30 companies of 100 men each including four sergeants, four corporals, and two drummers. There were to be a captain, two lieutenants, and an ensign for each company. The Crown was to appoint the field and staff officers, as well as one experienced lieutenant per company.

Most of the colonies eventually raised their quotas of Marines. New Hampshire, Delaware, South Carolina, and Georgia, however, sent none. The latter two colonies, with North Carolina, had participated in an expedition to St. Augustine while Delaware was included in the quota of Pennsylvania. The colonial forces raised were ultimately formed into a single regiment under the command of the Lieutenant Governor of Virginia, Colonel William Gooch; Spotswood had died at Annapolis on the eve of the expedition.

Admiral Vernon established his advanced base at Jamaica, and was joined there by the eight transports carrying the Marine force from North America in October 1740. Three months later, additional troops from England under Lord Cathcart, and later under General Thomas Wentworth, arrived. After a council of war was held in early January 1741, it was determined to proceed to windward in order to observe the motions of the French fleet at Port Louis in Hispaniola.

Vernon sailed from Port Royal during the last week of January. He made Cape Tiberon, the western point of Haiti's southern jaw early in February. After several days observing the French fleet, Vernon decided to attack the Spanish treasure city of Cartagena. The fleet sailed on 25 February from Haiti, and on 4 March anchored in Playa Grande Bay between Cartagena and Point de Canoa.

Cartagena could only be approached by way of the narrow passage of Boca Chica, eight miles south of the town. This entrance was defended on its northern shore (Tierra Bomba) by Fort St. Louis. The central work was supplemented by several small redoubts, St. Philip, St. Jago, and a small fort, Battery de Chamba. On the south side of the passage was a fascine battery, and in a small bay to its rear another. Facing the entrance on a small island stood Fort St. Joseph. From this island to the northern shore there was a boom, behind which stood four large Spanish warships moored with their broadsides covering the entrance. Beyond this passage lay the harbor of Cartagena. About midway toward the town the harbor grew narrow, being pinched by two peninsulas, the western peninsula crowned by Fort Grande Castillo, and the eastern bearing a horseshoe battery of 12 guns. East of the town was castle St. Lazar with its numerous heavy cannon.

The attack on Cartagena opened on 9 March with a heavy bombardment by the fleet of the smaller defenses on the passage's northern shore. Marines were landed on Tierra Bomba and soon occupied the redoubts, advancing to Fort St. Louis before reinforcements were thrown ashore. On the 19th the Marines occupied the southern entrance, and stormed Fort St. Joseph five days later. The remaining batteries were subsequently silenced as was the fortress of Grande Castillo. Only the outlying fort of St. Lazar remained.

Having cleared the way into the inner harbor, Vernon now decided to land a detachment on the western side of the bay for the purpose of attacking St. Lazar. However, a heated controversy over the methods and strength required to capture the fortress ensued between the Admiral and General Wentworth, postponing the attack for several days. On the morning of 9 April, 1,500 English Marines and grenadiers, accompanied by the Americans with scaling ladders, wool-packs, and hand-grenades, moved on the fort. The attack proved to be a ghastly failure. Under a hail of shot the ladder-carrying Americans fell back, stranding the grenadiers and Marines. Despite their terrible losses, a few succeeded in reaching the ramparts, but they too were obliged to retire. The failure of this attack broke the back of the British effort, even though they did attempt to bombard the town into submission.

By mid-April, it was decided that the fleet should return to Jamaica. The troops were tired, and sick with fever and dysentery. After the captured works were dismantled and destroyed, the expedition re-embarked and withdrew northward. Afterwards attempts were made upon St. Jago de Cuba (Santiago), Porto Cavallo, and La Guaira, but all were unsuccessful. The government's policy of offensive warfare against Spanish possessions in America was a failure. American losses during the expedition were heavy, and it has been estimated that not more than one-tenth ever returned home.[2]

By the end of the Seven Years' War in 1763, during which Americans again served as Marines in the British fleet, it would seem that numerous colonists possessed the training and experience which made them the best sort of raw material for an efficient Marine force. The lack of a true naval tradition, military discipline, and the poverty of the country, however, imposed limitations which, together with the overwhelming force of the enemy, seriously restricted the field of enterprise. Nevertheless at the opening of the American Revolution the patriotic cause was greatly aided by the activities of armed Marines afloat.

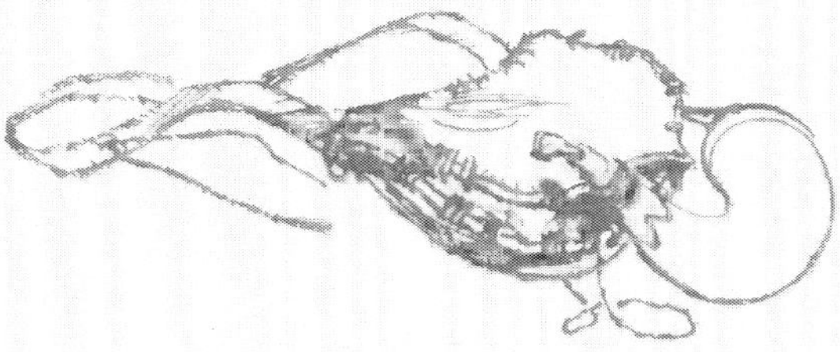

Willing

THE FIRST RECRUITS, DECEMBER 1775.
The painting recreates a scene which might have occurred on the Philadelphia wharves as the first Marine recruits were assembled, one month after a resolution was passed by the Continental Congress bringing the Marines into existence. Portrayed are Captain Samuel Nicholas and Lieutenant Matthew Parke, two of the first Marine officers commissioned.

CHAPTER I

Birth of the Continental Marines

10 November 1775

On Friday, 10 November 1775, Colonel Benedict Arnold stood on the banks of the St. Lawrence River and looked in frustration across a mile of storm-whipped water at the grand objective—Quebec. He and a thousand Americans, now ragged, had made a tortuous march of 350 miles through the Maine wilderness to get there. Now he could do nothing but shake his head and pray that the weather would clear so he could make the crossing before British reinforcements, known to be on the way, arrived.

Outside Boston on that same day General George Washington and his army of 17,000 were encamped at Cambridge in reasonable comfort. True, there were shortages of blankets, clothing, and powder. True, also, that to the men the siege looked endless and amongst the officers there was dissatisfaction with the plan proposed to the Congress to reorganize the army. But on the whole the American position looked promising.

In Philadelphia that Friday morning when the delegates to the Second Continental Congress gathered in the Assembly Room of the State House, much of their talk was about the situation and supply of the army at Cambridge. At ten the President of Congress, John Hancock, pounded the gavel and the daily session began. The first major item of business concerned the purchase and shipment of medicine to Cambridge, but as the day wore on discussion moved to a subject that had been tabled late the previous afternoon—Nova Scotia.

A week before, the Continental Congress received a petition from the inhabitants of Passamaquoddy, Nova Scotia, informing them that a Committee of Safety had been formed, and that they wished "to be admitted into the association of the North Americans, for the preservation of their rights and liberties." Acting upon the petition, the delegates on 2 November resolved that a committee of five be appointed to consider the matter and report the steps which it thought proper to take. Of the delegates considered, the five members finally settled upon were: Silas Deane of Connecticut, John Jay of New York, Stephen Hopkins of Rhode Island, John Langdon of New Hampshire, and John Adams of Massachusetts.[1] As was the custom, the first named would have been the chairman.

The committee began its work almost immediately. Often toiling late into the night, the five men debated possible courses of action and the reasons behind them. John Adams, though not a revolutionary expansionist, saw in the petition a chance not only to secure the liberties that would be won by the Americans for the Nova Scotians, but a chance to capture badly needed military stores and naval facilities for the American cause. Although of secondary importance when the discussions began, the prospects of taking the extensive British supplies and facilities in Nova Scotia, thus damaging ministerial naval designs in the Americas, became the major area of concern.

Discussions continued into the first full week of November, and by mid-week the committee was ready to present its proposals to Congress. Simple in its detail, the proposed plan for a naval expedition to Nova Scotia would be an ambitious undertaking. First, the scheme called for the creation of two battalions of Marines from the forces then under the command of General Washington. The two battalions, termed the First and Second Battalions of American Marines, would consist of one colonel, two lieutenant colonels, and two majors, with the remaining commissioned and non-commissioned officer corps structured along the lines of a Continental Army regiment. Excluding officers, each battalion would have 500 privates assigned to it. The two battalions would be further subdivided into ten companies of 50 privates plus officers. The reason given for this breakdown was "that in fitting out any ship of War one of these Companies would compleatly man a small Vessell and two of them make a large Proportion of Marines for the largest." An important requirement for both officers and men was that all should have served in the merchant service, or be acquainted with maritime affairs so as to be able to serve "to advantage by sea, when required."

Once raised, the two battalions of Marines would march overland from Cambridge to either Newburyport, Massachusetts, or Portsmouth, New Hampshire, there to rendezvous on 1 December with sufficient ships capable of transporting them plus three months' provisions to Nova Scotia. Meanwhile, a number of men would be sent ahead to gain information as to "the Temper and Disposition of the Inhabitants . . . with respect to the Present Struggle . . . and how far they may be willing and able to take an active Part in the present Dispute." Also two swift boats were to be sent to ply the waters of the Bay of Fundy specifically to learn of the British military posture in Halifax. After determining the situation, both throughout the province and in the town of Halifax, the Marines were to embark for Minas, located in the upper reaches of the bay.

Armed with flintlocks, long-handled hatchets, spears, and 32 rounds of ammunition per man, the Marines were to land on the western shore of the Nova Scotia peninsula. From there, they were then to "make a forced March [of about 40 miles] for Hallifax and possess themselves of that Town and of the naval and other Stores there and if practicable of the Shipping." Once the town was taken, the Marines were to destroy the docks and yards and carry off the military stores in their retreat. But if the force was able to capture ships in the harbor, they were to remain until driven out by a superior British force, whenever it could be brought to bear on them.

Debate on the committee's proposed plan by Congress consumed several days. Although no transcript or memorandum remains of what transpired, it is known that the five-man committee came into Congress well-armed for a spirited defense of their proposal. Should a member have inquired as to the British military response to such a force, the committee was ready with an example. Several months before, Colonel Arnold's expedition was supposed by the British to be destined for Halifax. To counter it, General John Burgoyne had ships and troops sent, but as the committee pointed out "not enough to make Resistence to two such Battalions" as those which were then being considered. Still, the most effective argument for adopting the proposal was that of the damage which would be done to the British. "Should the Expedition succeed, the Consequences will be of the Utmost Importance, nothing less than the greatest Distress, if not the Utter Ruin of the ministerial Navy in America." But if Congress should, "by any Accident," find the proposed expedition impractical the committee was ready to recommend that the two Marine battalions be retained, since they would be of "Utmost service, being capable of serving either by sea or Land."[2]

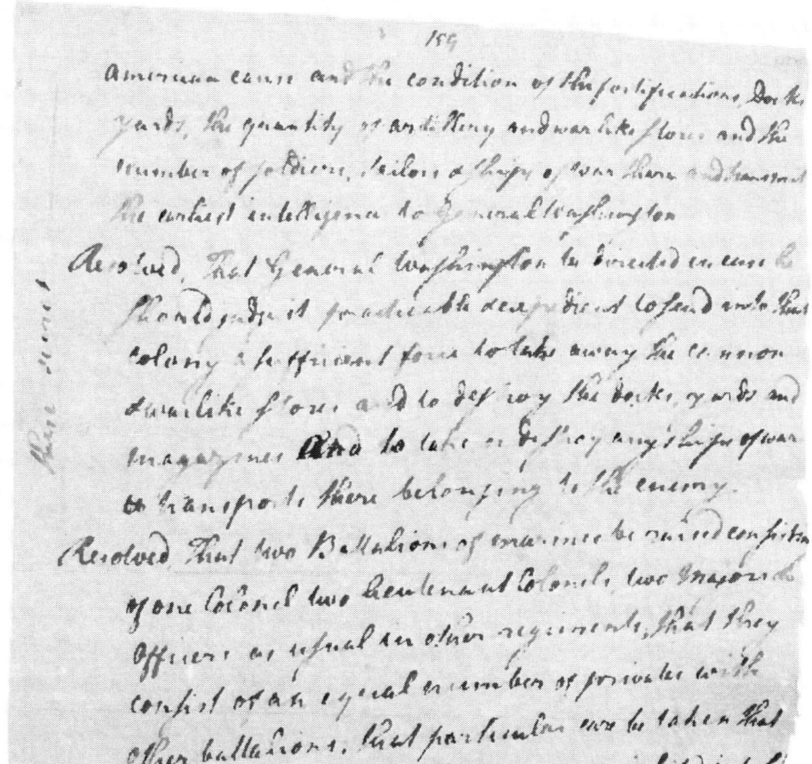
159

American cause and the condition of the fortifications, Dock yards, the quantity of artillery and warlike stores and the number of soldiers, sailors & ships of war there and transmit the earliest intelligence to General Washington

Resolved, That General Washington be directed in case he should judge it practicable & expedient to send into that colony a sufficient force to take away the cannon & warlike stores and to destroy the docks, yards and magazines And to take or destroy any ships of war or transports there belonging to the enemy.

Resolved That two Battalions of marines be raised consisting of one Colonel two lieutenant Colonels two Majors & Officers as usual in other regiments, that they consist of an equal number of privates with other battalions; that particular care be taken that no persons be appointed to offices or inlisted into said Battalions, but such as are good seamen, or so acquainted with maritime affairs as to be able to serve to advantage by sea, when required. That they be inlisted and commissioned for and during the present war between Great Britain and the colonies, unless dismissed by order of Congress. That they be distinguished by the names of the first & second battalions of American Marines, and that they be considered as part of the number, which the continental Army before Boston is ordered to consist of.

Resolved

Resolve of 10 November 1775

The committee, however, had little to worry about on that cold Friday afternoon, for the members of Congress voted to accept all the major items contained within its report. In addition to adopting the committee's recommendation for the sending of a number of persons to gather information, Congress resolved that the two battalions of Marines be created:

> *Resolved*, That two Battalions of marines be raised, consisting of one Colonel, two Lieutenant Colonels, two Majors, and other officers as usual in other regiments; and that they consist of an equal number of privates with other battalions; that particular care be taken, that no persons be appointed to office, or inlisted into said Battalions, but such as are good seamen, or so acquainted with maritime affairs as to be able to serve to advantage by sea when required: that they be inlisted and commissioned to serve for and during the present war between Great Britain and the colonies, unless dismissed by order of Congress: that they be distinguished by the names of the first and second battalions of American Marines, and that they be considered as part of the number which the continental Army before Boston is ordered to consist of.

The Marine battalions were authorized, but a third resolution passed that day left the fate of the Nova Scotia expedition in the hands of General Washington. Following the day's session, President Hancock transmitted the adopted resolves to Cambridge for Washington's information and comment.[3]

The Naval Committee

Closely paralleling the work of the Committee on Nova Scotia was that of the Naval Committee. Since late October the committee had struggled to establish a Continental Navy. The first suggestions for an independent naval establishment came from New England, where problems of port and coastal defense were of prime importance. Although there were many casual advances for such a naval force, the first formal movement in behalf of a Continental Navy came on 26 August 1775. Voting on a number of recommendations, the Rhode Island Legislature instructed its congressional delegates to use their influence during the coming session "for building at the Continental expense a fleet of sufficient force, for the protection of these colonies, and for employing them in such manner and places as will most effectually annoy our enemies."[4]

It was not until 3 October that the Rhode Island delegates presented their instructions to Congress. But debate on the proposals was postponed from time to time and it was several weeks before Congress took them under serious consideration. Final action on the Rhode Island plan for an American fleet would have to wait until mid-December when Congress ordered 13 frigates to be fitted out at Continental expense. Meanwhile, intelligence had arrived in Congress which indicated that two British brigs laden with arms and powder were on their way to Quebec. Realizing that the capture of the two vessels meant sorely needed arms and ammunition for the army at Cambridge, a motion was made that a committee of three be appointed to prepare a plan for intercepting the two ships.[5]

Opposition to the motion, according to John Adams, was "very loud and vehement." Delegates such as Edward Rutledge of South Carolina seemed to have realized that the motion would be the first step in the creation of the Continental Navy, a step which he and several other southern delegates were unwilling to take. Such an undertaking, Rutledge declared, was "the most wild, visionary, mad project that ever had been imagined. It was an infant taking a mad bull by his horns; and what was more profound and remote, . . . it would ruin the character and corrupt the morals of all our seamen." These arguments were answered by the motion's supporters who dwelt on "the great advantages of distressing the enemy, supplying ourselves, and beginning a system of maritime and naval operations"; all of which "were represented in colors as glowing and animating."[6]

After a long and lively debate the motion was carried by a small majority and a committee composed of Silas Deane, John Langdon, and John Adams was appointed. Later the same day, the three men recommended, and Congress resolved, that General Washington be directed to secure from Massachusetts two armed ships, and that the governors of Connecticut and Rhode Island be requested to cooperate in the project. Also included in the resolve was the recommendation that commanders appointed to the vessels encourage the enlistment of seamen and Marines. This mention of Marines is the first indication that the infant Naval Committee considered Marines both a necessary and integral part of any naval force assembled by the Continental Congress.[7]

Following the resolves of 5 October, the movement toward a Continental Navy quickened. On the 13th, Congress voted to fit out the two armed vessels and send them out to intercept British transports laden with military stores. To estimate the expenses involved, another three-man committee was appointed. Silas Deane and John Langdon were again designated as members, but John Adams was replaced by Christopher Gadsden of South Carolina, an apparent attempt to gain southern support. With congressional action on the second committee's report two weeks later, Congress simultaneously authorized the purchase of two additional ships. This time the objective was not limited to intercepting enemy shipping, but expanded to include "the protection and defence of the united Colonies." In addition to finally committing itself to a navy, Congress reconstituted the committee. Instead of three, the committee's membership was expanded to seven by the addition of John Adams, Stephen Hopkins of Rhode Island, Joseph Hewes of North Carolina, and Richard Henry Lee of Virginia. The committee, which had previously operated under several titles, now became known as the Naval Committee.[8]

By the beginning of November, four vessels had been authorized and the American colonies committed to a policy of naval warfare. But naval legislation did not slow, it moved more rapidly.[9] Under the able leadership of the aged ex-governor of Rhode Island, Stephen Hopkins, the committee set to work purchasing, equipping, and manning the four Continental ships. In this connection Silas Deane journeyed to New York and Connecticut, while John Adams wrote to James Warren and Elbridge Gerry in Massachusetts to inquire whether ships might be purchased, hired, or built in the province. Adams also asked whether there were suitable officers and men who could be enlisted in the navy and Marines. It is assumed that Adams' inquiries were not only connected with his work on the Naval Committee, but also with his duties on the Nova Scotia Committee. These initial contacts with the northern colonies, however, produced little assistance.[10]

In its search for suitable vessels, the committee had only to look as far as the Delaware River where four were located and purchased. The first ship procured was the *Black Prince*, a relatively new 300-ton merchantman owned by a group of Philadelphia businessmen. Shortly after purchase she was rechristened the *Alfred*, in honor of the founder of the British Navy. Within weeks the *Alfred* was followed by three other ships: the *Columbus* (formerly the ship *Sally*); *Cabot* (formerly the brig *Sally*); and the *Andrew Doria* (formerly the brig *Defiance*).

The *Alfred* and *Columbus* were large and clumsy in comparison with the *Cabot* and *Andrew Doria*. Since they were initially built to carry cargo, both ships required extensive internal bracing to bear the burden of their armament. Of the four ships, little is known about their dimensions and builders with the exception of the *Cabot*. Since she was captured in 1777 and taken into the Royal Navy, her dimensions have survived. British naval records show that the *Cabot* was a 14-gun brig with a deck length of 74'9½", keel length of 53'7", beam of 24'8", a hold depth of 11'4", and a displacement of 189 tons.[11] *

General Washington's Reply

As the Naval Committee proceeded with the outfitting of the four ships, the question of the Nova Scotia expedition and the raising of the two battalions of Marines again was considered. On 27 November, General Washington's thoughts concerning the resolves of 10 November were placed before Congress. Washington, although agreeing to send two men to Nova Scotia to gain intelligence, thought the decision to raise the Marine battalions from within his army impractical. The Continental Army at the time, the general pointed out, was in a period of realignment. Since the reorganization involved

* For a detailed description of the four Continental ships after renovation, see Chapter III.

the reconciliation of many different interests and the judging of certain field officers on their merits, the same difficulties would again arise with the appointment of a colonel to command the two battalions. The greatest problem, as Washington saw it, would occur in the selection of junior officers and enlisted men. Because Congress had placed a requirement that all officers and men be acquainted with maritime affairs, "they must be picked out of the whole army, one from this Corps, one from another," thereby damaging the whole system. Thus, the raising of the two battalions would not only cost the army time, anxiety, and pain, but would also weaken it. Due to the difficulties that would arise, Washington suggested that the two battalions of Marines be raised in New York and Philadelphia; "where there must be now numbers of Sailors unemployed?"

Washington not only questioned the advisability of creating the battalions from the Continental Army, but the necessity of an expedition to Nova Scotia. His available forces were low, and he needed all that could be mustered. Noting information had been received indicating additional troops had arrived to reinforce the British garrisons in Boston, and that the enemy would probably take advantage of the first bad weather to break the siege, General Washington asked whether this was the time to weaken American lines by employing his forces "on any other Service." Notwithstanding, Washington assured Congress he would "use every endeavour to comply with . . . [its] Resolve."[12]

On 28 November, Washington again wrote John Hancock. After making a number of inquiries amongst his officers, Washington informed the President that it would be impossible "to get the Men to inlist for the Continuance of the War, which will be an insuperable Obstruction to the formation of the two Battallions of Marines." Nevertheless, Washington said he would complete the reorganization of the army and then "enquire out such Officers & Men as are best qualifyed for that service, and endeavor to form these two Battallions out of the Whole."[13] Among the officers contacted in this respect was Colonel John Glover of Marblehead who supplied Washington with a list of 17 men whom he thought qualified to serve as captains in the Marine battalions. None were ever appointed.[14]

Although Washington remained concerned about raising the two battalions until late January 1776, Congress, on 30 November, relieved him of the responsibility. That day it ordered Washington to suspend the raising of the Marine battalions and ordered that they be created independently of the army.[15] Several days later Hancock informed Washington of the resolves and asked that he think of "proper Persons to command . . . [the] Corps, and give Orders for enlisting them wherever they may be found."[16] At present, only one name is known to have been submitted by Washington for consideration; this he did in the spring of 1780.[17]

Marine Officers Commissioned

The views expressed by Washington in his letter of 19 November seemed to have had the desired impression on Congress. For soon after, the idea of an expedition to Nova Scotia was abandoned and not heard again until late the following year. But Congress was not ready to abandon the two battalions of Marines. On 28 November it commissioned the first Marine officer—Samuel Nicholas.[18]

Little is known about this young Philadelphia Quaker prior to his appointment as captain of Marines. The only son of Mary (Shute) and Anthony Nicholas, a blacksmith, Samuel was born in 1744. Socially, Samuel Nicholas achieved early prominence. In 1760 he was admitted to the Schuylkill Fishing Company which, despite its commercial name, was an exclusive gentlemen's club devoted to the rod-and-reel and pleasures of the table. Six years later, he became one of the founders of the Gloucester Fox Hunting Club, whose membership was drawn from the leading families of Philadelphia as well as from the country gentry of Gloucester County, New Jersey. Through his association with these two clubs he became acquainted with such prominent Philadelphia gentlemen as John Cadwalader, Thomas and James Wharton, Robert and Samuel Morris, John Nixon, Clement Biddle, and Thomas

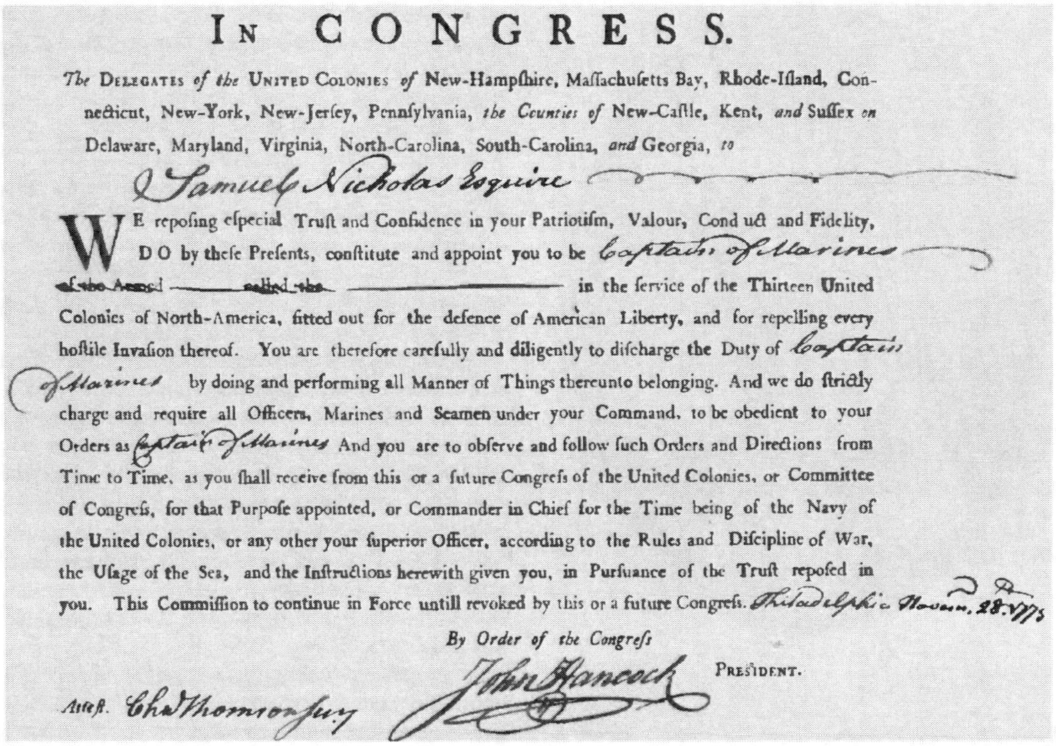

In CONGRESS.

The Delegates *of the* United Colonies *of* New-Hampshire, Massachusetts Bay, Rhode-Island, Connecticut, New-York, New-Jersey, Pennsylvania, *the Counties of* New-Castle, Kent, *and* Sussex *on* Delaware, Maryland, Virginia, North-Carolina, South-Carolina, *and* Georgia, *to*

Samuel Nicholas Esquire

WE reposing especial Trust and Confidence in your Patriotism, Valour, Conduct and Fidelity, DO by these Presents, constitute and appoint you to be Captain of Marines ~~of the Armed —— called the ——~~ in the service of the Thirteen United Colonies of North-America, fitted out for the defence of American Liberty, and for repelling every hostile Invasion thereof. You are therefore carefully and diligently to discharge the Duty of Captain of Marines by doing and performing all Manner of Things thereunto belonging. And we do strictly charge and require all Officers, Marines and Seamen under your Command, to be obedient to your Orders as Captain of Marines And you are to observe and follow such Orders and Directions from Time to Time, as you shall receive from this or a future Congress of the United Colonies, or Committee of Congress, for that Purpose appointed, or Commander in Chief for the Time being of the Navy of the United Colonies, or any other your superior Officer, according to the Rules and Discipline of War, the Usage of the Sea, and the Instructions herewith given you, in Pursuance of the Trust reposed in you. This Commission to continue in Force untill revoked by this or a future Congress. Philadelphia Novem. 28. 1775

By Order of the Congress

John Hancock President.

Attest. Chas Thomson Secy

Samuel Nicholas' Commission as Captain of Marines.

Willing—all of whom later played leading roles in the Revolution. The vocation he followed in the years before the Revolution is not definitely known, although there are unsubstantiated indications that he was an innkeeper and served in the colonial merchant service. A position in the latter, or an acquaintance with maritime affairs would certainly have qualified him for an appointment in the Marines. If, on the other hand, he had no experience in such affairs, he could possibly have achieved his appointment through a recommendation by one or more of his many prominent acquaintances.[19]

With Nicholas' appointment, the recruiting of men and the selection of their officers began. Among the first officers picked were Joseph Shoemaker, a member with Samuel Nicholas of the Fox Hunting Club, and Isaac Craig. A popular Irishman born in County Down, Craig emigrated to America with his brother James in 1765. Settling in Philadelphia he began as a journeyman carpenter and quickly achieved a master rating. Although he had no maritime experience, he joined the Marines in late November 1775 as its oldest lieutenant.[20] In December, Captain Shoemaker and Lieutenant Craig were joined by Captain John Welsh, and Lieutenants John Fitzpatrick, Robert Cummings, John Hood Wilson, Henry Dayton, Matthew Parke, and a Lieutenant Miller.[21]*

Virtually nothing is known about the background of the one captain and six lieutenants appointed in December with the exception of Matthew Parke. Lieutenant Parke was born to well-to-do parents in 1746 near Ipswich, England. Early in life he accompanied his paternal grandfather, who had been an aide to the Duke of Marlborough during the Battle of Blenheim and later governor of the Windward Islands, to Virginia. After several years in the southern colony, Colonel Parke returned to England leaving his grandson to make his way in the new coun-

* Lieutenant Miller either failed to accept his commission or resigned from the service before the fleet sailed, since he is not included on the Muster Roll of the *Columbus*.

Matthew Parke, by an unidentified artist.

try. As the revolution began Parke journeyed to Philadelphia where he later received an appointment as lieutenant of Marines.[22]

Recruitment of men was carried out by the three captains (Nicholas, Shoemaker, and Welsh), and the two ranking lieutenants (Craig and Wilson). Rendezvous were established early in December, probably in a number of Philadelphia public houses, and drummers with highly decorated drums paraded the streets to attract recruits. One such drum, upon which was painted a coiled rattlesnake about to strike, bore the motto "Don't Tread on Me." This design, as one observer noted, was probably the device intended to decorate the arms of North America.[23] *

One of the most successful recruiters was Lieutenant Isaac Craig. Beginning his efforts on 9 December, he succeeded in signing nine men the first day, four the second, and by 22 December he had enlisted more than 40 men. Of the men recruited only eight were born in America; a majority being from Great Britain and Ireland with a few from Germany, Switzerland, and Holland. They came not only from differing backgrounds, but also from widely separated trades and professions. The company included a doctor, jeweler, baker, wool comber, miller, breeches maker, butcher, and several carpenters. Statistically, the average age of Craig's recruit was 25.5 years, and the average height, 5 feet 5 inches.[24] This is in contrast with the present-day Marine recruit who is 18 years old and four inches taller than his Revolutionary War counterpart.[25]

Naval Regulations Adopted

As the five Marine Officers recruited their men, the Naval Committee of Congress worked diligently on legislation necessary for the organization of the Navy. On the same day that Samuel Nicholas received his appointment as captain of Marines, Congress adopted the "Rules for the Regulation of the Navy of the United Colonies." Formulated by John Adams and based upon British naval regulations, these brief rules dealt chiefly with discipline, pay, and rations, and were to apply to all men serving on board a Continental vessel.

The majority of the 44 articles concern conduct and are little more than a penal code. Should a sailor or Marine be caught swearing, he was to be punished by the wearing of a wooden collar, "or some other shameful badge of distinction." For drunkenness, an enlisted man would be put in irons, while an officer guilty of the same offense forfeited two days pay. The maximum punishment an officer might inflict on a seamen or Marine for a minor offense was "twelve lashes upon his bare back, with a cat of nine tails." For the crimes of quarreling, embezzelment, robbery, falling asleep on duty, mutiny, and desertion, which deserved greater punishment, a court-martial would be convened. All court-martials for capital offenses were to consist of 12

* The drum design was probably taken from the rattlesnake flag designed by Christopher Gadsden, and used as Esek Hopkins' personal flag as commander of the Continental Fleet.

officers; "at least three Captains and three first lieutenants, with three Captains and three first lieutenants of Marines." The sentence of such a court-martial could not be executed without the review and confirmation of the fleet's commander. An automatic death sentence was only to be imposed for murder and severe cases of cowardice in battle.

The regulations also fixed rations for those on board Continental ships for each day of the week. On Saturday, for example, the regulations called for one pound of bread, one pound of pork, one-half pint of peas, and four ounces of cheese to be issued each man. Variety was achieved by substituting potatoes, turnips, rice, pudding, and butter for the main staples. Each man was also given a half pint of rum per day, and a "discretionary allowance on extra duty, and in time of engagement."

In addition to establishing a penal code and determining the daily issuance of provisions, the rules prescribed the following monthly pay scale for Marines and Marine officers: captain, 26⅔ dollars; lieutenant, 18 dollars; sergeant, 8 dollars; corporal, 7⅓ dollars; fifer, 7⅓ dollars; drummer, 7⅓ dollars; and private, 6⅔ dollars. The regulations also established 18 naval ranks and their monthly payrate; the highest being that of captain, 32 dollars, and the lowest, able seaman, 6⅔ dollars.

Appended to the naval rules was a contract of enlistment which was to be read and signed by each member of the crew. According to the articles a bounty was to be deducted from the proceeds of prizes and paid to those men, or their heirs, who died, lost a limb, or were incapacitated in an engagement. The ship commander was to receive $400, captain of Marines $300, and inferior officers, seamen, and Marines, $200. In addition, the man who first sighted a ship which was later captured was to receive a double share of prize money, while the person who first boarded a prize was to receive three shares. The articles also prescribed that 10 prize shares were to be set aside for "such inferior officers, seamen and marines, as shall be adjudged best to deserve them by the superior officers."[26] The contract, however, failed to specify the normal distribution of prize shares. This would have to wait until early January the following year when Congress determined that proceeds from the sale of a prize would be divided into twentieths. A captain of Marines would share equally in three twentieths with naval lieutenants and masters, while a lieutenant of Marines would divide two and one half twentieths with the surgeon, chaplain, purser, boatswains, gunners, carpenters, master's mates, and the fleet's secretary. A sergeant of Marines would share three twentieths with the remaining noncommissioned officers, and privates (seamen included) would divide eight and one half twentieths. The lion's share of the prize money, of course, went to the fleet commander and ship captain.[27] *

As well as enacting rules regulating shipboard life and the distribution of prize money, Congress established the term of enlistment for seamen and Marines. On 5 December it resolved that they would be engaged for one year, ending 1 January 1777, "unless sooner discharged by Congress." This change, particularly in reference to Marines, was precipitated possibly by Washington's comment that few men in his army would consider serving in the Marines for the duration of the war. The change also brought the term of service into line with the customary length prescribed for Continental and state troops.[28] **

Naval Officers Appointed

Regulations of the Navy consumed only a portion of the Naval Committee's time, the remainder was spent equipping and officering the four Continental ships. To add to the Committee's work, Congress on 2 December authorized the purchase of two additional vessels and the renting of a third. The Rhode Island sloop *Katy*, later renamed the *Providence*, was leased on her arrival in Philadelphia, while two other vessels, the *Hornet* and *Wasp*, were purchased and outfitted in Baltimore, Maryland. Also on 2 December, Congress directed the Naval Committee to prepare commissions for the officers who would command the seven ships. Once these were prepared, the process of appointing suitable officers began.[29]

A measure of nepotism was obvious when, on 22 December, the Naval Committee laid before Congress for confirmation a "list of officers by them appointed." Heading the list was the 57-year-old Rhode Islander, Esek Hopkins, the brother of Stephen Hopkins, a member of the Naval Committee. Esek was designated Commander-in-Chief of the Fleet—a rank equal in status to that held by General Washington. He was, as General Henry Knox later observed, "antiquated in figure," but "shrewd and sensible," whom "I . . . should have taken . . . for an angel only he swore now and then." The choice of Hopkins was a promising one, for his background included privateer service during the French and Indian War. Two other Rhode Island men joined the Commodore on the list: his son John Burroughs Hopkins, who was given command of the *Cabot*, and a good friend, Abraham Whipple, who got the *Columbus*.[30]

Samuel Nichols, by an unidentified artist.

* The regulations adopted by Congress on 28 November remained in force throughout the Revolution.

** According to the resolve of 10 November Marines were to be enlisted "for and during the present war between Great Britain and the colonies, unless dismissed by order of Congress."

Heading the short captain's list was Dudley Saltonstall of Connecticut who was to command the *Alfred*. "A Sensible indefatigable Morose Man," Saltonstall owed his appointment to his brother-in-law Silas Deane, of the Naval Committee. Unlike Saltonstall, of the other two captains appointed, Nicholas Biddle seems to have achieved the command of the *Andrew Doria* by merit alone. Biddle, a Philadelphian, had an impressive naval background. Only 25 years old in 1775, he had seen service in the Royal Navy with the future Lord Nelson. When the war began he quickly joined the Pennsylvania State Navy and was appointed captain of the *Franklin* row-galley. But the duty was monotonous, and he resigned his commission in hope of getting a more substantial command in the infant Continental Navy.[31]

First on the list of lieutenants was John Paul Jones. A native of Scotland, Jones joined the sea service early in life to seek fame and fortune. By late 1775, he was employed in Philadelphia converting the *Alfred* from a merchant ship to a ship of war. It was Jones who, in the presence of the fleet's new Commander-in-Chief on 3 December, is said to have raised the Grand Union Flag to the top of the *Alfred's* jackstaff. Four days later he received his commission as lieutenant and acting captain of the *Alfred* through his good friend Joseph Hewes, after refusing the command of the *Providence* which soon after went to Captain John Hazard. He later regretted this action, since it meant a loss of seniority and a lieutenancy under a man whom he disliked. Also included on the lieutenant's list was Hoysteed Hacker who was subsequently given the command of the *Fly*.[32]

Two captains not on the 22 December list were those appointed later to command the two ships fitting out in Baltimore. They were Captain William Stone who previously commanded the *Hornet* and was to continue as such, and Captain William Hallock who was appointed to the schooner *Wasp*.

The Marines Prepare for Action

By late December the five companies of Marines had been recruited and each man paid a month's advance. The only thing that remained was to outfit them. Unfortunately, the Naval Committee of Congress did not have the means, and therefore the main burden of equipping the Marine companies fell to the Pennsylvania Committee of Safety. On 27 December Samuel Nicholas received from Robert Towers, the Committee's commissary agent, 100 provincial muskets, 100 bayonets and scabbards, and 12 iron ramrods.[33] But this was not enough to fulfill his requirements. Therefore, on 30 December an urgent plea from the Pennsylvania Council of Safety went out to the county committees for additional arms:

> The great demand for fire arms, in order to equip the Boats and Vessels employ'd in defence of the River Delaware, and to supply the Marines on Board the Continental Arm'd Vessels now ready to sail, has occasion'd the necessity of our collecting all the Arms belonging to the publick in every part of the Province. The Associations in this City have already deliver'd up all that were in their hands, and we hope those in the Country will cheerfully comply and deliver up all that are in their Custody. We hope before the Country can be exposed to danger, a sufficient number of Arms will be procured to furnish those who cannot supply themselves, and in the mean time most earnestly request you would exert yourselves to procure as fast as possible and send down to this Commee all the publick Arms that are in your County, to be employ'd in immediate Service.[34]

This appeal was partially answered two days later when Towers delivered 50 stands of arms to the Marines then quartered in the Philadelphia barracks.* Three days later, on 3 January 1776, another 86 muskets were ordered for delivery to Captain Nicholas' men. Still, the deliveries did not meet the number of arms required to fully equip the five companies of Marines. It was mid-January before this was accomplished.[35]

If the Marines were in need of arms, they were also in need of a uniform. But one was not forthcoming. At this time the army at Cambridge was the major recipient of all uniforms produced in the colonies or purchased in Europe. Therefore, Marines,

* A stand of arms was one musket, one bayonet, and, in some cases, a cartridge box and belt.

in most cases, wore the clothing they were enlisted in, or clothes they purchased with the pay they received. It was only later the following year that the Naval Committee prescribed a uniform, and even then it was difficult to procure.

As Marines in the five companies received their weapons they were sent to the Willing, Morris, and Cuthbert wharves. There they relieved details of the First Pennsylvania Battalion who since 2 December had stood guard over the six Continental ships and the piled military stores on board. On 3 January 1776, the Marines in turn were relieved of the responsibility as they boarded the ships they had guarded and set out for an unknown destination.[36]

DEFEAT ON LAKE CHAMPLAIN, 13 OCTOBER 1776.
Two days after the Battle of Valcour Island, Benedict Arnold beached his galley and four gunboats and ordered them burned in an effort to prevent their capture. This painting recreates the scene soon after the boats were grounded, when Marines were directed to jump overboard, ascend a bank on shore, and form a line to defend the vessels and flag against the British.

CHAPTER II

Outbreak of Hostilities: Early Marine Participation

Arnold's Champlain Fleet

In 1763 the Peace of Paris ended the Seven Years' War in Europe and with it the French and Indian War in America. The flush of victory was short-lived as Great Britain found her treasury almost drained, a huge public debt, and additional territories to administer. To sustain this burden and to meet ongoing expenses, the British Parliament saw no alternative but to lay taxes on the colonies, which existed, according to its view, for the benefit of the mother country.

For almost one hundred and fifty years the American colonies had enjoyed unusual freedom in the management of their own affairs. This was the result not so much of the granting of such freedoms by the British government, but stemmed from Great Britain's preoccupation with local interests, from indifference, inefficiency, and distance; in short, from benign neglect.

The immediate and sometimes violent objections of the Americans to their new taxes both baffled and angered the British. The colonists' position was that, while Parliament could legislate for the colonies, taxes could only be levied through their direct representatives—of which they had none in Parliament. Several British Parliamentarians, indeed, agreed with the argument. To circumvent direct taxation, the ministry imposed duties on certain items imported into the colonies, since it was acknowledged that Parliament had the right to regulate trade. When the Americans refused to import the goods so as to avoid the duties, England sent troops to protect against mob violence the customs collectors and those loyal merchants willing to accept the imports.

Noting that the revenue derived from the indirect taxes would be used to pay the salaries of royal officials and, viewing the presence of British troops as a threat to their civil liberties, the objecting colonists advanced the argument a step further. They now questioned Parliament's right to legislate for them at all. The ministry retreated and repealed all the duties except for the one on tea, not from any agreement with the argument advanced by the Americans but from practicality, since the revenue produced by the duties was insufficient. The tax on tea, which was retained to preserve the legal supremacy of Parliament, immediately led to mob action against a shipment of tea in Boston harbor in December 1773. Parliament retaliated by closing the port, annulling sections of the Massachusetts charter, and enacting several coercive laws.

The Americans now asserted their equality; they said that their colonial assemblies stood on equal footing with Parliament and thus they alone could legislate for the colonies, while Parliament legislated for Great Britain. Concerted action replaced the previously fragmented objections when a Continental Congress was called to meet at Philadelphia in September 1774 to consider formal protests and defensive policies.

It was at Boston where decisive action was taken. Parliament had declared Massachusetts to be in a state of rebellion, and the government's commander in chief in North American, General Thomas Gage, made up his mind to use maximum force to quash resistance. It was the eighteenth of April 1775 and the armed truce in the province was rapidly coming to an end. The commitment was made—a British force would make a thrust from Boston to Concord where the Provincial Congress met and where military supplies were said to be stored. Early the following morning rebel and redcoat met on the misty village green at Lexington. That night Gage's worst fears had been realized; the day's bloodletting had aroused rather than subdued the rebels. By the end of April rebel troops, ill-trained yet formidable, had encircled Boston and immobilized the largest concentration of British soldiers in North America.

As Gage peered fearfully from Boston's forts towards the growing rebel army, the Massachusetts Committee of Safety began planning for a secret expedition into the northern New York wilderness in hope of obtaining cannon for the siege. The expedition was the brainchild of an ambitious, 34-year-old merchant turned soldier from Connecticut, Benedict Arnold. Soon after learning of the events at Boston and the great need for siege guns, he assembled about 50 men and started out for the American lines. By the time he reached the camp at Cambridge, Arnold had formulated a plan whereby the British fort at Ticonderoga could be captured and its 80 heavy cannon, 20 brass guns, and 10 to 12 large mortars brought back to Boston. According to Arnold, "the place could not hold out an hour against a vigorous onset."[1] As a result of Arnold's promises to supply the besieging army with cannon, the Committee of Safety appointed him colonel for "a secret service" with authority to raise not more than 400 men. Once the force was enlisted, the committee instructed that they then march and take Fort Ticonderoga, leave a garrison, and return with the cannon and stores.[2]

Immediately the energetic Arnold set out to enlist his men, but soon learned to his surprise that another expedition with the same purpose was on the march against the fort. Organized by several prominent citizens of Hartford, Connecticut, the mission was given to a tall man with a roaring, bragging voice from the New Hampshire Grants, Ethan Allen, and his Green Mountain Boys. Fearing a rival to his command, Arnold hurried to the rallying point at Hand's Cove on the eastern shore of Lake Champlain. Around midnight on 9 May Arnold reached the rendezvous and immediately attempted to wrest command of the forces from Allen. In part he was motivated by a sincere belief that the capture of Ticonderoga was too important a task to be entrusted to a man with no real authority. Furthermore, Arnold was well aware of the considerable glory which would surround the capture and wished to be in on it. The possibility of Arnold assuming command so enraged Allen's forces that most were prepared to turn around and march home unless assured that they would be commanded by those with whom they had enlisted. A compromise was soon reached whereby Arnold and Allen would march together at the head of the column, and that Arnold would refrain from issuing orders.

The compromise was reached, but the fort still had to be taken. In two boats that were on hand, Allen and Arnold crossed the lake about three in the morning with the first contingent of 83 men. Reluctant to lose the element of surprise, Allen left the remaining 140 men on the eastern shore and he and Arnold led the men to the fort's gates. The lightly defended fort was quickly taken, as was the British outpost at Crown Point.[3]

Meanwhile, a detachment of Allen's men under the command of Captain Samuel Herrick had gone to Skenesborough (Whitehall) at the head of Lake Champlain. There they seized Major Philip Skene, a loyalist landowner, his family, and a small schooner. Built by Skene in 1774, the 40-ton schooner had a

Benedict Arnold, by Thomas Hart.

deck length of 41 feet, keel length of 31 feet, extreme width of 14 feet, 9 inches, and a hold depth of 3 feet, 1 inch. On 11 May the schooner, rechristened the *Liberty*, set out for Ticonderoga under the command of Captain Eleazer Oswald. Arriving on the 14th, she was quickly armed with four carriage guns and six swivels.[4]

Arnold immediately appreciated the importance of the seizure; after all there were no roads around Champlain, only the lake itself. The force controlling the lake could control the passage of all troops in the area. This was of vital importance as the fort at Ticonderoga could not be regained by the British without reinforcements from Quebec, and the lake was their only avenue of approach. For these watery exercises, Arnold would find it necessary to improvise a force of Marines.

Shortly after the capture of Fort Ticonderoga, a council of war was held from which Arnold emerged as commander of a proposed expedition to St. Johns, with hopes of surprising the British garrison there and capturing the *George III*, a large, well-armed sloop. Arnold gathered a force of 30 men and shoved off in a batteau on the 16th for the northern end of the lake. Later that day the *Liberty* sailed northward to meet Arnold; the sailors were commanded by Captain John Sloan and the soldiers by Oswald.

The *Liberty* handled poorly to windward and made slow progress until the wind shifted on the 17th. Meanwhile, Arnold scouted on ahead with his small force. With the wind aft, the *Liberty* overtook Arnold and he reboarded. Good progress was made until the wind died late in the afternoon on the 17th and the schooner was becalmed close to Pointe au Fer, some 30 miles south of the destination. Impatient to move on, Arnold put 35 men into the small boats and, leaving the *Liberty*, rowed all night, landing near the town at daybreak. A scout advised that the garrison was not alerted so Arnold moved at once. With surprise on their side, the Americans took both the fort and the sloop with little trouble. Along with the sloop, they captured nine boats and two brass cannon.

It was learned that the fort's commanding officer had departed the previous day after word of Ticonderoga's fall had finally reached him. Destined for Montreal, he was expected to return shortly with reinforcements, planning to board the *George III* and sail south to recapture the fort. With this intelligence, Arnold decided to depart at once, rather than fight a pitched battle on alien ground against a stronger foe. Burning five of the boats to prevent their recapture, he sailed south with the *Liberty*, *George III*, and four boats. Not far below the town he met Allen and his forces on their way to St. Johns. Allen decided, against the advice of Arnold, to proceed on to St. Johns where he set up camp on the opposite shore of the Richelieu River. The following morning he was attacked by the British force from Montreal and forced to retreat.[5]

Arnold's flotilla returned to Ticonderoga on 20 May. There the force set to work rearming and manning the *Liberty* and *George III*, which was renamed the *Enterprise*. While the *Liberty* retained her original armament, the *Enterprise* was eventually armed with six carriage guns and ten swivels. Manning the two vessels, however, was a more difficult job. Men with a facility or knowledge of seamenship were not to be had. On the other hand, there was an ample supply of soldiers who could act as Marines. In manning the *Enterprise*, for example, seven men of Captain Herrick's company were drafted by Arnold, or volunteered to serve as Marines. In all, the sloop carried 4 officers, 10 sailors, and 17 Marines.[6]

On the same day Arnold attacked St. Johns, the Continental Congress officially learned of the capture of the *Liberty* and the forts. Instead of directing offensive actions against the British forces to the north, Congress suggested that a strong garrison be established and that an inventory be made of the captured war material so that it might be returned when harmony between Great Britain and her colonies was again restored. Thus the lake forces were placed in a defensive status. A little over a month later Congress rewarded their efforts by placing them on the Continental payroll as of 3 May 1775.[7]

In the weeks following his attack on St. Johns, Arnold busied himself by leading almost daily reconnaissance cruises from the anchorage which he established at Crown Point. But in mid-June he found that his authority to command was in question. A special commission from the Cambridge Committee of Safety had arrived and began to delve into both Arnold's behavior and his apparent attempts to circumvent the orders of Congress. Not only had Arnold been aggressive in his actions, but he had gone so far as to suggest an invasion of Canada, which was in direct violation of the policy suggested by the Continental Congress. On 23 June, the commission informed Arnold that he was to conform to the orders of Congress and subordinate himself to Colonel Benjamin Hinman of Connecticut, or get out. Arnold coupled his refusal with his resignation.[8]

The *Liberty* and *Enterprise* continued to cruise the lake all summer for the protection of the outlying settlements and for reconnaissance. In late July the *Liberty*, "well man'd with both Sailors and Marines," began one of its short cruises. While proceeding up the lake, she made a number of stops to let Captain James Stewart and his lieutenant of Marines go ashore in order to obtain information as to British positions to the north. At one stop they were informed by several Indians and Frenchmen that British ships at St. Johns were ready to sail for Ticonderoga and Crown Point. With news of the imminent British attack, Stewart immediately sailed south for Ticonderoga where he transmitted his information to Major General Philip Schuyler, now commander of the Northern Department.[9]

British preparations at St. Johns to create a naval power, among other things, caused American military leaders on Lake Champlain to reconsider a continuance of the defensive posture. When Congress ordered General Schuyler to Ticonderoga it suggested that "if [he] finds it practicable, and that it will not be disagreeable to the Canadians," he should at once seize St. Johns, Montreal, and any other part of Canada that it might occur to him to take.[10] But Schuyler had doubts about the practicality of advancing into Canada and on 3 August felt that the "Enemy's naval strength . . . will prevent our getting down the Sorrel [Richelieu] River to St. Johns."[11] Two weeks later he departed Lake Champlain temporarily, leaving General Richard Montgomery in command.

General Montgomery, onetime member of Parliament and former British officer, had no such doubts. It was obvious to him that the British ships must be destroyed. On 28 August, a 1,200-man American force embarked on board the newly constructed gondolas *Hancock* and *Schuyler*, and several armed bateaux at Crown Point. Escorted by the *Enterprise*, *Liberty*, and several rowboats mounting a single 12-pounder, the straggling fleet got underway for St. Johns. After an initial attack and failure in mid-September, the tide of battle turned abruptly the following month, and the Americans occupied the British fort on the Richelieu River. By late November, Montgomery was in possession of Montreal and ready to sail for Quebec.

From his headquarters at Cambridge, General Washington initiated another expedition against Quebec. With Montgomery's force already enroute to Quebec by way of the St. Lawrence, Washington dispatched a small army under General Benedict Arnold by sea from Newburyport up the Kennebec River with orders to make himself master of Quebec

and divert General Sir Guy Carleton's forces from St. Johns, allowing Montgomery free passage. On 19 September, Arnold and his men sailed from Massachusetts, but it was not until 3 December that Montgomery joined him before Quebec.[12]

Activities in Canada developed rapidly in the winter of 1775–1776. Following the failure of the assault on Quebec, General Montgomery's death, and the arrival of decisive enemy naval power, the Americans were forced to retreat. Montreal and St. Johns soon fell in rapid succession as the British made their power felt. By the summer of 1776, the Americans found themselves back where they had started from a year before: Ticonderoga and Crown Point.

Little or no progress was made from November 1775 to July 1776 in augmenting the American fleet on Lake Champlain, although General Schuyler had initiated the construction of several gondolas in late May. In fact, there was little impulse for such a move since the lake was under American control. However, when the Americans under Benedict Arnold came fleeing out of Canada in June 1776, the prospect of being followed by British naval units from the north became painfully evident. The building of additional vessels was begun at once, and by late July the fleet included three schooners (*Royal Savage*, *Revenge*, and *Liberty*), one sloop (*Enterprise*), and several row-galleys and gondolas, finished or on the blocks.

Construction of the vessels was a relatively simple task, but manning them with seamen and Marines proved more difficult. On 21 July Lieutenant Colonel Thomas Hartly (6th Pennsylvania Battalion) informed Major General Horatio Gates (Commander of troops at Fort Ticonderoga) of the problem, specifically in connection with the enlistment of Marines: "Mr. [Robert] Hop[e]s, a young gentleman from York, in Pennsylvania, was wounded in the action at Trois Rivieres, where he behaved with the utmost bravery. . . . He was recommended to General Arnold, and was commissioned as a Lieutenant of Marines on board the large schooner [*Royal Savage*]. He was to have twenty-five men." But, as Hartly continued, the "vessel will be ready to sail in a few days. I would wish that you would be pleased to take such steps as to have him the complement of 25 Marines." To make up the contingent, Hartly suggested that "good men should be draughted from the Army as Marines."[13] *

In response to Lieutenant Colonel Hartly's suggestion, General Gates issued orders for the immediate enlistment of seamen and Marines from the Northern Army two days later:

> Whereas it is of the utmost consiquence that a well regulated body of Seamen and Marines should be immediately draughted from the several Brigades of this Army, to the end that the Arms of the United States may continue to support their Naval Superiority & Command of the Waters of Lake Champlain & for the encouragement of such Seamen & Marines as shall be so draughted for the important purpose aforesaid, the Hon'ble the Congress of the United States have ordered each non Com'd. Officer, while employed in the service aforesaid, Seamen & Marines, an additional pay of Eight Shillings per month, over and above the pay they are entitled to receive in the Corps to which they respectively belong.
>
> The following is the detail for the Non Com'd. Officers, Seamen & Marines to be draughted from each Brigade. They are to parade tomorrow [24 July] at eight o'clock & immediately to be delivered to the Officers appointed to command them.
>
> Each Officer will keep a Roll of their Names & Regiments & Companies that their pay abstracts may be regularly made up & adjusted once a month. . . .

Horatiŏ Gates, by Charles Willson Peale.

Details for the Seamen & Marines

	Serg'ts.	Corpls.	Drs.	Privates
First Brigade	6	4	4	64
2d.	6	4	4	62
3d.	6	2	2	60
4th.	10	6	6	102
	28	16	16	288 [14]

Following the issuance of the general order, Gates informed Hartly that Robert Hopes would be assigned to the *Royal Savage* where he hoped his future behavior would be such as to "merit further promotion." In addition, Gates noted that "service in the vessels will be a post of honour," and "those who distinguish themselves therein may depend upon my protection."[15] But Gates' congratulations were premature. On 21 July Hartly notified the general that although Hopes was commissioned, he had neglected to consult Hopes on his assignment. As a result, Hopes refused to accept his commission.[16]

Undaunted, Lieutenant Colonel Hartly recommended another officer from his battalion, Ensign James Calderwood:

> Mr. Calderwood, an officer in this battalion, who is a scholar, and has been three and a half years in the sea service, part of which as a Midshipman on board a man of war, is desirous of serving as an officer of Marines on this Lake. He is a person of courage, and will show himself worthy of an appointment. I beg leave to recommend him to your Honor to be appointed a

* Robert Hopes was commissioned ensign on 9 January 1776 in the company commanded by Captain Moses McClean —a unit of the 6th Pennsylvania Battalion.

The Royal Savage.

> First Lieutenant of Marines on board the schooner *Royal Savage*, in the room of Mr. Hop[e]s. I would be very glad that his commission should bear the same date with that to Mr. Hop[e]s.

Ensign Calderwood was accordingly commissioned a lieutenant of Marines on board the *Royal Savage*.[17] *

The drafting of seamen and Marines, and the appointment of appropriate officers, continued sporadically for the next three months. Although large at first—300 men in late July and an additional 300 in mid-August—the drafts by September were cut by 90 percent. On 4 September, for example, 33 men of Colonel Asa Whitcomb's regiment were ordered to the fleet as Marines under the command of Marine Lieutenant Calderwood, who then had charge of Marine recruitment for the fleet.[18] In all, the full fleet carried over 900 seamen and Marines.

Virtually nothing is known about the role and duties of Marine officers and men on board Arnold's fleet. Although they received approximately the same pay as Marines directly authorized by the Continental Congress, they retained the trappings of their former units in terms of uniform, weapons, and accoutrements. The only additional piece of equipment which might have set them apart from the ordinary soldier of the Northern Army were the brass blunderbusses they were issued.[19] As Arnold characterized the men of the fleet: "We have a wretched motley Crew; the Marines, the Refuse of every regiment, and the Seamen, few of them, ever wet with salt Water."[20]

By 20 August preparations to meet the enemy were about complete. The schooners *Royal Savage*, *Revenge*, and *Liberty*, the sloop *Enterprise*, and the gondolas *Boston*, *New Haven*, *New York*, *Provi-*

* James Calderwood was appointed quartermaster of the 6th Pennsylvania Battalion (Colonel William Irvine) on 9 January 1776, and then promoted to ensign in Captain Robert Adams' Company, 1 May 1776. On 1 August 1776, he received his commission as lieutenant of Marines. He subsequently raised an independent battalion and was mortally wounded in the Battle of Brandywine, 11 September 1777.

dence, Connecticut, Philadelphia, Spitfire, and the row-galley *Lee* were ready. The large, round-bilged galleys *Congress, Trumbull,* and *Washington,* however, required armament and crews. On the 24th Arnold set sail from Crown Point, secure in the belief that his fleet was far superior to anything the British could assemble at St. Johns. At Willsborough, half-way down the lake, the fleet encountered a heavy storm in which the *Spitfire* nearly foundered. Forced to weigh anchor, the fleet retreated to the shelter of Buttonmould Bay, where for three days it rode out the heavy gale. On 1 September Arnold again gave the signal to weigh anchor, and a fresh breeze carried the fleet down to Schuyler's Island and then to Windmill Point at the headwaters of the Richelieu River.[21]

At Windmill Point on the 6th, the fleet was joined by the row-galley *Lee* and the gondola *New Jersey.* Fearing that the British would erect batteries on the shore and rake his fleet then moored in a line across the lake, Arnold moved his ships southward to an anchorage off the Isle la Motte. Again on the 19th he shifted his position, this time to a point just above Cumberland Head, ten miles north of Valcour Island. Four days later he retired to a shielded anchorage between the island and the lake's western shore. There he moored his fleet in a curved line between the island and the New York shore. With the arrival of the three large galleys on 5 October, Arnold repositioned the fleet so that the *Washington* protected the right flank, the *Trumbull* protected the left, and *Congress* the center. There they waited.[22]

Sir Guy Carleton's fleet sailed from St. Johns on 4 October and proceeded up the lake cautiously in search of the Americans. Slowly it passed Isle La Motte, Grand and Long Islands, and by the morning of the 11th had rounded Cumberland Head. Two miles beyond Valcour Island Arnold's fleet was sighted. As the British swept back towards Arnold's position against a northerly wind, the Americans got their first look at the opposition's strength. The 300-ton, full-rigged ship *Inflexible* mounted 18 guns and the radeau, or floating battery, *Thunder,* 21. In addition, the British had assembled two schooners, four longboats, and 45 other smaller craft, into what probably appeared to the average Marine to be an immense squadron. The fleet was manned by over 70 experienced seamen and carried a total armament of over 90 guns. This in contrast to the American fleet which had 86 guns, a majority of which were much smaller in caliber.[23]

Seeing that the British were having a difficult time against a perverse wind, Arnold ordered the *Royal Savage* and the three galleys to get underway and commence the attack. Incompetently handled in the first but deadly exchange of broadsides, the schooner grounded and was quickly cut to pieces. As the galleys rejoined the original defensive line, the *Carleton,* a 14-gun British schooner, and several gunboats made their way up the channel and closed on the Americans. With no more than 350 yards separating the two opposing lines, the real battle began.

For more than five hours the British and Americans exchanged broadside after devastating broadside. On board the *Congress,* Arnold was everywhere, pointing and firing each gun, all the while shouting encouragement to his seamen and Marines. By late afternoon, the *Carleton* was out of action as were numerous gunboats. On the American side, the *Congress* and *Washington* were in bad shape.[24] The former had been hulled more than a dozen times, while the latter had lost almost all her officers, with the exception of Brigadier General David Waterbury and his lieutenant of Marines, Heathcote Muirson.[25] * In addition, the *Philadelphia* was so badly hit that she sank shortly after the action.

At dusk, the British concluded that the following day would see the complete destruction of the American fleet and thus withdrew their ships. That night Carleton anchored his fleet in a semi-circle about the beleaguered Americans. On board the *Congress,* the American captains gathered for a hastily called conference. Most agreed that although the fleet could not survive another day since ammunition was in short supply, there would be no surrender. Instead, they would attempt an escape. Close to midnight the battered fleet slipped silently passed the slumbering British and sailed for Ticonderoga.[26]

At dawn the following morning, the American fleet anchored off Schuyler's Island to effect repairs. There the *Boston* and *Providence,* found to be badly damaged, were scuttled. Shortly after noon the fleet again resumed its flight southward. Meanwhile, the British had discovered the Americans' escape and immediately started after them. It was not until the

* Heathcote Muirson, son of loyalist Doctor George Muirson of Suffolk County, Long Island, joined the Continental service against the express wishes of his family. On 13 October 1776 he was captured when the British took the *Washington,* and paroled in April 1778.

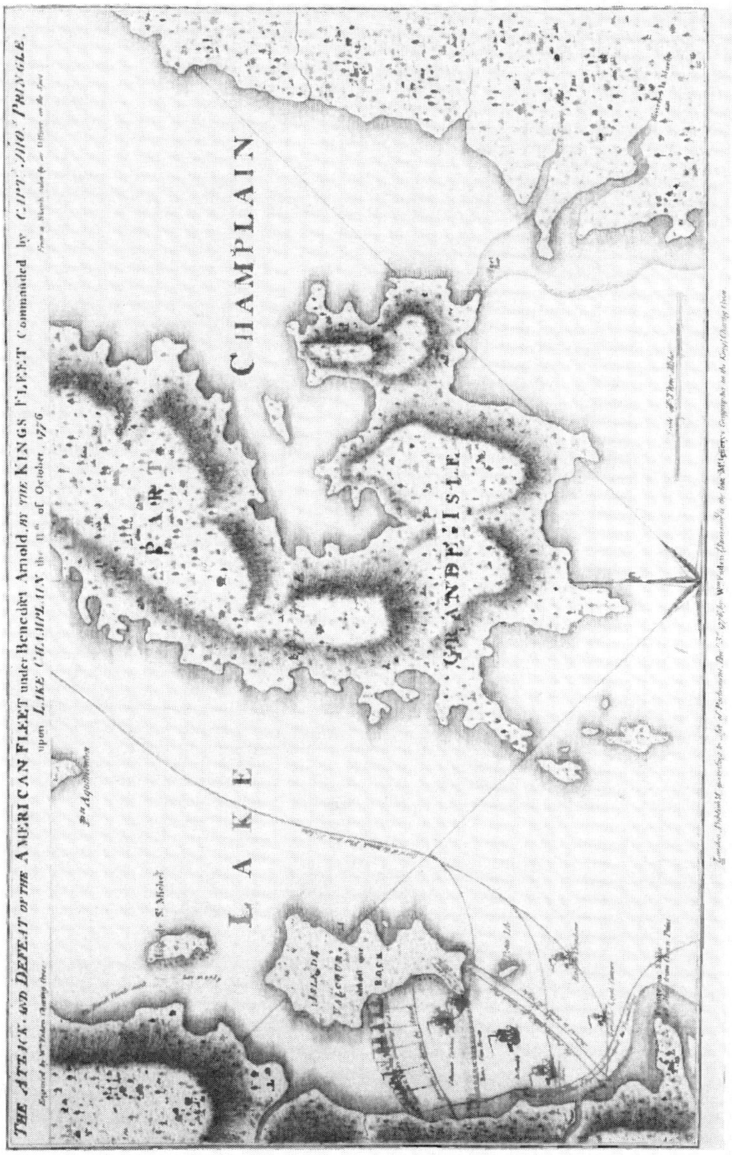
THE ATTACK AND DEFEAT OF THE AMERICAN FLEET under Benedict Arnold, BY THE KINGS FLEET Commanded by CAPT. THO.S PRINGLE,
upon LAKE CHAMPLAIN the 11th of October, 1776
LAKE CHAMPLAIN
GRANDE ISLE

morning of the 13th that the American fleet was again sighted and engaged. The first victim to fall to the British onslaught was the crippled *Washington*, which, after taking several broadsides, struck her colors. For two and a half hours the unequal combat raged. Then suddenly the *Congress* broke through the encirclement, and with four gondolas headed onto shore.[27] Joseph Cushing, who served as a sergeant of Marines on board Arnold's galley and was later appointed a brigadier general in the Massachusetts militia, describes the events which brought an end to the running battle with the British:

> He [*Arnold*] then ran his own gallery and four gun boats ashore in a cove, on the eastern coast of the lake, and set them on fire, but ordered the colours not to be struck; and as they grounded, the marines were directed to jump overboard, with their arms and accoutrements, to ascend a bank about twenty-five feet elevation, and form a line for the defence of their vessels and flags against the enemy, Arnold being the last man who debarked.[28]

With his fleet in flames, Arnold then headed through the woods towards Crown Point and Fort Ticonderoga.

In three days of fighting the Americans had lost 11 out of 15 ships. Only the *Revenge*, *Enterprise*, *Connecticut*, and *Trumbull* remained. Of the 700 men in the original force, only little more than half returned. Although Arnold had been utterly defeated, his gallantry and that of his seamen and Marines did not go unnoticed. On the day they returned to Fort Ticonderoga, General Gates sent "his thanks to Genl. Arnold & the Officers, Seamen and Marines of the Fleet, for the gallant defence they made against the great superiority of the Enemies Force." Gates further observed that, "such magnanimous behaviour will establish the Fame of the American Arms throughout the Globe."[29]

The British stayed in the area of southern Lake Champlain long enough to occupy the hastily abandoned fort at Crown Point. But with winter approaching, their fleet in need of repair, and their crews depleted by casualties, the British soon retreated northward. The conquest of Fort Ticonderoga and the invasion through central New York would have to wait until the following year, but the time purchased by the men of Arnold's fleet allowed regiments to be raised that in 1777 would meet and defeat General John Burgoyne's invading army at Saratoga.

Washington's New England Navy

Shortly after the fall of Fort Ticonderoga and the creation of the small fleet on Lake Champlain, General George Washington took command of the Continental Army around Boston. There, in late July 1775, he found his forces sprawled out in a great crescent of installations about the city. To the north, west, and south, the Americans looked down on the 6,000-man British army and precluded their movement by land. But to the east, however, Boston harbor opened toward the Atlantic. Through the narrow "guts" and between the myriad islands of the bay, British vessels sailed with reinforcements and supplies. They feared no molestation as the Americans possessed no navy, and the few privateer raiders and state naval vessels that occasionally appeared hardly posed a threat. Thus it was only logical that the Americans should consider mounting some sort of a naval defense.

Accordingly, Washington decided to create a small naval force to supplement his military operations ashore. Since he determined to use detachments from the army to officer and man the ships, no authority other than his commission as commander in chief was required. Like Arnold, the sea captains appointed by Washington would find it necessary to improvise detachments of Marines long before the Continental Congress took action.

The first of the vessels commissioned was the *Hannah*, a 78-ton coastal schooner. After supplementing her canvas, a draft on the regiments recruited in Marblehead, Beverly, and the other shore towns of Massachusetts provided the initial manning for the *Hannah*. The added inducement of a share in non-military prizes assisted the final recruiting drive and by late August, a crew of 50 officers, seamen, and Marines had been gathered up. Under the command of Captain Nicholson Broughton of Colonel John Glover's Marblehead Regiment, the *Hannah* sailed from Beverly on 5 September with detailed orders from Washington to seize all enemy vessels bound to or from Boston, laden with soldiers, arms, ammunition, or provisions for the British army.[30]

Unfortunately, this first cruise of the *Hannah* was to lead to almost every problem to which a navy is prone, including mutiny. After taking the sloop *Unity* on the morning of his second day out, Captain Broughton found to his great disappointment that his prize was a recapture and as such would have to be returned to her original owner, John Langdon of New Hampshire, rather than be auctioned. Following the return to port, a long and acrimonious debate ensued which ended with an armed revolt of the *Hannah's* crew. Only the arrival of a detachment from Washington's headquarters saved the situation, but the mutineers were subsequently tried and found guilty.

With a new crew from Glover's Regiment, the *Hannah* made many one-day cruises, but in each case, Broughton returned at dusk to the safety of port without captures. In early October, the *Hannah* pursued an unarmed transport but was unsuccessful in overtaking it before it reached the safety of Nantasket Roads. The British, outraged at the chase, sent out the armed sloop *Nautilus*. Shortly thereafter, intent on returning to Marblehead for the night, Broughton was passing among the islands of the middle bay when a lookout spied the *Nautilus*. The British sloop was much swifter than the *Hannah* so that, although the *Nautilus* was some five miles astern when Broughton started driving for home, once off Beverly they were neck-and-neck. Broughton decided to give up the chase, and ran the schooner aground on a sandbar in the cove. The *Nautilus* opened fire as the Americans fled through the shallows for shore. The *Hannah's* crew managed to strip several small cannon, and much to the surprise of the British, opened a rather accurate fire from shore. In attempting to flee, the *Nautilus* went aground and was very severely damaged before the incoming tide refloated her, permitting her to limp back to Boston.

In the aftermath of the *Hannah*, Washington developed a distaste for seamen that was to remain with him. It is clear that only the most pressing military necessity prevented Washington from abandoning the entire idea at once.

Concurrently with the demise of the *Hannah*, instructions arrived from John Hancock who informed Washington that two unarmed and unescorted British ordnance brigs were at sea, destined for Quebec. Hancock ordered Washington to approach the

Massachusetts Council of Safety for two armed vessels to intercept the brigs and give "proper encouragement to the Marines and Seamen that shall be Sent on this Enterprize."[31] As if in anticipation of Congress' request, Washington on 4 October instructed Colonel Glover to procure two craft at Continental expense. As a result the schooners *Hancock* and *Franklin* were hired, manned with approximately 70 officers, seamen, and Marines each, and then ordered to the mouth of the St. Lawrence under Captains Nicholson Broughton and John Selman, respectively.[32]

In early November, the *Hancock* and *Franklin* arrived in the neighborhood of the Strait of Canso, but did not enter the St. Lawrence. After making several unauthorized captures, the two ships turned their attention to the island of St. Johns (Prince Edward Island) and its capital Charlottetown. There they pillaged both public and private property and made prisoners of three prominent citizens, including the acting governor. On their return to Massachusetts, Washington immediately ordered the prisoners released and their property restored. Washington also severely reproved both Broughton and Selman for the unwarranted acts of violence and the failure of their enterprise.

Meanwhile, four other vessels, in addition to the

Hancock and Franklin, were put into service with Washington's fleet. Like the earlier vessels, the schooners Lee, Harrison, Warren, and the brigantine Washington were hired. Under the charter agreements signed by Washington's agents, the owner was responsible for expenses incurred in fitting out the vessel, while the agent took care of the expenses of operation, "officers, marines & Sailors Wages," and reimbursement of the assessed value should the ship be lost.[33] In late October the Lee, commanded by Captain John Manley of Marblehead, and Harrison, Captain William Coit of Connecticut, set sail. Early the following month the Warren, Captain Winborn Adams of New Hampshire, and Washington, Captain Sion Martindale of Rhode Island, followed them to sea in search of British transports. The Lee, Harrison, and Warren were manned by 50 officers, seamen, and Marines, and carried 4 carriage guns and 10 swivels, while the Washington carried 80 men and 10 carriage guns.[34]

The fleet of four ships had little success; more often than not they returned to port without prizes. As winter approached and the cold northerly winds increased, the seamen and Marines grew discontented. On 23 November Washington's agent at Plymouth, William Watson, found the Harrison's crew "an uneasy sett of fellows, who have got sowerd by the severity of the season," and on the 29th he observed "that the people on board the Brigantine Washington are in general discontent, & have agreed to do no Duty on board s[ai]d vessel, & Say that they Inlisted to Serve in the Army & not as Marines."[35] If his seamen and Marines were dissatisfied, General Washington was equally displeased with their conduct:

> The plague, trouble and vexation I have had with the Crews of all the armed Vessels is inexpressible; I do believe there is not on Earth a more disorderly set; every time they come into Port, we hear of nothing but mutinous Complaints . . . the Crews of the Washington and Harrison have Actually deserted them, so that I have been under the necessity of ordering the Agent to lay the latter up and get hands for the other on the best Terms he could.[36]

But the situation was not altogether disheartening. On 4 December, Washington learned of a fortunate capture made by Captain John Manley in the Lee. In mid-November, Manley received intelligence concerning the existence of an unescorted British transport expected momentarily in Boston. He immediately set sail from Beverly and on 29 November sighted the British brigantine Nancy. She was quickly overtaken and carried into Gloucester where a search proved her to be loaded with a large cargo of military ordnance. But the news of this valuable American prize was tempered somewhat by the loss of the schooner Washington, which was captured by HMS Fowey on 5 December.

At the end of December 1775, the enlistments of the seamen and Marines expired and new crews were recruited. In addition, all the vessels received new captains: Daniel Waters took the Lee, Samuel Tucker the Franklin, Charles Dyar the Harrison, William Burke the Warren, and John Ayers was assigned to the just-hired Lynch. To command the small fleet Washington appointed John Manley commodore, and assigned him the Hancock. On 20 January, three of the five captains received their commissions and instructions. It is noteworthy that, even though the naval committee of the Continental Congress had issued regulations governing naval affairs late the previous year, Washington's six vessels, naval officers, Marines, and seamen would be governed by army regulations.[37]

Throughout the early months of 1776, Washington's little flotilla continued to attack and seize enemy shipping. But in March, with the evacuation of Boston by the British and the movement of Washington to New York, immediate control of the fleet passed to General Artemas Ward. Soon after the evacuation, Captain John Manley was appointed to command one of the frigates being constructed by Congress and was replaced on board the Hancock by Samuel Tucker. James Mugford of Marblehead replaced Tucker as temporary commander of the schooner Franklin. Mugford was the man who brought Washington his largest and most valuable prize.

East of the entrance to Boston harbor on 17 May, Mugford and his crew of 21 seamen and Marines observed the large unescorted British transport Hope making for port, unaware that the British no longer controlled the city. Since she was heavily laden, Mugford chanced the she would not resist. His suspicions were correct. Under the guns of British men-of-war lying at Nantasket Roads, the Franklin soon overtook the Hope, and without opposition carried her into Boston. There she yielded 75 tons of powder and 1,000 carbines, a portion of which were later issued to Marines in the fleet.[38] Provoked by the loss of a valuable transport, the British attacked

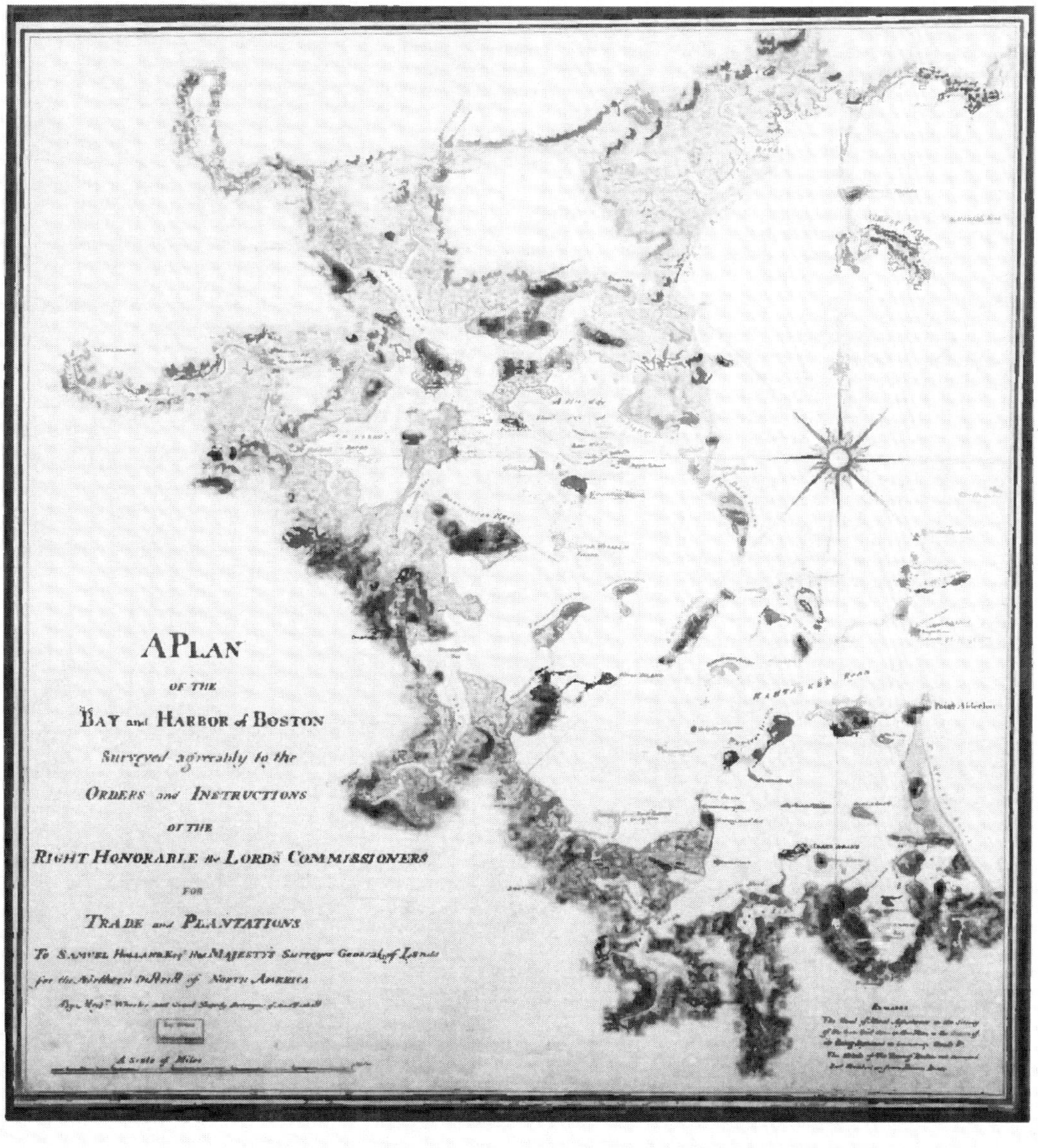

the *Franklin* two days later. In the fight Mugford and several of his men lost their lives.

The five ships of the fleet continued to cruise Massachusetts Bay throughout the remainder of 1776. On 16 June, two days after the last British vessel had abandoned Boston harbor, two British transports loaded with Scottish troops approached the port. Unsuspectingly they entered and were soon engaged and taken by the *Warren* and three other ships, in addition to the *Defence* (Captain Seth Harding) of the Connecticut State Navy. Two days later an additional transport was taken. Success, although spectacular at times, was sporadic, and in early 1777 the Marine Committee of Congress ordered the fleet disbanded and the officers and men taken into the Continental Navy.

Upon his arrival in New York in April 1776, General Washington, in concert with the New York Committee of Safety, created a small fleet similar to the one at Boston. Designed to supplement land operations and suppress the illicit trade with the enemy, the *General Schuyler*, *General Putnam*, *General Mifflin*, and the *Lady Washington* were commissioned and manned with seamen and Marines from the army. The fleet's existence was short-lived. In August, the British occupied New York, thus bringing to an end the unhampered movement of the fleet.

Despite the fact that Washington's fleet and the fleet established by Arnold on Lake Champlain did not operate throughout the American Revolution, both played unique, but different roles. Arnold's fleet, although first created for offensive operations against British forces to the north, became the only bulwark against British efforts to split the colonies in 1776. The fleet organized by Washington, on the other hand, concentrated on capturing British vessels and supplies, thereby withholding vital support to the enemy and at the same time providing the American army with badly needed clothing and equipment. Nevertheless, they provided a training ground for several officers and men of the Continental Navy and Marines.

Both, however, contained a common weakness, that of army control. In the two areas of operation, naval aspects of the combined operations were conducted by army officers under army direction instead of a competent naval officer, with full authority over naval matters. This error led to misunderstandings, jealousies, and discord, often evident at the lowest level. Similarly, since the two naval forces were both raised from the army, little attention was paid to the enlistment of able officers and men. It is no wonder that both Arnold and Washington complained, often bitterly, about the quality of seamen and Marines who served on board their ships. This might have accounted for Washington's particular distaste for Marines, and presumed feeling that they performed no mission that soldiers could not improvise, a feeling that seems to have persisted amongst Army generals and succeeding soldier-presidents.

Laying aside the roles and inherent weaknesses of both fleets, the need for Marines was there. Whether they served as riflemen on board ship, manned the guns, or served as part-time seamen, the distinction was made: there would be seamen and there would be Marines.

LANDING AT NEW PROVIDENCE, 3 MARCH 1776.
With hope of gaining sorely needed powder for Washington's army, 230 Marines and 50 seamen under the command of Marine Captain Samuel Nicholas landed on the island of New Providence early in March 1776. This painting depicts the moment Continental Marines stepped ashore from ships of the Continental Fleet, commanded by Commodore Esek Hopkins.

CHAPTER III

The New Providence Raid

The Continental Fleet

Heralding the departure of the Continental Fleet, a colonial newspaper reported that "the first American Fleet that ever swelled their sails on the western ocean . . . sailed from Philadelphia amidst acclamation of many thousands assembled on the joyful occasion."[1] A British informer, describing the same scene, failed to mention the great number of spectators but simply stated that the ships *Alfred* and *Columbus* accompanied by two brigantines sailed from the city of Philadelphia at one o'clock.[2] On that Thursday, the fourth day of January 1776, Lieutenant James Josiah on board the Continental brig *Andrew Doria* opened the logbook. There he penned the following austere entry: "At 2 P M Cast off from ye Warf In Company with ye Comodore Ship Alfred, Columbus & Cabot, Light airs from ye Westward & much Ice in ye River."[3]

Earlier that morning four of the six vessels of the Continental Fleet stood at the wharves below South and Water Streets ready to sail. On either side of the Willing and Morris Wharf were the *Alfred* and *Columbus*. Largest of the merchantmen purchased by the Naval Committee, the *Alfred* was conspicuous with her white bottom, a broad black band along the waterline, and bright yellow sides. Protruding from the newly installed lower deck gunports were twenty 9-pounders, while on the upper deck she sported ten 6-pounders. On her bow was a striking figurehead—a man in armor drawing a sword. The *Columbus*, second in size only to the *Alfred*, carried eighteen 9-pounders on the lower deck and ten 6-pounders on the upper. Her bottom was white, but unlike the *Alfred*, her sides were painted entirely black and she bore no figurehead.

Downstream, at James Cuthbert's Wharf, were the two brigs, the *Andrew Doria* and the *Cabot*. Named in honor of the Genoese admiral, the *Andrew Doria* carried sixteen newly installed 6-pounders. Alongside stood the *Cabot*, a small brig fitted out with fourteen 6-pounders. Her sides were painted yellow and a small female figurehead adorned her bow, distinguishing her from the *Andrew Doria's* black sides and no head.

The remaining two vessels of the Continental Fleet were the *Providence* and the *Fly*. The *Providence*, formerly the Rhode Island armed sloop *Katy*, was still undergoing alterations in order to install twelve 4-pounders on her decks. The *Fly*, a schooner carrying six 9-pounders and assigned to the fleet as a tender, had scraped sides which were covered with tar, this in contrast to the *Providence* which was entirely black.[4]

As the last of the supplies were loaded, the Marines stood ready to receive their orders of assignment. Captain Samuel Nicholas, with his two lieutenants, Matthew Parke and John Fitzpatrick, and the company of Marines (approximately 60 men) he had recruited the previous month, were ordered to Commodore Hopkins' flagship, the *Alfred*. Reporting to the *Columbus* was Captain Joseph Shoemaker, Lieutenants James Dickenson, Robert Cummings, and their 60 Marines. Lieutenant Isaac Craig and 38 of the 44-man company he had raised throughout December boarded the *Andrew Doria*. Ordered on board the *Cabot* were 40 Marines officered by Captain John Welsh and Lieutenant John Hood Wilson. The *Providence*, which was still being readied for sea, received 14 men raised by Lieutenant Wilson, and 6 of Lieutenant Craig's men. Once on board they were put under the command of Lieutenant

Esek Hopkins, by an unidentified artist

Henry Dayton.[5] No Marines were assigned to the schooner *Fly*. That afternoon the fleet put into the river.

Despite the casting off of lines, the fleet did not make ready for sea but moved to piers at Liberty Island, south of Philadelphia in the Delaware River channel. This shift not only lessened the danger of being caught in the ice forming around the South and Water Street wharves, but allowed each ship opportunity to gather those members of the crew who had not as yet not reported. Anticipating absences among those officers and crewmen who might have found it disagreeable to venture out in the midst of winter, the Naval Committee had a notice prepared and posted in the public houses throughout the city. The notice ordered "every Officer in the Sea and Marine Service, and all Common men . . . who have enlisted into the Service of the United Colonies on board the Ships now fitting out, to report immediately on board their respective ships." If they did not, they would be judged deserters. To facilitate the transportation of stragglers to the ships, boats were stationed at the Willing and Morris Wharf.[6]

The following day, the Naval Committee prepared two letters of instruction for the commodore, both of which probably reached him on the 6th. The first letter was of a general nature and concerned directions for the conduct of the fleet; these the commodore was at liberty to issue to his captains. In this letter, he was ordered to ensure that discipline, order, and peace were preserved amongst those on board the ships, and that "all men under your command be properly fed and taken care of when they are in health, as well as when they are sick or wounded." In addition to those orders pertinent to the discipline and care of personnel, Hopkins, as Commander-in-Chief of the fleet, was to give directions for the ship captains to follow in case of separation; appoint appropriate officers to command prizes; accord special attention to the care of arms and ammunition, making certain that they were "always fit for immediate service"; and guarantee that prisoners were humanely treated.[7]

Simultaneously, the Naval Committee issued to Hopkins a secret letter of instructions and sailing orders. The second letter, unlike the first, showed the effects of southern delegate pressure on the Continental Congress. It was they, who in October of the previous year, first dismissed the idea of a Continental Navy as the "madest idea in the world."[8] Now with the Navy established, and the first American fleet ready to sail, it was directed not against Canada or British occupation forces in and near Boston, but against British forces in the southern colonies. With the appearance of British ships off the southern coast in the autumn of 1775, the southern delegates finally endorsed demands of New Englanders for a Continental naval force. In return they received assurances that the first services performed by the fleet would be against enemy forces in that section of the colonies.

Not only reflecting southern pressure, but also the optimism of the Congress as a whole, the Naval Committee's secret orders expressed a desire to see that "our unnatural Enemies . . . meet with all possible distress on the Sea." To accomplish this task, Hopkins was first ordered to proceed immediately and in the most direct manner to the Chesapeake Bay in Virginia, and there "search out and attack, take or destroy all the Naval forces of our Enemies that you . . . [might] find there." John Murray, the Fourth Earl of Dunmore and Royal Governor of Virginia, had organized a small fleet the year before and was retaliating against rebellious segments of

his colony. On New Year's Day 1776, Lord Dunmore with his fleet attacked and burned Norfolk, Virginia. It was against Dunmore primarily, and then against his cohorts to the south that Hopkins was directed. Once Lord Dunmore had been dealt with, Hopkins was ordered to "proceed immediately to the Southward and make yourself Master of such forces as the Enemy may have both in North and South Carolina." This second phase accomplished, the commodore was then "to proceed Northward directly to Rhode Island, and attack, take and destroy all the Enemies Naval force" there.[9]

As the *Alfred, Columbus, Andrew Doria,* and *Cabot* lay tied up at Liberty Island in the thickening ice, final efforts were made to equip the separate Marine detachments. On January 5th, Pennsylvania's resourceful Committee of Safety delivered 37 muskets and 136 bayonets to Lieutenant Craig on board the *Andrew Doria.* In addition, the Committee issued to Captain Nicholas on the *Alfred,* 67 muskets, 60 bayonets, 50 scabbards, and 200 pounds of musket balls.[10] Still lacking in adequate arms, Nicholas again appealed to the Committee of Safety, and on 9 January they ordered the commissary agent to apply to John Cadwalader for all the muskets the Pennsylvania Militia colonel could spare.[11] While at Liberty Island Captain Nicholas received, in addition to a shipment of miscellaneous arms, a large supply of brilliantly colored hussar uniforms which included caps, coats, waistcoats, and trousers.[12] *

Although different in style and color from the uniform later prescribed by Congress, the Marines on board the *Alfred* achieved partial uniformity in appearance. The only other contemporary description of a Marine's clothing appeared in the January 9th edition of the *Pennsylvania Evening Post.* In a small, but conspicuous, advertisement, John Welsh, captain of Marines on board the *Cabot,* offered a two dollar reward for the return of Peter M'Tagart, one of his Marines. In addition to giving a rather complete physical description of the deserter, the advertisement included a list of clothing M'Tagart was wearing at the time he left the ship: "Light brown coat, white cloth jacket and breeches, blue stockings and new shoes."[13] Brown clothing was the common civilian and military attire of the period, and given a choice, a deserter probably left his post in unidentifiable clothing. So it can be assumed that there was a lack of uniformity in style and color among Marine uniforms of the Continental Fleet. This lack of uniformity is mainly attributable to the absence of a centralized supply network, and more specifically to the want of a prescribed uniform.

* The uniforms were probably blue and conformed in style with the English cavalry uniform of the period.

Six Weeks in Ice

The ice had thinned sufficiently by 17 January to allow the four original ships of the fleet plus the *Providence* and the *Fly* to drop down the Delaware River as far as Reedy Island. There, winter closed in again. For the next six weeks the fleet lay held in the ice and was unable to clear the river. This respite gave the ships time to take on additional quantities of water, wood, and other supplies brought by wagon from Philadelphia.

Later, John Paul Jones was to remember this period as one in which he and the other lieutenants stood the deck, watch after watch, day and night, in order to prevent desertions.[14] Exact numbers are unknown, but the rate of desertion while at Reedy Island did increase drastically. Many who had joined in an initial burst of enthusiasm soon lost their patriotic fervor when confronted with hard work, small-pox, frostbite, and the constant chill of the winter wind. Local townspeople and militiamen along the New Jersey side of the Delaware rounded up a number of deserters and returned them to their respective ships. Others were not as fortunate; they were caught making their way back to Philadelphia, jailed, and in several instances retrieved by detachments of Marines.[15]

The six weeks frozen in the ice at Reedy Island gave Commodore Hopkins a good deal of time to reread his instructions and ponder the situation. At least two important factors had an impact on his thinking as he re-examined his orders. When the fleet had sailed from Philadelphia the topic of conversation in official circles concerned the Continental Army's desperate need for gunpowder and other armaments. "Our want of powder is inconceivable," General Washington wrote to Joseph Hewes, a member of the Naval Committee, on 25 December 1775.[16] The commodore must have been aware of Washington's needs, for his brother Stephen was a member of Congress and sat on the Committee itself. Also he must have had knowledge of the intelligence report that reached the Secret Committee of Congress the previous November which reported large quantities of powder at New Providence Island in the Bahamas, and of the Committee's ensuing recommendation:

> *Ordered*, That the foregoing Committee take measures for securing and bring[ing] away the said powder and that it [be] an instruction to said Committee in case they can secure said powder to have it brought to the port of Philadelphia or to some other port as near Philadelphia as can with safety.[17]

The "foregoing Committee" was the Naval Committee which a month later authored the commodore's orders.

Of equal importance were reports reaching Philadelphia in mid-January which spoke of a number of well-armed ships being added to Lord Dunmore's fleet. Specifically, the addition of HMS *Liverpool*, a 28-gun frigate, now gave Lord Dunmore equal if not superior firepower in a possible confrontation with the American fleet. Fire superiority was not the commodore's only worry. His men were untrained in naval warfare, his fleet made up of converted merchantmen, and the numerous delays had taken their toll of both men and ships. Hopkins finally concluded, as he later testified before Congress, that the "alterd . . . Stacion of the Enemy Priticler as to the Strength of the fleet at Virginia," permitted him to use the clause in his orders which allowed him latitude should difficulties arise.[18] "If bad Winds or Stormy Weather, or any other unforseen accident or disaster disable" the fleet, Hopkins was then authorized to "follow such Courses as your best Judgment shall Suggest to you as most useful to the American Cause and to distress the Enemy by all means in your power."[19] This is what he attempted.

At 10 on the morning of February 11th, the six ships of the Continental Fleet again dropped down the river. Anchoring near Cape Henlopen two days later, the fleet was joined by the sloop *Hornet*, 10, and the schooner *Wasp*, 6, two vessels outfitted in Baltimore and ordered by the Continental Congress to join the main fleet in the Delaware Bay.[20] On the same day, a midshipman on board the *Andrew Doria*, one John Trevett, was promoted to lieutenant of Marines and transferred to the *Columbus* whose Marine company was shy an officer due to the absence of Lieutenant James Dickenson.

As the fleet awaited favorable weather, Hopkins busily drafted the general signals for the fleet as well as specific orders for the individual vessels. On February 14th, the commodore in a letter to Nicholas Biddle, captain of the *Andrew Doria*, ordered him to keep company with, and observe all signals given by the *Alfred* once the fleet was underway. Should the *Andrew Doria*, or any of the other ships, become separated from the remainder of the fleet and be unable to rejoin it in four days, Captain Biddle was then to make his way to the southern part of Great Abaco Island in the Bahamas and there wait a minimum of 14 days for the fleet. If the other seven ships failed to appear within that period, Biddle was then ordered to take the *Andrew Doria* and cruise in such places as "will most Annoy the Enemy."[21] By issuing these orders Hopkins had finally abandoned any thought of sailing against enemy forces in Virginia or the Carolinas. Instead, he seemed to favor a descent upon the town and forts of New Providence in order to secure powder and other armaments located there.

On the eve of departure there was some apprehension on the part of at least one of the captains of the fleet concerning the forthcoming cruise. In a letter to his brother James of Philadelphia on 15 February, Nicholas Biddle began with a short, rather positive report on the status of the *Andrew Doria*, but soon his thoughts turned to fears of personal failure and possible capture:

> May He who has Mercifully led me through the World keep My dear Brothers Cheek from ever being stained with a Blush for Me. I know well the Glorious Cause I am engaged in. And if ever I disgrace it May My Kind father who gave me being instantly Blast me in Mercy to me. . . . If in spite of my best endeavoir I should be taken, If Fortune Should frown on me I hope I shall bear up against it with . . . Fortitude Patience and Resignation.

Notwithstanding, Biddle concluded his letter with an air of self-satisfaction: "Never in my Life was I better pleased with a trip I was going to take than I am with this."[22]

Southward, Into the Atlantic

Whether Captain Biddle was pleased or not, final preparations were complete. About mid-day on Saturday the 17th, seamen on board the flagship unfurled the fore topsail. The commodore had given the signal to weigh anchor. With a strong wind blowing from the northeast, the eight ships of the first American fleet passed Cape Henlopen and moved out into the Atlantic.[23]

Soon after leaving the Delaware the fleet nearly met with disaster. Two days out, what had been a favorable northeast wind soon became a raging gale. In the midst of the heavy seas and thick weather, the fleet lost sight of the *Hornet* and *Fly*. Unbeknownst to the other six ships of the squadron, the *Hornet* and *Fly* had collided, forcing the *Hornet* to return to port and the *Fly* to stay behind to

The First Continental Fleet, 17 February 1776[24]

Vessel	Guns	Men: Seamen	Men: Marines	Naval Captains	Marine Officers
Ship *Alfred*	Twenty 9-Pounders Ten 6-Pounders	160	63	Commander of the Fleet Esek Hopkins Capt Dudley Saltonstall	Capt Samuel Nicholas 1st Lt Matthew Parke Lt John Fitzpatrick
Ship *Columbus*	Eighteen 9-Pounders Ten 6-Pounders	160	63	Capt Abraham Whipple	Capt Joseph Shoemaker 1st Lt John Trevett Lt Robert Cummings
Brig *Cabot*	Fourteen 6-Pounders	90	43	Capt John B. Hopkins	Capt John Welsh 1st Lt John Hood Wilson
Brig *Andrew Doria*	Sixteen 6-Pounders	65	39	Capt Nicholas Biddle	Lt Isaac Craig
Sloop *Providence*	Twelve 4-Pounders	62	21	Capt John Hazard	Lt Henry Dayton
Sloop *Hornet*	Ten 4-Pounders	70	—	Capt William Stone	Lt John Martin Strobagh
Schooner *Fly*	Six 9-Pounders	30	0	Lt Hoysteed Hacker	
Schooner *Wasp*	Eight 2-Pounders	43	5	Capt William Hallock	Lt William Huddle
	Totals	680	234		

make minor repairs. Though a constant watch was ordered on board the surviving ships, the sloop and schooner remained unaccounted for.[25]

Upon the *Hornet's* return to port, Marine Lieutenant John Martin Strobagh resigned his commission because he found "the service by sea to disagree with him," and thought he could be of "more use in the land service."[26] Apparently the severe Atlantic gale had much to do with his decision.

After the short storm, the remainder of the passage to the Bahamas was without major incident. It was not until the fleet had been at sea for two weeks that the first enemy vessels were sighted. With Great Abaco Island in the distance, the *Alfred* intercepted and captured two coastal sloops belonging to New Providence.[27] On the afternoon of 1 March, the squadron anchored in 12 fathoms off the southwest side of the island.[28]

In a Precarious State

For some time the British had considered the possibility of an invasion of the Bahama Islands by American rebel forces. As early as August 1775, General Thomas Gage warned Montfort Browne, Governor of the Bahamas, of intelligence indicating that ships were being readied throughout the colonies and that they were destined to "take Advantage of the State of the Island of Providence and make an attempt to Seize His Majestys Property there." He further stated that in order to ensure the integrity of the King's stores, he would send two transports and a Man-of-War to New Providence and remove all the artillery, ammunition, and military equipment.[29] One month later the provincial council made its reply. Noting that such action would expose the islands to an invasion by its neighbors, the French and Spanish, and cause "dangerous Insurrections among our Slaves," the council concluded that it would be imprudent for it to permit the removal of any military stores or equipment.[30] This matter and further speculation lay officially dormant for the next six months.

With the sailing of the first American fleet in early 1776, the British concluded that the squadron's ultimate destination was either New York or the relief of Boston, not the invasion of the southern colonies or England's Caribbean possessions. Disturbing reports continued to arrive in Nassau. Captain Andrew Law, "an officer in His Majesty's Land Service," arrived in New Providence on 25 February and brought with him news of a "considerable squadron" being assembled at the Delaware Capes whose destination was "against this island." Upon hearing the news the governor told Captain Law to keep the information secret until the council could be called into session.[31] The council was not summoned, since it was probably assumed that this was another of the many false rumors to reach the island. Thus the intelligence caused little stir, and no military preparations were made.

The town of Nassau was, as it is today, the administrative center of the Bahama Islands. In the past the town had been the headquarters for both piratical and Spanish invaders, and stories were told of the Spanish roasting the British governor over a fire after they had killed him. One of the few contemporary accounts of a visit to the Island of New Providence was given by the German doctor, Johann David Schöpf. Visiting the island in

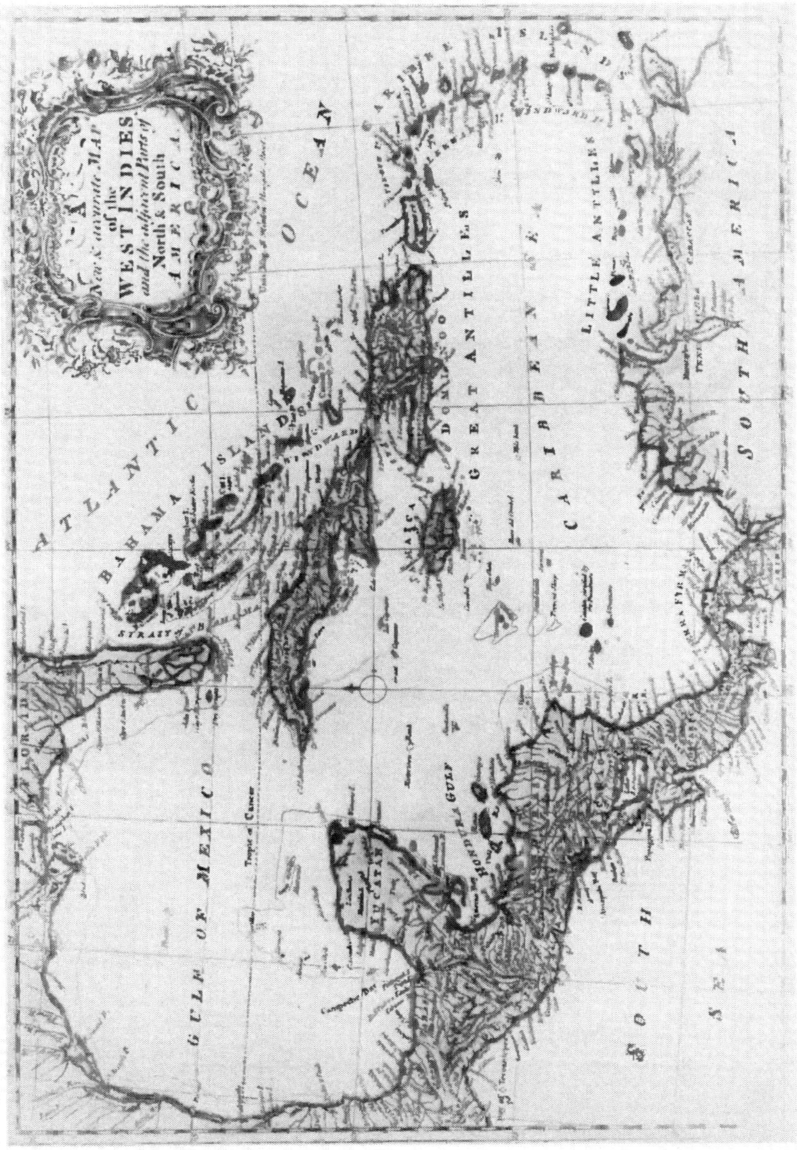
A New & accurate MAP of the WEST INDIES and the adjacent Parts of North & South AMERICA
ATLANTIC OCEAN
BAHAMA ISLANDS
GULF OF MEXICO
Tropic of Cancer
YUCATAN
HONDURAS GULF
GREAT ANTILLES
CARIBBEAN SEA
LITTLE ANTILLES
SOUTH AMERICA
SOUTH SEA

1784, Schöpf described in great detail the town, the buildings, and the inhabitants of this little remote colonial capital:

> The houses are of wood, all lightly built and of simple construction; according to the needs of the climate here, attention has been given only to roof, shade, space, and air. No chimneys are to be seen, and but few glass-windows. The houses stand apart, surrounded by trees, hedges, and gardens. . . . There is but one tolerably regular street, or line of houses, which runs next [to] the water. . . . A church, a gaol and an Assembly-house make up the public buildings of the town. . . . There is no pavement in the town, but none is needed, since the streets, like the whole island, are almost wholly stone. The inhabitants of the town of Nassau are a few royal officials, divers merchants, shipbuilders and carpenters, skippers, pilots, fishermen, and what laborers are needed, with several families who live on the returns from their lands and the work of their slaves. The real planters . . . live near to the town on their estates.[32]

With the American fleet at anchor approximately 50 miles to the north, the defensive posture of Nassau was in a precarious state. The recent recall of His Majesty's sloop of war *Savage*, 8, left only the armed schooner, the *St. John*, 4, in the harbor. Since the *St. John* was in desperate need of repair it could not hope to resist effectively an American attempt to invest the island. The main organized fighting force and the backbone of the island's defense had been a detachment of the 14th Regiment. As with the *Savage*, the detachment had been withdrawn several months before by order of General Gage and sent to America. Therefore, the island's defense devolved upon the provincial militia composed of about 300 men and commanded by Major Robert Sterling. Since a majority of the militiamen were engaged in fishing or other seafaring activities, the number actually present on the island at any one time fell far short of the 140 men needed to garrison the forts and protect the town.[33] For all intents and purposes the island of New Providence lay open to any one who wished to take it.

Not only was the island devoid of adequate defense forces, but it was also lacking in energetic and decisive leadership. A week after receiving the intelligence brought by Captain Law, Governor Browne was again made aware of the fleet's approach, and once more he did nothing. On the morning of 1 March, Captain George Dorsett arrived in New Providence with specific information as to the location and possible destination of the American squadron. While on a whaling cruise which took him near the island of Abaco the previous day, he spotted a fleet of seven vessels "standing in toward the Land from the North East."[34] Thinking the fleet headed for Nassau, he immediately returned to port and informed the governor of the developments. As before, Captain Dorsett was told to keep the information to himself until the council could be summoned. Once again the council was not called; "No Militia arrayed, No forts put in order, in short no preparation at all was made for the defence of the place."[35] Later the governor was to be severely criticized concerning his response to the two reports he had received:

> Martial Law might have been proclaimed; the Arms & Accoutrements of the Militia examined and their Defects remedied; the Forts put into a Condition of Defence and supplied with those Articles which, from a long peace, were deficient; and a Mode of Defence considered and determined on. . . . On whom this unaccountable Neglect is chargeable we need not mention.[36]

Whether Governor Browne had contradictory intelligence or whether he feared possible repercussions should he take decisive action is unknown. By his own actions and the state in which he kept the island of New Providence, he could not but have assisted the American fleet in achieving its goal.

On Saturday, 2 March, after learning of the defenseless state of the island from the captains of the two captured sloops, Commodore Hopkins ordered the fleet to begin final preparations for the attack. As seamen readied the guns and checked the rigging, Marines were issued muskets, ammunition, and broad swords and then transferred from their respective ships to the two sloops and the *Providence*.[37] Late that evening the fleet weighed anchor and sailed for New Providence. The planned frontal surprise attack worked out by the commodore called for the fleet, with the exception of the two sloops and the *Providence*, to stay just beyond sight of land until the last possible moment. Meanwhile the three vessels with the Marines hidden below deck were to sail into the harbor at dawn and once they "got in Close to the Fort . . . they were then to land Instantly & take possession before the Island could be Alarmed."[38]

At this time the island of New Providence had two forts. As Schöpf noted in his description of the island, "the harbor of Providence is formed and protected by a small island, called Hog Island, lying to the north." Because the island was so situated, "there are thus two approaches to the harbor, one to the east, and the other to the west."[39] Realizing that the harbor had to be protected from intruders, the Bahamians had erected a fort at each entrance. To protect both the town and the western entrance to the harbor they built Fort Nassau, a square palisaded stone fort with two bastions. Initially constructed in the late 17th century, the fort was completely destroyed by a combined French and Spanish force in 1703, but then rebuilt in 1744 by Peter Henry Bruce, an engineer of international note. Since the 1740s Fort Nassau had fallen into disrepair. By May 1775, Governor Browne was to report that even though the fort mounted forty-six 12 and 18-pounders, it was indefensible because if the guns were fired there was a great possibility that the walls would collapse. To forestall such an event, he had placed a 6-pounder at the fort's entrance to fire all the required alarms, salutes, and signals.

Fort Nassau not only suffered from decaying walls, but also from its position. Located at the waters edge, the fort was completely dominated by the high ground to the south. From these hills the musket fire of an attacking force could prevent defenders from serving the guns of the fort.[40]

Protecting the eastern entrance was Fort Montagu, a small square stone redoubt mounting seventeen 12 and 18-pounders. Also built in the 1740s by Peter Bruce, the fort's main purpose was to shut "the back door through which the place often had been surprised."[41] Writing to the Duke of Montagu in August 1742, Governor John Tinker observed that with the completion of the eastern fort the island of New Providence had become "the strongest possession in British America."[42]

Hopkins Blunders

In the early morning hours of Sunday, 3 March 1776, the ships of the fleet rendezvoused three leagues north of Nassau. Here Commodore Hopkins blundered. Instead of just the three sloops entering the harbor, the fleet suddenly sailed on and appeared "a little to the windward of the Bar" as the dawn broke.[43] The harbor pilot quickly awakened Governor Browne, who, without dressing, hurried to the door of Government House. Seeing the American fleet at the harbor entrance, he ordered the provincial council to assemble at Fort Nassau without delay. About a quarter of an hour later, with most of the council in attendance, the governor ordered an alarm sounded by three guns. Forebodingly, two of the three gun carriages collapsed. Following the alarm, the drummers beat the long roll to call the militia to arms. About seven o'clock, the most respected of the council members, Samuel Gambier, rode into the fort to find the panic stricken Browne still clothed in his night shirt. At the time the governor was considering shipping the fort's gunpowder with Captain William Chambers in the *Mississippi Packet*, which stood in the harbor loaded with lumber. Dispassionately, Gambier suggested that the gunpowder would be necessary for the defense of the island and that they should immediately prepare in the time that was available for that eventuality. He also proposed that Chambers be dispatched to reconnoiter the enemy and learn of their intention, and that a strong detachment of the militia be sent to secure Fort Montagu. His first suggestion was quickly adopted and Chambers was sent out, but adverse winds and swells forced him to abandon his mission and return without the requested intelligence. It was then proposed that the militia roll be called for the purpose of inspecting arms and ammunition. Fewer than 30 men had reported since the drums were sounded and a majority were without arms or possessed weapons unfit for service. Finally at nine, with the men on their way to Fort Montagu under the command of Lieutenant John Pratt, Governor Browne returned to Government House saying, "that he would just go home, & make himself a little decent."[44]

With the sound of the three-gun alarm, Commodore Hopkins realized that the fleet had been discovered and that he no longer possessed the element of surprise. The winds that had prevented Chambers in the *Mississippi Packet* from crossing the bar had forced the Americans further toward the island than originally planned. Any attempt now by the fleet to penetrate the harbor would have it become the easy prey of the fort's guns. With new orders, the fleet "suddently tacked and stood toward the East." At ten the fleet finally anchored six miles from Nassau in Hanover Sound.[45]

Commodore Hopkins at first proposed that the fleet sail around the west end of the island, land the Marines, and then attack the town from the rear. But this course was soon abandoned when it was learned that there was no suitable place to anchor, nor any road from that end of the island to Nassau, and that the time required to shift the fleet would give the islanders time to prepare an effective defense. With information provided by the impressed pilots, John Paul Jones suggested that the squadron might find anchorage six miles to the west at Hanover Sound. Though Hopkins first objected to being dependent upon the pilots, he finally accepted their assistance. Lieutenant Jones, as he later wrote, then took one of the pilots with him to the "fore-topmast-head" and from there guided the fleet through dangerous waters to a safe anchorage.[46]

With the fleet forced to move from its original landing site, new plans had to be formulated. After consulting with his captains, the commodore decided to land the Marines on the eastern shore of the island and attempt the capture of Fort Montagu. Since the "back door" to Nassau would be much more difficult, Captain Nicholas' 234 Marines were to be augmented by a force of 50 sailors under the command of Second Lieutenant Thomas Weaver of the *Cabot*, who was well acquainted with the area. As the rest of the fleet remained in Hanover Sound, Marines on board the *Providence* and the two captured sloops, covered by the *Wasp*, would make the first amphibious landing executed by Continental Marines.[47]

Ashore at Last

A short time after noon the Marines and sailors splashed ashore at "the Creek," two miles east of Fort Montagu. Nearby was the small village of New Guinea, inhabited by the free slaves and mulattoes of New Providence. At first they thought the invading Americans to be Spaniards and became alarmed, doubtlessly conjuring up pictures of being taken prisoner and again sold into slavery. They soon discovered otherwise. By two o'clock all the Marines and seamen had landed and formed marching ranks.[48]

The original force of 30 militiamen under Lieutenant Pratt sent to Fort Montagu earlier that Sunday morning was reinforced about ten o'clock by another detachment of equal size commanded by Lieutenant Burke, as enemy ships were sighted in the distance. Soon after the second detachment arrived at the fort, the whaleboats carrying American Marines and sailors were seen approaching the beaches two miles to the east. Immediately Lieutenant Pratt formed half his force into a scouting party under Lieutenants Burke and Judkin and ordered it down the beach "to reconnoitre and if possible prevent their landing."[49] But by the time the party reached the enemy beachhead there was little that could be done. The size of the assembled American force left only one alternative—gather whatever intelligence possible and then retreat. Therefore, a man was sent under a flag of truce to inquire as to the identity of the invaders and their purpose. On his return the militiaman informed Lieutenant Burke that "they were sent by the Congress of the United Colonies in order to possess themselves of the Powder and Stores belonging to His Majesty."[50] Seeing that he could not possibly forestall the Americans, Lieutenant Burke ordered an immediate withdrawal in the direction of Fort Montagu without firing a single shot.[51]

Meanwhile, word of the American landing was brought by courier to Fort Nassau, but Governor Browne still had not returned. In his absence, the council ordered Major Robert Sterling to march the main body of the militia (approximately 80 men) to Fort Montagu and if possible prevent the enemy from landing. As the militia was being formed into ranks the governor arrived, presumably now dressed, and apologized for his tardiness saying that he had been detained by "a violent fit of Cholick."[52] After learning of the situation, he then took charge of the militia and set out to reinforce Fort Montagu

Following his arrival at the eastern fort, Browne ordered a detachment of 40 men under the command of Captain-Lieutenant Walker and an ensign to advance toward the enemy beachhead and reinforce Burke and Judkin. As the second group advanced down the beach they met the first in retreat, and both returned to the fort. With the Americans advancing toward Fort Montagu and with his militia in full retreat, Browne realized that his situation was precarious. If he did not act quickly there was a possibility that he would be outflanked and cut off from the town. If on the other hand he stood his ground, all that protected the town was a fort garrisoned by two old men and a few militiamen. As the remainder of his retreating forces entered the fort, the governor looked around and withdrew several nails from his pocket. A local gentleman standing nearby asked him what the nails were to be used for. "For spiking up the guns of Fort Montague," Browne replied. "God," said the gentleman who asked the question, "if that is to be the Case I don't know what business we had here."[53] The governor then ordered an immediate evacuation of the fort, and all the cannon spiked with the exception of three which he instructed to be fired in the direction of the approaching Marines. Leaving two men to man the three guns, Browne mounted the only saddled horse and beat a hasty retreat to Government House where he remained for the next several hours.[54]

As Captain Nicholas and his force of 284 Marines and sailors moved along the cove trail, a "prodigious thicket on one side and the water on the other,'" the three guns at Fort Montagu were fired.[55] Many Bahamians later claimed that at this point Governor Browne lost his best opportunity for halting the American invaders. As one of the governor's severest critics, Merchant William Taylor, recalled, the coastal path "was the very place where our forces should have opposed them." Not only did the militiamen have adequate cover, a small sturdy fort, and sufficient "men to have Cut off a Thousand of them," but they were confronted with a poorly equipped and undisciplined enemy. In a letter to the American Secretary, Lord George Germain, Taylor paints a highly uncomplimentary picture of the American force:

> My Lord if you had seen the miserable figure the Enemies did Cut . . . for they had not so much as one field Piece, let alone Battering Cannon, nor a scaling Ladder, nor so much as an Ax to have made a gap in our Pallisades in order to have got under our great Guns, not one armed vessel had they steering along shore to cover them, and to seal the Woods in Case We had fired upon them, nor had they so much as Boats rowing along shoar to take them off if routed, tho' the path was not above a Stones throw from the sea side all the way they had to March, indeed they came more like Sheep to the Slaughter than men to fight.

Continuing, Taylor noted that "several of the Enemy declared afterwards that if they had met with a warm reception in the Wood, they would have sur-

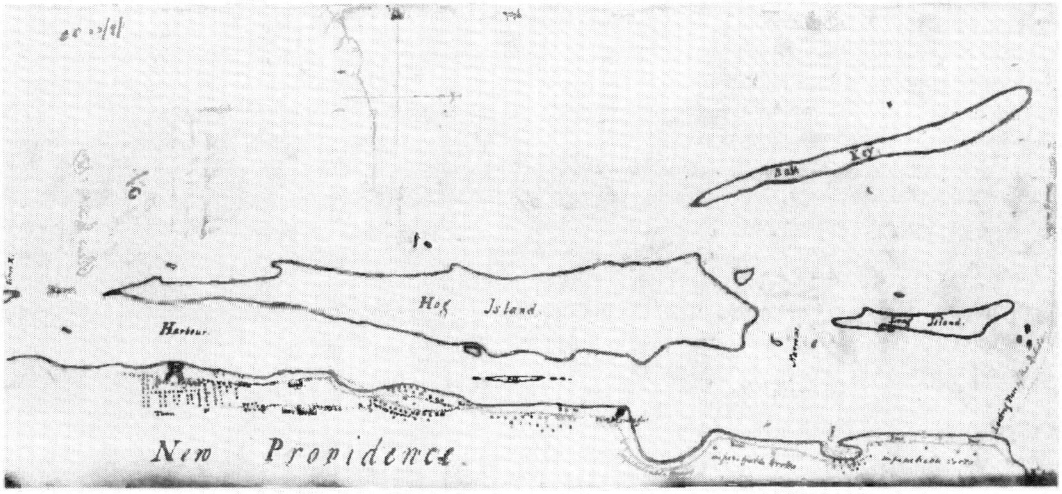

Landing Place of the Marines.

rendered."[56] Whether these observations were correct or not, he and others were at a loss to understand why Browne chose not to either organize Captain-Lieutenant Walker's or Lieutenant Burke's detachments into an ambush, or build defensive earthworks along the coastal trail.[57]

The suggestions that an ambush or breastworks be established along the trail were simple solutions to a more complex problem. Faced with a *fait accompli*, there was little the governor could do at Fort Montagu, since the possibility existed that the Americans would by-pass the fort and assault the town. Browne had lost the initiative when he failed to organize a strong and immediate response; neither an ambush nor the erection of defensive works at this late hour could regain it. Therefore, the only course of action open was to retreat.

Whether the governor's decision to retire to Fort Nassau was wise or not, the immediate results were disastrous. Each man in the militia must have realized that with the evacuation of the eastern garrison and the forces in retreat, the town was open to the ravages of the enemy. Therefore, his first thoughts were for the security of his family and possessions. Yet it was remarkable that after making sure their families were safe, over half of the militia reassembled at Fort Nassau.[58]

By three o'clock the governor had recovered enough to return to Fort Nassau and resume command. The militia was then put to work building defensive positions and assembling provisions for the expected battle. Still short of the required men needed to garrison the fort, Browne ordered that the drums again be sounded and a pistol be offered to any able-bodied Black who would join him in the fort. The plea had little effect. Although his forces were depleted by desertions, and few were willing to volunteer, the governor's most pressing need was for information as to the disposition and plans of the American Marines.[59]

Fort Montagu Taken

At the firing of the three cannon from Fort Montagu, Captain Nicholas cautiously halted the Marine column, and thinking the fort still occupied, consulted with his officers as to "what was best to be done."[60] It was decided that an emissary should be sent to acquaint the Bahamians with the American intentions. Marine Lieutenant John Trevett was then selected for the post. As he moved up the trail under a flag of truce. Trevett met Lieutenant Burke coming from the opposite direction.[61] Burke informed him that he had explicit instructions to "wait on the Commanding Officer of the Enemy to know his Errand and on what Account he had landed his Troops."[62] Captain Nicholas soon met with Lieutenant Burke and was told "that it was by the Governors order" that the three cannon were fired and that they had "spiked up the cannon, . . . abandoned the fort, and retired to the fort within the town." Nicholas informed Lieutenant Burke that he and his men were only after the military stores on the island and that he would be seeing the governor in the morning. With that the weary Americans marched on and took possession of the deserted fort. It was then decided that instead of continuing the attack, the force would spend the night at Fort Montagu. As Captain Nicholas later reported: "I thought it necessary to stay all night, and refresh my men, who were fatigued, being on board the small vessels, not having a convenience either to sleep or cook in."[63]

With the threat of an immediate assault somewhat relieved, Governor Browne leisurely contemplated his next move. Remembering that the high ground to the south completely dominated the fort, he dispatched Captain Hodgon and Ensign Barrett with 40 men to Govenment House to secure the vantage point.[64] His precautions were sound for one of Captain Nicholas' first objectives the following morning was to take possession of the strategic hill "with [its] two four-pounders, which commands the garrison and the town."[65] After the men were positioned around the governor's residence, he issued a call for all the militia officers and principal men of the community to assemble in Fort Nassau at eight o'clock that evening.[66]

Once the men were settled in for the night, Nicholas sent a courier to Commodore Hopkins on board the *Alfred* with information as to the day's events. Hopkins, upon learning of the rather peaceful developments, seems to have sensed that an opportunity for a bloodless conquest was at hand and he decided to issue a manifesto addressed to the residents of New Providence stating his intentions.[67] In the manifesto, Hopkins offered to spare the town and its peoples' personal property in exchange for nonresistance and noninterference.

> To the Gentlemen Freeman and Inhabitants of the Island of New Providence.
> The Reasons of my landing an armed force on the Island is in Order to take Possession of the Powder and Warlike Stores belonging to the Crown and if I am not Opposed in putting my design in Execution the Persons and Property of the Inhabitants Shall be Safe, Neither shall they be Suffered to be hurt in Case they make no Resistance.
> Given under my hand on board the Ship *Alfred* March 3rd 1776.[68]

The manifesto seemed to have the desired effect; for, as Lieutenant Governor John Brown later related, "this paper was handed about amongst the People to the Eastward of the town, . . . and induced several of the Inhabitants to refuse coming to defend the Fort & others to Join the Rebels."[69]

The reluctance of many of the inhabitants to fight the Americans was reflected by the militia officers and the more prominent residents as they met in the council of war called that evening by Governor Browne. The first question put to them by Browne was "whether from the Force the Rebels were supposed to be, and the Condition Fort Nassau was then found to be in, with what Strength was then in the Fort It was defensible or not?"[70] They replied, 14 to 10, that due to the poor condition of the cannon carriages and the lack of essential equipment, it was not. The six member provincial council then met in closed session to consider the governor's next question: What was to be done with the almost 200 barrels of powder located in the magazine at Fort Nassau?[71] Although Browne did not want the Americans to get hold of a single barrel, he feared that "sending away the whole of it might enrage a disappointed enemy, and induce them to burn the Town, and commit other depredations."[72] Concurring with the governor, the council ordered that the bulk of the powder be immediately shipped to safety.

Once again the council requested the services of Captain William Chambers. Responding like "a man truely attach'd to Government,"[73] Chambers unhesitatingly cast overboard the lumber in the *Mississippi Packet's* hold consigned for Jamaica. Around midnight, he began supervising the loading of 162 barrels of gunpowder brought by militiamen from Fort Nassau.[74] One hundred and nineteen barrels were placed on board the *Packet*, and another 43 were loaded on the *St. John*.[75] By two on Monday morning, the loading was completed and the two vessels sailed for St. Augustine with orders to deliver the powder to Patrick Tonyn, Governor of East Florida.[76]

The escape of the *Mississippi Packet* and the *St. John* with the bulk of the gunpowder can be attributed directly to Commodore Hopkins' failure to close the main entrance to Nassau harbor. For some unknown reason Hopkins held the main elements of the fleet (*Alfred*, *Columbus*, *Cabot*, and *Andrew Doria*) at a safe anchorage in Hanover Sound during the night while stationing the *Providence*, *Wasp*, and the two captured sloops off Fort Montagu at the eastern harbor entrance. John Paul Jones later wrote, with the benefit of hindsight, that the removal of the powder "was foreseen, and might have been prevented, by sending the two brigantines [*Cabot* and *Andrew Doria*] to be off the Bar."[77] There was nothing which prohibited this from being accomplished. Therefore, with the loss of the powder went the pri-

mary objective of the expedition, and Commodore Hopkins must take full responsibility for that loss.

Even with the two ships safely on their way to St. Augustine, Governor Browne could get little rest that night. The departure of the powder and the resultant defenseless state of Fort Nassau, "almost instantaneously produced a Desertion of three fourths of the men and Negroes." Still hoping to make some sort of stand, Browne asked the council whether or not it would be advisable to recall the detachment he had earlier sent to secure Government House.[78] When the council voted for their return, Browne and William Bradford, a council member, went up to Government House and brought back the 40-man detachment. After learning of the events which had transpired that evening, they, too, deserted the fort, saying that they could not defend it alone. The council then advised the governor that the fort should be immediately evacuated. Totally rejecting their opinion, Browne said that he would never willingly leave the fort "as long as a man would stand by him."[79] His dissent was to no avail. The situation worsened and by "day break the Governor, Council & Officers finding themselves almost entirely deserted, were at last obliged to leave the Fort."[80]

Fort Nassau and the Town Secured

Morning found Fort Nassau deserted, the governor in his residence, and members of the provincial council either at home or "hiding themselves amongst the Rocks."[81] Captain Nicholas had awakened his men before daybreak and by dawn they had covered the mile from Fort Montagu to the eastern edge of the town. There they met yet another emissary from the governor who desired to know their intentions. Nicholas again gave the same answer.[82] The messenger then told him that "the Western Garrison (F. Nassau) was ready for his reception, and that he might march his Force in as Soon as he Pleased."[83] After an hour's wait, Captain Nicholas moved his men into the town. As the Marines and sailors marched down Nassau's main street, Nicholas selected a guard and went up to Government House where he demanded and received the keys to Fort Nassau. Soon after he rejoined his men the Americans took possession of the fort without "firing a single shot." There they found 40 unattended "cannon mounted and well loaded . . . with round, langridge, and canister shot" for their reception.[84] The British colors were then hauled down and the Grand Flag of the United Colonies run up in their place.[85]

Soon after the fort and town were secured, a report was sent to Commodore Hopkins informing him that it was now safe to bring the fleet into the harbor. Only the *Alfred* and an escort entered port; the remainder of the fleet stayed behind at Hanover Sound until early Wednesday morning, 6 March, when they too entered and safely anchored off Fort Nassau.[86] At that time, the schooner *Wasp* was posted outside the harbor entrance "to serve as a lookout" until Saturday, when she too passed through the channel and anchored with the rest of the fleet.[87]

As the landing force awaited the arrival of the commodore, several Marines in Fort Nassau noticed the "Governor . . . and his council walking in his Piazza, and his servants waiting below with horses." Assuming that the governor was about to escape to the interior of the island, Captain Henry Dayton and Lieutenant John Trevett asked Captain Nicholas if they would "go and take him." Captain Nicholas pointed out that he as yet had no instructions from Hopkins concerning the governor, but they might do as they pleased. Trevett, Dayton, and "a young officer from Philadelphia" then went to Government House to bring Browne to the fort. In Trevett's words:

> [We] informed him he must go with us to the fort. He made reply it was beneath his dignity as Governor to go to the fort. We made him this reply he must go, then he says it must be by the force of arms. Then we told him it was by force of arms and he walked down with us to the fort.[88]

Why the governor did not chose to make an escape before the Americans took possession of Fort Nassau is unknown. His presence was unnecessary after the decision was made to abandon the fort. He had, as a later critic pointed out, "near five hours viz. from four oClock the time of our coming out of the

Fort with him 'till nine, as the Country was open and his Family out of Town." Thus he could have avoided being taken prisoner by the American Marines.[89]

Sometime after coming ashore, Commodore Hopkins met with Governor Browne. Presumably it was not a cordial meeting, for Hopkins had been informed that Browne had shipped almost all of the powder to St. Augustine the night before. After the meeting Hopkins ordered the governor confined for a short period in Fort Nassau, "in a place without food, water, bed, table, or chair," and then removed to Government House, for, as the governor later complained, "the better convenience of their officers and men." There he was held prisoner for four days while his 32-man Marine guard "used at descretion all my wines and other Liquors as they did everything else they had occasion for." On the following Sunday, 10 March, Hopkins sent a second Marine detachment to Government House with orders to bring the governor to the *Alfred*. After informing Browne of their orders, they told him he had a quarter of an hour to prepare himself. Browne replied that he would have nothing to do with them since they had violated the "proclamation issued by their admiral upon his landing," and therefore they "had forfeited every degree of Honor, faith, and Confidence." Whereupon he was, to use his own words, "seized, Collard, and Draged away like a fellon to the Gallows in the presence of a Dear wife and a aged Aunt . . . who were treated with such abuse, and such language," and put on board the *Alfred*. Later when he confronted the commodore and demanded to know for what crime he had been confined, Hopkins replied: 'for presuming to fire upon . . . [my] troops from Fort Montague.' The real reason, as Browne later thought, was surely for "daring to send away the powder."[90]

Along with Governor Browne two more men were imprisoned on board the *Alfred*. The first was Browne's personal secretary, a half-pay officer in the British Army, Lieutenant James Babbidge. The second man was Thomas Irving, counselor and collector of the King's quit rents in South Carolina and formerly the Inspector General of the King's customs for North America.[91] Both men accompanied the governor on the return trip, and all were later paroled.

Commodore Hopkins had more important matters to worry about during the two weeks following the initial landing than the loud protests of Governor Browne. Prior to Hopkins' arrival an inventory of the military stores in both forts had been prepared and was handed to him when he stepped ashore. Soon thereafter, Marines and sailors were tasked with loading the captured cannon and ball on board ships of the fleet.[92] The *Andrew Doria*, for example, received 4,780 shot and shells to replace her stone ballast which was put on shore.[93] Even though similar quantities were placed on board the larger ships of the fleet, there was still insufficient space for all the spoils. The arrival of the *Fly* on 11 March did not increase the fleet's carrying capacity, and therefore Hopkins contracted with Captain Charles Walker, master of the *Endeavor*, to carry the remainder of the captured material.[94]

In addition to supervising the loading of the military stores, Hopkins was concerned with the administration of the captured town and the activities of his own men. Although his relations with the townspeople were friendly, there were problems. The proclamation issued by the commodore which forbade "injury to private property," went against "the inclination of some of the leading Officers" to allow some pillaging. But Hopkins was a man of "humanity & influence"; a man not easily pressured into such action once he had given his word.[95] One of his first acts was to "Cause a Triangle to be set up upon the Parade, and if the Inhabitants made the least Complaints against any of his people, they were punished Immediately." His courts martial were "short, and decisive, the Number of Lashes few, but very severly given."[96] Although the governor later complained that the townspeople were overly hospitable, there is little reason to believe that the citizenry welcomed their conquerors with open arms, considering the number of complainants who besieged the commodore.

The health and well being of his own men, however, proved to be the commodore's greatest worry. The Marines and sailors had come ashore without provisions, and had no opportunity to procure any from the fleet during the first several days on the island. Therefore, an arrangement was made with a local merchant, Nathaniel Harrison, for food and other supplies to be sent to Fort Nassau during their stay.[97] But the lack of adequate provisions was not the only worry. With the warm weather, outbreaks of small-pox and fever became more widespread, so much so that by the time the fleet returned to

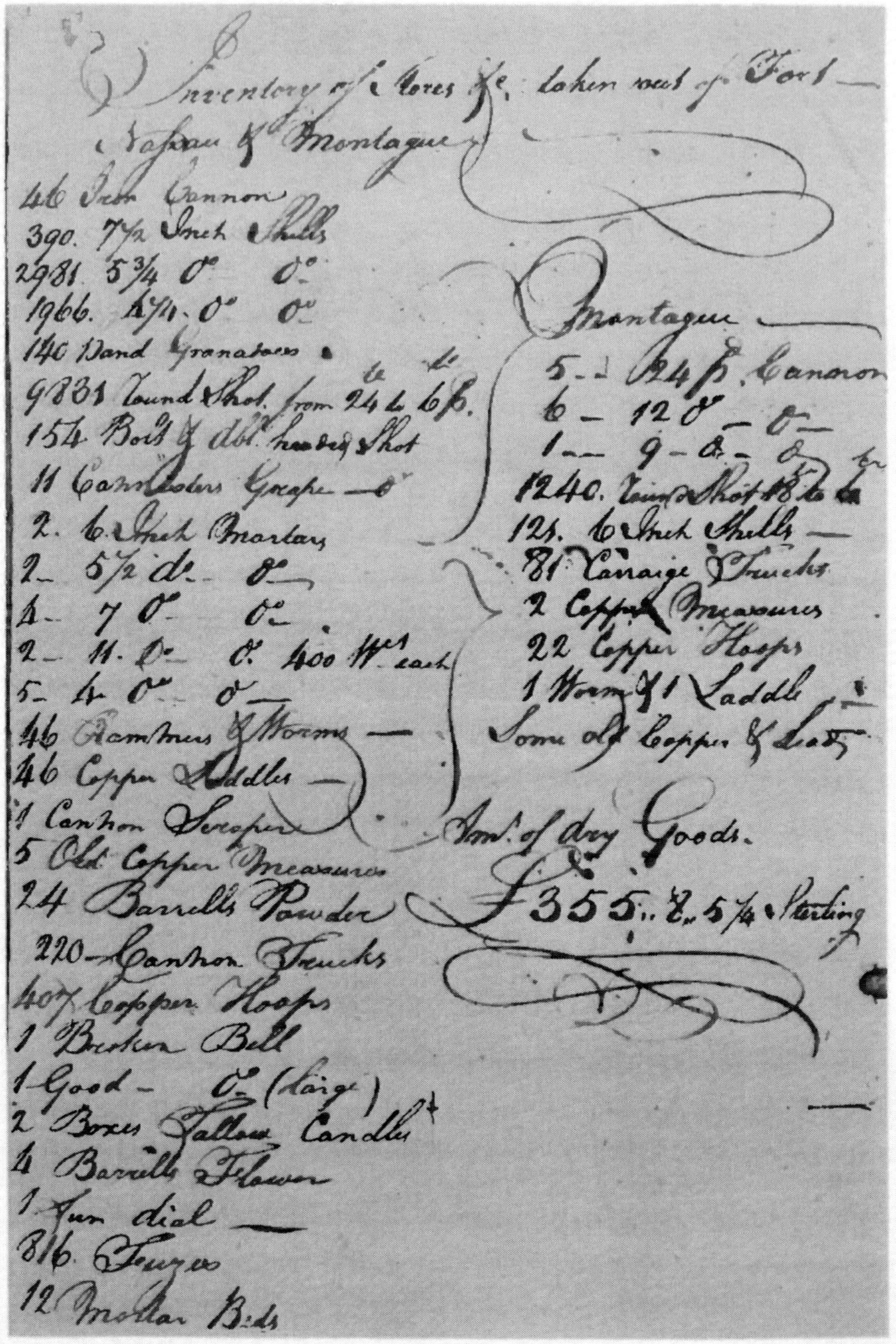

Inventory of Stores &c. taken out of Fort — Nassau & Montague

46 Iron Cannon
390. 7½ Inch Shells
2981. 5¾ D° D°
1966. 4½ D° D°
140 Hand Granadoes
9831 Round Shot from 24 to 6 lb.
154 Bar & dbl. headed Shot
11 Cannisters Grape — D°
2. 6 Inch Mortars
2 — 5½ D° — D°
1 — 7 D° D°
2 — 11 D° D° 400 Wt. each
5 — 4 D° D°
46 Rammers & Worms —
46 Copper Laddles —
1 Cannon Scraper
5 Old Copper Measures
24 Barrells Powder
220 Cannon Trucks
40 Copper Hoops
1 Broken Bell
1 Good — D° (Large)
2 Boxes Tallow Candles
4 Barrells Flower
1 Sun dial —
816 Fuzees
12 Mortar Beds

Montague —
5 — 24 lb. Cannon
6 — 12 D° — D°
1 — 9 — D° — D°
1240. Round Shot 18 to 6
121. 6 Inch Shells —
81 Carriage Trucks
2 Copper Measures
22 Copper Hoops
1 Worm & 1 Laddle
Some old Copper & Lead

Amt. of dry Goods.
£355.. 8.. 5¼ Sterling

Stores Captured at New Providence.

the continent in early April, several men had died and about 140 men had to be placed in hospitals.[98]

Meanwhile, Lieutenant William Grant in the *St. John* had arrived in St. Augustine and on 7 March gave Governor Tonyn a detailed account of the American landing on New Providence.[99] The following morning the governor dispatched a letter to Captain Andrew Barkley, commander of the Georgia squadron then in the Savannah River. In the dispatch, Governor Tonyn suggested that immediate steps be taken to intercept and destroy the American fleet.[100] When the governor's letter was received, a council of war was held on 14 March on board HMS *Scarborough*, the squadron's flagship. The council recommended that it would be inadvisable to leave the transports then loading provisions for British troops at Boston unguarded, and that "in all probability . . . [the Americans] have quitted Providence by this time." The council's recommendation and the governor's letter were then sent by packet northward, and nothing was done to intercept the American fleet.[101]

Home Without Powder

By Saturday, 16 March, the stores were finally loaded and secured below decks. The Marines and seamen who took part in the landing were then embarked and the order given to prepare to sail. Although Commodore Hopkins would not let Governor Browne leave the *Alfred* in order to bid farewell to the family he had been "tore away from . . . at a moment's notice," he apparently allowed someone on board the following day, for Browne, even though he was watched by "innumerable sentrys," was able to dispatch a letter to the Earl of Dartmouth noting the sacrifice he had made for the Crown.[102] When all was ready the commodore ordered that the signal to weigh anchor be given. As the billowing canvas caught the Sunday afternoon breezes, the American fleet made for the open sea.[103]

The American raid on the island of New Providence was a success on the whole, if not a brilliant one. The assault expanded the war for independence beyond the boundries of the thirteen colonies. No longer would it be viewed as just a localized rebellion by a few radical dissidents; the Americans had now internationalized the war. The Continental squadron left one British Caribbean outpost in a precarious state and challenged British supremacy on the sea. Now other British enclaves ranging from the West Indies to Canada found themselves open to a possible Marine amphibious assault.

A MARINE LIEUTENANT DIES, 6 APRIL 1776.
A month after the capture of New Providence, the first large sea battle of the American Revolution took place between ships of the Continental fleet and HMS Glasgow, off the coast of Rhode Island. This painting portrays the action, and the resultant death of Marine Lieutenant John Fitzpatrick on board the Alfred.

CHAPTER IV

Uncombined Operations, April-December 1776

The *Lexington* and *Reprisal*

A gentleman named Bernard, from Elizabethtown in New Jersey, came ashore in New York on 11 March 1776 after being released from His Majesty's ship *Phoenix*. Apparently disturbed over the capture by the British of his valuable brig, *Diligence*, several days before, he immediately informed the New York Provincial Council that Captain Hyde Parker of the *Phoenix* "was about fitting out his schooner Tender with carriage guns and other necessaries, to make her a vessel of force, to be stationed and cruise of Egg Harbour." Realizing the importance of the intelligence since the "Schooner may be very detrimental to the trade of this Colony and the neighboring Colonies, and at this critical juncture in particular, as ammunition, arms and other necessary stores, may thereby be seized and prevented from landing," the provincial council ordered its marine committee to consider the information and immediately advise the whole of what action should be taken.[1]

The following day, the committee informed the council of its suggested remedy. Basing its recommendation "on information received from Mr. Barnard . . . and from other circumstances known to them," the committee suggested that a letter be sent immediately to the Continental Congress requesting "that they order an armed vessel fitted out, to join one from this port for the protection of trade." Whether by accident or design, the committee extended the cruising ground of the *Phoenix's* tender 50 miles north and south to include "the Capes of the Delaware and Sandy Hook." This action was bound to have an effect on the delegates at Philadelphia, for the fleet raised late the previous year had put to sea and the lightest of British ships now might prey with impunity upon the city's trade. The letter, approved by the provincial council, was soon on its way by post-rider to John Hancock, President of the Continental Congress.[2]

Covering the 90 miles in less than a day, the post-rider arrived in Philadelphia around midday on the 13th. Considering the letter of primary importance to the Marine Committee of which he was a member, Hancock placed it before it for immediate consideration. After a short discussion, Robert Morris, Pennsylvania's member of the committee and one of Philadelphia's leading merchant-shippers, brought the information before Congress and introduced the following resolution which was adopted without debate: "*Resolved*, That the marine committee be directed to purchase the armed vessel now in the river Delaware, on the most reasonable terms, for the service of the continent, and that her destination be left to the said marine committee."[3] The resolve was marked secret, for the ship was to be sent out after the *Phoenix's* tender, and there were many spies.[4]

The vessel the Marine Committee was empowered to purchase was the Maryland armed brig *Wild Duck*, which had just arrived from St. Eustatius with a load of powder. As Captain James Tibbett unloaded his cargo, members of the Marine Committee went down to look the vessel over. They found the brig to be a little larger (length 86 feet; beam, 24 feet, 6 inches) than the brig *Cabot* purchased the previous year. Her sides were black with yellow

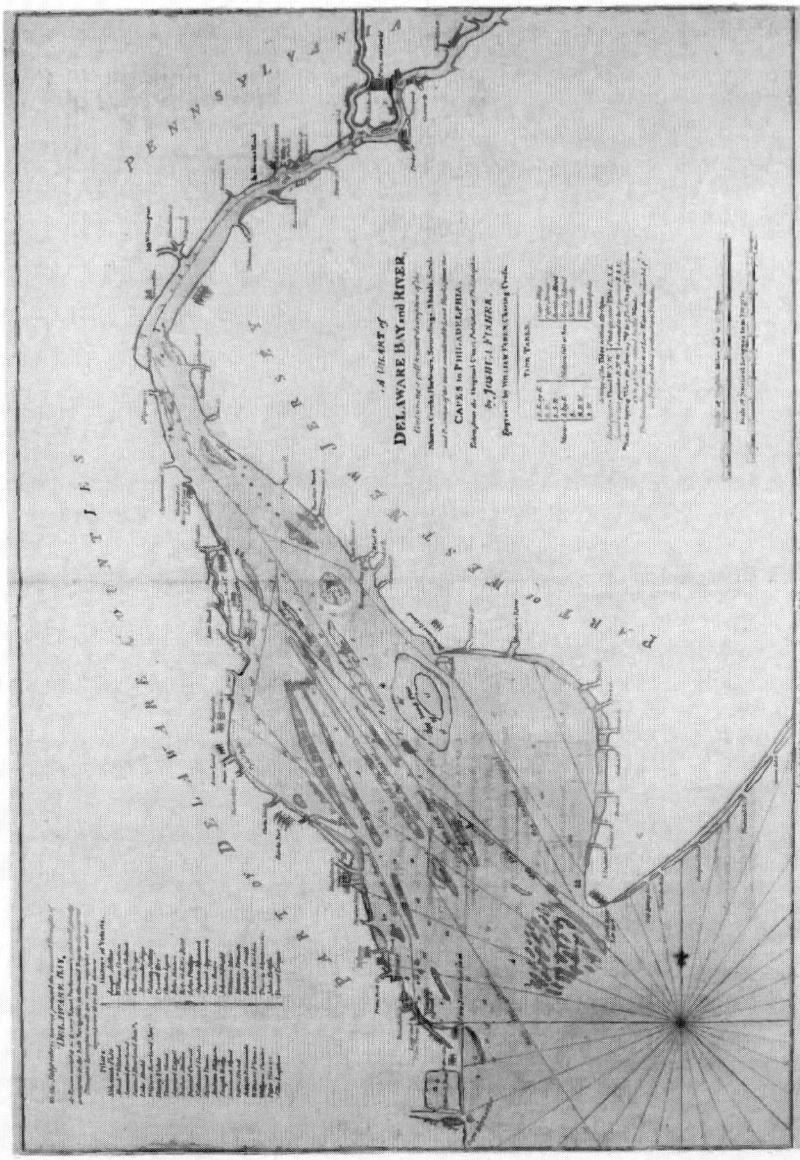
A CHART of
DELAWARE BAY and RIVER
CAPES to PHILADELPHIA
by JOSHUA FISHER

moldings, her low rounded stern was lead-colored, and she was armed with 16 guns. After arrangements were made with the Maryland delegates to Congress for the purchase, the *Wild Duck* was renamed the *Lexington*. Her reconditioning was assigned to Joshua Humphrey, a Philadelphia shipbuilder who previously had converted four merchantmen into vessels of war for the fleet commanded by Esek Hopkins. In nine days the work of strengthening her bulwarks, caulking, and repainting her sides was completed.[5]

The day after the *Lexington* was acquired her command was given to John Barry, a 31-year-old, Irish-born shipmaster. The balance of the commissioned officers were appointed by the Marine Committee several days later: Luke Matthewman, first lieutenant; Robert Scott, second lieutenant; and Abraham Boyce, captain of Marines. There was also a lieutenant of Marines appointed, but his identity remains unknown.[6]

As the *Lexington* awaited the delivery of arms and equipment from the Pennsylvania Committee of Safety on 26 March, an express from Lewes, Delaware, arrived in Philadelphia with disturbing news. Colonel Henry Fisher, the committee's watchdog at Cape Henlopen, indicated in a letter that a large sloop-of-war with a small tender had passed the Capes and entered Delaware Bay.[7] In response, four row-galleys of the Pennsylvania Navy were quickly dispatched the following day as far as Reedy Island with orders to cooperate with the *Lexington* in the destruction of all enemy vessels found in the river.[8] The *Lexington*, however, was not ready. Captain Barry had neither enough arms nor men. His Marine captain was still on shore attempting to round out the ship's Marine detachment, and Barry himself was pleading for more small arms to equip them.

As the four row-galleys sailed on the morning of the 27th, a second express from Fisher brought disastrous news. The vessel he had termed a large sloop-of-war two days before, now turned out to be a man-of-war of not less than 40 guns.[9] Later she was identified as the 44-gun British frigate *Roebuck*, which had been sent to open the Delaware and destroy all obstacles to navigation placed there by the "Inhabitants of Pensylvania in open Rebellion against the King."[10]

The new intelligence pressed the Pennsylvania Committee of Safety into action. That night the committee had an assortment of miscellaneous arms and equipment delivered to Barry on board the *Lexington*.[11] Although short-handed as to Marines, Barry decided that he had to sail in order to meet the enemy's newest threat. Therefore in the early morning hours of 28 March, the *Lexington* proceeded down the Delaware River leaving behind Captain Boyce and the Marines he had recruited.

Recessed on 27 March, Congress met the following day and was informed by the Marine Committee of the latest intelligence from the Capes.[12] Again Robert Morris rose and moved that a second vessel be purchased and fitted out for the protection of the Delaware. But unlike the resolve of 13 March, the resolve passed on the 28th specified that the mission of this new vessel was to join the *Lexington* and cruise in coastal waters between New York and Virginia. With little trouble the committee decided on the purchase of the brig *Molly* which was later renamed the *Reprisal*.[13] The *Reprisal* was a much larger vessel than the *Lexington* in terms of size, armament, and the canvas she carried. Her sides were painted black except for the white moldings on her quarters and a mingling of yellow and black on her stern. In contrast to similar vessels, the *Reprisal* had no quarter galleries; only her nine gunports on either side broke the line between wind and water.[14] To command the newly purchased brig, the Marine Committee settled upon a Maryland shipmaster, Lambert Wickes.

As Captain Wickes and his three lieutenants raised their complement of seamen, his newly appointed Marine officers, Captain Miles Pennington and Lieutenant John Elliott set about the same task. But delays in the recruitment of both seamen and Marines soon arose, for the *Wasp* and *Hornet* had returned to port in early April and the manning and refitting of these two vessels took precedence over the *Reprisal*. Thus both Captains Wickes and Pennington were forced to leave Philadelphia and seek recruits up and down the Delaware. Indications are that Captain Pennington recruited as far south as Wilmington where he signed up the first black Marine, John Martin or "Keto." The Marine captain enrolled at least 2 sergeants and 24 privates for service on board the brig *Reprisal*.[15] *

* John Martin or Keto was the "property" of William Marshall of Wilmington, Delaware, and was enlisted in April 1776. Martin remained with the Reprisal and died when the ship foundered off the Newfoundland Banks in October 1777.

As men were recruited and the *Reprisal* made ready, news of the *Lexington's* victory over the British sloop *Edward* arrived. Soon after leaving Philadelphia, Barry took the *Lexington* down the Delaware and past the row-galleys at Reedy Island. Not wishing to engage in an unequal contest with the *Roebuck*, Barry slipped silently past Cape May and headed out to sea. Proceeding up the Jersey shore he put into Egg Harbor on 3 April. There he learned that the *Phoenix's* tender was on her way northward, probably having already returned to New York. Since the tender was out of his reach, Barry then decided to turn southward and try his luck along the eastern shore of Maryland and Virginia.

With a stop at the Capes to take a small sloop and to convoy a group of merchantmen out of the Delaware Bay, the *Lexington* was abreast of Cape Charles, Virginia by 7 April. There in the early morning hours a strange sail was sighted to the southwest. Immediately Barry called the crew to quarters and ordered the guns made ready. As the two ships closed, the opposing vessel was identified as being the tender assigned to HMS *Liverpool*. With no more than several hundred yards separating the *Lexington* and the *Edward*, Barry opened fire and for an hour the two vessels exchanged broadsides. The end came when the guns of the wrecked sloop fell silent and the *Edward's* captain allowed his ship to be boarded by a prize crew. Following the battle, Barry penned a terse report to the Marine Committee in which he noted that his raw seamen and Marines "behaved with much courage."[16]

After the *Edward* was secured, Barry headed northward. On 8 April as both ships neared the Delaware Capes, the *Edward* veered off and made for Philadelphia where she was later condemned as a legal prize.[17] The *Lexington* continued up the coast and the following day put into Egg Harbor for repairs. While there, orders arrived from the Marine Committee which directed the *Lexington* to Cape May where she was to meet and convoy the sloop *Betsy* carrying Silas Deane, Congress' first emissary to France. Also, Barry was instructed to pick up Captain Abraham Boyce and his Marine recruits who would be landed at the Cape by the *Wasp*.[18]

Repairs to the *Lexington*, although slow, were made and she soon sailed for Cape May. Anchoring there on the 15th, a boat was sent ashore to bring back Captain Boyce and the few Marines he had been able to sign-up on his far-flung recruiting mission. As soon as the boat returned, Boyce reported to Barry that the *Betsy* had cleared the Cape two days before, and that the *Wasp* had returned to Philadelphia. Since Barry's recent orders from the Marine Committee made it mandatory that he safely convoy the *Betsy*, Barry decided to make fast sail in order to overtake her.

Although the *Lexington* was rigged for full sail and had a brisk wind across her stern, Barry was unable to catch the sloop. Not far from Bermuda he gave over the chase and again turned northward. As dawn broke on 26 April, a fleet of 17 ships was sighted to the southeast, and it was not long before Barry learned that he had fallen in with a portion of the British fleet destined for Charleston, South Carolina.[19] Immediately, the British admiral dispatched the frigate *Solebay*, 28, to the northwest after the American sail,[20] but the *Lexington's* luck held out. The frigate soon ceased the pursuit and *Lexington* was able to make it to the Capes and up the Delaware to Philadelphia without a major incident.

The Defense of the Delaware

When Barry returned to Philadelphia with the *Lexington* on 5 Bay, he found a nervous city. Although his appearance sparked a moment of excitement, this was soon dispelled by the arrival of another dispatch from Colonel Fisher. Two British frigates had started up the river, reported Fisher.[21] With the *Lexington* under repair, the Pennsylvania Committee of Safety was called into immediate session the following day to plan for the city's defense. After hasty discussions orders went out to Captain Thomas Read, commander of the Pennsylvania fleet at Fort Island, "to call every Boat & Soldier to their stations, & each to prepare for immediate action." Several fire ships and two chains of fire rafts were also ordered to stations near the fort.[22]

Hours of uncertainty passed as the ships at Fort

Island awaited news. Close to midnight a second express arrived—this time from Port Penn, some 20 miles south of Wilmington. "Two Ships of War, a Top Sail Schooner & three small Vessels, supposed to be tenders," were sighted, the report noted. Early the following morning, the committee again met and "gave orders to Capt. Reed to order down the thirteen armed Boats under the command of the Senior Officer, with directions to proceed down the River, when they must concert the best method to attack, take, sink, destroy, or drive off the said Vessels if possible," without "exposing any of the Boats to Capture or destruction." A shallop was then ordered loaded with shot and ammunition from the city's stores for the fleet's support. From the Marine Committee, an order was secured directing Captain Wickes of the *Reprisal* to "supply the Boats with as many Voluntiers as may offer."[23]

At 1230 on 7 May, Thomas Read received his orders from the Committee of Safety and immediately called a conference of his captains. The orders were read aloud, and then Captain Read asked for comments. In response, the galley captains voiced a concern about their depleted crews. Through minor adjustments, the galleys would be brought up to strength. The meeting then concluded with an agreement that the galleys would rendezvous that night below Hog Island, just south of the fort, and then proceed down the river the next morning.[24]

Soon after the council of war adjourned, Captain Lambert Wickes joined Read on board the Pennsylvania state ship *Montgomery*. The provincial commander began by reading the part of his orders which applied to the Continental captain, but Wickes interrupted saying that he had "received no orders and could not let any of his Men go without it, tho' [he] wish'd at the same time to do it, and would Immediately on Receipt of the orders." Later in the day Captain Read delivered to Wickes a copy of the orders, which the *Reprisal's* captain "comply'd with." The following morning Wickes dispatched a party of men to the boats: to the two pilot boats went First Lieutenant Robert Harris and 10 seamen; and to the two fire boats went Captain Miles Pennington, two sergeants, and 24 Marines, "equip'd with Small arms, all in great Spirits."[25]

Captain Pennington seemed to have sensed that he would soon be sent into battle, for only a week before he had made out his last will and testament. Witnessed by his lieutenant, John Elliott, he bequeathed to his wife all "Wages and Prize money which shall be due me from said Ship [*Reprisal*] or any other Vessel." The making out of a will prior to a cruise was widespread at the time, and ensured that a man's family would receive the monies they were entitled to.[26] In Pennington's case, however, the will ensured nothing, for his wife Catherine was still petitioning for the settlement of his prize money as late as 1785, eight years after her husband's death.[27]

The 13 row-galleys of the Pennsylvania Navy rendezvoused at Hog Island in the early morning hours of 8 May, and then proceeded to make their way through the chevaux-de-frise and down the Delaware. The *Reprisal*, however, was unable to pass the obstruction and thus remained above the barrier with the *Montgomery*, forming a second line of defense. As both ships settled in for the remainder of the night, an order from the Committee of Safety arrived directing both ships to get through the chevaux-de-frise and follow the galleys. Poor weather and adverse tides forbade the attempt. At daylight they would try again.[28]

In Philadelphia that morning, the Committee of Safety worked feverishly on the city's defense. Ammunition was supplied to the city's militia, who were ordered to assemble later in the day on the common.[29] At noon, word arrived from Wilmington which placed the *Roebuck* and *Liverpool* at the mouth of Christiana Creek.[30] Undaunted by the latest intelligence, the committee continued its preparations. Meanwhile, the Marine Committee had the *Hornet* made ready and then ordered her to Fort Island. As the *Lexington* was still under repair, the Committee decided to make use of her crew. Thus an order went out to Captain John Barry to:

> Collect your officers & Men and repair down to the Provincial Armed Ship Capt Reed, and Supply him with as many People as he may want to completely Man that Ship for Immediate action, you will also spare any of them that may be wanted on board the Floating Battery, or on board the Ship *Reprisal* and in short We expect the utmost exertions from you, your officers & Men in defending the Pass at Fort Island, and to prevent their coming up to this City.

Barry and his crew could, the Committee concluded, "go down in the sloop *Hornet* Capt Hallock."[31]

Near two o'clock in the afternoon, suspense was heightened by the rumble of distant cannon. Immediately drums called men to arms and a number of boats were dispatched down river "to assist as the report was that the *RoeBuck* of 40 & the *Liverpole* of 28 guns with tenders were got above New Castle standing up river."[32]

The British expedition up the Delaware River was not an attempt to bombard or capture the city of Philadelphia, but to clear the river of obstructions and open it to navigation. Soon after the two frigates entered the Delaware on 5 May, the schooner *Wasp* was sighted and given chase. By later afternoon on the 7th, the frigates were off the mouth of Christiana Creek, having forced the *Wasp* up the creek in order to seek refuge. There they anchored for the night, preventing the *Wasp's* escape and allowing crews to refill the water casks and make minor repairs.[33]

The fog-shrouded forenoon of 8 May found the two British frigates still at anchor off the creek. As the fog cleared around one o'clock, a fleet of 15 vessels was sighted coming down the river. Composed of the 13 row-galleys, each carrying a single heavy gun (either an 18, 24, or 32-pounder), a floating battery armed with ten 18-pounders, and a sloop "fitted as a fire ship," the 15 ships appeared neither formidable nor impressive. However sluggish and unwieldy the fleet might have appeared, the two frigates immediately cleared for action. With no more than a mile separating the two forces, the galleys opened up but failed to inflict any damage on the British ships. Under sail the frigates moved closer, brought their broadsides to bear, and began a thunderous bombardment. For two hours the combatants exchanged shot until the *Roebuck* went aground and the galleys retired, failing to pursue the advantage.[34]

As the battle raged in the river, spectators and militiamen lined the Delaware shore, filling every house. Colonel Samuel Miles, who with his battalion of militia watched the contest from shore, later remarked that, after two hours of engaging at long range, "a great deal of amunition has been wasted." He, however, noted that "our Boats fire much better than the other Vessels, but in my opinion ingage at two great a distance." And although he believed that "three or four hundred shot have past between them," there was no damage done to either the galleys or the frigates.[35]

After the battle was broken off, the fleet retired to Chester where it remained until the following morning.[36] On board the *Roebuck*, the night was spent guarding the ship against a surprise attack. Although numerous attempts were made to refloat the frigate, it was not until four o'clock the next morning that success was achieved. Back in the channel, Captain Andrew Hamond anchored to await the dawn.[37]

Above the chevaux-de-frise, the *Montgomery* and *Reprisal* awaited news from the galleys. It had been several hours since the firing had ceased and Captain Read wondered about the day's events. At two-thirty the following morning, an express arrived bringing a demand from the galleys for more powder and shot. Shortly thereafter, Read dispatched a boat southward with all the powder and shot that could be spared from the magazines of the *Montgomery* and the floating battery *Arnold*. But lacking definite information, he was reluctant to abandon the second line of defense at Fort Island, although he promised to "take the first opportunity to get throo" the barrier. Soon after the powder supplies were sent down the river, the *Hornet* arrived with Captain Barry, his seamen, and Marines.[38] Following a look at the situation, Barry wrote a hasty note to Robert Morris urging that the *Lexington* be quickly fitted out, "for the More there is the better."[39]

By noon on 9 May, the *Reprisal* had made it through the chevaux-de-frise, but the *Montgomery* was not as fortunate. Her pilot ran her ashore, and in warping off she jammed into the *Reprisal*, carrying away Wickes' jib boom. Finally straightened out—"Just below the Chevaux De frize"–Captain Read protested to the Committee against proceeding down the river: "I do think and I believe that our Best station is above, to guard the pass and to get there as soon as the tide will admit, abreast of the Battery." Although his officers agreed in the opinion, Read consulted Captain Wickes.[40] There is no record of the conference, but it is evident that the Continental captain opposed the plan to return through the barrier. Therefore, both vessels remained anchored at a point just below the obstruction.

Down river meanwhile, the galleys left their anchorage at Chester and slowly approached the two British frigates.[41] At the first indication of movement, the two frigates weighed and with the wind in their favor started after the galleys. But after two hours, little headway, and a fresh breeze from the southwest, the *Roebuck* and *Liverpool* headed down stream in hope of drawing the galleys into deeper water where they "could run near them, and have a better chance of destroying them." The galleys immediately followed, keeping "up a smart fire, but cautiously remained at their usual distance." By sunset the galleys had pursued the frigates as far as New Castle where the cannonade ceased, it being too difficult to sight the guns. The two opponents, however, continued to drop down the river—the galleys finally anchoring for the night at New Castle and the frigates several miles below.[42]

The damage done to the *Roebuck* and *Liverpool* during the second day's engagement was extensive. Several shot had pierced the *Roebuck's* hull just above the water line on both her sides and stern. Likewise, an 18-pound shot had entered an upper port, destroying a 9-pounder, killing one man, and wounding two more. The rigging, sails, and spars of the 44-gun frigate suffered considerably. The *Liverpool* had "several Shot through our Sails, some few in our Hull, and one in our bowsprit." Damage suffered by the galleys was minimal.[43]

The following morning the two frigates got underway and stood down the river.[44] Expecting the battle to resume, the galleys "Spread across the River" in a battle line. But to the galley's surprise, the *Roebuck* and *Liverpool* continued down the river and

Scrimshaw Carving Portraying a Continental Marine.

moored at the upper end of Reedy Island. With some relief a council of war was called where it was determined that the galleys would remain off Newcastle "untill we Can be Amply furnished with Amunition of Every Kind." Once the resupply had taken place, only then would they "make a general Attack upon the Enemy." Ammunition returns indicated that the course chosen was correct, for the galleys had only 20 rounds of powder per gun, and a few 24-pound shot: scarcely enough to reengage, let alone sustain a lengthy cannonade.[45] It was not until 11 May that the Committee of Safety ordered the ammunition replenished, and by the time it was filled, the British frigates had abandoned the anchorage at Reedy Island and dropped down to Delaware Bay.[46]

On 15 May the *Roebuck* and *Liverpool* reached the Delaware Capes. Noting that little could be done "without more Ships, a Bomb brig, and a body of Troops to act with them," Captain Hamond stood to sea the next morning—the *Liverpool* with orders to cruise off the Delaware Capes, and the *Roebuck* to proceed southward.[47]

The defense of the Delaware against the British frigates *Roebuck* and *Liverpool* marked the first instance of Continental Marine cooperation with state naval and Marine forces. Although their contributions were minimal in terms of the number of men involved, this action anticipated further cooperation in subsequent land battles and amphibious expeditions.

Under Cape May

Following three days of battle on board the provincial galleys, Captain Miles Pennington and his Marine company returned to find the *Reprisal* ready for service. Unlike other vessels, the *Reprisal* had not been emptied of powder and shot to supply the galleys, for the Marine Committee refused to allow such drafts for ammunition on its ships. Although Captain Wickes had maintained his supplies intact, his seamen were not as fortunate. Several had become violently ill and required immediate hospitalization, but were refused on the grounds that the hospital at Fort Island was just for provincial use. Only by invoking the aid of the Marine Committee were the men admitted.[48]

The Marine detachment of the *Lexington* returned to Philadelphia to find their ship in the same condition as when they departed—still under repair. The excess time, however, was not wasted. Marine recruiting parties scoured the city, and had little trouble enlisting men for the next cruise. Among those enrolled was another John Barry, who appears on the brigantine's muster roll as "John Barry, Serjant," born in Ireland. Of this sergeant of Marines, who was recruited on 13 May, little is known.[49] But he, and others recruited during the same period, swelled the ship's contingent to well over 100 officers, seamen, and Marines.

On 17 May the *Lexington*, far better armed and manned than she had been in March, joined the *Reprisal* and *Hornet*, then at Chester. Under orders from the Marine Committee, the two brigs and sloop were to convoy merchantmen idled by the recent battle, down the river and into open sea. Soon after obtaining pilots, the three Continental vessels dropped down the river and came to anchor under Cape May on the 25th. There they found both friend and foe.[50] In mid-channel stood the *Liverpool*, while in the upper reaches of the bay lay the *Wasp*. The following day the British frigate unexpectedly put to sea, only to reappear on the 27th, give the four vessels chase, and then retire beyond the Capes.[51] For the next several weeks the two brigs, sloop, and schooner played tag with the frigate *Liverpool*, and then upon her return to New York, with her replacements, the frigate *Orpheus* and sloop *Kingfisher*.

Early in June, Captain Wickes wrote the Marine Committee requesting that they permit the *Reprisal* to cruise. Apparently believing his crew ready, he considered offensive action of more value than lying in wait behind the Capes. The time he chose for his request was propitious, for Congress had directed the Committee of Secret Correspondence to secure a Continental ship and send it to the West Indies to procure muskets and gain intelligence on French naval and troop movements.[52] To carry William Bingham, the Continental Agent picked for the mission, the Marine Committee singled out the *Hornet*. But she proved unseaworthy, and the *Reprisal* was selected as her substitute. On 13 June the *Wasp* arrived at Cape May bearing Mr. Bingham, and soon after he was transferred to the *Reprisal* for the trip to Martinique.[53] Events at the Cape, however, postponed his departure.

Impatient to put to sea soon after William Bingham had been brought on board, Captain Wickes found his way blocked by the British frigates lying off the Capes. For two weeks the *Reprisal* and the 13 merchantmen she was to convoy to the West Indies waited. Then at dusk on 28 June, the American brig *Nancy* was spotted east of Cape May. Chartered by the Pennsylvania Committee of Safety, the brig carried arms and powder from St. Croix and St. Thomas. Immediately the *Lexington* and *Wasp* were dispatched to assist, but with darkness upon them they were unable to locate her. The following morning the *Nancy* was sighted, but finding her being pursued by the frigate *Orpheus*, the two Continental ships retired.[54]

As the *Nancy* approached the Capes, Richard Wickes—brother of Lambert and a lieutenant on board the *Reprisal*—gathered a party of seamen and Marines and set out to assist in her defense. Once on board the brig and seeing that she could not adequately defend herself against the 32-gun frigate, Wickes ran her ashore. Joined by Captain Barry and a portion of his crew, Wickes immediately began unloading the brig's valuable cargo. All but 100 barrels of powder had been removed before the *Orpheus* and *Kingfisher* anchored abreast of the *Nancy* and opened fire. The Americans quickly set fire to the remaining powder and retired. Several boats from the *Kingfisher* were dispatched to take over the brig, but five minutes after they boarded the powder blew.

The battle to save the *Nancy* left Captain Wickes' brother dead, but allowed the *Reprisal* and the merchantmen safe passage. On 3 July, after the British frigates had withdrawn up the coast, the *Reprisal* and her convoy set out to sea leaving the *Lexington*, *Hornet*, and *Wasp* behind to maintain the watch.[55]

Voyage to Martinique

Until late afternoon on 3 July, the *Reprisal* and the merchantmen under her care maintained a close formation. Then "a good distance from the Land," the convoy scattered, each to make its own way. Sailing southward, the *Reprisal* on 11 July took the 240-ton *Friendship* and several days later the schooner *Peter* and brigantine *Neptune*. Prize crews composed of seamen and Marines were put on board the three captures, and they were dispatched either to Philadelphia, or some other convenient port.[56] The remainder of the cruise southward was relatively uneventful and on 27 July the *Reprisal* rounded Pointe a la Mare, sighting St. Pierre, Martinique's principal seaport.[57]

As the *Reprisal* approached the harbor, the British sloop *Shark* was sighted lying at anchor. Not wanting to endanger the life of his valuable passenger in a sea fight, Wickes immediately ordered a boat lowered to row Bingham ashore. Shortly thereafter, the *Shark* sailed out of the harbor to look over the strange sail with "Colours which I was unacquainted with (being red and white striped, with a Union next the Staff)." As the entire population of the little French seaport turned out to watch the outcome, the *Shark* hailed the *Reprisal* twice and ordered the American brig to lower her boat.[58] In reply, Wickes told the British to "come and do it themselves."[59] The British then fired a single shot across the *Reprisal's* bow, and ordered her to heave to. Three times the *Shark* fired warning shots, but the *Reprisal* did not answer. Then on the last hail the American brig responded with a broadside—one shot of which ripped through the *Shark's* quarter, wounding a Marine.[60] The half hour engagement ended when the French shore battery fired twice at the *Shark*, forcing her to sea.[61]

Captain Wicke's request for protection was granted by the French, who allowed the *Reprisal* permission to enter St. Pierre harbor for necessary repairs. Captain John Chapman of the *Shark*, who demanded of the French that they permit him to take possession of the *Reprisal*, was told that the French were neutral in the affair and that he was forbidden to engage any ship in the French roads or under their forts.[62]

As it happened French policy in the West Indies had been defined several days before the *Reprisal* arrived. D'Argout, the French military head of Martinique, had received orders by a frigate from France the preceding Sunday which ordered him "to give all possible assistance and protection to the American vessels." D'Argout also was to permit American prizes to be sold or disposed of in the ports of Martinique.[63] In early August a second demand that the *Reprisal* be turned over arrived, this time from Vice Admiral James Young. Again the French refused.

By late August the *Reprisal*, then at Fort Royal, was ready to sail. In her hold were muskets, powder, and other merchandise for the American cause. Refitted and newly painted, the *Reprisal* left Martinique on 26 August, arriving in Philadelphia on 13 September, after a passage of nineteen days. In October the *Reprisal* was again placed at the disposal of the Committee of Secret Correspondence and ordered to take Benjamin Franklin, who had been appointed a commissioner to France, to his post. She sailed about 1 November and anchored in Quiberon Bay a month later, taking two small prizes enroute.[64]

The *Lexington* Goes to Sea

While the *Reprisal* was in Martinique, the British blockade of the Delaware eased allowing the *Lexington* opportunity to escape. On 2 July, the Marine Committee had decided that the *Lexington* was of little use "cooped up at Cape May," and therefore directed Captain Barry out to sea.[65] In concert with the Continental sloop *Sachem* (reconditioned prize *Edward*), Barry cleared the Capes on 10 July. Since the *Kingfisher* was enroute to New York, and the frigate *Orpheus* was at anchor some distance southeast of Cape Henlopen, the Americans' departure went unobserved. Soon thereafter, the sloop and brig separated, each vessel seeking its own fortune.

Skirting the Virginia coast for two weeks, the *Lexington* failed to sight a sail. Then on 27 July, a lookout reported a vessel downwind. Immediately the *Lexington* gave chase. After a series of skillful maneuvers and carefully placed broadsides, the sloop *Lady Susan* hove to and was boarded. A second prize (*Betsy*) was taken several weeks later, after which the *Lexington* headed for the Delaware Capes. On 26 September, the American brig anchored off Philadelphia and two days later John Barry relinquished his command. The Marines under the command of Abraham Boyce, however, remained with the vessel.[66]

In October the brig *Lexington* (Captain William Hallock) was placed under orders of the Secret Committee of Congress—as distinct from the Committee of Secret Correspondence—and sent to the West Indies for military supplies. On her return in December, she was captured off the Delaware Capes by the British frigate *Pearl*. Captain Hallock and the remainder of the naval officers were transferred to the British frigate and a small prize crew of seven put on board the *Lexington*. Later the same day the American crew, under the able leadership of Marine Captain Abraham Boyce, recaptured the ship and took her into Baltimore where she remained until February the following year.[67]

Return of the Continental Fleet

While the *Lexington* and *Reprisal* were fitting out at Philadelphia in late March, the Continental Fleet under Esek Hopkins departed the island of New Providence on 16 March 1776 and sailed northward. At two in the afternoon of the third day out, the *Alfred's* barge came alongside the *Andrew Doria* and delivered a letter from the commodore to Captain Biddle.[68] In it were Hopkins' brief set of sailing orders the fleet was to follow on the return voyage:

> You are to keep Company with the Ship I am in if possible—but should you Separate by accident you are then to make the best of your way to Block Island Channel, and there to Cruise in 30 fathom Water South from Block Island six days in order to join the Fleet.

If a ship was unable to rejoin the fleet, it was then ordered to cruise in such places as will "most annoy the Enemy," or enter the nearest port.[69]

Later that afternoon Captain Biddle ordered his men to clear the ship for action and exercise the guns. Despite the fact that a majority of the crew were down with the fever, he was not going to be unprepared should the fleet encounter enemy vessels. What they met with, however, was a severe storm which began on 22 March and continued for the next three days.[70] During the gale the fleet lost sight of the *Wasp*, although she later made her way alone to the Chesapeake Bay.[71]

Just over a week's sailing from New Providence the first unidentified sail was sighted. The *Andrew Doria* and the *Providence* immediately gave chase, but she was found to be a Carolina schooner bound for France, and she was allowed to proceed. The next day a second sail was sighted and again the *Andrew Doria*, this time in company with the *Fly*, took up the chase. After an eight hour pursuit she turned out to be a French schooner bound for Cape Francois from New London, Connecticut. But before she again proceeded on her voyage, her captain gave the fleet good news; Boston was now in American hands.

The welcomed news brought a moment of joyous relief to the seamen and Marines on board the fleet, but only a moment; for a large portion of the men were still fighting the fever and small-pox. At least one death per day was commonplace. The ship's surgeons could do little, since they too were taken ill. On 3 April, as the fleet neared the rendezvous point off Block Island, approximately 10 miles off the Rhode Island coast, Isaac Craig, captain of Marines on board the *Andrew Doria*, lost his sergeant, Thomas Vernon Turner. But by then the weather had cooled, thereby alleviating much of the sickness.

As the fleet approached Block Island on the evening of 3 April in a thick fog, the *Andrew Doria* and *Columbus* veered out of sight, but reappeared the following afternoon. When they came up with the rest of the squadron, they informed the commodore that each had taken a prize—a small sloop, and the schooner *Hawke*.[72] Again on 5 April another British ship was secured. Mounting 8 carriage guns, 10 swivels, and carrying 48 Marines and seamen, His Majesty's bomb brig *Bolton* was indeed a major prize.[73] Several other smaller vessels were captured that day as the fleet cruised south of Block Island. Hoping to pick up additional vessels, the commodore formed the fleet into two parallel columns. In the easterly column was the *Cabot*, with the *Alfred* approximately 100 yards astern. About a quarter mile to the starboard were the *Andrew Doria* and *Columbus*. Taking up the rear were the *Providence*, *Fly*, and the captured British vessels. As luck would have it, the fleet's good fortune was to change that evening.

An hour into the midnight watch two unidentified sails were sighted to the southeast. From their silhouettes it was surmised that one was a large frigate and the other possibly its tender.[74] The immediate call for 'all hands to quarters' rang out on each of the twelve ships which then composed the fleet. On board the *Alfred*, Captain Nicholas was roused out of bed and his company of Marines ordered to assemble. Once collected and outfitted for action, First Lieutenant Matthew Parke took the main body of the company and boarded the barge on the main deck, while Nicholas and Second Lieutenant John Fitzpatrick took the remainder to the *Alfred's* quarter deck. With the *Alfred* astern and the *Andrew Doria* and *Columbus* not far away, the *Cabot* bore off toward the distant sails.[75]

As soon as the *Cabot* came within shouting range she was hailed by the unknown ship, who identified herself as the *Glasgow*, Tyringham Howe commanding, a 20-gun British frigate. The *Glasgow* then de-

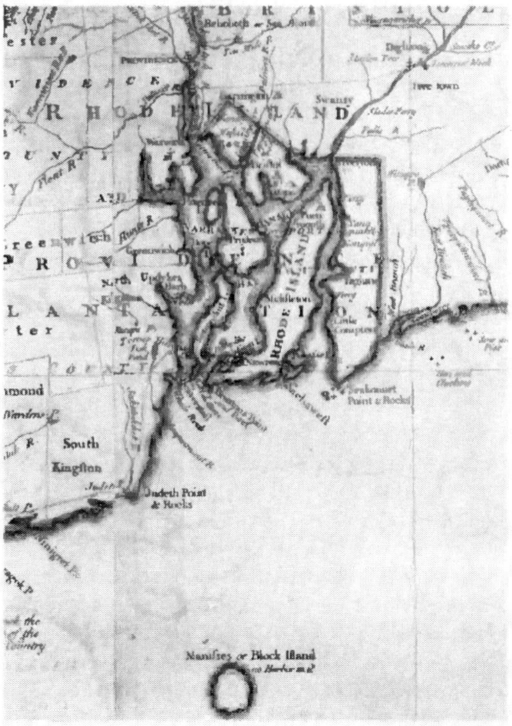

Block Island and Narragansett Bay.

manded to know who the other ships with the *Cabot* were. "The *Columbus* and *Alfred*, a two and twenty gun frigate," replied Captain John B. Hopkins.[76] As if to reinforce Hopkins' answer, a grenade was thrown from high atop the *Cabot's* mainmast, which exploded harmlessly on the *Glasgow's* deck. The *Cabot* then opened up with a broadside from seven of her 6-pounders, but the *Glasgow* returned it twofold, killing the *Cabot's* master (Sinclair Seymour), two Marines (Patrick Kaine and George Kennedy), and wounding eight men including Captain Hopkins and Marine Lieutenant John Hood Wilson.[77] As a result of the *Glasgow's* two 9-pound broadsides the *Cabot* was forced to retire and make repairs.

Reeling away under the weight of the *Glasgow's* batteries, the *Cabot* almost ran afoul of the *Andrew Doria* who was about to make her pass at the British frigate.[78] By the time the *Andrew Doria* was clear, Captain Saltonstall had brought the *Alfred's* twenty 9-pounders and ten 6-pounders into action. Presently the two ships were hurling broadside after broadside at each other. In one of the first exchanges, Captain Nicholas' second lieutenant, John Fitzpatrick, was felled by a musket ball through the head. As Nicholas was later to eulogize: "in him I have lost a worthy officer, sincere friend and companion, that was beloved by all the ship's company."[79]

After several more broadsides, a lucky shot from the *Glasgow* carried away the *Alfred's* wheel block and the lines which led to the tiller. Immediately, she became unmanageable, and it was several minutes before the *Alfred's* crew could bring her under control. In the meantime, the *Andrew Doria* closed on the *Glasgow's* port quarter, but she too veered off after taking several shots which damaged her hull and rigging. The *Columbus* only got off a few shots at the *Glasgow's* stern during the initial engagement, before she was becalmed. The sloop *Providence* throughout the battle seemed content to tack back and forth out of range.

The battle continued for an hour and a half during which time the *Columbus* managed to work her way back into the fight, but without much success. On board the *Glasgow* Captain Howe, expecting that he was about to be boarded, ordered the dispatches thrown overboard and a new course set for Newport. The *Glasgow* then bore away with the *Alfred*, *Andrew Doria*, and *Columbus* "within musket shot, on . . . [her] Quarters and Stern."[80]

The pursuit continued until daybreak with the three American ships occasionally bringing their guns to bear on the shattered ship. Even though much cut up, the *Glasgow* was able to outrun the Americans who were loaded down with the spoils captured at New Providence. As they approached Newport harbor after a chase of nearly seven glasses (three and one half hours), Hopkins, apprehensive about engaging the British fleet stationed there, signaled the ships to give up the chase and steer to the southwest. The *Glasgow* had made good her escape, but she was so badly damaged that she was later sent back to England for repairs. As one observer noted, the *Glasgow* entered Newport harbor "under all the sail she could set, yelping from the mouths of her cannon (like a broken leg'd dog) in token of her being sadly wounded."[81] Yet all that the Americans received for the night's work was the *Glasgow's* tender, while sustaining moderate damage.

Casualties were extremely light considering the fierceness of the battle. The *Glasgow* had one man killed and three wounded—all by American Marine musketry. Of the American ships, the *Alfred* lost five killed and seven wounded; the Cabot four killed and seven wounded; and the *Columbus* and *Andrew Doria* both had one man wounded.[82]

By noon on April 6th, the fleet had reassembled and signals were given for the squadron to head for New London, where the spoils could be unloaded and repairs could be made in safety. At sunset they were northwest of Block Lsland when a thick fog rolled in from Long Island Sound and engulfed the fleet. Dawn the following day found only the *Cabot* and the *Andrew Doria* together and the rest of the fleet nowhere in sight. Shortly after noon the two passed New London lighthouse, entered the Thames River, and anchored. That night they were joined by the *Alfred* and the remainder of the fleet with the captured British vessels in tow. The following morning, 8 April, the full American fleet moved up the river and anchored off the town of New London, Connecticut.[83]

At first there were congratulations. John Hancock expressed the feelings of Congress when he wrote to Esek Hopkins after receiving the Commodore's operational report: "I beg Leave to congratulate you on the Success of your Expedition. Your Account of the Spirit and Bravery shown by the Men, affords them the greatest Satisfaction; and encourages them to expect similar Exertions of Courage on every future Occasion." Yet Hancock, as if to foreshadow future events, noted that the escape of the *Glasgow* was regretted, but hastened to add that it was not due to any misconduct on the part of any officers or men.[84]

The loss of what would have been a valuable prize, however, later mushroomed into a full-fledged controversy in which only a few would be spared the indignity of public censure. Criticism of the commodore's conduct could be heard in the streets and taverns of New London soon after the fleet's arrival. Eventually it reached the halls of the Pennsylvania state house where Congress instituted an investigation. After many drawn-out discussions, the Marine Committee recommended that he be censured.[85] Agitated by the vote, the political storm continued to swirl about the commodore until late March 1777 when Congress voted to suspend Hopkins from his command.[86] Such an able commander deserved better.

Demise of the Continental Fleet

Soon after the American fleet put into New London on 8 April, it was found to be in shambles. Of a total complement of approximately 1000 seamen and Marines, one-quarter were sick with "some New Malignant Fever," and required hospitalization.[87] Resignations and desertions also taxed the already depleted crews. To supply manpower, Commodore Hopkins requested and received a draft of 170 men from the forces under General Washington. But the draft was insufficient. Therefore both seamen and Marines were shifted from one vessel to another. Although inadequately manned, the fleet sailed from New London on 19 April.[88]

The fleet arrived at Newport harbor on 25 April, following a minor mishap off Fisher's Island, and then moved up the river to Providence.[89] There Hopkins received an order from General Washington requesting that he return the men who had been loaned to him, as they were needed in New York.[90] This unexpected request caused problems, for the recruitment of additional seamen and Marines in the Providence area was impossible due to the overwhelming attractions of privateering. But the commodore acquiesced, although as he pointed out, "we Still continue to be Sickly on board the Fleet so that it will be impossible to go to Sea with the Fleet before we get recruited with hands which will not easily be done."[91]

To compound the problems of the fleet even further, Hopkins received information from the Marine Committee concerning two small British fleets in southern waters—one in Virginia, and the other at Wilmington, North Carolina. While the committee did not directly order the fleet against the two naval forces, it suggested that "there is no Service, from the present Appearance of things in which You could better promote the Interest of your Country."[92] Later convinced of the weak condition of the fleet due to the lack of personnel, the Marine Committee abandoned the scheme as impractical.

By early May, three of the vessels were fitted out and ready to sail. The command of the *Providence* was given to John Paul Jones, who on the 12th was ordered to New York with the soldiers loaned to the fleet by Washington.[93] The same day, Hopkins ordered the *Andrew Doria* and *Cabot* on a month's cruise to "such places as you think will most annoy the Enemy."[94] The other three vessels of the Continental fleet remained behind, the *Alfred* and *Columbus* at Providence waiting for fresh recruits while the *Fly* patrolled the entrance of Narragansett Bay.

Having taken Washington's troops to New York, the *Providence* returned to Newport harbor and on 13 June was ordered to convoy vessels between Narragansett Bay and Long Island Sound.[95] Before Jones departed, he received on board a new lieutenant of Marines, Alpheus Rice. The previous month Rice had petitioned Commodore Hopkins for a position in the fleet:

> That your Petitioner desireous to be of Service to his Country has Quit'd the land Service with a design to get into the Fleet under Your Comd. & as your Petir. is well versed in the Small Arm exercise—would be greatly obliged to your Honr. if a Vacancy should present that You Would Consider your Petir. as he is now & has ben for some Time upon Expences.[96]

On 12 June, Rice was appointed lieutenant of Marines and ordered on board the *Providence*. The 22 Marines, whom he had recruited from the Rhode Island Brigade, joined him two days later, doubling the Marine contingent.[97]

After several unsuccessful attempts to convoy the *Fly* and other vessels past Fisher's Island, the *Providence* returned to Newport.[98] There Jones, on 18 June, received new orders which directed him to proceed instead to Boston.[99] Two days later, the *Providence* sailed for Boston where a group of transports loaded with coal were collected and subsequently conducted safely into Delaware Bay, arriving 1 August.[100]

The *Andrew Doria* and *Cabot* sailed on a short cruise to the eastward on 19 May. Before they sailed, however, both received new Marine lieutenants. Assigned to the *Cabot* was John Carr (or Kerr), who took the place of James Hood Wilson, killed in the fight with the *Glasgow*. Kerr's immediate superior remained John Welsh, captain of Marines.[101] To reinforce Lieutenant Isaac Craig and his depleted group of 12 Marines on board the *Andrew Doria*, Hopkins appointed Lieutenant John Trevett.* With Lieutenant Trevett came 17 additional Marine privates who were taken from the *Alfred*.[102]

Soon after getting to sea, the *Andrew Doria* and *Cabot* fell in with the British frigate *Cerberus* and separated. The *Cabot* bore away to the eastward, while Captain Biddle turned the *Andrew Doria* southward. By nightfall on 20 May, the *Cerberus* gave over the chase and retired. Unable to regain contact with the *Cabot*, the *Andrew Doria* continued southward toward Nantucket Island. The following morning, several miles south of the island, the brig took a sloop loaded with rum, sugar, and salt. A prize crew was put on board and the sloop immediately dispatched to Providence.[103] For the next eight days the American brig sailed northwestward in hope of encountering British transports bound for Halifax.[104]

Dawn on Wednesday, 29 May, found the *Andrew Doria* approximately 600 miles southeast of Boston.[105] As the brig's lookout gazed toward the northern horizon, he sighted two ships. Immediately Captain Biddle swung the *Andrew Doria* to the north and began the pursuit. Within a hour Biddle captured what turned out to be two unarmed British transports. Bound for Halifax, the transport *Oxford* carried a 100-man company of the 42d Royal Highland Regiment, while the other transport, *Crawford*, had a company of the 71st Regiment, likewise about one hundred strong. As part of a 32-ship convoy, the two transports had become separated in an Atlantic storm.[106]

The capture of the *Oxford* and *Crawford* immediately posed a problem. The crews and passengers of the two ships outnumbered the *Andrew Doria's* crew by three to one. Captain Biddle's solution was simple: separate the sea officers from their crews and the land officers from their companies. Therefore, instructions went out to transfer all the troops to the *Oxford*, and the officers to the *Crawford*. In addition, all arms and equipment belonging to the two companies were ordered moved to the *Andrew Doria* along with the British seamen. Lieutenant James Josiah was appointed prize-master of the *Crawford*, while Lieutenant John McDougall was picked for the same task on board the *Oxford*. To

* Isaac Craig is shown as a captain of Marines on the muster roll of the *Andrew Doria* (dated 10 May 1776), but he was not appointed to such a position by Congress until 22 October 1776.

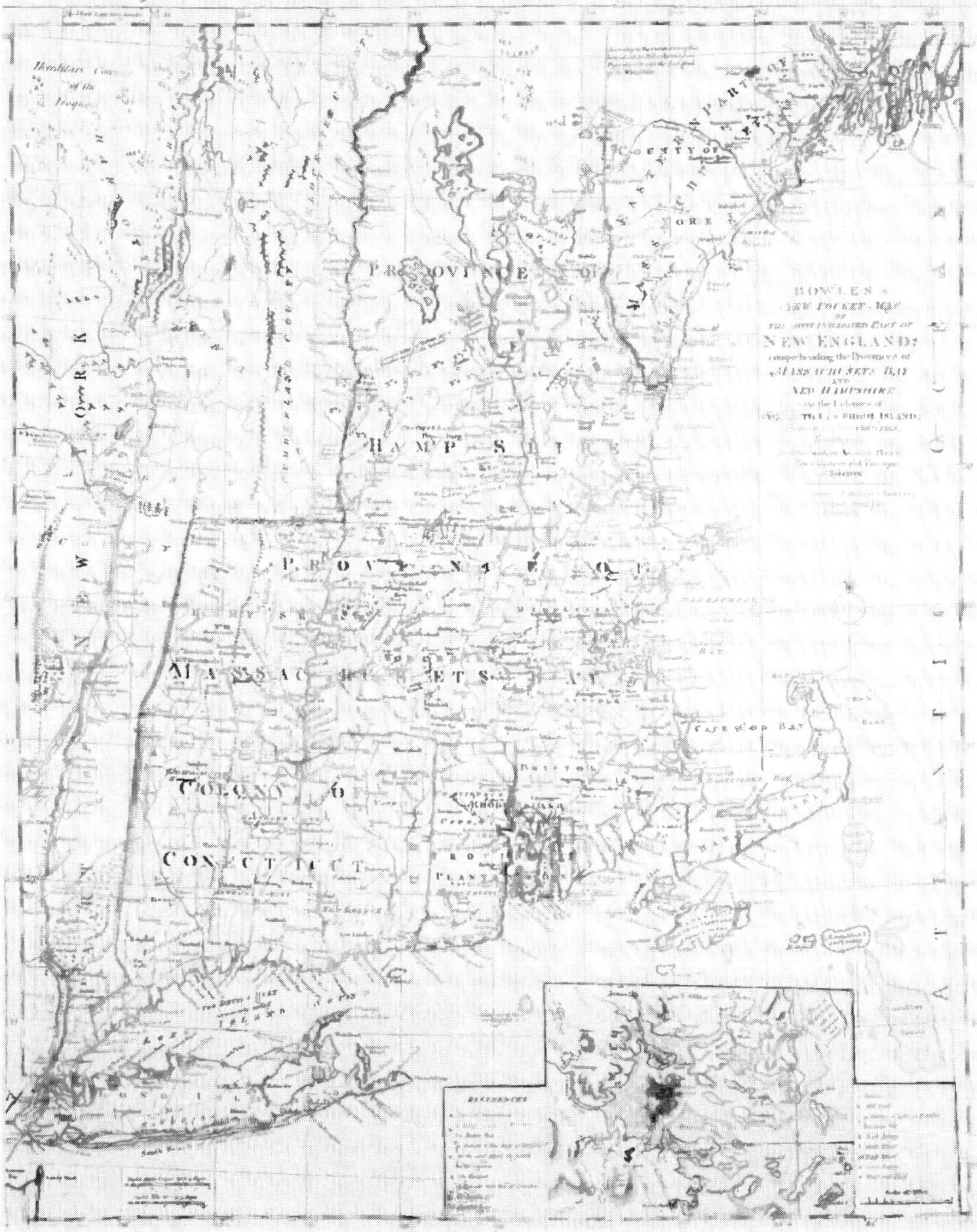
NEW ENGLAND
NEW HAMPSHIRE
MASSACHUSETTS BAY
CONNECTICUT
CAPE COD BAY
REFERENCES

join Lieutenant McDougall as prize-master's mate, Biddle chose Marine Lieutenant John Trevett.[107]

When the brig and her two prizes got underway, they headed westward, back to Providence. During a majority of the return voyage, the sea was smooth and the horizon empty. But on 11 June their luck turned. Some 90 miles south of Martha's Vineyard, five enemy sail were sighted.[108] Without delay the three ships separated; the *Andrew Doria* to the east, the *Crawford* to the south, and the *Oxford* to the west. The *Andrew Doria* successfully eluded the British fleet, but she unfortunately lost sight of her prizes and returned to Newport harbor empty-handed.[109] The *Crawford*, however, was not as successful. Early the following morning she was taken by the *Cerberus*, and ushered to Long Island by the frigate's tender. But on 19 June, off Fire Island, she was recaptured by the *Schuyler*, a sloop of Washington's New York fleet.[110]

The *Oxford*, although successful in eluding the enemy fleet, had its own problems. Before the prize was out of sight of the *Andrew Doria*, on 11 June, the 300 prisoners on board rose and overpowered the 11-man prize crew. Of the event Marine lieutenant John Trevett merely remarked: "I could not blame them, for I would have done the same." Instead of remaining in the area, the captors decided to head for Hampton Roads where they expected to join up with Lord Dunmore. By good fortune, for they were without a competent navigator, the *Oxford* made it to the Virginia Capes nine days later. There they inquired about the whereabouts of Dunmore and were told he was then some 40 miles up the James River. The information, however, was incorrect; the fleet under the command of the Virginia governor lay northward in Chesapeake Bay.

"After giving three cheers," according to Trevett, the unsuspecting British weighed and stood up the James River. There they met two armed boats of the Virginia Navy who retook the *Oxford* and carried her up to Jamestown. As soon as the prize was secured, a company of frocked Virginia riflemen—"the finest I ever saw," commented Trevett—marched the British troops off toward Richmond. Lieutenants McDougall and Trevett accompanied the riflemen as far as the Virginia capital at Williamsburg, where they were treated "very politely" and given expense money for their return to Rhode Island.[111]

At Newport harbor, meanwhile, the *Columbus* was made ready for a projected cruise. Like other vessels of the fleet, Captain Abraham Whipple's ship suffered from an immediate and partial crew loss following its return from New Providence. Hardest hit by sickness, desertion, and resignation was the Marine company. The *Columbus* lost all three of her Marine officers by resignation shortly after her return to New London.[112] No reason has been found for the resignations of Captain Joseph Shoemaker and Lieutenant Robert Cummings, but Lieutenant John Trevett relinquished his position in order to take a similar post on board the *Andrew Doria*.[113] Lieutenant Matthew Parke was transferred from the *Alfred* and promoted to captain to replace Captain Shoemaker. To fill one of the vacant lieutenancies, Whipple elevated Sergeant Edward Burke.[114] Trevett's position, however, was left empty for want of a qualified officer.

On Tuesday 18 June, the *Columbus* sailed out of Newport harbor. As she slowly emerged from the East Passage of Narragansett Bay, the *Cerberus* was observed standing in from Block Island. Immediately the *Columbus* cleared for action and discharged a broadside. Her opponent, however, responded with three in rapid succession and then withdrew. Crippled, the *Columbus* returned to Newport with the aid of the *Andrew Doria*, *Providence*, and two Rhode Island galleys. The encounter with the *Cerberus* postponed for several weeks the *Columbus'* cruise.[115]

In early August, the *Columbus*, with the *Andrew Doria*, again headed out. The two ships separated and on the 10th, the *Columbus* took five sugar ships, only two of which made it into port. After two months at sea without much success, Whipple returned to Providence on 29 September—the *Columbus* in need of repair. Soon after returning, the British blockaded the entrance to Narragansett Bay, forcing the *Columbus* to remain at anchor until February 1777.

When the *Andrew Doria* returned to Newport in mid-June, she found new orders awaiting her. Under the new instructions Biddle was to take the black brig to sea as soon as possible, "and Cruise till your Provisions are out or you have weaken'd your Brig by Manning Prizes so as to make it unsafe to Cruise."[116] Shortly after assisting the *Columbus* back into Newport following her encounter with the *Cerberus*, the *Andrew Doria* in concert with the *Fly* set sail. Hugging the Rhode Island and Connecticut shores to avoid the British frigates, the two vessels put into New London on 26 June. There they found the *Cabot*, just in from the cruise begun in company with the *Andrew Doria* more than a month before.[117]

While in the Connecticut port, Biddle took the opportunity to replenish his supplies. Five dozen shirts and as many pairs of shoes were requisitioned for his seamen from Nathaniel Shaw, Jr., the newly appointed Continental agent. In addition, Shaw provided 22 pairs of sorely needed trousers for Lieutenant Craig's company of Marines on board the *Andrew Doria*.[118]

On 30 June, the *Andrew Doria* cleared New London and in a thick fog headed out into the Atlantic. For almost two weeks nothing was sighted. Then on 11 July, a large vessel was spied off to the northwest. In less than an hour, Biddle overtook the lumbering merchantmen, *Nathaniel and Elizabeth*, and put a prize crew on board. Without delay she was dispatched into the closest Continental port with her valuable cargo of sugar and rum. With provisions running low, the *Andrew Doria* returned to Newport after several more days of profitless cruising.[119]

Soon after his return, Biddle was informed by the Marine Committee of his appointment to the frigate *Randolph* with orders to proceed on a cruise that would terminate at Philadelphia.[120] While preparations were being made for this last cruise, Lieutenants McDougall and Trevett arrived and rejoined the *Andrew Doria*. In company with the *Columbus*, the black brig departed Newport once more. Within a few days of separating from Whipple's ship, the *Andrew Doria* took her first prize, the *Molly*. Twenty-four hours later a second prize, *Lawrence*, was picked up. In the next two days the *Andrew Doria* intercepted two more merchantmen. On the first, Biddle placed Marine Lieutenant Trevett in command and ordered him to return to Providence. Lieutenant McDougall was put on board the second and ordered to the same port.[121]

On the morning of 17 September, the *Andrew Doria*, with six prizes to her credit, entered Delaware Bay and by nightfall had anchored off Chester. Several days later the brig was turned over to the Continental shipyard at Philadelphia for repairs.[122] The *Andrew Doria's* Marines remained attached to the brig since their enlistments had not yet expired. Isaac Craig, however, left the ship and was appointed captain of Marines on board the Continental armed galley *Champion*.[123]

Almost a month before the *Andrew Doria's* arrival, Captain Jones in the *Providence* sailed for Rhode Island, but not before he had replaced his Marine officer, Alpheus Rice. Prior to entering the Delaware, there had developed a heated argument between Rice and William Hopkins, the master; over what it is uncertain. As a result, Lieutenant Rice was confined. At Chester on 9 August, Rice wrote to Captain Jones and apologized for his misconduct:

> I have given Mr. Hopkins satisfaction, & there now subsists, a good understanding between us.—I am conscious, that I was culpable.—but have this, to say in excuse—that I was in a passion.—am sorry for it.—& pray that you would overlook it.[124]

Jones, however, could not overlook quarrels between his officers that might harm the operation of the *Providence*. Therefore, Rice was discharged on 12 August and replaced by Edmund Arrowsmith.[125] Alpheus Rice remained in Philadelphia and was appointed by Congress on 29 August to the *Hampden* as lieutenant of Marines. Rice, however, failed to assume his new post.[126]

Providence departed Delaware Bay on 21 August and within a week had captured the brigantine *Britannia*, which was soon sent off to Philadelphia. Dropping down to the Bermuda latitudes, the *Providence* fell in with the British frigate *Solebay* on 1 September. After a fast chase which lasted four hours, the frigate came within musket range and fired "several broadsides of double-headed and grape shot." But the *Solebay's* marksmanship was poor, and not a single shot touched the *Providence*. Exploiting his ship's nimbleness to the fullest, Jones escaped by a "Hairbreadth."[127]

After several additional weeks of unsuccessful cruising, Jones decided to run north for Nova Scotia.

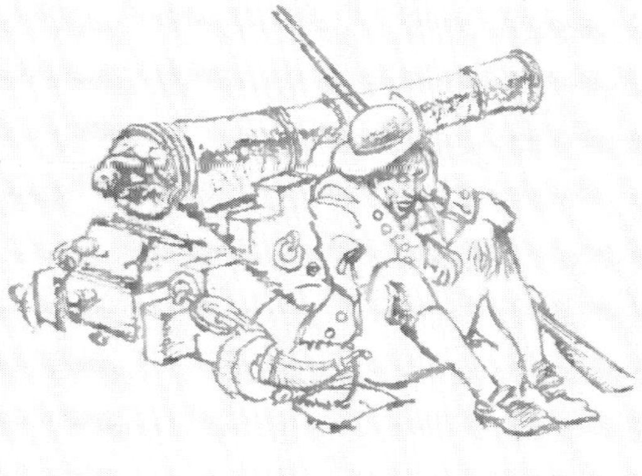

Although short of water and wood, he "had besides a prospect of destroying the English shipping in these parts." On 20 September, as the *Providence* hove to in order to permit the crew a chance to fish between Cape Sable and the mainland, His Majesty's frigate *Milford* of 28 guns was sighted to the windward. Without delay the frigate bore down on the *Providence*, but when she came within cannon range Jones made full sail, gaining the advantage. For eight hours the *Providence* gave the *Milford* what Jones later described as a "wild goose chase," which tempted her "to throw away powder and shot." By nightfall, the day long "mock engagement" had so excited Jones' contempt for the frigate, that when she "rounded to, to give his broadside," Jones ordered Lieutenant Arrowsmith "to return the salute with only a single musket."

The next morning, the frigate was seen "standing to the westward"; *Providence* made for Canso. There Jones got water and wood, and also several recruits from the local fishing population. About three fishing boats were seized and either burned or sunk. Having received intelligence at Canso that a group of Jersey merchantmen were anchored at Ile Madame, Jones dispatched a tender and his boat, both full of armed seamen and Marines, to investigate. The intelligence proved correct, and the Jerseymen quickly surrendered. Three of the vessels were released and as many were destroyed. On the afternoon of 26 September, following a violent storm, the *Providence* put to sea with her recent prizes. And on 8 October, entered Narragansett Bay after a cruise of 49 days, 16 prizes, and the destruction of the fishing fleets at Canso and Ile Madame.[128]

Back in Newport, Jones proposed that an expedition be sent to the West Coast of Africa where he was sure a large number of English merchantmen could be taken.[129] Commodore Hopkins, however, had determined to send a small fleet to Cape Breton in order to capture the British coal fleet, release American prisoners working in the mines, and destroy the Newfoundland fisheries. Therefore, he ordered Jones to take command of the *Alfred* (Captain Saltonstall having been reassigned), and in company with the *Hampden* (Captain Hoysteed Hacker) raid Cape Breton.[130] But before Jones embarked on his mission, he managed to increase the *Alfred's* complement by taking everyone out of the *Providence*. Like the seamen, Lieutenant Arrowsmith—subsequently promoted to captain to fill the new post—went on board the *Alfred* with his Marines. There he joined Lieutenants William Hamilton and Alexander Neilson, both of whom had recently been elevated from positions as Marine sergeants.[131]

As preparations were made for the expedition, the brigantine *Lawrence* with Lieutenant Trevett at the helm arrived in Newport. After his prize was secured, Trevett joined the *Hampden* as its Marine officer, Lieutenant Alpheus Rice having refused the post.[132]

The expedition to Cape Breton was begun on 27 September, but before the two ships could make it out of Narragansett Bay, the *Hampden* struck a sunken ledge, so damaging her that she could not proceed any further. Her officers and men were then transferred to the *Providence*, and the expedition made a fresh start on the first day of November. The *Alfred* and *Providence* first called at Tarpaulin Cove in the Elizabeth Islands off Cape Cod, where they took the Rhode Island privateer *Eagle*.[133] Since Jones suspected that the sloop harbored deserters from the Continental Navy, he dispatched Lieutenants Trevett and Arrowsmith with several armed Marines to the privateer. After "examining the Ships Articles," Trevett ordered his men to impress "all . . . [the men they] could," about 25, and put them on board the *Alfred*.[134] This action later caused problems for Jones and contributed to his dislike for Commodore Hopkins who did not fully support him in the matter.

On 11 November, the *Alfred* took her first prize, the brigantine *Active*, bound from Liverpool to Halifax with a load of dry goods. The following morning a second prize was taken, the 350-ton armed transport *Mellish*. She carried a cargo, according to Trevett, of "10,000 suits of Soldiers Clothing ready made 1 set light horse accoutrements with carbines and a valuable invoice of Medicine chests! . . . trunks of Silk gowns and dry goods suitable for Gen. Burgoynes army at Quebec." The suits of clothing—"every article complete for a Soldier from the hat, to the shoes"—were later sent to General Washington's army, which at the time was withdrawing through New Jersey.[135]

Several days after taking the *Mellish*, the *Providence* parted company with the *Alfred*, there being a disagreement between Captain Hacker and his crew over the condition of the ship. On 22 November, from off Canso, Jones sent in armed boats which burned a grounded transport, destroyed an oil warehouse, and carried off a small, fast schooner, to take the place of the *Providence*. In addition, the boats brought back a gentlemen who informed Jones that three British frigates were searching for him. Off Louisbourg, two days later, three strange sails were sighted, but fortunately they were not the frigates. The *Alfred* easily captured all three.

With five prizes in tow, Jones decided that "for the interest and Honor of the Service" he would return to port. However, high winds and a stiff gale slowed progress. On 7 December, as they crossed St. Georges Bank east of Cape Cod, the ships encountered a British frigate which proved to be HMS *Milford*. That night Jones sent all but one of his prizes ahead, and then tried to lure the frigate away. Unsuspectingly, the frigate followed. The next morning found the *Milford* nine miles astern of the *Alfred* and her consort, *John*. Later in the day, after a short chase, the three ships closed and Jones fired off four cannon at the Englishman. The *Alfred* then made sail as the *Milford* again gave chase. In the early afternoon, the Englishman gave up his pursuit of the *Alfred*, tacked, and bore down on the *John*. Unlike the *Alfred*, the prize vessel was no match for the 28-gun frigate. The *Milford* soon forced her to heave to, and sent a boarding party to recapture the prize. After shaking the *Milford*, Jones entered Massachusetts Bay and anchored at Boston on 14 December.[136]

The New Frigates

Throughout 1776, the main concern of the Marine Committee was the construction and manning of the 13 Continental frigates authorized by Congress in December 1775. Unlike Commodore Hopkins' fleet of converted merchantmen, these frigates were to be built as warships from the keel up. Without regard to political influence, the Committee assigned the ships to the colonies possessing a competent shipbuilding industry. Thus they decided that two were to be built in Massachusetts, two in Rhode Island, two in New York, four in Pennsylvania, and one each in New Hampshire, Connecticut, and Maryland. Once the plans were approved, the Committee entrusted the selection of a builder and the overall supervision of the frigate's construction to the Marine Committeeman from the colony in which the ship was to be built. An additional responsibility of each committeeman was the selection and recommendation of competent naval and Marine officers.

Under the direction of John Langdon, naval agent for New Hampshire, the keel of the frigate *Raleigh*, 32, was laid at Portsmouth on 21 March 1776. Two months later she was launched. As a close personal friend of Langdon, Thomas Thompson was nominated and quickly approved by Congress as the frigate's commander. The selection of her Marine officers, however, took much longer. After a number of names were advanced, all were rejected in favor of three men from New York, New Hampshire, and Massachusetts, whose appointments were approved by Congress on 22 July.[137]

Appointed Captain of Marines for the frigate *Raleigh* was George Jerry Osborne of Exeter, New Hampshire. Born 15 March 1732, Osborne had been a school teacher in Portsmouth for several years before joining the Revolution as a captain in the artillery. Selected to serve as Osborne's lieutenants were Stephen Meads, of New York, and Nathaniel Thwing of Boston. Captain Osborne joined the *Raleigh* on 12 August, seven days before First Lieutenant Meads.[138] Second Lieutenant Thwing boarded on the 21st. The *Raleigh* carried approximately 90 Marines.[139]

Down the coast in neighboring Massachusetts, two more frigates were being constructed. Under the watchful eye of Thomas Cushing, the *Boston*, 24, was launched on 3 June 1776, and the *Hancock*, 32, slightly over a month later. Like Langdon, Cushing received a number of requests for positions as Marine officers on board the two frigates. Among them was the petition of Massachusetts militia Captain John Ford, of Chelmsford, "a distinguish'd officer in the Bunker hill battle [who] has offer'd himself as Capt. of Marines and says, he will bring all his men with him."[140] For unknown reasons, Captain Ford was rejected. Under pressure from John Hancock in early July, Thomas Cushing recommended Richard Palmes of Boston as captain of Marines on board

Seth Chapin, by an unidentified artist.

the *Boston* (Captain Hector McNeill) and Benjamin Thompson as his lieutenant. Palmes' commission was approved by Congress on 23 July, and Thompson's the following day.[141] Cushing, however, delayed until October 1776 in recommending a Marine officer for the *Hancock* (Captain John Manley), even though John Hancock threatened to appoint one himself.[142] On the 10th of that month, Seth Baxter of Braintree, Massachusetts, was appointed captain of Marines on board the *Hancock* upon Cushing's probable recommendation.[143] Early the following year, William Bubier, a life-long resident of Marblehead and veteran of Colonel John Glover's Regiment, joined the frigate as lieutenant of Marines.

The two Rhode Island frigates, *Providence*, 28, and *Warren*, 32, were built at Providence. Although Stephen Hopkins was authorized to supervise their construction, he turned over the building of the two frigates to an 11-man committee of prominent merchants and ship owners. By 15 May, the committee had launched the *Warren*, and three days later would launch her sister ship, the *Providence*.

The selection of Marine officers to serve on board the two frigates was begun early. In April, William Jennison, a member of Captain Nightingales's Rhode Island cadet company, journeyed to Providence and appeared before the committee, who subsequently appointed him first lieutenant of Marines for the *Warren* (Captain John B. Hopkins).[144] Throughout May and the early part of June, Jennison tried to recruit his alloted 25 Marines with the 60 pounds provided by the committee. By mid-June, he found that he was unable to accomplish the task, and therefore quit the service to join a company of volunteers being raised for duty with Washington. On 14 June, the Rhode Island Committee approved John Grannis as captain of Marines for the *Warren*, and Barnabas Lothrop as second lieutenant. To replace William Jennison as first lieutenant, the committee selected George Stillman.[145]

Disturbed over the fact the William Jennison had accepted their commission, taken their money, and then instead chose to volunteer for service with Washington, the Rhode Island committee established recruiting quotas for Marine officers appointed to the *Providence*. A captain of Marines was to enlist 40 "good men," a first lieutenant, 33, and a second lieutenant, 27. This enlistment quota was to be met by each officer before he was allowed to accept his commission. On 20 June, the committee approved Silas Devol as captain of Marines, and two days later selected Seth Chapin of Mendon, Massachusetts, as second lieutenant. The position of first lieutenant was filled with the appointment of Avery Parker on 26 June.[146]

Construction of the 28-gun frigate *Trumbull* was carried out 20 miles up the Connecticut River at the Chatham shipyards. Under the supervision of Silas Deane, and his brother Barnabas, the frigate was completed and launched on 5 September 1776. In the selection and appointment of a commander for the frigate and a captain of Marines, nepotism was the rule. On 22 August, Congress appointed Deane's brother-in-law, Dudley Saltonstall, to the *Trumbull's* command, followed shortly thereafter by the appointment of Dudley's brother, Gilbert, as captain of Marines.[147] Two months later, on 16 October, Jacob White was selected for the post of lieutenant of Marines on board the frigate.[148]

The two frigates built by Lancaster Burling on the Hudson at Poughkeepsie, were the *Montgomery*, 28, and the *Congress*, 24. Although the frigates were never completed, a captain and a lieutenant of Marines were selected. In response to the wishes of Congress, the New York Committee of Safety, on 4

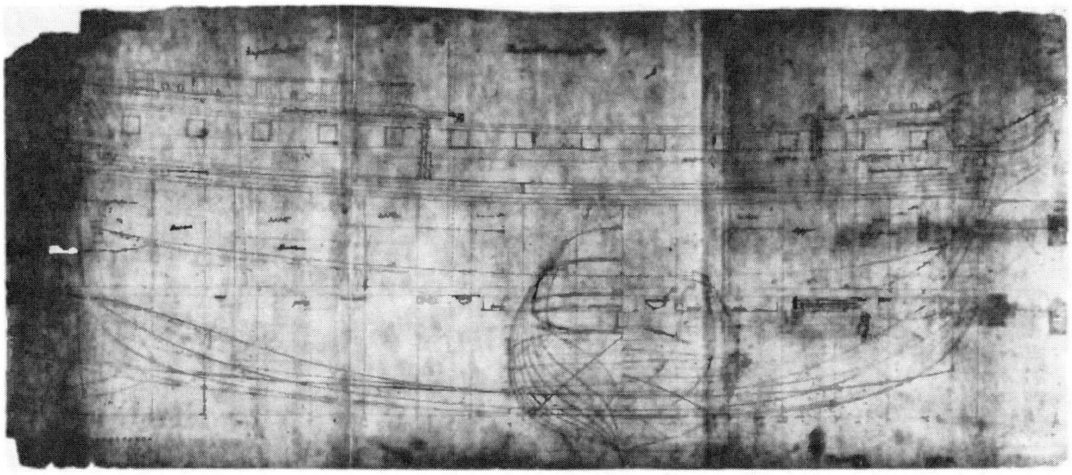

Draft of the Frigate Randolph.

December 1776, nominated Victor Bicker, Jr., as lieutenant of Marines for the *Congress*, gave him 200 dollars, and directed him to enlist 30 Marines, who would receive "Continental Bounty Pay and Rations." The Marines were assigned to guard the two frigates and stores "belonging to the Continent" at Poughkeepsie during the winter.[149] Lieutenant Bicker and his 30 men served with the *Congress* until 6 October 1777, when the frigate was burned to prevent her capture.

Sometime early in 1777, the New York Committee of Safety appointed William Matthewman, captain of Marines for the frigate *Montgomery*. A veteran of three month's service with the 4th New York Regiment, Matthewman, like Bicker, remained with the frigate until 6 October 1777, when she, too, was destroyed in order to prevent her capture by the British.[150]

Under the supervision of Samuel Purviance, Jr., an energetic Baltimore merchantman, the frigate *Virginia*, 28, was constructed in the shipyards of Mr. George Wells at the Maryland seaport. Although the ship was launched in the fall of 1776, the Maryland delegates to Congress proposed to the Marine Committee in late June, that John Stewart be appointed captain of Marines, Thomas Pownal first lieutenant, and Richard Harrison second lieutenant.[151] The commission of all three men, previously officers in the Maryland militia, were confirmed by Congress on 25 June 1776. Captain Stewart and Lieutenant Harrison, however, were later removed and replaced by James Disney (22 October 1776), and Thomas Plunkett (9 December 1776), respectively.[152]

The *Randolph*, 32, *Washington*, 32, *Effingham*, 28, and the *Delaware*, 24, were built at or near Philadelphia. Their construction, unlike the other nine frigates, was directed by the Marine Committee itself.*

The Marine Committee hoped that at least some of the new frigates would cruise before the end of the year. But for a number of reasons, none got to sea in 1776. The major obstacle was the enlistment and retention of seamen and Marines. Commodore Hopkins summed up the problem when he stated: "The whole attention of merchants and seamen at present seems to be on Privateers." To counteract the attractions offered by privateers, Hopkins urged that Congress liberalize the allotment of prize money.[153] His suggestion was accepted on 30 October. After 1 November 1776, officers, seamen, and Marines of Continental ships would be entitled to one-half the value of merchantmen, transports, and store ships taken, and the whole value of warships and enemy privateers.[154] This revision had little effect. On numerous occasions throughout the remainder of the war, Continental vessels were laid up due to the lack of seamen and Marines, but privateers suffered from no such deficiency.

In addition to offering liberalized prize shares, Congress raised the pay of all ranks in the navy, but only two in the Marines. The pay of a captain of

* The appointment of Marine officers to the four frigates built at Philadelphia will be detailed in Chapter V.

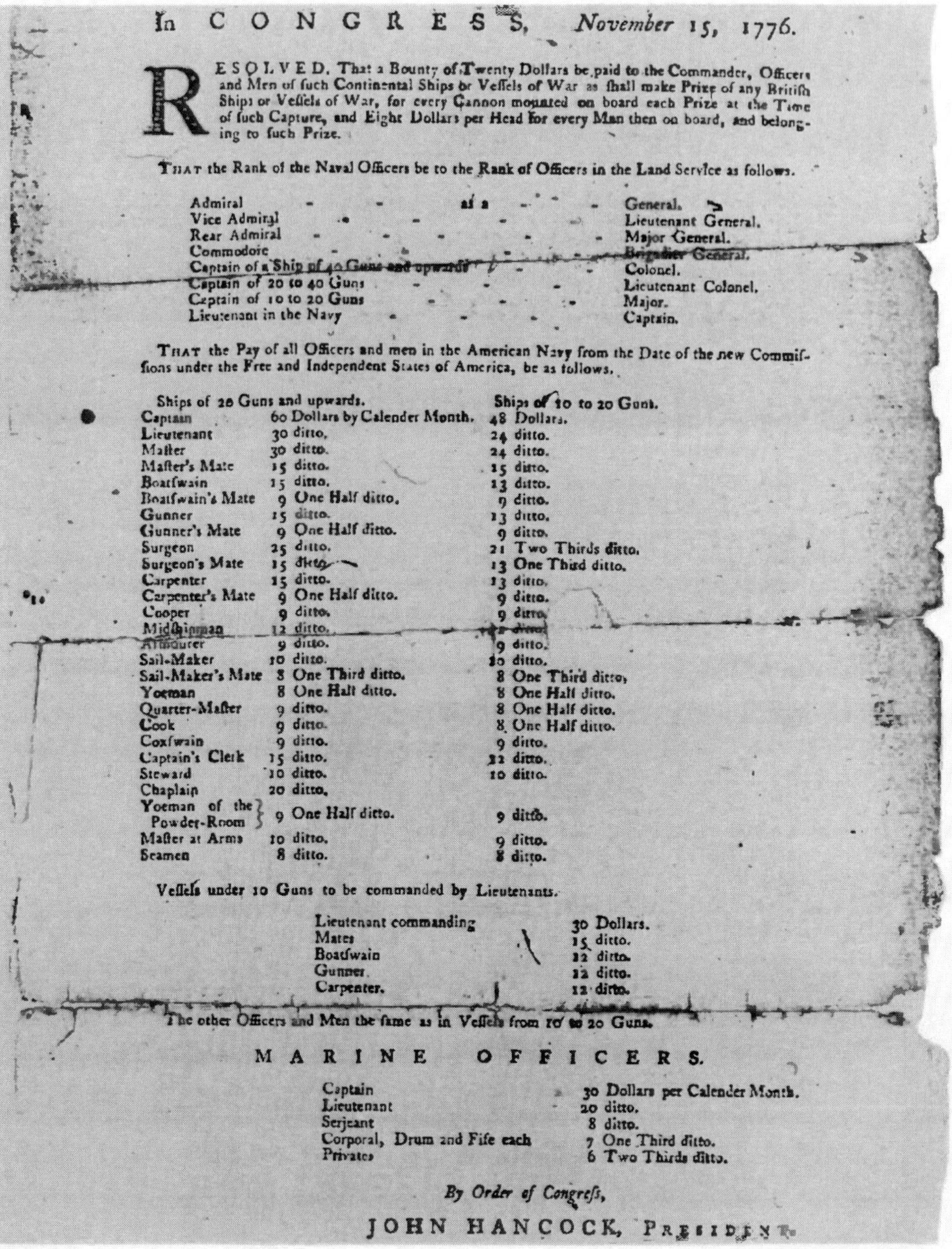

In CONGRESS, *November* 15, 1776.

RESOLVED, That a Bounty of Twenty Dollars be paid to the Commander, Officers and Men of ſuch Continental Ships or Veſſels of War as ſhall make Prize of any Britiſh Ships or Veſſels of War, for every Cannon mounted on board each Prize at the Time of ſuch Capture, and Eight Dollars per Head for every Man then on board, and belonging to ſuch Prize.

THAT the Rank of the Naval Officers be to the Rank of Officers in the Land Service as follows.

Admiral	as a	General.
Vice Admiral		Lieutenant General.
Rear Admiral		Major General.
Commodore		Brigadier General.
Captain of a Ship of 40 Guns and upwards		Colonel.
Captain of 20 to 40 Guns		Lieutenant Colonel.
Captain of 10 to 20 Guns		Major.
Lieutenant in the Navy		Captain.

THAT the Pay of all Officers and men in the American Navy from the Date of the new Commiſſions under the Free and Independent States of America, be as follows.

Ships of 20 Guns and upwards.		Ships of 10 to 20 Guns.
Captain	60 Dollars by Calender Month.	48 Dollars.
Lieutenant	30 ditto.	24 ditto.
Maſter	30 ditto.	24 ditto.
Maſter's Mate	15 ditto.	15 ditto.
Boatſwain	15 ditto.	13 ditto.
Boatſwain's Mate	9 One Half ditto.	9 ditto.
Gunner	15 ditto.	13 ditto.
Gunner's Mate	9 One Half ditto.	9 ditto.
Surgeon	25 ditto.	21 Two Thirds ditto.
Surgeon's Mate	15 ditto.	13 One Third ditto.
Carpenter	15 ditto.	13 ditto.
Carpenter's Mate	9 One Half ditto.	9 ditto.
Cooper	9 ditto.	9 ditto.
Midſhipman	12 ditto.	12 ditto.
Armourer	9 ditto.	9 ditto.
Sail-Maker	10 ditto.	10 ditto.
Sail-Maker's Mate	8 One Third ditto.	8 One Third ditto.
Yeoman	8 One Half ditto.	8 One Half ditto.
Quarter-Maſter	9 ditto.	8 One Half ditto.
Cook	9 ditto.	8 One Half ditto.
Coxſwain	9 ditto.	9 ditto.
Captain's Clerk	15 ditto.	12 ditto.
Steward	10 ditto.	10 ditto.
Chaplain	20 ditto.	
Yeoman of the Powder-Room	9 One Half ditto.	9 ditto.
Maſter at Arms	10 ditto.	9 ditto.
Seamen	8 ditto.	8 ditto.

Veſſels under 10 Guns to be commanded by Lieutenants.

Lieutenant commanding	30 Dollars.
Mates	15 ditto.
Boatſwain	12 ditto.
Gunner	12 ditto.
Carpenter.	12 ditto.

The other Officers and Men the ſame as in Veſſels from 10 to 20 Guns.

MARINE OFFICERS.

Captain	30 Dollars per Calender Month.
Lieutenant	20 ditto.
Serjeant	8 ditto.
Corporal, Drum and Fife each	7 One Third ditto.
Privates	6 Two Thirds ditto.

By Order of Congreſs,

JOHN HANCOCK, PRESIDENT.

Congressional Broadside, 1776.

Marines was raised from 26⅔ dollars per month to 30, while a lieutenant's pay was increased from 18 to 20 dollars per month. The monthly pay for a sergeant, corporal, drummer, fifer, and private, however, remained the same as established by Congress in November 1775. The pay of all Marine "Non-commissioned officers and soldiers" was based on the wage scale for those of equal rank in the "land service."[155]

To further separate Continental naval and Marine officers from those in the private service, the Marine Committee prescribed a distinctive uniform for each in September 1776. Opting for the traditional blue color of British naval uniforms with minor modifications in color and cut, the committee settled upon a blue coat with red lapels and cuffs and yellow metal buttons, a red waistcoat, blue breeches, white stockings, and the usual tricorn hat for American naval officers. For Marine officers, the color chosen was green:

> A Green Coat faced with white, Round Cuffs, Slash'd Sleeves and Pockets; with Buttons round the Cuff, Silver Epaulett on the right Shoulder—Skirts turn'd back, Buttons to suit the Faceings.
> White waistcoat and Brieches edged with Green, Black Gaiters & Garters.
> Green shirts for the Men if they can be procured.[156]

Why the color green was selected remains a mystery, although there are numerous theories. Some attribute the selection to an attempt to imitate the green uniforms of John Cadwalader's company of Philadelphia militia who wore "green short coats faced with white, white vests & breeches, black knee bands & epaulet, dashes of cloth buttoned at the sides, extending to the calves of the legs & white silk stockings."[157] Others attribute it to the availability of green-dyed material. Still others suggest that supplies of green cloth might have come from the depots of captured enemy clothing. For whatever reason lay behind the choice, the colors green and white were to dominate Marine clothing throughout the Revolution.

At the end of 1776, the Continental Marines completed their first year of operations. It was a year which saw the disintegration of centralized control. No longer would the Marines be as closely knit as the group envisioned by Congress in November 1775. With the construction of the 13 new frigates, and the addition of several reconditioned vessels to the Continental Navy, Marine companies were raised independently of the original battalion structure written into the 10 November resolve. What began as a small expeditionary force of 234 men in January 1776, by the end of the year had grown into a fragmented force of over 600. The developments of 1776 ensured that the Continental Marines would never again achieve the status of an independent "corps," which had been their position in late December 1775.

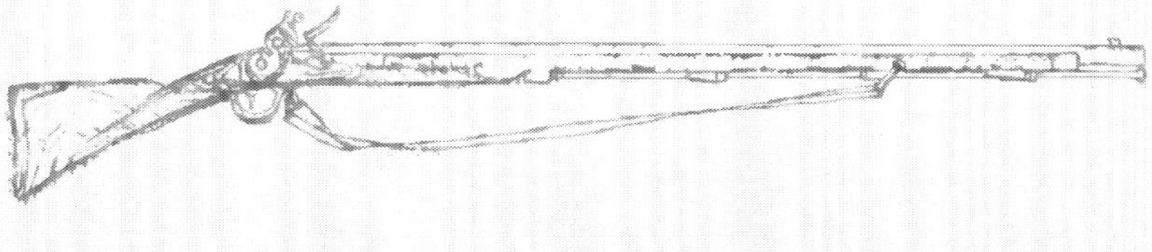

MARINES WITH WASHINGTON AT PRINCETON, 3 JANUARY 1777.
The first attempt by Continental Marines and Philadelphia militiamen to dislodge the British before Princeton failed. This painting portrays General George Washington rallying them for a second attempt in which they succeeded in scattering the defenders and halting the enemy campaign.

CHAPTER V

The Trenton-Princeton Campaign

Four New Companies

Late spring 1776 was a period of inactivity for Marine Captain Samuel Nicholas. Following the Continental Fleet's return from the Bahamas, Captain Nicholas remained on board the *Alfred* in hope that she would soon put to sea. As spring turned to summer, the ship was virtually stripped of seamen and Marines to man the smaller vessels of the fleet. Nicholas was the fleet's senior Marine captain and thus not eligible to command one of the smaller Marine detachments. His prospects for a cruise in the near future were dim, and so he asked that he be permitted to return to Philadelphia on personal business. Commodore Hopkins granted his request in mid-June, and sent him off to the seat of government with dispatches for Congress; considering him "very Capable of giving . . . an impartial history of our whole proceedings."[1]

Upon his arrival at Philadelphia, Captain Nicholas immediately waited on John Hancock, who informed him that Congress had conferred a majority upon him several days before. Hancock then directed him to report to the Marine Committee, which at the time was beginning its inquiry into the conduct of Commodore Hopkins during the New Providence raid and complaints against Captains Dudley Saltonstall and Abraham Whipple. Nicholas' appearance before the Committee was short, and after completing his personal business, he asked to be returned to the fleet in Rhode Island. Instead of complying with the request, the Marine Committee detached him from the *Alfred* and ordered him to remain in Philadelphia to organize and discipline the "four Companies of Men being then raising for the Frigates on the Stocks."[2]

On 25 June, Congress appointed officers to head the Marine detachments of the four frigates being built at or near Philadelphia. Benjamin Dean of Philadelphia was designated as captain of Marines on board the 32-gun frigate *Washington*. Two other Philadelphia residents, Peregrine Brown and Abel Morgan, were to serve as his first and second lieutenants, respectively. Samuel Shaw became captain of Marines on board the second 32-gun frigate, *Randolph,* and Franklin Reed his first lieutenant. The position of second lieutenant was filled in September with the selection of a Frenchman from the West Indies, one Panatiere de la Falconniere. Heading the list of Marine officers appointed to the 28-gun frigate *Effingham* was the young Philadelphia schoolmaster, astronomer, and mathematician, Andrew Porter. Daniel Henderson was appointed his first lieutenant, and James McClure served as second lieutenant. The smallest of the four frigates was the *Delaware* of 24 guns; its Marine detachment was officered by Captain Robert Mullan, a Philadelphia innkeeper. Selected as his subordinates were David Love, first lieutenant, and Hugh Montgomery, second lieutenant.[3]

Two other men were also appointed Marine officers on the 25th. Joseph Hardy was selected as captain of Marines. At the time he was ship's clerk on board the *Columbus* and continued as such throughout the remainder of the year. William Gillmore was commissioned second lieutenant, but to

Andrew Porter, by Edward Marchant.

which ship he was assigned remains unknown. Indications are that he served as an extra lieutenant for the four companies until early December when he resigned his commission and joined the army.[4]

Recruitment of men for the four Marine companies was begun almost immediately. On the day Robert Mullan was appointed he enlisted two brothers, Collin and Peter York, whom he designated as the company's drummer and fifer.[5] Mullan then established a rendezvous at his inn, Tun Tavern on the east side of South Water Street, between Chestnut and Walnut Streets, from which his two lieutenants made frequent recruiting forays. The city of Philadelphia proved to be the major source of recruits for Mullan's company, but not the only one, for Lieutenants Love and Montgomery roamed as far west as Lancaster and as far south as Kent County, Maryland.[6]* Officers of the other three companies followed the same pattern in recruitment as did Robert Mullan; the only exception being Andrew Porter. While his lieutenant, Daniel Henderson, scoured the countryside for recruits, Porter concentrated on his school where about 100 students were enrolled. It is probable that a number joined their headmaster and volunteered for service with the Marines.[7]

Throughout July and August recruitment progressed slowly and, as in previous periods, men often regretted their decision. On 14 August, Franklin Reed, first lieutenant in Captain Shaw's company, reported the first deserter, one "Angus Cameron, born in Ireland, about 5 feet 8 or 9 inches high, dark complexion, with short black hair, curled behind, a little pock marked, about 30 years of age." Lieutenant Reed offered 30 shillings reward and a payment of "all reasonable charges," but he had no takers. The advertisement was repeated the following month, again without success.[8]

In late August, while Marines were still being enlisted, Nicholas requested arms. Congress complied with the request by directing the Secret Committee on 22 August to deliver to Major Nicholas a number of muskets sufficient to arm the Marines.[9] But it was only through the influence of Nicholas and the pressure of his friends that he succeeded in retaining these arms, for on the 29th, Congress directed that the Secret Committee provide the German Battalion with "proper arms," and if it could not otherwise do so it should use for "this purpose the arms ordered to be put into the hands of the marines."[10]

Activities of the four companies of Continental Marines stationed in Philadelphia consisted mainly of guard duty. On 16 September, at the request of Pennsylvania's Council of Safety, Congress ordered Major Nicholas to provide daily guards for each of the four frigates, in addition to the state prison, and the powder stores belonging to the Continent.[11] One week later Congress partially reversed its previous order and directed that two of the companies march immediately to Fort Montgomery on the Hudson River, there to guard Continental stores, and the frigates *Congress* and *Montgomery*.[12] But because of enemy threats to the city, the two Marine companies remained at their posts in Philadelphia.

In early October, speculation was that British General Sir William Howe would soon send a portion of his army from New York to invade Pennsylvania. Committees were organized by both the Continental Congress and the Pennsylvania Council of Safety to work out a joint plan of defense. Pennsylvania's

* Among the men recruited by Mullan were two blacks, Isaac Walker, and "Orange Negro." Both received the same pay and allowances as the white enlisted man, and it is assumed that they performed the same duties.

three-man committee, on 16 October, submitted its recommendations to the Continental Board of War.[13] In addition to two Virginia battalions and the German Battalion, the committee suggested that the four companies of Continental Marines be:

> Continued in this State, or at Trenton or Billingsport, in New Jersey, as Necessary defence of this City, not only against the British Troops, but the growing party of disaffected persons, which unhappily exists at this time; also for the protection of the vast quantity of Stores belonging to Congress that are now in the public Magazines; and at the same time to carry on such Works of defence as have been or may be thought necessary.[14]

The recommendations of the Pennsylvania committee were adopted, but with British troops seemingly engaged in Westchester County, New York, for the near future, the threat of an imminent invasion quickly passed and was forgotten for the moment.

Major Nicholas, by early November, had a well-organized and well-disciplined battalion of Marines housed in comfortable barracks on Second Street in the Northern Liberties, just above the city's limits. Their health was well taken care of for on 4 November Congress directed the celebrated Dr. Benjamin Rush "to take them under his care, and see them properly provided for."[15] Nevertheless, the desertion rate was high, and few deserters ever returned to duty. One who did was Henry Hassan, a private in Robert Mullan's company. On 24 November his court-martial was held. Facing a board of five officers headed by Mullan, Hassan pleaded guilty to the charges of desertion and quitting his guard post without being properly relieved. The sentence was severe: "fifty lashes for Desertion & Twenty one Lashes for Quiting his guard," to be "on his bare Back well laid on at the head of his Company."[16] Private Hassan was returned to duty, but a little over a month later he deserted a second time.[17]

Philadelphia, although it had considered the threat of a British invasion in October of no real alarm, was disturbed in mid-November with a report that a large enemy fleet was moving southward from Sandy Hook. In response, Congress directed the Marine Committee to arrange its naval forces in the Delaware "as will best conduce to defeat the designs of the enemy."[18] One of the measures adopted by the Committee was to have the *Randolph* made ready for sea. Captain Samuel Shaw's company of Marines was ordered from the barracks to the frigate, and a rendezvous was opened for the immediate enlistment of a crew. Even though the rumored enemy fleet proved to be a group of British merchantmen returning to England, preparation of the *Randolph* for sea duty continued.

Fort Washington fell on 16 November and Fort Lee on the 20th; the British were in New Jersey and a real threat to Philadelphia. The defense of the capital, therefore, took precedence over the manning of the *Randolph*.

General Washington and his army were in a precarious situation by late November 1776. After the fall of New York, Washington had been forced to withdraw to Harlem Heights, then to White Plains, and then to Peekskill in upper Westchester County. He crossed the Hudson on 13 November and began

his painful retreat to Hackensack, to Newark, to Elizabethtown, and then to Brunswick. There, on 30 November, Washington penned a letter to John Hancock in which he pleaded for reinforcements: "Their arrival is much to be wished, the situation of our Affairs being truly alarming and such as demands the earliest aids."[19] But major reinforcements were slow in coming. With a confident British force of 10,000 in pursuit, Washington's weakened army of 3,000 abandoned Brunswick on 2 December, marched through Princeton, and then halted on the east bank of the Delaware River at Trenton.

Off to Trenton

News of the British occupation of Brunswick and their design to march toward Philadelphia, sent the city into near panic on 2 December. Townsfolk quickly loaded wagons with belongings and prepared to forsake the city for shelter with relatives in the countryside. On orders of the Council of Safety, shops were closed, schools shut, and all citizens and militia ordered out to defend the city. By late afternoon, as one diarist recorded, Philadelphia took on "a martial appearance."[20] Among the troops called out were the three companies of Continental Marines, Captain Shaw's company being on board the *Randolph*. Muskets were inspected and bayonets and cartouch boxes issued. All was ready by nightfall, and the three companies, less the sick and a skeleton guard for the frigates, boarded gondolas for the trip up the Delaware River to Trenton where they would assist Washington.[21]

John Cadwalader, by Edward Bawers.

The situation in Philadelphia remained chaotic following the Marines' departure. The Associators (militia) composed of clerks and shopkeepers of the city and the Northern Liberties were ordered to assemble and be formed into a brigade on the 3rd, which was placed under the command of Colonel John Cadwalader.[22] The brigade was divided into three battalions: the first commanded by Colonel Jacob Morgan; the second by Colonel John Bayard; and the third by Lieutenant Colonel John Nixon.[23] In all the brigade numbered approximately 1,200 men. Colonel Cadwalader with a portion of the brigade—well clothed, although poorly armed—marched northward on 5 December to join Washington; the remainder were to follow the next day.

After a rather cold and snowy march along the Bristol Road, Colonel Cadwalader assembled his troops opposite Trenton and then crossed the Delaware. General Washington, although he first contemplated sending Cadwalader to reinforce the rear guard at Princeton, ordered him on 7 December to remain at Trenton and await the arrival of the rest of his brigade. In the meantime, Washington requested that he take the seamen from the frigates *Delaware* and *Washington* under Captains Charles Alexander and Thomas Read, and Major Nicholas' Marines in his charge, "till a further disposition of them can be made, if necessary, letting me know in the meanwhile if they came out resolved to act upon Land or meant to confine their Services to the Water only."[24]

General Washington's Retreat

Somewhat puzzled at the week-long inactivity of the British at Brunswick, Washington started out for Princeton with about 1,200 men to reinforce General Lord Stirling (William Alexander) later on the 7th. His march was halted by the receipt of news that the enemy was advancing and that an attempt would be made to cut off the troops at Princeton. Instead of proceeding farther, Washington ordered an immediate retreat to Trenton and then a withdrawal across the Delaware. In the process, everything of value that could be used by the British to continue the chase was either destroyed or removed to safety.[25]

General Howe's pursuit of Washington across New Jersey was slow and cautious. Reaching Princeton on the 7th, he divided his forces into two corps; the first under Major General James Grant was to advance southward, directly to Trenton, and the second under Major General Charles Cornwallis was to march to Maidenhead, halfway between Trenton and Princeton. The next day the vanguard of Grant's troops reached Trenton, just as the last of the Americans were crossing the river into Pennsylvania. But as Washington had ordered the removal of all boats from the east bank of the river and the positioning of cannon on the west, the British advance was abruptly checked on the 8th. Cornwallis' troops searched the east bank as far as Coryell's Ferry, 15 miles above Trenton, but no boats were to be found. Once Howe learned that Grant and Cornwallis had lost momentum, he abandoned the planned invasion of Pennsylvania and established a line of armed camps stretching from Hackensack to Burlington on the Delaware. As soon as the camps were organized and the troops moved into quarters, Howe departed New Jersey for New York City, there to spend the winter months.[26]

Washington remained convinced that Cornwallis had been stopped only temporarily by the river. Word was received on the 10th that the British were repairing the bridges below Trenton and as the general wrote Congress, it "seems to indicate an intention of their passing lower down."[27] Responding to the British moves, Washington requested that the Pennsylvania Council of Safety order Commodore Thomas Seymour of the state navy to send one of his galleys up the river as far as Dunk's Ferry in order to gain intelligence on enemy movements in the area.[28] In addition, he ordered Cadwalader's third battalion to Dunk's Ferry to guard the crossing, and a scouting party of Pennsylvania riflemen across the Delaware, near Bordentown.[29]

Charles Cornwallis, by Thomas Gainsborough.

Two hours before daylight on 11 December, Colonel Carl E. U. von Donop of the Jaeger corps of Hesse-Cassel left Trenton with a detachment, passed over Crosswicks Creek, and proceeded southward to take possession of Bordentown and Burlington. Some resistance was offered by the local militia, but the Hessians reached Bordentown and moved on to Bustleton. A short distance from Bustleton, they encountered the American scouting party and forced it to withdraw across the Delaware. Upon his arrival at Burlington, von Donop was informed that his presence would subject the town to a cannonade from several galleys under Commodore Seymour, which had recently anchored off the town. He urged that a delegation of townspeople confer with the commodore, and then he and his officers retired to await Seymour's answer. Meanwhile, Hessian guards were seen patrolling the streets of Burlington and immediately the fleet opened fire. Commodore Seymour then informed the town that if Hessians had indeed occupied it, he would have no mercy on the place. Upon hearing of Seymour's answer, von Donop concluded that it was unwise to continue his operations, and he withdrew northward. The following day, Marines of the Pennsylvania privateer *Hancock* under Captain William Shippin, "a smart little fellow," occupied the town.[30]* A foothold within striking distance of British forces had been gained.

Reports from the scouting party and Commodore Seymour confirmed Washington's suspicions. Based on the reports, General Washington on the 12th began to deploy the remainder of his forces along the Delaware River to the south. Brigadier General Philemon Dickinson was to secure Yardley's Ferry and place his troops along a front extending to a point two miles below the ferry where Brigadier General James Ewing's sector began.[31] General Ewing was to make his headquarters at a ford near Hoop's Mill on Biles Creek and occupy a front which stretched as far as the Bordentown Ferry.[32] Colonel Cadwalader took over the ground south of Ewing down to Dunk's Ferry, with his headquarters at Bristol. All three officers received somewhat the same orders: "Spare no pains or expence to get Intelligence of the Enemy's motions and intentions"; defend against a British attempt to cross the river; and if routed to withdraw to "the strong ground near German Town."[33]

Cadwalader's forces moved into Bristol on the 13th and the following day quarters were assigned. The Marines were boarded in the town's Quaker Meeting House, while Major Nicholas was quartered at Bessonet's Tavern. Accommodations for the remainder of the brigade were found in several public houses and private homes scattered about the town.[34]

The *Randolph* Readied

As General Washington set up a defensive line along the western bank of the Delaware, General Israel Putnam took charge of the defense of Philadelphia. But before much could be done, an express rode in from the Delaware Capes with news that HMS *Roebuck* had anchored off the bay, blocking exit to the sea.[35] Congress immediately became panicky. The Marine Committee was directed to send fast sailing vessels to cruise off the capes and warn inbound merchantmen of the danger. General Putnam was given complete charge of the city's defenses and was to have combustibles prepared to burn the frigates and other Continental vessels should they be in danger of falling into enemy hands. Congress next turned its attention to the two vessels about ready for duty—the *Randolph* and the sloop *Hornet*. Both were directed to observe Putnam's orders "in preventing the enemy from passing the Delaware." Congress then made a unique bounty offer to the officers, seamen, and Marines of the *Randolph*. When the "General has no further occasion for the use of the frigate *Randolph*, for the defence of this city," resolved Congress, and if Captain Biddle succeeds in getting to sea, "this Congress will reward him and his people with a present of 10,000 dollars." Congress then adjourned to Baltimore and gave to General Washington dictatorial powers in "all things relative to the department, and to the operations of war."[36]

Congress left behind one of its most able and energetic members, Robert Morris. Determined not to put Continental vessels to the torch, Morris proposed to General Putnam that the *Randolph* and *Hornet* be sent to sea immediately; Putnam "readily consented." Morris then proceeded to supply the *Randolph* with powder and ammunition as Captain Biddle obtained additional hands from the city prison. Believing that the frigate had "Seamen & Marines sufficient to work the Ship," and that it was better to send her to sea shorthanded than "to let her remain & be destroyed," Morris ordered the *Randolph* out of port on the 13th:

> You are therefore to repair on bd the *Randolph* Frigate of wch you are Commander and proceed with

* Several days later Shippin and his small company of Marines were attached to Cadwalader's 2d battalion.

> all possible diligence to Sea. We shall not at this time prescribe to you where you are to Cruize nor form any plan whereby to distress our Enemies—the first Object is to get the Ship well manned and for this purpose you may Cruize where you think is the best Chance of meeting Merchantmen or Transports without Encountering Frigates &c we find by [expe]rience the Seamen taken in prizes are in Gneral very [wi]lling to enter into our Service we think you will be able [in] this Manner to compleat your Number admitting [all] such as enter freely & induce them thereto show [the va]rious Encouragement given by Congress .

After the *Randolph* was sufficiently manned, Biddle was to return and cruise the coast for enemy shipping destined for New York.[37]

Instructions to Captain James Nicholson of the *Hornet*, who had recently arrived with a group of seamen and Marines from Baltimore, varied from those issued to Biddle. Since the *Hornet* was sufficiently manned, she was directed to proceed to South Carolina after clearing the capes, and from thence to Martinique where supplies might be obtained for the army.[38]

The *Randolph* and *Hornet* cast off from Fort Island with the tide on 14 December and dropped down to the chevaux-de-frise below Hog Island. The following day an express overtook them with letters for the *Hornet* and a request that the two ships put into Chester to await a merchantman destined for France. While anchored at Chester, a pilot boat arrived from Philadelphia bearing an urgent dispatch from Morris recalling both ships. According to the latest information, the *Roebuck* had recently been reinforced by the armed sloop *Falcon* and two bomb ketches, therefore both ships would be in great danger should they proceed farther down the river. "My labours appear to be lost," Morris wrote Congress two days later, "and sorry I am for the disappointment." He, however, was hopeful that "the first smart northwester" would drive away the blockaders and open the way for the two vessels to get to sea.[39]

The safety of the *Randolph* and *Hornet* was not Morris' only concern. On the 13th he wrote General Washington requesting his cooperation in endeavoring to save another frigate, the *Delaware*. Specifically he asked that her captain (Charles Alexander), seamen and Marines be permitted to return to Philadelphia in order that the frigate could be taken to sea as soon as possible.[40] Washington immediately forwarded the letter to Colonel Cadwalader, under whose command they had been placed. Upon receipt of Morris' urgent request, Cadwalader ordered Captain Alexander and his seamen back to Philadelphia and directed that a group of Marines be assembled. Major Nicholas responded by taking a draft of men from each of the three companies and forming a 20-man detachment which he placed under the command of Lieutenants Daniel Henderson and David Love.[41]

The formation of the detachment of Marines for the *Delaware* called for an immediate shift among officers. To replace David Love, Sergeant James Coakley was elevated to the temporary rank of first lieutenant and Peter Bedford, a friend of Robert Mullan, replaced Hugh Montgomery, who had recently been assigned the duty of quartermaster for the Marines. Second Lieutenant James McClure filled the vacancy left by Daniel Henderson, but a new lieutenant was not appointed by Captain Porter to replace McClure. In addition, about this time a

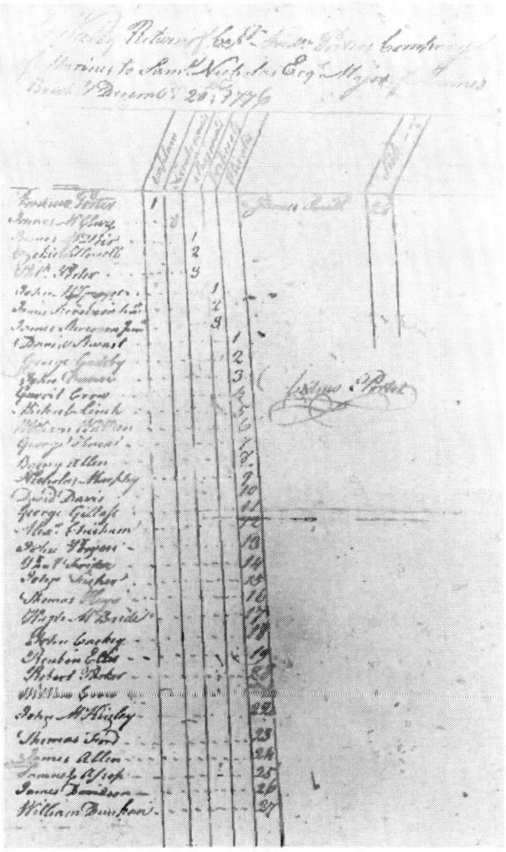

Captain Porter's Roll, December 1775

familiar Marine captain, Isaac Craig, joined Nicholas as adjutant.[42]

The loss of 20 men from Major Nicholas' Marines seems to have had little effect upon Cadwalader's command at Bristol, for the British had gone into winter quarters on 14 December and all was quiet for the moment. The Marines, even though they now numbered approximately 130 officers and men, were required to contribute their original share of the brigade's main guard: one corporal, eight privates, and a company commander on a rotating basis as captain of the guard. The function of the main guard at Bristol and smaller sector guards posted along the brigade's share of the Delaware River was not only to defend the western bank, but to gather information and harass the enemy since there was always the possibility of a British attack. It was this possibility that continually plagued General Washington.[43]

Washington's Important Stroke

At no time since he established the thin defense along the Delaware's western bank was Washington free of the apprehension that an enemy crossing was planned. He was also aware that if such a plan were carried out in the near future, his troops could not hold, and their collapse would prove disastrous to both the army and possibly the cause. And there was also 31 December, the day when his whole army might practically disappear due to expiration of enlistments. Therefore, the only solution to his many problems was to take the initiative and attack.

As early as 14 December, Washington expressed the hope that if reinforced by troops under General Lee, he might "under the smiles of providence, . . . effect an important stroke."[44] News of the withdrawal of General Howe's main force to New York and the scattered disposition of the remaining troops was encouraging. A decision had to be made and a definite plan evolved, for such an enterprise would risk the whole army. By the 24th, Washington had formulated a plan which he presented to his commanders. After short discussions, the plan was adopted.

The primary object of Washington's "important stroke" was Trenton. The river was to be crossed in three places by three separate divisions. Brigadier General Cadwalader was to command one, composed of his own brigade, Colonel Daniel Hitchcock's brigade of Rhode Island Continentals, and a company of Delaware militia under Captain Thomas Rodney. This division was to cross the river near Bristol, and together with Colonel Samuel Griffin's troops, who were already in New Jersey, move northward and join Washington at Trenton.*

To the north of General Cadwalader was the second division under Brigadier General Ewing. Chiefly composed of Pennsylvania militia, Ewing's division was to cross the river at Trenton Ferry, just below the town. Once on the Jersey side, the plan called for Ewing to seize the Assunpink Bridge to prevent the Hessians in Trenton from escaping the attack launched by the principal division.

* Washington felt obliged to promote Cadwalader to the rank of Brigadier General of the militia since his command was to include a Continental colonel.

The third and principal division was commanded by Washington. It was made up of about 2,400 men from the forces which had been involved in the initial retreat across New Jersey. General Washington's force was to cross at McKonkey's Ferry, about nine miles above Trenton, and then march to the town. Once his own and Ewing's divisions had taken Trenton, and after Cadwalader had joined them, Washington planned to push toward the British posts at Princeton and Brunswick.

Washington Crosses the Delaware

On Christmas day in Trenton, the 1,400 Hessians under Colonel Johann Gottlieb Rall celebrated the holiday in the usual German manner with drinking and feasting, but in the American camp across the river, the day was spent assembling the troops which would cross the Delaware that night. As the day progressed, a snow storm set in and the wind picked up from the northeast. At six in the evening, the first of Washington's troops loaded into boats and put across the river. Unit after unit followed, but by midnight the river ice had so slowed progress that the force was three hours behind schedule. It was not until four the following morning that the division formed and began to march.

About four miles from McKonkey's Ferry, Washington halted his column to give the men time to adjust to the intensified storm. When the order to march was given, the division split into two corps; the first under General Nathanael Greene swung left and took the Pennington Road, while General Arthur St. Clair proceeded down the River Road. Within a mile of Trenton, Greene's force divided once again, forming a rough semi-circle about the town. After driving in some Hessian pickets north of town at seven, the main attack on Trenton began.

When orders were given to advance, Captains Thomas Forrest and Alexander Hamilton of the artillery trained their six field-pieces down King and Queen Streets and immediately opened fire. The shots tore into the ranks of alerted Hessians who gave two volleys and then withdrew in confusion. Driven to the edge of town, Colonel Rall reformed his broken lines and started back, intending to make a bayonet charge. But as his men marched back accompanied by the regimental band, it was already too late; the Americans were everywhere. From each window and doorway they picked off the returning Hessians. Colonel Rall ordered a charge and then a retreat, but the Americans now held Assunpink Bridge and there was no escape to the south, so Rall signaled a withdrawal to an orchard at the southeast end of town. He had hardly given the order when two bullets struck him in the side and he was carried into a nearby house where he later died.

Without their commander, Rall's men continued their flight to the orchard and then north, seeking a way out, but the situation was hopeless and they surrendered. Meanwhile, a second group of Hessions under Major von Dechow tried to cross the Assunpink, but they were met and driven back. They, too, were finally forced to lower their colors and ground their muskets.

Washington's losses during the two-hour battle were light: one killed and three wounded. Hessian losses, on the other hand, were heavy: 22 killed, 83 wounded, and 891 captured. Only about 40 percent of Rall's force managed to escape, either to Bordentown or Princeton.

Cadwalader Fails

While Washington's force had fulfilled its assignment, the two planned crossings to the south were complete failures. "About dark," General Cadwalader moved his troops out of their quarters at Bristol toward Neshaminy Ferry where they planned to cross the Delaware. Colonel Timothy Matlack's battalion of Pennsylvania riflemen led the march followed by the light infantry and militia; Continental troops brought up the rear. Major Nicholas' Marines accompanied the militia.[45]

Upon the division's arrival at Neshaminy Ferry, Cadwalader found the river choked with ice and the crossing was delayed. Finally, the decision was made to abandon Neshaminy for Dunk's Ferry several miles below, in hope of finding a better crossing. There Matlack's battalion embarked in several large boats to cover the landing of the remainder of the division. They landed on the Jersey shore with great difficulty, having traversed about 150 yards of ice before reaching the river bank. About 200 yards inland, the advance party formed and awaited the others. Shortly after Matlack's battalion was ashore, the 1st and 3rd battalions of Philadelphia Associators started to cross. As the last of the Associators landed, the storm increased and the ice began to thicken. Not wanting to risk his fieldpieces on the newly formed ice, Cadwalader ordered his men back to the Pennsylvania side.[46]

General James Ewing was much less daring than Cadwalader. After surveying the river conditions at Trenton Ferry, he thought a crossing impossible, and thus did not attempt one.

Following the swift and complete victory at Trenton, Washington recrossed the Delaware into Pennsylvania and immediately dispatched his prisoners to Philadelphia where Robert Morris was again trying to get the *Randolph* to sea. The enemy ships were still in Delaware Bay, and Captain Biddle was more reluctant than ever to proceed without a full ship's complement. At year's end, Morris expressed his diasappointment with the *Randolph's* captain and crew in a letter to Richard Henry Lee:

> You cannot conceive how I am vexed & mortified to find after the deal of pains & trouble I have taken that the *Randolph* Frigate is still at the Piers & Ice making in the River, but the Officers of that Ship show great reluctance to go away without being compleatly manned & that is not possible, she might have been at Sea before now had they exerted themselves for that purpose but they had constantly in View to wait for more Men, this has its foundation in a noble principle which has hindered me from complaining to the Marine Committee, altho I have scolded the Officers like a Bitter Whore for their dilatoriness; they say they wish to Fight & not to run, I tell them they must run until they can fight.[47]

With the victory at Trenton, the immediate threat to Philadelphia was removed, and thus it was no longer necessary to send the *Randolph* to sea.

Into New Jersey

Late on 26 December, Washington, at his headquarters at Newtown, Pennsylvania, received a letter from General Cadwalader explaining the reasons for his failure to complete the crossing of the Delaware. Having not yet been advised of Washington's present position, Cadwalader added that he planned another attempt to cross the river the following morning.[48] Washington hurriedly dispatched a letter to Bristol informing Cadwalader that he had returned to Pennsylvania and asked that he defer his "intended Operations till you hear from me."[49] Cadwalader failed to receive the letter in time, and at sunrise on the 27th "proceeded about 2 Miles above Bristol & embarked."[50]

By one in the afternoon, all except Colonel Hitchcock's regiment had crossed the river. As his initial contingent was about to embark, a messenger arrived and announced that Washington had returned to Pennsylvania. Hitchcock thereupon decided against crossing. When Cadwalader received the news he immediately called a conference of his officers, and after considerable debate it was decided to remain in New Jersey until further orders. The discussion then turned to possible courses of

George Washington, by Charles Willson Peale.

action: "some were for proceeding to Mount Holly to attack the Hessians," while others were inclined "to march on to Bordentown which might be expected to be weakly provided with Troops." An alternate suggestion that the force proceed to Burlington, however, was adopted and ordered.[51]

Cadwalader's forces reached Burlington about nine in the evening on the 27th, but found no enemy troops. Colonel Joseph Reed, who had scouted ahead, sent back word that the way to Bordentown was clear. At four the following morning, Cadwalader moved his forces to Bordentown where they took possession of a large quantity of abandoned Hessian stores, but found no food. After surveying the devastation wrought by the enemy in their retreat, Cadwalader pushed on to Crosswicks, four miles to the northeast, where food for his men was procured.[52]

Washington had planned to return to New Jersey anyway, but Cadwalader's crossing seemed to have forced his hand, therefore he devoted the day before the scheduled crossing to the subject of reenlistments. Fortunately, Congress had voted him the authority to offer each man whose enlistment was about to expire, a bounty of ten dollars if he decided to remain an extra six weeks. Washington approached each Continental regiment with the bounty offer in turn, but only 1,500 out of an original force of 2,400 agreed to the additional service. On the 29th his depleted army started to cross the Delaware, but it was two days before they entered Trenton.

Meanwhile at Crosswicks, Cadwalader was determined to push to Allentown, but his officers balked, declaring "it was madness to attempt it; for it would use up all our brave men not one of whom had yet given out but were dreadfully fatigued."[53] Cadwalader, however, remained resolute and on the evening of the 28th sent a party to Allentown and then on to Cranbury to gather intelligence. The remainder of the division stayed on in Crosswicks where quarters and food were obtained. During the next three days Cadwalader maintained his position by sending out small patrols into the countryside. On one such partol, Nicholas and his Marines captured a British sympathizer who informed them that the ex-sheriff of Monmouth, Colonel John Lawrence, was collecting men at the Monmouth Court House. According to the prisoner, Lawrence had thus far enlisted 70 men and imprisoned 20 Americans for refusing to join his band of Tories. Nicholas immediately requested permission to go after Lawrence, but Cadwalader refused, noting that "it was not an object at this time."[54] Washington, although he left the decision up to Cadwalader, thought that if "it is not too far distant from your Army, it will certainly have a good effect to quell such Insurrections."[55]

The last day of 1776 found Washington again busily engaged in persuading the Continental troops to stay an additional six weeks. One by one the regiments were paraded and formed into ranks. Washington then addressed them:

> 'My brave fellows, you have done all I asked you to do, and more than could be reasonably expected; but your country is at stake-your wives, your houses and all that you hold dear. You have worn yourselves out with fatigues and hardships, but we know not how to spare you. If you will consent to stay only one month longer, you will render that service to the cause of liberty and to your country which you probably never can do under any other circumstances.'[56]

By New Year's Day, Washington was able to muster a total of 3,335 men.

With the question of enlistments settled, Washington began planning future operations. A council of war was called on the evening of 1 January where it was decided that Cadwalader's troops at Crosswicks and General Thomas Mifflin's forces at Bordentown should be brought to Trenton. Further, there is ample evidence that the council determined to make an attack on Princeton by way of the Quaker Road, instead of via the Trenton-Princeton Road which might prove costly.[57]

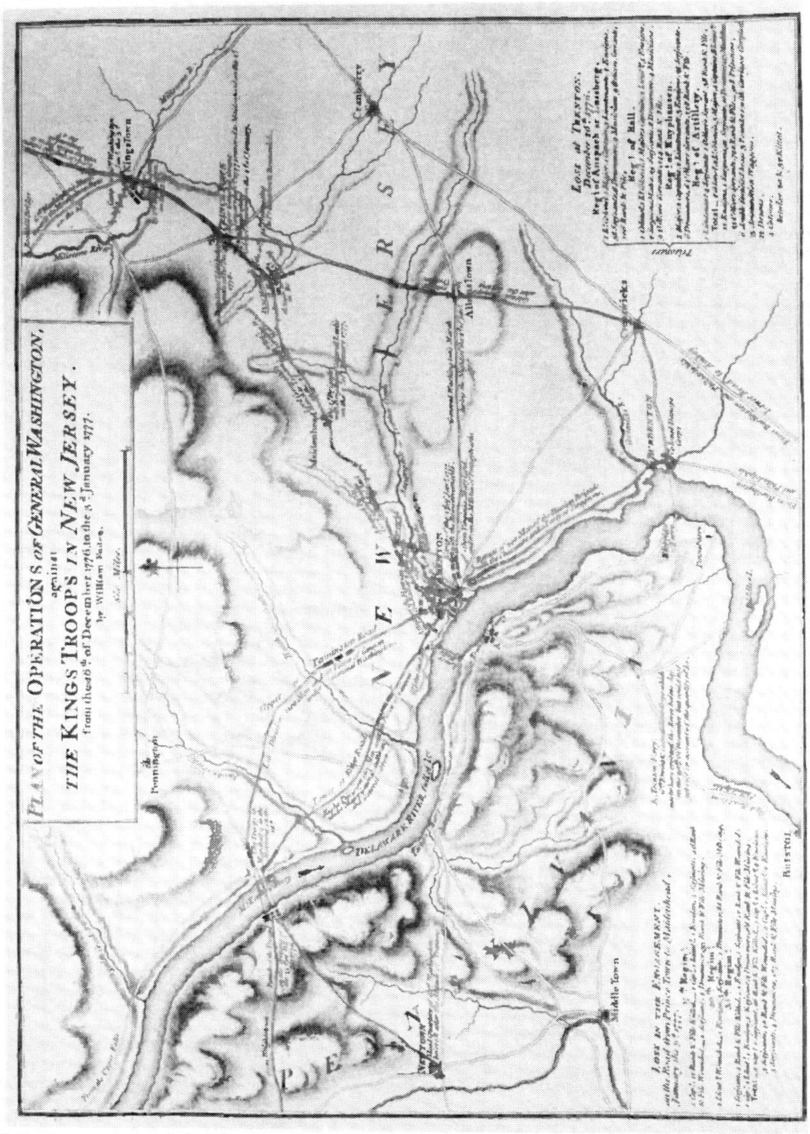
PLAN OF THE OPERATIONS OF GENERAL WASHINGTON,
against
THE KINGS TROOPS IN NEW JERSEY.
from the 26th of December 1776, to the 3d January 1777.
by William Faden.

Assunpink

As Washington made preparations for future attacks, a British retaliatory force of 8,000 was on the march. At dawn on New Year's Day, Major General James Grant moved the 4th Brigade, dragoons and Hessian Grenadiers out of Brunswick. Six hours later Grant's vanguard entered Princeton. The main British force followed. Throughout the remainder of the day he ordered the troops to be alert and ready to march at a moment's notice. Late that evening, Major General Cornwallis arrived from New York and immediately called a conference of the officers where he announced his intention of marching on Trenton the following morning. He then ordered all troops, except the 4th Brigade and 16th Dragoons who were to garrison Princeton, prepared for the march at daybreak.

At one in the morning on 2 January, Cadwalader's troops started out for Trenton. The roads were muddy and the march tedious, but by sunrise the bulk of Cadwalader's force had reached the town. Soon after quarters were assigned and a meal begun, the call to arms was sounded. Without delay the brigades were assembled and moved to prepared positions on the southern bank of Assunpink Creek. General Mercer's brigade was stationed at Phillip's Ford, about two miles up the creek. Below him was General Cadwalader's command, positioned in an open field about a mile from town. General St. Clair's brigade was posted on the high ground east of Assunpink Bridge.[58]

The British advance from Princeton began at eight on the 2nd. In three columns, Cornwallis' force passed through Maidenhead about noon and shortly thereafter approached Shabakunk Creek. There they met troops under Colonel Edward Hand which had been sent forward to dispute the advance. For three hours the Americans held until forced to retreat to previously prepared positions at the northern end of Trenton. There Hand again held off the British, this time for an hour. As the enemy pressed forward, Hand was forced to yield ground a second time to prevent being outflanked. At this point, Washington ordered Hitchcock's brigade across the Assunpink to cover Hand's retreat to the south side. Hit in the front by Hessian Grenadiers and on the right by British Light Infantry, Hitchcock's men suddenly broke ranks and raced back to the bridge.

As the New Englanders spilled across the bridge, Washington observed the British Light Infantry approaching the creek below the bridge with the apparent intention of crossing. Immediately Hitchcock's brigade was ordered reassembled and sent down, but American artillery intervened and drove the enemy infantry back. The Hessian grenadiers then attempted to force the bridge and Cadwalader's light troops—Nicholas' Marines among them—were ordered "to fly to the support of that important post." Again the enemy was repulsed.[59]

On to Princeton

After the Hessian withdrawal, both sides continued the artillery duel until mid-evening when the two camps settled in for an uneasy night's rest. At the British camp, some two miles northeast of Assunpink Creek, Cornwallis decided to begin the main attack at Phillip's Ford the following morning. In the American camp, Washington called a hasty council of war. Should the fight at Trenton be continued or should they withdraw southward? Both questions were heatedly debated. Then someone observed that since the enemy's main forces were at Trenton, there should only be a small garrison at Princeton and that an attack there would not only gain sorely needed supplies but would avoid any appearance of a retreat. The suggestion was immediately taken up and a plan formulated. An attack would be launched against Princeton, and if successful, they would push on to Brunswick.[60]

Preparations for the march were begun at once. Washington "ordered the guards to be doubled, a strong fatigue party to be set to work on an intrenchment across the road near . . . [Henry's] mill, within distinct hearing of the sentinels of the enemy," and the heavy baggage removed to Burlington. He then directed that the campfires be heaped with fence rails and kept burning until daylight. An hour after midnight the troops were "silently filed off by detachments" and the march to Princeton commenced.[61]

"The night," recorded Major James Wilkinson, "although cloudless, was extremely dark, and though calm most severely cold, and the movement was so cautiously conducted as to elude the vigilance of the enemy." The course followed by the American army first took them through the woods paralleling Mirty Run and then to Quaker Road which ran from Princeton to Allentown.[62] About midway through the march "there was a cry that they were surrounded by the Hessians and several corps of [Mifflin's] Militia broke and fled towards Bordentown." The rumor was false and the main portion of the column "remained firm and pursued their march without disorder." At sunrise, the army crossed Stony Brook bridge on Quaker Road and halted in an open field beyond. Here Washington divided his army into two columns. Mercer's and Cadwalader's brigades under General Greene were to march up Quaker Road and secure the bridge on the main road to Trenton. The second column headed by Major General John Sullivan was to take Saw Mill Road and enter Princeton from the east.[63]

As Washington's ragged force of 6,000 stood less than three miles from Princeton, an unsuspecting British force of 700 under Lieutenant Colonel Charles Mawhood began its march to Trenton. Mawhood's brigade consisted of the 17th and 55th

The Battle of Princeton, by Wiliam Mercer.

Regiments of Foot, and one troop of the 16th Dragoons; the 40th Regiment remained behind to guard Princeton. Shortly after crossing Stony Brook, Mawhood sighted a group of armed men emerging from the woods to the south. Supposing them to be part of the American army fleeing Cornwallis, Mawhood dispatched a mounted dragoon to reconnoiter. He then ordered the 17th Regiment to withdraw across the bridge to an orchard on the high ground above the stream and the 55th to return to Princeton with the supply train.

Mawhood had observed what proved to be the vanguard of Mercer's brigade on its way to the bridge. Mercer, thinking the enemy a small scouting party sent from Princeton, pushed up the hill to his right in hope of cutting off their retreat. As his troops crested the hill and descended through the orchard, Mawhood's men who were entrenched behind a bank, "rose and fired" at the Americans. Their first shots, according to an American sergeant, "passed over our heads, cutting the limbs of the trees under which we were marching." The 17th then retreated a short distance to their packs and established a new line of battle. As the Americans got off their first rounds, the British charged with bayonets and soon overpowered them, mortally wounding General Mercer and others. The remainder of Mercer's decimated brigade fled in disorder.[64]

General Cadwalader, who had halted his column, heard the firing in the distance and without delay marched toward it. Moving his troops to the high ground he observed Mercer's brigade in full retreat and immediately ordered the column to form into divisions from the right; Morgan's battalion taking the left flank, and Matlack's riflemen and Nicholas' Marines the right. He then directed the troops to advance in formation and fire at will, but their first rounds were fired at such a distance that they had no effect upon the enemy. Seeing the situation, Cadwalader ordered the divisions to double up and reload as they advanced. But it was too late; his

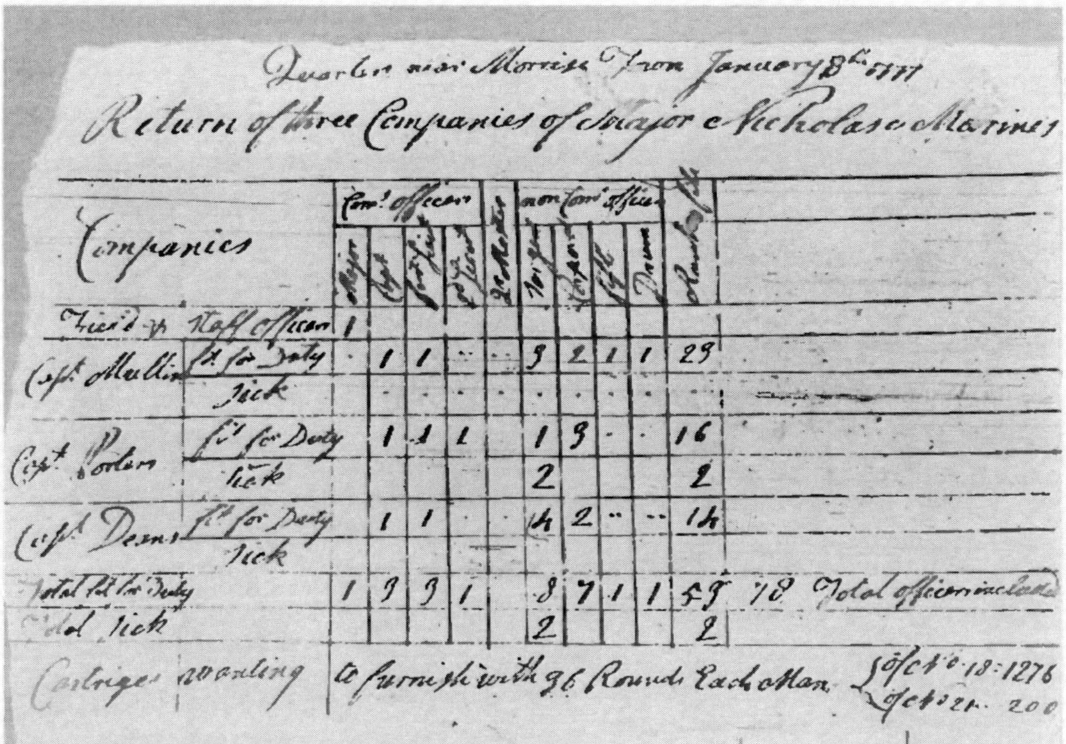

Quarters near Morristown January 8th 1777

Return of three Companies of Major Nicholas's Marines

Companies		Com.d officers: Major	Capt	First Lieut	2d Lieut	Qr Master	Non Com.d officers: Serjeants	Corporals	Fife	Drum	Rank & File
Field & Staff officers		1									
Capt. Mullin	fit for Duty	·	1	1	··	··	3	2	1	1	29
	Sick	·	·	·	·	·	·	·	·	·	·
Capt. Porter	fit for Duty		1	1	1		1	3	·	·	16
	Sick						2				2
Capt. Dean	fit for Duty		1	1	·	·	4	2	··	··	14
	Sick										
Total fit for Duty		1	3	3	1		8	7	1	1	59
Total Sick							2				2

78 Total officers included

Cartriges wanting to furnish with 36 Rounds Each Man. Lost Octo 18: 1276 Lost Octo 21 200

Return at Morristown, January 1777

troops had approached to within 50 yards of the British who raked the oncoming Americans with several devastating volleys. As a result, Cadwalader was driven back 40 yards and obliged to abandon one artillery piece. About 100 yards from the enemy's line Cadwalader tried, but failed to regroup his men. At this point Washington appeared and attempted to rally the troops, but to no avail. Meanwhile Captain Joseph Moulder's two-gun battery atop the hill continued to pound the 17th. The situation was resolved with the arrival of reinforcements from Sullivan's division.[65]

Gradually the scattered American forces were reformed and ordered to advance. Colonel Hand's riflemen went against Mawhood's left, while troops commanded by Cadwalader and Major Israel Angell pressed the center. Under the renewed attack, the British line gave way and the 17th was forced to retreat. The Americans continued to press the British regiment until it scattered. Lieutenant Colonel Mawhood and his staff tried to join the 40th and 55th Regiments in Princeton, but they too were forced back and with remnants of the 17th he later reached Cornwallis' rear guard at Maidenhead. Meanwhile, elements of the 55th Regiment advanced a short distance from Princeton, but were met and routed by Sullivan's Continentals. As the Americans entered the town, the balance of the 55th withdrew to prepared positions surrounding Nassau Hall where it surrendered without a fight.

Washington estimated the British losses in the battle of Princeton at 500 men, of whom 100 "were left dead in the Field."* American casualties were set at "6 or 7 Officers and about 25 or 30 Privates" slain.[66] Among the officers killed were General Mercer, Colonel John Haslet, Captain Daniel Niel, Ensign Anthony Morris, Jr., and Pennsylvania Marine Captain William Shippin.[67]

To Morristown

Brunswick had been part of Washington's original plan, but after two days of marching and fighting his men were too exhausted to travel 17 miles farther. A short distance from Princeton along Millstone River, Washington halted his troops and called his officers into conference to plan their next move. Many argued for an immediate attack on Brunswick, while others warned of the dangers of such an enterprise. The debate was soon resolved by Cornwallis who awoke that morning to find that the Americans had abandoned their positions on the Assunpink. He broke camp immediately and hastily marched for Princeton. Since he could not risk a general action, Washington decided to give up Brunswick and move at once for Morristown, where a base for the army had already been established.

At dusk on the 3rd, the Americans reached Somerset Courthouse where they bivouacked for the night. "Our army was now extremely fatigued," wrote Captain Thomas Rodney of the Delaware militia, "not having had any refreshment since yesterday morning." At daybreak the following morning they were once again on the march. At Pluckemin they halted to "await the coming up of nearly 1000 men who were not able through fatigue and hunger to keep up with the main body, for they had not had any refreshment for two days past," and had been "obliged to encamp on the bleak mountains whose tops were covered with snow, without even blankets to cover them." The army rested at Pluckemin for two days, and on 6 January marched to Morristown.[68]

Situated on a high triangular plateau backed against Thimble Mountain, Morristown was an excellent defensive position. Approachable from the east only through the rugged Watchung Mountains the village was nearly equidistant from Newark, Amboy, and Brunswick, the main British posts in New Jersey. Morristown not only provided a secure encampment, but a launching point from which to initiate offensive operations against British outposts and movements along the roads connecting New England with Philadelphia.

Shortly after their arrival, the Marines seemed to have been separated from Cadwalader's brigade and

* Actual British losses were 270 of all ranks.

marched to "Sweets Town," some two miles from Morristown.[69] There Nicholas' 80 men reduced by transfers, desertions, and deaths from the original complement of 131, found lodging. Like the remainder of the army, Marines found it difficult to procure food on the locally inflated markets, therefore Nicholas was obliged to apply to his superiors for additional money to provision his men. General Cadwalader responded on 11 January by providing Nicholas with the sum of ten pounds; hardly enough to sustain 80 men for any length of time.[70] In addition to food, the Marines also suffered from a shortage of clothing. When they volunteered for the campaign in early December, the Marines had expected it to be of relatively short duration and thus were unprepared to withstand the long, cold New Jersey winter. On 16 January, Cadwalader provided them with 21 pairs of shoes and 18 pairs of mittens, and again on the 19th he had delivered another 28 pairs of shoes and mittens.[71] Like the money to purchase provisions, they were hardly sufficient to fulfill their needs.

Their Separate Ways

The winter felt by the three Marine companies at Morristown also touched the fourth; Captain Shaw's company on board the *Randolph*. Since mid-December his men had borne the brunt of the cold northern winds that whistled across the half deserted decks to stand guard over the frigate moored at Fort Island. Inactivity soon led to dissatisfaction, and on 15 January four Marines deserted. Three days later another Marine, with two landsmen, slipped away from his post. Captain Biddle offered a reward of £5 for each, or £35 for all seven, but whether any of them were apprehended is not apparent.[72]

Meanwhile, news had come from the capes reporting that the bay was clear of British ships, but the ice-choked Delaware still posed a threat. A sudden shift in the wind during the last days of January brought warmer weather and the ice began to break. Taking advantage of the unseasonably mild weather, Robert Morris ordered the *Randolph* to sea.[73] In concert with the *Hornet* and *Fly*, Captain Biddle was directed to convoy several merchantmen safely to sea and then cruise southward to the Virginia Capes, there to seize enemy warships "interrupting the commerce of the United States."[74] The *Randolph* cast off from Fort Island on 3 February and three days later passed Cape Henlopen—the first of the 13 Continental frigates to put to sea.

The six weeks extended enlistments had expired by the end of January and many Continental troops (including several artillery companies) headed home from Morristown. Since these artillery companies were required to turn in their cannon and ammunition, there was an excess of fieldpieces and a deficiency of men to work them by the end of the month.[75] To help meet this manpower shortage, the three companies of Continental Marines were transferred to the artillery on 1 February.[76] Captain Isaac Craig later recorded: "In February we were ordered to join the Artillery and learn that Duty which Orders I gladly received As I had already acquired a considerable Knowledge both in the Theory and Practice of Gunnery and Projectives—Arts, I always delighted in."[77] Following their transfer, Nicholas' Marines were moved from "Sweets Town" back to Morristown and placed under the control of General Henry Knox.

Of the three, Captain Mullan's company of Marines served the shortest period as artillerymen. Colonel George Weedon, Washington's acting Adjutant General, ordered Mullan on 20 February to take charge of 25 British prisoners and conduct them safely to Philadelphia where he was to await orders of the Pennsylvania Council of Safety respecting their disposition.[78] Upon his return, the Council seemed to have no further use of Mullan's company and thus returned its control to the Marine Committee. Since the frigate *Delaware* was still unfit for sea, Mullan was then directed to procure accommodations and await further orders.

Captain Mullan obtained barracks space for his company at the leased home of Abraham Wilt on the east side of Second Street, between Mulberry (now Arch) and Sasafras (now Race) Streets. It appears that his company kept the barracks in wretched condition, as the report of Peter Ozeas, Deputy Barracks Master, illustrates:

> A Return of Abraham Wilts house in Second Street Capt. Mullens Compy of Marines, a dirty yeard, Burned the whole stairs in the Kitchen, dresser & mantle piece, the whole yard full of human dung till to the door. I desired them 5 days agon to clean it but they have done nothing yet.

Mullan, however, failed to put the house in proper repair before the quarters were re-inspected, and all damages were charged to the company's account.[79]

After 1 April 1777, Mullan's company of Marines seems to have disappeared. Although there were a number of Marines who continued to serve with the *Delaware*, it is not definitely known whether Robert Mullan remained as their commander.

In mid-March, Isaac Craig and Andrew Porter resigned their commissions as captains of Marines and accepted commissions with the same rank in the artillery. Craig joined the Pennsylvania artillery regiment being formed by Colonel Thomas Proctor, while Porter accepted an appointment in the Continental artillery under General Knox. Craig's commission was dated 14 March, and Porter's two weeks later. Porter's commission, however, was later made retroactive to 1 January 1777, thus precipitating a conflict between the two officers as to seniority.[80]

Captain Benjamin Dean's company of Marines remained as artillerymen until 1 April. He returned with his company to Philadelphia about that date and went on board the frigate *Washington*. The frigate remained in port for several months and when the time of his men expired in late June, Dean resigned from the Marines. His first lieutenant, Abel Morgan, remained on board and took command of the Marine detachment.[81]

The incorporation of Nicholas' three companies into the army had the superficial effect of dispersing both Marine talents and personnel, but far beyond this, the action crushed the idea of an independent corps of Marines, finally reducing them to a series of ship's detachments. The New Jersey campaign of 1776–1777 had a similar effect on Major Samuel Nicholas, who no longer functioned as the head of the Marines but remained essentially a high-ranking officer without assignment.

THE EVACUATION OF BILLINGSPORT, 2 OCTOBER 1777.
With the British increasing their hold on the Delaware River early in October 1777, it was decided to abandon the fort at Billingsport, New Jersey. This painting portrays a landing party of Continental Marines from the brig Andrew Doria covering the evacuation of the garrison to Fort Mifflin in the face of oncoming British soldiers.

CHAPTER VI

New Frigates in Action and the Debacle on the Delaware, 1777

Two Escape

Lieutenant of Marines John Trevett returned to Providence during the last days of December 1776 to find the sloop *Providence* ready for sea. After a month's leave he was again eager to take up the fight against those who recently had driven his family from their Newport home. As he took his post on board the sloop, word was brought in that one of the blockading British frigates, HMS *Diamond*, had run aground near Warwick Neck in upper Narragansett Bay. It was a golden opportunity which Trevett thought should not be lost.

On board the frigate *Warren*, Commodore Esek Hopkins saw the grounding not only as a possible way to break the British stranglehold on his fleet of five ships, but an opportunity to make amends for his past shortcomings. Should he succeeding in capturing the *Diamond*, surely his reputation, so badly damaged before Congress the previous year, would be refurbished. To avoid endangering his flagship in the bay's shallow waters, Hopkins took a portion of his crew to man the sloop *Providence*. Captain Abraham Whipple was directed to take charge of the sloop, and also instructed to detail a part of the *Columbus'* crew for the expedition.

When the *Providence* arrived on the scene in the early afternoon of 2 January, it found the *Diamond* still caught on the soft bottom of the shoal, but yet upright. Several passes were made within musket shot of the stranded frigate, and answered by the *Diamond's* stern guns. Meanwhile, two field pieces on shore were brought within range and joined the *Providence* in the attack. At sunset the *Diamond* careened and ceased firing, but the *Providence* continued until darkness made it difficult for the crew to sight the guns. Shortly after dark, the commodore made two trips to the beach to confer with the local militia commander. On his last, the launch that was to carry him to the sloop failed to return and he therefore spent the remainder of the night ashore. While Hopkins was away, the *Diamond* freed herself on the incoming tide and all hope of taking her faded. The *Diamond's* sides were pierced several times and her rigging cut, but her crew remained unharmed. Damage and casualties on board the *Providence* went unreported.[1]

Commodore Hopkins had missed his chance. Soon after the abortive attempt to seize the *Diamond*, the officers of the *Warren* sent a petition to Robert Treat Paine, speaker of the Massachusetts House of Representatives, accusing Hopkins of blasphemy against God and the Continental Congress, of inhuman treatment of prisoners, and the "blamable manner" in which he handled himself during the engagement with the British frigate. Prominent among the signers was John Grannis, captain of Marines.[2] In late February, Grannis, as spokesmen for the crew, took a second petition to Philadelphia and placed it before the Marine Committee.[3] As a result of the petition and the committee's examination of Grannis, Congress resolved on 26 March, that "Esek Hopkins be immediately, and he hereby is, suspended from his command in the American navy."[4] A year passed before the commodore was formally dismissed from Continental service.[5]

Early in April, Captains John B. Hopkins, Abraham Whipple, and Dudley Saltonstall were ordered to make every effort to get the frigates *Warren*, *Providence*, and *Trumbull* to sea in search of enemy merchantmen and transports.[6] Due to the blockade of both the Connecticut River and Narragansett Bay, all three frigates remained idle throughout the rest of the year. Of the seven Continental ships locked in the harbors of Connecticut and Rhode Island, only one escaped.

Following the engagement with the *Diamond*, command of the sloop *Providence* was temporarily given to Lieutenant Jonathan Pitcher who successfully guided her through the blockade and out to sea toward the end of February. Before reaching open sea, however, the *Providence* passed so near a British frigate in the darkness that the crew could be heard talking on board, and at sunrise she was becalmed within sight of the enemy at Newport. Short of men, Pitcher immediately put into New Bedford, Massachusetts where the sloop's complement was filled-out with pressed seamen.

The *Providence* put to sea again and sailed eastward. Off Cape Breton she engaged an enemy brig laden with soldiers and stores destined for the British army at Quebec. After an hour of intensive fire, the brig surrendered and Marine Lieutenant John Trevett was sent on board to take possession. What he found was depressing. "In the Cabin the Flours," observed Trevett, "was Spread as full of Wounded Men, as You Could seasley find Room to put you foot and I found tha ware some of them Irish as tha Cry out for Jeses Sake to Spar's thare Lives tha ware Very Badly Wounded." On board the *Providence* the situation was about the same: the sloop was shattered and Lieutenant Pitcher badly hurt. After the soldiers were transferred to the *Providence*, a prize crew was put on board the brig and the cruise given up. "That Night the wind blew Hard with Squarls," and by dawn the following morning the brig had vanished. Alone the sloop sailed for Nantucket where Captain John Peck Rathbun replaced the wounded Pitcher.

In June, the *Providence* once again put to sea, destined to cruise in the vicinity of New York. Off Sandy Hook a 16-gun enemy merchant ship in company with a brig, schooner, and sloop was sighted. Captain Rathbun decided to engage the ship, for if taken, the others could be captured leisurely. After two days of bitter fighting, in which the three smaller vessels coordinated their attack with the larger ship, Rathbun had to settle for the schooner. The *Providence* continued to cruise the area for two more months before returning to New Bedford in August where the sloop was repaired and new recruits obtained.

Scraped and caulked, the *Providence* sailed southward in November. After weeks at sea the sloop put in for Charleston, but as she was about to cross the bar shortly after midnight a British privateer was sighted. The Britisher ordered "the Dam Yanke Bugers to Hall Down the Cullers" as he opened for action. The fire was answered by "a yankee welcome," with "A handsome Brard Side." Throughout the night the *Providence* chased the privateer and by morning she was so close that a lieutenant standing on the roundhouse of the privateer sniped at the sloop. In response, Marine Lieutenants Trevett and Michael Molten broke out muskets, went to the bow, and after three shots killed the lieutenant. The Americans boarded the privateer and carried her into Georgetown, South Carolina where they remained until the end of the year.[7]

Providence, Rhode Island had been the center of naval activity in 1776, but the following year it was supplanted by Boston. This was the result of the British occupation of Newport and control of Narragansett Bay, the authorization in April of the Navy Board of the Eastern Department which was to be located in Boston, and the suspension of Esek Hopkins. In addition, Boston possessed a fine harbor which was under American control.

Moored at Boston were the untried Continental frigates *Hancock* and *Boston*, plus the ship *Alfred* and brig *Cabot*, both of which had recently returned from successful cruises. Under Captain Joseph Olney, the *Cabot* was the first of the four ships to sail from Boston in 1777.

On 15 January, Olney received orders to repair to Boston, take command of the *Cabot*, and cruise for enemy transports. John Bradford, Continental agent, supplied the brig with provisions, and by 27 February she was ready to get underway, but a severe snow storm, however, caused some damage and prevented the sailing. On 9 March Olney was still in port, but shortly thereafter put to sea in company with the Massachusetts state ships *Tyrannicide* and *Massachusetts*.[8]

Off the Nova Scotia coast, following several days of uneventful sailing, an enemy ship was sighted rapidly bearing down on the brig. She proved to be the *Milford* of 28 guns. The *Cabot* immediately turned and crowded sail. Cut off from his consorts, Olney was forced to beach his brig near the mouth of the Chebogue River after two days of running.[9] So close was the *Milford* that Olney and his crew had barely enough time to abandon ship let alone destroy the brig. The British version of the beaching is somewhat disdainful:

> Olney, and his motley Crew (including what they call a Captain, two Lieutenants and a numerous party of Marines) fled into the woods in the greatest hurry and confusion, carrying their small arms and ammunition with them but could not *spare time* in their flight to set fire to their vessel, which they might very easily have done. Olney himself was the first man in the boat, having *jumped* into her whilst they were hoisting her out.
>
> A party of 20 seamen and about as many marines, were detached from the *Milford* as soon as possible in quest of them, but the Rebels having started first, proved too nimble of foot to be overtaken. Marching along shore to the northward of the Cape, they seiz'd upon a light schooner, plundered the adjacent houses for provisions, and set off for N. England, to blaze abroad their heroic exploit.[10]

Upon returning to Boston a court of inquiry was held which cleared Olney of all blame in connection with the loss of the *Cabot*.*

The *Fox* Captured and Lost

The orders of the Marine Committee sending the ships to sea, were premature. Although the *Boston* was fitted with cannon fairly rapidly, the arming of her counterpart at Portsmouth was delayed. When the *Boston* finally stood down the bay on 20 January she was not in company with the *Raleigh*, but the frigate *Hancock*, Captain John Manley. For three months the two frigates remained at anchor while additional crewmen were sought.[11]

The frigate *Boston* had been moved from the docks to the harbor channel "in order to keep her Crew on board." Desertions plagued both the ship's crew and its Marine company under Captain Richard Palmes, and Lieutenants John Harris and Robert McNeill. As in the previous year, privateering proved

The actions of the *Milford* in New England waters prior to the capture of the *Cabot* were cause for concern among the commercial interests of Boston and Portsmouth. Her captures had depressed commerce and finally provoked the assemblies of Massachusetts and New Hampshire to offer all necessary assistance to complete the fitting out and manning of the Continental frigates *Boston*, Hector McNeill, and *Raleigh*, Thomas Thompson. The offers of aid were accepted by the Marine Committee which ordered the two frigates against the *Milford*.

* Lieutenant John Carr may have been the *Cabot's* Marine officer at this time for he joined the brig *Lexington* on 8 May 1777 from the *Cabot* and was discharged from the *Lexington* on 22 May 1777.

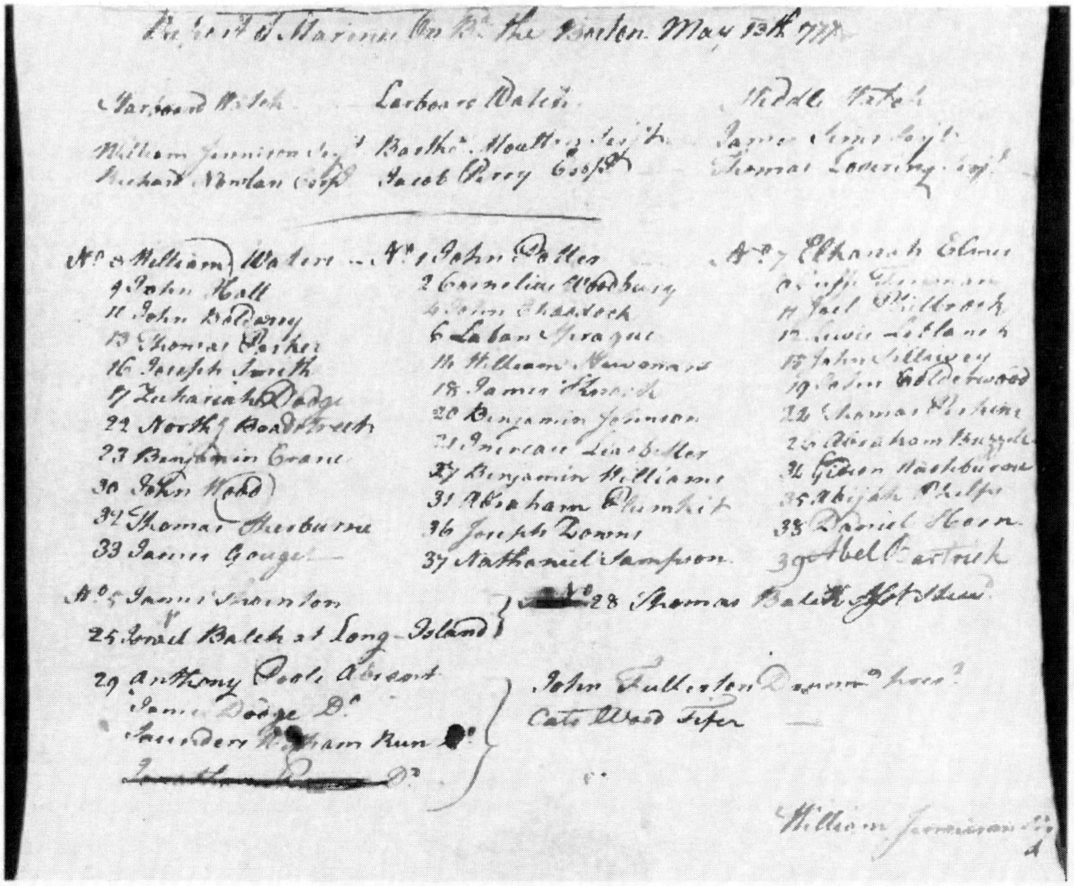
Return of Marines On Bd. the Boston May 13th 1777

Starboard Watch	Larboard Watch	Middle Watch
William Jennison Serjt	Bartholomew Moulton Serjt	James [illegible] Serjt
Richard Nowlan Corpl	Jacob Perry Corpl	Thomas [illegible] Corpl
No 8 William Waters	No 1 John Potter	No 7 Elkanah [illegible]
9 John Hall	2 Cornelius Woodbury	10 [illegible]
11 John [illegible]	4 John Chadwick	11 Joel [illegible]
13 Thomas Parker	6 Laban Sprague	12 Lewis [illegible]
16 Joseph Smith	14 William [illegible]	15 John [illegible]
17 Zachariah Dodge	18 James [illegible]	19 John [illegible]
22 North [illegible]	20 Benjamin Johnson	24 Thomas [illegible]
23 Benjamin Crane	21 [illegible]	26 Abraham [illegible]
30 John Wood	27 Benjamin Williams	34 Gideon Washburne
32 Thomas Sherburne	31 Abraham [illegible]	35 Elijah Phelps
33 James Gouge	36 Joseph Downs	38 Daniel [illegible]
	37 Nathaniel Sampson	39 Abel Partrick

No 5 James [illegible]
25 Israel Balch at Long-Island } No 28 Thomas Balch Sick ashore

29 Anthony Poole Absent
James Dodge Do
Saunders [illegible] Run Do
[illegible] Do } John Fullerton Drummer [illegible]
Cato Wood Fifer

William Jennison [illegible]

Marine Watch List

to be the major villain. Of the problem Captain Hector McNeill asked:

> With what Spirit can an officer advance Monies to Cloathe the Naked Objects, who offer themselves, as willing to serve in their several Capacity's, if the next moment those Men may with impunity go away in a Privateer, or enter into any other Corps, either by Sea or Land? and run no risque by being detected? [12]

Both Manley and McNeill went heavily into debt by advancing money to their crews for clothing and other necessities.[13]

The two frigates in concert with nine privateers set sail to the eastward at the end of May. Within six days the privateers had parted with the frigates, due either to bad weather or a desire for prizes. On the 29th, a small brig laden with duck and cordage was taken and sent into Boston. The following day the frigates fell in with four large transports bound for New York and escorted by the *Somerset* of 64 guns. According to McNeill; "Capt. Manley was not convinced of the size of our Opponent untill she was within Shott of him, when very Luckily for him the *Hancocks* Heels saved his Bacon." The *Somerset* left her charges and gave chase to the *Hancock* and *Boston*. Throughout the remainder of the day the *Somerset* pursued the two American frigates until darkness forced her to retire to the convoy.

Shortly after the encounter with the *Somerset*, Commodore Manley ordered a change in course to the northeast in hopes of finding less protected enemy transports. Nothing was sighted except for a "few miserable Fishermen" until the morning of 7 June when they encountered the *Fox*, a 28-gun British frigate commanded by Captain Patrick Fotheringham.[14]

The Englishman at first made preparations to engage, but then turned away and made a run for it. After several hours in chase, the *Hancock* drew alongside. "A Spiteful Short Action Ensued," wherein the *Fox* "pegg'd Mr. Manley's ribbs so well that he had his pumps going" when the *Boston* closed 45 minutes later. Shortly thereafter, Fotheringham hauled down his colors and surrendered to the Americans.[15]

The capture of the *Fox* again opened an old disagreement between Captains Manley and McNeill. Although the *Boston* was tardy in arriving on the scene, McNeill was the first to reach for the fruits of victory. He sent his first lieutenant on board to take command of the prize, and his two lieutenants of Marines, John Harris and Robert McNeill, to maintain order among the enemy crew. This action provoked Manley to demand that Browne be withdrawn, for the *Hancock* had borne the brunt of the battle. McNeill acquiesced, but the Marine lieutenants stayed. The loss of Harris and McNeill on board the *Boston* prompted Captain McNeill to promote William Jennison to the post of acting lieutenant of Marines.[16] *

Soon after a portion of the *Fox's* crew was set free, Captain McNeill suggested that they now turn southward for Charleston, South Carolina, there to join Captain Nicholas Biddle in the *Randolph* and cruise for the West India Fleet. Manley at first agreed, "but in a few days alter'd his Mind and his Course . . . Nevertheless," McNeill noted, "I follow'd him as the Jackall does the Lyon, without Grumbling except in my Gizard." For the next several weeks the two American frigates and their prize wandered about the area encountering "Nothing but Spaniards and French Men."[17]

In mid-July, Manley ordered the squadron to stand for Massachusetts. Eastward of Cape Sable on the morning of the 6th, the *Hancock* captured a sloop laden with coal out of Louisbourg. Captain Manley took the sloop in tow and the fleet headed toward the southwest. Late the same afternoon, three ships that appeared to be in chase were sighted astern. They were the British ship *Rainbow*, 44, the frigate *Flora*, 32, and the 18-gun brig *Victor*. Captain Manley, "far from taking notice of the Signal made by the *Fox* who was then astern," continued towing the sloop throughout the night, forcing the *Boston* and *Fox* to shorten sail. Sunrise the following morning found the *Flora* ahead of the *Boston*, the *Rainbow* to her lee, and the *Victor* to the stern. With a fight inevitable, Manley burned the coal sloop and prepared to engage.[18]

Flora, the lead British ship, was the first to engage the *Boston*, the rear American frigate. Commodore Manley, mistaking the *Rainbow* for a 64-gun ship-of-the-line, tacked and stood to the southward. After the first broadsides, the Americans scattered; the *Boston* to the northward, the *Fox* to the eastward, and the *Hancock* to the southward. The *Rainbow* immediately followed the *Hancock*, but for some unexplained reason Manley shifted weight forward thus slowing his vessel down. At about four in the morning of 8 July, the *Rainbow* opened up with her bow chasers and several broadsides of round and grape. Four and one half hours later, Manley was ordered to strike if he expected quarter. The American commodore took a few minutes to consider the ultimatum, but as there was a fresh breeze he made a last effort to outrun the Englishman. Commodore Sir George Collier responded with yet another broadside. After a chase of more than 39 hours Manley struck to the *Rainbow*, first of the American frigates to do so to a British ship.[19]

Captain of Marines Seth Baxter and Lieutenant William Bubier accompanied Manley as prisoners to the *Rainbow*. Later the two Marine officers were carried into Halifax from whence they were exchanged in late 1777. Captain Manley was transferred from Halifax to New York and exchanged the following year.[20]

While the *Hancock* was engaged, the *Fox* with her American crew was run aground after heavy action with the *Flora*. The two Marine officers on board, Lieutenants Harris and McNeill, were captured and sent into Halifax. In January 1778, they were exchanged for British prisoners of equal rank. The frigate *Boston* luckily escaped capture and sought sanctuary in the mouth of the Sheepscott River on the Maine coast. There she rested for several weeks while British ships paraded off the coast waiting for her to emerge.[21]

About the end of July, the British reduced their vigil off the river's mouth, allowing the *Boston* time to sneak out. Avoiding British warships, she ran along the coast and entered Casco Bay about 1 August. There McNeill again waited while four Royal frigates patrolled off the harbor.

* William Jennison joined the *Boston* as a "volunteer" in January 1777 after serving in the New York campaign.

A NEW MAP of
NOVA SCOTIA,
and
CAPE BRETON ISLAND
NEW ENGLAND and CANADA.
GULF OF ST LAURENCE
ATLANTIC OCEAN

Palmes Arrested

The presence of enemy ships and the inactivity of his own were not McNeill's only concerns, for friction and dissension pervaded the *Boston's* ward-room. Richard Palmes, captain of Marines, seems to have been singled out by McNeill as the leading cause. On 10 August Captain McNeill placed Palmes under arrest for "unofficer like behaviour" and repeated breach of orders with the warning: "At your Perril break your Arrest, in which case I shall treat you as you deserve."[22] The following day a stronger order to Palmes was issued:

> You may thank your own folly and impertinence for what has now befallen you. I dispise your insinuations of Cruelty, as indeed I do Every thing Else you can say of me consistent with truth. You may go to the house of Office as offten as Nature calls, provided you return immediately to your berth and keep your Tounge Still as you pass and repass.[23]

Lieutenant of Marines William Jennison was tasked with maintaining the guard over the senior Marine officer.[24]

The cause of conflict between Palmes and the *Boston's* captain is unclear, but it happened at a time when unity on board the ship was necessary. Again during a lull in British vigilance, the *Boston* slipped out of Casco Bay and moved down the coast to Portsmouth. Several days later she made a final dash for Boston.[25]

McNeill's reception in Boston was less favorable than he had expected. Opinion held him responsible for the loss of the *Hancock* and the imprisonment of the popular John Manley. McNeill considered himself blameless and therefore in a letter to the Marine Committee placed full responsibility with Manley. His officers, however, felt differently. Had McNeill assisted Manley, not only would the *Fox* have been saved but the *Rainbow* and *Flora* might have been taken. Differences between McNeill and his officers were further aggravated by the captain's refusal to pay the officers their share of the prize money.[26]

Marine Captain Richard Palmes, who had literally cast his commission into the fire, was still under arrest and had to be dealt with.[27] Once in Boston, Palmes showed signs of behaving properly and therefore McNeill returned him to duty. Their differences, however, were irreconcilable and McNeill was forced to ask the Eastern Navy Board to bring him to trial for "misaplication of the Ships Stores, Neglect of duty, disobedience of orders, and attempts to Excite Murmuring and Mutiny among the Ships Company."[28]

While under arrest Palmes journeyed from Boston to Providence where he presented his case before two members of the Eastern Navy Board—John Deshon and William Vernon. Not only did he convince the gentlemen of the validity of his position, but they in turn awarded him his share of the prize money and immediately transferred him from the *Boston* to the *Warren*, frigate. Stunned by their action, Captain McNeill turned to the Marine Committee for an affirmation of his position:

> Never was a man taken from under Arrest and preffer'd to any other Employment without first undergoing a Court Martial. Nor is it possible that ever good order should Exist in armys, or fleets, without takeing care to punnish, cashier, or repremand, such as shall on due trial be found guilty of such offences. If precedents of this kind be permitted once to take place, farewell Discipline and good Order, farewell Honour, and honesty. The Service will then become a recepticall for unclean birds who will hereby be Encouraged to take Shelter there, and all men of good principals will totally forsake it.[29]

The Marine Committee, however, sustained the Eastern Navy Board's decision, brushing aside the complaints of McNeill.[30] Here the matter was officially dropped. Captain McNeill thought the decision by both bodies unsatisfactory, but he let the question of Palmes rest.

Although he was willing to let the question of the arrest and trial of Richard Palmes die, McNeill was unrestrained in letting higher authority know his opinion of the necessity of Marine officers on board Continental frigates:

> I must now beg leave to give my Opinion respecting Marine officers for such Ships as ours, so much hampered for want of room. I think in concience a Subeltern is Enough, three Marine officers takes up so much room to accomodate them that we are pinch'd beyound measure to afford it. Then they have Little or no duty to do, are allways in the way and apt to disagree with the Sea officers so that it takes much trouble to mannage them, then they run away with so much of the prize money from Officers who are realy useful, that 'tis painfull to hear the murmerings it Occasions. Might it

> not be proper to Lessen their Number down to one on board the frigates and give what the other two did Enjoy between the Chaplin and Surg'n.

As an addendum to his suggestion that the Marine officers complement of Continental frigates be composed of one instead of three, McNeill recommended the name of a young man whom he thought qualified for the single post.[31] Congress rejected both his views and his choice.

The Marine Committee began to take notice of the increasing number of complaints against McNeill. Coupled with McNeill's recent criticism of Manley, the Committee on 12 November suggested to the Eastern Navy Board that it would be advisable to suspend him "till his conduct respecting his last Cruize is properly enquired into."[32] Due to Manley's absence, however, the hearing was deferred until his release the following year. Meanwhile McNeill was suspended from command. Shortly after the *Boston* received a new captain and Richard Palmes was allowed to resume his post as captain of Marines on board the frigate.

Osborne's Recruits

At Portsmouth the problem of recruiting Marines for the *Raleigh* proved a stumbling block to the frigate's departure. Due to the shortage of qualified men, Captain George Jerry Osborne of the Marines was forced to enlist 11 men out of Colonel Pierse Long's Regiment—then under orders to march to Fort Ticonderoga. Unhappy over the loss, Long wrote the New Hampshire Committee of Safety and demanded their return.[33] The committee, agreeing with the colonel's request, ordered Osborne to "release them immediately," since the Marines were not considered a part of the 88 battalions authorized by the Congressional resolve of 16 September 1776 which permitted recruitment from existing units.[34]*

Thomas Thompson, captain of the *Raleigh*, was the first to receive the order and then passed it onto Osborne. In his reply, Thompson reminded the committee that:

> The men have entered as Marines not Mariners, & Marines by a Vote of Congress are on the same footing as soldiers in the land service; Notwithstanding we can keep them & are not accountable to the Committee of Safety of any State for such conduct; yet rather than any reflection should fall on us or that we should be the means of retarding the too long neglected march of Colo. Long, shall agree to give them up, if he again desire it.

Thompson also warned the committee of the always present competition of privateers and that they had failed to observe the "Embargo" against allowing private ships to be manned in deference to Continental ships.[35]

Captain Osborne's letter of reply to the committee covered somewhat the same ground as Thompson's: "I think, Gentlemen, if my Company is to be ruled & commanded by the Regulations of the army, I must in consequence reckon my Department as part of the same." He then noted that he had received "orders from the Major of Marines to fill my company equal to the Land service."* Osborne then noted, as did Thompson, that the men would be returned should Colonel Long request them.[36]

Both Thompson's and Osborne's reply to the committee's order had no effect. On 8 February, the committee again wrote to Captain Osborne, restating their argument:

> The Committee desirous of promoting the publick service, and in consequence of Col. Long's solicitations wrote you their opinion of the Impropriety of Enlisting men belonging to said Col. Long's Regiment; and are still of the same opinion; as you cannot pretend your Company is included in the Eighty-eight Battalions expressly named in the Resolve of Congress as having liberty to recruit out of the Regiments in Service.[37]

With the committee's reply, the matter was dropped and the men returned. The problem, however, was to arise again.

The *Raleigh* lingered on in port shorthanded and inadequately armed for several more months. Early

* The resolve authorized 88 new battalions for the Continental Army. In creating these battalions, enlistment from existing units was also authorized.

* This short statement is highly significant for it has been previously assumed that Major Samuel Nicholas exercised no control over Marine companies other than those assigned to the four frigates built in Philadelphia. This is the first, and only known instance of such control.

A View of Portsmouth, New Hampshire.

in August she was joined by the *Alfred* under the command of Elisha Hinman. By the end of the month the *Raleigh* was ready and the two Continental ships sailed out of the Piscataqua River and put to sea, destination France.

Three days out a small schooner bound for Halifax was taken. She was in ballast except for 20 barrels of flour and carried over 4,000 dollars in counterfeit Congressional and Massachusetts state currency, which Thompson said he would "commit to the Flames saving a Sample." The schooner was then burned. On 2 September a snow was taken, and from the passengers it was learned that on the previous day the ship had straggled from the Windward Island Fleet which was under the convoy of four British ships—the *Camel*, *Druid*, *Weazel*, and *Grasshopper*. After taking copies of the convoy's sailing orders and signal book, the prize was sent off to the first safe American port. The next morning the fleet was sighted from the mastheads and by sunset the Americans were near enough to discern 60 sail bearing east by north.

Thompson immediately hailed and informed Hinman of his intention to run into the fleet and attack it at sunrise. He ordered the *Alfred* to keep close to the *Raleigh's* stern until they were alongside the English commodore's ship, then they would both attack. During the night the wind shifted and by morning the Americans found themselves to the leeward—a distance of two or three leagues. The *Raleigh* soon closed the gap, but the *Alfred* "being Tender-sided could not carry Sail & therefore fell a great Way to Leeward and a Stern." Since it was evident that the *Alfred* could not make up the distance, Thompson "determin'd . . . to make Sail & stand into the Fleet and take . . . [his] chance alone."

The *Raleigh* hauled into the midst of the fleet using the captured signals. With his ports down and guns housed, Thompson maneuvered the *Raleigh* to within pistol shot of the 14-gun sloop *Druid*, and then "up Sails and out Guns, hoisted the Continental Colours & told them to strike to the Thirteen United States." The action threw the *Druid* into complete confusion. The *Raleigh* then unleashed her first broadside. While Marines in the tops kept up a constant fire with their muskets, a dozen more broadsides were fired. In danger of being surrounded by other escort vessels, Thompson soon broke off the attack and ran down to the *Alfred*. The three British warships pursued the two ships until nightfall when they were forced to return to the convoy. For the next three days the *Raleigh* and *Alfred* hounded the fleet, but the ships kept close together and refused to engage the Americans. On the fourth day the Continental ships gave over the chase and reset their course for France.[38]

Fitting-out the *Ranger*

Early in 1777 Robert Morris, in the name of the Marine Committee ordered John Paul Jones to take command of a squadron composed of the *Alfred, Columbus, Cabot, Hampden*, and sloop *Providence*, and make an ambitious cruise southward. First they were to capture the British island of St. Kitts where Morris said they would "make a considerable booty." Thence, the fleet was to proceed to Pensacola and take British sloops-of-war which were becoming a nuisance in the area. Next they were to cruise off the mouth of the Mississippi and capture valuable British river transports, "give them an alarm at St. Augustine," show the flag in Georgia and the Carolinas, and finally cruise the West African Coast or attack the British slavers bound for the West Indies. The main object of the expedition was to deliver several strategic blows to British outposts. As Morris stated:

> It has long been clear to me that our Infant fleet cannot protect our own Coasts; and the only effectual relief it can afford us is to attack the enemies' defenceless places and thereby oblige them to station more of their Ships in their own Countries, or to keep them employed in following ours, and either way we are relieved so far as they do it.

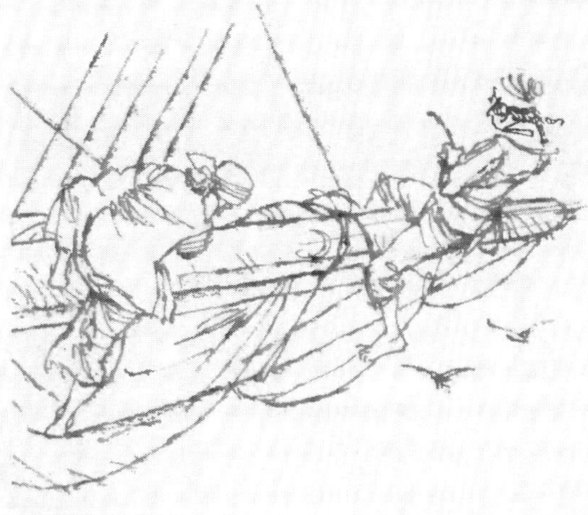

To accomplish the orders, Morris suggested that Jones carry as many Marines as possible "for they will be usefull & necessary in all your Land Excursions."[39]

The Marines, however, would never be used. The immediate opposition of Esek Hopkins and his suggestion that there would be numerous difficulties encountered in manning the ships quickly quashed the whole project.

John Paul Jones soon recovered from the disappointment of having lost the command of a squadron. In March he was appointed to command one of the three vessels which Congress had ordered to be purchased at Boston. But again the command failed to materialize. In May Jones was given a third command, but that, too, was abandoned when the ship's captain refused to turn over the vessel to Jones. Finally on 14 June 1777, Jones was given command of the new 18-gun ship *Ranger*, just launched at Portsmouth.

Although neither plan nor drawing of the *Ranger* has survived, her measurements have been preserved. Rated as a sloop-of-war, her gun deck measured 97 feet, 2 inches, keel 77 feet, 9 inches, beam 27 feet, 8 inches, and hold depth 12 feet. The *Ranger's* bulwarks were originally pierced for 20 guns, but her weight prohibited her from carrying

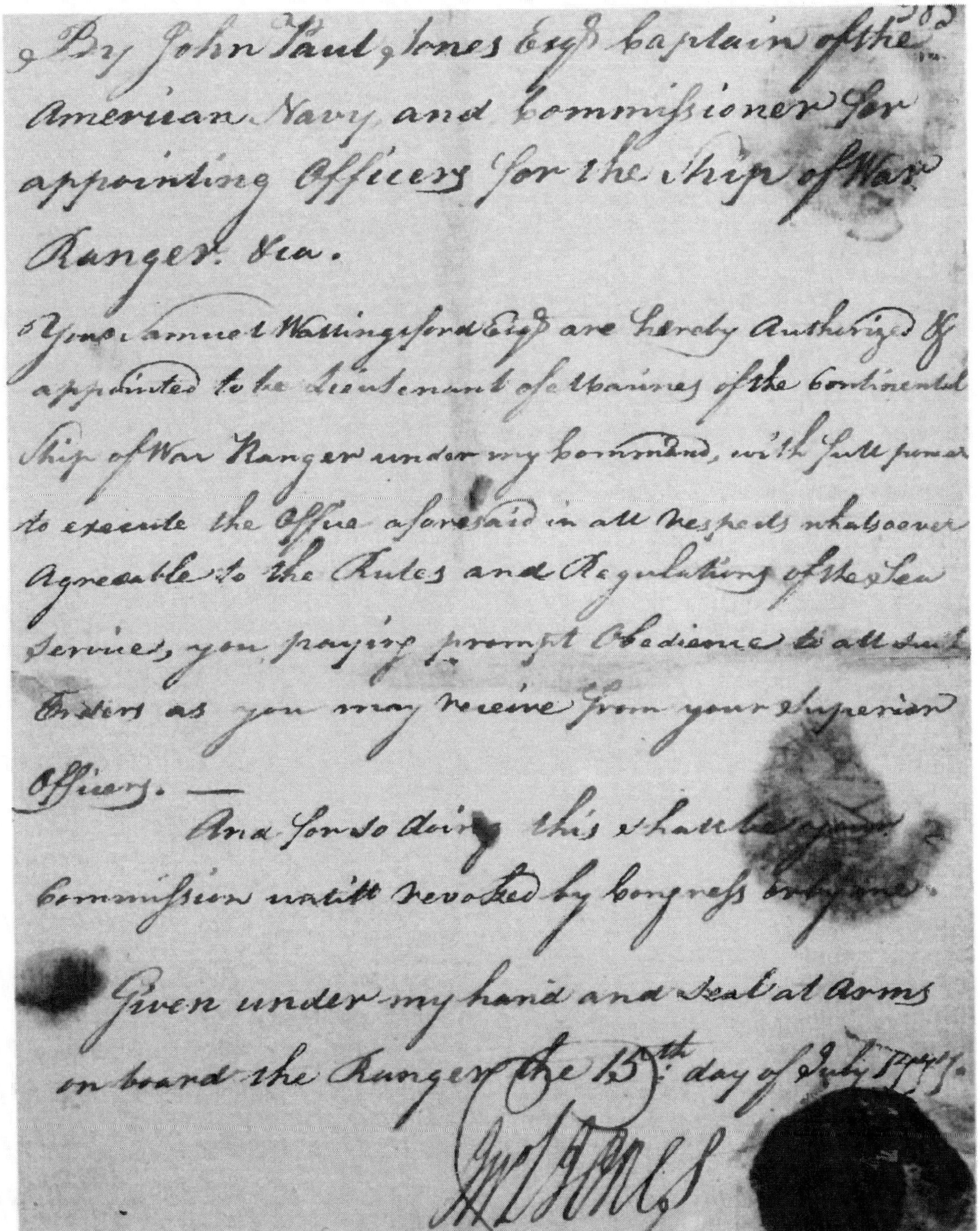

By John Paul Jones Esqr Captain of the American Navy and Commissioner for appointing Officers for the Ship of War Ranger &ca.

You Samuel Wallingford Esqr are hereby Authorized & appointed to be Lieutenant of Marines of the Continental Ship of War Ranger under my Command, with full power to execute the Office aforesaid in all respects whatsoever Agreeable to the Rules and Regulations of the Sea Service, you paying prompt Obedience to all such Orders as you may receive from your Superior Officers. —

And for so doing this shall be your Commission untill revoked by Congress or by me.

Given under my hand and Seal at Arms on board the Ranger the 15th day of July 1777.

JnoPJones

Samuel Wallingford's Commission as Lieutenant of Marines

more than 18. Her sides were painted black with a broad yellow stripe, and she had three fully rigged masts. From contemporary descriptions it is possible to infer that the sloop had a fine, sharp bow, and an undercut stern which made for speed in a good wind. "The Ranger taken altogether will in my Judgement prove the Best Cruizer in America," wrote Jones to Captain Matthew Parke of the Marines. "She will Always be able to fight her Guns under a most excellent cover; and nothing can be better calculated for Sailing fast or making Good Weather. . . . Since the establishment of the Navy none have had so good, do fair a prospect before then as ours is at present."[40]

Immediately after his appointment, Jones' primary concern was the outfitting of his ship. But once that was underway, the problem of manning the *Ranger* arose. As was the case with other ships, the selection and appointment of officers was carried out by a local committee; thus nepotism, favoritism, and friendship took precedence over qualifications. For first lieutenant, the *Ranger* received Thomas

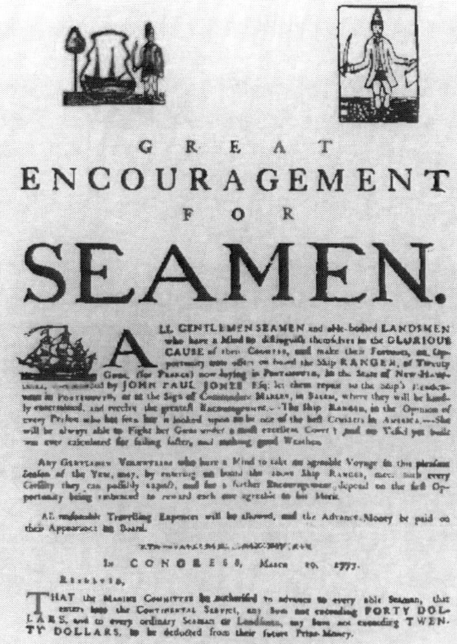

Recruiting Broadside, 1777.

Simpson, brother-in-law of John Langdon, Continental agent concerned with the construction and outfitting of the *Ranger*. Elijah Hall and David Cullam, appointed second lieutenant and master respectively, were prominent Portsmouth residents and well acquainted with the sea. Appointed lieutenant of Marines was Samuel Wallingford, also a local resident. Although he had no maritime training, Wallingford did possess extensive military experience as both a lieutenant and a captain in the New Hampshire militia. The only exception to those locally selected was captain of Marines Matthew Parke, the one friend Jones was allowed to appoint.

John Langdon and William Whipple were not of any assistance in obtaining a crew; Jones was delegated that responsibility. In procuring men for the *Ranger*, Jones wrote to Captain Parke who was then at Providence and ordered that he and Marine Captain Edmond Arrowsmith enlist both seamen and Marines:

> As you are confident that a number of good men will enter with you at Providence to sail under my Command You are directed to enter as Many as Possible—but especially good Seamen—as I mean to Appoint some Petty Officers from among them—you are to call on Captain Arrowsmith of the Marines and order him to assist and co-operate with you in this Enlistment.—for this purpose let him go round with a Drum, Fife and Colours as Often as may be proper, and it is my express desire & direction that both of you exert yourselves [in] this Bussiness to the Utmost of your skill and Abilities.[41]

Parke and Arrowsmith managed to sign 21 men, but soon lost them to Captain John B. Hopkins of the *Warren*.

Likewise Jones sent Lieutenant Wallingford out to beat the drum for recruits. Wallingford was given posters and ordered not only to enlist Marines but all qualified seamen as well, since the *Ranger* was competing with the frigate *Raleigh* for men. To make service on board the sloop-of-war more attractive, the Marine lieutenant was authorized to grant all "reasonable Travelling expenses," and a bounty of 40 dollars for every man who appeared at the ship.[42] Although some time was spent in Portsmouth, most of his recruiting took place about Somersworth, his home. There he enlisted both kinsman and boyhood neighbors.

Towards the end of July 1777, Jones began to worry about the progress of enlistments for the *Ranger* was nearly ready for sea. Thus on the 27th he ordered Captain Parke to leave Providence and go to Boston where Jones hoped more men would be found. Should he find crewmen there, Parke was to enter them as volunteers or midshipmen since no other vacancies existed, and, as Jones noted "the prospect of promotion is, I think, an Ample inducement."[43]

The recruiting efforts of Parke and Wallingford proved not to be of much success. Men often signed the roll and after receiving an advance wage, deserted. At the end of August, Jones was in such dire need of men that he petitioned the New Hampshire Committee of Safety for permission to enlist "matrosses" (cannoneers) from the forts and garrisons of Portsmouth harbor.[44] His request elicited little support, and a month later he again appealed to the New Hampshire government, but in much stronger terms: "United as the Continent is its interest must take Precedence of all private concerns."[45] Once more no action was taken. By 11 October Jones was so exasperated that he penned an angry letter to John Langdon in which he stated that Congress had authorized him to "Equip, Officer and Man the *Ranger* as well and as soon as possible," and he expected his orders "to be duely Honored."[46] Langdon paid no attention.

Despite John Langdon's neglect, the *Ranger* by the end of October had about 150 officers and men on board. Her canvas was finally completed, the ship rigged, and her Marines were collected and properly armed. At last, on 1 November, a moderate northwest wind blew down the Piscataqua River and the *Ranger* put to sea for France with the news of General Burgoyne's surrender at Saratoga. Thirty-two days later, the *Ranger* anchored at Paimboeuf on the Loire River.[47]

Two Foul Cruises

Another vessel that would be ordered to France in 1777 was the Continental frigate *Randolph*, Captain Nicholas Biddle. After putting to sea from Philadelphia in February, she encountered a strange sail which proved to be a French ship. A boarding party found her papers in proper order and let her go, but not before lieutenant of Marines Panatiere de la Falconniere had "Begged a large Jug of Wine of the Capt." This action was later cited by other officers of the *Randolph* in a petition to Captain Biddle demanding the Marine lieutenant's removal.[48]

Captain Biddle's intention to continue cruising southward came to an abrupt end after the *Randolph* suffered a split mast in a heavy gale and was compelled to put into Charleston, South Carolina, for repairs. Before arriving there a mutiny broke out among English sailors on board. Captain Samuel Shaw's Marines were quickly assembled and armed on the quarter-deck, but were not pressed to use their weapons, for the sight of the armed Marines promptly quashed any ideas the mutineers might had had to seize the ship.[49]

In mid-March the *Randolph* made Charleston harbor after what Captain Biddle described as "one of the most disagreeable Passages that ever I experienced."[50] There sick crewmen were put ashore and carpenters set to work replacing the mast and damaged spars. The repairs took longer than expected and it was not until early June that the *Randolph* was again ready to sail. Several days after being brought down from the shipyard to the harbor, a bolt of lightning from a late spring storm shattered the new main mast forcing the frigate to remain in port a month longer.

Toward the end of July a new mast had been put into place and the *Randolph* was again made ready. Before the frigate could cross the bar, a summer squall swept in from the east and lightning against struck the ship's main mast, splitting it from cap to deck. Disheartened, Captain Biddle set about replacing the mast for the third time.[51]

During the interval between the damaging of the third mast and its replacement, the problem of Panatiere de la Falconniere had to be settled. The other officers of the frigate had had their fill of the Frenchman from the West Indies and wanted no more of him. At the end of August they filed a formal complaint with the captain, accusing the second lieutenant of Marines with conduct "unbecoming an officer," and of being "a disgrace to the *Randolph*."

"As there is not Officers sufficient to form a Court

of inquiry," the petition began, "We are sorry to be under the disagreable necessity of addressing you." The officers then set forth the evidence upon which they had based their charge:

> He has suffered himself to be beat on those without resenting it—
> He gave an Order for Eighty Pounds on Mr. [Edward] Blake, with intent to defraud the Person to whom he gave it—
> He always when on shore Associates with the worst of Vagabonds, and those houses no Gentleman ever go to—
> He was one Night put under the Town Guard, for stabbing a Soldier, but shewing his Commission and the wound not like to prove mortal, was let go—
> He has behaved so as to be most Effectually hated and despised by every one on Board, the Vessel and by all who knowed him on shore—
> He is the most Obseene talker, and greatest Reprobate and blasphemer we ever heard—
> He has been heard to Damn the Trinity in the most shocking expression
> He has frequently declared that if any Man affront him, he would Assasinate him for it if it was 20 Years afterwards

Several additional charges were made, and the complaint concluded:

> For these and a Thousand more instances we could mention (if necessary) of the same nature we think him not only unworthy of holding a Commission in the Randolph, but a nussance to the ship. And therefore beg you to Rid us of him.[52]

Captain Biddle's reaction to the complaint was characteristic and swift. "Lieutt Falconier of Marines," he wrote to the Marine Committee, "is so exceeding Troublesom and has behaved in so many instances unworthy the Charracter of an Officer that I can not hesitate a moment to comply with the Request of the Officers which I inclose under cover to you and which contains the principle Charges against him I have ordered him to proceed to Congress."[53] Whether Falconniere ever reported to the Marine Committee or Congress is unknown.

With a stiff breeze blowing out of the west on the first day of September, the *Randolph* stood to sea, but after months in port the frigate moved rather sluggishly through the water under full sail. By the 3d, however, the *Randolph* had gained the latitude of St. Augustine and had turned seaward in hopes of sighting enemy merchantmen running between Jamaica and New York. At twilight that night her hopes were rewarded—several sails were sighted over the port bow. Throughout the night the frigate shadowed the vessels and in the following morning's light the masthead lookout made out two ships, two brigs, and a sloop.

Captain Biddle immediately called the crew to quarters as the *Randolph* bore down on the convoy. Although the distance was great, one of the ships—*True Briton*—opened fire while the others crowded sail in an effort to escape northward. The Americans, however, withheld their fire.[54]

Captain Shaw of the Marines was a tall and remarkably thin young man, and always thought himself too skinny a target for an enemy cannon ball. When a shot carried away the mizzen shroud at his back that morning, "he then thought himself in some danger."[55]

As the *Randolph* approached within pistol shot, Biddle ordered a single gun to be fired. That was sufficient, for the *True Briton* immediately hauled down her colors. Soon after a prize crew was put on board, the *Randolph* continued the pursuit of the remainder of the enemy convoy. Three of the merchantmen were overtaken, but the fourth slipped away to the southwest. Gathering the four rum and sugar-laden prizes together, Biddle concluded the cruise and made for Charleston.[56]

The arrival of the *Randolph* and her prizes electrified Charleston. But Biddle had more to worry about than his good fortune; his ship was rotting. Soon after anchoring, the frigate was moved into the yards where workmen began scraping and resurfacing her bottom.[57] Meanwhile, from Pennsylvania, expresses brought both good and bad news. From the Marine Committee Biddle received orders sending the *Randolph* to France, but from news dispatches he heard of the disastrous fighting around and the capture of Philadelphia.[58]

Fight For the Delaware

A View of Philadelphia, 1777.

The citizens of the American capital had recovered somewhat from General Howe's threats of invasion and capture by late spring 1777. The Continental army in New Jersey still stood between the British and the city. But General Washington was apprehensive about their inactivity. On 31 May he wrote to Virginia's Governor, Patrick Henry, of the expected sailing of a large enemy fleet—estimated at 100 ships—from New York. What Howe's immediate object was could only be a matter for speculation, but Washington believed he had two possible purposes: either to possess the Hudson River or attack Philadelphia by way of the Delaware. Washington's intelligence was premature, for the fleet would not sail until the end of July.[59]

Of the three Continental Marine companies which participated in the New Jersey campaign, only two returned to Philadelphia in the spring. Captain Robert Mullan's company was tasked in late February with escorting a group of British prisoners to the capital. On its return, the company moved into temporary barracks on Second Street and awaited the completion of the frigate *Delaware*. Marines under the command of Captain Benjamin Dean served as artillerymen until 1 April when they, too, returned to Philadelphia in order to assume posts on board the frigate *Washington*. Captain Andrew Porter's company, on the other hand, continued with the artillery until 28 April when the company was disbanded and a new artillery company formed. Of the 35 Marines in Porter's original company, only nine, including Lieutenant James

McClure, chose to remain with their commander. Porter's Marines received pay for services as artillerymen until 1 July 1777.[60]

At the Philadelphia docks on 5 April, the frigate *Delaware* was about ready to sail. The last of the pig iron to be used for ballast was delivered, and the Marines under Captain Mullan finally settled on board.[61] Three days later Captain Charles Alexander received his sailing orders from the Marine Committee:

> You are hereby directed to proceed to Cruize for the enemies Transports and Provision Vessels that may be coming from Europe to New York. You will proceed so far to the Eastward and so far from the Coast as will be the least likely to fall in with the Enemies Ships of war until your Men are well practiced and fitted to engage with Strong Ships. When that is the case you are then to draw nearer to the Coast of North America, and Cruise along the said Coast from Block Island to the Capes of Virginia doing your true endeavour to take, burn, sink, or destroy as many of the enemies Ships and Vessels of every kind as you may have the good fortune to fall in with.[62]

On the 9th, even though inadequately manned, the *Delaware* cast off and dropped down with the tide to Fort Island.

Like the frigate *Boston*, the *Delaware* remained anchored in the channel off Fort Island to prevent her crew from deserting. So great was the propensity of seamen and Marines to escape southward once they had signed the roll and received an advance wage, that on 10 April the Continental Navy Board at Philadelphia wrote to the Pennsylvania Committee of Safety and asked that they "issue orders to the four Ferries on the *Schulkyll* that no Sea-men or Marines be suffered to cross without a pass from such Authority as you may think proper to appoint."[63] The problem eased as the days passed.

Throughout the spring and into the summer the *Delaware* remained above the chevaux-de-frise as final manning and outfitting took place. This inactivity soon resulted in personal quarrels, one of which took place between her two Marine lieutenants and erupted into a controversy that was carried on in the open press. It began on 25 June when Second Lieutenant Alexander Neilson entertained several guests on board the frigate.* Wanting to provide them with refreshments, Lieutenant Neilson broke open the sea chest he shared with Lieutenant Daniel Henderson and borrowed some of his liquor. Upon returning, Henderson charged Neilson with theft and demanded a court martial. A court martial was held and Neilson was honorably acquitted, but ordered to make restitution. Henderson disagreed with the findings of the court and the evidence given. The matter seems to have been laid to rest after Lieutenant Neilson restored all that he had taken from Henderson.[64]

While the *Delaware* stood off Fort Island, the frigate *Washington* remained anchored at Philadelphia. With his ship still uncompleted, Captain Dean had a difficult time in retaining his Marines. A number deserted because of the inactivity and enlisted on board ships of the Pennsylvania State Navy.[65] In addition, his Marines often became the target of rowdy bands of soldiers who frequented the docks in search of easy money.[66] As a result of this inactivity, Captain Dean apparently resigned his commission and passed control of the *Washington's* Marines to his lieutenant, Abel Morgan.[67]

On 23 July Howe's fleet of more than 260 warships and transports, laden with fifteen to eighteen thousand soldiers and an impressive array of military equipment, set sail from Sandy Hook. Seven days later the lead ships of the fleet were sighted by the Pennsylvania Committee of Safety's lookout, Henry Fisher, at Lewes.[68] Messengers were sent immediately to Philadelphia with the news, and then on to Washington at Coryell's Ferry who dispatched his troops across the Delaware to the capital. On 2 August, Fisher wrote that the fleet had withdrawn.[69] At the Falls of Schuylkill near Germantown Washington halted his army to await sure intelligence of its destination. On the 10th, news arrived that the fleet had been sighted off Sinepuxent Inlet in Maryland, 30 miles below the Delaware Capes, heading south. Four days later the fleet rounded Cape Charles. It would take 11 more days to reach its landing at the Head of Elk in upper Chesapeake Bay.

News of the British fleet sent Philadelphia into near panic. To protect the city against an assault by water, the frigate *Delaware* was immediately ordered down river with several state galleys to reinforce the *Andrew Doria* and *Fly*, which had been dispatched in April.[70] By 20 August the frigate and her charges had passed through the chevaux-de-frise at Billingsport and anchored off Port Penn. The following

* Alexander Neilson (Nelson), former second lieutenant of Marines on board the *Alfred*, seems to have accompanied John Paul Jones to Philadelphia in early 1777. Following David Love's resignation, Daniel Henderson was advanced to first lieutenant in Mullan's company, and Neilson assigned as second lieutenant.

morning a British frigate and two schooners were sighted abreast of Bombay Hook.[71] The Americans gave chase and the enemy withdrew. Several days later, three more galleys and two fire ships were ordered to assist the force in defending the river.[72]

On 25 August, General Howe landed near the Head of Elk and on 11 September engaged Washington's Continentals at Brandywine. This struggle over, another two weeks were consumed in marching from Chadd's Ford to Philadelphia, due to the many skirmishes having to be fought with the harassing parties which hung on the flanks of the victorious British army. On 26 September, at half past eight in the morning, Lieutenant General Cornwallis, with 3,000 British troops entered Philadelphia. By 10 the city was theirs.[73]

Three days earlier an order went out from Lewis Nicola, commander of the city, to remove all shipping anchored at Philadelphia either up or down the river.[74] The following day Lieutenants James Coakley and Abel Morgan gathered their small Marine guard detachments and repaired to their respective frigates, the *Effingham* and *Washington*.* Later, both frigates moved up the river to berths at Burlington. All other ships were ordered down river to join the state fleet under the command of John Hazelwood.

Soon after the British gained control of the American capital, their engineers were tasked to construct two batteries along the banks of the Delaware "to prevent the Enemy's fleet repairing up the River or annoying the city."[75] These batteries were begun on the evening of the 26th and completed the following morning. The lower battery was to be the most powerful—"four medium 12 pounders and two 5½ inch howitzers"—since it was from the south that an attack most probably would come.[76]

John Hazelwood, by Charles Willson Peale.

For fear that an attack might occur at any moment while the platforms were being laid, two 12-pounders and one howitzer were placed on the bank and armed.

In an effort to unify the American naval command on the Delaware, the Marine Committee, then sitting in Lancaster, Pennsylvania, sent a letter to Captain Alexander on the 26th informing him that he and the rest of the Continental captains would now be under the command of Commodore Hazelwood of the Pennsylvania Navy. The committee in its letter also reiterated the importance of maintaining control of the river. They noted that if the Delaware is "bravely and properly defended the possession of Philada. will probably turn out to be the ruin of the British Army."[77]

An American Frigate is Captured

Off Fort Mifflin, early the following morning, the captains of the combined state and Continental fleet gathered to discuss what was to be done concerning the defense of the river. The subject of passage up the river arose during the conference and it was decided that Captain Alexander should proceed to Philadelphia with the *Delaware*, state ship *Montgomery*, and Continental sloop *Fly*. Once off the city he was instructed to do everything in his power to "annoy the Enemy should they attempt to Come in to our City, & should you see them preparing any works for Cannon or Hauling Cannon near the river,

* Lieutenant James Coakley returned to Philadelphia with Robert Mullan's company of Marines. Since the uncompleted *Effingham* was denuded of Marines by the continuation of Porter's company in the artillery, Coakley and several privates seemed to have been assigned to guard her.

you are in that case to send a Flag a Shore & warn them if they do not desist in making any preparations of Fortifying any where, that you will in that case fire on the City." Should the situation be other than that expected, the conduct of the expedition was left to Alexander's "prudent & effectual management."[78]

Soon after sunrise Alexander collected the *Montgomery*, *Fly*, and five galleys, and set out from Gloucester Point for Philadelphia. As the Americans approached the lower battery, the *Montgomery* came into the wind some distance from shore, but the *Delaware* moved closer. According to the commander of the British battery, the frigate came within "400 yards, and in another tack or two would have been alongside our guns." The battery in an attempt to drive off the frigate fired once, disobeying issued orders not to fire until fired upon.

The *Delaware* returned the fire and was soon joined by the other American vessels. Thereafter, the frigate attempted to push farther upriver. In doing so she received "some shot, and a shell which set her on fire, the people were thrown into great confusion, neglected the management of the sails, and she ran aground within 250 yards" of the battery, at a point off the lower end of Windmill Island. Several minutes later she struck to the British. With the frigate aground in mid-channel and the *Fly* driven ashore on the New Jersey bank, the *Montgomery* and galleys returned down-river.[79]

On board the grounded frigate, Marines and sailors scurried to lower boats in an attempt to escape to the opposite bank. Marine Lieutenant Neilson and a group of men were able to get away before a detachment of British Marine grenadiers boarded and took control of the frigate. The rest of the officers and crew were taken prisoner and immediately hustled off to the city jail. Several days later they were transferred to the state house where many of them remained until the end of British occupation in 1778. The *Delaware*, much to the chagrin of the Americans, was taken into the British fleet.[80]

Marines to the Rescue

The advance squadron of the British fleet, led by the *Roebuck*, sailed up the river as far as Chester on 28 September. There at midday on the 29th, it prepared to ferry elements of the 10th and 42nd Regiments under Lieutenant Colonel Thomas Stirling across the river. Destined to attack the weakly garrisoned American fort at Billingsport, they did not complete the crossing until late in the afternoon of 1 October.[81] Meanwhile, Colonel William Bradford of the Pennsylvania militia and his 112 men at the fort wondered about the strength and intentions of the British landing party. At Woodbury, six miles away, Brigadier General Silas Newcomb of the New Jersey militia estimated the British strength from intelligence reports at 400, somewhat equal to American strength in the immediate area. During the night of 1 October, Newcomb assembled about 300 men and two or three field pieces, and marched off to engage the enemy who had not moved from the landing place at Raccoon Creek.[82]

As General Newcomb and his force began their march south, Colonel Stirling and his force advanced northward. At nine in the morning on the 2nd, the two forces met and engaged in a short, but heated fight. During the first engagement Newcomb realized that he had greatly underestimated British strength, which he now conceived to be about 1500. Under constant fire, Newcomb was forced to withdraw northward. After a second sharp engagement at Mantua Creek, the British withdrew and turned their attention toward the fort.

With news of the New Jersey militia in retreat, Colonel Bradford ordered an immediate evacuation of the Billingsport garrison to Fort Mifflin. Guard boats officered by Marine Lieutenants Dennis Leary and William Barney of the Continental brig *Andrew Doria* worked feverishly until all the ammunition and men were transferred to safety. A few Americans remained behind to spike the guns and set fire to the fort's buildings. Not long after, in spite of the fire from American galleys, the fort was taken.[83]

With the fort in their hands, the British procedeed to remove the lower chevaux-de-frise and finally succeeded in cutting away part of it, affording a

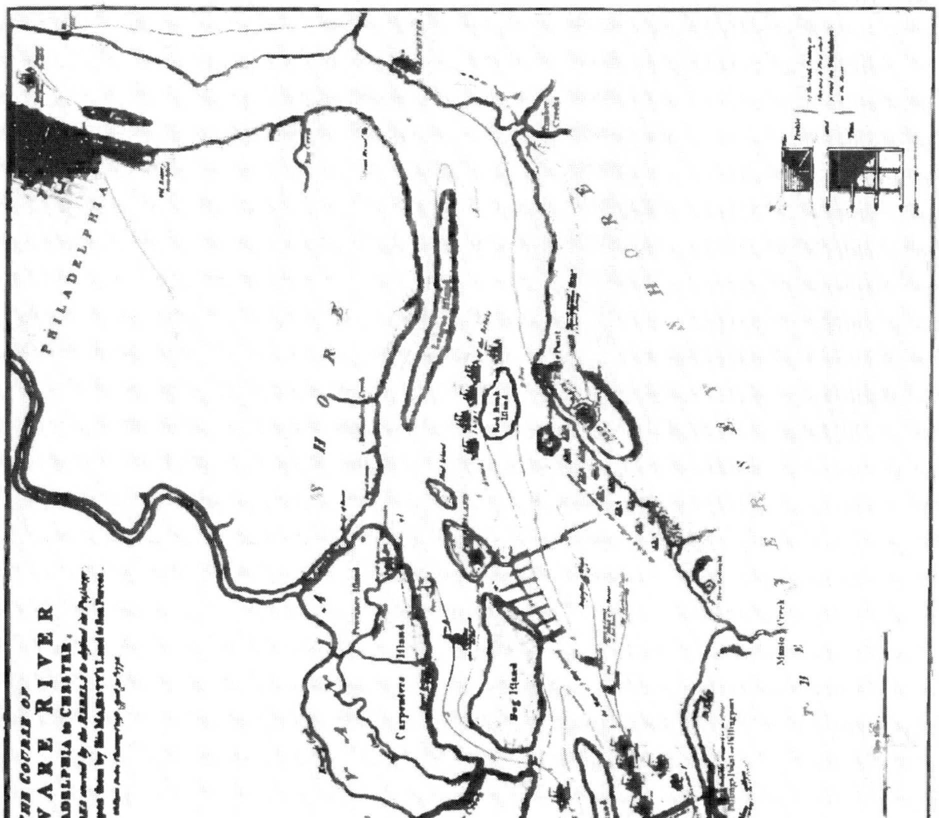
THE COURSE OF
DELAWARE RIVER
from PHILADELPHIA to CHESTER,
PHILADELPHIA
CHESTER

William Barney, by Charles St. Memin

passage for their ships. As the work continued the Americans repeatedly endeavored to prevent the dismantling of the obstruction by sending down galleys and fire rafts. Nevertheless, on 14 October the passage was widened to about 100 feet, and two days later it was buoyed. By the 22d the British fleet had warped through. They were too late to prevent the costly repulse of three Hessian battalions by the Americans who held Fort Mercer at Red Bank.

Checked by the American fleet, the British now attacked Fort Mifflin instead of renewing the attack against Fort Mercer. Early on 23 October, the British fleet and batteries thrown up on the Pennsylvania bank of the river combined to bombard the small fort on an island in mid-channel. Aided by the effective fire of the American fleet, the fort successfully weathered the first assault. About 10 in the morning, the British frigate *Augusta* began to burn after running aground the night before. At noon the fire ignited her powder store, and she blew up. The *Merlin* which had got aground at the same time was set afire and also destroyed. The first assault on Fort Mifflin had been costly.

Following this repulse, the British installed more powerful batteries opposite the fort and a second attack was made on 10 November. Under a tremendous bombardment from the fleet and shore batteries, the stronghold held out until the 15th when it was evacuated to Fort Mercer. A few days later Fort Mercer was also abandoned. The American fleet was now entirely without shore protection. On 21 November several galleys and smaller vessels ran by the city under intense fire and escaped up river. The larger Continental vessels and fire ships finding "no wind . . . were set on Fire and burnt, which made a most terrible conflagration, to the great joy our our cruel & wicked enemies."[84] By early December 1777, the British completely controlled the Delaware River from Philadelphia to the capes.

In 1777 the British brought the war to America. Narragansett Bay, the Connecticut River, New York, and the Delaware River were either blockaded or in British hands. The enemy for the most part controlled the American coast. Nevertheless, Continental Marines survived, as did the Army and Navy. So flexible were the Marines, that 1777 also saw an expansion of Marine activity into European waters.

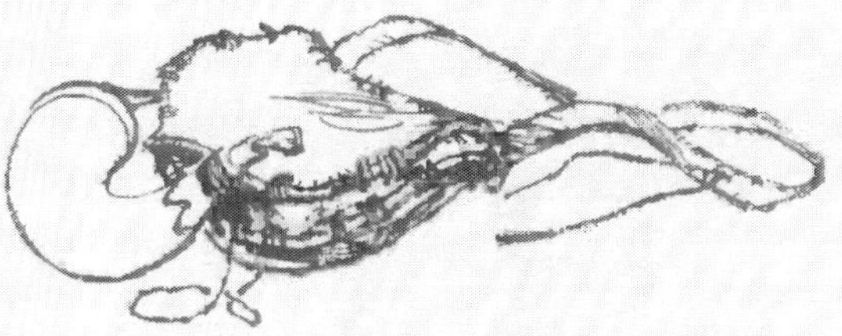

LAUNCHING OF THE WHITEHAVEN RAID, 22 APRIL 1778.

During the spring of 1778, John Paul Jones sailed into the Irish Sea determined to raid the English seaport of Whitehaven. This painting portrays Jones (center) about to launch the raid with two boats filled with 30 volunteer Marines and seamen.

CHAPTER VII

Across the Atlantic, 1777-1778

Franklin's Naval War

When the American Revolution broke out, the French Government was careful to adopt a policy in accord with its own interests and consistent with the letter, if not the spirit, of agreements worked out with the Government of England. Louis XVI's advisors, although they disapproved of the subjugation of the colonies by force, were reticent to show their hand prematurely, lest England and her American colonies make up and celebrate their reconciliation by an attack on French possessions in the West Indies. So France allowed the civil war to continue and awaited an opportunity to intervene, while England tore herself apart.

It was essential, wrote the French Foreign Secretary, Comte de Vergennes, on 12 March 1776, that the intentions of France and Spain not only be pacific but friendly, in order that the British government might entangle itself too deep to retreat. The courage of the colonists, on the other hand, would have to be sustained "by some secret favours" from France. They should be supported furtively with arms and money, and informed that, while it was below the French King's dignity to associate openly with insurgents, he was disposed to recognize them as allies if they would renounce their allegiance to Britain and declare themselves an independent nation.[1]

Such was the position of the French Government when Silas Deane, the first commissioner sent by the Continental Congress, arrived in Paris in June 1776. The prospects of cordial relations between the American Republic and the Court at Versailles were most encouraging. Failures seemed to follow in the wake of the diplomatic methods employed by colonial leaders until the astute statesman, Benjamin Franklin, was appointed American commissioner in late 1776. From the time he assumed his post his influence was apparent throughout all departments of the French administration. Millions in loans began to flow, secretly purchasing military stores for Washington's army and equipping American ships which preyed upon English commerce.

Since the French were not yet ready to declare openly on the American side, Franklin decided to initiate a second front on his own. As soon as she could refit, after her arrival in France, the *Reprisal* sailed for the English Channel in late January. Within a matter of days, the yellow and black brig captured three enemy merchantmen which were sent off for L'Orient. Continuing northward on 5 February, Captain Lambert Wickes encountered the *Swallow*, a 16-gun royal mail packet destined for Lisbon. Soon the sound of the *Reprisal's* 6-pounders and the *Swallow's* 4-pounders thundered across the channel mouth. Broadside after broadside were fired, but neither vessel sustained heavy damage due to a rough sea. On board the American brig, after a half hour, one man had been killed and naval Lieutenant Robert Harris had had his left arm torn off above the elbow by the bursting of one of the 6-pounders.[2]

On the *Swallow* two sailors had been burned with gunpowder, "their cartridges having blown up by accident," and the flames spread along the deck. Nevertheless, as the *Reprisal* moved closer, musket and small arms fire from the packet swept the brig's deck.[3] Marine Lieutenant John Elliott and several of his men were wounded, the lieutenant receiving a musket ball in the wrist.[4] The vessels closed, and American boarders swarmed over the side. Just 45

Benjamin Franklin, by Johann Martin Will.

minutes after the engagement began, the packet's captain surrendered and the British flag was hauled down.[5]

Once the prisoners were on board the brig and a prize crew placed on the *Swallow*, the *Reprisal* turned southward for L'Orient. The return voyage, although ardous due to high seas, proved profitable.[6] It was during this period that Wickes took a fifth and final prize—the ship *Betty*, British owned and bound from Bordeaux to Londonderry with "Brandy Claret & Hoops." She was 180-tons, and largest of the prizes. A prize crew was placed on board, and her cargo, crew, and passengers transferred to the *Reprisal* which then continued southward.[7] Proceded by two of her earlier prizes, the American brig arrived off the French port on 13 February.[8]

The seizure of the *Swallow* and the four British merchantmen immediately brought wild protests from the British Ambassador in Paris, Lord Stormont. The French Foreign Minister cooled things off by ordering Wickes and his prizes out of French waters within 24 hours. But by this time the damage had been done—all the prizes had been sold and their cargoes auctioned. In his report to the Committee of Secret Correspondence of Congress, Franklin admitted that Wickes had given "some trouble and uneasiness to the [French] court," but he added, "We have ordered him to make another cruise before he returns to America."[9]

Before Wickes put to sea again, Franklin found two even likelier candidates to carry out his program of naval warfare against Britain—American naval Captain Samuel Nicholson, and an Irish seafarer named Gustavus Conyngham. The younger brother of the Navy's senior captain, James Nicholson, Samuel found his way to France in early 1777, and was directed by Benjamin Franklin to purchase a cutter "suitable for the purpose of being sent to America."[10] In pursuance with these instructions, Nicholson secretly purchased the cutter *Rochefort* in Dover and then had her brought to France where she was renamed the *Dolphin*. Since the cutter was originally intended to be used as a packet, her French crew refused to continue on board when they learned that she would be used as a vessel of war. Not pressing the point, Nicholson discharged the crew and began the tedious process of enlisting a new one at Nantes in April.[11]

Gustavus Conyngham, a hot-headed daredevil from County Donegal, found himself stranded in Europe after British protests to the Dutch had forced him to abandon his ship, *The Charming Peggy*, loaded with military supplies for the colonies, in Dutch waters. Corrupt Dutch officials then seized it and sold the ship for next to nothing. Through a friend, Conyngham met William Hodge, a Philadelphia merchant employed by the commissioners to purchase vessels at Dunkirk for the Continental naval service, who recommended the Irishman to his superiors as a captain for one of these ships. Franklin accordingly filled out a blank commission provided by Congress, and ordered Conyngham to return to Dunkirk where he assumed command of an English-built lugger, renamed the *Surprise*.[12]

The *Surprise* was fitted out, manned with seamen and Marines, armed with 10 guns, and got to sea about 1 May. Heading along the coast of Holland under Continental colors, Conyngham made his first captures within a few days. On 3 May the packet *Prince of Orange*, enroute from Harwick to Helvoetsius, was intercepted. On the following day the 110-ton brig *Joseph*, bound from Messina to Hamburg, with a cargo of wine and fresh fruit, was taken. With these two prizes the *Surprise* returned to Dunkirk.[13] The British ambassador in Paris immediately protested, and the French were obliged to yield to his demands.[14] The lugger was seized, her captain and crew imprisoned, and the prizes returned, all in an apparent attempt by the French to avoid a rupture with Britain until the time was ripe.[15]

A Squadron Gathered

Gustavus Conyngham, by an unidentified artist.

Meanwhile, the Continental brig *Lexington* had arrived at Bordeaux and her captain, Henry Johnson, at once departed for Paris. There the commissioners informed him of their plan for a bold attack by the *Reprisal*, *Lexington*, and *Dolphin* on the Irish linen fleet out of Dublin. Under orders to join Nicholson and Wickes at Nantes, Johnson returned to his ship.[16] Before he could raise anchor 17 seamen he had recruited from the two prizes taken on the crossing requested discharges. This left the *Lexington* with only 45 crew members on the eve of the projected cruise to the Irish Sea. The problem was quickly solved during the first week of May with a large influx of French hands. In addition, two Americans were secured, George Thayer and John Carr, formerly of the Continental brig *Cabot*. Carr had been languishing in a British prison for months, having been captured on one of the *Cabot's* prizes in October 1776. Escaping to France, he signed up with Johnson as a lieutenant of Marines on 8 May.[17]

Eleven days later, the *Lexington* sailed from Bordeaux, passing down the Gironde River and out into the Bay of Biscay. Captain Johnson observed a strange sail soon after emerging from the river's mouth and later identified her as a British man-of-war. Falsely thinking his pursuer was the ship *Foudroyant* of 80 guns, he called for full sail and quickly outran the lumbering ship-of-the-line.* Forced off her course momentarily, the *Lexington* continued on her way, and on 22 May entered the Loire—much to the relief of Wickes and Nicholson who had begun to fear for her safety as the days passed.[18]

Within hours of the *Lexington's* arrival, Lieutenant of Marines John Carr was discharged, after serving only 14 days. Carr's discharge left only two other Marines on board the brig, Lieutenant James Connelly and Sergeant John Barry. It is probable that a number of the 32 French landsmen and able seamen recruited at Bordeaux were assigned to them as Marines.[19] The *Dolphin* suffered somewhat from the same deficiency, while the *Reprisal* did not. Of a total crew of 35 on board the cutter, only eight were Marines.[20] The *Reprisal* had a full complement of about 40 men under Captain Miles Pennington and Lieutenant John Elliott.

The orders for the cruise were issued on 23 May by Wickes, who was the senior officer. If at all possible the vessels were not to part from each other, but in such an event they were to continue to "Cruize through the Irish Channel or to the North West of Ireland," until they arrived "off the Isles Orkney, and there Cruize 5 or 6 Days for the Fleet to Come up." Should the remainder of the fleet fail to appear, they were to sail southward "for Bilboa, or St. Sabastian & there Refit as fast as possible for Another Cruize." Prizes were to be sent into French and Spanish ports, but to maintain a mask of neutrality, "the Prize Master must not Report or Enter her as Prize, but as an American Vessel from a port that will be most likely to gain Credit according to the Cargo, she may have on board." Continuing in the same vain, Wickes ordered that "if you meet a Dutch, French, Dean, Sweed, or Spanish Vessel when you have a Number of Prisoners on board, I think it would be well to put them on board any of those Vessels, giving as much Provision and Water as will

* The *Foudroyant* was then in Portsmouth. The ship sighted was probably HMS *Burford*.

serve them into Port." Wickes' concern for the security of his prizes and ship was a direct result of the imprisonment of Conyngham's crew and other indications that France would avoid entangling itself in a conflict with Britain at that point.[21]

The American squadron set sail from Nantes on 28 May. Two days out, Wickes and his little fleet encountered what was again thought to be the 80-gun *Foudroyant*, but in fact turned out to be the 74-gun ship-of-the-line *Burford*. She fired several shots at the *Lexington* before the squadron outran her, setting its course for the northwest. The linen fleet was missed, but the Americans continued to cruise the mouth of the Irish Sea and English Channel, making 14 captures within a five day period, a total of 18 prizes in all. Off Ushant on 27 June, the *Burford* was again sighted and this time her captain, George Bowyer, had no intentions of permitting them to slip through his fingers. Following a chase which consumed the greater part of the day, the Americans were compelled to separate; the *Reprisal* and *Dolphin* putting into St. Malo, and *Lexington* into Morlaix.[22]

While Wickes' squadron readied itself in early May, the American Commissioners procured the release of Conyngham and his crew. Fearing that the crew would disperse, the Irishman was given a second commission and he and his crew were ordered immediately on board a cutter which Hodge had recently purchased at Dunkirk. The vessel was aptly named the *Revenge* and carried 14 guns. Quickly standing out to sea in mid-July before she could be detained in port, the cutter cruised for several months, seizing and sometimes destroying British ships in the North Sea, Baltic, and on all the coasts of England and Ireland. Conyngham sailed completely around the British Isles, even landing at one point on the coast of Ireland for water, before putting into the port of Ferrol in Spain. The *Revenge* cruised successfully throughout the remainder of the year, sending a large number of prizes into Spanish ports.[23]

The cruises of the *Reprisal*, *Lexington*, *Dolphin*, and *Revenge* caused Lord Stormont to renew his protests. Furious, the British Ambassador threatened to resign and thus break off diplomatic relations with France if something was not done about American vessels. Still officially uncommitted, the French Government reproached the American Commissioners for failure to keep their ships away from French ports. To indicate further displeasure with the Americans, Vergennes ordered the arrest and confinement in the Bastile of William Hodge, and directed that the *Reprisal*, *Lexington*, and *Dolphin* be sequestered and detained until sufficient security could be given that they would return directly to America. Hodge was soon released, but the French would entertain no excuses concerning the three American vessels.

After being driven into port at the end of their cruise around Ireland, Captains Johnson and Wickes were engaged for several weeks in refitting their damaged vessels. The *Dolphin* was soon converted into a packet, the original purpose for which she was purchased, and Captain Nicholson and his crew were assigned to a frigate then being readied at Nantes. Upon learning of Nicholson's appointment, and desiring to advance his junior Marine officer, John Elliott, Wickes strongly recommended him as commander of the Marines for the new frigate. "I take this Oppertunity," wrote the *Reprisal's* captain, "to Recommend my Lieut. Marines as a Very proper Person for Captain of Marines for Capt. Nicholson, as I think him entitled to Perferment in Prefference to any other Person, as has been in the Service from the Commencement of the Reprisal's being Armed & behaved himself Very Well."[24] With the recommendations of both Captains Nicholson and Wickes in hand, the Americans Commissioners granted John Elliott a commission of captain of Marines on board the frigate, later named the *Deane*.[25] Captain of Marines Miles Pennington, Lieutenant Elliott's senior on the *Reprisal*, saw his brother Marine officer leave the ship with regret, but little imagined that Lieutenant Elliott not only secured a promotion thereby, but also saved his life.

Stormont's demands became so insistent that they could no longer be evaded, therefore the Commissioners acceded to the French demand that the *Reprisal*, *Lexington*, and *Dolphin* proceed directly to America and cruise no longer in European waters. By mid-September all three ships were ready for sea. Wickes wished to make the return voyage in company with Captain Johnson, but they did not meet. The *Reprisal* and *Dolphin* sailed forth from St. Malo on 14 September, while the *Lexington* departed Morlaix three days later.[26]

Two days out of Morlaix, the *Lexington* fell in with the fast sailing royal cutter *Alert*, of 10 guns. Only after a bitter fight, during which Johnson exhausted his ammunition, did the *Lexington* surrender. Lieutenant John Bazely of the cutter described the action:

> I gave chace at five in the Morning, and came up with him at half past Seven, had a close Engagement till Ten when He bore up and made Sail, as soon as I got my Rigging to rights, again gave chace and came up with him, at half past one, renewed the Action till half past two, when he Struck.
>
> I have been so fortunate as to have had only two Men killed and three wounded, one of which is since Dead with my Mast, Rigging and Sails much cutt and Damaged.

The *Lexington* lost seven killed and 11 wounded. Among those who died in the engagement was James Connelly, lieutenant of Marines.[27]

Captain Johnson and a portion of the *Lexington's* crew were transferred to the *Alert* which then made for Plymouth. Soon after their arrival, the captives were confined at Mill Prison from which only a few made their escape; among those who did were Captain Johnson and his sergeant of Marines, John Barry.[28]

Off Ushant, the *Dolphin* and *Reprisal* separated; the cutter heading southward while the brig sailed westward toward her destination, Portsmouth, New Hampshire. Two weeks later, and within days of her arrival, the *Reprisal* foundered on the Newfoundland Banks. All of her crew except the cook were lost.[29]

The ship *Alfred* and frigate *Raleigh* arrived at L'Orient in November 1777. But times were not propitious for the cruising of Continental ships in European waters. Again Lord Stormont, the British Ambassador, protested the presence of American ships, and again the French government acquiesced. On 25 November, the American Commissioners ordered both vessels to return to America; suggesting a circuitous route by way of Africa and the West Indies where unescorted British merchantmen could be taken. Early the following year the *Alfred* and *Raleigh* were homeward bound.[30]

From Portsmouth to Solway

The sloop *Ranger*, arriving from Portsmouth, worked her way down the coast of Lower Brittany in the early morning hours of 2 December; by late evening she had made her way up the Loire and was anchored off Paimboeuf, the deep water port for Nantes.[31] Soon after his arrival, Captain John Paul Jones journeyed to Paris where he conferred with the American Commissioners about his future command. Prior to his departure from Portsmouth, Jones had been promised the command of *L'Indien*, a large frigate then building in Amsterdam. All he received from Franklin, Silas Deane, and Arthur Lee were evasive promises, since very delicate negotiations were underway in the French capital.[32]

With news of the American victory at Saratoga, Comte de Vergennes became more fearful that if an alliance was not concluded soon, some sort of peace without American independence might be reached. His spies informed him that Lord North, the British Prime Minister, had been in contact with both Franklin and Deane. Worried that an agreement might have been proposed to settle the dispute, the French foreign minister immediately convened the Council of Ministers early in January, and warned them that it was time to act if the colonies were to be kept in the war. The council voted for an alliance, and within a month it was signed.

Now, with no further problems about French neutrality, Franklin turned his attention to Jones and the *Ranger*. As to *L'Indien*, the American Commissioners concluded that it would be politically advantageous to transfer the frigate to the French government, rather than to take possession. Deprived of what he thought would be a fine command, Jones then proposed a cruise in the *Ranger*. On 16 January Franklin and Deane issued the desired orders. They instructed him, "after equipping the *Ranger* in the best manner for the cruise," to "proceed with her in the manner you shall judge best for distressing the enemies of the United States, by sea or otherwise, consistent with the laws of war, and the terms of your commission." Should he make "an attempt on the coast of Great Britain," continued the orders, Jones was advised not to return to France immediately since the French were still nominally neutral; the alliance having been agreed to, but not yet signed. Finally, the two American Commissioners cautioned him against giving offense to the subjects

The Continental Ship Alfred.

of France or any other neutral power, lest he destroy the diplomatic framework which had taken them months to construct.[33]

Jones had already stated his ideas on American naval policy to the Commissioners upon his arrival. Sharing the views expounded by Robert Morris nearly a year before, Jones noted that the most effective employment of American vessels against the enemy was "to surprize their defenceless Places and thereby divide their Attention and draw it off from our Coasts," rather than attempt to cope "with their Navy as their numbers and Force is so far Superiour to ours."[34] But what specific plan Jones had in mind when he returned to Paimboeuf in late January, is not known. Soon after his return, however, he spent several weeks outfitting the *Ranger*. New sails were cut, swivel guns mounted in the fighting tops, and the bow ports altered so the guns could be trained ahead in a chase. Quantities of clothing and provisions were purchased; stocks of match, and barrels of gunpowder were provided; and dozens of pistols, cutlesses, and blunderbusses were obtained. The *Ranger* was ready.[35]

Early in February, the *Ranger* dropped down the Loire with several merchantmen to Quiberon Bay where a French fleet under Chef d'Escadre (Commodore) La Motte Piquet lay.[36] On board the sloop trouble was brewing among the officers; they wanted to get to sea, not take part in one of Jones' crazy schemes in which there would be no prize money. They had received promises in Portsmouth that they would "make their Fortunes," but so far not a penny was forthcoming from the sale of the two prizes taken during the Atlantic crossing. Their mutterings and rumblings soon took the form of a complaint against the *Ranger's* captain of Marines, Matthew Parke. Captain Parke, they protested, had "embarked on board the Ranger, with a view of acting according to his Rank, in a Frigate which you were to take the command of, on our arrival in France." Since "no Captain of Marines is allow'd to any Ship or Vessel, under twenty Guns, we take it as a hardship *peculiar to us*, that a person *in his Capacity*, should remain in the Ship, to take the fourth part of the three twentieths which are the Shares belonging solely to *us* (as Lieutenants and Master of the Ship) of any Prize money to be divided for her Officers and Men." Therefore they requested Jones

John Paul Jones, by the Comtesse de Vendahl.

"to dispose of Captain Parke in such a manner, that he may not Interfere with us, in any future division, a late instance of that kind has been *very disagreeable.*"[37]

Captain Parke unhappily wrote to Jones five days later, 19 February 1778, and requested a discharge. "On our arrival in France," he noted, "I expected to have the Happiness of going on board the Frigate you was to have had the command of, this has failed, and these Officers are very much dissatisfied at my continuing on board any longer in my Station." Continuing, Parke wrote: "considering my future happiness and the welfare of the Service, I do request a discharge from this Ship (tho' very disagreeable) that I may return in the Frigate Deane to America."[38] Reluctantly, Jones accepted his resignation, though he thought his lieutenants and master selfish, and their petition "Illiberal and Unjust."[39]

After days at anchor near the head of Quiberon Bay, Jones sailed for Brest on 3 March. The voyage took longer than expected due to bad weather, but on the 8th the *Ranger* put into the little port of Cameret, about eight miles below Brest.[40] There Jones dismissed Captain Parke, and temporarily replaced him with Lieutenant Jean Meijer of the Swedish Army.[41] While in the northern French port, Jones communicated his plans to "end the barbarous ravages perpetuated by the English in America" to Lieutenant-Général le Comte d'Orvilliers, commander of the fleet at Brest. "I proposed," Jones later related, "to descend on some part of England and there to destroy merchant shipping. My plan was also to take someone of particular distinction as a prisoner to hold him as hostage to guarantee the lives and exchange of Americans then imprisoned in England." The poor treatment of captured American seamen was a sore point with officers and sailors alike, and Jones hoped to change British practice, if not policy in this area.[42] But against what part of Britain Jones would direct his attack, he had not decided.

Soon after her arrival, the *Ranger* was moved up the Bay to Brest, and the crew allowed liberty after almost a month at sea. The town offered many allurements for the average seamen or Marine, since it was esentially a military port, and several took the opportunity to desert. A few were captured and returned to the sloop under guard where they were confined in irons. Early in April, the *Ranger* again put down the river for Cameret, there to be careened and graved once more before sailing. The time consumed scraping and caulking the sloop's bottom allowed the junior Lieutenant of Marines Samuel Wallingford and a number of other officers a chance to view the town's fortification and yards, in addition to touring a new fort being constructed nearby.[43] The *Ranger* was refloated on 4 April and the following day she and the French frigate *Fortunee* sailed down the bay. Poor weather prevented them from reaching open sea and the two vessels returned to port. Awaiting favorable winds, Wallingford had his Marines exercised in small arms drill, while naval Lieutenant Thomas Simpson worked with seamen on the sloop's eighteen 6-pounders. The weather looked promising on the 8th, and the two ships put to sea.[44]

The French frigate escorted the *Ranger* for four days and then unexpectedly returned to Brest. On 10 April, between Scilly and Cape Clear, a brigantine carrying flaxseed was taken and sunk, and on the 17th, off Dublin, a 250-ton ship loaded with merchandise and English port was captured and sent back to Brest with a prize crew. From Wicklow Head, the *Ranger* stood north into the Irish Sea.

About 10 in the morning on the 18th, Jones was off Point of Ayre, the northern cape of the Isle of Man, ready to cross the channel to Whitehaven, when he was challenged by the 8-gun British revenue cutter, *Hussar*. After a chase of several hours, the cutter escaped through the shallows of Luce Bay, and the *Ranger* turned back.[45] "Had the Captain have permitted the Marines to fire on them when they first came under our lee Quarter," recorded the sloop's surgeon, "[we] might have taken Her with great Ease."[46]

During the next few days a schooner and sloop were taken near the entrance of Firth of Clyde; both were sunk to the dismay of the *Ranger's* prize conscious crew. Jones then reentered the North Channel and on 20 April, off Carrickfergus, detained a small fishing vessel. From the crew he learned that HMS *Drake*, a 20-gun sloop-of-war, was anchored a short distance up Belfast Lough. Immediately, Jones "ordered the ship to be put about in order to go in and cut her out," but the sloop's crew refused. They consented instead to surprise the *Drake* that night by entering the lough and anchoring just to her windward, thus exposing the enemy's decks to the *Ranger's* musketry. The plan failed, but Jones made a safe return to the channel in hope of coming back the following night.[47] The weather turned and the crew was "very much fatigued," so the attack was abandoned and the sloop took shelter under the south shore of Scotland.[48]

Whitehaven and St. Mary's Isle

April 22d dawned fair and cold; snow covered both sides of Solway Firth and the Isle of Man—the "three kingdoms," as Jones observed. He now decided to carry out his planned descent upon the English coast. Of the numerous seaports which dotted the inlets and coves, the *Ranger's* captain settled upon the port of Whitehaven, partly because he was familiar with it, having sailed from there for Virginia at the age of thirteen, and partly because of information which placed a large number of vessels in the port's harbor. The crew saw no point in the raid since it brought neither prize money nor military advantage.[49] Lieutenants Thomas Simpson and Elijah Hall encouraged the dissent; as Jones later commented: "they were poor; instead of encouraging the morale of the crew, they excited them to disobedience; they persuaded them that they had the right to judge whether a measure that was proposed to them was good or bad." When the captain called for volunteers, both lieutenants were the first to decline since they were "ill with fatigue."[50]

As the *Ranger* crossed Solway Firth, the wind grew faint, so that by midnight the sloop was still miles away from Whitehaven. Two boats were then lowered, and 30 volunteers ordered over the side. Jones took command of one, with Lieutenant Meijer as his second, and Marine Lieutenant Wallingford officered the other, with Midshipman Benjamin Hill as his mate. At midnight the two boats, loaded with men and combustiles, left the ship and, after several hours of hard rowing, they arrived at the harbor's outer pier just as dawn was beginning to break. Unconcerned over the possibility of being discovered, Jones dispatched Wallingford's boat to the northern end of the harbor with orders to set fire to an estimated 150 merchantmen anchored there. Meanwhile, he and a few men scaled the walls of the port's southern battery, spiked the guns, and secured the four sentinels found asleep in the guardhouse.

Shortly after the fort was taken, Jones returned to his boat expecting to see the shipping in the north ablaze. Instead he "found the Boat under the direction of Mr. Hill and Mr. Wallingsford returned, and the party in some confusion their Light having burnt out at the Instant when it became necessary." By the strangest coincidence, his own party was in the same situation. Undaunted, Jones obtained a light from a nearby public house, and went a short distance to the southern end of the harbor where about 70 to 100 ships lay grounded. There he had a fire built in the steerage of the largest, the collier *Thompson*. A barrel of tar was obtained after a search and poured on the flames in hopes that it would spread.

It was now a full hour after sunrise, "and as sleep no longer ruled the world," Jones thought it time to retire. Besides, hundreds of townsfolk were swarming down to the water's edge after being awakened by a deserter who had run from house to house with the alarm. The two boat crews then reembarked after releasing a number of American prisoners, and began to row out to the *Ranger*. Presently the inhabitants brought up a few unspiked cannon and began to fire. Their shots, however, did no damage; instead, it "afforded some diversion which my People could not help shewing by discharging their pistols &ca. in return of the Salute."[51] By seven Jones and his men were on board the *Ranger*, and the raid was over.[52]

With the night's work at Whitehaven done, Jones now proposed to complete his plan by a descent upon St. Mary's Isle in Kircudbright Bay across Solway Firth. There he hoped to seize the Earl of Selkirk and carry him to France where he would be held as a hostage for the better treatment of American prisoners in England or to exchange him for a number of them. Unfortunately, his plans were to be frustrated.

By ten on the same morning as the raid on Whitehaven, the *Ranger* had crossed the Firth and was abreast of Little Ross, the island which marked the entrance to the bay.[53] The sloop's cutter was then lowered into the water, and Jones, Lieutenant Wallingford, Master David Cullam, and a dozen sailors and Marines embarked. Thirty minutes later the cutter grounded at the point of the isle. Leaving a guard, Jones, his two officers, and the squad of well-armed crewmen marched up the path to the manor house. Along the way they encountered the head gardener who was told that the party was a press gang seeking recruits for the Royal Navy. Questioning the man, Jones learned to his dismay that the earl was absent. He then turned about and began to walk back to the cutter—his mission a complete failure. Cullam and several others stopped him and demanded that they should be allowed to loot the house, since they had returned empty-handed from Whitehaven, and that New England homes had not been shown the same "delicacy" by British sailors. Jones then made a quick decision. Wallingford and Cullam with some of the men could go to the house, demand the family silver, and carry it off, but they were not to search the house or demand anything else.[54]

Shortly before noon Lady Selkirk, who had just finished breakfast with her overnight guests, observed a group of strange men surround the house. The children and guests were immediately told to go upstairs, while she and her butler would remain below to face the intruders. Stationing a man armed with a musket, bayonet, two pistols, and cutless at each door, Cullam and Wallingford entered the house and spoke to the countess. They told her that they were from a frigate belonging to the United States, and that they had been instructed to demand her silver. Lady Selkirk quickly decided to do as they wished, since there was no one to defend her or the house. She entered the pantry where she found her butler trying to hide some of the plate in a maid's apron, but told him to stop and hand it all over to them. The officers called for sacks and began filling them with the spoil. David Cullam the senior officer, the countess observed, was clad in blue clothes which to her did not resemble a uniform. He had "a vile blackguard look," and "seemed by nature a very disagreeable & one may say a bad man." Marine Lieutenant Wallingford, on the other hand, "was a civil young man, in a green uniform, an anchor on his buttons which were white," over which he wore "a blue great coat.[55] He "seemed naturally well bred & not to like his employment."[56]

When the sacks were filled, Cullam called for an inventory of the plate. Lady Selkirk produced it and the *Ranger's* master glanced at it for a second and then asked: "Where is the teapot & coffeepot." The butler brought the two pieces, but Cullam was not satisfied; he insisted there was more. The Countess assured him otherwise. After a short walk through the house, the two officers prepared to leave.[57] Before they reached the door, however, Lady Selkirk requested a receipt for what they were about to carry away. Lieutenant Wallingford began to write it, but got no further than "this is to cert. . . ." when he "found fault with the pen & said it was no matter for everybody would soon know they had been there."[58] Both officers then accepted a single glass of wine, took their leave, formed up their men, and marched quickly off to the cutter where Captain Jones was waiting. The party rowed out to the *Ranger*, which promptly made sail for the channel.[59]

The damage done by the *Ranger* in her two raids was inconsequential. British estimates of losses at Whitehaven ranged from £250 to £1,250, while the plate was valued at 3,000 livres or 600 dollars in France.[60] The effects of the raids, however, were tremendous, as the *London Chronicle* reported:

> A number of expresses have been dispatched to all the capital sea ports in the kingdom where any depredations are likely to be made; all strangers in this town are, by an order of the magistrates, to be secured and examined; similar notices have been forwarded through the country, &c. and, in short, every caution taken that the present alarming affair could suggest.[61]

Other newspapers printed letters condemning the defenseless state of the coast and laxness of the government. The country was thoroughly aroused. In response, the Admiralty ordered HMS *Stag* sent in pursuit of the "Yankee pirates"; however, her chances of intercepting the sloop were slim.

Drake Captured and Wallingford Lost

Instead of returning to France around the western coast of Ireland, Jones decided to remain in the area. By daybreak on the 24th the *Ranger* stood off Carrickfergus, the entrance to Belfast Lough.[62] The plan was to enter the lough a second time and attack the enemy sloop *Drake* "in plain day." The crew, however, thought the project "by no means right, because it involved honor more than self-interest, their only motive."[63] Spurred on by the ship's two naval lieutenants, they took this occasion to mutiny. Lieutenant Meijer happened upon the plan in time and warned Jones, who, when Master Cullam rushed him as a signal for the mutiny to begin, put a pistol to his head and quashed the plot.[64] Meanwhile, the *Drake*, whose captain had been warned of the *Ranger's* recent activities, prepared to stand out into the channel. Close to Kilroot Point a strange vessel was sighted entering the lough. Uncertain as to her identity, the *Drake's* captain ordered a boat sent to investigate.

Jones kept his now-quieted crew below decks, his guns masked, and the sloop's stern toward the approaching boat. Soon the gig drew alongside and the British lieutenant in charge came on board, where he was immediately informed that he and his crew were now prisoners. "This trifling success had such an exhilarating effect on my crew," Jones later wrote, "that they no longer objected to giving battle." Taking the boat in tow, Jones then decided to draw the *Drake* into the channel where he would have more room to maneuver.[65]

The *Drake* worked slowly out of Carrickfergus Bay against an unfavorable wind and flood tide. After weathering Kilroot Point, Jones allowed her to approach within hailing distance. As she did, her captain ordered the English colors hoisted; the *Ranger* responded by displaying the Stars and Stripes. "What ship is this?" demanded the Englishman. "The American Continental ship Ranger," replied Master Cullam, "we have been waiting for you, come on."[66] At this point Marine Lieutenant Wallingford appeared on deck dressed in his regimentals, "looking vary pale." One of the officers nearby asked him "why he came on deck for a mark to shoot at that he had better return and dress in sailers dress which he did."[67]

"The sun was now little more than an hour from setting, it was therefore time to begin," observed Jones. "The Drake being then astern of the Ranger, I ordered the helm up." This placed the sloop directly in front of, and across the *Drake's* bow. From that position, he fired a broadside of grape

The Irish Sea.

which swept along the enemy's deck. "The action was warm, close, and obstinate," Jones reported; "it lasted an hour and four minutes." Then the enemy called for quarters; her captain and four others dead, her lieutenant and 19 others wounded, and her rigging cut to pieces. The men killed on board the *Ranger* were Samuel Wallingford—who received a musket ball in the head—and two seamen; five were wounded.[68] Sergeant of Marines John Ricker was immediately given charge of the ship's Marine detachment until a replacement could be found.[69]

The day after the battle, 25 April, crews of both ships were employed in repairing the damage. Before noon a large brigantine from Whitehaven, bound for Norway happened to sail so close to the *Drake* that Jones was obliged to take her. A prize crew was put on board and the vessel ordered to Brest. That evening a solemn service was held on deck and the body of Marine Lieutenant Samuel Wallingford committed to the deep, "with the Honours do so brave an Officer." The following morning repairs were completed, but the wind suddenly shifted, so Jones decided to return to France by way of the west coast of Ireland rather than the southern channel.

By the 4th of May, the *Ranger* and her prize had made their way down the Irish coast and were approaching their destination. The next morning, off Ushant, a strange sail was sighted to the north. Jones, eager to please his crew with another prize, turned the sloop about and ordered Lieutenant Simpson, in command of the *Drake*, to do the same. Simpson, however, held to his southerly course. After ascertaining that the vessel was a neutral Swede, Jones turned south and caught up with the *Drake*. Furious with his lieutenant for deliberately disobeying his orders, Jones placed him under arrest and sent Elijah Hall on board to take over command of the *Drake*. The two sloops then headed for Brest, arriving there safely on the evening of the 8th.[70]

On his arrival in France Jones was plagued with numerous problems. One, however, was quickly solved, that of filling the vacancy left by the death of Lieutenant Wallingford. On 26 May, Jonathan Williams, American agent at Nantes, wrote to Jones and recommended William Morris, a man whom he thought would fill the vacancy well. Morris was the type of man Jones would like—he was a man unencumbered by the desire to enrich himself.[71] As Morris himself wrote: "my intentions are intirely to Serve my Country and wish to enter into the Servise with no other View."[72] Jones agreed with Williams' assessment and Morris was appointed lieutenant of Marines upon his arrival at Brest in June.

The major problem Jones had to contend with upon his return was that of the arrest of Thomas Simpson. The trouble concerning the naval lieutenant was an outgrowth of the unfortunate state of affairs on board the *Ranger*. For numerous reasons there had arisen a feeling of discontent among the crew; a feeling that was encouraged by Simpson, who, Jones charged, had gone so far as to incite a mutiny before the battle with the *Drake*. On that occasion, according to Jones, Simpson "held up to

the crew that being Americans fighting for liberty, the voice of the people should be taken before the Captain's orders were obeyed."[73] Jones also held his lieutenant responsible for the failure at Whitehaven, therefore, the only thing Jones could have done, which was consistent with Navy regulations and discipline, was to place Simpson under arrest. To his crew that was tyranny since the lieutenant was extremely popular with them, while the opposite was true of Jones.[74]

The American Commissioners in Paris regretted the whole affair. They begged Jones to be more flexible in the matter, and informed him that it was their desire to see Simpson released on his parole to go to America and face a court-martial.[75] Jones complied with their wishes and on 10 June he freed Simpson.[76] Five weeks later Simpson was released from custody and put in command of the *Ranger* which was to return to America. On 27 July, the lieutenant went on board the sloop, "to the joy and Satisfaction of the whole Ships company," according to Surgeon Green. Not long afterwards the *Ranger* sailed from France in company with two other American ships, one of which was the frigate *Boston*.[77] Jones, however, remained behind at Paris in hopes of receiving a new command.

With Adams to France

The *Boston* was brought to France in April by Samuel Tucker who succeeded Hector McNeill in command of the frigate after the latter's suspension. Three months earlier, however, the ship had been anchored in Boston harbor refitting for a new cruise. Her new captain was on board and her captain of Marines, Richard Palmes, had been reinstated. Towards the middle of January the frigate was about ready. Lieutenant of Marines William Jennison had returned from a successful recruiting trip to Douglas, as had Palmes from Boston. On the 20th, Captain Tucker issued orders to Palmes which were to guide him in the regulation of his men:

> You being Commanding Officer of Marines on board the Ship Boston under my Command—My Orders are that the Commissioned & Non commissioned Officers are to go on board with the Men under your Command & reside there constantly at their Duty—
>
> You are to be particularly careful that a commissioned Officer does constantly lie on board every Night—
>
> You are to obey such Orders as you shall receive from me, or the Commanding Officer of the Ship in my absence—
>
> The Commanding Officer of the Ship is not to give any Marine Leave to go on shore upon their own Occasions, if you have any reasonable Objection thereto
>
> The Marines are to be exercised in the Use of their Arms as often as you shall think proper, and are to be employed as Centinels, and upon any other Duty & Service on board the Ship which they shall be capable of—Such as not on Duty that 24 hours as Centinels, therein to be subject to the Officers of the Ship—but they are not to be obliged to go aloft, or be beat or punished for not shewing an Inclination to do so—being assured the ambitious will do it without driving—Those Marines not on immediate duty are to be turned on Deck by a Serjeant to do the Duty as above—
>
> The Marine Arms, Cartooch Boxes, Drums, Fifes & Accoutrements are under the immediate Inspection of you or your Officers—You are to be careful that the Arms be kept in good Order, & keep a sufficient number of Cartridges & Balls in readiness for Action—You are welcome to go on shore when you please, leaving such Orders with your Officers, as for the Safety & good Order of the Ship—[78]

There orders are the first indication of the exact duties and responsibilities Marines and their officers were to have on board a Continental ship.

The *Boston* weighed on 1 February, dropped down to Nantasket Road, and anchored. The following day her allotted amount of powder was loaded as final preparations were made for the upcoming cruise.[79] On the 10th, the Eastern Navy Board informed Captain Tucker that he had been selected to carry John Adams to France, where he was to replace Silas Deane as one of the three American Commissioners.[80] Accompanying Adams would be his eleven-year-old son, John Quincy Adams, and the sons of Silas Deane and William Vernon, all on their way to Europe for an education. Three days later, the *Boston*'s barge called at Braintree and returned with the distinguished charges.[81]

The frigate once again weighed on 15 February; coming to anchor in Marblehead harbor shortly before noon. There Tucker signed on his final contingent of officers and crewmen. As Adams noted in his diary, the ship still was in need of men, for those already on board were unruly and inexperienced: "he [Tucker] has very few Seamen indeed. All is as yet Chaos on board. His Men are not disciplined. The Marrines are not. The Men are not exercised to the Guns. They hardly know the Ropes."[82] Tucker prepared to sail the next morning, but adverse winds and heavy snow held the frigate at Marblehead for another day. Shortly after dawn on the 17th, the *Boston* departed the Massachusetts port "firing seven guns for Salute."[83]

Four days out, the *Boston* was hit by a storm of hurricane force. The wind raged and the sea churned for three days. "No Man could keep upon his Legs," recorded Adams, "and nothing could be kept in its place an universal Wreck of every Thing in all Parts of the Ship, Chests, Casks, Bottles &c."[84] During the storm a bolt of lightning happened to strike the captain of the mainmast, burning "a place in the Tope of his head about the bigness of a Quarter Dollar." The man lived another three days and then died raving mad.[85]

In fair weather again, the *Boston* made better than eight knots, covering about 200 miles in 24 hours. On 11 March, Tucker requested and received permission from Adams to chase a vessel which had been sighted to the windward. Coming within range a short time later she was found to be a 14-gun British letter of marque.[86] Suddenly she opened up with three guns, but quickly silenced them after absorbing the *Boston's* first broadside. The prize was the ship *Martha* out of London for New York with a cargo of provisions and stores valued at 84,000 pounds sterling. Tucker placed naval Lieutenant Hezekiah Welch on board as prize master and dispatched the vessel to Boston.[87] *

At Bordeaux

The *Boston's* turbulent voyage across the Atlantic came to an end on 1 April when she glided past the Tower of Cordovan and anchored just below Bordeaux.[88] On the 4th, Marine Captain Richard Palmes, accompanied by Adams and his young charges, departed for Paris carrying Tucker's instructions and signals.[89] Meanwhile, the *Boston* dropped down the Gironde River to Lormont where the frigate was careened, thoroughly cleaned, and a new mainmast stepped in. Returning up river on the 10th, the frigate was moored opposite the city's exchange and the crew given liberty. The officers, however, went about their duties. The following day Lieutenant Jennison submitted a clothing requisition for the ship's Marine company:

> Wanted for the Use & Service of Marines belonging to this Ship 40 Green Coats faced with white—
> 40 White Waistcoats &
> 40 White Breeches—
> The Buttons for the whole to be a plain White
> Coats to be opened sleeved, & a Belt to every Waistcoat.

The completed uniforms were delivered and distributed to the men one month later.[90]

Captain Palmes returned to Bordeaux on the 27th carrying with him secret orders from the three American Commissioners in Paris.[91] Under the orders, the *Boston*, once refitted, was to make a voyage "towards the entrance of the Baltic, or some other distent Seas, where the Ships Company might have an opportunity of making ample Profits to themselves, as well as acquiring the honour of serving

* During this engagement John Adams is supposed to have joined the Marines in a moment of intense patriotism, as the following extract of a letter purportedly authored by Captain Samuel Tucker and published in the *Daily National Intelligencer* on 16 September 1826, indicates:

> We fell in with a very large armed ship, though not a cruiser. She, however, soon appeared in a posture of engaging, and having our ship in readiness, with the men at their quarters, it became my duty to give Mr. Adams such information as was necessary. He followed me on deck where we conversed a few minutes on the subject of taking the ship, and after listening a moment or two to my entreaties for his safety, took me by the hand with a *God bless you* and descended the gangway ladder into the cockpit. I stepped after, and coming alongside the ship, hailed. His answer was a broadside and he immediately struck his colors before I could, to *good advantage*, discharge a broadside into him. Being very near, and in such a position that the smoke blew directly over our ship, while looking around on the quarter deck, observing the damage we had sustained from his fire, I observed Mr. Adams among my Marines, accoutred as one of them, and in the act of defence. I then went to him and said, 'my dear sir, how come you here?' With a smile he replied, 'I ought to do my share of the fighting.' This was sufficient for me to judge of the bravery of my venerable and patriotic friend Adams.

their country."[?] The exact destination was left to Captain Tucker's discretion. Concerned about the reduced compliment of seamen and Marines, the commissioners further recommended that Tucker "engage as many [men], at Bourdeaux by honorable means, as possible." This order was to distress not only Captain Tucker, but also his Marine officers.

The *Boston* stayed at Bordeaux for better than two months while recruiting parties scoured the French countryside. But problems in the enlistment of Marines soon arose. Early in May a Frenchman named Jean Morel volunteered as a Marine; his wife, however, objected. In a complaint lodged with the port's Commissary of the Marine, she said her husband had been forced on board the *Boston* and was being kept there against his will.[92] News of the complaint, and the supposed ill-treatment of Frenchmen by officers of the ship spread, causing naval Lieutenant Musco Livingston to remark that the reports have "disgusted them [the French] much against our ship."[93] Nevertheless, Marines were recruited, as the following petition for payment submitted by Cazaneuve, a French sergeant of Marines, indicates:

> Captain Tucker told me he was in want of Marines to complete his Crew, I answered that if he was permitted to raise me I should endeavour to procure him fit and proper subjects; In consequence thereof he sent me to the Commanding officer of the Admiralty of Bordeaux, to get sd. permission—I returned in a very short time with a satisfactory Answer And by his orders in Consequence of sd. answer I raised from sd. day first April till twenty first May following more men than were requisite to complete said company at my Cost and Expense. . . . Remark if you Please that not one received a farthing bounty and that I was put to many trifling Costs on the Account of their business.—

Pointing out that the green Marine uniform would attract more young men, he then continued by describing the process by which he obtained one:

> I had a green coat which I could have turned if he [Captain Tucker] would agree to furnish the facings and other Articles wanting on which he told me to go to his tailor to have it made and that he would pay for it; afterwards he told me to have Breeches and Waist coat suitable to the Coat.[94]

Later, unfortunately, the French sergeant and the men he had recruited about Bordeaux proved both troublesome and mutinous.

A Short Cruise and a Sullen Crew

By 6 June the *Boston* was ready for sea once again and sailed with several vessels under convoy of a French frigate.[95] Off Isle de Dieu, three days later, Captain Palmes issued orders to his lieutenants concerning the regulation and conduct of Marines should the frigate engage the enemy:

> Whenever All hands are called to Quarters, they [Sergeants and corporals] are to muster all the Marines on the Quarter Deck, there to await further Orders.—
>
> As soon as they are paraded, the Serjeants of each Division, are to apply for the Keys of the Arm chest—in order that each man may be supplied with a Gun & Cartridge Box filled with Cartridges, or any other proper Weapon for Defence—
>
> After they are supplied with their armour, they are to be paraded & marched to their particular Quarters, there to be kept as closely covered from the sight of the Enemy as possible—
>
> No one is allowed to speak, but the officers Commanding, or the Serjeants, or in their Absence, the Corporal, are most particularly to see that every man's Gun, & other accountrements are in the best order possible—
>
> If any thing should be wanted, they are immediately to inform the commanding Officer therewith—
>
> The Serjeants & Corporals are to parade with the Main Body as soon as in Gun Shot of the Enemy—

Following the engagement, each man was to return his weapon in "as good Order as when delivered to him," it having been thoroughly cleaned with oakum. Palmes then concluded with the warning that any man who deserted his post in time of action will be "shot upon the Spot"; however, he hoped that all would "act worthy of the Glorious Cause in which they are Engaged."[96]

After a month-long cruise during which the *Boston* took four valuable prizes, the frigate put into L'Orient on 3 July.[97] Once in port, however, trouble immediately arose among the Frenchmen who had been enlisted as Marines at Bordeaux. On the 8th, Captain Palmes reported that eight French Marines had refused to do duty as sentinels and as a result he had ordered that they forfeit their wages and prize money. Palmes then suggested two further courses of action, "for the peace & safety of the Ship." First, he recommended that the Frenchmen be moved ashore since they "are not disposed to serve us." Second, he proposed that only American Marines be given duty as sentinels, and that they be exempted from all other responsibilities on board ship. Palmes also pointed out that the Americans constituted only a relatively small portion of the company (one sergeant, two corporals, one drummer, and twelve privates), and therefore shifts would have to be lengthened.[98]

As Captain Tucker studied Palmes' suggestions, the eight Marines in question were allowed liberty. Once on shore, they created such a stir by complaining of the ill-treatment they had received at the hands of the *Boston's* officers that a French general went on board the frigate and asked all Frenchmen whether they wished to remain with the ship. All replied that they had no such desire. The general then ordered the 24 French Marines and 23 French sailors ashore, ignoring both Captain Tucker's loud objections and evidence in the ship's book that they were not pressed but had signed on as volunteers.[99] Tucker's letter of protest to Paris drew a noncommittal reply from the commissioners who understandably wanted to do nothing which would disturb the new ally.[100] Having disposed of his three prizes, Tucker disgustedly weighed anchor on 1 August and sailed for Nantes, there to join the *Providence*.[101]

To Sea at Last

The frigate *Providence*, Captain Abraham Whipple, was still under orders in early 1778 to escape her extended confinement in the Providence River, but after a prolonged period of inactivity, the frigate had neither sufficient officers, nor an adequate number of men to run the British blockade. Therefore, the Navy Board of the Eastern Department directed that new officers be appointed and that additional money over and above that allowed by Congress be offered to every man entered on board.[102]

The original Marine officers appointed to the *Providence* at the time of her launching had, by early 1778, resigned. As Sergeant of Marines Stephen Earl later stated:

> [I] served about six months on board the said Frigate in the capacity of Serjeant at which time [January 1777] the Second Lieutenant of Marines was broke and this deponent advanced to the place of the Second Lieutenant, the name of said 2d Lieut was Seth Chapin—that this deponent served in the capacity of said 2d Lieutenant for about four months at which time Avery Parker, the first Lieutenant of Marines on board said Frigate was broke and this deponent was then advanced to the first Lieutenant of Marines on board said Frigate and served as such for the term of four Months— at the expiration of the said four months the Captain of said Marines, by the name of Silas Devoll, was appointed Second Lieutenant on board the said Frigate whereupon he this deponent was advanced to the Captaincy of Marines in which capacity he this deponent served until the month of April 1778.[103]

Only one officer, Zebediah Farnham of Windham, Connecticut, had been commissioned during the interim, therefore Captain Whipple's first task was to locate replacements. Whipple, by early March, had selected a man whom he thought well-qualified for the position of captain of Marines to replace Stephen Earl who had temporarly assumed the post; his name, William Jones. A former captain in the Rhode Island Regiment, Jones was then recommended in glowing terms to the Navy Board at Boston: "He is a man well acquainted with the service having been in it most of the War, and has great interest in raising a Compy. of Marines. His appearance and Character is such I apprehend as will do

Abraham Whipple, by Edward Savage.

honor to the service.'[104] Shortly after Jones' appointment, a young man from New London appeared at Providence with 20 men he wished to enlist as Marines on board the frigate. In return, Jonathan Woodworth received a Marine lieutenant's commission.[105] The appointment of William Waterman of Cranston, Rhode Island, on 29 April gave the frigate her final Marine officer.[106]

Even though undermanned, the *Providence* made sail on the night of 1 May. Destined for France to obtain cannon and other supplies, the frigate moved swiftly down the river in a brisk wind. Upon entering Narragansett Bay she discovered the British frigate *Lark* lying in wait, and a severe running battle ensued. Whipple claimed to have disabled the enemy frigate and to have sustained damage in his own rigging before passing out of the bay. Narrowly evading yet another British ship-of-the-line the fourth day at sea, the *Providence* continued on course. A passage of 26 days brought her to Nantes in company with a British brig laden with wine.[107]

Marine Captain William Jones, like his counterpart on board the *Boston*, was ordered to Paris with dispatches upon the frigate's arrival in France. There he was received by the three American Commissioners who informed him of their wish that the frigate return to America with a load of clothing and arms for Washington's army, as soon as she was ready. Meanwhile at Paimboeuf, several miles below Nantes, the *Providence* was careened, and repairs begun on her damaged mast and hull.[108] The work had progressed little when Jones returned in mid-June, and he was ordered to take charge of the repairs. After several weeks at his new post, he caustically remarked:

> Methinks that if the almighty had been as Long making Such a number of frenchman (As there are) as that they are finishing our mast, he might have had Business for ages yet to Come, and they would have been Somewhat Like other Parts of his workmanship. It would Give me infinite Satisfaction if I could make them understand what I would Say, whereas talking has no Effect. May he who Superintend the universe Deliver us from the hands of those infernal Souls.[109]

By mid-July, the repairs were completed and the loading of copper, cork, arms, ammunition, and clothing begun.[110]

William Jones, by Marcus Waterman.

Off to America

The frigate *Boston* anchored at St. Nazaire on 3 August. Three days later, the *Providence* came down the Loire River to join her sister ship; each ship complimented one another with a 13-gun salute. They then sailed in company and arrived at Brest on 14 August to meet the *Ranger*. "Thank God, there will be two frigates and a sloop of war belonging to the thirteen United States together, and I hope Heaven will send us success in the cruise, and that we all may return to America, plentifully loaded with his divine goodness." These were Captain Samuel Tucker's reverent hopes that their combined strength would be blessed with enemy ships and rich cargoes.[111]

The three Continental ships sailed together from Brest on 22 August carrying materiel for the American cause. Three prizes, two brigs and a snow, were taken on the passage. Under the able command of Captain Whipple the squadron made Portsmouth, New Hampshire, on 16 October. Six days later Whipple and Tucker sailed for Boston, leaving Simpson behind to make repairs on the *Ranger*.[112]

Conyngham's End

The Continental cutter *Revenge* under Captain Conyngham cruised from Spanish ports with much success during 1778. The friendly attitude of the government and population towards the cause of the colonists aided the bold Irish captain, often to the chagrin of British officers. "Judge of the situation of our spirited commander," wrote an officer of the British ship *Monarch* on his return from Cadiz, "when during the time we lay there (seven days being detained by the wind) we had the mortification to see the usual honours paid to two Dutch frigates, and above all the Revenge American privateer, commanded by Cunningham, who came swaggering in with his thirteen stripes, saluted the Spanish admiral, had it returned, and immediately got product; the Spaniards themselves carrying on board wood, water, fruit, and fresh provisions; all which we were eye-witnesses of, as he anchored directly under our stern, within two cables length."[113] The British made many attempts to capture the American cutter, but Conyngham could not be caught; even in sight of British men-of-war he continued his depredations, and on one occasion, off Cape St. Vincent, he actually brought to and burned the tender to

HMS *Enterprize* under the very guns of the frigate.[114]

The friendly feelings of the Spanish government were short lived. By August all ports in both Bourbon kingdoms had been closed to the ship, due to the crew's insistence that the captain treat neutral vessels with the same "respect" he treated those of the enemy. The *Revenge* was allowed to refit, and in September Conyingham set sail for the West Indies. There he found better fortune; as Conyngham himself put it, the cutter "kept the British privateers in Good order in those seas."[115]

Following the successes in the Caribbean, the *Revenge* set her course for the Delaware Capes. Her passage was uneventful and she reached Philadelphia on 21 February 1779.[116] There Conyngham delivered a precious cargo of arms and munitions to the Continental Board of War. Not long after, the *Revenge* herself was turned over to the Marine Committee and sold at public auction.[117]

The two years, 1777 and 1778, which saw the introduction and expansion of Continental Marine activities in European waters, were years of mixed blessings. Marines on board the *Lexington* had been captured and imprisoned in England, while their brothers on board the *Reprisal* met their fate on the Newfoundland Banks. Yet during the two-year period, Marines did participate in many sea captures and engagements, as well as the two important raids conducted by Captain John Paul Jones at Whitehaven and St. Mary's Isle, Scotland. Although these accomplished little in terms of material gain or military advantage, both again proved that Marines were an integral part of any broad program of naval warfare.

FLAG RAISING AT NEW PROVIDENCE, 28 JANUARY 1778.
Marines under the command of Captain John Trevett captured and held the island of New Providence for a second time in January 1778. This painting recreates the moment at which the thirteen Stars and Stripes was raised over the island's main fort by Trevett and his men.

CHAPTER VIII

A Year of Landings and Leg Irons, 1778

"and take What we Pleased"

It was evident that the British, by the beginning of 1778, were finding the subjugation of their revolting colonies a serious undertaking in the light of the American victory at Saratoga. The French alliance with the United States not only increased this feeling, but emphasized the need for more energetic measures to quash the rebellion. On 18 February Lord George Germain, the British Secretary of State for the Colonies, ordered Sir William Howe to open the year's campaign by destroying shipyards as well as ships from Connecticut to New Hampshire. Subsequently, these orders were expanded. On 8 March he told Sir Henry Clinton, Howe's replacement, that, if unable to defeat Washington at once, he was to abandon further offensive operations ashore and spend the summer raiding the coasts of New England. Later, when the Franco-American alliance was made public, the British completely revised their strategy. The war with France was to have priority over that with the colonies, troops were to be withdrawn from Philadelphia, ordered to fall back on New York, and if necessary, mount attacks on the French West Indies and protect British possessions in America; and frigates were to be sent home to bolster the channel fleet. The safety of England herself now became the major objective.

Meanwhile, the Americans strove to make the best possible use of their meager naval resources. A second expedition to New Providence was undertaken, this time by a single ship, the sloop *Providence*.

Shortly after his arrival at Georgetown, South Carolina, in November 1777, Captain John Peck Rathbun was informed by a merchant captain who had just returned from the Bahamas that the *Mary* had put into Nassau for repairs. The news immediately brought back memories of his brief encounter with the enemy brig off New York the previous June. In the short but heated battle, Rathbun's well-liked sailing master, George Sinkens, was killed. Now, wrote his captain of Marines, John Trevett, "we ware. . . . Determined to Take Fort Nassau and then we Could Have Command of the Town and Harber and take What we Pleased."

Captain Nicholas Biddle, who was then in Charleston, thought the scheme so presumptuous that he attempted to persuade Rathbun of its futility, requesting that his good friend Captain Trevett be transferred to the frigate *Randolph*. Rathbun, however, was confident and determined that his plan would succeed. He needed Trevett and would not let him leave, even if it meant a disappointment for the Marine officer. But by the time the *Providence* set sail in early January, Trevett had reconciled himself and was encouraged by the undertaking: "I have had A Long time to think of What I am A Going to undertake but I an Very well satisfied that we Are in a Good Cause & we are a fiting the Lords Battel."

The sloop was scarcely out to sea before she was chased by three enemy vessels. To lighten his ship, Rathbun ordered all his water, wood, and much of his provisions cast overboard, but his pursuers continued to gain. By nightfall the *Providence* was ahead of the enemy ships, but not yet out of reach. As soon as darkness closed, the sloop's sails were taken in, and all lights put out. Within a few hours the enemy passed under full sail and continued on their way without seeing the American. Having safely eluded the British ships, Rathbun spread his canvas and continued southward by a somewhat different course, anchoring at Abaco Island several days later.

A portion of the crew replenished the sloop's water supply at Abaco, while the remainder went to work constructing a scaling ladder for use in assaulting the New Providence forts. Late in the afternoon of 27 January, the *Providence* set sail for Nassau, some 40 miles to the southeast; her topmast down, her gunports closed, and her crew below deck. Shortly before midnight the sloop dropped anchor off Nassau harbor, and her barge was lowered into the water. Twenty-six Marines, "All smart," filled their pockets with extra ball cartridges and prepared to go ashore. One reluctant fellow, however, tried to make excuses. "I can not Run," muttered the lame Marine. "You," said Trevett, "are the Man I should Chuse."

The sloop's barge would not accommodate the complete landing party, so Marine Lieutenant Michael Molten went in first with 15 men, and returned the barge for Trevett and the other 11 Marines.[1] The party landed about a mile west of Fort Nassau where final preparations were made for the attack.[2] Meanwhile, the *Providence* remained at anchor off the western point of Hog Island.

Fort Nassau, situated between the town and the harbor, was a small semi-star-shaped stone fortification which included within its walls a barracks, officers' quarters, kitchen, casements, and a powder magazine. Surrounding the fort on three sides was a row of high, pointed, wooden pickets which formed a palisade. Five days before the landing, John Brown, chairman of the island's council, wrote to Lord Germain and assured him that Fort Nassau and the other fort to the eastward, Fort Montagu, could withstand any assault directed against them: "I now deem the Forts sufficiently fitted to keep out any Force the Rebels may ever think of sending this Way." Upon Nassau's walls, Brown continued, were mounted four 18-pounders, four 9-pounders, four 6-pounders, and four 4-pounders; all in such a state of readiness that he was "under no great apprehension of Danger from the Rebels should any of their Vessels again attempt to come into the Harbour."[3]

Trevett, who had taken part in the first expedition to the island in 1776, remembered that he had removed one of the pickets on the west side of the fort during the two-week stay. Cautioning his men to remain silent, he made his way alone to the spot and found to his delight that the picket had not been replaced. As he crawled through the opening in the fence and quietly approached the wall he heard one of the sentinels cry, "All is well." Soon after he was answered by two similar calls; one from the other side of the fort, and one from a ship lying nearby.

Waiting a few minutes for the sentry to complete

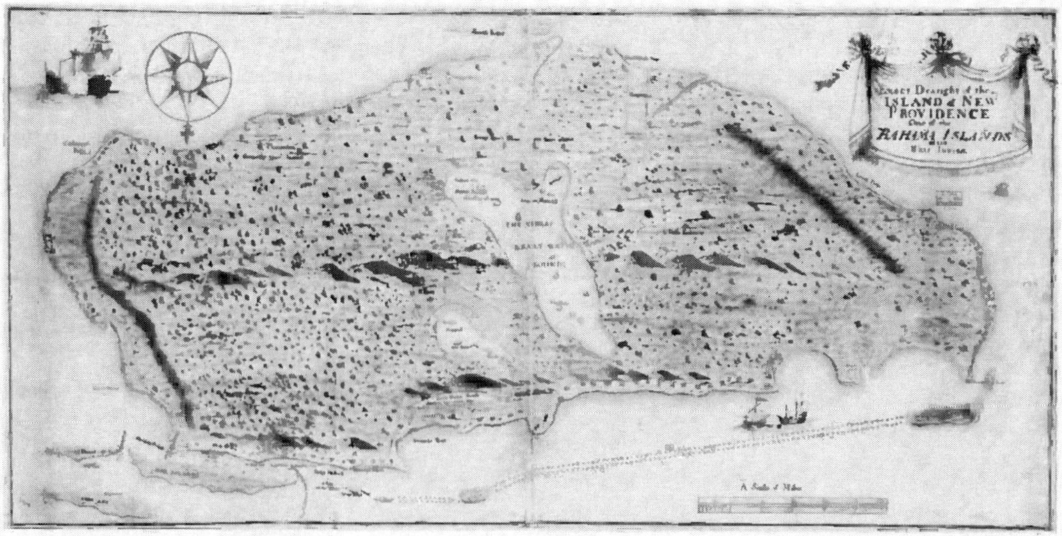

his rounds, Trevett returned to his men and then led them back to the fort through the opening in the palisade. Having assumed that the signal was given every half hour, he ordered the Marines to crouch beneath the wall and wait. About thirty minutes after Trevett heard the first call, the sentry again appeared and cried, "All is well." After he had received his reassuring replies and walked back into the fort, Trevett raised the scaling ladder near one of the embrasures, and led the Marines up the ladder and over the wall.

Once inside the fort, Trevett reminded his men of his previous order for silence, since a pistol shot was sure to alarm the town. With his men behind, the Marine captain rounded the corner of a barracks building and suddenly met one of the British sentinels "full Butt." He was quickly overpowered and forced into the first open doorway. "For God sake What have I Dune!" cried the terrified sentry. In a reassuring voice, Trevett asked the prisoner if there were others in the fort. Only one, responded the prisoner, and he was at the other end. The second sentinel was quickly seized, and the two men questioned separately by Trevett, who learned to his amazement that the two were the only guards detailed to secure such an important fort. Asking the reason for this apparent lack of security, Trevett was informed that they could summon 500 armed men within minutes merely by firing one or two of the loaded cannon.

Somewhat frightened by the prospects of such a confrontation, the Marine captain immediately alerted his men. Some were put to work filling cartridges with gunpowder from the fort's magazine, while others moved some of the heavy cannon to point them at strategic streets in the town and at the several vessels moored in the harbor. A few men were stationed at intervals along the walls to warn of approaching visitors. Every half hour throughout the remainder of the night a Marine on the seaward side shouted, "All is well," and another on the opposite side gave the responding signal. They were answered each time by sentries on board at least two ships in the harbor, who little realized that the enemy was in control of the fort, and that the situation was far from well.

At daybreak, Trevett had the American flag (thirteen stars and stripes) hoisted over the fort. Having landed without food or drink, he then sent one of his men with a note to James Gould, a merchant who lived nearby. Gould, a former resident of Newport, Rhode Island, was known to be sympathetic to the American cause, and in making contact with him instead of with the island's lieutenant governor, Trevett hoped to rally others also favorably inclined.

In response to the Marine captain's note, Merchant Gould rushed to the fort, and Trevett climbed down the scaling ladder to meet him. After the two exchanged greetings, Gould requested to know to which fleet the Americans belonged. To the fleet under Commodore Nicholas Biddle now anchored at Abaco Island, Trevett responded. They had been sent to Nassau to take off the ship *Mary* and all warlike stores, but were not to touch any private property. Continuing, Trevett also told Gould that 200 men and 30 officers were in the fort, and that he had plenty of provisions for the men. But, he said, "I must have Breckfast Gott for My thirty officers." Gould agreed to supply all that was requested, and then quickly returned to the town. In a short time he sent back ample quantities of bread, butter, and coffee. "A Very Good Breckfast & Dubel Allowance" we had, recorded Trevett.

Meanwhile, Captain Trevett had sent Lieutenant Molten with two men to Fort Montagu with orders to give the two-man garrison "no time to parle." Quickly covering the four miles which separated the two forts, Molten entered Fort Montagu and informed the two guards that 230 Americans now occupied Nassau and that they had better surrender. Both agreed, and Molten's party took possession.

Shortly thereafter, Trevett thought it time to take the 16-gun ship, *Mary*. Accompanied by a midshipman from the *Providence*, four freed American prisoners commandeered a boat from one of the nearby wharves and rowed out to the ship, lying a short distance from the fort.* The *Mary's* captain, Henry Johnson, was sick ashore, and his lieutenant at first refused to let the Americans on board. Trevett, watching the whole proceedings from the fort's wall, then took a direct hand in the matter: "I was All Redey to Settel the Affair with them I Haled [from the Fort] with hard Langege & some hard Names."[4] Trevett's threats soon convinced the stubborn lieutenant, and the Americans were allowed to take possession of the ship without firing a shot.

* The four prisoners were among 20 Americans captured by the *Gayton* and sent to Jamaica. From there they had made their way to New Providence where they hoped to obtain passage to America.

The midshipman remained on board while the four sailors escorted the *Mary's* crew and arms to the fort. In the ship's hold was found a valuable cargo of sugar, rum, and coffee.[5]

During the remainder of the morning, boarding parties made up of Marines and the released American sailors were sent out to take the four other vessels in the harbor—all captured by an English privateer weeks before. Among the vessels were the brig *Washington*, loaded with rice and indigo; the sloop *Tryal*, with a small cargo; and two coastal schooners.[6]

About mid-afternoon, the *Providence* sailed into the harbor after a night's anchor off Hog Island. An hour later another sail appeared off the horizon and was identified as the 16-gun British privateer *Gayton*, commanded by Captain William Chambers, the very ship which had captured the four vessels then anchored in the harbor. Since the *Gayton* was no match for the *Providence*, Rathbun and Trevett decided to attempt a ruse. They ordered the American flags lowered both on board the sloop and at the fort, and then waited for the *Gayton* to approach. Once she was within range of the fort's cannon, she could either be captured or sunk. Loyal islanders saw through the plan and moved to block it. Men, women, and children turned out upon the hills behind the town and began waving hats, coats, and aprons to warn the ship away. Chambers, apparently interpreting the actions as a greeting, continued on course. Finally, several islanders rowed out in boats and managed to warn Chambers of the trap.

As the privateer bore away, Trevett ordered the flag hoisted and gave the command for three 18-pounders to fire. One shot pierced the *Gayton's* hull but apparently did little damage. Chambers then circled Hog Island and reentered the harbor through the eastern passage. Anchoring close to Fort Montagu, he hurried ashore to learn what had transpired.

Trevett, foreseeing the possibility that he could give little help to Lieutenant Molten in such a situation, ordered him to spike the guns, break the rammers and sponges, throw the powder into the sea, harass the enemy, and then withdraw to Fort Nassau. By darkness, Molten and his two men had accomplished their mission and returned to the fort.

With the *Providence* moored abreast of the town by springs on her cables to bring both sides to bear should trouble develop, and Fort Nassau in Marine hands, the Americans settled down to a grand dinner of turtle meat—prepared and served on china dishes by servants of the island's first ladies. Loyal Bahamians, shocked by the day's events, began moving their personal property inland, seemingly fearing the worst despite assurances to the contrary.[7] Late that evening the island's council gathered to consider the situation. Like the council meeting conducted on 3 March 1776, the members unanimously decided, "that any Attempt to recover the Forts would be useless, and might be attended with the disagreeable consequence of the destruction of the Town." This decision was made as a result of the lack of arms and ammunition and American guarantees that neither persons nor their property would be injured.[8]

The Americans spent a peaceful, though watchful, night, and the next morning began preparing the captured vessels for sea, but the islanders, unlike the night before, had now become restless. Encouraged by Chambers, large numbers began to collect around the governor's house and upon the hills behind the fort—"thare Guns &c Glesining" in the sun. This gave Trevett little worry for his men were all in "Good spirits as Many of them is Very Ugent for Me to fire on What tha Cal Black Berds." But Trevett wanted to avoid bloodshed if at all possible.

Before noon, a large number of the inhabitants approached, and halted "within Half a Pistel shot" of the fort. One of the group immediately recognized Trevett from the 1776 attack, and shouted: "thare is that Dam Buckerer Come Again that Carred away Governor Brown." Trevett paid no attention. A few uneasy minutes passed before a three-man delegation headed by Robert Hunt, a council member, broke through the crowd and drew near the fort. The Marine captain immediately ordered the scaling ladder put over the wall, and climbed down to meet them. At the direction of the governor, John Gambier, they had come to inquire as to American intentions, Hunt said. Trevett responded by again repeating the orders and assurances he had given Gould the day before, and closed with a demand that supplies needed to refit the four vessels be brought to the fort, and that the people be directed to disperse. With that he turned and reentered the fort.

Trevett waited an hour for a reply to his demands, and then ordered the ladder again lowered. He would go to see the governor himself. If, on his way through the crowd he should be attacked, Lieuten-

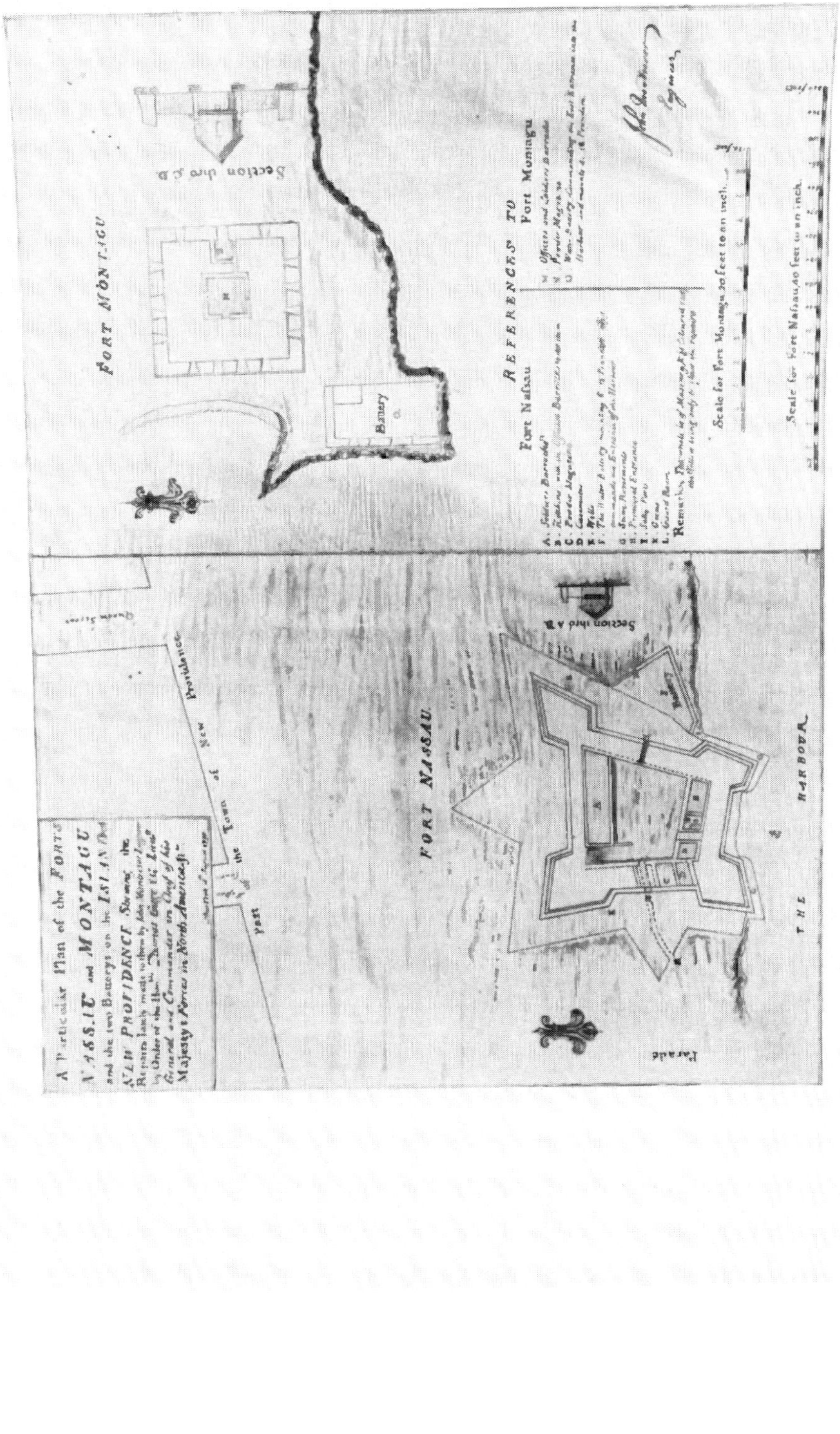
A Particular Plan of the FORTS NASSAU and MONTAGU and the two Batterys on the ISLAND of NEW PROVIDENCE
FORT NASSAU
FORT MONTAGU
Section thro A.B.
Section thro C.D.
Battery
Town of New Providence
Parade
THE HARBOUR
REFERENCES TO
Fort Nassau
Fort Montagu
Scale for Fort Montagu 20 feet to an inch
Scale for Fort Nassau 40 feet to an inch

ant Molten was directed "to not mind Me," but give them two or three 18-pounders loaded with grape. Sword in hand, the Marine captain climbed down the ladder and set out for the governor's house. On his way he met Hunt who informed him that Gambier had agreed to his demands, and suggested that they go for the supplies. By darkness, the materiel had been moved to the fort, and the vessels in the harbor ordered readied by the following evening.

After a peaceful night, Trevett was joined by Captain Rathbun early the following morning. Upon his arrival, the naval captain informed his Marine officer that the *Providence* would be ready for sea the next day and that three pilots would be needed to conduct the vessels out of the harbor. In order to procure the pilots, who would not come forward of their own accord, Trevett suggested they auction off 23 barrels of rice a few hours before sailing. This would entice the pilots out of hiding to a place where they could be easily identified, and spirited out to the ships. Rathbun thought it an excellent plan and ordered Trevett to put it into effect the following morning.

That afternoon Trevett dined with his good friend James Gould. While sitting at the table covered with an assortment of the island's delicacies, he happened to glance out the window and observed one of his men running to the house. He immediately excused himself and went out to meet the fellow. The excited Marine told Trevett that Captain Chambers was landing his men. The Marine captain hurridly returned to the fort where he also learned that the islanders had placed several cannon on Governor's Hill, and that they planned to attack the fort that evening.

Meanwhile, Gould assembled his belongings and started out with his family for the safety of Hog Island. As he passed the fort, Captain Trevett came out and smilingly said: "I Hope You are not Afraid." "No," replied Gould, "I am not Affraid but I Do not Expect to see You any More, for tha . . . have All the Privaters Crew att the Goveners House and [are] Making Every Preparation to Attack you to Night; for tha have Discoverd Your Strength; and If I thout . . . that You had nothing But that Damd Bumboat I would fight Aganst You My self." Trevett smiled and told Gould that if he happened by the following morning he would still find the Marines in control of the fort. With that the two friends parted.

As soon as Trevett returned to the fort he asked John Scranton if he could shinny up the flagpole. The young Marine from Rhode Island muttered a few harsh words and said that he could. Trevett then told him to fetch a hammer and nails, and to wait until he had prepared a note to Governor Gambier. In the letter, Trevett ordered the governor to disperse the men who had gathered on the hills overlooking the fort. If he did not within 15 minutes, the Americans would cannonade the town, and "he May Rely on itt that . . . [we] will Give no Quarter nor take none." At the same time the note was delivered, Scranton climbed up the flagpole and began nailing the Stars and Stripes to the pole—a gesture signifying an intent to fight to the last man. The threats served their purpose for within 20 minutes the hills were deserted.

During the evening the town was like a graveyard; the British being more disposed to sleep than fight.[9] This was not the case, however, for Captain Chambers, who now was in possession of Fort Montagu. With the aid of a few loyal inhabitants and several hundred blacks, he planned to launch a joint land-sea attack on Fort Nassau after dark. According to his plan, the land force would march on the fort while the *Gayton* would run down the bay and engage the *Providence*. At 11 that night, the *Gayton* started off but an unskilled pilot ran her aground. With the privateer of no use, the scheme was abandoned.[10]

At Fort Nassau, meanwhile, the Marines feverishly prepared for the morning's sailing. Powder and cartridges were moved out of the fort's magazine and put on board the *Providence*. The guns were spiked, and the ladders, sponges, rammers, and other warlike equipment destroyed. By daybreak all was in readiness.

Soon after sunrise, Trevett had the 23 casks of rice rolled out in front of the fort. He then took his drummer and marched by beat of drum to the market square where he told a number of people who gathered around that he had a quantity of rice which he would sell cheaply. If it were not sold, he would dump it into the sea. By eight, a large group had collected about the fort and Trevett began breaking open three casks whose contents he distributed among the children. All this time the three pilots needed to conduct the ships out of Nassau harbor were being selected. Once chosen and informed of their task, Trevett had them forceably removed to the *Providence*. His plan having succeeded, the Marine captain gave away the remaining barrels of rice.

By 10 the vessels were manned and ready for sea. Only Trevett and a few Marines remained ashore. While they waited for the sloop's barge, an officer from the *Gayton* appeared with an invitation from Chambers addressed to Captain Rathbun and his officers. Promising that there would be no treachery, he asked that they join him in a bowl of punch at a local tavern. Trevett took a cynical view of the invitation, and sent word to Chambers to bring the *Gayton* out and "take the sloop Providence then I would take some Punch."

As soon as Trevett and his men were on board, the *Providence* and her captives put to sea. Close to sunset, they anchored off Abaco Island and made final preparations for the passage northward. The Marines were given a fine dinner and ordered to rest, "as not one of our Men nor My self Had any sleep from the Time we landed until now for When we Relived gards we took them from ond part of the fort and put them on duty att A Nother so as to Deceve the inhabitens and Make them think we had a grate Number of Men in the fort."

The *Providence* and the *Mary* (under the command of Captain Trevett) made their separate ways to New Bedford, Massachusetts, surviving a bitterly cold voyage during which at least one man froze to death and others suffered severe frostbite.[11] Once in home waters their troubles began anew. The Eastern Navy Board ruled the *Mary* a merchant ship and informed Rathbun and his crew that they were entitled to only one-half the proceeds of her sale. Immediately a bitter argument arose between Trevett, the crew's agent, and the Continental Agent in Boston. A court of inquiry was convened and it held in favor of the government. Trevett, however, was unsatisfied with the decision and journeyed to York, Pennsylvania, determined to plead his case before the Marine Committee of Congress. Unable to appear before the committee, he left Pennsylvania after a short stay and returned to New Bedford where he learned, to his disappointment, that the *Mary* and her valuable cargo had been burned by a British raiding party.[12]

Meanwhile, the sloop *Providence* was extensively rebuilt, and by mid-May she was ready for sea. About that time, Rathbun was issued instructions not to leave port until orders were received from the Committee of Foreign Affairs. The captain was not informed of the reasons for his detention because the Marine Committee had decided to send the *Providence* to France with secret dispatches for the American Commissioners. These orders, however, were never issued, and the sloop remained in port.[13]

Faced with the loss of 15 seamen and Marines because of Trevett's refusal to distribute the prize money due them, Rathbun spent most of the late spring recruiting new hands. By mid-June, this was accomplished with little difficulty.[14] About that time Trevett returned to Massachusetts to find that he no longer commanded the Marines on board the *Providence*; the Navy Board having directed William Vernon, a member, to find a replacement for the Marine officer now out of favor with the Board.[15] No record remains as to who Vernon chose to succeed Trevett.

The *Providence* sailed in late July or the first week in August on a short cruise to the east of Nova Scotia. On 7 August, off Louisbourg, she fell in with an enemy convoy of 30 armed transports destined for Halifax. The American sloop immediately singled out a transport loaded with Highland troops and engaged her. The battle began at sunset and by midnight the *Providence* had to withdraw and take refuge in an inlet along the Nova Scotia coast to effect repairs.[16] Before returning to Boston in early September, the sloop took a snow from Scotland.[17] The *Providence* was out again in mid-November, sailing with the 18-gun Continental ship, *General Gates*.[18]

"fighting at a most infernal rate"

The frigate *Randolph*, after a brief but successful foray into the Atlantic in September 1777, returned to Charleston, South Carolina, for extensive repairs. By early December she had been thoroughly graved, rerigged, and moved from Hobcaw to the main harbor.[19] Naval Lieutenants John McDougall and Joshua Fanning were placed in charge of enrolling a crew and within a short time swelled the frigate's complement to more than 250 officers and men. The frigate's Marine contingent still suffered from a lack of leadership. Lieutenant Franklin Reed had resigned his commission almost a year before, and Lieutenant Panatiere de la Falconniere had been sent under arrest to Congress in September. Only Captain Samuel Shaw remained.

Meanwhile, the British renewed the blockade. Four enemy ships, recently released after convoying Howe's transports safely up Chesapeake Bay to the Head of Elk, now stood off the Charleston bar. From the viewpoint of South Carolina Privy Council President, John Rutledge, their arrival could not have been more ill-timed. Large numbers of merchantmen destined for Europe were ready to clear the harbor, while several ships were soon due from France with military supplies and clothing consigned to the state government. There was little chance of any vessel getting into or out of port with the area beyond the bar patrolled by fast-sailing British frigates.[20]

Nicholas Biddle, by James Peale

The task of driving away the blockaders was far beyond the capability of the South Carolina Navy, which in November 1777, consisted of two small brigs, a schooner, a sloop, and several pilot boats.[21] Small wonder then that President Rutledge began to pin his hopes on the *Randolph*. In a conference with Captain Nicholas Biddle in mid-December, he proposed that the *Randolph*, accompanied by three ships engaged by the state, clear the coast of all enemy ships of war. In order to facilitate this task, he noted that the Privy Council would authorize several detachments of the Continental regiments to serve on board the fleet as Marines.[22] Anxious for action, Biddle eagerly accepted the command.

Rutledge convened the Privy Council the following day, presented his plan, and informed the council that Captain Biddle had agreed to take command. The council assented. The ships *Volunteer* and *General Moultrie* were then ordered into the service of the state on terms agreeable to both the owners and the government. Immediate instructions were sent to Edward Blake, first commissioner of the state navy, advising him that the ships had been engaged, and that the council intended to order the state brig *Notre Dame* on a cruise. Blake was asked to "have those Vessels equipp'd with the utmost Expedition, & furnished with such Articles as may be necessary for this Cruize," and to appoint a proper person to determine the value of the two private vessels.[23]

The council, in addition, resolved to offer a bounty of 10 dollars and a monthly pay of 21 pounds to seamen entering on board the three vessels. Once this information was conveyed to the state navy board, the council turned its attention to Rutledge's promise for Marine detachments from the Continental regiments. Since the troops were not under the authority of the state, here the council could only request, not demand.[24] Rutledge therefore appealed directly to Major General Robert Howe, commander of all Continental forces in the Southern Department:

> The trade of this port being likely to suffer great injury from the vessels of war, which have for some days past, been in sight of the town. In order to clear the coast and protect the trade, Capt. Biddle has agreed to go on a cruize with the Randolph and several other vessels, engaged by the state, to be put under his command, but it being thought expedient that a number of marines should be embarked in the vessels: the council have advised, that you should be desired to order as many of the continental troops under your command as Capt. Biddle may apprehend to be necessary for this service, to be detached upon it. As I do agree in opinion with the council, I do, therefore, and in pursuance of their advice, request that you will be pleased to give the necessary orders for this purpose.[25]

Upon receipt of Rutledge's request, Howe instructed Brigadier General William Moultrie to call a council of war the next day to determine whether Continental troops could be spared and, if so, how many and for what length of time. The council listened to the proposal, debated, and then decided. While seeing no impropriety in sending the troops requested, the council was of the opinion "that we have not men enough to defend the state should it be properly attacked."[26]

The findings were forwarded to General Howe, but he returned them with a request that the officers reconsider, as he was "certain the military would be highly censured for not complying with the requisition of the Governor and Council." A second council was held, and again they reasoned that they could not in good conscience order Continental troops on such an expedition, but they concluded, "they are ready and willing to obey orders."[27]

With this second pronouncement, General Moultrie waited upon Rutledge. The president listened impatiently and then expressed his displeasure with the military's decision. Angrily, Rutledge said that only "150 men would be needed for the expedition." An astute soldier, Moultrie acquiesced; if he refused he was afraid the state would take it upon themselves, "as they have so earnestly requested the detachment." The expedition, he later wrote to General Howe, "may be attended with good consequences."[28]

In comparison, the naval authorities moved swiftly to implement the desires of the Privy Council. The two private ships were valued, and their captains ordered to begin recruiting the 120 men needed to man each vessel. Letters of marque were granted Captain Philip Sullivan of the *Volunteer* and Captain Jacob Johnston of the *General Moultrie*. In addition, the Privy Council resolved to add two more vessels to the fleet: the brig *Fair American*, Captain Charles Morgan, and the brig *Polly*, Captain Hezekiah Anthony. Both commanders were directed to enlist men; Morgan to enroll 80, and Anthony, 90. Recruitment for the four vessels did not go well and by early January the *Volunteer* was released to her owners and what hands had been recruited were transferred to the *General Moultrie*.[29]

Orders for the Continental troops to be sent to the fleet to serve as Marines were issued on 11 January; one company each from the First and Second Regiments of the South Carolina Line, a half company from the Fourth Regiment, and a platoon from the Fifth were chosen. Their instructions in part read:

> Each Regt. to provide their men with 18 Rounds per man to be put into a Military Chest on board the Vessels they go in . . . The officers Commanding parties are to take Care to keep Good order and Discipline with the Sailors, & assist the Capts, of the Vessels to the utmost of their power, in attacking the enemy.[30]

The following morning the captains and subalterns met at Charleston's new barracks to draw for vessels. The coveted assignment to the *Randolph* was won by the company from the First Regiment commanded by Captain Joseph Ioor, with Lieutenant George Gray as his subaltern.[31]

Early on Tuesday morning, 13 January, the captains of the four ships that were to accompany the *Randolph* appeared before the commissioners of the South Carolina navy. Their ships could not possibly be ready for sea, they said, until the next spring tides, as "the Men on board each Vessel Ought to be Stationed & Trained some Little time before they sail."[32] Captain Biddle was informed of the delay that afternoon and asked not to sail without the vessels. The spring tides, to his disgust, were over a month away.

From the deck of the *Randolph*, at anchor in Charleston harbor, in the early morning hours of 15 January, sailors and Marines alike observed flames and smoke billowing aloft from the city. Charleston was burning. Started in a bakehouse in Union Street just below Queen, the fire was fanned by a strong winter wind blowing out of the northeast. The flames spread unhampered to neighboring houses, while sparks ignited roofs of dwellings in adjacent streets. All Charleston southwest of Queen and East Bay streets was doomed to destruction as the fire mounted in intensity. Throughout the day citizens, troops, and a landing party of seamen and Marines from the *Randolph* battled the blaze and, as the wind died down, succeeded in bringing it under control, but 252 houses had been destroyed, and a major portion of the city lay in ruins.[33]

Faced with this calamity, final arrangements for the naval expedition had to be postponed.[34] Stores intended for the enterprise had been destroyed, but the principal delay resulted from the attention paid to the homeless and destitute, those victims "worn down with fatigue, standing to watch over their property, covered with blankets, and shivering with cold."[35]

A week after the fire, however, the four captains appeared before the navy board and reported that their ships would be ready to sail within two weeks.[36] Continental troops destined to serve as Marines, therefore, were called up. On 23 January, from his headquarters in Charleston, General Moultrie issued the orders. "1 Capt. 1 Subaltern, 2 Serjts & 48 Rank and file from the 1 Regt. to go on Board the Randolph tomorrow morning as was order'd before," the directive read. "The Boats will be ready at the market wharf for them." Three days later the remainder of the detachments were ordered on

board the state vessels, and the ships themselves ordered down into Rebellion Road to join the *Randolph*.[37]

Only the lack of clearance over the bar and adverse winds held the fleet at Rebellion Road, for the enemy blockaders had not been sighted for almost a week. The way, therefore, was free to proceed to sea without having to engage while navigating the channel. On Thursday morning, 12 February, the winds turned and, one by one, the vessels moved out. Beyond the bar before noon, the pilots were discharged, and the fleet stood to the eastward, but remained in sight of the coast.[38] Lookouts strained to sight the enemy frigates, but to no avail for they had found it impossible to maintain the blockade after a flurry of prize taking.[39]

With no enemy frigates to disperse or destroy, the *Randolph* and her consorts left the coast and made for the southeast. In two weeks the fleet was windward of Barbados—within the path of trade between Britain and her Caribbean possessions. No British merchantmen were to be seen; all that were sighted, overtaken, and then examined proved to be French or Dutch. This conspicuous lack of enemy traders could only mean one thing, thought Biddle, news of their presence had reached the enemy and they had sent heavy ships out in quest of the American fleet.[40]

On the afternoon of Saturday, 7 March, 60 leagues due east of Barbados, a sail was sighted. As the ship drew closer, Biddle ordered the crew to quarters, and the *Randolph's* four consorts to stand after the flagship. Near sunset, the enemy ship was identified as the 64-gun ship-of-the-line *Yarmouth*, Captain Nicholas Vincent, out of Antigua in search of the Americans. The *Randolph* immediately hove to and waited for her opponent; the *General Moultrie* promptly followed suit. The other three vessels tried to do the same, but all were carried off course toward the west.[41]

An hour after darkness, the *Yarmouth* hauled abreast of the *General Moultrie*, fired a warning shot, and demanded to know the ship's identity and from whence it came. Thinking her to be a merchantmen, the 64-gun ship then stood for the *Randdolph*. The American frigate in reply to her hail hoisted her colors and gave the *Yarmouth* a broadside.[42] Thereafter the roar of their thunderous discharges pierced the night—a terrific engagement in which, according to one observer, "the Randolph appeared to fire four or five broadsides to the Yar-

The
HARBOUR
of
CHARLES TOWN
in
South Carolina
with
A VIEW of the TOWN
from the SOUTH SHORE of Ashley River.
ASHLEY RIVER
LONG ISLAND

mouth's one."[43] The two ships were, said another, "so near as to throw their hand grenades from their tops on each other's decks."[44]

The *Randolph's* fire was rapid and accurate. Her broadsides soon shot away the *Yarmouth's* mizzen topmast, bowsprit, and cut her rigging to pieces.[45] On the frigate's forecastle and quarter deck her Marines, including the South Carolina infantrymen, directed their musket fire onto the enemy's decks and into her tops, accounting for a portion of the five men killed and the wounding of 12 others on board the *Yarmouth*. The American, as an awed observer recorded, was "fighting at a most infernal rate."[46]

A quarter hour after the opening broadside, Captain Biddle was wounded seriously in the thigh. A chair was immediately ordered up to the quarter deck, and the ship's surgeon began dressing the wound. As the captain, propped up in his chair, continued to encourage his crew to victory, an enemy shot found its way to the *Randolph's* powder magazine and blew her apart.[47] Only four sailors, of a crew of 305 seamen, Marines, and infantrymen, survived. The remainder of the fleet scattered soon afterward and all made their way safely into port.[48]

The *Alfred* Lost

The loss of the frigate *Randolph*, Captain Nicholas Biddle, and his gallant young Marine officer, Samuel Shaw, was a severe blow to the Continental Navy and Marines. Equally serious was the capture of yet another Continental ship, the *Alfred*, and the subsequent imprisonment of her crew.

The *Raleigh* and *Alfred*, having made the voyage to France in the fall of 1777, set sail together on 29 December for America via a circuitous route calculated to take some prizes. Off the bar at Senegal, Africa, they captured a wine-laden sloop before turning westward toward the Windward Islands.[49]

By 9 March 1778, they had reached latitude 16° 31′ north, longitude 55° 40′ west, and at six in the morning sighted two sails to the northwest. Captain Thomas Thompson of the *Raleigh* hove to for the *Alfred*, and ordered Captain Elisha Hinman to run down and observe the sternmost ship. By ten, being within five or six miles, it was clear to the Americans that they had encountered two armed enemy ships instead of merchantmen. The *Raleigh* and *Alfred* hauled on the wind, and held the same course as the two Englishmen. Then, Thompson reported:

> On seeing us haul the wind headmost ship tack'd, then the other tack'd; by this we found that they were trying to work up and get our wakes; the Raleigh I found sailed nearly as well as they, but the Alfred dropped a-stern and to leeward withal.—As the weathermost ship pass'd under the Alfred's lee, standing to southward on the third tack, Captain Hinman hoisted his colours and fired several shot, which were returned under English colours; they were then two miles apart, and the other ship four miles to leeward to her consort, the Alfred was about three miles a-stern of us. I had determin'd to tack to the Southward and on seeing the firing I ordered the master to put the ship in stays immediately and to stand towards the Alfred that we might close together, when the weathership stood to the Northward again, which would have been a favorable opportunity to attack her if she chose to come up, as the other ship to leeward could not have fetch'd up to his assistance in less than two hours. The instant our helm was put a-lee (without waiting any signal from me) I saw the Alfred right before the wind, and set studding-sails and every other light sail she could set, it was then half past twelve, wind E.N.E. light breeze. I had not determin'd in my own mind what was best to be done, as I knew not their real force, but I imagined either was an overmatch for the Alfred; which I suppose Capt. Hinman discover'd being near them, and knowing the Alfred's best sailing was on one mast, thought that it was his only chance to escape; he might likewise think the ships might pursue us, being the largest ship . . . I am sorry he tried to effect his escape at that time as I plainly saw the leewardmost ship bearing then S.W. would cut him off before he could pass her or I could give any assistance. As I saw the Alfred bear away I did not go about just then, but ordered the courses to be hauled up, thinking that would tempt the weathermost ship to stand on for us, but found in a few minutes they were determin'd to act more cautiously, as they both made towards the Alfred; I then ordered the master to veer and make sail toward the Alfred, and run between her and the other ship to take off her fire and give the Alfred an opportunity to escape, who I thought gained upon them, but in a few minutes the two got up and began a furious fire, which was returned by the Alfred as fast as they could; just as we had got our studding-sails hoisted we had the modification to see the Alfred haul down her colours; it was then one o'clock, the firing lasted about ten minutes; we were then within three miles of the ships.[50]

In light breezes and a smooth sea, the British quickly took possession of the *Alfred* and then turned to the *Raleigh*. Faced with what he evaluated as a "superior force," Thompson altered course to effect his escape. He ordered lumber and spare gear to be thrown overboard to lighten the ship, and other weight, about 35 tons, shifted from "right to midships" to improve sailing. Slowly the *Raleigh* pulled ahead, and after 19 hours in chase, the enemy turned away and stood southward toward her prize.[51] The "superior force" which captured the *Alfred* and put the *Raleigh* to flight was composed of the *Ariadne*, 20 guns, and the *Ceres*, of 16 guns—the Continental ships jointly mounted 56 guns.[52]

After surrendering the *Alfred*, Captain Hinman and his officers, including Captain of Marines John Welsh, First Lieutenant William Hamilton, and Second Lieutenant Nathaniel Richards, were taken on board their captors who sailed for Barbados. There all American officers, except Nathaniel Richards, were removed to the *Yarmouth* for passage to England. On 18 July, the *Alfred's* officers were confined to Forton Prison, Portsmouth; a week later Captain Hinman made his escape.[53] His two Marine officers, however, remained in confinement for several months, despite the pleas of Hinman for their exchange, before making their escape to France, and then to America.[54]

Marine Lieutenant Richards, unlike his compatriots, was given a parole. "On our arrival at Barbados after capture," he later recorded, "I was recognized by Capt. Thompson of the British Ship Yarmouth a 74 [64], who knew me when a child & was intimate in my Father's family many years before. By his influence with Capt. Pringle of the Ariadine, & by the intercession of Capt. Hinman, I was permitted to return home on parole. Capt. Hinman was particularly desirous for me to return that I might give an account, as early as possible, to the Navy Board, of the manner of our capture and of the desertion of our consort the Frigate Raleigh." But Richards was to encounter a delay. "From Barbados I proceeded in a Cartel for Martinique where I took passage for America in the Brig Charming Sally, E. Hammond Master. I was again taken by his Majesty's Frigate Ambuscade, Mr. Cartney Commander—carried into Halifax and imprisoned." Lieutenant Richards was released on 7 July 1778, and sailed in a cartel for New London, Connecticut, where he arrived on the 28th, "after an absence of about fourteen or fifteen months."[55]

Upon returning to Boston with the *Raleigh* in early April, Captain Thompson found himself censured for not supporting the *Alfred*. William Story, clerk of the Navy Board at Boston, wrote to William Vernon on 29 April: "The doctr. of the Alfred has been at the Board and gives a particular Accot. of Capt. Thompson's behaviour. He is Condemned by every One and they are Crying out why don't your board turn him out and hang him &c &c. I am Sorry the Service Suffers by the Misconduct of the officers in the navy."[56] Two weeks later, the Marine Committee of Congress wrote to the Eastern Navy Board, and recommended that Thompson be suspended "until a full and fair enquiry can be made into his Conduct."[57] An inquiry was held, resulting in Captain Thompson's court martial and dismissal from the service.

Escape From the Providence River

The vessels shut up in the Providence River for more than a year made efforts to run the British blockade in the spring of 1778. First among them was the Continental frigate *Warren*, commanded by John B. Hopkins. The long period of inactivity had taken its toll, especially upon the frigate's Marine company. Of the original officers appointed and men enlisted, not one remained by late December 1777. Captain of Marines John Grannis and his lieutenants, George Stillman and Barnabas Lothrop, had resigned their commissions in order to seek more active employment, and few Marines elected to remain once their enlistments had expired. Therefore, a search was begun in December for qualified men to command the frigate's Marine detachment.

In January 1778, Congress commissioned Elihu Trowbridge a captain of Marines and assigned him to the frigate *Warren*. He was an experienced officer, having served as one of General Washington's Life Guards, and as a second lieutenant in Colonel Charles Webb's Second Connecticut Regiment until 4 December 1777 when he was discharged due to poor health.[58] The identities of Trowbridge's first and second lieutenants are unknown.

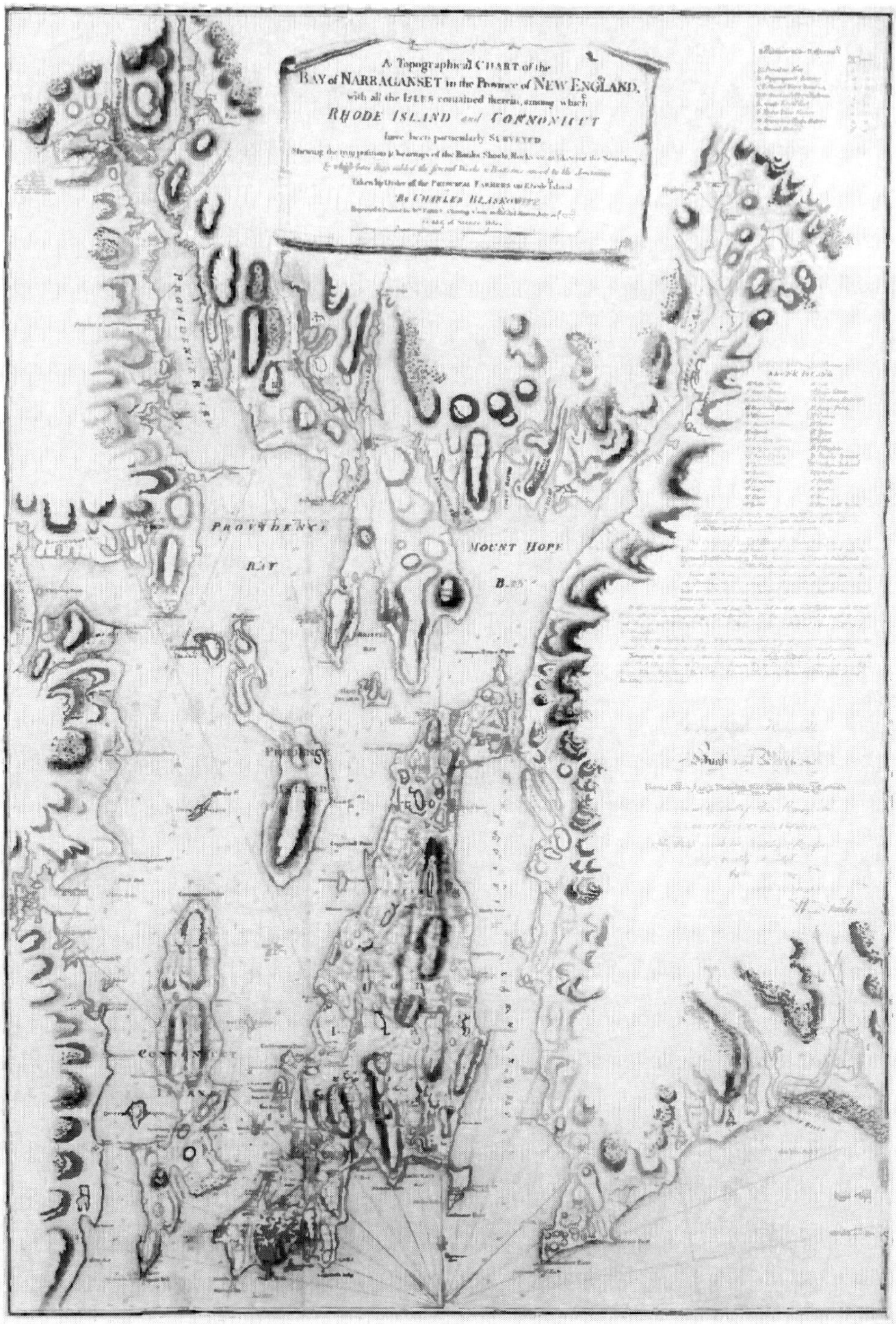
A Topographical CHART of the
BAY of NARRAGANSET in the Province of NEW ENGLAND,
with all the ISLES contained therein, among which
RHODE ISLAND and CONNONICUT
have been particularly SURVEYED
Taken by Order of the PRINCIPAL FARMERS on Rhode Island
By CHARLES BLASKOWITZ
PROVIDENCE RIVER
PROVIDENCE BAY
MOUNT HOPE BAY

Once the Marine complement had been recruited, and the frigate provisioned, Captain Hopkins moved down the Providence River. As he neared the blockading enemy ships off Newport, favorable circumstances combined to aid the *Warren* in her escape. The night was exceedingly dark with little wind until the critical moment when she passed closest to the British force, then the wind suddenly shifted to the northwest, carrying Hopkins on his way.[59] The *Warren* was chased a short distance and sustained a moderate amount of damage. Her mizzen yard was shot away, main yard damaged, and she was hulled several times; only one man was slightly wounded, but after a year cooped up in the river, she was now free.[60]

Hopkins was under orders to reach New London if at all possible. The wind was unfavorable and the weather extremely cold and severe. Many of the *Warren's* 170-man crew did not possess as much as a change of clothing and Hopkins felt compelled to cruise southward toward a warmer climate.[61]

At latitude 24° off Bermuda, the *Warren* made her first prize, the ship *Neptune* from Whitehaven for Philadelphia, loaded with salt and dry goods. Out of her Hopkins took 120 bales of duck and some other articles, possibly clothing for his crew. Soon after, a snow from St. Eustatius bound for Ireland was taken. Both prizes were manned and ordered to a safe port in New Hampshire or Massachusetts. The *Warren* then turned northward and sailed for Boston, arriving on 23 March.[62]

For months the *Warren* lay at Boston "in a most destitute and forlorn Situation" for want of a crew.[63] Throughout the spring and early summer months, both Hopkins and Trobridge scoured the New England countryside for men to replace those whose enlistments had expired. By late July, each man had accomplished his task, and the frigate was then prepared for sea.[64] The Marine Committee proposed that the *Warren* and other Continental frigates then at Boston join the French fleet under Admiral d'Estaing off New York.[65] Hopkins did not follow the proposal, but set out instead on a short cruise in search of the cork fleet.[66]

The *Warren* took her first prize some 400 miles east of Cape Charles on 5 September, more than a month after the frigate had sailed. She was the schooner *Thomas* from Jamaica bound for Halifax laden with Molasses. The prize was manned and ordered into a safe port in Massachusetts. After taking several more prizes, the American frigate turned northward and headed for Boston, arriving there early in December.[67]

The *Columbus* Aground

The next vessel to attempt to run the British blockade was the ship *Columbus* under Captain Hoysteed Hacker. Since the expectation of success was not too bright, precautions were taken. Five miles below Providence all her stores, except four guns, were unloaded, which "brings the Ship to a light set of Ballast and much facilitate her Sailing, besides if she falls into the Enemies hands they will gain only an Old Ship and Twenty five Men wch. we think sufficient to run the Ship to N London."[68] Upon reaching that port, her guns and stores, which were to be transported overland, would be placed back on board the ship. Only a favorable wind and thick weather were awaited. During the night of 27 March, Hacker made the attempt.[69]

By chance, a merchant brig sailed the same evening as the *Columbus*. The brig was detected by an enemy ship whose signal caused the British frigates *Maidstone* and *Sphynx* to get underway immediately. As the two frigates approached the entrance to Narragansett Bay, they sighted the *Columbus* in midchannel. Finding escape impossible, Hacker ran the Continental ship aground near Point Judith.[70] He, his Marines under Captain Joseph Hardy, and a large number of townspeople spent the night stripping the ship of her sails, rigging, guns, and the remaining stores of powder and small arms.[71] The following afternoon a party of British sailors and Marines was driven off by musket fire from the beach as they attempted to attack the stranded ship. The enemy tried again just before nightfall and succeeded in setting the *Columbus* afire.[72]

The following day, wrote Captain Hacker, "Came on a strong Gale, which parted her [the *Columbus*] into four pieces; as soon as the Gale abated, we got to work upon her and set the parts on fire, the more readily to Come at the Iron, of which we savd as near as can be estimated, betwixt five & six Tons. . . . The Anchors lay in such a manner, as to render it extremely difficult, if not impracticable, to get them a shore till Warm Weather." Soon after the salvage operation was completed, Hacker discharged all but eight of his crew, "reserving them to row the Barge to what ever place she may be Orderd, to dry the sails or whatever may be most necessary."[73] The next vessel to attempt the hazardous dash from the Providence River was the frigate *Providence*, and she was successful.

The Connecticut frigate, *Trumbull*, Captain Dudley Saltonstall, remained in the Connecticut River the whole year, being unable to pass over the bar at the mouth of the river. In March, when the tides ran high, Saltonstall was furnished extra men and 4,000 dollars for more equipment, but to no avail.[74] Displeased with the failure, John Deshon of the Eastern Navy Board traveled to Connecticut determined that "she must be intirely stript of her Yards and Top Mast, and all her Story even to a Swept Hole that if possible to bring her to 9 or 10 feet Water." But his astute supervision also failed and the frigate remained locked in her river-prison.[75]

Lieutenant Pownal in Irons

The frigate *Virginia*, Captain James Nicholson, had waited nearly a year for a chance to run the British blockade of the Chesapeake Bay. Several attempts had been made, but all proved fruitless. By December 1777, officer resignations, expiration of enlistments, and desertions had become commonplace. The frigate's Marine company, however, seems to have suffered little. Captain James Disney did resign, but he was quickly replaced by Second Lieutenant Thomas Plunkett. William Barney, who had journeyed from Bordentown with his brother Joshua and a group of replacement seamen from the late *Andrew Doria*, succeeded Plunkett as second lieutenant.[76] Thomas Pownal, appointed in June 1776, still retained rank as first lieutenant.

In late March 1778, Nicholson decided to make yet another attempt at running the blockade; this time against the advice of the Maryland Council of Safety who thought the "Chance of getting through so small."[77] On the 30th, at eight in the morning, the *Virginia* hurriedly weighed anchor, and sailed from Annapolis in company with a Maryland brig. In the rush to make her escape, she left behind her tender with 19 hands under Marine Captain Plunkett and naval Lieutenant Fanning. Throughout the day the wind blew hard toward the northwest, making progress difficult. At three the following morning the frigate struck a shoal, and as she pushed over she lost her rudder. Leaking badly, Nicholson thought it advisable to anchor until daylight when repairs could easily be made. At daybreak two British frigates were discovered nearby—one within "two gunshots." Immediately, Nicholson ordered his barge hoisted out, and called for volunteers to join him in abandoning the frigate. Without pausing even long enough to destroy valuable papers, supplies, or weapons, he ordered his nine oarsmen to pull for shore. The captain's hasty departure left naval Lieutenant Joshua Barney in command. Barney's first thought was to try to save the frigate. He suggested cutting the anchor cable in the hope that the frigate would drift ashore on Cape Henry, out of reach of the enemy. Barney's fellow officers vetoed his suggestion and they, along with some 300 crewmen, watched as the British drew near. With the captain gone and capture imminent, it was difficult to keep order on board especially after some thirsty crewmen broke into the ship's liquor stores. The British, however, seemed in no great hurry to take the *Virginia*, and it was not until ten in the morning that a boat from HMS *Emerald* drew alongside to take possession.

As for Nicholson, he made it safely to shore. The next day he came out to the frigate under a flag of truce to negotiate a possible parole for his men and to pick up his personal effects. Being unable to obtain the parole, he returned ashore.[78] Captain Nicholson was later brought before a court of inquiry, but acquitted of all blame for the loss of his ship.[79]

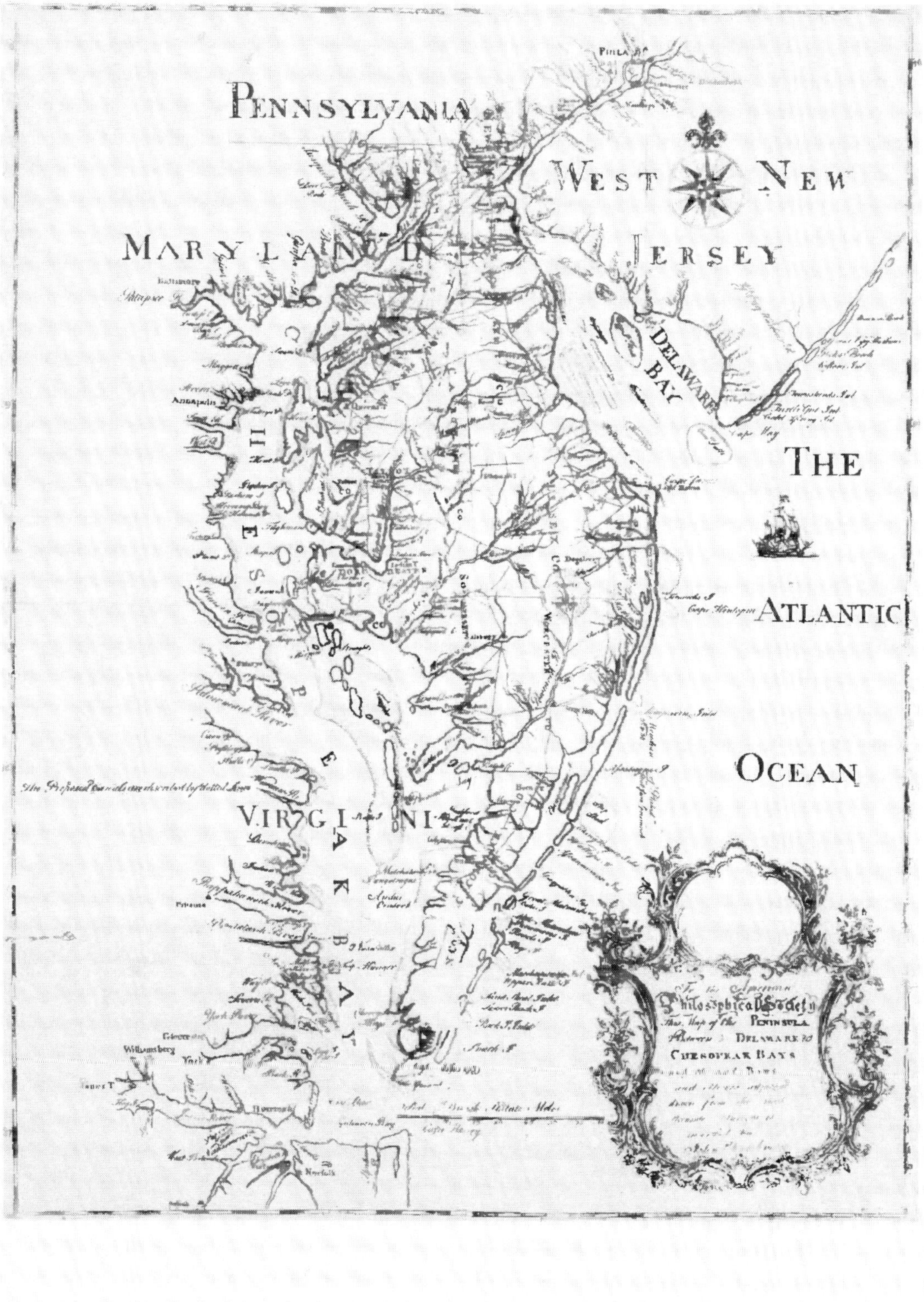
PENNSYLVANIA
MARYLAND
WEST NEW JERSEY
DELAWARE BAY
THE ATLANTIC OCEAN
VIRGINIA
CHESAPEAK BAY
Baltimore
Annapolis
Williamsburg
York
Norfolk
Cape May
To the American Philosophical Society This Map of the PENINSULA Between DELAWARE & CHESOPEAK BAYS

The captured officers and crew of the American frigate were distributed among the enemy's Chesapeake squadron. Marine Lieutenant William Barney was exchanged immediately. Lieutenant Pownal was sent on board the *Emerald,* where he and naval Lieutenant Joshua Barney were "treated with every mark of attention and kindness."[80] By early summer the British squadron had become so crowded with American prisoners that it was decided to send a majority to New York. Lieutenant Pownal was among those sent northward. Soon after his arrival, he was transferred to a prison ship, and then to Admiral John Byron's 64-gun flagship, *Ardent.*[81] In late August, Pownal was exchanged for an officer of equal rank from the captured British frigate *Mermaid.*[82]

Barges on the Delaware

The frigates *Effingham* (Captain John Barry) and *Washington* (Captain Thomas Read) were blockaded in the Delaware River above Philadelphia shortly after the British occupied the city. Fearing that the enemy would come up the river and capture or destroy these vessels, General Washington advised that they be stripped and sunk. This was accomplished late in November 1777. The following month several attempts were made to refloat the two vessels, but to no avail.

The few Continental Marines that remained after the scuttling of the two frigates and the destruction of the combined Continental and Pennsylvania fleets gathered at Bordentown, the meeting place of the Navy Board of the Middle District. Of the original officers assigned to the four Continental vessels in the river, only five were left: Captain Dennis Leary and Lieutenant William Barney of the brig *Andrew Doria,* Lieutenant Abel Morgan of the frigate *Washington,* Lieutenant Alexander Neilson of the frigate *Delaware,* and Lieutenant James Coakley of the frigate *Effingham.* The Marine crews, likewise, were but skeletons of the original detachments that had been recruited 18 months before, a majority having since deserted, while a few had been captured.

By the beginning of January 1778, a number of officers and men were chafing at the inactivity at Bordentown. To most of the others, who had passed through the strenuous two-month fight, the inactivity was enjoyable. Nevertheless, plans were afoot to avenge the defeat. On the 29th, the Marine Committee ordered Captain John Barry to organize a boat expedition down the river for the purpose of annoying the enemy, capturing or destroying their transports, and cutting off their supplies and diverting them to the use of the Continental army, then in desperate straits at Valley Forge.[83] Two barges belonging to the frigates *Effingham* and *Washington* were immediately procured and armed with a single 4-pounder. Manning the two boats with volunteer seaman and Marines proved to be more difficult. After numerous attempts, about 40 men were secured. Towards the first week in February all was in readiness and Captain Barry took command of one boat with Marine Lieutenant James Coakley as mate, while naval Lieutenant Luke Matthewman officered the other.[84] It is unknown whether any of the other Marine officers joined the expedition.

Hugging the Jersey shore, the two boats, with oars muffled, slipped silently past Philadelphia; below they were joined by five other boats, half-manned by Pennsylvania seamen and privateers. Meanwhile, General Anthony Wayne's tattered brigade pushed into Wilmington, destined for New Jersey in search of cattle and hay for the Continental army. On the morning of the 19th, the two forces joined as Barry's craft were ordered to ferry the brigade from Wilmington across the Delaware and up to Salem. While the boats lay tied up in Salem Creek, Wayne's men spent the next several days rounding up the few cattle that could be found nearby. By late afternoon on the 23d, about 1,000 head had been collected. How to get them to Pennsylvania became a problem. After much discussion it was determined that the cattle would be driven through central New Jersey, and across the Delaware above Philadelphia. While this was done, Barry's little fleet of boats would proceed up the river as far as Mantua Creek and then return, burning all the hay stored along the New Jersey shore. This would divert enemy attention from Wayne's crossing at Burlington.[85]

On 7 March, after the short but successful hay

burning expedition, Captain Barry's little fleet captured two ships and a schooner off Port Penn. The ships were transports, each with a crew of 14 men, bringing forage and supplies from Rhode Island to the British army in Philadelphia; the schooner was manned by a crew of 35. Two days later a number of British vessels came up Delaware Bay, forcing Barry to burn the transports to prevent their recapture. He attempted to run upriver with the schooner, but being hard pressed he was compelled to run her aground off New Castle. All cargoes of the vessels, however, were saved and ultimately found their way to the Continental army. Throughout the balance of March and April, Barry's boats remained idle at Wilmington due to British dominance of the river.[86]

Early in May the expected British expedition moved up the Delaware. Composed of 700 men, the British force landed at Whitehill without opposition. There the half-submerged frigates *Effingham* and *Washington*, together with a brig and sloop, were set afire. The troops then took possession of Bordentown where several more vessels were destroyed. The punitive expedition destroyed 40 vessels; a staggering loss for the Continental government and State of Pennsylvania.[87]

More misfortunes befell the Continental Navy and Marines during the months of March, April, and May 1778 than any other similar period throughout the war. The effect was depressing and naturally caused loss of confidence. Submerged in the gloom, members of the Marine Committee became almost fatalistic as they learned of each new disaster:

> The Enemies ships do indeed swarm in the Seas of America and Europe; but hitherto only one of our Frigates hath been captured on the Ocean. Two have been burned in North River, two sunk in Delaware, one captured there, and one in Chesapeak. The Alfred we are just informed was taken on her passage home by two frigates in sight of the Rawleigh. The particulars of this capture and why she was not supported by the Raleigh we are ignorant of. I hope Capt. Thompson is not culpable. I entertain a high opinion of him. The Columbus is a trifling Loss, and I should not much lament the Loss of the Alfred if her brave Captain, Officers and men were not in the hands of a cruel enemy. Our little fleet is very much thinned. We must contrive some plan for catching some of the Enemy's Frigates to supply our Losses; but we must take care not to catch tartars. It is reported that Capt. Biddle of the Randolph, in an engagement with a sixty gun ship, was blown up. We have been so unfortunate that I am apt to believe almost any bad news; but this report I cannot believe.[88]

Small wonder that courts of inquiry were held to determine the culpability of those responsible for the loss or capture of Continental vessels.

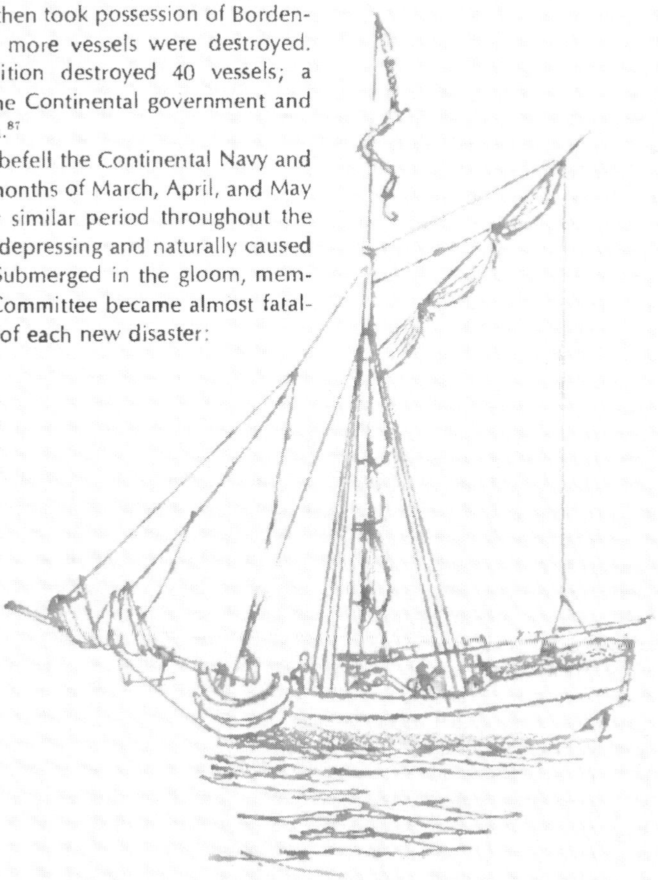

Another Misfortune

In the midst of numerous disasters, the achievements of John Barry with the barges on the Delaware stood out as the one major success. In reward, the Marine Committee appointed him commander of the frigate *Raleigh*, following Captain Thompson's suspension, and ordered him to repair to Boston.[89] When he reached the northern port in late June, he found his ship to be the site of a series of courts martial being conducted by the Eastern Navy Board. For about two weeks he waited while the board deliberated as to the guilt or innocence of Captains Hector McNeill, Thomas Thompson, and Hoysteed Hacker. Finally, with the trials concluded, he took possession of the ship.[90]

Throughout the balance of July and early August, the frigate's new captain strove to recruit a crew. By mid-August, the *Raleigh* boasted a complement of 235 officers and men, including about 50 British prisoners who were signed on as Marines, and a large detachment of seamen from the blockaded frigate *Trumbull*.[91] Of the original Marine officers appointed in 1776, only two remained: Captain George Jerry Osborne and First Lieutenant Stephen Meads. Replacing Nathaniel Thwing as second lieutenant was Jabez Smith, Jr., of Groton, Connecticut, a veteran of service with the Continental brigantine *Resistance*.[92]

Captain Barry's fear that he would have to await the refitting of the French Fleet before he could put to sea ended early in September with the receipt of instructions from the Marine Committee. He was to sail on a cruise between Cape Henlopen and Ocracoke Inlet, North Carolina, for the express purpose of destroying "certain armed Vessels fitted out by the Goodriches, or any other of the enemies Vessels that may be infesting that Coast." The cruise was to be so managed as to permit a call at Hampton, Virginia once a week for further orders. The Continental brigantine *Resistance* was to accompany the frigate, the committee being unaware that the vessel had sailed several weeks before.[93]

At six o'clock in the morning on 25 September, the *Raleigh* weighed anchor and stood out of Boston harbor with a brig and sloop under convoy. At noon two strange sails were sighted at a distance of eight to ten leagues to the southeast. The *Raleigh* hauled to the north and prepared for action. The vessels, which were the British 50-gun ship *Experiment* and the *Unicorn* of 22 guns, followed. The chase continued nearly two days before Barry, under the mistaken impression that he had shaken off his pursuers, changed course. Steering east-southeast and then south by east, "in order to keep clear of Cape Sable," the two enemy ships suddenly reappeared about half-past nine on the morning of the 27th. Barry promptly "wore ship, and hauled our wind to the N. westward," under full sail.[94]

With a fresh breeze blowing, and the *Raleigh* making better than eleven knots, it looked as if his pursuers might be left behind. But by noon the wind had moderated and the *Unicorn* was gaining rapidly while the *Experiment* held her position astern. Marine Captain Osborne continues the narrative:

> About half past four P.M. tacked and stood to the S. westward, in order to discover the headmost ship's force; at the same time saw several islands, but could not tell the name of either. Our ship being cleared for action, and men at their quarters, about five P.M. crossed the headmost ship [*Unicorn*] to windward, athwart her fore foot; on which we hoisted our colours, hauled up the mizen sail, and took in the stay sails, and immediately the enemy hoisted St. George's ensign, she appearing to be pierced for twenty eight guns; we gave her a broadside, which she returned; the enemy then tacked and came up under our lee quarter, and the second broadside she gave us, to our unspeakable grief, carried away our fore topmast and mizen top gallant mast; he renewed the action with fresh vigor, and we, notwithstanding our misfortunate, having in a great measure lost the command of our ship, we determined for victory. He then shot ahead of us, and bore away to leeward. By this time we had cleared our ship of the wreck. The enemy plied his broadside briskly, which we returned as brisk; we perceiving his intentions were to thwart us, we bore away to prevent his raking us, and; if possible, to lay him aboard; which he doubtless perceived, and having the full command of his ship, prevented us by sheering off, and dropping astern, keeping his station on our weather quarter.
>
> Night coming on, we perceived the sternmost ship gaining on us very fast; and being much disabled in our sails, masts and rigging, and having no possible view of escaping, capt. Barry thought it most prudent, with the advice of his officers, to wear ship and stand for the shore, if possible, to prevent the ship's falling into the enemy's hands, by running her on shore; the engagement continuing very warm, about twelve midnight saw the land, bearing N.N.E. two points under our lee bow. The enemy, after an engagement of seven

> hours, thought proper to sheer off, and wait for his confort; they showing and answering false fires to each other.[95]

Four men, including the first lieutenant of Marines, had been killed on board the *Unicorn*, and a large number wounded.[96]

The *Experiment* soon came up and poured in three broadsides in rapid succession. The *Raleigh* responded in kind. A moment later the frigate grounded. For a quarter hour, both British ships maintained a heavy fire; then they bore away to await the dawn, and an easy conquest.[97]

Shortly after the cannonade died down, Barry launched his boats. Evacuation of the frigate was difficult, but within two hours all 220 men were landed on barren Wooden Ball Island, just south of the mouth of Penobscot River. Fifteen men were unaccounted for and probably dead. Since the desolate rock could not be adequately defended, Barry decided to destroy the *Raleigh* and get his men to the New England coast.[98]

A boat was ordered back to the frigate with instructions to set her afire and then row westward toward the mainland. Meanwhile, Barry and Marine Captain Osborne would take about 50 men, including the wounded, and try to reach the coast. Marine Lieutenant Jabez Smith, naval Lieutenant David Phipps, and 132 men would remain on the island. Should the mainland prove to be a few miles away, Barry planned to evacuate the rest of the crew by dawn, but the scheme failed due to treachery.[99]

On the morning of 28 September, the two British ships opened fire on the *Raleigh*. Midshipman Jeacocks, who had been ordered to set the frigate afire, promptly hauled down the colors. A flag of truce was sent to the island and the remaining crew forced to surrender. The *Raleigh* was then hauled off the shoal at high tide and manned with a British prize crew. Captain Barry with those of his crew who escaped found their way back to Boston.[100] The *Raleigh* was taken into the British service and the frigate's captured officers, including Jabez Smith, paroled and later exchanged.[101]

New Companies For New Ships

Several newly built ships and others purchased under Congressional authority were put into service to offset losses at sea during 1778. One vessel obtained in New England was the *Industrious Bee*, an 18-gun brigantine, later rechristened the *General Gates*. On the recommendation of John Bradford, Boston prize agent, John Skimmer was appointed captain.[102] To command the 15-man Marine detachment, the Eastern Navy Board selected Richard McClure, Jr., an experienced artilleryman from Boston.[103]

Early in the summer of 1778, the brigantine left Boston on her first cruise. Two brigs were captured in short order and sent to Boston where they arrived at the end of July with their cargoes of molasses and fish. Two more prizes were taken in August, the schooner *Polly* and the letter of marque brig *Montague*. During the hour-long engagement with the *Montague*, the enemy privateer expended all her "Shott and every piece of Iron of all Kinds that could be ramm'd into the Tube of the Cannon, they at last put in their Jack Knives & the Capt. roll'd up his Speaking Trumpet & fir'd it away." In the action Captain Skimmer was killed.[104]

After the death of Skimmer, command of the brig was given to Daniel Waters, who had served in Washington's fleet. In mid-November she sailed with the sloop *Providence* for the West Indies and was at sea until after the first of the year.

During the year, a frigate of 32 guns was added to the Continental Navy. Built on the Thames River at Norwich, Connecticut, the *Confederacy* was launched on 8 November, and towed down the Thames to be completed at New London. Recruitment of a crew proved difficult in the face of enticements offered by privateers. Enlistments were generally for one cruise, and discipline was lax. The Continental navy and Marines, on the other hand, paid low wages, demanded long enlistments, and divided prize money between the government and the crew. Small wonder that the frigate's captain, Seth Harding, had to send out patrols in search of deserters.

Prior to the frigate's launching, the Eastern Navy Board appointed Harding's subordinate Marine offi-

Gurdon Bill, by an unidentified artist.

cers. Selected as captain of Marines was Joseph Hardy, who had been languishing in New London after the ship *Columbus* was run aground and burned in March:

> As a Captain of Marines in the American Navy you are hereby required to [go] on board the ship Confederacy under the command [of] Seth Harding Esqr. there to do & perform the duty of a Captain of Marines for which this shall be your Warrent.[105]

Appointed as Hardy's first lieutenant was Gurdon Bill of Norwich, Connecticut, grandson of Major Joshua Huntington, agent in charge of the *Confederacy's* construction. Recommended by Huntington, and later commissioned second lieutenant of Marines, was Ephraim Bill, Gurdon's younger brother.[106]

During the remaining months of 1778, Hardy and the Bill brothers concentrated on the arduous task of winning recruits. Captain Harding's excellent reputation was of little help, despite Governor Jonathan Trumbull's assertion that Harding could "expeditiously" man the ship.[107]

A second Continental vessel went into commission during 1778, the 32-gun frigate *Alliance* built at Salisbury, Massachusetts. Little is known about the design of this vessel, even though a contemporary report states she was 151′ 0″ on the lower deck, 36′ 0″ beam, and 12′ 6″ depth of hold. She was approximately the same type of vessel as the *Confederacy*, but slightly smaller. Authorized by Congress more than two years before, the *Alliance* encountered the same difficulties and delay in getting ready for sea as the Connecticut-built frigate. Like her sister ship, the *Alliance* did not cruise until the following year.

Selected to command the new frigate was Pierre Landais, a strange, half-mad Frenchman. A member of an old St. Malo seafaring family, Landais had served in the French Navy since youth, had been wounded in the Seven Years' War, commanded a small vessel during Bougainville's circumnavigation of the world, and had risen to the rank of *Capitaine de Brulot* (fire-ship) by 1773. In 1775 he accepted an appointment as Lieutenant of the Port of Brest, not knowing it to be a dead-end command, and was discharged from the French Navy. In 1777 he persuaded American Commissioner Silas Deane to send him to America in command of a supply ship, with a letter of introduction to Congress recommending him for a commission in the United States Navy. Upon his arrival at York in April 1778, the Marine Committee promptly appointed him captain of the *Alliance*.[108]

The appointment of Landais' subordinate Marine officers was done by the Eastern Navy Board instead of the Marine Committee. Named captain of Marines was Matthew Parke, who had resigned from the *Ranger* in February because of a dispute with other officers of the sloop. He was allowed to leave the ship and voluntarily returned to the United States on board the frigate *Deane*.[109] As first lieutenant, the Navy Board selected one of its own, James Warren, Jr., the son of James Warren, Sr., a member.[110] Thomas Elwood of Connecticut was chosen to serve as second lieutenant.[111]

In addition to the *Confederacy* and *Alliance*, two other vessels joined the Continental Navy: the frigate *Deane* built at Nantes and the *Queen of France*. Under the command of Captain Samuel Nicholson, the *Deane* arrived at Boston early in May with a cargo "consisting of 12000 suits of cloaths for the army, compleat, a large quantity of copper, tin and lead, besides many other valuable articles."[112] Since only one Marine officer had been appointed (Captain of Marines John Elliott) in France, the Navy Board soon filled the vacancies with the selection of Samuel Pritchard of Boston as first lieutenant, and Samuel Gamage, also of Boston, as second lieutenant. Although appointed, the two new Marine officers did not see active service during 1778, for the frigate remained tied up at Boston throughout the remainder of the year.[213]

The old 28-gun frigate *Queen of France* was purchased in France for the navy by the American Commissioners. Under Captain John Green of Philadelphia and with a French crew, she sailed into Boston harbor carrying a valuable cargo of clothing and supplies in May 1778. Upon her arrival, the French officers and men were paid off, but the frigate would have to wait another year before she received her American officers and crew. Selected to officer her future Marine detachment would be Captain Edmund Arrowsmith and Lieutenant Peter Green.

The Continental Marines, already greatly reduced, were further depleted in 1778 by the loss of Marine detachments which had served on board the frigates *Washington*, *Effingham*, *Randolph*, *Virginia*, *Raleigh*, and the ships *Alfred* and *Columbus*. Of the numerous detachments raised, only those on board the frigates *Boston*, *Warren*, *Providence*, *Trumbull*, *Deane*, and sloops *Ranger* and *Providence*, remained. To these were added those of the *Alliance*, *Confederacy*, and brig *General Gates*. The Marines, therefore, showed a gradual decline and their condition at the end of 1778 was by no means satisfactory, even though they had participated in a number of naval successes.

WILLING'S MARINE EXPEDITION, FEBRUARY 1778.
Naval Captain James Willing took 35 Marines down the Ohio and Mississippi River during the winter of 1778 in an effort to quash British influence along the banks of the lower Mississippi. This painting depicts the loading of spoils on board the armed boat Rattletrap from one of the many loyalist plantations raided.

CHAPTER IX

Marines on the Mississippi, 1777-1779

Willing's Marine Expedition

In February 1778, James Willing, who fashioned himself a naval "Captain in the service of the United Independent States of America," raided loyalist settlements along the lower Mississippi with a company of Marines in Continental pay. Not much is known about this man before he officered the expedition. He was born 9 February 1751 to a prominent Philadelphia family, his elder brother, Thomas, being a business partner of Robert Morris and a member of the First Continental Congress. Upon receiving his patrimony, he moved to Natchez where he unsuccessfully engaged in trade and rapidly squandered his fortune in dissolute living. "Should Mr. Willing's out standing Debts come in by Next Aprill," wrote one of his New Orleans creditors, "you may depend on a Remitance of the Gent's. Acct. of About three or four Thousand Dollars, but as I am Apprehensive there is little dependence on a great Amt. of his Debts to be Colected by that time would not advise to lay any great dependance."[1] When news of the revolution reached Natchez, he became one of its most ardent followers, trying unsuccessfully to arouse support for the cause.

Willing returned to Philadelphia on business in 1777 and, through the influence of his brother, succeeded in meeting with the Commerce Committee of Congress early in the fall. As no record of the meeting's substance exists, it is assumed that he painted a vivid picture of strong loyalist influence in the Natchez area, warned of a possible attempt to close the Mississippi to American shipments of war supplies north from New Orleans, and advanced a plan for an expedition to the lower Mississippi to enlist or compel the support of West Florida. Undoubtedly the Committee had a similar plan in mind, although in July Congress had rejected George Morgan's plan for a sizeable American attack on Pensacola and Mobile which had been strongly supported by Benedict Arnold. By late summer, however, the Committee had decided upon a less ambitious expedition, and commissioned James Willing a captain in the navy for the purpose.[2] The commission, it seems, was granted without the general knowledge of Congress.

Willing's instructions, although lost, can be reconstructed. His letter of introduction to General Edward Hand, commander of Fort Pitt, for example, states that he is charged with some dispatches for New Orleans, and that "it is of great Importance that Mr. Willing get speedily down, and that he has truly faithful People with him."[3] A letter of the same date to Oliver Pollock, Continental agent in New Orleans, indicates that Willing was not only entrusted with dispatches, but was also expected to bring stores, which Spain had agreed to deliver at New Orleans, up the Mississippi and Ohio Rivers to Fort Pitt.[4] "Mr. James Willing will return with these Boats and to him you may commit your dispatches, also the superintending Care of the Goods." This is further corroborated by a request Hand received from Willing prior to his departure from Fort Pitt:

> Tho I expect to bring at least five Boats from New Orleans laden with dry Goods and Navigated by 20 or 25 Men each I request one favour of you to give the Necessary Orders for a Sufficient Quantity of Flour to be lodged for me at the Arkansaws not less than Sixty

or Seventy Barls. of 250 lb or 300 lb each as the Men must have 1½ Rations per day also Twenty Bls. of Pork, so as to be there the beginning of April.[5]

In addition to carrying dispatches and conveying supplies up the river, Willing claimed several years later that he was also instructed "to capture whatever british property he might meet with."[6] This claim is substantiated by Willing's letter to Oliver Pollock of 30 May 1778:

> In the first place to Begin with my instructions; the following Extracts will serve to specify their Tenour—After being ordered to make prize of all British property on the Mississippi River I was instructed to apply to the Governor of this Province for Liberty to make Sale of them. that obtained I am again instructed to pay one moiety [share] of the Net proceeds into Your hands as agent for the Congress.[7]

Whether this order pertained only to the destruction of enemy vessels on the river, or included all British property along the shore as well, is unknown. But Willing, it seems certain, was expected to seize British property in conjunction with his other duties.

Armed with his instructions, Willing proceeded to Fort Pitt, arriving there on 12 December 1777. With a burning desire for action equalled only by the zeal of his reckless nature, he immediately entered upon the duties of his assignment. On the 21st he sent the following list of "sundrys" needed for the expedition to General Hand:

List of sundrys for Expedition to New Orleans—
1 Lieutenant or Ensign
24 regulars
1 Sergeant
1 Corporal

27—1 Boat to row 12 or 14 Oars—
Besides these I shall want to take Eight or Ten Volunteers, who must sign the Shiping articles—
30 Kegs of Flour
3500 or 4000 lb. of Biscuit
15 Blls of Beef of 200 lb each
15 Blls of Pork
2 Bushes. of Salt
100 Glls. of Wiskey
A Ship Carpenter—on Act of the Boats that are to Come up—
2 Swivels
40 Stand of Arms, compleat Bayenet &c—Some of these riffles
250 lb of Gunpowder
Ball in proportion
100 lb of Iron Ball for Swivels—Matches—flints
Cartridge paper & some writing do
Hides to Cover the Boat
Potts to coock in and fallg Axes, Tomihaws, some Kettles
Cloathing for the Men if to be had—

Concluding, he suggested that the men be drafted, "that they may get ready being anxious to get off the Season being far advanced."[8]

The armed boat *Rattletrap* was assigned to Willing's command and officers of various regiments directed to solicit volunteers for a company of Marines.* The effect of their call was spontaneous. Escape from the tedium of garrison duty and the promise of action on a distant front, where riches awaited the victor, were the compelling inducements that resulted in a mad scramble for enlistment in the first Marine unit to ply the western waters. The eagerness with which the troops responded is best evidenced by the number of volunteers who surrendered their places in the ranks of the 13th Virginia Regiment for assignment in the new company. Twenty men from the regiment voluntarily joined the expedition. An additional 14 men were enlisted from the other regiments garrisoning Fort Pitt. Upon the recommendation of General Hand, Robert George and Robert Elliott were appointed first and second lieutenants, respectively.[9]

Great was the stir at Fort Pitt as Captain Willing assembled his expedition. Strategy for the campaign had been determined during the Philadelphia conference, but it would seem that the naval captain's progress did not meet with the approval of General Hand. Writing to Richard Peters, secretary of the Board of War, on 24 December, Hand had the following observation:

> Capt. Willing had arrived here a few days before my return from Fort Randolph [at Point Pleasant]. I have in the best Manner supplied him with such things as he wanted.—but am afraid the river will be shut up before he gets away.[10]

With matters of supply and organization settled, Willing got away from Fort Pitt during the night on 10 January 1778, "with very Good Water."[11]

As the expedition made its way down the Ohio, Willing and his Marines immediately achieved considerable notoriety. A short distance below the point where the Wabash empties into the Ohio, they seized the Becquet brothers and their large bateau laden with pelts, as well as a Mr. La Chance and his cargo of brandy. The Frenchman, Rocheblave, com-

* The boat *Rattletrap* was purchased from John Gibson at a cost to Congress of 300 pounds Pennsylvania currency. The craft is supposed to have been of the galley type with about ten oars and stern sweep. Her armament, if Willing's request was honored, consisted of two ¾-pound swivel guns.

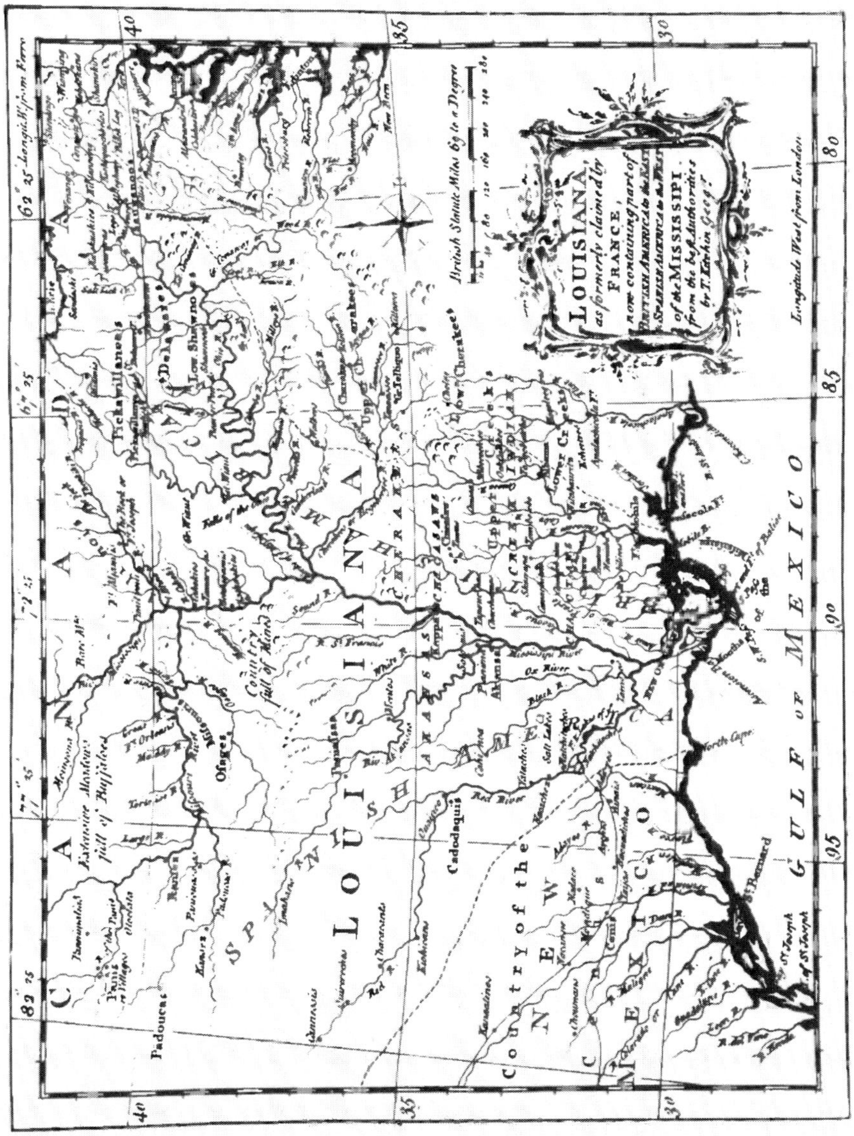
LOUISIANA
FRANCE
GULF OF MEXICO
British Statute Miles 69 to a Degree
Longitude West from London

mander at Kaskaskia, believed the expedition directed against Illinois, and interpreted the seizures as an indication of what might be expected should the Americans enter the area in greater force. In fact, when British Lieutenant-Governor Henry Hamilton at Detroit, the "hair buyer," heard of George Rogers Clark's capture of Kaskaskia several months later, he supposed the captors to be from Willing's command, aided by the Spaniards. But the expedition's objective was not Illinois.[12]

After the seizure of the Becquet boats, the *Rattletrap* continued to drift down the Ohio. Somewhere along its course, Willing intercepted a dispatch from Francisco Cruzat, Spanish Lieutenant Governor at St. Louis, to George Morgan, and endorsed it: "Recvd and Examined this 17 Jany 1778 J. Willing."[13] By the time the expedition reached the Mississippi, two canoes and ten recruits had joined the force. Among those who volunteered was George Girty, subsequently appointed a second lieutenant in the Marine company. Girty was the youngest member of the four "Girty Boys."[14] Following the example set by two of his notorious brothers, James and Simon, he later deserted to the British.[15]

The next record of the Marine expedition is from the Spanish post at the mouth of the Arkansas, where 14 American settlers petitioned the authorities for protection:

> The Petition of Sundry American Families arrived since the departure of James Willing Esqr. Captain in the Army of the United States of America Shewith, that your Petitioners upon their arrival at the English side found the Place called Concord abandoned and understood that Captain Willing left Orders that nobody should remain there: In consequence where of as well as from the fear they have of being disturbed by Savages in the Royal Interest of Great Britain and also not chosing to go further down the River until certain intelligence be had from below, Your Petitioners humbly request that you will take them under your Protection.[16] *

Willing's flotilla reached the plantation of loyalist Anthony Hutchins, a short distance above Natchez, on 19 February. Hutchins was taken prisoner, his slaves and all valuable propertly seized.[17] Before proceeding further down the river, Willing decided to divide his Marine force; two canoes, one commanded by Robert George and the other under Thomas McIntyre, would scout ahead. Late that Friday, 19 February, the two canoes arrived at Natchez, a thriving farm community populated by American, English, and French settlers, and until recently the home of Willing himself. "Dressed in hunters Frocks and armed," the small force immediately went ashore to "take all the inhabitants and secure their properties," as one distraught loyalist later related to British authorities.[18] At daybreak the following morning the main body arrived. The American captain's activities at Natchez are best described by the settlement's inhabitants who thought it in their best interest to capitulate rather than resist:

> [He] sent orders to all parts for us to convene in order that at the same time that we should be made prisoners of war to the United States he might take possession in their name of this jurisdiction. The inhabitants, mindful of the unfortunate state of their very great remoteness from protection and fearful of the confiscation of their property, felt it necessary to go to said Captain Willing to propose measures . . . and the inhabitants unanimously delegated four of their number to treat in their name for a capitulation, which should be formed in the best terms possible.

In their meeting with Willing, the four-man committee proposed the following:

> That we will not in any fashion take arms against the United States of America, nor help to supply, nor give any assistance to the enemies of said States.
>
> That our persons, slaves and other property of whatever description shall be left secure, and without the least molestation during our neutrality.

On 21 February, Willing approved of the proposals for the town's capitulation and future neutrality:

> Approved in full by me in the name of the United States of America (every public official of the crown of Great Britain who holds property in this District being excepted) and those who have held commission and signed the oath of neutrality contained in the preceeding articles.
>
> The property of all British subjects who are resident of this District is likewise excepted as belonging to enemies of the before mentioned States.[19]

Once guaranteed protection, the people of the Natchez district were somewhat secure in their persons and property. Willing only demanded provisions for his men. He did, however, order "all the Single Men in the Natchez to join him and the married People who were well affected to him to remove over to the Spanish side within 15 days." Whether this order was enforced or not is unknown, but a number of district men did decide to join him. Among them was Richard Harrison, a settler, whom Willing made a lieutenant.[20]

Willing's guarantee of protection applied only to Natchez. South of the district he carried out a ruthless campaign in which the crops of British planters

* Captain Willing's branch of service is obviously in error.

were destroyed, their houses burned, their cattle and hogs killed, and their slaves carried off. Many planters crossed the river and took refuge in Spanish territory. Others were made prisoner. One of Willing's victims, William Dunbar, later wrote:

> The Party was commanded by James Willing of Philadelphia, a young man who had left this Country the year before; perfectly & intimately acquainted with all the Gentlemen upon the river at whose houses he had been often entertained in the most hospitable manner & frequently indulged his natural propensity of getting Drunk—This was the Gentleman our friend & acquaintance, who had frequently lived for his own conveniency for a length of time at our houses, I say . . . this was the man who it seems had solicited a Commission by which he might have an oppy of Demonstrating his gratitude to his Old Friends—Several of his boats are already arrived at the Fort, where they make no secret of their intentions: they have taken from the Natchez Col. Hutch[ins] with his Negroes; they plundered Harry Stuart's House and have seized the negroes & other property at Cuming's Plantation; they divide the property at Castles', taking one half for his Partner's share, but leaving the Other unhurt for himself—Presently two of the Batteaux passed the land at Mr. Walker's plantation, where above 100 shots were discharged in half an hour's time, which were afterwards found were employed in wantonly killing the Hogs & other stock upon the plantation, in the afternoon the rest of Willing's boats pass down & lastly about sun sett the Genl. himself dropt down & put ashore at Walker's, where the scene that followed markes well the nature of the man,/I had almost said the Brute/ The Houses were immediately rumaged & every thing of any value secured for the Comodore's use, after which the Heroick Captain ordered his people to set fire to all the houses & indigo works, which was accordingly done & they were quickly consumed to ashes—Twas not enough to pillage & plunder the man at whose house he had been often most hospitably entertained, his ruin must be compleated by a piece of wanton cruelty, from which the monster cou'd derive to himself no advantage . . . the Houses of the English Gentlemen on the British side were plundered & among the rest mine was robbed of every thing that cou'd be carried away—all my wearing apparell, bed & table linen; not a shirt was left in the house—blankets, pieces of Cloth, sugar, silver ware, In short all was fish that came in their nett, they destroyed a considerable quantity of bottled wine, tho' they carried away no liquor; . . . the orders given by their head were to drive down my negroes & if opposed by any one to shoot 'em down . . . in the whole I was plundered of £200 sterg. value—

Continuing his rather lengthy diary entry, Dunbar speculated as to Willing's motives for plundering his former friends:

> About this time it was discovered that Willing had left Fort Pitt with a Batteau & only 25 men with orders to proceed to N. Orleans & take charge of such stores as were prepared for him by Pollock & return with the same up the Ohio, & also if he found himself able, to make Captures of British property on the River but no directions were given to disturb the peacable Inhabitants on Land—Notwithstanding of these orders, Cap. Willing conceived the design of making his fortune at one Coup upon the ruin & destruction of his Old Friends & Intimates—His chief reason was that he had by his folly squandered a fortune upon the river & twas there he ought to repair it. In order to effect this, his hellish purpose, he recruited & collected together on his way down, all the vagabonds & rascalls he met with, of which kind the river is always full; Engaging them upon the aluring Expecta[tion] of enriching themselves by plunder; and his numbers, upon his arrival in this settlement amounted to 200, much more than sufficient to accomplish his design, if we consider that perhaps one half of the Inhabitants were in the American Interest, which circumstances being well known to the Loyal part of the people, was the means of tying up their hands & preventing their attempts to oppose the Bantitti, not to mention report of their great numbers, which were at first much exagerated—[21]

Another victim, Frederick Spell, wrote of the seizure of a boat in which Stephen Shakespear was transporting his property to safety in Spanish territory:

> The Batteau being fastened Head and Stern to the Spanish Shore, and we had laid our selves down to rest, we were awakened by a sudden Noise which we found to be occasioned by one Calvert and some armed Men under his command on the Spanish Shore casting off the Ropes, who immediately ordered me down into the Cabin and not speak a word. I not immediately complying Calvert ordered one of his Men to Blow my Brains out, on which the Man presented his Gun and said he would blow all hell through me, upon which I haveing obeyed they pushed off from Shore for New Orleans.[22]

Although excess and wanton destruction often accompanied the seizures of loyalist property, the plundering was not indiscriminate. Friends of the American cause were not molested.

At Manchac, on 23 February, an advance party of 18 Marines under Lieutenants McIntyre and Harrison captured the British sloop *Rebecca*, mounting sixteen 4-pounders and six swivels. Sent up the river to frustrate the plans of the Americans, the seizure of this vessel ended, for a time, British control of the waterway. Making "prisoners on their parole of all the Inhabitants in Manchack Settlement," the two lieutenants then carried the *Rebecca* down to New Orleans where she was later fitted out as a cruiser and renamed the *Morris*. After a short stay in the Spanish city, the party pushed towards the Gulf.[23]

Learning of Willing's presence up river, Pollock began to organize supplementary forces and laid plans for future American operations. His nephew, Thomas Pollock, and a Captain Lafitte, with 41 men, were ordered up stream to assist Willing, and to bring back British property. Instead, the 26 French and Spanish boatmen enlisted by Lafitte floated down river by night and joined the two canoes manned by a slightly larger group of Americans under Lieutenants McIntyre and Harrison. Continuing downstream, they came upon the British brig *Neptune*, destined for Jamaica with lumber. The Americans tried to persuade the French and Spanish boatmen to join in an attack upon the vessel; two refused and left the party. The others joined the Americans in seizing the enemy vessel. Five passengers were put off into a boat manned by a few Americans and 16 or 17 boatmen, who, according to a description, were "armed with Cutlasses and wore either Cockades or Deer Tails in their Hats." Later the *Neptune* was brought up to New Orleans where she was detained as a Continental prize.[24]

News of Willing's exploits spread rapidly throughout the British province of West Florida. The uncertainty and uneasiness which had prevented any armed resistance to the Americans now prompted an emigration to the safety of Spanish territory. Plantations were abandoned and slaves, cattle, and valuables transported across the river, or put on board barges for New Orleans. Spanish officials, eager to remain neutral in the struggle between Britain and her colonies, welcomed the refugees and extended their hospitality. The number of West Floridians who availed themselves of the Spanish offer of refuge is not known.[25]

Trouble at New Orleans

The welcome extended to Willing and his party was comparable to that accorded to British refugees. This cordiality was due in part to the respect and influence Oliver Pollock had with Governor Bernardo de Galvez and his predecessors. Largely because of his personal friendship with the governor, Pollock succeeded in obtaining freedom of the city for the Americans. In addition, a public building was assigned to them as a barracks and Pollock was permitted to auction off their plunder which consisted mostly of slaves, although Spanish law forbade commerce with foreigners.[26] The exact amount of profit secured from the sale is unknown, but reliable estimates placed it around $60,000, a rather meager sum considering the value of buildings burned, livestock and crops destroyed.[27]

Although the British were most appreciative of the hospitality shown them by Galvez, they strenuously protested the welcome given Willing and especially the official approval of the sale of plunder in New Orleans. By mid-March, the arrival of petitions detailing the damage done by Willing and claiming that the Americans had seized British property illegally, became almost a daily occurrence. To consider their merits, the governor appointed a three-man commission, but the petitions themselves caused Galvez little personal alarm until the British sloop *Sylph*, commanded by John Fergusson, arrived in the Mississippi. On 14 March, Fergusson vigorously took up the cause of Willing's victims:

> Having the honor to command one of His Britannic Majesty's Ships in this River, and having information, that your Excellency has received into your Government, a Body of Armed Men, Enemies to my Sovereign, and that you have suffered them, from the Spanish Territory, to commit depradations on this River, by forcibly seizing upon the Vessels, Property, and Persons of British Subjects, in violation of the Treatise of Peace, the Law of Nations, and the Rights of Men. I cannot help looking on such conduct on your part, as a tacit, if not an open declaration of war, against the King, my Master.[28]

The same day, Galvez replied to Fergusson's charges with equal vehemence:

> I do not know, how you can take my mode of proceeding as a declaration of war, Violation of the Treaties of Peace, Law of Nations and Rights of Men; when at the same time I do not think, that I have taken a single step, which does not indicate a religious observation of the same Rights, which you accuse me of having violated. . . .
>
> If I have received the Americans upon the Territories, which I command, it was out of regard of the same rights of Men, which you . . . say, I do not observe. . . .
>
> You are at liberty to put what construction you pleased upon my way of proceeding: my ideas are certainly no others, than to remain neutral in the present War, according to the commands of the King, my Master: and should you (persisting [in] your opinion of my way of acting) pretend to commit the least hostility from Manchac to Balise towards the Subjects of my Sovereign, or any one under his Protection in this Colony you will find me as resolute to repress you by force of arms as I am willing to preserve the friendship & good understanding of the two Nations.[29]

In the course of the correspondence which passed between Governor Galvez and Captain Fergusson, the points at issue were those raised by Willing's victims in their petitions: the release of all British subjects and their property, the restoration of all prizes made on the Mississippi, and the latitude allowed Willing in New Orleans. In his reply to British concerns about the impropriety of harboring Willing and his men, Galvez declared that his protection of the Americans was in consonance with the position taken by the leading powers of Europe. To the demand that he restore and release all British property and persons, he asserted that Spain was under no obligation to protect British subjects on British soil or off the shore of British territory, and that the seizures made by Willing would not be restored until the completion of the official investigation. As to the demand that all prizes made on the Mississippi be returned, the governor disclaimed all responsibility for any seizure made above Manchac, since both Spain and British shared control of the river. Below, Galvez agreed that no violence was permissible either by the British or Americans, therefore certain boats and cargoes which had been captured below Manchac would have to be returned to their owners.[30]

Bernardo de Galvez, by an unidentified artist.

Galvez' decision that much of the captured property ought to be restored was a blow to the Americans. On minor items such as the return of one or two slaves they acceded to the wishes of the governor, but on major items, such as the *Neptune* and Stephen Shakespear's boat, they protested at length. The capture of the brig *Neptune* was legal, contended Willing, for she was seized on the open Mississippi, not in a Spanish port or under Spanish protection. Citing the judgment of the Vice Admiralty Court at Pensacola in the case of American vessels taken in the Mississippi that "Captures so made came under the Denomination, as that of prizes made on the High Seas," he claimed the *Neptune* a "Lawfull Prize." In similar fashion, he laid claim to Shakespear's boat. It had been captured first in British territory, then was "Stolen off in the night by Mr. Shakespear contrary to his Faith pledged," and recaptured floating in the Mississippi, "tho' fastened to a Log of Wood near the [Spanish] Shore."[31] But Galvez remained inflexible. The seizures must be returned, and he added that he would give the same reply if the whole United States Navy were on hand.[32]

While controversy raged over other captures, one prize taken by the Americans seemed secure, the sloop *Rebecca*. Soon after her arrival at New Orleans she was renamed the *Morris*, and Pollock appointed Willian Pickles to her command. Selection of her other officers took more time. On 1 April, Pollock chose Robert Elliott to be her captain of Marines, and Daniel Longstreet, his first lieutenant.[33] Then the arduous process of converting the sloop to a cruiser began. Carpenters were put to work removing her upper deck to make room for the sixteen 6-pounders that had been ordered from Havana. In addition, her bow and quarters deck were to be altered and strengthened for the "2 Bow & 2 stern Chacers 8 four pounders . . . Swivels Cohorns &ca." When completed and fully armed, and with a crew of 150, Pollock hoped to send the *Morris* out to convoy ships to Atlantic ports, to cope with any British vessel in the area, and thus protect upriver shipments of American goods.[34]

Early in April, the sloop *Hound*, under Captain Joseph Nunn, replaced the *Sylph* off New Orleans. Like Fergusson, Nunn continued to champion the demands of British subjects and press for their resolution.[35] Galvez, likewise, continued to maintain a bold front toward both the British and Americans, while at the same time realizing his predicament, the protection and favors he had shown to Willing and his Marines now placed him in a "critical situation with his neighbors." Afraid the British would retaliate, he requested reinforcements and began preparing New Orleans for a possible attack. In addition, he required every British and American subject in the city to take an oath of neutrality or leave.[36] A few Englishmen chose the latter course, but the majority complied. The Americans accepted the oath unanimously. Galvez may have relieved his own anxiety through the administration of these oaths, but he certainly incurred the displeasure of British authorities in West Florida who considered them another indication of his support for the American rebels. Yet they were reluctant to embroil themselves in a war with Spain as a result of Galvez' protection of Willing's party.[37]

British authorities showed no reluctance in challenging American control of the river. In March, they sent a force of 50 men to Manchac. Before dawn on the 19th, they surprised the small American garrison, killing two men, one woman, and wounding between eight and ten. Fourteen were taken prisoner. When Willing learned of the attack, he immediately sent another detachment northward which recovered the post without opposition.[38]

In April, Willing's concern for the control of Natchez prompted him to send Marines under Lieutenant Harrison to see that the oath of neutrality was being observed. Loyalist Colonel Anthony Hutchins, learning that men would be sent, broke his parole at New Orleans and hastened to his home where he "excited the inhabitants of Natchez to take up arms, by declaring on OATH to the people, that this detachment was coming up with the determination of robbing the inhabitants of their property without exception." Alarmed, the settlers armed themselves and organized a defense at White Cliffs, preparing to resist any further American attempts at plundering. Harrison was warned of the danger by John Tally, a few miles below the ambush, and he sent Tally ahead to inform the group that the Americans intended no hostile action. The American lieutenant then came up and crossed the river toward the Natchez settlers. As he approached, he discovered that he had entered a trap. Harrison then shouted for all "friends to the United States" to stand clear; the reply came for all "friends to the Natchez" to fall below the gunwales. A shot was fired, by whom it is uncertain, but each group later claimed that the other fired first. Then the firing became general. The Americans, outnumbered and more exposed than the settlers on shore, got the worst of it. Five men were killed, several were captured, and a few returned to New Orleans with Lieutenant Harrison.[39]

Again in May, Willing was frustrated in another attempt to renew control of the river. He blamed his failure to get possession of Manchac this time on the "unexpected appearance of the Enemy in the Lakes," and on the discontent of his own men whose accounts had not been settled by Pollock, but he continued to hope that "a party may yet get up [in] time enough to open the Levy, drown the Country,

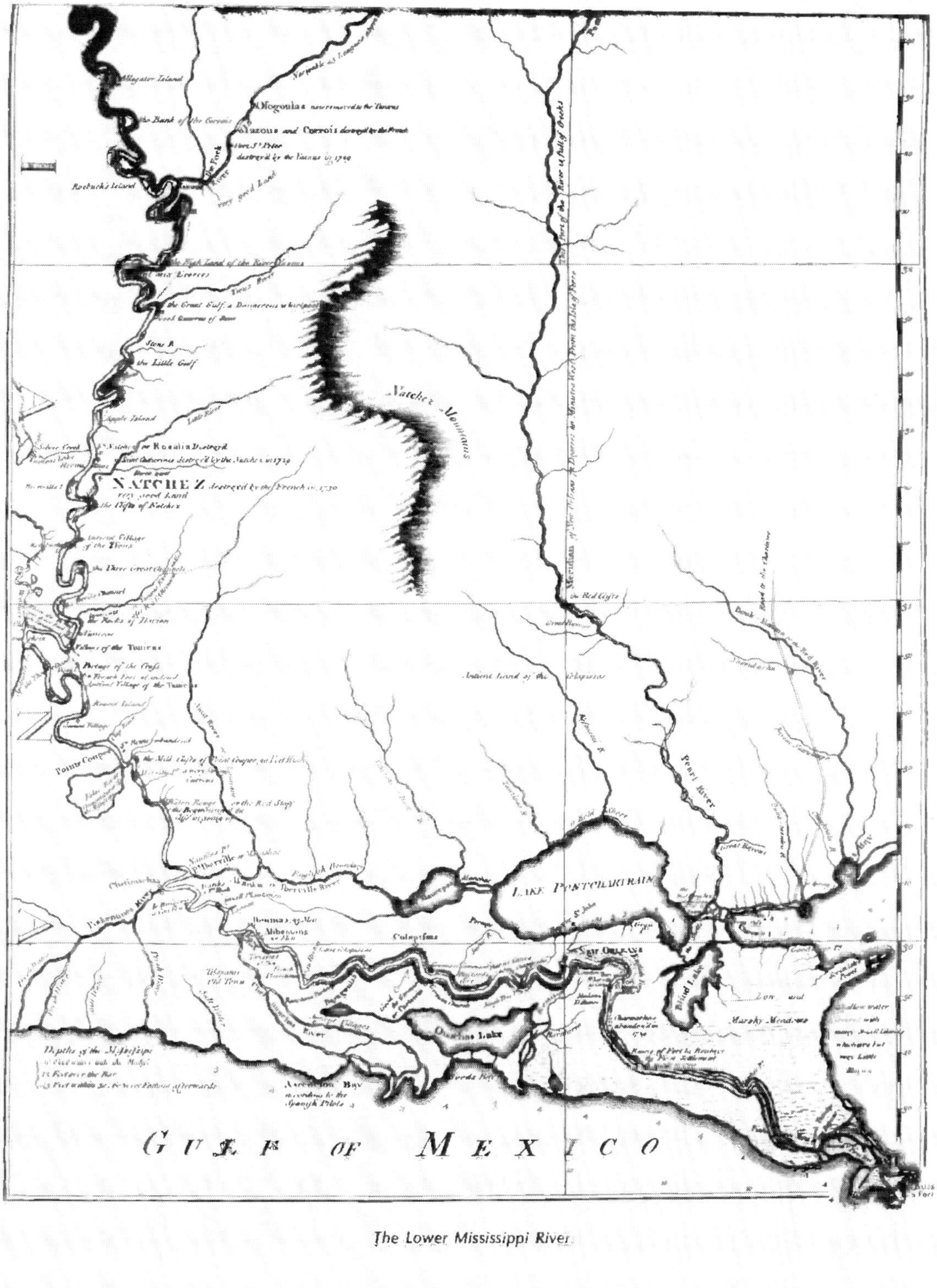

The Lower Mississippi River

Felle some Trees in the Baiu and by Burning and destroying the Buildings and other materials, put a stop to their Operations untill such Time as sickness or ye arrival of a Reinforcement might effectively prevent their fixing themselves Solidly."[40] Willing was not able to carry out this desire. Instead Governor Peter Chester of West Florida encouraged the settlers to return to the province "and restore yourselves to that full allegiance & fidelity which you owe to Your Sovereign & to your Country—otherwise Your Refusal will be Considered as a Criminal neglect of that duty, which is due from you as faithful subjects, to the Best of Kings."[41]

Shortly after Willing's final attempt to wrest control of the river, Manchac was garrisoned by a British force sent from Pensacola, made up of 40 regulars, 60 rangers, and a small party of Carolina refugees. Defending the river was the sloop of war *Sylph* with a crew of 150. Two hundred men protected Natchez. As a result, communications and the shipment of supplies between New Orleans and Fort Pitt were interrupted, if not halted altogether.[42] Disillusioned, Willing wrote of the loss of the English bank as "an unlucky circumstance to me and to the States, although it has given me an opportunity of knowing my friends from my enemies."[43]

Having thoroughly aroused the British in West Florida, Willing found that he was unable to leave New Orleans by the Mississippi. As his stay in New Orleans stretched into months, both Galvez and Pollock soon found it difficult to do business with the American officer because he lacked judgment and was inclined to dissipate in New Orleans rather than attempt to return to the states. The Spanish governor's displeasure was incurred early when Willing issued a proclamation at New Orleans to his paroled prisoners. Galvez considered this action a disregard of Spanish sovereignty and called the American captain to task for taking advantage of the hospitality extended him.[44] Willing apologized and explained that his offense had been unintentional and that he thought this the "least objectionable method" of communicating with his paroled captives.[45] But the longer Willing stayed the more uncomfortable Galvez became.

The friction between Galvez and Willing was exceeded only by the open break which developed between the American captain and Pollock. By the end of May, the rift was so great that Willing felt it necessary to address Pollock in writing "to prevent any Verbal altercation." After quoting his instructions from Congress and condemning Pollock for his poor administration of the finances of the expedition, Willing wrote:

> But it is my Business at present to Insist upon a total change of Procedure—
>
> My Men and Officers are discontented, myself displeased and the Governor himself highly disatisfied with Your Conduct and what is of the most serious consequence My Men are deserting and the American *Bank* as it is termed is become proverbially Ridiculous In a word the Service suffers and our Enemys rejoice—This therefore is to *insist* that you forthwith make out all your accounts so that the one half belonging to me and the Men be instantly divided and that you have the Ballance that is due on that Score ready to pay into my Hands on Monday or Tuesday next. . . .
>
> This being done I am well convinced there will be no further discontent and that the Men will chearfully return to their Duty—I need say no more than that the Service and honor of the States requires your immediate complyance, and that every ill effect which may arise, will be entirely owing to your Neglect—I have a Copy of this to lay before The Congress and Governour if needfall.[46]

However vehement the letter, Willing was entirely correct as to desertions. The plunder-seeking recruits had departed for home. The remaining force, held together by the efficient conduct of Lieutenants George and Harrison, consisted of the greater portion of the original 35-man Marine company.

In response to this "very Extraordinary and Unexpected Letter," Pollock suggested that had Willing given himself "a moment Reflection you would never have sign'd." He promised a full and speedy accounting, but laid the blame for the delay on Willing's failure to place a value on the *Rebecca*. Expressing full confidence in his own conduct, should Willing lay his "unsupportable Grievances" before Congress "In order to throw the Odium upon me," and explaining the governor's displeasure was due to Willing's military maneuvers rather than his own actions, Pollock concluded by warning the American captain "to be more cautious whom you touch as some how or other the Enemy is acquainted with Your Proceedings."[47]

Pollock grew increasingly anxious to be rid of him as Willing's stay lengthened. His reports to Congress indicate his annoyance. From his letter of 6 July:

> The Small Party you sent here under the Commander of Captn. James Willing without any Order or subordination, has only thrown the whol river into confusion and created a Number of Enemies and a heavy Expence which would not have happened had they been otherwise Governed and a Proper number sent, however the onlv Remedy for what has past is a speedy dispatch.[48]

Again on 11 August:

> What his next Pretence for tarrying here will be God knowns, but as there is a clear Passage for him and his Party to go up, part by Land and Part by Water through the Spanish Territories by way of the Appelousaa & Nachetosh and join Colo Clark, I am determined to stop all Supplies in order to get him away.[49]

Wishing for Willing's early departure was easier than bringing it about. First Pollock attempted to route Willing and his men by way of the Mississippi. The increased British military movements in West Florida frustrated this plan. His next recourse was the sea route. On 16 June, he asked Galvez for permission to fit out the *Morris* in order that the vessel might be used to carry Willing and his men home. Even though granted, delay followed delay. Meanwhile Willing asked the governor for permission to go north through Spanish territory and under Spanish protection, thus escaping the fury of the Natchez settlers.[50] Galvez at first refused, suspecting that Willing's request was merely a pretext for new ravages on the English settlements. Nevertheless, on 14 July, he acquiesced and authorized passage through Spanish territory, but the venture was considered too hazardous and quickly abandoned.[51]

A month later the plan was revived by Lieutenant Robert George who requested permission to lead the men, "lately commanded by Willing," northward.[52] Galvez replied that he would grant such permission only if the Americans would swear "to follow the route directly, and not to offend or bother during the journey any English subjects, neither their possessions nor their persons, but on the contrary to treat them with the same consideration as if they were Spanish subjects."[53] George and Harrison immediately furnished the required oath and Galvez issued a formal pass for the party by way of Opelousas, Natchitoches, and the Arkansas.[54] Shortly thereafter George and his party left New Orleans. By

June the following year they had completed their trek, and had joined George Rogers Clark at Kaskaskia where the Marine company was disbanded, and a portion enlisted in a new artillery company commanded by Captain Robert George.[55]

In mid-November Willing, accompanied by Lieutenant McIntyre, departed New Orleans for Philadelphia with dispatches from Galvez and Pollock to Congress. But his journey was an unfortunate one. The sloop in which he took passage was captured off the Delaware Capes by the British privateer *Columbus*, and he was taken prisoner to New York.[56] On 5 July 1779, his exchange for Nicholas Ogden was proposed and then rejected. Toward the end of the year he was exchanged for Lieutenant-Governor Henry Hamilton of Detroit.[57]

Longstreet and the *West Florida*

In August 1778, the *Rattletrap* was "condemned not being fit for service," which left one American vessel of war on the Mississippi, the *Morris*.[58] For almost a year, Pollock labored to complete the conversion of the sloop and by late July 1779 the job was almost finished, as he reported to the Commerce Committee:

> In my Letters . . . of the 10th April, I wrote you particularly respecting the Ship Morris, and at that I expected the proper Guns from the Havana, but now find there are none to be got for Love or Money; in conse-

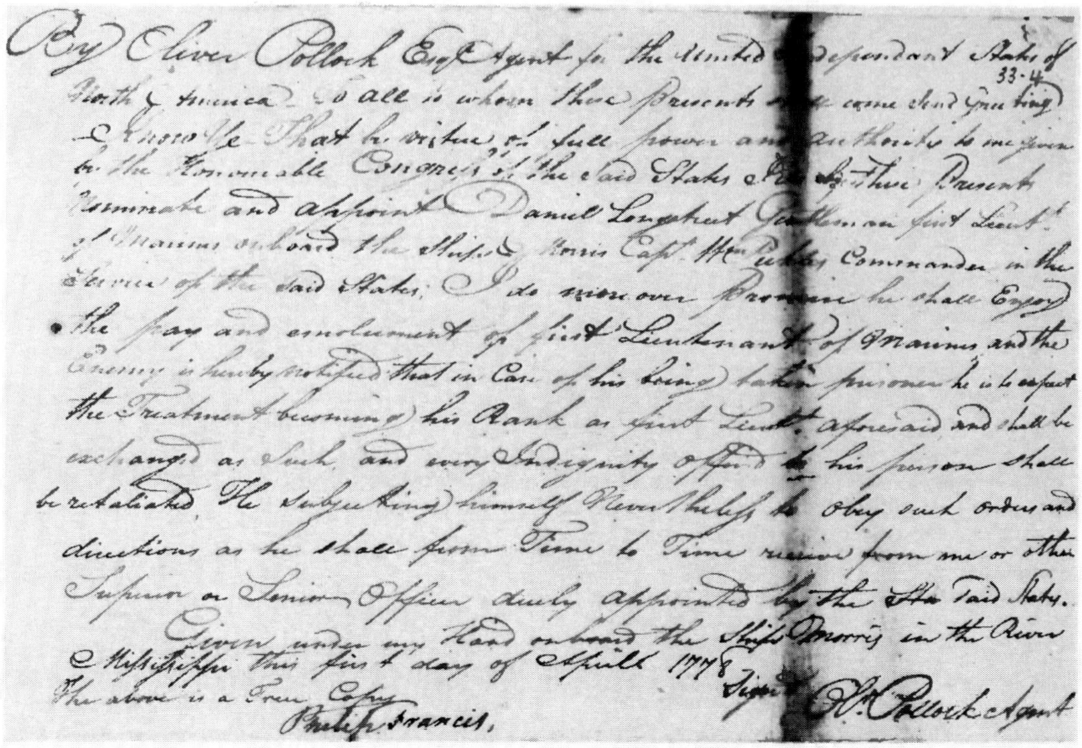
By Oliver Pollock Esqr Agent for the United Independant States of North America To all to whom these Presents shall come Send Greeting 33-4

Know ye That by virtue of full power and authority to me given by the Honorable Congress of the said States I do by these Presents nominate and appoint Daniel Longstreet Gentleman first Lieut. of Marines on board the Ship Morris Capt. Wm Pickles Commander in the Service of the said States; I do moreover Promise he shall Enjoy the pay and emolument of first Lieutenant of Marines and the Enemy is hereby notified that in Case of his being taken prisoner he is to expect the Treatment becoming his Rank as first Lieut. aforesaid and shall be exchanged as Such, and every Indignity offered to his person shall be retaliated. He subjecting himself Nevertheless to obey such orders and directions as he shall from Time to Time receive from me or other Superior or Senior Officer duly appointed by the said States.

Given under my Hand on board the Ship Morris in the River Mississippi this first day of April 1778

Signed Ol. Pollock Agent

The above is a True Copy
Philip Francis.

Daniel Longstreet's Commission as Lieutenant of Marines.

> quence I have got the 8 Six Pounders with 10 of her own four Pounders all ready mounted on the main Deck, and 6 Ditto [4-pounders] on her quarter Deck, besides swivels, with 76 Men and the English Deserters coming on board daily from Manchak.[59]

Pollock's only problem now was orders. Since he had received none, Pollock, "for the present defered sending her out and determined to shift a little longer, in hopes every moment to hear from" the Committee.[60]

While Pollock waited for the expected orders, a hurricane swept over New Orleans and along the river with such destructive force that the whole province of Louisiana was left in ruins. Up and down the river, for a distance of 40 miles, houses were wrecked, crops destroyed, and livestock killed. Unfortunately for the American cause, the *Morris*, now ready for sea, was sunk. Of this blow, Pollock wrote:

> I am extremely sorry to inform you of the Misfortune of loosing the Ship Morris at this Levy by a Hurricane that happened here the 18th Inst. which has destroyed great part of this Town, and sunk several of the Kings Galleys and other vessels in this River, in short the scene was too distressing for me to describe on paper let it supply. I and my Family as well as many others had a narrow Escape for our Lives. I saved all my Books & Papers which were the only Articles in my House I could keep dry all the Ship's Crew (except Eleven who were unfortunately drowned) got saved on the Wreck and were picked up about three Leagues below this Town. This is a most unlucky circumstance for this Government, and particularly to myself, after all my trouble, vexation of Mind, and expence in fitting out this Ship, and just now, when I had her ready and a Channel opened to commence Hostilities again the English at this place, to loose her in the critical Moment.[61]

Among the survivors were Marine Captain Robert Elliott and Lieutenant Daniel Longstreet. Elliott, apparently dissatisfied, later made his way to Kaskaskia and then to Philadelphia.[62] Lieutenant Longstreet, however, chose to remain behind and take command of the *Morris'* Marine detachment in hope of again putting to sea.

In the midst of the desolation wrought by the hurricane, Governor Galvez determined to renew preparations for attacking English posts on the river, as Spain and Britain were now at war. Plans were drawn up for operations against Manchac, Pensacola, and Mobile. By late August 1779 all was ready, and Galvez set out for Manchac with a force of nearly 600 Spanish, American, and Indian troops. At that time, there were a number of English vessels, loaded with military stores destined for Manchac, on Lake Pontchartrain, as well as an English ship mounting 16 guns said to be laying at Ship Island. In order to protect his force, Galvez ordered an armed schooner, "which was saved down the River where they had little of the gale," transferred to Pollock as a replacement for the *Morris*. The officers and crew of the ill-fated ship were then removed to the schooner and ordered to engage the enemy wherever possible on the Lake.[63]

With little trouble the schooner moved out on the Lake, and on 10 September, "after a very hot dispute of about twenty Minutes," captured the British armed sloop *West Florida*. The 70- or 80-ton vessel under the command of Captain James Payne mounted four carriage guns and carried a crew of 30. Captain Pickles lost three and had five wounded during the engagement, two of whom later died. "The taking of this Vessel is an infinite advantage to this place [New Orleans]," wrote a friend of Pollock, "as she prevented all communication by the Lakes to Manchack and Pensacola."[64]

Of significance also was the capture of British settlements on the north side of Lake Pontchartrain. Captain Pickles, then on board the sloop *West Florida*, landed several seamen and Marines under the command of Marine Lieutenant Daniel Longstreet between "Bayou La Combe and the River Panchipaho" on the 21st and ordered them to take possession of British settlements in the area. Shortly after this was accomplished, Pickles left a number of men to protect the inhabitants against "Indians and others," and withdrew to New Orleans. The articles of capitulation acknowledging the subjection of the settlers to the United States were submitted by Pollock, and later by the Congress, in support of United States claims to West Florida, and the right of free navigation of the Mississippi River.[65]

Returning to New Orleans after the successful capture of Manchac, Galvez proceeded to marshal his resources for an attack on Mobile, and then Pensacola. While the plan provided for a coordinated movement from New Orleans and Havana, Navarro, Governor of Cuba, was slow to respond, as he favored a combined attack on St. Augustine by a joint American and Spanish force instead. After much delay, Navarro finally agreed to send a force for the Mobile expedition. Confident of success, Galvez embarked his force of 750 regulars, militia, and freed slaves and dropped down the river on 14 January 1780.[66]

Provisioned and equipped for 60 days, the *West Florida* was directed to join the Spanish fleet. Captain Pickles, under Pollock's orders, was to take part in the reduction of Mobile and Pensacola, "for the space of twenty Days or longer if necessary, or requested by the Commander in Chief of the Spanish Fleet." This accomplished, he was to proceed to Philadelphia, having taken on a cargo of taffia and sugar at Havana. In the event that an enemy vessel was taken "that will suit you better than the sloop," Pickles was authorized to "dispose of the said sloop to the best Advantage and ship your Men on board the captured Vessel and proceed [on] your Voyage in her as already directed."[67]

The Spanish fleet of over 13 vessels waited at Berleas for the expected Havana fleet until 4 March. When it failed to appear, Galvez put to sea and, on the 6th, was overtaken by a strong gale between Dauphin Island and Mobile Point. Four ships were grounded on a sand-bar, but the troops on board were safely put ashore. Galvez landed his artillery on 15 March "about a mile below Mobile & hauled them through the woods to a Road that led about of Mobile Town." On the 17th at eight in the morning,

as an American observer wrote, "he Opened a Battery of nine 18. pounders a bout a Qt. of a mile from the fort [Fort Charlotte]. At 4 in the afternoon a white flag was hoisted in the fort soon after they capitulated."[68]

In the attack upon Mobile, the *West Florida* captured a small sloop, and was "of great service" to Galvez.[69] After the surrender, she and her prize were taken to Philadelphia and sold, her crew having been relieved, paid, and possibly assigned to other Continental vessels.[70]

Willing in his address to Congress, a few years after the Marine expedition, reported that "his operations in that quarter created a considerable diversion by drawing off a part of the enemys force from their main army and employing a number of armed vessels to protect the navigation of the Missisippi."[71] His statement, although correct in part, had a much different consequence than Willing himself implied. The expedition, and Willing's inhumanity and lack of restraint, instead crystalized loyalist feeling and opposition, thus for a time losing the province of West Florida to the American cause. The conclusion must be that Willing and his Marines did more harm than good and that such results as were beneficial to the United States should be credited to the industriousness of Oliver Pollock, rather than Willing and his Marines. What Willing failed to accomplish, was made up for by the activities of Lieutenant Daniel Longstreet and his small Marine detachment on board the Continental sloop *West Florida*.

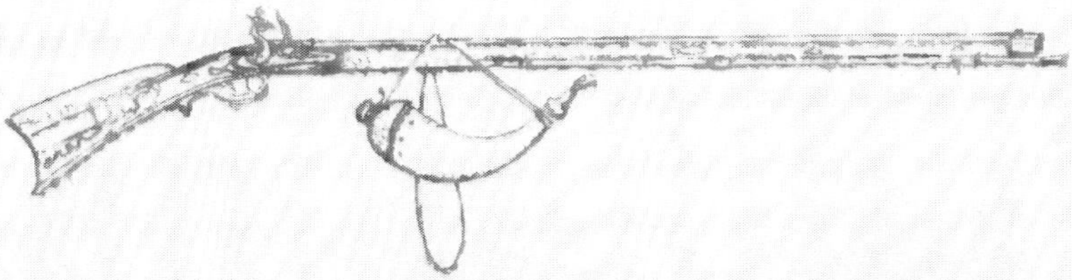

ASSAULT AT PENOBSCOT, 28 JULY 1779.
In the spring of 1779, the British established a strong outpost on the tip of the Bagaduce Peninsula in Penobscot Bay, Maine. This painting recreates the assault of Marines under Captain John Welsh up the steep, heavily forested bluff on the peninsula's western face, initiating a drive that was intended to rid the area of British troops.

CHAPTER X

Success at Sea and Disaster at Penobscot, 1779

Gratifying Achievements

"There remain in this harbour Six Ships," advised the Eastern Navy Board at the close of 1778. Only one problem prevented the Board from predicting when any or all of them would be ready to sail:

> We mean the manning of the Ships this alone delays them every Ship here might sail in fourteen days if they could be manned it is not easy to describe to you the difficultys we are under on this account & the losses the Continent sustains by the Conduct of the Privateers who are always seducing by every art the Men from the public Service after they have received the bounty & clothing & become largely in debt these men they secreet & send privately to some out ports where they go to take them in & elude the search of our officers.[1]

What the remedy should be was left to Congress, but for its part the board urged an embargo on privateering until Continental ships were manned.[2]

The six ships at Boston were the frigates *Alliance*, *Deane*, *Queen of France*, *Warren*, *Providence*, and *Boston*. If one adds to these the *Ranger* at Portsmouth, the *Confederacy* and *Trumbull* in Connecticut, and the sloops *Providence* and *General Gates* in the West Indies, one has a complete list of ships carrying detachments of Continental Marines at the beginning of 1779.

Captain Samuel Nicholson of the *Deane* had his ship in complete readiness; her wood, water, and provisions were on board, plus 200 men. The *Queen of France* had most of her stores, and 136 officers and men. All that was required was the completion of her gun carriages and some other last minute repairs. Needing much the same was the *Alliance*, with 180 officers and men on board. Captains Whipple, Tucker, and John B. Hopkins had their vessels alongside Boston wharves all graved, but in various stages of fitting and ballasting. The *Warren* was being repaired and the *Providence* was fitting a new foremast. The *Boston* had but "a few" men, the *Providence*, 11, and the *Warren* none.[3]

As refitting progressed, Marine officers from each of the ships received money and then went out into the countryside to appeal for recruits. Most returned with what they could scrape up from the far-flung towns, others did not. Marine Captain Elihu Trowbridge of the *Warren* was supplied with 390 pounds and sent to Springfield to recruit Marines, but as the Navy Board reported to their counterparts in Philadelphia, both he and the funds were missing:

> Three days past a Man who he had engaged as a Sergent of Marines brought down Six men with him whom Capt Trowbridge left at Springfield to Enter men promising him to return in a few days & pay off his Expences &c. waiting many Weeks and hearing Nothing from him borrowed money to discharge the board of those men and some others that would not come down with him arrived here as above & Entred with his Men on board the *Warren* . . . he Informs that by the best Intelligence he could get Trowbridge gamed away the public money & deserted the Service.[4]

Trowbridge, apparently a very crafty fellow, made his way to Philadelphia, and on 8 February appealed to Congress for more money:

> Your Memorialist begs leav humbly to Represent to Your Honours—that the great expenses which attend the Recruiting service, the frequent repitition of it arising from the shortness of the Term, the men were engaged for, the difficulty which Attend Recruiting, and the small allowance granted by the public, therefore brings a most intollorable burden upon Your Memorialist—he has already expended the whole of his private fortune in the service of the public, and is unable to account for all the public Moneys which has gone through his hands unless your Honours will be pleased to grant him an Allowance for the necessary Expenditures in the Recruiting

> Business in which he has been so frequently engaged, beyond what is allowed by the present regulation.[5]

Taken in by his plea, the Navy Board of the Middle District advanced Captain Trowbridge another 200 dollars. Soon thereafter Trowbridge again disappeared.[6]

Sometime late in January 1779, the Navy Board at Boston appointed Captain Richard Palmes to replace Trowbridge on board the *Warren*. Palmes, unlike his predecessor, put his recruiting funds to good work by going out into Boston with drum and fife, and placing appeals in the local newspaper:

> The Continental Frigate Warren, John B. Hopkins, Esq; Commander, will sail on a Cruize in Six Days from the above Date; all that are willing to enter as Marines, Seamen or Landsmen (Friends to the Independent States) are desired to repair on board; All Marines that was formerly under my Command will be prefer'd.[7]

The response, however, was slow, and by mid-February the *Warren* and *Boston* had but 70 men each, and the *Providence* only 30.[8]

The sloop *Providence* arrived at New Bedford on 11 January from a cruise begun the previous November in company with the *General Gates*. Plying the lucrative, well-traveled shipping lanes, Captain Rathbun had taken five prizes, all of which reached New England ports safely. The first capture, a Quebec schooner laden with flour and bound for New York, was made in concert with the *General Gates* before the two sloops parted in a heavy gale. Another of the prizes, the ship *Nancy* from Glasgow, had been taken by an American privateer off Barbados, retaken by the British, and then captured again by the *Providence*.[9]

While the *Providence* was repairing her ripped copper sheathing at New Bedford, Rathbun asked to be detached from his command because of poor health. The Eastern Navy Board appointed Captain Hoysteed Hacker to replace him; "a man," commented William Ellery, "of whom I have not the highest opinion."[10] This was Hacker's second tour in command and he lost no time in refitting and manning the sloop.

Shortly after the arrival of the *Providence*, the frigates *Deane* and *Alliance* sailed from Boston. The *Alliance* was bound to France and parted with her consort on the third day out. The *Deane*, provisioned for a four-month cruise, headed southward and soon captured the ship *Viper*, "a letter of marque, fitted out at Liverpool, mounting 16 guns and 75 men." Later "Capt. Nicholson took and burnt a ship belonging to London in ballast from New York to Cadiz for a load of brandy, wine and fruit, for the use of the army and navy at New-York."[11]

The *Deane* cruised about the West Indies for four months before stopping at Martinique where she took on dispatches for Congress. Early in April, the frigate appeared off the Delaware Capes, but was prevented from going up the river by poor weather. Eager to forward William Bingham's dispatches to Philadelphia, Nicholson sent them up to the city in a packet boat with Captain of Marines John Elliott, "who in the passage being covered with the sail, and asleep, by some accident rolled overboard and drowned."[12] On 17 April, the *Deane* sailed into Philadelphia.

Captain Thomas Simpson brought the *Ranger* around from Portsmouth on 25 February, and moored her in Nantasket Road near the *Warren* and *Queen of France*.[13] The three ships lay in the road awaiting crewmen and wind until the morning of 13 March when they came to sail in company in order "to Chastice the Insolence of those Small Cruisers upon the Coasts of Virginia & the Carolinas."[14] Orders to all three captains were exactly the same:

> Sweep in the first place this coast from the Southward of Cape May to the Bar of Charles Town, and afterwards to Cruize in such Latitudes and Longetudes which are best calculated to give the greatest aid and protection to the Trade of Delaware, Chesapeake and Charles Town, and as often as circumstances and the safety of your Ships will admit of it, you are to enter the mouths of Delaware & Chesapeake for the purpose of destroying the small armed Vessels from New York that lurk about the Capes to the certain destruction of almost every Merchantman that sails . . . when the state of your Provisions requires, you will return into the Port of Philadelphia or some convenient one in the Bay of Chesapeake.[15]

For almost a month "Strong Gales and excessive hard Squals and a furious Sea" battered the three ships as they moved southward. Once into the warmer latitudes, seamen and Marines were brought on deck to exercise with cannon and muskets, firing at selected targets.[16] The practice came in handy, for on 6 April, off Cape Henry, Virginia, the first prize was taken. At four in the morning the following day two fleets were sighted, one to leeward of 10 sail and another to windward consisting of nine sail. Captain Hopkins of the *Warren* made signal for all vessels to give chase to the windward fleet.

The Englishmen were a fleet of transports and store ships bound for Georgia from New York under

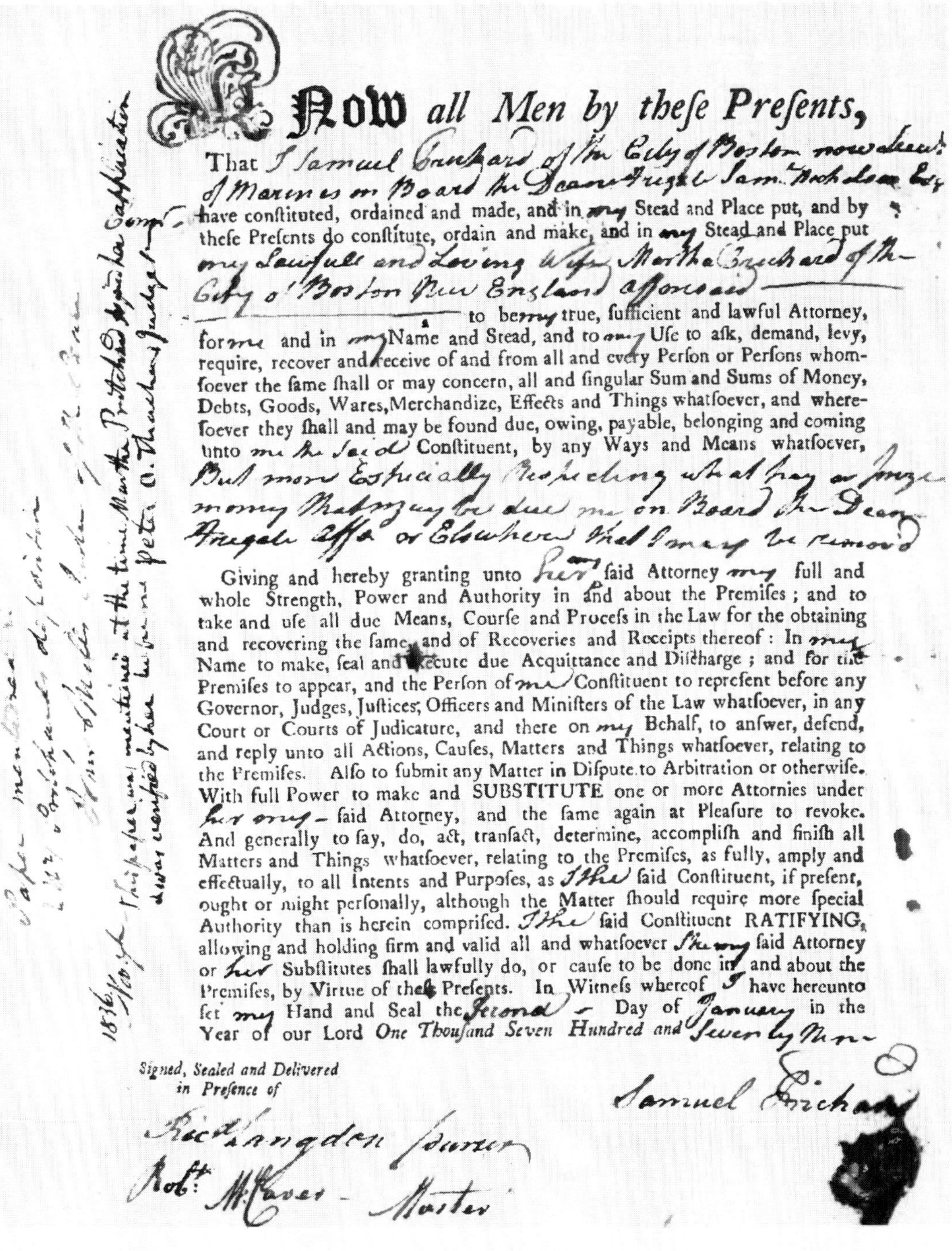

Know all Men by these Presents,

That I Samuel Prichard of the City of Boston now Lieut. of Marines on Board the Deane Frigate Sam. Nicholson Esq. have constituted, ordained and made, and in my Stead and Place put, and by these Presents do constitute, ordain and make, and in my Stead and Place put my Lawfull and Loving Wife Martha Prichard of the City of Boston New England aforesaid — to be my true, sufficient and lawful Attorney, for me and in my Name and Stead, and to my Use to ask, demand, levy, require, recover and receive of and from all and every Person or Persons whomsoever the same shall or may concern, all and singular Sum and Sums of Money, Debts, Goods, Wares, Merchandize, Effects and Things whatsoever, and wheresoever they shall and may be found due, owing, payable, belonging and coming unto me the said Constituent, by any Ways and Means whatsoever, But more Especially Respecting what pay or prize mony that may be due me on Board the Deane Frigate affs or Elsewhere that I may be removed

Giving and hereby granting unto her my said Attorney my full and whole Strength, Power and Authority in and about the Premises; and to take and use all due Means, Course and Process in the Law for the obtaining and recovering the same, and of Recoveries and Receipts thereof: In my Name to make, seal and execute due Acquittance and Discharge; and for the Premises to appear, and the Person of me Constituent to represent before any Governor, Judges, Justices, Officers and Ministers of the Law whatsoever, in any Court or Courts of Judicature, and there on my Behalf, to answer, defend, and reply unto all Actions, Causes, Matters and Things whatsoever, relating to the Premises. Also to submit any Matter in Dispute to Arbitration or otherwise. With full Power to make and SUBSTITUTE one or more Attornies under her my said Attorney, and the same again at Pleasure to revoke. And generally to say, do, act, transact, determine, accomplish and finish all Matters and Things whatsoever, relating to the Premises, as fully, amply and effectually, to all Intents and Purposes, as I the said Constituent, if present, ought or might personally, although the Matter should require more special Authority than is herein comprised. I the said Constituent RATIFYING, allowing and holding firm and valid all and whatsoever She my said Attorney or her Substitutes shall lawfully do, or cause to be done in and about the Premises, by Virtue of these Presents. In Witness whereof I have hereunto set my Hand and Seal the Second Day of January in the Year of our Lord One Thousand Seven Hundred and Seventy Nine

Signed, Sealed and Delivered in Presence of

Samuel Prichard

Richd Langdon Gunner

Robt. McCaver — Master

The Last Will of Samuel Prichard.

convoy of the 20-gun ship *Jason*. By two on the same afternoon, the American squadron had taken seven enemy sail including the escort vessel. The transports carried no troops, but were richly laden with provisions, dry goods, and accoutrements for a regiment of light dragoons which the British intended to raise among southern loyalists.[17]

After manning the prizes, Hopkins ordered the squadron and its prizes to stand to the northward, "having had intelligence of a large number of armed vessels being off Chesapeake and Delaware Bays." Off the Georgian Banks in a thick fog on 11 April, the *Warren* became separated from the rest of the fleet, and sailed into Boston harbor alone on the 16th.[18] The *Queen of France* and six of the prizes straggled in several days later, while the *Ranger* took two of the captures with her to Portsmouth.[19]

When the *Warren* dropped anchor in Nantasket Road, Hopkins sent Marine Captain Richard Palmes up to Boston to report to the Navy Board. Although elated over news of the success, the Board nevertheless was cognizant of the fact that once tied up to a city pier, the ship's company would scatter. Palmes was instructed to return to the ship with orders for Hopkins to remain in the Road and not come up. Palmes replied that it was too late for when he left the *Warren* was already under sail. Hopkins himself appeared before the Board the following morning and explained that he could not get his officers and men to sail until a settlement of the prize money was made. It was their contention that they had signed on for a cruise, and it was over. In less than a week the crew evaporated, and the *Warren* lay idle, unmanned, and useless.[20]

To prevent the same thing from happening to the *Queen of France*, the Board sent word to Captain Olney to anchor in Nantasket Road, keep his officers and men on board, and submit a list of the ship's needs. Olney ignored the order, brought the ship up to town, and allowed his crew ashore. The *Queen of France* was soon as deserted as the *Warren*.[21]

The reason for this conduct on the part of both Hopkins and Olney became clear to the Board when it learned that these two commanders had, by threats and coercion, been designated as prize agents for their crews. Unethical, if not illegal, this meant that Hopkins and Olney would not only receive a captain's share but also a commission on each man's entitlement. Disgusted, the Board wrote to the Marine Committee:

> We must inform you by Appearances and their Conduct that they are more Attached to their own Interest & Emoluments then to the honor & benefit of the United States we may pronounce this without doing them any Injustice no men were ever more asiduous in Search of gain than those two Captains have been.[22]

After much deliberation, both commanders were suspended and ordered to appear before a court of inquiry, or court martial if necessary. In the interim the Marine Committee named John Peck Rathbun to the *Queen of France*, and Dudley Saltonstall to the *Warren*.[23]

Hopkins also precipitated the downfall of his captain of Marines, Richard Palmes. Three days after the *Warren's* arrival, Hopkins sent Palmes to Philadelphia with a letter for Congress, contrary to the expressed orders of the Navy Board.[24] Since Palmes had "left his Duty at a Critical time without the leave & against the Judgement of Mr. [William] Vernon," the Board suspended Palmes, and appointed Captain John Welsh in his place.[25] With the *Alfred's* former captain of Marines came William Hamilton, his former lieutenant on board the same ship and cell-mate in Forton Prison at Portsmouth, who was appointed a lieutenant of Marines.

The brig *General Gates* returned to Boston on 13 April 1779. After parting with the sloop *Providence* the previous December, Captain Daniel Waters fruitlessly cruised about the West Indies several months before putting into Martinique for repairs. On his passage northward he fell in with the Massachusetts brig *Hazard*, and together they took an enemy brig loaded with fish.[26] When a survey showed that the *General Gates* was not worth repairing, she was sold.[27] This left Marine Lieutenant Richard McClure without a command.

The same day the *General Gates* arrived in port, the frigates *Providence*, *Boston*, and sloop *Providence* sailed. For more than two months, manning the two frigates at Boston was painfully slow. With Palmes' departure from the *Boston* early in February, recruiting duties fell to Marine Captain Seth Baxter who had been unemployed since his exchange in October 1777. In conjunction with Captain William Jones, Lieutenants Zebediah Farnham and William Waterman, Marine officers of the *Providence*, Baxter scoured the city for recruits, often to the dismay of army officers who concluded that it was useless for them to recruit men "if the Marines Officers may take them from us." On 28 March, William Cooper joined the *Boston* as second lieutenant of Marines, and with his help Baxter completed the frigates' Marine contingent early in April.[28]

The three Continental vessels were sent out at the request of the Massachusetts Council to sweep the Bay for ten days in search of small enemy cruisers reported to have left New York with the purpose of ravishing the Massachusetts coast.[29] The frigate *Providence* was to return to Nantasket Road within several days to await an important assignment from the Marine Committee.[30] The *Boston* and sloop *Providence,* once the local task was completed, were directed to proceed to Delaware Bay to take on a supply of bread and await further instructions from the Marine Committee.[31]

No enemy vessels were discovered in the waters off Massachusetts, and the *Providence* frigate returned to Boston.[32] The *Boston,* in compliance with orders, headed for Philadelphia, while the sloop lingered in the New York area. Off Sandy Hook on 7 May, the *Providence* sloop, "after an obstinate engagement," took the 12-gun brig *Diligent* which had left New York for the Chesapeake with Commodore Collier's fleet and had strayed from the main body. Before the Englishman struck, she absorbed several devastating broadsides and accurate volleys of Marine musketry resulting in eight killed and 10 wounded. Captain Hacker had two killed and 12 wounded, two of whom, "Mr. [James] Rogers, sailing master, and Mr. [John] C[h]ilton, Lieut. of marines, two valuable officers, are since dead, much regretted."[33]

Captain Hacker carried the *Diligent* and a recaptured schooner into New Bedford. Wishing to forestall repetition of the Hopkins-Olney affair, the Navy Board reminded Hacker that he had been directed to cruise for three or four months, and not to harbor any idea that quick success and a return to port after only a few days meant the cruise was ended.[34] To make certain that the sloop's crew would not have a chance to desert, Hacker was ordered out into the bay in pursuit of enemy vessels together with his prize and two Massachusetts state ships which were to join him off Nantucket.[35] Before sailing from New Bedford Hacker had the vacant post of lieutenant of Marines filled. Chosen to take Chilton's place was Robert Davis, a Massachusetts-born former Continental army officer.[36]

The *Providence* and the *Diligent* entered Boston harbor on 10 June after retaking "two Nantucket wood Sloops & a small French Sloop of no great value."[37] The *Diligent* was purchased by the Board for the Continental Navy at the unheard of price of £26,000. Hacker was allowed to choose command of either ship, but he preferred to stay with the *Providence.* Command of the *Diligent* was given to Lieutenant Philip Brown, first lieutenant of the *Providence.*[38]

In the meantime, Captain Samuel Tucker had brought the *Boston* up to Philadelphia. Toward the end of May, he was joined by the new Connecticut-built frigate *Confederacy* which had been delayed for several months at New London. Manning the frigate with seamen proved to be a major problem for Captain Seth Harding, but for Marine Captain Joseph Hardy the problem was equipment. Early in March he appealed to Major Joshua Huntington for finished arms, cartouch boxes, flints, ball, and all available empty cartridges, so that his men could begin filling them.[39] A month later he needed more cartridge thread, ball, and a set of drums and fifes. If the latter was not sent immediately, he wrote Huntington, he would be obliged to purchase them wherever he could find them.[40] By mid-May all was ready, and the *Confederacy* cast off from New London for the Delaware Capes. The frigate rounded Cape Henlopen several weeks later and sailed up the Delaware as far as Lewestown, where she was to await further instructions.[41]

The Marine Committee intended that the *Boston* and *Confederacy* cruise off the coast between latitudes 35° and 40° and take, burn, sink, or destroy as many enemy ships as possible. In particular they were to meet and protect American merchantmen from the West Indies, and escort them safely into Delaware and Chesapeake ports. Both frigates were to return to the capes within three weeks.[42]

During the first week of June the two frigates sailed past Cape Henlopen and out into the Atlantic. Within weeks two merchantmen had been shepherded safely into port, and three prizes taken, the most significant of which was the 26 gun British privateer *Pole.* Marine Lieutenant William Cooper of the *Boston* was put on board as prize master, and the vessel sent off to Philadelphia.[43]

Returning to the lower Delaware Bay, the *Boston* and *Confederacy* were met by the *Deane* which had spent more than two months at Philadelphia. During this time repairs were made, but apparently no one was appointed to fill the vacancy left by the death of Marine Captain John Elliott in April, although several vied for it.[44] The Marine Committee ordered the *Confederacy* to proceed up the river as far as Chester,[45] and the *Boston* and *Deane* to sea. Tucker and Nicholson were directed first to the Chesapeake and, if no British ships were to be found, to cruise in such places considered best for intercepting enemy transports destined for New York and homeward-bound West India ships. Both ships were ordered to remain at sea until September and then put into Boston.[46]

The *Deane* and *Boston*, after disposing of the enemy, sailed out of Chesapeake Bay on 29 July in company with two ships of the Virginia State Navy and a convoy of 13 merchantmen, from whom they soon parted. The successful five-week cruise netted eight prizes, including the sloop of war *Thorn* of 16 guns. The most important prize was the *Earl of Glencarron*, a Glasgow ship bound for New York with a valuable cargo of dry goods. Also found on board were plates, type, and a large quantity of paper for counterfeiting Continental money. It was the British intention to further depress an already depreciated and discredited currency by flooding the states with bogus bills. The frigate returned to Boston on 6 September with two of the eight prizes.[47]

The *Ranger* stayed at Portsmouth from April, when she returned from the cruise with the *Warren* and *Queen of France*, until 13 June. Under persistent urgings by the Navy Board, Captain Simpson brought the sloop to Boston. By mid-June, the *Providence*, *Queen of France*, and *Ranger* were ready to put to sea on a joint cruise. Orders were issued to Captain Abraham Whipple, senior officer of the group, to proceed without delay and cruise to the south of the Newfoundland Banks for the purpose of intercepting enemy transports and merchantmen destined to or from New York and the West Indies. When provisions were exhausted Whipple was allowed the discretion of returning to a convenient Continental port or replenishing in the West Indies and cruising in that area during the winter.[48] The *Providence*, *Queen of France*, and *Ranger* stood out on 17 June.[49]

While wrapped in a Newfoundland fog on 18 July, the sound of ships bells was heard. As the overcast began to lift, Whipple's squadron found itself in the midst of a Jamaica fleet of about 60 sail under convoy of a British 74 and several smaller vessels. The *Queen of France* was virtually alongside a large merchantmen. Under the pretense of being a British frigate, she sent a boat to the English ship and quietly took possession of her. Without arousing suspicion, the Americans took 10 ships in the same manner before retiring at nightfall.[50]

Eight of the 10 prizes safely reached port; two were retaken. The captured merchantmen were heavily laden with rum, sugar, coffee, and cotton, later valued at over a million dollars. Upon the squadron's return to Boston on 21 August, all hands shared handsomely in the good fortune. Seaman Andrew Sherburne, for example, received:

> One ton of sugar, from thirty to forty gallons of fourth proof Jamaica rum, about twenty pounds of cotton, and about the same quantity of ginger, logwood and alspice, and about seven hundred dollars in paper money, probably worth fifty dollars in specie.[51]

The squadron's Marine privates each received approximately the same share as did Seaman Sherburne.

At the time Whipple's squadron left on its cruise in June there remained anchored in Boston harbor three Continental ships: the frigate *Warren*, Captain Dudley Saltonstall; sloop *Providence*, Captain Hoysteed Hacker; and brig *Diligent*, Lieutenant Philip Brown. Before the summer was out, the Marine detachments on board each of the three ships would be severely tested at Penobscot, Maine, and their conduct found to be exemplary.

The Penobscot Expedition[52]

By early 1779, the improving naval situation in America was beginning to tell on British shipping. Privateers, especially those operating from Boston, were becoming increasingly expert in intercepting Crown shipping between New York and Halifax, Nova Scotia. These brazen attacks on military transports forced the British into a convoy system, siphoning off warships to protect the convoys while at sea. The long distance between New York and Nova Scotia prompted the British to seek out a protected anchorage, accessible to the sea lanes, from which aggressive patrols could be mounted against the raiders.

The coast of Maine had several advantages. First, its rocky shoreline was pierced by a number of estuaries, many offering suitable harbors. Secondly, it was believed that the eastern provinces of Massachusetts Bay, as the current state of Maine was then known, contained a large percentage of loyalists among the population. It was hoped that this population would assist the British in establishing a site, or at very least, not actively oppose them. Finally, command of the Maine coast would deny the colonists access to the forests upon which they depended not only for naval timber, but also for firewood and construction lumber. Although not absolutely necessary to the British naval effort, these resources were more accessible in Maine than at many other sites and could therefore be economically exploited for the Crown, while denying them to the colonies. Indications are that another element in British thinking was the possibility of establishing a new colony in Maine as a refuge for loyalists from the colonies. This new colony, subsequently referred to as New Ireland, was to occupy the area between the Penobscot and St. Croix Rivers.

Orders were relayed from London through the commander-in-chief of British forces in the colonies, General Sir Henry Clinton, to Brigadier General Francis McLean, commander of military forces in Nova Scotia. McLean was specifically instructed to establish, on the Penobscot River, "a respectable Work capable of containing 3 or 400 Men, with Magazine & Barracks compleat."[52] Clinton suggested a force of approximately 500 men, but left the final determination to McLean.[53] Clinton enclosed instructions for delivery to whichever officer McLean might designate to command the expedition. Among the provisions was one under which the inhabitants of the area were to be offered grants to lands they then cultivated, in return for taking a loyalty oath to the Crown.

The British force was initially to be made up of 400 men from the 74th Regiment (Argyle Highlanders), and 100 men from the King's Orange Rangers, the latter unit composed of New Jersey Loyalists.[54] By late May, however, McLean had changed his mind on two counts. First, he decided to enlarge the force to 640 men. The 74th would provide 440 under Lieutenant Colonel John Campbell of Barbreck. Since the Orange Rangers could not provide a total of 200, they were detached from the force and the 200 drawn from McLean's own regiment, the 82d. Secondly, McLean decided to command the force himself, and made provisions with the comamnder of British naval forces at Halifax, Captain Andrew Barkley, for the necessary transports and escorts. The convoy departed on Sunday, 30 May 1779, under the watchful eyes of the *Blonde*, 32, *North*, 14, *Nautilus*, 16, *Albany*, 14, and *Hope*, 6.

The ships arrived in Penobscot Bay on Saturday, 12 June, after an uneventful voyage.[55] McLean spent the next three days reconnoitering the banks of the river to locate the best site for the fort. Finally he settled upon the peninsula known variously as Magebiguiduse, Biguiduce, or simply Bagaduce. The peninsula extends into the bay from the eastern shore, approximately as far as the northern tip of Isleboro, then Long Island. Connected to the mainland by a very narrow neck, it measures two miles east to west, and about one mile north to south. At the time the British arrived, it was almost completely covered with a dense, virgin evergreen forest, largely fir and pine. A protected bay opened to the south, below the present site of Castine, its far shore being formed by the northern coast of a neighboring peninsula. McLean picked a high plateau near the middle of the peninsula from which cannon could command the bay as the site for the fort. The western (river) side of the peninsula was obscured by forest, but it dropped precipitously to the shore some 50 to 75 feet below.

By the time the force had been ashore a week and a half, all supplies had been unloaded and rolled up steep hills from the bay to the fort site. Once the transports were unloaded and ready to return to Halifax, a long, and rather acrimonious debate developed regarding the size of the naval fleet to be left with McLean. Captain Barkley, commander of the convoy, wished to retire with the transports and all warships except the *Albany*, Captain Henry Mowat. His withdrawal assumed a degree of urgency when it was learned that several American frigates were cruising in the area and might be heading for Halifax. That port was largely defenseless during the absence of Barkley's fleet. McLean disputed the matter with Barkley, but with little immediate success.* Ultimately, the *Hope* sailed on 26 June to carry dispatches to New York, and the *Blonde* departed for Halifax on 28 June. The sloops *Albany*, *North*, and *Nautilus* remained behind at Bagaduce with the transports. With the departure of Captain Barkley in the *Blonde*, Captain Mowat assumed command of British naval forces at Bagaduce.

Word spread quickly about the British landing. Apparently the first word to reach the Massachusetts Bay Council was in a letter from Reverend John Murray of Georgetown, Lincoln County. Writing on 18 June, Murray passed on the best information he had at the time, indicating that an army of at least 800, and maybe as many as 1,500, soldiers had landed.[56] Writing from Pownalborough on the 19th, Brigadier General Charles Cushing of the Massachusetts Militia noted the landing and suggested that the Lincoln and Cumberland County militia might be utilized to expel the British, but only on condition that the state council supply all provisions.[57] Within days several other similar letters arrived at Boston.

A Fleet Formed

It was logical that the demand for action should center in Boston. Not only was the affected area part of Massachusetts Bay, but that state's trade and privateering were all directly threatened. The Maine coast had, for decades, offered convenient harbors to escape storms, access to lumber products, fishing waters, and, more recently, a haven for privateers preying on British shipping. It was unthinkable that a state as maritime-oriented as Massachusetts Bay would let such a challenge go unanswered.

Among the first steps taken by the Council of Massachusetts Bay was to appeal directly to the Navy Board of the Eastern Department for assistance in routing the British. The Navy Board in turn advised the Marine Committee, and recommended that Continental ships be assigned to the attempt.[58] Without waiting for Marine Committee concurrence, the Navy Board wrote to the Massachusetts Council saying the Board did "most heartily concur in any proper measures for dislodging the enemy from Penobscot and are preparing the Warren, Sloop Providence, and Brige. Diligent for that purpose."[59] As would be expected, manning the vessels was difficult. At least one hundred experienced seamen were needed for the *Warren*, later to be designated flagship for the expedition's naval commander, Captain Dudley Saltonstall. The sloop and brigantine also needed crewmen.

* The intensity of the two commanders' feelings may be gaged by the careful preparation of a series of letters exchanged in which each verbal discussion is preserved in writing.

On 29 June, the Council of Massachusetts formed a small committee to inform the state of New Hampshire of the force being raised. A letter was sent on the same day to New Hampshire Governor Mesech Weare asking that his council be informed and that appropriate assistance be dispatched.[60]

New Hampshire agreed to send the 20-gun ship *Hampden*, under the command of Captain Titus Salter. The *Hampden* was a privateer, whose owners contracted her to the state especially for the Penobscot expedition. Armed with 6- and 9-pounders, she carried a complement of 130 men.

Also included in the flotilla were three vessels of the Massachusetts State Navy. Originally designed as a sloop, the brig *Tyrannicide*, commanded by Captain John Cathcart, was built at Salisbury, Massachusetts, in the spring of 1775. She was a veteran of several successful cruises, during one of which she was nearly captured, thus requiring a change of rigging upon returning to port. With a crew of 90, she carried twenty 6-pounders. The brig *Hazard* of 18 guns, added to the Massachusetts navy in the fall of 1777, carried a crew of 100, and was also armed with 6-pounders. Lastly, the *Active*, 16, was a British-built brig which had been captured by the *Hazard* off St. Thomas on 16 March 1779. She was, in turn, purchased by the state and, under Captain Allen Hallet, joined the state navy.

These seven Continental and state naval vessels were augmented by 12 privateers chartered by the state:

Ship	Number of Guns	Captain	Crew
Black Prince	18	Nathaniel West	100
Charming Polly	6	John Palmer	20
Charming Sally	20	Alexander Holmes	200
Defence	14	John Edmonds	90
Hector	20	John Carnes	130
Hunter	20	Nathan Brown	130
Monmouth	20	Alexander Ross	100
Pallas	14	James Johnston	80
Putnam	20	Daniel Waters	130
Renown	14	Robert Adamson	95
Sky Rocket	16	William Burke	120
Vengeance	18	Thomas Thomas	140*

In addition, a large number of merchantmen were chartered to carry supplies from Boston, and militia from York, Lincoln, and Cumberland Counties. Among the transports were the following:

Transport	Type	Captain
Abigail	Sloop	William McGlathery
Allen	Schooner	Levi Tower
Bethaiah	Sloop	Edward Freeman
Brittania	Sloop	Robert Johnson
Centurion	Sloop	William McLellan
Defiance	Sloop	Daniel Mitchell
Dolphin	Sloop	( ) Kent
Fortune	Sloop	David Drinkwater
Hannah	Sloop	Meto Sampson
Industry	Sloop	William Young
Job	Sloop	Jonathan Sprague
Nancy	Sloop	( ) Grafton
Pigeon	Sloop	Luther Little
Race Horse	Sloop	( ) Turner
Rachel	Schooner	John Wylie
Safety	Sloop	William Kent
Sally	Sloop	Daniel Carver
Samuel	Ordnance brig	James Brown
Sparrow	Sloop	Samuel Drinkwater
Unity	Schooner	( )**

* This is a composite list, taken from a number of sources, several of which are in conflict. Most agree that the number of chartered privateers was 12, but several list the *Revenge* instead of the *Renown*. The confusion is not easily dispelled as only two of the above ships were not lost. The *Pallas* left before the British arrived, while either the *Renown* or *Revenge* apparently returned to Boston shortly after the convoy arrived off Bagaduce. Because no claim was made for loss, records are incomplete in this case.

** This list is drawn primarily from Volume 145, Massachusetts Archives, as substantiated by and amended by secondary accounts.

Plans called for a force of Continental and state Marines to be augmented by up to 1,500 militia. The militia were to be called from the three southernmost counties of the current state of Maine. Jeremiah Hill, the Adjutant General, went to Cumberland County where he oversaw the call-up and assembly of units from that region. Hill noted in subsequent testimony that the troops "were collected with the greatest reluctance."

Not only was the call-up short of its goal of 600 men, but the quality of those rounded-up left a good deal to be desired. Hill noted, "some sent boys, old men, and invalids, if they belong to the Train Band or Alarm List they were soldiers, whether they could carry a gun, walk a mile without crutches, or only *compos mentis* sufficient to keep themselves out of fire and water."[61] Nevertheless, Hill saw to it that the 433 rank and file who answered the call were marched from Falmouth to Townsend (now Boothbay harbor) where the land forces were to rendezvous with the transports for the final leg of the voyage to Penobscot.

The contingents from York and Lincoln Counties were also understrength. At Townsend, Brigadier General Solomon Lovell, a militia officer of considerable reputation within the state and commander of the land forces on the expedition, was able to count only 873 effectives instead of the expected 1,500. Timing was of great importance—the Americans wanted to get to Bagaduce before the British had time to build a strong fort. Therefore, Lovell opted to maintain the time schedule, sailing from Townsend with the understrength militia units, while instructing their respective colonels to make up the deficits and send them on to Bagaduce as soon as possible.

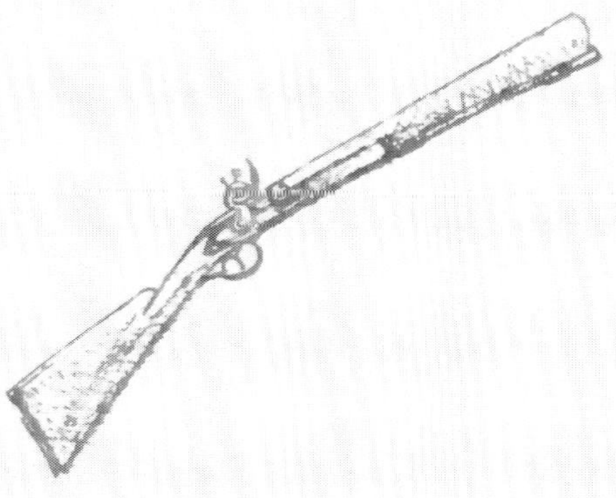

On to Penobscot

In mid-July, small groups of transports and privateers had rendezvoused in Nantasket Roads and sailed for Townsend. By 23 July all units were at anchor off Townsend and men were going on board. The fleet sailed on 24 July. That evening, it was becalmed and anchored off the Fox Islands (now Vinalhaven) at the mouth of Penobscot Bay. Meanwhile, the *Tyrannicide* and the *Hazard* had been ordered ahead as scout vessels. A short distance up the bay, Captain Williams of the *Hazard* sent Second Lieutenant William Cunningham ashore with a party to find any inhabitants of the area who might offer recent intelligence on the enemy. Cunningham returned with three men, all of whom were in turn sent to the flagship *Warren* where Commodore Saltonstall could talk to them. Another boat went into Camden where it picked up Captain Mitchell, an inhabitant of Belfast, who was supposedly well acquainted with the area in which the British were operating.

Shortly after dawn on Sunday, 25 July, the fleet started up the bay toward Bagaduce. At mid-morning, Lieutenant Brown of the Continental brig *Diligent* received orders from Saltonstall to run on ahead of the fleet to reconnoiter the banks and search out the British fleet. Near Bagaduce, Lieutenant Brown's attention was drawn to three men who ran down to the water's edge, waving at the ship. Suspecting a trap, Brown armed a boat crew before sending them toward the men. As it developed, one of the three had recently been near the British lines and estimated the enemy strength at 450 to 500. He also stated that the fort was not even half finished at the time. The men were sent to the *Warren* where Saltonstall was informed of their news. Brown advised the Commodore that the ships should prepare to move in immediately, indicating also that he felt the fort could be easily carried. Displaying, for probably the first time on the expedition, a caution which would later cause the attempt to founder, Saltonstall retorted that "none but a madman would go in before they had reconnoitered, and it would be the height of madness to attempt it."[62]

Saltonstall may be forgiven, in part, because the situation which presented itself at Bagaduce apparently bore little relation to that foreseen in the plans. As noted, the river (west) end of the Bagaduce Peninsula was a steep precipice; the bay (south) side was protected by cannon mounted on the fort and others on Banks (Nautilus) Island in the bay; the other approaches were inaccessible. The basic assault plan, as developed while enroute, indicated that the planners were not prepared for the rugged terrain and hostile warships. The plan indicated that:

> Colo [Joshua] Davis on the Arrival of the Transports at the place of Rendezvous will immediately throw out the flat botom'd Boats and man them in order that the Troops may be landed with Dispatch whenever such Orders shall be given
>
> Major [Daniel] Littlefields Detachment [York County Militia] will land First to serve as an advance party & to cover the Flanks of the first Line and are to be commanded by Colo [John] Tyler, Colo [Jonathan] Mitchells Regt will land next and form Immediatly in the same Order in which they were review'd taking care that the Compy & Boats are properly arranged previous to landing that the Troops may be less liable to Confusion in forming in Case of Opposition. The first Line will be supported by one Field Piece landing next after them, if the Ground permit, and Colo [Samuel] McCobbs Regt landing in the same manner as above mentioned with one Field Piece will form the second Line, he will detach a small party from each flank for a Cover. The first Line and light Corps to be commanded by Brig Genl [Peleg] Wadsworth. The 2nd Line by Colo McCobb. The rest of the Train will be in readiness for landing more Artillery, or acting as they shall receive Orders. as soon as the Troops are landed the Boats will retire to the Transports, so as to be out of danger from the shore in order to supply the Troops with Water and Ammunition in case of an Action. Colo Davis will give particular Attention to conducting the Boats and Transports, in the mean Time the Officers will be careful that every man is compleatly equipt in Arms and Ammunition and that they have drink in their Canteens and a Morcel for their Pockets. The Artillery order'd to the first and second Line of the Army will advance as near the Centre of each Line and as close in the rear as the nature of Things will admit. 'Lt. Col. [Paul] Revere will order a suitable Number of Officers and men to the Pieces order'd on Shore and see them well supply'd with Ammunition, he will apply to Colo Davis for a flat botom'd Boat, for Landing the Artillery and Stores, and have them ready at the shortest notice.' Major [William] Lithgow is appointed to act as a volunteer Aid de Camp to Genl Lovell and is to be respected and obey'd accordingly. The Adjt of each Regt Detachment or Corps is orderd to attend the Hd Quarters at 8 OClock in the morning and Six in the Evening for Orders. The Genl flatters himself that should there be an Opportunity he will have the utmost exertions of every Officer and Soldier not only to maintain, but to add new Lustre to the Fame of the Massachusetts Militia.[63]

By three on the afternoon of Sunday, 25 July, the

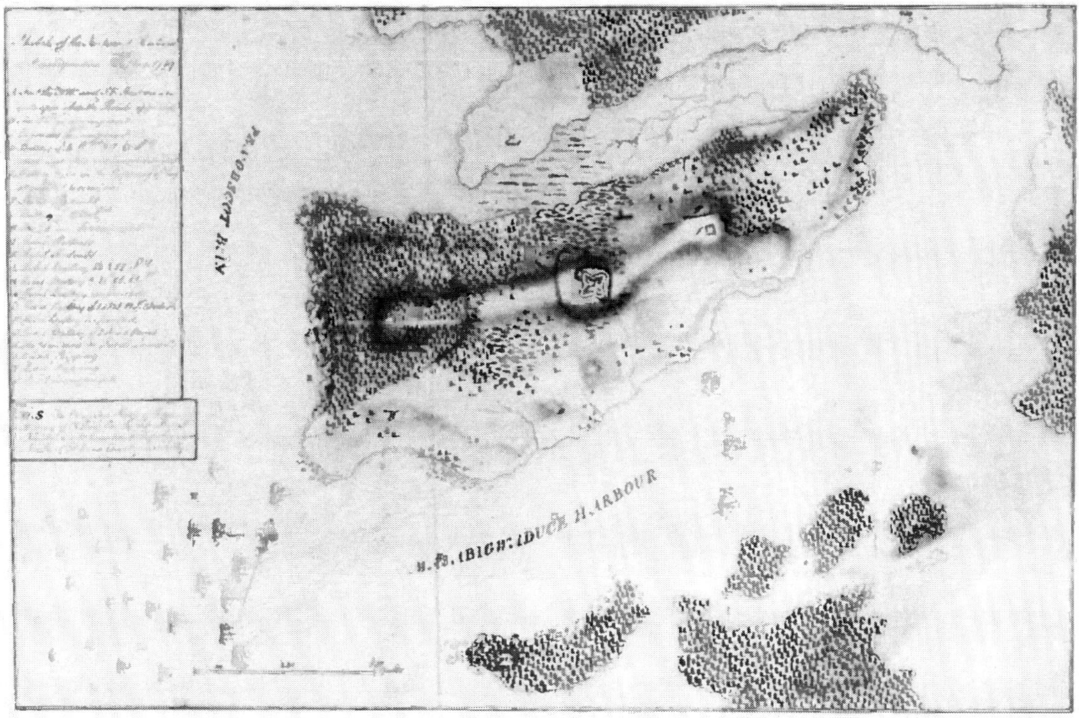

harbor. The transports were kept in the background, American fleet arrived off the mouth of Castine effectively screened by the naval vessels and privateers. The British warships were found to be arrayed in a line across the entrance to the harbor, protecting the approaches to the fort while screening their fleet of transports which rode at anchor farther up the harbor.

American reconnaissance of the fort itself indicated that it was a fortress in name only. "The breast-work or fort," wrote Marine Sergeant Thomas Philbrook some years later, "was rather a rough looking concern, built with logs and dirt, and not more than three or three and a half feet high, which our long-legged militia-men would have straddled over without much difficulty."[84] Nevertheless, the British exploited the manpower on board their ships in the harbor, and utilized, some say impressed, local labor so that shortly after the arrival of the Americans, the site had been strengthened. The walls had been raised sufficiently to permit firing in barbette, while two cannon had been mounted to support the muskets. The lines were closed and chevaux-de-frise defenses had been started on the ramparts of the fort. Finally, the British had stripped the cannon from the starboard side of their vessels, since they were arranged in line across the harbor with their port sides facing outward. These cannon were emplaced at various sites on Bagaduce and on nearby islands and peninsulas.

In preparation for the major assault of the 28th, the Americans undertook a number of probing actions, generally designed to feel out the British defenses. Intermittent, but occasionally brisk, cannonades were exchanged. Under cover of the *Hazard*, *Tyrannicide*, and *Charming Sally*, a landing was apparently attempted at dusk of the first day. The assault was unsuccessful owing to a stiff breeze which sprang up, requiring the recall of the boats. All day Monday, 26 July, shots were exchanged with the three British ships. Little damage was suffered by either side, but the British vessels retired up the harbor a short distance, establishing another line.

At about six Monday evening, a landing was made on Banks Island where the British had erected a position containing several cannon. A council of war held on board the *Warren* earlier in the day had concluded:

That the Marines under the Command of Capt. Welsh should land & take Possession of the Island in the intrance of the River; who are to be Covered by Arm'd Vessels: One of which upon the Lodgment of the Marines, will Hoist English Colours at Main Top Gallant Mast Head—and in case they want assistance to a white Flagg at Ditto—& the attention of the Enemy be diverted at the same time by a Feint of the shipping to enter the Harbour—And the first Division of the Land Force feigning to Land with the Marines; but as soon as the Marines shall appear to be Lodged on the Island, they are to fall up the River, and Land the opposite side of the Peninsula. Further Operations are to depend on the success of the above-mentioned Plan.[65]

The landing was covered by the *Providence*, *Pallas*, and *Defense*. The number of Marines involved is unknown, but one account by a participant, Marine Sergeant Thomas Philbrook, relates that he landed with 30 Marines from the *Providence*, "with as many more from the brig, all under the command of Capt. R. Davis of the continental army."[66] The brig is unidentified, but was probably one of the Massachusetts naval vessels: *Active*, *Hazard*, or *Tyrannicide*.

Upon seeing the approaching American boats, the force of about 20 British Marines on the island retired, leaving behind four light field pieces. The Americans may not even have been aware that the British were actually on the island just before the landing. This is supported by the fact that the Americans worked silently through the night to set up two 18- and one 12-pound cannon. They fired these early the next morning, believing that this was the first knowledge that the British had of the Americans being ashore.

Although there were no casualties on either side during the actual landing, a round fired by a British cannon early on Tuesday landed near the American position, killing two and wounding three others. Later on Tuesday, another chance round from a British battery landed in a boat carrying Major Littlefield, commander of the York County Militia, and two other men, killing all three.

On Tuesday, after the island was secure, the main force was withdrawn, although, Marines from the *Pallas* and *Providence* remained ashore, developing the cannon site and occasionally firing on the British. Indications are that the *Providence*, *Pallas*, and *Defense* remained off the southwest coast of Banks Island for several days, supporting the Marines ashore. A contingent of Marines remained on Banks at least through the landings on the 28th, keeping up a diversionary cannon fire on British positions. By occupying the island, the Marines successfully denied its use to the British and forced them to move their vessels still farther up the bay to escape the battery's fire.

By Tuesday evening, a fundamental disagreement seems to have developed between the naval and army commanders. Basically, the issue revolved around the order in which the British forces would be destroyed. The army officers favored an attack by the armed ships on the three British vessels in the harbor. It was felt that, if these three ships could be destroyed, a safer and more advantageous landing could be made on the harbor side of the peninsula where the ground sloped somewhat more gently to the water. Such an attack on the British ships could be mounted by the numerically superior American fleet, assisted by the battery on Banks Island. The naval officers, on the other hand, especially Commodore Saltonstall, refused to endanger their ships by attacking the British fleet under the very guns of the fort. Saltonstall's position was that the army and Marines should land and overrun the fort, after which the fleet could then safely destroy the British vessels. The problem with this arrangement was that the army and Marines would have to make their assault at the base of precipitous cliffs, hardly a desirable spot. The land forces seemed to have lost the argument as the naval captains refused to discuss it further, saying they "think the Question out of their Province; and therefore leave the Matter to the Officers of the Army, and Capt. Welsh."[67]

At the same time, another disagreement was

growing between the senior naval commander, Commodore Saltonstall, and his subordinates on the various ships. Apparently several of the privateer captains were growing impatient over the delay and so formulated and circulated a petition. Addressed to the Commodore, the petition stated:

> That we your Petitioners strongly Impressed with the importance of the Expedition, and earnestly desire to render our Country all the Service in our power—Wou'd Represent to your Honour, that the most spedy Exertions shou'd be used to Accomplish the design we came upon, We think Delays in the presant Case are extremely dangerous: as our Enemies are daily Fortifying and Strengthning themselves, & are stimulated so to do being in daily Expectation of a Reinforcement—We dont mean to Advise, or Censure your past Conduct, But intend only to express our desire of improving the Present Opportunity to go Immediately into the Harbour, & Attack the Enemys Ships—however we humbly submit our Sentiments to the better Judgment of those in superior Command—therefore wait your Orders Whether in Answer to our Petition, or otherways—[68]

The petition contained the signatures of 32 officers from 11 ships.

At a council of war held on board the *Warren* on 27 July it was decided to proceed with the landing before attacking the British ships. Approximately 850 militia would be available to support 227 Continental and Massachusetts Marines. Eighty of Colonel Revere's cannoneers would also be available. Scheduled to begin at midnight, 27 July, the landing would be supported by all available naval and privateer vessels. The assault force was divided into three units or divisions. The Marines were to land on the right under the command of Captain John Welsh, senior Continental Marine officer. The second division was composed of Colonel McCobb's Lincoln County Militia with Revere's artillerymen, now armed with muskets, in reserve. The third division, on the left, was the Cumberland County Regiment under Colonel Mitchell.

Extended delays were experienced in transferring the landing force to the flat-bottomed boats. The situation was aggravated by the fact that many men had had little or no sleep the previous night. Captain Philip M. Ulmer's company of Knox County Volunteers, assigned to Colonel McCobb's sector, was kept standing for hours in boats so crowded as to preclude sitting down.

By three in the morning, 28 July, several hours behind schedule, the naval vessels were arrayed in position off-shore just out of musket range, each offering a broadside to the enemy. Aligned directly off the beaches were the *Hazard*, *Active*, *Tyrannicide*, and *Charming Sally*. Once in position, the order was given "to begin to fire into the Woods with an Intent to scower them of the Enemy, which was Immediately obey'd."[69]

Storming the Heights

As the four ships opened up, the *Warren* immediately engaged the British ships from long range, forcing them to move still farther up the harbor. Other ships commenced firing with round and doubleheaded shot; the loud report of the guns was echoed by the crashing of trees and the fall of branches.

At first light, shortly after five, and under partial cover of fog, the boats approached the shore. British defenders still on the heights initiated a heavy fire of muskets, the balls falling all around the boats. The attackers leaped ashore and milled about in some confusion prior to seeking cover. The soldiers had been instructed to assemble in line of battle once ashore, but the heavy fire from above, together with the nearly perpendicular precipice, prevented such a formation. Moving in small groups, they started to climb, hand over hand, unable to fire their weapons as their hands could not be freed during the climb.

Undoubtedly the stiffest opposition, by both American and British accounts, was that faced by the Marines on the right. The heights on the center and left were defended by relatively green troops of the 82d Regiment and were overrun with little difficulty by the Americans, once the precipice was scaled. On the right the Marines faced a rear guard of 20 men protecting the harbor battery. This group, under the command of Lieutenant (later Sir) John Moore, seemed at first to be about to retreat with the others. Rallied by the young lieutenant, they held and returned the Marines' fire, finally at very close range. Among the first Marine casualties were

Captain Welsh, apparently killed by a single musket shot, and his lieutenant, William Hamilton, who lay at the bottom of the cliff severely wounded.[70]

At last, threatened by encirclement, Moore's troops withdrew toward the fort, seven men having fallen in the exchange. The Americans then regrouped and counted casualties. In addition to Welsh and Hamilton, 32 had either been killed or wounded, among them eight Continental Marine privates from the *Warren*.

Several Marines under Lieutenant William Downe, commander of the Massachusetts Marines on board the *Tyrannicide,* advanced to the edge of the woods near the fort, but were stopped as they had not enough support. The British seemed to have expected that the Americans would exploit their victory by immediately attacking the fort, for General McLean stood at the flag pole, ready to lower the colors after offering at least the semblance of opposition to the numerically superior Americans. Such was not to be.

The naval vessels should, at this point, have entered the harbor and attacked the British ships, but they failed to do so, largely due to the hesitancy of Commodore Saltonstall. Without removing the threat of reinforcement and flanking fire from the ships, the ground commanders were reticent about trying to carry the fort. As a result, the Americans halted their advance at the top of the precipice, working quickly to set up defense lines around it. The initial thrill of victory was soon dispelled by a confusion of orders and drills. Orders went out to land the cannon and this project occupied a good number of the troops ashore. Late in the afternoon, the *Warren* and three brigs started toward the entrance of the harbor, there to fire on the British ships, but they seemed to have hauled up prematurely. In any case, during a sharp exchange, the expertise of the British gunners was clearly demonstrated. The *Warren* suffered two rounds through the mainmast, one in her bowsprit, and another through the forestay. This damage, suffered in less than half an hour, led to Saltonstall's decision to back away and anchor off shore. Thus, after a brave and spirited assault, against almost inaccessible enemy positions, the Americans bogged down, desparately in need of a forceful commander.

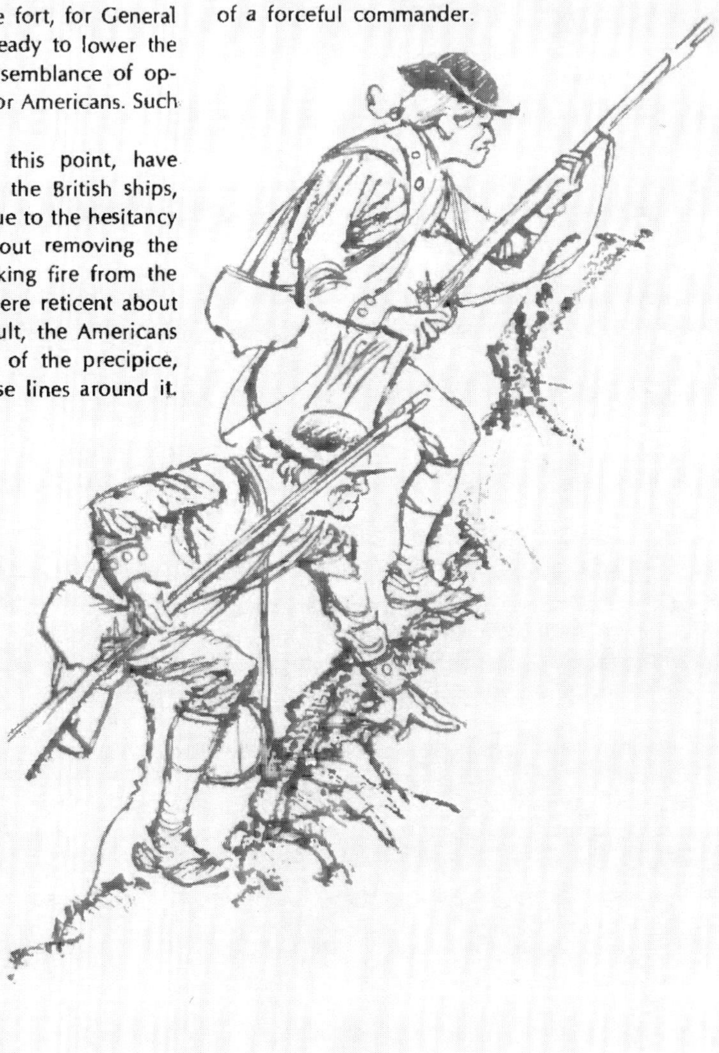

Prospects of Victory Fall Away

The next few days were generally uneventful. The Americans landed their cannon and set them up atop the hill. They entrenched their lines in order to be protected from the direct fire of the British cannon in the fort. Occasional reconnaissance parties went out to survey the British positions and probe their lines. Throughout the period, the American side adopted a defensive posture.

During a council of war on 29 July, it was decided to erect a fort on the ground held by the Americans, there to better cannonade the British. The work was to be put under the command of Captains Salter of the *Hampden* and Thomas of the *Vengeance*, with eight men and one officer from each ship to do the work. This may, in part, have been an attempt to keep the men occupied, as there are clear indications that the men were becoming unruly and hard to control. Most were only peripherally acquainted with military discipline; essentially civilians, they were not accustomed to military life. As early as 28 July, the same day as the assault, General Lovell had to remind the soldiers and Marines ashore that none were permitted to leave the lines without his personal written permission. Two days later, General Lovell's orderly book notes:

> The General is much alarm'd at the loose and disorderly inattentive Behaviour of the Camp since the Arrival of the Troops at this place and is sorry to think that their Honor so lately acquired is already endanger'd by scattering abroad from the Lines to prevent the impending Danger, he positively orders that no non-commission'd officer or Soldier presume to be more than twenty Rods absent from his Lines on any Pretence whatever without particular Permission from the Commanding Officer of the Detachments to which he belongs . . . all non commission'd Officers and Soldiers are strictly forbid to fire their Guns in such a loose unsoldierlike manner as has been practiced of late . . . The Commanding Officers of Picquits will give such Orders to his Centerioes as to prevent all non commission'd Officers and Soldiers from going below them.[71]

By 5 August the situation had further deteriorated. Morale was low, no plans were in effect to take the offensive, and the possibility of British reinforcements arriving became greater by the day. In an attempt to get things moving again, General Lovell addressed a written request to Commodore Saltonstall in which he asked whether the ships would enter the harbor and destroy the British warships. Saltonstall held a council of war on board the *Warren* on 6 August at which it was decided to answer Lovell in the negative. The ships would attack only when the general agreed first to storm the fort.[72] Lovell was informed of this at once and he announced it at a council held for the army officers. This council unanimously agreed that the fort would not be carried under these circumstances.[73]

Lovell had been keeping the authorities in Boston informed of the progress, or lack thereof, a practice which Saltonstall had been ignoring. The Navy Board wrote the Commodore on 12 August, noting his silence and urging him to act immediately "as a Reinforcement are probably on their passage at this time."[74] The Massachusetts Council, on the other hand, took more decisive action by requesting reinforcements from General Horatio Gates, then at Providence, but it was at most a gesture of desperation as the time involved in moving a body of troops would have been prohibitive.

At this point in the siege, events seemed to have assumed a course of their own, bound inexorably toward tragedy. Councils of war were held daily after 9 August, but they invariably recessed with no agreement. Relations between the land and naval forces deteriorated to a point where the two groups frequently held separate councils.

By Friday, 13 August, the siege had been in effect for almost three weeks and the British side seemed to be getting stronger, at least the fort itself was growing more formidable by the day. The last in a series of assault plans was drawn up, but, before it could be put into effect, a heavy fog lifted and enemy sails were detected entering the lower bay. The discovery was made by the *Active* and *Diligent*, both of which had been cruising off the bay in order to detect just such an intrusion. The *Diligent* immediately ran up the bay to sound the alarm.

Word of the American expedition to Penobscot had been received at New York on 28 July. A powerful naval squadron under Sir George Collier consisting of the 64-gun ship *Raisonable*, the *Blonde* and *Virginia*, both 32-gun frigates, one 28, the *Grayhound*, two 24s, the *Camillia* and *Galatea*, and the 14-gun sloop *Otter* departed 3 August for Maine. Although numerically inferior to the American fleet, the squadron was heavily armed and its 1,530 men experienced.

Luckily for the Americans, a rain squall came up, followed by more fog and then darkness. During this brief respite, all men and cannon were taken off the shore and loaded on board ships of the fleet once again. At daybreak the Americans tried to get underway in order to form a fighting line, but the tide was against them. As the first British line drew closer, the American ships broke and ran. Most of the armed vessels turned upstream, trying to seek safe haven in the small inlets along the upper river. The transports also turned upstream, but many were intercepted before they could get very far. By nightfall, most American ships had been either destroyed by their own crews, or were captured by the British. On the morning of 15 August, the few remaining vessels were blown up by their crews as British search parties neared their hiding places.

The disaster was complete by the 16th. Marines, sailors, and militia were fleeing through the Maine wilderness toward home. The entire American fleet was smoldering on the banks of the Penobscot River. During the first week of September, the weary participants began filtering into Boston.

The destruction of the fleet was a devastating blow to the State of Massachusetts. Except for the three Continental ships and the one from New Hampshire, the state had agreed to indemnify the owners for losses suffered. In addition, the state lost its entire navy. Total charges to the state for shipping losses exceeded one million pounds sterling. This was but a fraction of the total cost for the expedition, as provisions, payrolls, and other costs were not included. Governor Weare of New Hampshire subsequently estimated the cost to the Continental government at four million pounds, while that of Massachusetts was set at seven million.[75]

Courts marital were held in Massachusetts after the fiasco, and an attempt was made to determine what went wrong and why. Generals Lovell and Wadsworth were exonorated of all blame. Commodore Saltonstall was tried by a board consisting of the Navy Board commissioners, as his commission was in the Continental Navy. Found guilty, he was declared unfit to again command a Continental vessel. Paul Revere's conduct was examined in some detail but he seems to have cleared himself to the satisfaction of his peers.

On at least one occasion in subsequent years, the issue of sending another fleet to Penobscot was discussed, but serious consideration was never given the idea. The British remained at Penobscot for some years, finally evacuating of their own accord.

In retrospect, the expedition seems a disaster for most of those concerned, if only for the material losses incurred. Total American casualties are indeterminate but exceeded 100 and may have approached 400. Overall, it was a demoralizing defeat with a direct impact on subsequent amphibious operations. It was not until the War with Mexico that Marines would again attempt to launch a sizeable amphibious operation. The reasons for the defeat were manifold: divided command, poor communications and planning, ill-trained forces, and insufficiently aggressive leadership. Despite these shortcomings, victory was within grasp on at least one occasion, only to be lost through indecision.

The number of Continental Marines was small, and they were never in a position to determine the outcome themselves. Nevertheless, indications are that they performed admirably when called upon. The successful occupation of Bank's Island and the attainment of the heights at Bagaduce attest to their bravery and determination. Unfortunately, their valor was not sufficient to compensate either for the timidity of certain officers, or for the poor training of the militia.

A Coppy of Return Given In to Major Samuel Nichlas of Men Enlisted by Capt. Robert Mullan, Since Aug. 9th 1779 to Jan 1st 1780.

	Mens Names	[illegible]	[illegible]	[illegible]	[illegible]	Mens Names	[illegible]	[illegible]
1	Abram Lewis Serj.	1				William Warner	1	
2	Jeremiah Mallon		1			Andrew Anderson	1	
3	Richard Kisby	1				James Carroly	1	
4	Francis Dempsey	1				Thos. Paul	1	
5	Edward Butler	1				James Murdock	1	
6	John Smith		1			John Kean	1	
7	John Henry	1				John Coil	1	
8	Andrew Ellice	1				John Woodford	1	
9	Samuel Everfelt	1				John Mackearny	1	
10	John Fife	1				John Miller	1	
11	John Egan	1				Patrick Higgin	1	
12	Joseph Hardunck	1				John Radle	1	
13	Thos. Conner	1				Nichos. Ultzman	1	
14	James Plowmer	1				John Barbro	1	
15	Peter Gurney	1				Thos. Smith	1	
16	Joseph Desaign	1				William Baker		1
17	James Deacon		1					
18	James Bryan		1					
19	Joseph Craft	1						
20	Theodora David	1						
21	James Crampton			1				
22	Robert McGinnis	1				on Board Confederacy with Capt. Boyce		
23	Charles Stephenson	1						
24	Thos. Harden							
25	Nathan Marchal				1			
26	Thos. Mathews				1			
27	Thos. Sappington				1			
28	James Nicholas				1			
29	Thos. Peckworth				1			
30	Richard Coats				1			
31	John Smith				1			
32	John Dougherty				1			
33	Richard Babington				1			
34	John McCashion				1			
35	Thos. Sellords				1			
36	George McCray				1			

Captain Mullan's Muster Roll.

"have a Company of Marines raised"

In the late summer of 1779, the number of British naval prisoners at Philadelphia was growing at an alarming rate. With no facility to hold them, Congress resolved that the Marine Committee be directed "herewith to cause the crews of vessels captured from the enemy, to be confined on board prison ships, and supplied and treated in all respects in the same manner as the crews of vessels belonging to these United States and captured by the enemy, are supplied and treated."[76] An appropriate vessel was quickly selected and readied, but guards to man her proved more difficult to provide. After applying to the Board of War "for a guard which could not be furnished," the committee on 5 August "ordered a Company of Marines to be raised for that purpose."[77] Two days later, by order of the Navy Board of the Middle District, Major Samuel Nicholas was directed "to have a Company of Marines raised at this place [Philadelphia] to consist of fifty Men including non commissioned Officers, to be inlisted during the war on the usual Terms and this to be done as expeditiously as possible."[78]

Major Nicholas, who had been assisting the Continental Army as a muster master and quartermaster since the fall of 1777, called upon Marine Captain Robert Mullan to aid in the task. Despite numerous hardships, Mullan raised 28 men by the end of August. Each man was given a bounty of between 100 and 200 dollars, a uniform, and immediately ordered on board the vessel.[79] For the next two months Mullan attempted to recruit the remainder from his quarters at the corner of Pine and Front Streets. Twenty-four additional men were enlisted, but by mid-October 28 out of a total of 52 had deserted. Excluding those men confined, dead, and sick, Mullan was only able to muster 15 guards.[80]

"Altho proper endeavours have been made," reported the Marine Committee, "very few Men have been recruited, nor is there a possibility of being able to Inlist in any reasonable time a sufficient number to guard the P. Ship. Under these circumstances the Committee beg leave to submit this business to the consideration of Congress."[81] Congress, although slow to act, on 12 November empowered the Committee "to discharge the vessel which they were directed to take up and employ as a prison ship."[82] Shortly thereafter, the ship was released and the remaining 15 Marines paid-off.[83] Nicholas and Mullan retired temporarily to private life.

At Year's End

While the two Marine officers were raising the company at Philadelphia, the frigate *Confederacy* lay in the Delaware River at Chester awaiting directions from the Marine Committee. Orders came down from the capital on 17 September informing Captain Seth Harding that his ship had been designated to carry home M. Gerard, French Minister to the United States. Harding was directed to avoid engaging "any vessel of equal or superior force" on the passage, and to make for any port Gerard thought proper.[84] A month later, Harding was informed that in addition to the French Minister he would have yet another distinguished passenger, John Jay, who had been appointed American Ambassador to Spain.[85]

After months of preparation, during which time Harding augmented his crew by impressing seamen and Marines to the dismay of both Continental and Pennsylvania governments, the *Confederacy* cleared the Delaware Capes on 26 October.[86] Smooth seas and pleasant weather were the daily routine for the first 12 days. Between five and six in the morning of 7 November, the *Confederacy* was making nine knots in a brisk breeze and rough sea when suddenly "her bowsprit and all her masts gave way in less than three minutes." After moving injured men below, all hands were put to work in an attempt to clear away the wreakage and rig jury masts.[87]

The next morning, closer examination of the frig-

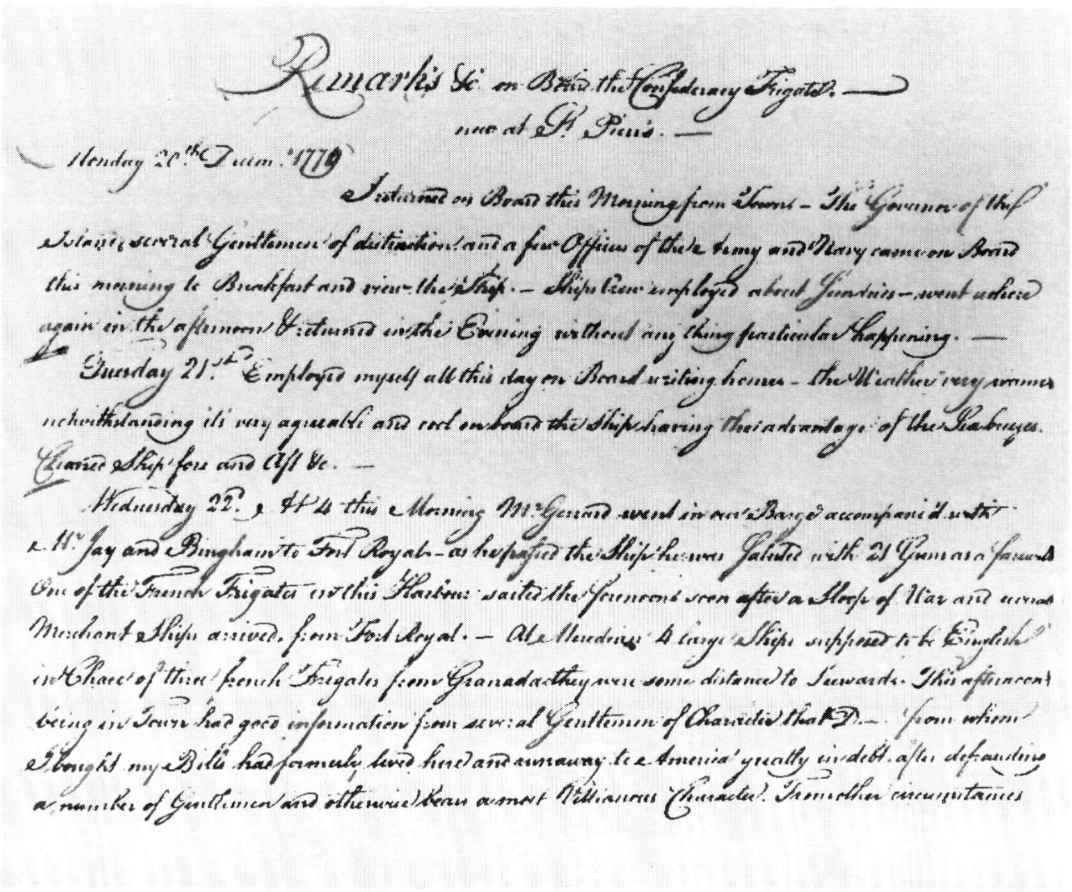

Remarks &c on Board the Confederacy Frigate —
now at St. Piere. —
Monday 20th Decem. 1779
I returned on Board this Morning from Town – The Governor of this Island, several Gentlemen of distinction and a few Officers of the Army and Navy came on Board this morning to Breakfast and view the Ship. – Ships Crew employed about Sundries – went ashore again in the afternoon & returned in the Evening without any thing particular happening. —
Tuesday 21st Employed myself all this day on Board writing home – the Weather very warm notwithstanding its very agreeable and cool on board the Ship having the advantage of the Seabreezes. Cleaned Ship fore and Aft &c. —
Wednesday 22d, At 4 this Morning Mr. Gerard went in our Barge accompanied with Mr. Jay and Bingham to Fort Royal – as he passed the Ship he was Saluted with 21 Guns as a farewell. One of the French Frigates in this Harbour sailed the forenoon soon after a Sloop of War and several Merchant Ships arrived from Fort Royal. — At Meridian 4 large Ships supposed to be English in Chace of three french Frigates from Granada they were some distance to Leewards. This afternoon being in Town had good information from several Gentlemen of Character that D— from whom I bought my Bills had formerly lived here and run away to America greatly in debt after defrauding a number of Gentlemen and otherwise bears a most Villianous Character. From other circumstances

Extract from Joseph Hardy's Journal.

ate revealed a "greater misfortune." The shank of the rudder had twisted and split, causing the frigate to toss and drift aimlessly with the wind and current. Temporary repairs were made, and after discussions with Gerard, Jay, and his own officers, Harding made for Martinique.[88]

After 53 days at sea, the *Confederacy* came to anchor at St. Pierre on 18 December, much to the relief of Marine Captain Joseph Hardy and his lieutenants, Gurdon Bill and Samuel Holt. Captain Harding began refitting the battered frigate at St. Pierre, while the two diplomats resumed their passage to Europe on board a French vessel.[89]

Following successful summer cruises, the *Boston, Deane, Providence, Queen of France,* and *Ranger* were back in New England, again being readied for sea. The *Ranger,* as usual, was at Portsmouth rather than Boston. By late September, the *Queen of France* and *Providence* were in Nantasket Road "with their Guns and great part of their Water & provisions on board Entering men very fast."[90] Within a month and a half, the *Boston* and *Ranger* were in Nantasket Road, fully manned, ready to proceed to sea, and needing only orders to get underway. As the four-ship squadron rode at anchor, Captain of Marines Seth Baxter of the *Boston* resigned his commission and was replaced by Captain Richard Palmes, who for six months had been in disfavor with the Navy Board.

Growing increasingly impatient with delays, by mid-November the four naval captains prepared a joint petition to the Navy Board requesting orders. The drain on provisions and the hazards of holding completely manned ships in port were noted.

Within a day of submitting the petition, orders were received from Philadelphia which directed the squadron to "proceed immediately to Charleston in South Carolina, and there persue the orders of the Commanding Officer at that place."[91] On 24 November the four Continental vessels sailed out of Boston for the last time. Behind in Nantasket Road lay the frigate *Deane*, in want of men and provisions, her two Marine officers having requested discharges, Samuel Gamage for reasons of "infirmities contracted in the service," and Samuel Pritchard because of a desire to engage himself elsewhere.[92]

Continental Marines had had an extremely active and rich year in 1779. At sea they participated in more captures than in any previous year. At Penobscot, however, they received a major setback. Although the Marines performed admirably, the fruits of victory were wrenched from their grasp repeatedly by the lack of aggressive leadership on the part of both the land and sea commanders. The disaster in Maine was to have a profound effect upon their activities for the next half century. Amphibious operations with sizeable Marine participation would no longer be attempted.

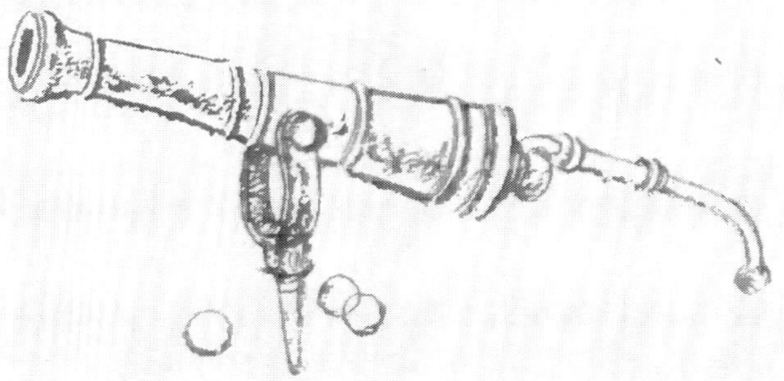

JOHN ADAMS REVIEWS JONES' MARINES, 13 MAY 1779.
This painting recreates John Adams' review of John Paul Jones' Marines after dining with the captain and several others at a stylish L'Orient inn. In truth, the men were part of a contingent of the French Regiment of Walsh who had volunteered for service as American Marines on board the Bonhomme Richard.

CHAPTER XI

Sailing Against England, 1779-1781

Passage to France

The frigate *Alliance*, under Captain Pierre Landais, was brought down the Merrimack from Salisbury to Newburyport and then to Boston at the beginning of August 1778. There the Eastern Navy Board ordered her prepared to receive the Marquis de Lafayette and his party who were to be carried to France. Problems among the captain and his officers prevented the frigate's early departure. Chiefly due to the unwillingness of American officers to sail under a French captain, Landais was forced to request that a number of them be replaced. Captain Matthew Parke of the Marines, as spokesman for the officers, warned Landais that if he ever replaced one he would have to replace them all. Landais backed down. This animosity was to continue until the naval captain's relief the following year.

The ill-feeling which existed between the French captain and his officers soon spread to the few crewmen on board and had an immediate and profound effect on the recruitment of others. Rumors were circulated that Landais had treated his crew with disdain, causing enlistments to falter. Faced with the superior inducements offered by privateers, they dropped still further. Short of the needed men to work the 36-gun frigate, 20 seamen were drafted from the *Boston* as were 30 sick Frenchmen from Admiral Comte d'Estaing's squadron. To further bolster sagging enlistments, a number of British prisoners who wished to escape confinement were taken on as Marines.

By early January 1779, the *Alliance* was ready; Lafayette and his party were on board, and the needed crew members obtained. On the 14th the frigate sailed with the *Deane*, which parted company two days later. The voyage across the Atlantic was relatively smooth, but not without incident. Two Swedish vessels were captured and the frigate lost her main topsmast in a storm. During the early morning hours of 2 February, mutiny was discovered among the English and Irish crewmen. All hands were immediately called on deck and held while their belongings were thoroughly searched for weapons and other evidence of the conspiracy. Nothing was found. A court of inquiry was then convened to question the supposed ringleaders, among them John Savage, master-at-arms, and William Murray, sergeant of Marines.[1] Murray in his first appearance before the court refused to acknowledge the existence of a plot, but later recanted. "Savage and he, with 70 more," he told the court, "had agreed to take the ship and carry her into some part of England or Ireland, and force one of the Lieutenants to take command of her." Continuing, the sergeant described the elaborate plans to seize control of the frigate, and the punishments to be meted out to the officers:

> They were to divide themselves into four divisions, the first to take the magazine, the three at the same time to force the cabbin, wardroom, and quarter-deck, then to take command of the arm-chests, and in case of opposition, they were to point the fore-castle guns aft and fire them, the guns being 9 pounders and all loaded. The party that was to go to the magazine were to kill the Gunner, Carpenter and Boat-swain; the other punishments for the other officers and French gentlemen were thus: Captain Landais was to be put in irons and sent in the cutter, without victuals or drink; the Lieutenants were to walk overboard on a plank from the ship side,

unless they would take charge of her and navigate the ship into England; the marine officers and the Doctor were to be hanged, quartered, and hove over-board; the sailing Master was to be tied up to the mizenmast, scarriefied all over, cut to pieces, and hove over-board.

Lafayette was to be put in irons and sent to England. As a result of Murray's disclosures, 38 of the mutineers were placed under arrest, and a heavy Marine guard mounted about the frigate to prevent further outbreaks.[2]

The *Alliance* anchored at Brest on 6 February after a passage of 23 days.[3] There she waited nearly a month for much needed repairs on her mast and new sailing orders. Early in April, John D. Schweighauser, American commercial agent at Nantes, ordered Landais southward to the Loire River port where John Adams had arranged for an exchange of British and American prisoners.[4] While at Nantes, Adams boarded the frigate in hope of returning with her to America, but new orders came for Landais. The *Alliance* was to join John Paul Jones at L'Orient.[5]

With a fair wind, the American frigate sailed from the Loire on 10 May. The following day she anchored off Isle de Groix after experiencing a difficult night navigating the French coast. On the 12th, the *Alliance* warped past the fortress at Port Louis and came to anchor off L'Orient. Soon after their arrival, Landais and Adams called upon Jones who was then on board his new ship, the *Bonhomme Richard*. There they learned that the *Alliance* had been placed under Jones' direction, a disappointment for Landais who desired an independent command, and for Adams who expected to return to America on board the frigate.[6]

An Elegant Dinner

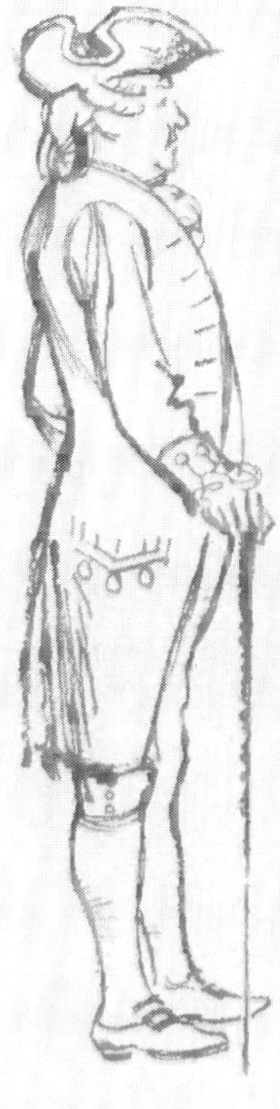

The following day Adams, Landais, and 16 other officers and gentlemen were given an "elegant Dinner" by Jones at L'Epee Royale on the L'Orient waterfront. The conversation, according to Adams, was "not very instructive," with the exception of one good piece of advice which he noted down in his diary, that there were two ways of learning French, "take a Mistress and go to the Commedie." Apparently this observation came from Jones who seems to have had a taste for both forms of entertainment. Surgeon Brooke of the *Bonhomme Richard* jokingly asked Adams for his opinion as to which method he preferred. In the same vein, Adams replied that he favored both, but "the Language is no where better spoken than at the Comedie."

After dinner Adams and Landais were escorted outside the inn by Jones to view the commodore's Marines. According to the American Commissioner, they were "dressed in the English Uniform, red and white. A Number of very active and clever Serjeants and Corporals are employed to teach them the Exercise, and Maneuvres and Marches &c." Apparently Adams left unimpressed because the Marine uniforms were not in accordance with those "established by Congress . . . red and white instead of Green."[7] However, the Marines, particularly the officers, were wearing their proper uniform of a red coat with white waistcoat and breeches. They were neither American nor French, but Irish, being members of the Infanterie Irlandaise, Regiment de Walsh-Serrant.* The dinner and review also appeared to have reinforced his opinion of Jones whom he had considered impertinent and ambitious almost a year before. The most trying year of the commodore's life seemed to have changed him little, thought Adams.[8]

A Ship at Last

Captain Jones spent more than a year on shore, beset with every sort of vexation and disappointment after his return to Brest in May 1778. His drafts on the American Commissioners for the support of his crew and the refitting of the *Ranger* were not honored due to the lack of funds. Therefore Jones himself had to borrow heavily in order to pay these necessary expenses. Also during this time he began the long and weary wait for another and larger ship. He had, as we have seen, tried to obtain *L'Indien* before he set sail for Whitehaven. Following his return he renewed these efforts with Franklin in Paris and at Passy. Franklin assured him on 1 June that he positively would have the frigate. On the 10th he wrote that all was "settled."[9] Yet the Dutch government, anxious to maintain its neutrality in the hostilities between Britain and France, refused to allow *L'Indien* to leave Holland.

John Adams, by Charles Willson Peale.

Other schemes were proposed, among them the command of a squadron of French ships at St. Malo destined to cruise the Baltic, but due to the natural jealousy of French officers, and other difficulties, every plan fell through. After months of diplomacy, exasperation, and hard work, Jones himself determined to go to Versailles and personally urge his own cause, taking the advice offered in *Poor Richard's Almanac:* 'If you want your own affairs to prosper, go yourself, if not send someone.'[10] Because of Jones' personal intervention, the French Minister of Marine, de Sartine, at last took an interest in his affairs, and the result was the purchase of an East Indiaman called the *Duc de Duras* in January 1779.

The vessel was 14 years old, unsound, and a dull sailer, having made several voyages to China and India under the Compagnie des Indes, the Crown, and private owners. But Jones, eager to command, accepted the *Duras*, and at once began the arduous task of converting her into a man-of-war. His first concern was to find enough guns of sufficient caliber to make the *Duras* a proper fighting ship, therefore he journeyed to foundries of Angoulême where he contracted for the needed ordnance: six 18-pounders, twenty-eight 12-pounders, and six 9-pounders. Meanwhile, with the consent of Sartine, and in honor of his good friend and benefactor, Franklin, the old East Indiaman was renamed the *Bonhomme Richard*.[11]

In fitting out the *Richard* there was much to be done, for his first orders had intended her to be the flagship of a proposed amphibious expedition

* The uniform of the Walsh Regiment was a red coat with yellow collar, blue cuffs and lapels, white waistcoat and breeches.

with Lafayette against the English city of Liverpool. To accommodate the 1,500 infantrymen expected to participate, the decks were strengthened and enlarged, and a roundhouse built on the quarter deck to house the general and his staff. Before long, Jones' command was enlarged by the addition of several more vessels. The new American-built frigate *Alliance* was assigned to Jones by Franklin, soon after her arrival in Europe. The French government gave him three other armed ships. The most important was the frigate *Pallas*, armed with twenty-six 9-pounders, and commanded by Captain Denis-Nicolas Cottineau de Kerloguen. Next in importance was the brig *Vengeance* (Lieutenant Philippe-Nicolas Ricot), lightly armed with twelve 4-pounders, and last, a smart, fast cutter named *Le Cerf* (Ensign Joseph Varage), armed with two 8-pounders and sixteen 6-pounders. All three French officers were given commissions in the United States Navy, but their ships continued to be maintained by the French government at no expense to America.

In May the planned descent upon Liverpool by Jones and Lafayette was abandoned. Instead an invasion of England was proposed, for which a large French and Spanish fleet was collected. For his part, Jones was to make a diversionary operation against Scotland, thus pinning down enemy forces that might oppose the main landing in southern England. This scheme too was given up, and it remained now for the American squadron to cruise independently.[12]

With a large ship to man, Jones was driven to near exasperation by the problems inherent in enlisting a crew. Not so with the selection of officers. Jones had good men to choose from among the naval lieutenants and privateer officers stranded in France or released from English prisons, but for officers to command his Marines he went to the Infanterie Irlandaise, Regiment de Walsh-Serrant in the French Army. There he engaged two lieutenants who were delighted to take on the British. Recommended by Captain Fitz-Maurice of the regiment, Lieutenants Eugene Robert MacCarthy and Edward Stack thought themselves the proper officers to command and discipline Jones' heterogenous group of Marines.[13] In late June, another lieutenant from the same regiment, James Gerald O'Kelly, joined his two

A View of L'Orient, France.

Infanterie Irlandaise, Regiment de Walsh-Serrant

comrades on board the *Richard*.[14] All three were issued Continental Marine Commissions by Franklin, signed by John Hancock.

Although Jones was willing to go outside the circle of American officers to appoint Marine lieutenants, he was at first unwilling to do so when it came to the selection of a Marine captain. Appointed, but not commissioned, was Alexander Dick, who John Adams described as of a "good family and handsome fortune in Virginia."[15] But as Captain Dick later wrote American Commissioner Arthur Lee:

> I could never brook the insults to which a Marine Officer is exposed; Capt. Jones I believe is a brave & experienced officer, but sea officers in general contract such an insul[t]ing manner of behavior that it is impossible for an Gentlemen of Spirit to serve under them; I boast myself to be a Virginian & it shall never be said I le[t] any man insult me with impunity: I have theref[ore] left the Poor Richard.[16]

Selected to replace Dick were two French lieutenant colonels, Paul de Chamillard de Varville and

Antoine-Felix Wuibert de Mézieres.[17] Both were invaluable to the commodore, but neither was granted a Continental Marine commission.

The Marine officers of the *Bonhomme Richard*, accordingly, were a more homogenous group than the company they commanded, and, as volunteers accepted by Jones, they were personally loyal to the commodore.

As for the crew, including the Marines, Jones himself observed that they were "one of the worst crews ever found on a vessel . . . . They were generally so mean that the only expedient I could find that allowed me to command was to divide them into two parties, and let one group of rogues guard the other."[18] Yet the crew was to be one of the best he ever commanded. Instead of the crew of spoiled, hometown boys he found on board the *Ranger*, he now had a crew of seamen and Marines who hungered for a fight. They numbered, as the commodore wrote Sartine, "380 officers, men and boys, inclusive of 137 Marine soldiers, 36 Landsmen and 32 Boys."[19]

Preliminary Cruise

Jones' squadron was ready to sail by early June, but before starting on his diversionary operation against England, he was ordered by Sartine to escort French merchant vessels to Nantes, Rochefort, and Bordeaux. On the 17th, the *Bonhomme Richard* and her four consorts sailed out of L'Orient and anchored off Isle de Groix. Two days later the squadron moved southward to perform the temporary escort duty. The first night out, when only a few miles off the coast, the *Richard* and the *Alliance* ran afoul of each other, carrying away the *Richard's* jibboom and the *Alliance's* mizzen-mast. Lieutenant Robert Robinson, the officer of the deck at the time, was court-martialed and dismissed for the accident, but the real culprit, according to Jones, was Captain Landais of the *Alliance* who refused to yield right of way to the flagship.[20]

The next evening the cutter *Cerf* captured a 14-gun sloop, but was obliged to abandon her with the approach of a superior enemy force. On 29 June, the *Bonhomme Richard* fell in with two British frigates off Isle de Groix. Jones immediately put about, but the two warships "fled with precipation." During this encounter, Jones discovered to his chagrin that the *Richard* was not fast enough to force an enemy vessel to fight against its will. He did, however, have "flattering proof of the martial spirit of . . . [his] crew."[21]

When the squadron returned to L'Orient on 1 July, Jones found new orders from Sartine awaiting him. Countersigned by Franklin, the directive authorized the commodore to conduct a commerce-destroying cruise north of the British Isles, and about the middle of August to put into the Texel, in Holland, from whence he was to convoy merchant vessels to France. Jones, as expected, promptly protested the poor use to which his squadron was to be put.[22] Franklin, somewhat irritated by Jones' complaints, replied:

> For as the Court is at the Chief Expence, I think they have the best right to direct. I observe what you write about a Change of the Destination. But when a thing has been once considered and determined on in Council, they don't care to resume the Consideration of it, having much Business on hand, and there is not now time to obtain a Reconsideration.[23]

So that was that, although Jones retained the notion that the small American Navy should make lightning strikes against the enemy's coastline instead of meeting them on the high seas where the British had the advantage.

While Jones pondered his new orders, trouble was brewing on board the *Bonhomme Richard*. The Marines and sailors came to blows,[24] and an incipient mutiny was discovered among the English prisoners who had signed on to escape confinement. As a result, a series of courts-martial were held. The ringleader, Quartermaster Robert Towers, got 250 lashes "on his bare back with a Cat of nine Tails at the Gangway."[25] Others were given lighter sentences. At the request of Jones, the 100 remaining English prisoners who had been enrolled were discharged as untrustworthy and replaced by American prisoners lately released from English prisons and an assortment of Portuguese, Danish, and Dutch sailors.[26]

A few days before sailing, Jones issued instructions to his captains requiring them to pay special attention to his signals and obedience to his orders. They were to maintain their stations, and never "loose Company with [the] Squadron." Sealed orders were given to each ship captain appointing a rendezvous in case of separation.[27] In addition Jones, at the request of Le Ray de Chaumont and against his own better judgment, executed a Concordat between himself and the other four naval captains. It gave each a degree of independence and provided for the control and distribution of prize money.[28] This last provision of the agreement created trouble later because the French, who were providing all expenses, thought they should have a say in the disposal of prizes.

An Ambitious Cruise

With the addition of two French privateers, *Le Monsieur,* 38 guns, and *La Granville,* 12, the squadron sailed from the Groix Roadstead on 14 August. For the first several days they encountered nothing but pleasant weather and moderate winds. On the 16th, a vessel from London bound for Madeira was intercepted but not detained.[29] On the 18th, the *Monsieur* took a prize. According to the Concordat it should have been shared by the whole squadron, but Captain Guidloup refused and the next day deserted the fleet. A large ship was chased on 19-20 August, but she eluded capture. On the 21st, the brigantine *Mayflower,* bound from Limerick to London with beef and butter, surrendered after two shots and was sent to L'Orient with a fiye-man prize crew.[30]

Mizen Head, Ireland, lay abeam at eight in the morning on 23 August, and with "Pleasant Weather & a fine Breese of Wind," the squadron stood to the northwest. At noon it was five miles SSW. of Great Skellig Island, which marks the southern entrance to Dingle Bay. Here began a series of accidents which cost the squadron the 18-gun cutter and two boats. At four in the afternoon, the brig *Fortune* was sighted and taken by two of the *Richard's* armed boats with little problem. She was bound from Newfoundland for Bristol with a cargo of fish-oil. A prize crew was placed on board and she was sent to France. Later the same day, the wind fell and the currents began to carry the *Bonhomme Richard* so close to shore that Jones ordered his boat lowered in order to tow the ship clear of danger. Manned by an irate coxswain and a group of homesick Irishmen, the boat soon veered away for the Irish coast. The jolly boat was immediately manned with several armed Marines and lowered away. The boat pursued the deserters with such zeal, according to Jones, that they got lost in a thick fog which rolled over southwest Ireland. The cutter *Cerf* was sent to look for the two boats, but she too disappeared.[31] Unable to overtake the boats and rejoin the squadron, the *Cerf* returned to France early in September after springing her mainmast and being chased by an English cutter.

A northwest gale on the night of the 26th compelled the *Bonhomme Richard* to stand off from the rest of the fleet. The following morning only the *Vengeance* was sighted, the *Alliance* and *Pallas* having fallen behind. For the next few days the wind

blew fair and brisk. By early afternoon on 30 August, the *Richard* and her single consort sighted the Flannan Islands off the Outer Hebrides, some 800 miles from L'Orient. Early on 1 September, the *Bonhomme Richard* took the *Union*, a valuable letter of marque bound from London to Quebec with a cargo of clothing for British troops in Canada. Late that same afternoon, off Cape Wrath, the *Alliance* with a prize (ship *Betsy*) rejoined the squadron.[32] Jones allowed Landais the privilege of manning both the *Betsy* and *Union*, but contrary to the commodore's orders the two rich prizes were sent to Bergen, Norway, instead of France.[33] There they were returned to the British consul by Danish authorities.[34]

The *Pallas* reappeared during the night of 2 September and rejoined the squadron. For the next several days the four ships cruised between the Orkney and Shetland Islands, taking several unimportant prizes. On the 4th, the weather turned so that land was not sighted again until the 13th when the Cheviot Hills loomed up from off Dunbar. During the nine days of tempestuous weather, the *Alliance* again disappeared. On the 14th Jones took two colliers loaded with coal bound from Leith to Riga.[35]

Up the Firth of Forth

"Knowing that there lay at anchor in Leith Road an armed Ship of 20 guns, with two or three fine cutters," wrote Jones in his report to Franklin, "I formed an expedition against Leith [seaport of Edinburgh] which I purposed to lay under a Large Contribution or otherwise to reduce it to ashes." The captains of the *Pallas* and *Vengeance* were summoned on board the flagship to discuss the commodore's plan. Numerous objections were raised by Captains Cottineau and Ricot, but countered by Jones who held out the prospect of squeezing a ransom of 200,000 pounds out of the town. At last both agreed to the project. So much time had been spent in discussing the plan, that by the time the conference concluded the wind had changed. Nevertheless, Jones was determined to proceed.[36]

As the squadron worked its way up the Firth of Forth, about 130 Marines were called to arms,[37] and provided with red British uniforms. Several were placed on board the two captured colliers, "so that the two ships looked like transports."[38] To complete the ruse, all officers of the fleet were ordered to don English naval uniforms, previously provided by Jones for such an occasion.[39] Meanwhile, the commodore composed an ultimatum "To the Worshipful the Provost of Leith," which Lieutenant Colonel Paul de Chamillard of the Marines was to present, and articles of capitulation for the town fathers to sign. Half the ransom was to be paid upon demand, while the remainder would be secured by six hostages, or Leith would be laid to waste.[40] The ransom, de Chamillard was to explain, should be regarded as a contribution toward the indemnity "Which Britain owes to the much injured Citizens of America. Savages Would blush at the Unmanly Violation and rapacity that has marked the tracks of British-tyranny in America from Which Neither Virgin innocence nor helpless age has been a plea of protection or Pity."[41]

Lieutenant Colonel de Chamillard never had a chance to demand the capitulation of Leith. For as Sir Walter Scott, who was then in the city as a young lad, wrote in the introduction to *Waverly*, "a steady west wind settled the matter by sweeping Paul Jones and his vessels out of the Firth of Forth." The wind, according to Jones, was so severe that one of the colliers taken on the 14th capsized, but without loss of life.[42]

Following the abortive attempt to extort ransom from Leith, Jones found it difficult to excite the interest of his French captains in other schemes. They were becoming increasingly uneasy at his remaining so long on the coast. If they tarried longer, warned Captain Cottineau, "We Should all be taken." Therefore the squadron sailed southward and in the course of the next several days a number of prizes were taken. By 21 September they were off Scarborough where two vessels were captured and a fleet whose masts had been sighted over the southern horizon was chased. About one the following morning, Jones forced one of the vessels of the fleet ashore between Flamborough Head and Spurn and took an English brigantine enroute from Holland.

THE
BRITISH ISLES,
Comprehending
GREAT BRITAIN AND IRELAND:
with
THE ADJACENT ISLANDS
by
SAMUEL DUNN,
Mathematician
SHETLAND
THE GERMAN
OR
NORTH SEA
IRISH SEA
THE CHANNEL

The Battle off Flamborough Head

At dawn on the 22nd, the squadron was off Spurn, the northern cape of the Humber estuary. Wishing to continue the chase begun the previous evening, Commodore Jones had the signal for a pilot hoisted atop the fore-topgallant masthead. Soon two pilots appeared. Thinking the *Bonhomme Richard* to be British, they informed Jones that "a King's frigate lay there in sight at anchor Within the humber Waiting to take under Convoy a number of merchant ships bound to the Northward." The wind being too light to risk the *Richard* inside the estuary, Jones tried to entice a number of the vessels anchored in the Humber out, but to no avail. So, during the evening, the *Richard* turned north toward Flamborough Head.

Shortly before midnight two strange sails were sighted. All hands were called to quarters and three signal lanterns hoisted. At daybreak the two ships were made out to be the *Alliance*, which had not been seen for over a fortnight, and the frigate *Pallas*. Counting the *Vengeance*, Jones now had four fighting ships under his command.

The squadron sailed slowly toward Flamborough Head before light winds on the morning of the 23d. About three in the afternoon, Jones sent one of the captured pilot boats with a lieutenant and 20 Marines to take a brig sighted to the windward, which he suspected was the vessel he had run ashore the previous day. A hour later, a merchant fleet of 41 sail appeared off the Head, bearing north-northeast. Realizing that his long-sought-after opportunity had arrived, the commodore abandoned the brig and hoisted the signal for a general chase.

Upon sighting the American squadron, all the merchant vessels crowded sail toward the shore, but the two ships escorting the fleet "steered from the land and made the disposition for the battle." Later they were identified as HMS *Serapis* and HMS *Countess of Scarborough*. The *Serapis*, commanded by Captain Richard Pearson, was a new, copper-bottomed frigate rated at 44 guns. Actually she carried 50; twenty 18-pounders, twenty 9-pounders, and ten 6-pounders. The sloop-of-war *Countess of Scarborough* carried 20 guns.

As the sun was setting, Jones made signal for the squadron to form a "Line of Battle." The *Alliance*, *Pallas*, and *Vengeance* paid no attention. Captain Landais, in the lead, hauled the *Alliance* to the windward, leaving the *Richard* alone to engage the *Serapis*; the *Pallas*, under orders, sheered off and made sail for the *Countess of Scarborough*; the *Vengeance*, being astern, simply sailed about, refusing to engage either ship.

By half past six, the commodore and crew were at general quarters, ready to do battle.[43] Jones was on the quarterdeck with Lieutenant Colonel Wuibert, while on the poop was Chamillard, Lieutenant James O'Kelly, and 20 Marines. First Lieutenant Richard Dale had charge of the gun deck and the main battery, and Marine Lieutenant Eugene MacCarthy, with several men, the powder magazine. The main top, commanded by Marine Lieutenant Edward Stack, was manned by 15 Marines and four sailors; the foretop by one midshipman, 10 Marines, and three sailors; and the mizzen top by one midshipman, six Marines, and two seamen. Armed with swivels, coehorns, blunderbusses, and muskets, the Marines and sailors stationed in the masts were ordered to clear the enemy's tops before turning their fire upon his decks.[44]

Presently the two ships were side by side on the port tack heading west, the *Bonhomme Richard* to the south and windward of the *Serapis* the wind being from the southwest. "Just as the moon was rising with majestic appearance, the weather being clear, the surface of the great deep perfectly smooth, even as in a mill pond," Captain Pearson of the *Serapis* hailed, "What ship is that?" The answer from the Richard was: "Come a little nearer, and I will tell you." Next, Pearson demanded to know what type of cargo the *Richard* was carrying. "Round, grape, and double-headed shot," came the reply. Instantly, both ships fired a broadside, and the battle began. At the first or second salvo, three of Jones' 18-pounders burst, killing "most of the men stationed at them,"[45] slightly wounding Marine Lieutenant MacCarthy,[46] and blowing away a portion of the deck above.

"The battle being thus begun," wrote Jones, it "Was Continued with unremitting fury." Each captain strove to maneuver his ship across the other's bow or stern in order to rake. The *Serapis*, being much more manageable, gained an advantageous position several times "in spite of my best Endeavours to prevent it," admitted Jones. After exchanging several broadsides, the commodore thought it more prudent to close with the Serapis rather than engage in a gun-to-gun duel which would be fatal for the *Bonhomme Richard*. Therefore Jones dropped sail, rammed the *Richard* into the *Serapis'* starboard quarter, and attempted to board, but the English sailors repulsed the boarders and Jones sheered off.[47]

The *Serapis* then attempted to cross the *Bonhomme Richard's* bow to rake her, but without much success. After several more maneuvers, the enemy's bowsprit "came over the . . . [*Bonhomme Richard's*] poop by the mizen mast," and Jones "made both ships fast together in that situation Which by the action of the Wind on the Enemie's sails forced her stern close to the . . . [*Bonhomme Richard's*] bow, so that the ships lay square alongside of each other, the yards being all entangled and the Cannon of Each Ship touching the opponent's side." Outgunned from the start, Jones knew that his only chance for victory was to disable the *Serapis'* rigging and kill off her crew by musketry and hand grenades. Boarding would be difficult because of the enemy's two covered gun decks. Captain Pearson, on the other hand, had to shake off the *Bonhomme Richard's* deadly embrace in order to bring his superior fire power to bear. Therefore he ordered the grappling hooks cut. But the *Richard's* sharpshooting Marines in the fighting tops picked off the English seamen who tried to carry out Pearson's orders.

Deprived of his 18-pounder battery by the bursting of the guns, and of his main battery of 12-pounders, under the command of Lieutenant Dale and later Lieutenant Colonel Wuibert, by the *Serapis'* 18-pounders, Jones had no cannon left except two 9-pounders on the quarterdeck. In addition, "Colonel De Chamillard Who commanded a party of 20. Soldiers on the poop had abandoned" through cowardice "that station after having lost some of his men." Therefore Jones' only advantage lay in the good marksmanship of the Marines and sailors on deck and in the fighting tops. Owing to their fast and accurate shooting, they cleared the enemy's tops, took control, and then fired down on the *Serapis'* deck and even into her hatches.

While the *Bonhomme Richard* desperately clung to the *Serapis*, the cutter *Vengeance* leisurely maneuvered at a safe distance, not daring to assist her consort. The *Pallas* hotly engaged the *Countess of Scarborough* and Landais in the *Alliance* behaved like a madman. During the early part of the battle, he did not help either embattled vessel. On the contrary, he assisted the adversary by raking the *Richard*, killing several sailors and driving others from their stations. Later, as he crossed her stern, Landais poured a broadside into the *Richard's* port quarter. Passing once more, the *Alliance*, despite numerous attempts to hail, gave the flagship a third and fatal broadside, killing several of Jones' "best men" and mortally wounding "a good officer on the forecastle,"[48] Marine Lieutenant James O'Kelly.

It was now about ten in the evening. "My situation Was really deplorable," Jones later reported. "The . . . [*Bonhomme Richard*] received Various shots under Water from the Alliance, the Leak gained on the pumps, and the fire increased so much on board both Ships. Some officers persuaded me to Strike; of Whose Courage and good Sense I Entertained an high opinion." But the commodore was not ready to call for quarters.[49] Not with such men in the masts as William Hamilton. An enterprising seaman, Hamilton took a basket of hand grenades and a live match out to the end of a yardarm that hung directly over one of the *Serapis'* open hatches and dropped a grenade through it, which ignited a quantity of powder cartridges.[50] At least 20 men were killed.[51]

Almost immediately after the big explosion, the mainmast of the *Serapis* began to tremble. Captain Pearson, seeing that the situation was hopeless, "lowered his flag and asked for quarter."[52] Naval Lieutenant Dale headed a boarding party which then took charge of the shattered British frigate. Pearson was brought on board the *Richard* and introduced to Jones, into whose hands he formally delivered his sword. The two captains then retired into Jones' wrecked cabin for a glass of wine. "Thus ended this ever memorable battle, after a continuance of a few minutes more than *four hours*."[53]

Meanwhile the *Pallas* had pursued and closely engaged the *Countess of Scarborough*. They exchanged broadsides for almost 20 minutes, when, for some unexplained reason, the *Pallas* dropped astern out of range. Captain Thomas Piercy of the *Scarborough* then sailed over to support the *Serapis*, but decided not to engage because he could not "distinguish one ship from the other." Cottineau then forced him to renew the action. After two hours, the disabled British sloop surrendered.[54]

The fight between the *Bonhomme Richard* and the *Serapis* had lasted nearly four hours. Casualties were heavy. Jones estimated his losses at 150 killed and wounded of a total crew of more than 300. At least 100 of the 400-man crew on board the *Serapis* had lost their lives.[55] Both ships were in a deplorable condition. The *Serapis'* spars, sails, and rigging were cut to pieces. The *Richard* was on fire in several places. Her rudder was missing, her stern frames and transoms were almost entirely shot away, the quarterdeck was "mangled," and at least five feet of water flooded the hold. As Jones observed in his report to Franklin: "a person must have been an Eye Witness to form a just idea of the tremendous Scene of Carnnag, Wreck and ruin that Every Where appeared. humanity Cannot but Recoil from the prospect of Such finished horror and lament that War Should produce Such fatal Consequences."

For the next two nights and a day the survivors worked valiantly to save the *Richard*. The fires which ravaged the ship were not quenched until ten the following morning. But all efforts to save her proved fruitless:

> After the Carpenters as Well as Captain De Cottineau and other men of Sense had Well Examined and Surveyed the Ship (Which Was not finished before five in the Evening) I found Every person to be Convinced that it Was Impossible to Keep the . . . [*Bonhomme Richard*] afloat so as to reach a port if the Wind Should increase it being then only a Moderate breeze. I had but Little time to remove my Wounded, Which now became unavoidable and Which Was effected in the Course of the night and next morning. I Was determined to Keep the . . . [*Bonhomme Richard*] afloat and if possible to bring her into port. for that purpose the first lieutenant of the Pallas continued on board With a party of men to attend the pumps With boats in Waiting ready to take them on board in Case the Water Should gain on them too fast. The Wind augmented in the Night and the next day on the 25 so that it Was impossible to prevent the good old Ship from sinking. They did not abandon her till after nine Oclock. The Water Was then up to the Lower deck, and a little after ten I saw With inexpressable grief the last glimpse of the . . . [*Bonhomme Richard*].[56]

Seaman John Kilby more aptly described the dignified end of a noble vessel:

> Altho Every man that could be spared from Every ship in the fleet was assisting on to Keep her above water: O heavens, it was Enough to bring tears from the heart of the most unthinking man. She went down head foremost with all sail set, Studensails, Top gallonsails, Royals, Sky Scrapers, and Every sail that can be put on a ship, Jack, Pedent [Pennant], & that Beautiful Insign, that she so gallantly wore while in action & when we Conquered. alass she is gone never more to be seen.[57]

To the Texel

The situation of the squadron on the British coast was becoming dangerous; each day that passed increased the possibility that enemy ships in search of the American fleet would find their mark. Before flight was possible, a vast amount of work had to be done repairing the *Serapis* to make her seaworthy. With the assistance of crewmen from the *Alliance*, her rigging was refitted and a temporary mainmast lifted into place. By the morning of the 30th, the squadron was ready to sail and the commodore signalled for it to stand to the eastward.[58]

Commodore Jones' squadron, now comprising the *Serapis*, *Alliance*, *Countess of Scarborough*, *Pallas*, and *Vengeance*, sailed straight across the North Sea. Notwithstanding the original instructions which named the Texel as the port of destination, Jones wished to put into Dunkirk where he hoped to place

John Paul Jones, by Richard Brookshaw.

his prizes and prisoners under French jurisdiction. The other captains would not have it; they insisted on following the original orders to the letter. Jones was forced to follow or to proceed alone to Dunkirk; he chose the former alternative. The passage took several days because of the crippled condition of the *Serapis*, but on 3 October all anchored safely in Texel roadstead.[59]

After receiving leave to repair and replenish his ships from a reluctant Dutch government, Jones' first objective was to land the sick members of his crew and his 500 British prisoners. In this effort he was frustrated by Dutch authorities who stiffly refused, lest they increase the already strained relations with England. Nevertheless, by late October arrangements were made to land only wounded prisoners and to house them in a fort on Texel Island. Given the responsibility of guarding and caring for the wounded was Lieutenant Colonel Wuibert and a group of his Marines.[60]

Meanwhile, the majority of British prisoners remained quartered in the cold, rat-infested hold of the *Serapis*. A number, however, chose to join the United States Navy and bide their time until circumstances allowed them to escape. At two in the afternoon on 14 October, while the officers and men were at dinner, 26 of the former British prisoners thought their time had arrived. Armed with knives, they commandeered a boat and made for shore. The alarm was given by a Marine sentry at his post on the gangway, who fired a number of rounds after the deserters. Marine Lieutenant Stack immediately armed a group of Marines and ordered them over the sides into a waiting yawl. Close to shore the deserters abandoned their boat, formed up with drawn knives, and waited for Stack and his men. "But as Soon as they heard the balls Whistle about their ears," a number surrendered; others still tried to escape. Only four men succeeded; the rest either drowned or were captured.[61] Among the latter was Andrew Paton who explained that it was a dangerous attempt, "but I was ready for anything, either to lose my life, or make it in some shape more comfortable; for here I was naked, eat up by vermin, and by cold got a disorder in my leg and foot that it was proposed to have it cut off."[62]

While the squadron was being refitted, Jones' Marines went about their duties. Guards were mounted about the small fort on the Texel and on board the *Serapis*.[63] Those not on duty were tasked to assist in the repair of the former British frigate; no shore leave was permitted. As days wore into months, his Marine officers became impatient with the numerous delays. On 4 November, Lieutenant Stack requested a leave of absence to visit his uncle, but Jones refused, saying that Stack could not "leave the Squadron without dishonor."[64] Two weeks later, however, Stack and MacCarthy were ordered by Francois Jacques, Comte de Walsh-Serrant, to rejoin the Second Battalion, which the King had ordered embarked at Brest for service in the Antilles.[65] Since both were on leave from the Walsh Regiment, Jones had no other choice than to comply with the order. Thus Stack and MacCarthy hurried off to Brest by way of Versailles, where each was promoted to captain in the Regiment de Walsh-Serrant, and granted 400 livres for their conduct "with Commodore Paul Jones in the fight against an English frigate of superior force that he mastered."[66]

In December, Lieutenant Colonel Wuibert requested that he be allowed to "return by the way of St. Eustatia to his Duty as an Engineer in America."[67] Jones, with regret, acquiesced and sent Wuibert on his way highly recommended. Of the five original Marine officers on board the *Bonhomme Richard* when she sailed, only one remained, Lieutenant Colonel Paul de Chamillard, who had recently returned from Passy with dispatches.

Off in the *Alliance*

In consequence of the charges brought against him by the officers of the squadron, Captain Pierre Landais was ordered to Paris by Franklin on 15 October.[68] Supplied with detailed accounts of his mad actions, attested to by a majority of the officers in the squadron, and even by a number of his own, Franklin temporarily suspended Landais from command. More than a month after Landais' departure, Jones took command of the *Alliance*, since the French government had decided to take responsibility for the entire squadron except for the *Alliance*. To avoid rupturing the delicate relations between France and Holland, Jones shifted his flag, all American officers and crewmen, and a large amount of supplies from the *Serapis* to the American frigate.[69]

Completely disgusted with the Dutch, the commodore sought the first favorable wind to depart. Ready to sail on 1 December, he waited nearly four weeks for an opportunity. On the 13th, he wrote to Franklin:

> We hear that the enemy still keeps a squadron cruising off here; but this shall not prevent my attempts to depart whenever the wind will permit. I hope we have recovered the trim of this ship, which was entirely lost during the last cruise; and I do not much fear the enemy in the long and dark nights of this season. The ship is well manned and shall not be given away . . . . I need not tell you I will do my utmost to take prisoners and prizes in my way from hence.[70]

At last, with a favorable east wind, the *Alliance* sailed on 27 December 1779 from the Texel and succeeded in running the British blockade.[71]

Hugging the Flemish shore to hold her wind, the *Alliance* ran through the Straits of Dover in full view of a British fleet anchored in the Downs. During the night of 28 December several enemy vessels were sighted, but none seemed interested enough to engage the American frigate. The next morning, as a precaution, Jones set general quarters by ordering "up all Hammocks" and directing that the crew be exercised on the guns. The *Alliance* was then off the Isle of Wight, in view of yet another enemy fleet at Spithead. By New Year's Day 1780, the frigate had cleared the English Channel, and was standing off Ushant.[72]

During this period the *Alliance* was not a happy ship. Quarrels broke out among the original officers and those brought on board by Jones from the *Bonhomme Richard*. The supposed cowardice of Captain Landais in the action off Flamborough Head was constantly argued and discussed. "Our wardroom," recorded Midshipman Nathaniel Fanning, "exhibited nothing but wrangling, jangling, and a scene of discord."[73] Jones, who liked his orders carried out smoothly and quietly, forbid the loud brawling, directing his officers "to pass and repeat the Word of Command without Confusion," and his crewmen "*to execute orders* and never to speak loud except in Case of Necessity but to observe as much Silence as Possible."[74] His orders seem to have been obeyed.

After clearing the Channel, the *Alliance* sailed south to the latitude of Cape Finisterre in the hope of taking prizes and prisoners, but almost every ship she encountered bore a neutral flag. Only one small English brig, bound from Liverpool to Leghorn, was taken and sent off to America.[75]

On 16 January, Jones decided to put into Coruña, the celebrated seaport of northern Spain. There he took on water, provisions, and a second anchor to replace one lost at the Texel. There, too, an incipient mutiny was arrested. The crew, asserting that Jones had promised to sail directly to L'Orient where they would be paid wages and prize money, refused to work.[76] The next day, however, "the People were satisfied by the Captain, so as to appear chearfully to Duty." Under Jones' direction they careened the frigate, scrubbed her bottom, and performed "sundry other Jobbs."[77]

Captain Jones sailed again on 28 January for another cruise off Cape Finisterre. Meeting with little success, he put into Groix Roads on 10 February. Several days later the *Alliance* was moored at L'Orient, alongside the *Serapis* which was awaiting condemnation.[78] The *Serapis*, *Countess of Scarborough*, and *Pallas* had sailed from the Texel to Dunkirk, from whence the *Serapis* proceeded to L'Orient. There she was eventually sold, as was the *Countess of Scarborough* at Dunkirk.[79]

Jones' first objective upon his arrival at the French port of L'Orient was to repair, refit, and make extensive alterations to the *Alliance*, the extent and cost of which were the subject of a series of letters between the captain and Benjamin Franklin. The American Commissioner, who positively forbid the sheathing of the frigate's bottom with copper and the purchase of new canvas and cordage, thus adding to the extraordinary expenses already incurred by Jones in Holland, concluded with an appeal for economy: "For God's sake be sparing unless you mean to make me a bankrupt, or have your drafts dishonoured for want of money in my hands to pay them."[80] Notwithstanding this appeal, from 19 February to 12 June, the crew of the *Alliance*, assisted by French workmen, exerted themselves to the utmost in repairing and refitting the frigate. By April, the essential repairs were completed, and Jones could record that "she had the reputation in Europe of being one of the best frigates of her time."[81]

Meanwhile, Franklin had received orders from the Congressional Board of Admiralty to send the *Alliance* back to America with large supplies of arms and clothing urgently needed by Washington's Army. On 19 February he wrote to Jones, urging him to sail quickly after taking on 15,000 stands of arms as ballast, and 120 bales of uniform cloth purchased by John Ross, United States Commercial Agent at Nantes. Jones was also told that Samuel Wharton, a Philadelphia merchant, and two returning diplomats, Arthur Lee of Virginia and Ralph Izard of South Carolina, would accompany him as passengers.[82] Repairs on the *Alliance* were incomplete, but Jones assured Franklin he would finish the task as soon as possible.[83]

Before Jones returned to America, the question of prize money had to be settled. His crew was becoming more and more discontented. Franklin in his letters to Jones referred to the question, and stated that the vessels he had taken were to be valued, but that the muster roll of the *Bonhomme Richard* was wanting in order to regulate the proportions given to each ship. At the same time, Franklin authorized Jones to draw on him for 24,000 livres[84] from which "the People Received one Months Pay" on 22 April —the first they had received since leaving America almost a year earlier.[85] In addition Franklin stated that Chaumont had authorized his banker at L'Orient to advance 100,000 livres for the Americans of the *Alliance* and *Bonhomme Richard*, on account.[86] The sum, however, was never paid.

Landais Takes Command

The prize-money question was a troublesome one, inducing Jones to journey frequently to Paris, ostensibly to hasten the adjustment of the claim. In his absence, Captain Pierre Landais, acting under the advice of Arthur Lee and Commodore Gillon of the South Carolina Navy, plotted to regain command of the *Alliance*. By argument and stealth it was not difficult for Landais to work upon the mutinous spirit of the crew by charging Jones with neglect of their interests regarding prize money due them, to such an effect that they addressed a letter to Franklin, signed by 115 officers and men, declaring that they would not raise the anchor, nor depart L'Orient, until their wages and prize money had been paid, and "until their legal captain, P. Landais, was restored to them." Captain of Marines Matthew Parke, Jones' 'favourite" according to Landais, refused to sign; Marine Lieutenant James Warren, Jr., did likewise. First Lieutenant of Marines, Thomas Elwood, however, went along with the conspirators.[87]

On the morning of 12 June, Jones, upon his return

Pierre Landais, by Charles St. Memin.

to L'Orient, assembled his crew and asked them if they had any complaints. No one answered, and the captain went ashore convinced that they were satisfied. That afternoon, Pierre Landais seized control of the *Alliance*. Captain Matthew Parke describes the event:

> June 12th about 2 OClock P M I was sitting in the ward room with Mr. T Lee & Mr Brown (two of the gentlemen passengers who came with us) & a few french officers that came aboard to pay Cap Jones a visit & see the ship we heard 3 chears Mr. Lee asked me what that meant—I answered him I did not know; upon which he run upon deck & immediately returned with the news that Cap Landais had come on board & asked me if I did not know of his coming I answered no. Soon after Mr. Lynd came down with a message from Cap Landais for me to go into the cabin—I immediately waited on him & welcomed him on board all the officers of the ship were likewise called into the cabin. After they had all come in Cap Landais asked them if they knew him to be their Captain by resolve of Congress they all answered Yes; upon which he had his Commission and orders from Congress read which was not objected to. A piece of writing was drawn up for the Officers to sign certifying they knew no other Commander but him. They all signed it but me, upon which Cap Landais asked me if I woud sign it. I answerd No sir—and said my commissn. and the orders I recd. from the honble. Navy Board at Boston was sufficient & untill I had further orders to the contrary I must own him to be my commander, or, untill an order was passed to suspend him; this piece of writing was destroyed upon my objecting to sign it he then went upon deck and ordered all hands to be called, when they were all come, he orderd his commission and orders from Congress to be read in answer three chears were given from the whole of the ships company.[88]

Those officers and men who refused to acknowledge Landais as commander, and all officers of the *Bonhomme Richard*, were sent on shore.[89]

That evening, in an apparent move to consolidate his command and insure its integrity, Landais ordered Parke's Marines armed. Guards with fixed bayonets were stationed at the gangway, and should "any armed people . . . attempt to come up by force, the beyonets are to be plied to them, and in case they should fire, or come with drawn swords, they are to be fired at."[90] Later, Parke passed on orders to his sergeant, John Farnam:

> It is my possitive orders that Centinels on the gangways, do not suffer any Boats to come alongside, before they acquaint the commanding Officer on the Quarter Deck, and that they suffer no boats to come alongside after the gun is fired in the evening till after the morning gun. All boats passing or re-passing must be hailed by the Centinels as soon as she comes within hail; and if any should attempt to force on board, you are immediately to acquaint me.—The Sergt. or Corporal having the Guard, is to be constantly on deck, and see that his guard is there with him, that they may be ready upon all calls.—If any musket &c should be lost or broke by accident or carelessness he is to report it to me in writing.[91]

No attempt was made by Jones or his men to regain control of the *Alliance*. Instead, the captain set out for Paris to seek renewed authority from Franklin.[92]

Jones returned from Paris within a week to learn that the *Alliance* had been warped from L'Orient harbor to nearby Port Louis. To get to sea she would have to pass through a narrow strait guarded by French batteries. In response to Jones' pleas for assistance, French authorities caused a boom to be moved across the frigate's only possible exit, and issued orders to the Port Louis batteries to fire on the ship should she attempt to pass. Similar orders were given to three French warships then in port, and a gunboat protecting the boom. In addition, armed boats with French Marines were readied should it become necessary to board the *Alliance*.[93]

At this point Jones made a final plea to the frigate's officers and crew by way of his old friend Matthew Parke:

> The Critical Situation in which the Alliance and her Company is involved at this Awlful moment awakes all my humanity and obliges me to lay before you and them the fatal Consequences of persisting in this Causeless Mutiny—Open your Eyes, I beseech you and do not suffer your sober Reason to be further Misled by the Misrepresentations of a Malicious party.[94]

Without waiting for a reply, Jones, to the astonishment of Parke and the relief of Captain Landais, gave up. He interceded with the French authorities to have their orders reversed and the boom removed. The possible loss of life, and creation of disagreements between allies to the profit of their common enemy, were Jones' reasons for this sudden change of heart.

> As the English could have considered this circumstance as arising from disagreements between France and America, and as in any case these ill effects would have dishonored the American flag, I was glad to show my moderation, not being able to consent to the spilling of American or French blood in order to give me a command.[95]

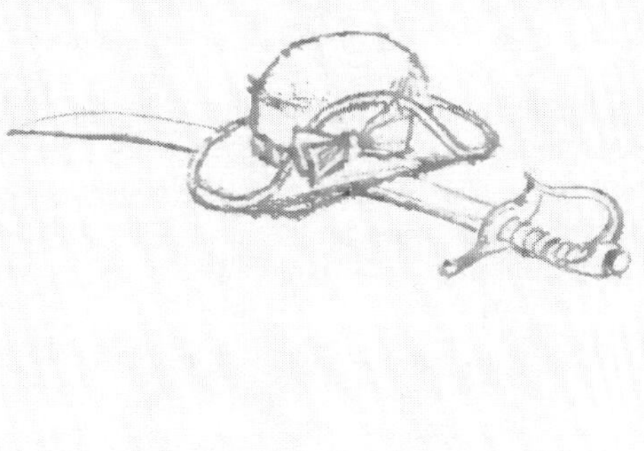

A Madman at the Helm

As soon as French authorities removed the boom, the *Alliance* warped through the narrow strait and into Groix roadstead. On 8 July, Landais sailed for America, but not before Captain Parke was arrested and confined 11 days for refusing to take an oath to obey and defend the captain no matter what happened. From this point, the captain's relationship with the officers and crew began to deteriorate. The remaining crewmen of the *Bonhomme Richard* who refused to obey their new captain were clapped in irons and thrown into the hold. On one occasion Landais suddenly ordered the ship's officers and passengers supplied with water from the common cask located on the quarter deck, an obvious insult to both officer and passenger alike. There were violent quarrels about fishing on the Grand Bank and about killing livestock which had been brought on board for the passenger's use. Finally on 11 August, the crew, with the cooperation of the officers and passengers, revolted. As seamen John Kilby related:

> Landas' conduct was such that, in the opinion of [Arthur] Lee, & all the passengeres, as well as the Officers on bd. proved Compleatly that he was in a measure beside himself. owing as was thought on account of his Conduct in the action [between the *Bonhomme Richard* and the *Serapis*] & Else where. It therefore was requested by all the passengers, & officers, Mr. Lee at the head, to draw up a paper which was signed first by Mr. Lee & all the passengers, to take the Ship from Landas. It was done & at the 4 o'clock watch it was to be carried into Effect. the plan was this, a quarter master, by name, James Prat[t], a Very Large Robust Rough seaman, Was to first take Charge of the Ship. It Really was Laughable to see the actions of the then Commodore Prat. as the watch was Relieved, Prat mounts the quarter Deck, Calls in the 2 Lieuts. of marines, To wit, 1st Lieut. Eldridge [Elwood], 2nd. Dr. Warran [Warren], the son of the great Doctr. Warran of Boston. they both Came to Prat with their hats in their hands, & asked what was his Command. Prat ans. & sd. take a file of soldiers & Confine Capt. Landas to his Caben. they had 12 soldiers with musquets Ready. Orders was given by Eldridge to do so. Landas at first Refused. Prat then ordered them to Charge bayonets. These were Charged, and Landas, put below under a guard of 4 Soldiers at the Cabin Door, after which Eldridge & Warran Came to Prat. Pray, sir, have you any further Command. Prat ans. yes, go down in the ward Room & Bring to me Lieut. [James A.] Digs [Degge]. in a Very small time Digs Came with his hat under his arm. addressing Prat, in those words, To wit, pray sir what is your oder. Why sd. Prat, I now give you Charge of this Ship & you are to observe one thing, that you are to Carry her into Boston & that Should you be ingaged by any Ship what Ever, you are to sink before you Strike the Colours. Digs ans: yes Sir your Command Shall be obey'd; after this Remark Prat became quartermaster again.[96]

Upon the frigate's arrival at Boston, the Navy Board ordered Captain John Barry to relieve Landais. The Frenchman refused to resign, and three stout Marines under Matthew Parke had to drag him, fighting and cursing, out of his cabin and ashore. At the court-martial even his friends testified that in their opinion Landais was insane. The court found, after a short deliberation, that he was guilty of allowing private goods to be shipped in place of public stores, incapable of handling a ship, and should be "broke and rendered incapable of serving in the American Navy for the future."[97]

The *Ariel*

After the departure of the *Alliance*, Jones was occupied in preparing the *Ariel* for sea. A 435-ton ship of war built for the British Navy in 1778 and captured off the Carolinas by the French, the *Ariel* was loaned to Franklin in order to assist in transporting the large amount of clothing and military supplies destined for the Continental army. Franklin hoped that Jones would load her promptly and depart, but she did not leave for four months. Surviving crewmen of the *Bonhomme Richard* formed the nucleus of the *Ariel's* crew, while the remainder was made up of released Americans, English prisoners, and a detachment of 37 Marines provided by the French government. Chosen to command the latter was Chevalier William Nicholson, brother of La Comtesse de Nicholson, Jones' latest paramour. Nicholson, a veteran of the Scots regiments in the French Army during the last war, was joined sometime later by a M. Joriss whom Jones selected as his lieutenant of Marines. Both were granted Continental Marine commissions by Franklin.[98]

In mid-July the *Ariel* began loading and toward the end of the month Franklin forwarded his dispatches for Congress in anticipation of her early departure. July passed and then August; Jones attributed the numerous delays to contrary winds; however, the *Ariel's* log indicates that the wind was from the east a good part of the time.[99] Probably the real reason was to await the loading of the two brigs which the *Ariel* was to escort. Meanwhile, Jones organized a series of grand and extravagant parties to which the cream of L'Orient's nobility was invited.[100]

Preparations completed and with moderate breezes from the east, the *Ariel* departed L'Orient on 5 September and anchored in Groix Roads. There she remained for over a month, largely due to a succession of gales which battered the coast of Brittany. On Saturday, 7 October, she put to sea in "fine Pleasent Weather," accompanied by the brigs *Luke* and *Duke of Leinster*. The next day, however, the winds increased to such a violent tempo that the *Ariel* was driven close to the rocks of the Penmarch Peninsula, between L'Orient and Brest. As the ship neared the rocks, Jones dropped anchor but still could not bring the ship into the wind. As a last resort he ordered the foremast cut away. "This produced the desired effects and *L'Ariel* immediately came into the wind; but the heel of the main mast would not stay in place and its movement could be compared to that of an unsteady drunkard." Jones ordered it too cut away, but in falling the mizzenmast and bridge were brought down with it. Completely dismasted, the *Ariel* now came into the wind.[101] For two days and nights she rode in that position while Marines manned the pumps.[102] On the 11th, when the weather began to moderate, a temporary mizzenmast was stepped into place and the *Ariel* set sail for L'Orient.[103]

Unfortunately, the damage sustained in the gale required another postponement, this time for over two months. Jones did his best to obtain a better ship, the French frigate *Terpsichore*. He resumed writing letters, not only to those in important positions at the French Court but to anyone who could or might assist him in the effort. Even Marine Captain Nicholson journeyed to Paris in an attempt to solve his captain's problem, but if Jones wanted the frigate, he wrote, he must come to Paris himself. Jones refused to leave his ship and therefore failed in his attempt to replace the *Terpsichore* for the *Ariel*. On 18 December he sailed for America.

Loaded with military supplies and carrying important dispatches from the French government, Jones had no desire to fight his way across the Atlantic and therefore he took the southern route along the northern edge of the tradewinds. Notwithstanding his precautions, a large sail was sighted a few hundred miles northeast of the West Indies. Jones tried to escape under the cover of darkness, but the ship was on top of him the next morning. Since he was unable to flee, Jones decided to stand and fight. He cleared for action, but kept the *Ariel's* gun ports closed and his French Marines below deck, all the time working up to a position for a close fight.

With his usual nerve, he moved into hailing distance and bluffed the English captain into giving a full account of his ship and himself. His name, he said, was John Pindar, and his ship, the *Triumph*, a British privateer of 20 guns. Jones even tried to persuade the captain to come on board the *Ariel*; when he refused Jones "hauled up the American flag and commenced firing." Taken completely by surprise, Captain Pindar made a feeble attempt at resistance and after several minutes struck his colors and called for quarters. "As a result," Jones recorded, "I ceased fire, and as usual after a victory there were many huzzas and cries of joy on L'Ariel, but a minute later the captain of the Triumph deceitfully set sail and fled." This was John Paul Jones' last battle under the American flag. On 18 February 1781, Jones brought the *Ariel* safely up the Delaware River and anchored at Philadelphia.[104]

While the *Ariel* lay at Philadelphia, Jones spent most of his time meeting old friends and cultivating delegates in the hope of obtaining a new command. On earlier occasions, Jones had exerted himself to the utmost on behalf of his crew to see that they were properly paid, but now he neglected them. The Board of Admiralty informed him on 19 May that the officers were demanding their pay, which the Board could not give them without an official pay roll.[105] Two weeks later the Board directed Jones "to exhibit a complete Payroll of the Ariel's Crew with all Possible dispatch," as "the reputation of the board, and the Interest of the public suffer exceedingly by its not being exhibited."[106]

After this was done, Captain Jones recommended that William Nicholson's "appointment as Capt. of Marines might be confirmed" by Congress. This request immediately prompted both the Board and Congress to consider the validity of all commissions issued by United States ministers aboard.[107] On 16 July, a committee composed of Theodoric Bland, Daniel of St. Thomas Jenifer, and Roger Sherman, made their report: "Commissions heretofore Granted by the Ministers of these United States at foreign Courts or any of them having been granted for tempory purposes are to be considered as temporary Brevets and nor giving rank in the Navy or Marine of these States and that therefore all such Commissions be and are hereby revoked." In a second resolution, the committee recommended that Nicholson be appointed a "Lieutenant of Marines in the service of the United States."[108]

Apparently this demotion in rank disturbed Nicholson enough so as to prompt him to refuse the commission and return to France. On 25 August a committee of Congress recommended that the paymaster of the Navy be directed to pay both Nicholson and Louis De la Vallette, who had replaced M. Joriss as lieutenant of Marines on board the *Ariel*, a total of 681.72 Continental dollars, being what was due them as of the 25th.[109]

With Jones' return to America, after an absence of more than three years, Continental Marine amphibious operations in Europe came to an end. No longer would Englishmen see the green-uniformed Marines of the *Ranger* step ashore, or the adopted Marines of the *Bonhomme Richard* ranging up and down their coasts. Jones had had a single plan of action: to strike at the enemy's homeland rather than engage him on the high seas where he was far superior. The amphibious role of Marines was always a crucial factor in this scheme. With Jones' exit, Marines reverted to their traditional duties on board Continental ships plying European waters.

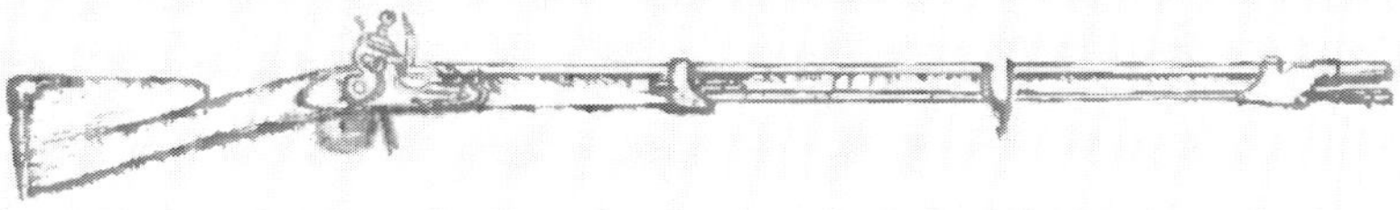

CHAPTER XII

Charleston, 1780

A Fleet to the Southward

The gallant American defense of Charleston in 1776 so discouraged the British that they attempted no further military operations in the south for more than two years. The failure of their efforts during 1777–1778 to achieve final victory in the north forced them to turn their attention again to the south. In December 1778, they took Savannah and during the next year the British were determined to capture the whole lower south. To frustrate this design, the Americans requested aid from the powerful French fleet under Admiral d'Estaing. Although he had been ordered to return to France, he appeared off the coast of Georgia early in September 1779 and joined Major General Benjamin Lincoln in an attempt to recapture Savannah. Through numerous delays the opportunity was lost and their assault when finally made was unsuccessful. D'Estaing then sailed for France, and Lincoln fell back to Charleston.

Following the failure of the joint American-French effort to relieve Savannah, the British and their American Tory friends grew ever bolder. Rather than move back into Rhode Island, it seemed to Sir Henry Clinton that the time had arrived to attempt the subjugation of the southern states by a campaign that should begin at Charleston, the site of his defeat in 1776. His plan was first to capture the city, establish a base there, and then proceed inland. Everything seemed favorable.

The same could not be said for General Lincoln's position. His garrison at Charleston was desperately weak. To hold the South Carolina capital, the commander of the Continental Army in the Southern Department had about 2,500 regular troops. Fear of an outbreak of smallpox, should they be cooped up in the city, kept a majority of the state's militiamen away.[1] Including reinforcements from North Carolina, they numbered somewhat less than the Continentals. To aid in the defense, four Continental ships were ordered south from Massachusetts.

Commodore Abraham Whipple's squadron arrived off Charleston on 18 December 1779 after a difficult 27-day passage from Nantasket Road. In a heavy gale several hundred miles northeast of Bermuda the *Providence* and *Ranger* sprung their mizzenmasts, the *Boston* her mainmast, and while the *Queen of France* received no damage, Whipple said, "I believe if the gale had continued twelve hours longer she would have foundered." On 5 December the squadron fell in with and captured the 12-gun privateer brigantine *Dolphin* from St. Augustine, sending her to Charleston with a prize crew.[2] On board the *Boston* three days later, Captain Samuel Tucker found time amidst the "blowing weather" and heavy seas to pen orders to Richard Palmes, his captain of Marines:

> As commanding officer of Marines, on board the Ship Boston, under my command, I desire you would give your Centry at the cabin door strict orders to keep the gangway under the half deck, clear to the cabin door, and allow no disturbance, under the half deck, and that no person come into the cabin before he sends word excepting the commission'd officers, Master Surgeon & purser—The Centrys in the quarter deck, allow no person to open the Arm Chests except, a Marine Officer, or a Sergeant by their orders.
>
> You are to give your Sergeants, orders, to turn up the after guard in their watch and keep them on the quarter deck until reliev'd by the next watch. The Sergeant never to [be] off the deck in his watch.
>
> You are not to allow the Centrys to converse with any person while on his post.

> You are strictly ordered to give your Sergeants orders, in writing, and in case they break them, make report to me, and they shall be broke.
>
> In case of misdemeanors, in your men, make report, and they shall be punished according to their offences and your desires.
>
> You are to discipline [drill] your Marines as often as the weather and ships business will permit.
>
> You will take particular care to place Centrys when order by the commanding [officer] on board.

Concluding, Tucker exhorted Palmes to exercise vigilance in executing the orders, and in maintaining order on board the frigate throughout the remainder of the voyage.[3]

Upon their arrival, Whipple informed Governor Rutledge and General Lincoln of the squadron's presence and requested pilots to take his ships over the bar. Lack of wind held the ships out for several days, but on 23 December, with a strong wind and flood tide, they came to anchor before the town. Work was begun immediately on repairing or replacing the damaged masts.[4] Captain Rathbun surveyed the *Queen of France* and found her in such poor condition that a "Jury of Carpenters" was called to decide whether the ship was worth repairing or not. On 4 January they "Condemned her as unfit for Cruising."[5]

As Marines and seamen on board the frigates *Boston* and *Providence* labored to clean and repair the two ships, the *Ranger*, accompanied by the schooner *Eagle*, was ordered out on 6 January to render assistance to a Spanish ship from Havana which had run aground about 20 miles north of the Charleston harbor.[6] Back in port five days later, Captain Simpson reported that after doing all in his power to help the Spaniard, he had sent the schooner ahead to give chase to a sail sighted to the eastward. Lieutenant Page, commanding the schooner, came up with the ship, spoke to her, and was informed that she was a British troop transport from New York bound for Georgia. Although not captured, she did make signals when pursued which led Simpson to conclude that she was part of a larger fleet from which she had separated.[7] This intelligence, when forwarded to the governor and General Lincoln, caused a considerable stir.[8]

On 17 January, Lincoln directed Whipple to dispatch two of his ships on a short cruise to the north and south of Charleston in order to give early warning should the enemy approach.[9] Whipple, without hesitation, decided upon the *Providence* and *Ranger*, thereby confining the *Boston* in port which William Jennison "considered as a precedent yet unknown in the British Navy."[10] The two ships dropped down into Rebellion Road on 21 January where they were becalmed for three days. On Sunday morning the 23rd the weather was thick, and a brig was spotted outside the bar slowly making her way along the coast. The commodore ordered the *Eagle* pilot boat out to "reconnoitre her" and report back as soon as possible. By convincing her master that he was approaching Tybee, the entrance to the Savannah River, the brig was decoyed into Charleston harbor and captured. She carried 14 guns and "17 Men part belonging to the New Levies of royal New Yorkers." The vessel was later taken into the Continental service, renamed the *General Lincoln*, and given to Captain Hoysteed Hacker to command.

Shortly after sunrise on 24 January, Whipple and Simpson put to sea and headed south. Off Tybee lighthouse on the 26th, they took two sloops carrying a portion of the loyalist Bucks County Light Dragoons from whom they gained the latest information on enemy movements in the area. Three days later both were chased back into Charleston harbor by two British "fifties & one frigate."[11] This was the last sortie made by the four Continental ships; thereafter, they concentrated on the defense of the harbor and town.

On 26 December 1779, General Sir Henry Clinton turned over his command in New York to Lieutenant General Wilhelm von Knyphausen and sailed south with Cornwallis as second in command. His fleet of 90 transports and supply ships carried eight British infantry regiments, five Hessian, several corps of loyalists, and detachments of artillery and cavalry—about 7,600 troops.[12] It was escorted by five ships of the line, six frigates, and several smaller vessels under the command of Vice Admiral Marriot Arbuthnot. Ironically, two of the frigates included in the naval squadron were the former Continental vessels *Raleigh* and *Virginia*.

The voyage south was a stormy one. Off Hatteras heavy gales almost destroyed the expedition; several of the transports were lost, most of the cavalry's horses perished, and provisions were badly damaged. A month after leaving Sandy Hook the first ships appeared off the mouth of the Savannah River where the expedition regrouped. After repairs to the fleet were completed, they sailed north for Charleston. On 10 February, the ships entered Edisto Inlet, and landed the troops the following day on Johns

Island, 30 miles south of the South Carolina capital.[13] Clinton's plan for the capture of the city envisioned a land siege combined with a strong naval assault from the seaward side.

At Charleston, meanwhile, Governor Rutledge placed the State Navy comprising "the Ships Bricole & Truit, the Briggt Notre Dame and the Galleys" under Whipple's command.[14] Two French ships were also added. Working parties made up of seamen and Marines from the four Continental ships were sent to refit the new additions "for Defence."[15] Rutledge further empowered the commodore to impress all idle seamen into service and, if necessary to forcibly enter and search houses suspected of harboring deserters or "draft dodgers."[16]

To determine the feasibility of mooring ships parallel to the channel entrance in a position to rake any vessel attempting to enter, Lincoln requested Whipple to have the inside of the bar and adjacent shoals sounded and buoyed.[17] Captains Hacker, Rathbun, Tucker, and Simpson were immediately ordered to accomplish this task. On 1 February they reported their unanimous findings:

> [We] Do declare upon due deliberation that it is our opinion impracticable Our Resons are. that when an Easterly wind is blowing, & the flood making in (such an opportunity as the enemy must embrace for their purpose) there will be so great a swell in Five Fathom Hole, as to render it impossible for a Ship to ride, moored athwart which will afford the enemy's ships, under full sail the advantage of passing us, should they effect that, the Continental Ships cannot possibly get up to Fort Moultrie as soon as the Enemy's.[18]

The general was disturbed by the captains' opinion since he was convinced that Charleston's only salvation lay in confining the enemy attack to the land. Although he acknowledged the possibility of losing ships should they be stationed near the bar, he nevertheless was "fully convinced that the probable services, which they will render there, should the enemy attempt to come over the Bar, and the evils consequent on their getting into this Harbour that the attempt ought to be made, and that the measure thereby can be justified." Therefore, he ordered Whipple to place the *Providence*, *Boston*, *Bricolle*, *Truite*, and several galleys in a position to command the channel entrance.[19]

Destruction of Fort Johnston

In order to further prepare for the expected enemy naval assault, Governor Rutledge ordered Whipple to "destroy all the Ranges, such as the Beacon and Lighthouse."[20] Chosen to accomplish the defensive destruction were 20 hand-picked Marines and seamen under Captain Simpson and an equal number under Captain Tucker. About midday on the 14th, the two groups set off for the lighthouse, and at two that afternoon the navigational aid was blown up.[21] In a letter to General Lincoln the next day, Rutledge ordered more destruction:

> I think Commodore Whipple's people, who were so dextrous in blowing up the Light house, yesterday, would, very readily & effectually demolish the walls of Fort Johnston—If you think this is a material service, I c[oul]d wish you w[oul]d be pleased to order it performed, immediately.

Later that evening, Whipple issued specific orders to Captains Simpson and Tucker for the fort's demolition:

> I have to direct that you immediately proceed and by the most effectual ways and means possible, are to level destroy and totally demolish the remains of Fort Johnston.

Benjamin Lincoln, by Charles Willson Peale.

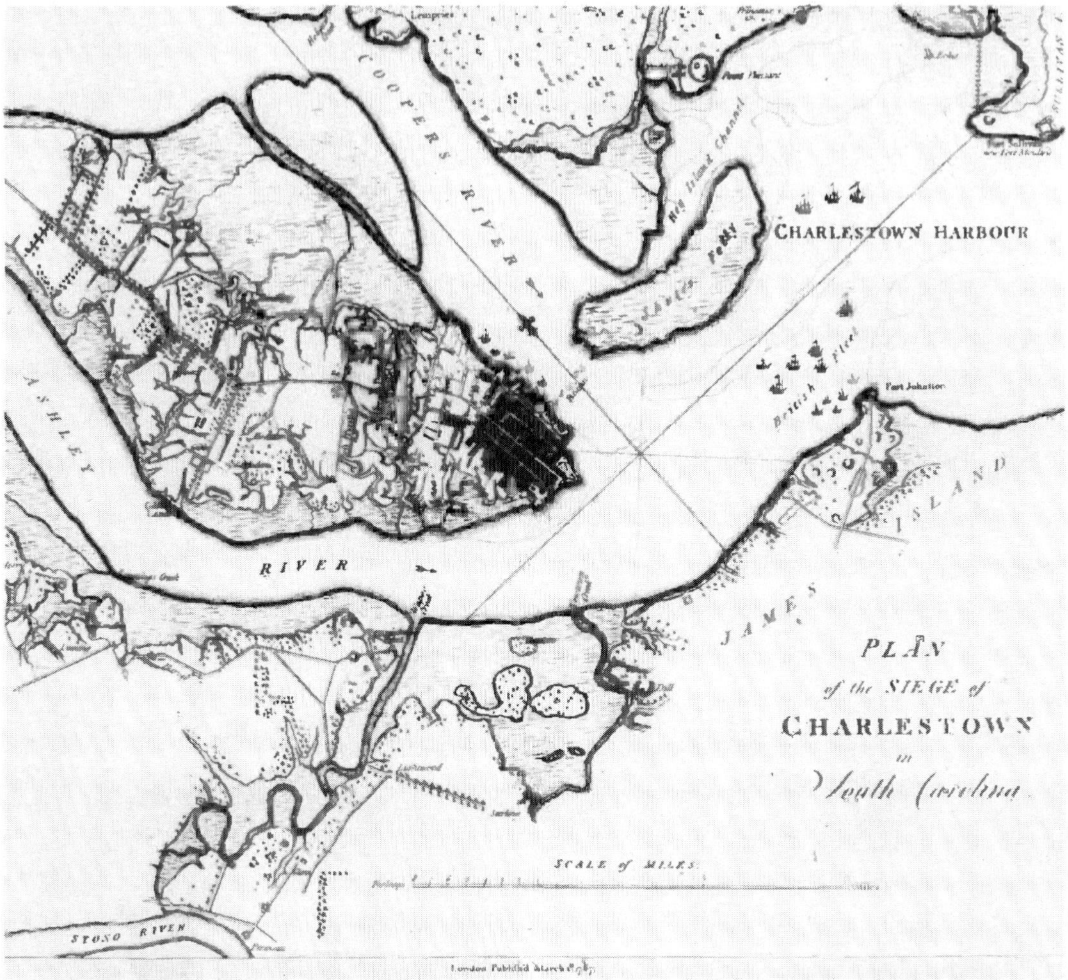

> As it may happen that a part of the enemy may attempt to surprize or attack you in the execution of these orders you will shew these orders to the Captains or commanding officers of Marines on board the Providence, Boston and Ranger and give such directions as may be necessary for the landing of all their marines in good order ready for immediate action or to secure a Retreat— As it is highly needful to be careful of our Men, you will reflect on the necessity of advanced Guards, and take every precaution in your power to hinder the enemy from taking any advantage.[23]

Early Wednesday morning, 16 February, Continental Marines were landed on James Island and ordered to surround Fort Johnston. Inside, meanwhile, sailors planted explosive charges at strategic points about the fort and then set them off. Once the fort was leveled and of no advantage to the enemy should they occupy it, the Marines and seamen hurriedly withdrew across the harbor.[24]

In mid-February, Captain Rathbun was ordered to haul the *Queen of France* up the Ashley River opposite Wappoo Creek. On the appearance of the enemy he was to signal the alarm—a gun fired three times at one minute intervals. In addition to supervising the positioning of state ships and galleys along the river, Rathbun was also ordered to maintain a constant row-guard along the southern shore and inlets. All was in an apparent effort to prevent Sir Henry Clinton's troops from crossing the river and attacking the town from the south.[25]

In compliance with General Lincoln's order of the 13th, the *Boston*, *Providence*, and *Ranger* were taken down and moored just inside the bar on 21 February.[26] Still the commodore and his captains remained unconvinced of their new position's effectiveness, and again recommended that "the ships can do most effectual service for the defence and security of the Town [if they were] to act in conjunction with Fort Moultrie."[27] This time the general acquiesced and allowed Whipple to station his ships off the fort.[28] Thus the defense of the city and its harbor now rested upon the combined fire power of Fort Moultrie and ships of the Continental Navy stationed nearby.

Immediately upon landing on Johns Island, Clinton sent part of his fleet to blockade the harbor. Leisurely moving inland he first seized Stono Ferry connecting Johns Island with adjoining James Island; then he occupied James Island itself and crossed Wappoo Creek which separated it from the mainland, on 7 March. He then erected batteries on the southern bank of the Ashley River opposite Charleston, forcing the *Queen of France* and other ships in the river to haul down and join the other defenders near Fort Moultrie. On 29 March he prepared to cross the Ashley and mount the siege.

Meanwhile, Admiral Arbuthnot's powerful fleet passed over the bar on 20 March without much op-

position and anchored just inside Charleston harbor. Seeing that his position off Fort Moultrie was untenable, Whipple, with General Lincoln's permission, shifted his entire force from the fort area up the Cooper River near Gadsden's Wharf.[29] A line of ships was then sunk across the river from the town to Shutes Folly. Among the eleven ships sacrificed to construct the chevaux-de-frise was Rathbun's decrepit *Queen of France*; the rest of the fleet was anchored in the river behind the boom.[30]

Marines to the Batteries

After Whipple's retreat up the Cooper River, the Marines' role on board ship came to an end. In company with the guns, powder, provisions, and the rest of the crew, they were ordered ashore to bolster frontline artillery units.[31] Captain Palmes' Marines were divided between Craven's Battery and the Granville Bastion, both located on Charleston's Cooper River front. Sandwiched between Palmes' troops were the *Providence's* Marines, who like the *Boston's*, were divided among the Exchange Battery and Broughton's Battery. Marines under Captain Arrowsmith were assigned to Gibb's Battery, situated on the Ashley River side of the city.[32] "In case of alarm," the separate detachments of Continental Marines were to assemble and join with the "James Island Company, the Detachment of the Berkeley County Regiment, . . . the Troops of Horrys dismounted Dragoons, and Col. Moultrie's Corps," and "occupy that part of the Lines between Mr. Livingstons house and Gibbs Wharf and to be under the Command of Col. [Francis Marquis de] Malmady."[33] This order was apparently never put into effect, for the Marines seemed to have remained at their assigned batteries throughout the remainder of the siege.

On 29 March Clinton crossed the Ashley River in force and moved to within 2,000 yards of the American works erected on Charleston Neck. On 9 April the British fleet ran past Sullivan's Island under fire from Fort Moultrie and anchored between James Island and the town. Charleston was almost completely surrounded. Fort Moultrie's small garrison surrendered without resistance on 5 May in the face of a combined assault from ship cannons and British Marines. Five days later, the beleaguered defenders of the South Carolina capital capitulated.

Surrender terms required that all American shipping in the harbor be turned over to the victors, and that all Continental troops, sailors, and Marines be made "prisoners of war until exchanged."[34] Five days following the surrender, Whipple, Hacker, Rathbun, Tucker, and Simpson addressed a joint plea in behalf of themselves and all other naval and Marine officers to Admiral Arbuthnot requesting paroles in order to return "to our friends and connexions in the New England States where we severally promise to remain inactive till a proper exchange shall have taken place."[35] Parole was granted, and the officers sailed north to Philadelphia in a cartel. Being excluded from the request, Continental Marines and naval enlisted men had to wait until June before the British chartered three cartels and sent them home.[36]

To Sea at Long Last

The capture of the three Continental ships at Charleston left only five Marine detachments afloat, those on board the frigates *Alliance*, *Confederacy*, *Deane*, *Trumbull*, and the sloop of war *Saratoga*. Since her launching in 1776, the frigate *Trumbull* had remained in the Connecticut River, unable to pass over the bar at the river's mouth. Shortly after Dudley Saltonstall's appointment to the frigate *Warren*, care of the *Trumbull* was given to Captain Elisha Hinman, who, by adopting an unspecified but new plan, reduced the frigate's draft to nine feet of water and passed her over her bar.[37] She was then taken to New London and fitted out for sea. There, besides James Nicholson of Maryland who was appointed to command the frigate, the only other officers to join the ship were David Bill of New London, Jabez Smith, Jr. of Groton, and Daniel Starr of New London, appointed lieutenants under Captain of the Marines Gilbert Saltonstall. Late in May 1780, the Board of Admiralty which had succeeded the Marine Committee in the administration of naval affairs, ordered the *Trumbull* to sea.[38] At that time she had a Marine contingent of about 30 men, a majority of whom had taken part in the ill-fated Penobscot expedition a year before.

Trumbull vs. *Watt*

The *Trumbull* had not been long at sea when she fell in with and captured the British 40-ton privateer schooner *Queen Charlotte* off Bermuda.[39] After sending the schooner to Philadelphia under a prize crew, the frigate continued southward until she sighted a large sail at half past ten on the morning of 1 June. She later proved to be the 36-gun British letter of marque *Watt*. In a letter to his father, Marine Captain Gilbert Saltonstall gave an account of the engagement between the two ships which turned out to be one of the hardest-fought naval actions of the war:

> Lat. 36.N., long 63 W. we saw a sail from mast-head directly to windward; as soon as she discovered us she bore down for us; we got ready for action, at one o'clock began to engage, and continued, without the least intermission, for five glasses, within pistol shot.

It is beyond my power to give an adequate idea of the carnage, slaughter, havock, and destruction that ensued. Let your imagination do its best, it will fall short. We were literally cut all to pieces; not a shroud, stay, brace, bowling, or any other of our rigging, standing. Our main top-mast shot away; our fore, main, mizzen and jigger masts gone by the board; two of our quarter deck guns disabled; thro' our ensign 62 shot, our mizzen 157, mainsail 560, foresail 180, our other sails in proportion; not a yard in the ship but received one or more shot; six shot through her quarter above the quarterdeck, four in the waste; our quarter, stern, and nettings full of langrage,* grape and musket ball. We suffered more than we otherwise should, on account of the ship that engaged us being a very dull sailer: Our ship being out of command, she kept on our starboard quarter the latter part of the engagement. After two and a half hour action, she hauled her wind, her pumps going—we edged away; so that it may fairly be called a drawn battle.[40]

In a second letter to his father on 19 June, Saltonstall elaborated:

Though we were cut to pieces she [*Watt*] has nothing to boast of should she get in—her sides were damaged as much as our sails and rigging—her maintop mast was hanging over her side just ready to go as well as her mainmast—her sails and rigging were not damaged as much as ours—as we fired principally at her hull, whence we concluded we did more executiong than they, though they did enough. Had we a sufficiency of langrage and aimed at her rigging instead of her hull, I think we should have carried her—our wads set her nettings on fire on her larboard quarter, which they cleared themselves of by cutting part of their nettings away. We saw them heave sundry of their men overboard during the action. Their wads set our nettings afire on our starboard bow—our main and quarter-deck guns expended 388 rounds, 86 of which were fired on the quarter-deck; the marines fired pistols during the engagement exclusive of which they fired near 1200 rounds.

Upon the whole there has not been a more close, obstinate and bloody engagement since the war. I hope it wont be treason if I dont except even Paul Jones—all things considered we may dispute titles with him.[41]

Obviously Captain Saltonstall was referring to the engagement between the *Bonhomme Richard* and *Serapis*.

The Marines Suffer

Casualties were exceedingly high on board the *Trumbull*. Among the 13 officers and men killed were Marine Lieutenants Bill, Smith, Starr, and Marine Sargeant Ezekial Hyatt.** Lieutenant Daniel Starr, as Captain Saltonstall later wrote:

Was wounded the later part of the engagement with a grape shot, which went in just above the right hip bone and was cut out behind. He lived until 4 o'clock Monday morning following [5 June 1780], when he died without a groan or struggle. I was with him most of the time after he was wounded till he died. The day after he was wounded he was out of his head and so continued until his decease. I suppose his bowels was mortified as he was insensible of pain.

Lieutenant Bill died instantly during the first few minutes of the engagement when a piece of langrage struck him, taking "off the upper part of his head." About 30 men were wounded, among them Gilbert Saltonstall. "I had eleven different wounds," he wrote, "from my shoulder to my hip; some with buck-shot, others with the splinters of the after quarter-deck gun. I had some shot through the brim of my hat; but was not so disabled as to quit the quarter-deck till after the engagement."[42]

The frigate's troubles were hardly over after the engagement:

The next day, the 2d it blew a heavy gale of wind, which soon carried away our main and mizzen masts by the board—the foretopmasts followed them, and had it not been for the greatest exertions our foremast must have gone also, it being wounded in many places, but by fishing and propping it was saved. In this situation no person would envy us—torn to pieces—our masts gone—at the mercy of the elements—the cries of the wounded—our decks lumbered with the destruction of the enemy the day before and the storm raging—our men beat out with fatique—a hull—a prize to anything that should happen to come athwart us of force—a beast not totally bereft of sympathy, of common humanity, must have sympathised with us in our condition.

We remained in this situation till the next day, the 3rd, our men having got a little over the fatigue of the engagement and the duty of the ship—the gale abating we got up jury masts and made the best shift. In the night the gale increased again, and continued from that time till we got soundings on George's Banks in 45 fathom of water the 11th instant. We got into Nantasket the 14th, the day following into the harbor.[43]

The *Watt*, greatly shattered, got into New York on 11 June.[44]

* Langrage: a kind of shot consisting of bolts, nails, etc., fastened together in a case, used for damaging sails and rigging in battles at sea.

** Smith died of wounds received, according to his gravestone, at Boston on 28 June 1780—aged 29 years. His space on board the *Trumbull* was filled some time during the summer with the appointment of William Fielding.

From Martinique to Philadelphia

Since December 1779 Captain Seth Harding had the frigate *Confederacy* at Martinique repairing the splintered rudder and trying to obtain new masts and spars. While repairs were under way at the island's naval base, French Admiral Piquet visited the frigate and reduced the size of the crew by demanding between 40 and 50 French sailors, and borrowing 30 Americans, amongst whom were "a Serjt. & 15 Marines," for a 20-day cruise. Marine Captain Joseph Hardy saw this as no great loss, on the contrary, it was "a good riddence of Lubbers."[45] He did, however, view the sufferings of his men due to the lack of adequate clothing with much more concern. On 29 January he addressed this problem in a letter to William Bingham, Continental Agent:

> Some time ago Captain Harding presented you my Indent of Marine cloathing, which was then much wanting and be assured it now increases [ ] necessity, the Marines in general not having Cloaths sufficient to cover their nakedness. Notwithstanding in that miserable condition, they are obliged to go thro' their Duty, Day and Night, in all the changes & inclemencies of the Weather, in this Country which had thrown the greatest part of them into Fevers, and now I have not Men enough, capable of going thro' the necessary Ships duty. Therefore you'll procure the Cloathing as soon as possible which will much Oblige me and render the Ship a great piece of Service.
>
> We have several good Taylors on board, who can assist in making up the Uniform, and will save some Expence.[46]

Like Captain Harding, Hardy received little satisfaction, for good clothing like new masts and spars

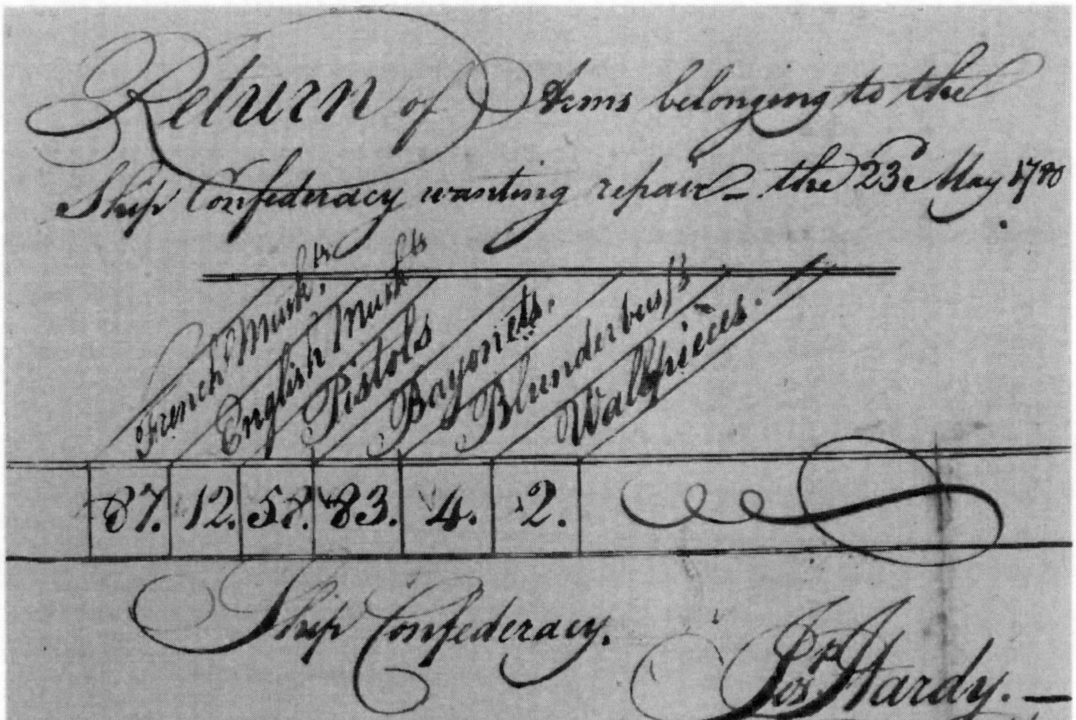

Return of Arms belonging to the Ship Confederacy wanting repair — the 23d May 1780

French Muskts	English Muskts	Pistols	Bayonets	Blunderbusss	Wallpieces
87.	12.	58.	83.	4.	2.

Ship Confederacy.

Jos Hardy. — Captn M.

Entered

Return of Marines' Arms.

was "very difficult to obtain, and some impossible to be had at this place."[47]

It was not until the end of March that the *Confederacy*, fitted out with makeshift masts, was again ready for sea. Captain Harding had decided to join a large French fleet under Comte de Guichen in an attack against the English at St. Lucia. Just as he was getting underway an express arrived from American agent Bingham with orders to make for Philadelphia. On Thursday morning, 30 March, the *Confederacy*, loaded with sugar and accompanied by five merchantmen, sailed from St. Pierre.

One day out, several men came down with the dreaded smallpox. To prevent further spread of the disease, the men were isolated in the cutter which was hung from one of the booms.

On 5 April, in a heavy sea, it was discovered that the maintopmast was badly sprung in two places. This was an unfortunate accident, but as Hardy expressed it, "when we recollect the Distresses and disagreeable events of our late Passage we can or ought to look upon this as a trifling circumstance." Trifling or not, Harding had to treat his jury rigged masts gingerly lest he find himself once again dismasted. Therefore prospective prizes were allowed to pass unmolested.

The remainder of the voyage to Pennsylvania was uneventful; sailors were exercised on the guns while Marines worked with small arms, powder, and made musket cartridges in their spare time. On 25 April, in squally weather, the *Confederacy* dropped anchor inside Cape Henolopen. The following day Second Lieutenant Thomas Vaughan took six men in a yawl and headed for Lewestown to secure a pilot, but once ashore Vaughan and the sailors disappeared. Not until the 27th did a pilot come on board the *Confederacy* and guide her up the Delaware to a safe anchorage off Philadelphia.[48]

Abraham Van Dyke

As spring turned to summer, the frigate lay "at Philada. and Chester.—Careening, Rigging, Manning &c."[49] For the enlisted Marine it was a period of inactivity, but for Marine Captain Joseph Hardy it was a time to put the company's affairs in order. Damaged weapons were repaired, and a quantity of excess equipment discarded.[50] Still he found time to converse with his counterpart on board the new Continental sloop *Saratoga*.

Launched from the Humphrey shipyards at Philadelphia on 10 April 1780,[51] the sloop *Saratoga* was quickly outfitted by the Board of Admiralty, even though the board was in dire financial straits. By June she was almost complete and the appointment of her officers was begun. About this time Abraham Van Dyke, a 61-year-old Dutchman from one of New Jersey's oldest families, appeared in Philadelphia with a letter of recommendation from General Washington, who felt that a captaincy of Marines would be of "some compensation for the hardships and losses which he had experienced on account of his steady adherence to the cause of his country."[52]

Van Dyke, although advanced in years, was a man who possessed the experience missing in his youthful counterparts. He was born in 1719, the son of Jan Van Dyke and Anna Verkerk Van Dyke, a stalwart Dutch couple who resided on the banks of the Raritan River in Middlesex County, New Jersey. During his youth, he had been a lieutenant of Marines on board a privateer in the French and Indian War.[53] At the outbreak of the Revolution he raised a company of militia in New York City, and with them was taken captive in September 1776, when Washington's Army fled Manhattan.[54] Elias Boudinot, American Commissioner of Prisoners, who had visited New York in 1778, reported that "Captain Vandyke had been confined 18 months for being concerned in setting fire to the City, when, on my calling for the Provost Books, it appeared that he had been made Prisoner & closely confined in the Provost 4 days before the fire happened."[55]

"He was treated with uncommon rigor during the whole time of a long captivity," Washington related to the Board of Admiralty, "owing, as is supposed, to the influence of those persons who remained voluntarily in the town, and to whom he was particularly obnoxious on account of his opposition to their measures from the commencement of the dispute. The little property he was possessed of fell into the hands of the enemy, and as no provision has been made for him, either in the Army or in the state, he had been reduced to a condition truly distressing."[56]

To please the general, Francis Lewis, a board member, drew up a captain's commission and sent it to Congress for approval. The commission was returned a few days later unsigned, for a sloop-of-war under 20 guns rated only a lieutenant of Marines. Lewis then went back to Congress with a compromise: "that . . . [Van Dyke] should be appointed to act at present and receive pay as lieutenant of marines," but be appointed a captain of Marines.[57] Again Congress rejected the captaincy recommendation. Beaten, Lewis returned on 24 July and urged that Van Dyke be commissioned a lieutenant of Marines instead of captain. Congress not only approved the appointment, but authorized that the commission be back-dated to 7 July.[58] Pleased with the appointment, Van Dyke joined Captain John Young on board the *Saratoga*.

Enlisting a crew for the *Saratoga* and recruiting personnel for the frigate *Confederacy* proved difficult, even though there were "a considerable number of Seafaring men in port."[59] After several weeks of futile effort, Captains John Young and Seth Harding went before the Board of Admiralty where they pointed out that the pay of officers and men "is now by Depreciation become so nearly reduced to nothing as to be Considered by Seamen as no reward for Pass'd services, or inducement to engage for the future."[60]

Out of Harding's and Young's appeal grew an Admiralty Board resolution to Congress that pay for officers and men be in specie, that subsistence pay for officers also be in specie, that a bounty of 20 dollars be paid every able seamen and 10 dollars to each ordinary seamen or landsman who would enlist for a year, and that 40 Continental dollars be considered the equivalent to one dollar in specie. Congress approved these recommendations on 11 July, but only as applied to enlisted men.[61]

Even with more lucrative incentives, it was difficult to enlist the needed men, therefore both Young and Harding turned to the city jail where receivers of stolen goods, thieves, and persons convicted of treason were recruited. In all cases, the enlistee's choice was between further imprisonment or service on board either the *Confederacy* or *Saratoga*. Of nearly 20 felons enlisted, only a few were taken on as Marines.[62]

As Harding and Young labored for recruits, the Continental Board of Admiralty continued to develop plans for a joint cruise of the frigates *Trumbull*, *Confederacy*, *Deane*, and the sloop *Saratoga*. But the *Confederacy* and *Trumbull* were in need of extensive repairs, and the *Saratoga* in want of men; only the *Deane* was ready.

"Come then, my brave Boys"

The frigate *Deane*, after much delay, had sailed on a short cruise to the eastward in mid-February. Returning to Boston eight weeks later with four valuable prizes to her credit, Captain Samuel Nicholson had the frigate caulked, graved, and a new suit of sail and rigging put into place.[63] Early in May, the frigate was ordered, with the *Trumbull*, to sail for the Delaware where they would join the other two Continental ships. As in previous sailings, willing seamen and Marines were difficult to find. "That lucky and fast-sailing Continental Frigate DEANE is now preparing for Sea," Captain Nicholson advertised on 25 May. All able-bodied seamen and Marines "have now the best of Chances for making their Fortunes; and those who miss this gallant Opportunity, may, perhaps, never have another:—Come then, my brave Boys, to the Rendezvous, . . . where you will find your jolly Companions, and receive greater Advantages than in any private Ship whatever."[64] The rousing appeal did nothing to encourage enlistments; even a state ban on privateer enlistments accomplished little.[65] By mid-July the *Deane* was only able to muster 175 men, far short of the desired complement of about 300 men and officers. Short handed, she and the *Trumbull* sailed for the Delaware on 27 July in pursuance with the Board of Admiralty's May orders, which had since been revived.[66]

The long-expected reinforcement from France, consisting of 5,000 troops under General Rochambeau convoyed by seven ships of the line commanded by Admiral de Ternay, arrived in Newport on 12 July. It was intended by the Board of Admiralty that the *Confederacy, Deane, Trumbull,* and *Saratoga* should cooperate with Ternay's fleet in accordance with the wishes of General Washington. Unfortunately Ternay was soon blockaded in Newport by a superior British fleet under Arbuthnot, who had returned from his victory at Charleston and the union of the two fleets was postponed.

While awaiting the repair of the *Confederacy* and the arrival of the *Deane* and *Trumbull*, the *Saratoga* was ordered to shepherd the *Mercury* packet with its important passenger, Henry Laurens, to sea.[67] On 13 August she dropped down the river, through the chevaux-de-frise, past Chester, and on the 15th anchored off Port Penn. There, two large vessels were sighted coming up the bay—the *Trumbull* and *Deane* at last. Short of water, both frigates soon sailed up the river, promising to return without delay. After four days of impatient waiting, the *Saratoga* and her charge put to sea. The *Mercury* was convoyed a short distance by the sloop which then turned southward. The packet continued on its way, but off the Banks of Newfoundland she was captured by a British frigate and taken to England where Laurens was confined in the Tower of London for almost a year. The *Saratoga*, meanwhile, cruised a short time before returning to the Delaware with a rich prize laden with rum.[68]

The *Trumbull* and *Deane* sailed for a three weeks' cruise on 2 September, and returned "without having taken anything worth naming, which is a great disappointment to the Admiralty Board."[69] The *Saratoga*, in an intervening voyage, took four more valuable prizes; however, all were recaptured. By mid-October, the *Trumbull, Deane,* and *Saratoga* lay at Chester and the *Confederacy* at Philadelphia, all still under repair. The ships remained in port for nearly a month preparing for sea under difficulties which caused endless delays. The difficulties, as might be expected, were mostly financial and not only hindered repairs to the vessels, but caused near mutiny on board the *Trumbull* whose crew wanted their pay and liberty to depart.[70]

Masts For the Frigates

Financial problems were not all that worried the Board of Admiralty during the fall of 1780. Sometime during the previous spring the board had made an agreement with James Wilson that he provide "masts for the Navy of the United States to be cut up Schuylkill, and consented that Captain [Dennis] Leary a Captain of Marines should superintend the workmen who were to be employed in the business."[71] Leary recruited a small company of Marines in Philadelphia and proceeded to Reading, a flourishing young settlement located in the Blue Mountain foothills. From there Marines and workmen moved 30 miles up the Schuylkill to "near the Gap of the second Mountain,"[72] and began cutting "Masts suitable not only for our ships, but for the ships of our illustrious Ally [France]."[73]

On Sunday, 27 August, Captain Leary was "alarmed with an account of an attack make by the Indians at a House about a mile" from his camp on the river. He immediately marched to the house with four Marines, where he viewed a scene of sad destruction. The man of the house and two of his three children lay dead, the third, a little girl, had been carried off by the Indians in their retreat. Stopping only to bury the dead and make inquiries, Captain Leary started in pursuit of the murderers the following day with 10 Marines. Captain Balty and Colonel Lindenmuth later reinforced him with about 50 men. The search was continued without luck until the 31st and then given up.

"Since the first attack a House & Barn have been burned on little Schuylkill, & two Horses taken a little Boy Son of one Shurr is also missing since Tuesday last," wrote Leary to William Moore, Vice President of the Pennsylvania Council on 1 September. As a result, he had "left about sixty men at the Different Settlements" to protect the inhabitants against any further depredations.

One Sunday about this time, a man and his wife appeared at the Marine camp and stated that their name was Hamilton and that the man was a carpenter. A few days later it was ascertained that the man's name was not John Hamilton, but Niel Tye, a notorious half-breed. To add to the mystery, Tye's wife sent a message to Leary begging him to guard himself against her husband who had "determined to scalp" him. A guard of Marines soon located the blood-thirsty Niel Tye and imprisoned him in the Reading jail. Captain Leary's only comment concerning this unusual affair was that it seemed "a little extraordinary."

All these incidents indicated to Leary that he needed additional men to perform adequately the mission assigned to him. Therefore he requested assistance from both the Board of Admiralty and the Council of Pennsylvania "in forwarding such Supplies of men & provisions as will be necessary for defending the post where we are, at least, if not the rest of the Frontier."[74] Leary's request and vivid report brought the desired aid. Although the Board of Admiralty asked Pennsylvania authorities for about 50 or 60 men,[75] a call went out for militiamen and about 300 were sent.[76] The arrival of this reinforcement seemed to have had the desired affect, for no further incidents against the workmen or inhabitants were reported by Marine Captain Dennis Leary.

The *Confederacy* and *Saratoga* Lost

Only the *Deane* was ready for sea by mid-November, and Congress proposed sending her to Cape Francois to secure much needed clothing and military stores.[77] Since Samuel Nicholson apparently had no desire to go to the West Indies, a leave of absence was granted him and a successor appointed for the cruise—his younger brother John. By the 20th the *Deane* had sailed; the *Trumbull*, *Confederacy*, and *Saratoga* remained. Upon the latter two Continental vessels, the Board of Admiralty lavished its full attention.

During the remainder of November, Harding and Young applied themselves assiduously to the multitudinous duties involved in getting their vessels refitted. Most important of Young's duties was to replace the officers who had been taken in the four prizes and those who wished to retire. Among the latter was Abraham Van Dyke who resigned because of advanced age and the overly strenuous life on board the sloop. His place was filled by Lieutenant Hugh Kirkpatrick, about whom nothing is known.

As November drew to a close, the *Confederacy* and *Saratoga* were at last ready to sail. To help round out their crews, felons were pardoned, private vessels were boarded and likely seamen and Marines were impressed. After a week of impressment a mighty howl arose from Pennsylvania's Supreme Executive Council. The "abuses of our Trade, Oppression of our Citizens & other Iregularities of Capt. Harding & his Officers" would have to stop.[78] Although Congress promised to remedy the ill and then did nothing, the practice stopped when the *Confederacy* dropped down the river on 5 December. Several days later, the *Saratoga* followed suit and joined Captain Harding off Reedy Island.

On 20 December, the two vessels got under way with a fleet of 13 merchantmen which they were to conduct to the West Indies, but the *Confederacy's* pilot decided it was too rough to venture out into open sea. Consequently, the responsibility for convoying the merchantmen fell to the *Saratoga* which cleared the capes just before dusk.[79] Enroute to the Caribbean, the sloop captured the 16-gun British privateer *Resolution* after firing a single broadside.[80] Shortly thereafter, she engaged the letter of marque ship *Tonyn* for a full hour. Musket fire of Kirkpatrick's Marines on the quarterdeck contributed to the enemy's distress before capture.[81] On 27 January Cape Francois, her destination, appeared off the sloop's bow.

Just before noon the *Saratoga* negotiated the tricky, narrow channel, and dropped anchor in the broad roadstead of the cape. Not far away lay the frigate *Confederacy*, which had taken only one prize on her rather short cruise southward. The cargo of military stores with which the two ships were to return was not ready for shipment, so both vessels lay in the harbor for several days while their crews enjoyed the varied life of the French settlement.

On 1 February, the *Confederacy* convoyed a group of four merchantmen bound for America and then proceeded on a two weeks' cruise with the French brig *Cat*, in compliance with the orders of the island's governor. Before her departure Captain Hardy's force was strengthened by the addition of 31 French Marines; for what purpose it is not known. The only vessels overtaken in their two week search for prizes proved to be American, and on the 16th Harding returned to Cape Francois empty-handed. Meanwhile, the frigate *Deane* had arrived after spending eight profitless weeks scouring the West Indies for rich prizes.[82]

After a short cruise by the three Continental vessels in late February, during which the enemy merchant ship *Diamond* was taken, they returned to port. Again they waited several weeks, while a sizeable merchant fleet—half bound for America and half destined for France—was assembled and loaded. Finally, on 15 March, this large fleet of over 80 ships stood forth from the harbor of Cape Francois. The fourth day out those vessels bound for France parted company while the sloop *Saratoga* veered off in chase of two enemy merchantmen. The first, a lightly armed snow, was easily captured, but while in pursuit of the second the *Saratoga* disappeared and was never heard from again.[83]

Off the Delaware Capes, on 14 April, Harding sighted a large frigate bearing down upon the *Confederacy* and its fleet of merchantmen. Alone, the *Deane* not in sight, the Connecticut captain lay back to give his convoy an opportunity to escape, and then ordered the frigate cleared for action. As he anxiously watched the approaching vessel, a second appeared. They proved to be the 44-gun British frigate *Roebuck*, and her consort the frigate *Orpheus* of 32 guns. When the lead ship came within range and ran out a second tier of guns, Harding "prudently" struck his color and "submitted without any resistance."[84] The British frigates then triumphantly carried the captured American into New York where she was later taken into the Royal Navy under the slightly altered name *Confederate*.

Captain Harding and a few officers, among them Marine Lieutenant Samuel Holt, were immediately paroled and sent to New London in a cartel. The crew was confined to the *Jersey* prison ship until exchanged or released. Marine Captain Joseph Hardy, Lieutenant of Marines Gurdon Bill, "& [the] greatest Part of Petit Officers" were sent to England for imprisonment.[85] But "by accident," the ship carrying them to England put into an Irish port where both Hardy and Bill made their escape. Later both got to France and from there they made their way back to America.[86]

Only Three are Left

The loss of the *Confederacy* and *Saratoga* left only three Marine detachments on board ships of the Continental Navy: that on board the *Trumbull*, still being fitted out at Philadelphia; that with the *Deane*, which escorting the New England contingent of the merchant fleet, had put safely into Boston on 16 April; and the detachment on board the frigate *Alliance*.

As the trial of Pierre Landais dragged on through the fall and winter of 1780, the *Alliance* was made ready for yet another voyage to France. There was the usual delay and difficulty in recruiting a crew, but in the *Alliance's* case the problem was exacerbated by the investigation. As the Eastern Navy Board informed the Board of Admiralty on 21 December: "she mans but slowly," and therefore they had been forced to send the officers into adjacent states to seek recruits.[87] The problem became so acute that the Navy Board made application to the state government for authority to impress seamen and to enlist soldiers. The request was denied, but permission was obtained to enroll volunteers from the guard at Castle William.[88] The recruiting officers found a deaf ear at the castle garrison, and it was again necessary to impress a considerable number of British prisoners. Some of the officers who had sailed previously in the *Alliance* were willing to continue under her new captain, among them all the Marine officers—Captain Matthew Parke and Lieutenants Thomas Elwood and James Warren—and newly joined lieutenant of Marines, Samuel Pritchard of Boston. Other officers who felt that Landais had been unduly abused left in disgust.

Seven weeks, including postponements, had been consumed with the court-martial of Pierre Landais. On Monday, 8 January, the trial of Lieutenant James A. Degge began. Unlike that of Landais, the Degge trial was concluded in only three weeks, but the verdict was the same. The former senior lieutenant of the *Alliance* was ousted from the Continental naval service.[89] For more than two months, the process of getting the frigate to sea had been sadly and needlessly impeded.

By early February, the *Alliance* had on board 226 officers, men, and boys—"quite enough," thought the Navy Board, "to make a good defense Against the enemy." Including the passengers—Colonel John Laurens, Thomas Paine, Viscount de Noailles, Major William Jackson, and their aides—she mustered 250 men as she dropped down to Nantasket Road, and there waited for a good wind. On 11 February the frigate stood to sea.[90]

"On our passage through the Gulf Stream we had a hard blow . . . and we had some difficulty in handling our main top sail, it being wet and frozen."[91] Once in mid-Atlantic, favorable winds wafted the *Alliance* steadily on her course. Early in March, as she approached the French coast, a small schooner and a tall ship were sighted. A shot from the forecastle gun soon brought both vessels to heel. The schooner was a British privateer, the *Alert*, mounting 10 guns; the ship, a Venetian merchantman, the *Buono Campagnia*, taken by the *Alert* a few hours before. Since the merchantman was neutral and did not carry contraband, Captain Barry allowed her to resume her voyage.[92] Shortly after noon on 9 March, the *Alliance* and her prize dropped anchor north of Isle de Groix "because the wind was high, and not fair, and the harbor [L'Orient] Difficult to enter."[93] There they remained for several days before entering the port.

"My particular orders," Barry wrote to Franklin, are "to wait here four or five weeks should there be any Public Stores offered in that time, I am to take on board as much as Convenient and proceed with them to Philad. and to Convoy any Public Ship or Ships that may have public Stores on board."[94] Within three weeks, Barry's orders were filled, and on 29 March the *Alliance* got under way for her return voyage in company with a large French letter of marque, the *Marquis de Lafayette*, loaded with military stores.[95]

Soon after sailing, a mutiny was discovered among a little group of British seamen. Mate John Kessler described the events:

> On 30th, of March evening an Indian, one of the forecastle men, gave Capt. Barry information of a combination among the crew for the purpose of taking the ship and pointed out three who had strove to prevail on him to be concerned therein. The three men were immediately put into irons and all the officers with such of the crew as could be confided in were armed and required to remain all night on deck.
>
> On the next morning all hands were called and placed on the forecastle, booms and gangways, excepting the officers and such part of the crew in whom Capt. Barry confided, who armed strongly guarded the quarter deck the steerage and main deck to keep the remainder of the crew togather on the forecastle and boom. The three designated men were brought out of their irons on the quarter deck, and being stripped and hoisted by the thumbs to the mainstay, underwent a very severe whipping before either would make any confession, the names of twenty-five of their accomplices were obtained from them before the whipping was discontinued as their accomplices were disclosed, they were called on the quarter deck, stripped and tied to the ridge rope of the netting and the whipping of all continued until it was thought all was disclosed that could possibly be obtained, and which finally proved to be. That it was intended to take the ship on her passage out by killing all the officers, in the middle watch of the night, except the second Lieutant P. Fletcher, who was to navigate her to some port in Ireland, or on failure to be also destroyed.[96]

Although brutal, Barry's actions were completely justified by all the laws of the sea in the Eighteenth Century. He was dealing with desperate men.

Through a cloudy dawn on 2 April, the masthead reported two sail to the northward. Signaling his consort to join in the chase, Barry ordered his crew to quarters. Instead of fleeing, both enemy vessels bore down upon the *Alliance*, and poured a single, but damaging broadside into the American frigate. Barry immediately struck back with two broadsides, cutting away most of the enemy's rigging. Heaving to amid tangled rope and canvas, both enemy ships struck. The larger of the two brigs proved to be the privateer *Mars* out of Guernsey. Considered one of the fastest sailers in Europe, she bore twenty 12-pounders and a crew of 111 men. The smaller brig was the *Minerva* out of the Isle of Jersey, mounting eight 4-pounders and carrying 55 men. Both were later manned by prize crews and ordered off for Philadelphia. A month later two other prizes—a brig and snow loaded with sugar from Jamaica—were taken.[97]

The weather created difficulties for the *Alliance* in May. Heavy gales, accompanied by lightning, became almost a daily occurrence. During one of them, on the 12th, the two prizes were driven from sight and did not return. Five days later, amid rolls of thunder and a heavy sea, a bolt of lightning struck the main topmast, carrying away the mainyard, springing the foremast, felling more than a dozen men, and burning the skin off several. Continuing her course with a shattered mainyard and temporary top and fore masts, the *Alliance* came across His Majesty's sloops-of war *Atalanta* and *Trepassey* on 27 May.[98]

Although rated as sloops, the *Atalanta* and *Trepassey* were a ship and brig respectively. The former carried sixteen 6-pounders and a crew of 125 men. She was commanded by Captain Sampson Edwards. The *Trepassey* mounted fourteen 6-pounders and was manned by a crew of 80. Her commander was a Captain Smith.

At dawn on the 28th, the English sloops hoisted colors, and their drummers beat the crew to quarters. John Barry in the *Alliance* did likewise. The sea was mirror-smooth and "by little puffs of wind we were enabled to get within short hailing distance at eleven o'clock." As the distance between the opponents dwindled, the *Alliance* opened the engagement with a broadside. "Unfortunately," recorded Midshipman Kessler, "there was not wind enough for our steerage way and they being lighter vessels, by using sweeps got and kept athwart our stern and on our quarters, so that we could not bring one-half our guns out astern to bear on them and thus laying like a log the greatest part of the time."[99]

The *Alliance's* quarter deck, swept by grape, and peppered by musket ball, was an inferno. There was not much protection for the Marines on deck and in the fighting tops, whose muskets sought a mark through the opaque clouds of smoke. Lieutenant of Marines Samuel Pritchard was mangled by a 6-pound shot and was carried to the cockpit mortally wounded.[100] Lieutenant James Warren received a severe wound in the right knee; later, the leg had to be amputated.[101] Sergeant of Marines David Brewer, son of Colonel David Brewer of Boston, died at his post, a musket ball through the head.[102]

About two in the afternoon, Captain Barry was wounded in the shoulder by grape shot and as he disappeared below deck, the *Alliance's* colors were carried away. "It so happened," Kessler continued, "that at the same time such guns as would bear on them had been fired and were then loading and which led the enemy to think that we had struck the colors and manned their shroads and gave three cheers." Their jubiliation was short-lived. Using a mizzen-brail for a halyard, a new Continental ensign was hoisted, "and our firing again began." Shortly before three a wind sprang up which slowly swung the *Alliance* around. The entire starboard battery was now brought to bear upon the enemy. The *Trepassey's* colors came down after one blistering broadside. The *Atalanta* still showed fight, but one more broadside ended her resistance. The battle had lasted more than three hours and had cost the *Alliance* five dead and 22 wounded. The *Atalanta* and *Trepassey* sustained a combined loss of 11 killed and 25 wounded.[103]

Through the late afternoon which continued hot and fair, men on board the *Alliance* were employed with repairs to the frigate, and taking possession of the two prizes. Captain Edwards and a few other officers were sent on board the *Alliance*, but Barry did not dare add 200 more prisoners to those already on board. He decided to send the *Trepassey* as a cartel to Halifax with all his prisoners, there to be exchanged for Americans. The *Atalanta,* which had been dismasted during the engagement, was fitted with jury masts and put in charge of naval Lieutenant Hezekiah Welch as prize master who was ordered to take the ship to Boston.[104] Some weeks later the vessel was recaptured off Cape Cod and sent into Halifax.[105] The *Marquis de Lafayette* which had parted with the *Alliance* in April was also un-

Frigate Alliance, Capt. John Barry, Passing Boston Lighthouse From Sea in 1781, by Matthew Parke.

lucky. She encountered the enemy's Jamaica fleet and was taken after three hours of hard fighting.

On the morning of 6 June, the battered *Alliance*, missing a mainyard, sails perforated with innumerable holes, came into Boston harbor—69 days from L'Orient. In spite of his wounds, Barry's first order of business was to urge the Board of Admiralty to sheath the frigate's bottom with copper. "It will not Cost so much to put it on the Ship as it will be to Clean her," he told the Board. "If you would order the Alliance to be sheathed with it you may Keep her the whole War, if not you may be assured that whenever she is Catched at Sea foul that you will loose her. . . . Sheathing the Ship with Copper will render an assential Service to the Country."[106] The Board agreed and the work was begun. The job of sheathing, cleaning, and repairing the frigate would continue into late fall.

The frigate *Trumbull* spent the first half of 1781 slowly fitting out at Philadelphia for a cruise, under the usual difficulties imposed by lack of funds and scarcity of seamen and Marines. The repairs were finally completed by early August. With an American privateer, letter of marque, and a convoy of 28 merchantmen, the frigate took her departure from the Delaware Capes on 8 August. At sea the same day, three sail were discovered to the eastward, two of which gave the convoy chase. That night, in squally weather, the *Trumbull's* foretopmast and maintopmast were carried away, forcing her to run before the wind thus leaving the rest of the fleet. By the time the wreckage was cleared away, one of the three ships sighted came alongside. "Immediately all hands were called to quarters," but as Captain James Nicholson reported:

> Instead of coming, three quarters of them ran below, put out the lights, matches, &c. With the remainder and a few brave officers we commenced an action with the Iris for one hour and thirty-five minutes, at the end of which the other ship came up and fire into us. Seeing no prospect of escaping in this unequal contest, I struck, . . . My crew consisted of 180 men, 45 of whom were taken out of the new goal—prisoners of war; they through treachery and others from cowardice betrayed me, or at least prevented my making the resistance I would have done. At no time of the engagement had I more than 40 men upon deck.[107]

The British 32-gun frigate *Iris*, formerly the American frigate *Hancock*, towed the *Trumbull* into New York where she was left to rot.[108] As in the case of the *Confederacy*, the officers and men of the frigate *Trumbull* were either paroled or confined on the *Jersey* prison ship where they languished until they were either released by their captors or died.

Crowns to Save the Continent

In the late summer of 1781 the Board of Admiralty, hampered by half-hearted cooperation on the part of Congress and a continual lack of funds, ceased to exist after only a year and a half of work. Although Congress voted to appoint an Agent of Marine to take charge of naval affairs, no one was to be found and the office was never filled. Therefore, on 7 September, it placed the management of these affairs under the care of Robert Morris, who, as Superintendent of Finance, had gradually accepted direction as the Board of Admiralty became increasingly impotent. Three days after he assumed the office, Morris summoned Major Samuel Nicholas and Tench Francis, a Philadelphia merchant, to his office where he informed them that the French frigate *La Resolve* had arrived at Boston. What seemed on the surface as routine information was in fact most secret, for the vessel had brought more than two and a half million crowns, a loan from Louis XVI of France. Nicholas and Francis were to bring this money to Philadelphia.[109]

During the course of the long conversation, the burly, 47-year-old Philadelphia merchant specified in great detail just what he wanted done. First, he wished as much of the amount as possible to be invested in good bills of exchange at Boston, both in the interest of safety and of weight. "With Respect to the Remainder," he continued, "I wish it to be brought here in the following Manner. Have small square Boxes made strong of Oak Boards to contain about fifteen hundred to two thousand Crowns each and have large strong Chests made of thick Oak Plank of such Shape as may be most proper and of Size to contain about twenty of the lesser Boxes."[110]

After the lid was nailed down, each chest was to be further secured by a heavy iron strap welded as well as bolted to the axle and tongue of a stout ox cart from which the body had been removed. Four oxen and a horse were to be purchased at Boston to draw each cart, and "the Oxen," as Morris shrewdly commented, "will be worth more Money here after performing the service than their first Cost." To drive the carts, sober, honest, and industrious teamsters were to be hired in Boston and each armed with a musket and bayonet. "I shall apply to the Board of War," he continued, "to order such an Escort of dragoons from Boston to Philadelphia as they shall deem sufficient to write to the commanding officer at Boston to supply an ample Guard whilst there also to supply Arms and Ammunition in defence of the Treasure should they be attacked on the Road, and to the honble General [William] Heath to supply such a Guard of infantry whilst this Money is travelling through the dangerous part of the Road and to take such other Measures as he may judge will render it perfectly safe against every attempt of the Enemy and secure it from all Thieves whatever."[111] Finally, Morris indicated the route Nicholas, Francis, and the money were to return to Philadelphia by: "From Boston to Worcester to Springfield to Greenwood to Salisbury to Fishkill to New Windsor or Newburgh to Sussex Court House or Newton to Easton to Philadelphia."[112]

With the explicit instructions in hand the two men set out immediately for Boston where after the purchase of bills of exchange, more than a million crowns remained to be transported back to the capital. The circuitous route proved to be tiring, but uneventful, and on 6 November the treasure-laden ox carts creaked into Philadelphia.[113]

Tench Francis was rewarded with a position in Morris' Bank of North America. As for Major Nicholas, he went back to the quiet life of a Philadelphia innkeeper. Although called to serve on a board which was to inquire into the loss of the frigate *Trumbull*, the Boston expedition was the last real duty he performed as a Marine officer.[114]

Before the end of 1781, the strength of the Continental Marines afloat was reduced to its lowest point during the war. Seven vessels had been lost within the past two years and with them nearly 200 Marines. Those on board the frigates *Alliance* and *Deane* now constituted the entire strength of the Continental Marine establishment.

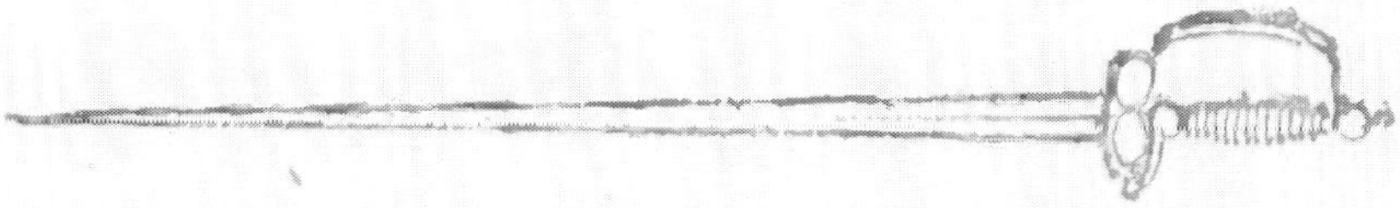

CHAPTER XIII

State and Privateer Marines

Maryland and Virginia

Thus far this volume has dwelt on the activities of Continental Marines, to the virtual exclusion of two other types of Marine forces which served during the American Revolution: state and privateer Marines. Only in those instances such as the expedition to Penobscot Bay and the defense of the Delaware River and the city of Charleston, where they cooperated with Continental Marines, have they been mentioned, yet they, with their naval counterparts, bore the major brunt of America's eight-year naval war with Great Britain.

With the possible exception of New Jersey and Delaware, each of the original thirteen states owned one or more armed vessels during the Revolution. Massachusetts, Connecticut, Maryland, Virginia, and South Carolina had the largest fleets. New Hampshire with one ship and Georgia with four galleys narrowly escaped being put in the same class as New Jersey and Delaware. The navy of no one state was as large as that of Congress', however, the total number of state vessels greatly exceeded those of the Continental Navy. Since the states alloted less than did Congress for naval purposes, they were unable to build large vessels. Then, too, the chief impetus for a navy on the part of each state was to defend its seaports, coastline, and trade routes. For such service small craft were demanded. A few of the states fitted out larger and stouter sailing craft for deep-sea navigation, but offensive warfare on the part of the state navies was secondary in importance, consisting mainly of commerce-destroying ventures conducted along the ocean paths of British trade.

The Maryland State Navy and corps of Marines were typical of the naval forces of the other states. They were local forces, built with local defense needs in mind. The Maryland Navy consisted mainly of galleys or barges which usually carried one or two fair-sized guns, and a crew of 65 to 80 men. The largest ship was the *Defence*. Built in Baltimore under the supervision of Captain James Nicholson, the ship had a keel length of 76 feet with a gun deck of 85 feet. Her draft was expected to exceed 12½ feet when fully loaded with four months' provisions, and 180 men. Her planned armament consisted of twenty-two 6-pounders, besides swivels, but as the time for commissioning drew near, the armament apparently was modified to eighteen 6-pounders on the main deck and two 4-pounders on the quarter deck.

Before the ship was launched, the Maryland Convention meeting at Annapolis had, on 14 January 1776, authorized a company of Marines. The Marines' monthly pay was somewhat less than that paid to Continental Marines:

> Every Drummer and Fifer—6 [Dollars]
> Every Private—5⅓ [Dollars]

Officers pay was to be settled later. For uniforms it was provided that "companies of . . . marines, each of them be yearly found and allowed, at the expense of this province, a new hat, short coat, waistcoat, pair of breeches, and hunting-shirt; out of their pay there may be reasonable stoppages for necessary shirts, shoes, and stockings." It was further provided that the uniform of the land forces and Marines be hunting shirts. Different colors were assigned to the various companies; those of the Marines were to be blue.[1]

It appears that the first four Marine officers appointed for the authorized company were Captain

Garrett Brown, and Lieutenants Thomas Walker, Joseph Smith, and William Morris. The exact dates of appointment are unknown, but Joseph Smith, in a letter to his brother on 20 March 1776, states: "I was just on the point of returning to New England when the Convention appointed me a Lieut. of Marines on board the ship *Defence*."[2]

The *Defence* was "into the river" on the morning of 9 March 1776, although the full complement of Marines intended for her does not seem to have been on board. Samuel Purviance, Jr., a well-known Baltimore merchant, in a letter to Daniel of St. Thomas Jenifer on the 8th, noted that Captain Samuel Smith's company had gone on board the ship as volunteers, apparently to augment the ship's Marine detachment. Smith's company was an independent infantry company from Baltimore that later became a unit of Smallwood's Maryland Regiment. As of 10 April 1776, they were dressed either in a white coat and waistcoat, mixed bearksin breeches, or brown coat and waistcoat, both lined with osnaburg and made short. A bonus was offered and, as Purviance continued, "I found this bold Stroke absolutely necessary as the Ship was really poorly off for Sailors & the Capt seemed to have no great depend[en]ce on his Marines."[3]

Captain George Cook took command of the *Defence* late in 1776, and sailed to the West Indies, capturing five small prizes laden with logwood, mahogany, indigo, rum, and sugar. Available muster rolls indicate that Marines on board numbered 4 officers, 3 sergeants, 3 corporals, 1 drummer, and 34 privates.[4] Shortly after returning from this cruise, Captain Cook, with about 60 or 70 of his crew, left Baltimore on the night of 17 December for the Head of Elk on the way to Philadelphia where they were to guard Continental stores and vessels.

The construction of galleys for the Maryland Navy and the need for men to crew them led to the organization of small detachments of Marines for galley service. Recommendation of John Stevenson of Harford County led the Council of Safety to appoint him a lieutenant of Marines for the row galley *Independence* (Captain Bennett Mathews) on 25 January 1777. The galley *Baltimore* received two Marine officers, Second Lieutenant John Crapper, appointed 17 May 1777, and First Lieutenant James Boyle on the following day. James McCabe was commissioned lieutenant of Marines for the *Chester* galley on 4 June 1777. Efforts to get the galleys into service were not particularly successful as the Council reported on 31 August 1777: "We have but two of the Row Galleys fitted. The Enemy have spare Men of War, enough to block them up separately; they may eventually, with their 120 Men, be serviceable."

Before the close of 1777, the General Assembly passed a resolution authorizing the sale of the ship *Defence*. The Marines were allowed either to enter on board the galleys or to join a matross (artillery) company. Rules were drafted to cover the procedure for having their names stricken from the rolls of the *Defence*.

By late March 1779, the chairman of the Maryland Council was forced to report to the General Assembly that serious problems were frustrating attempts to maintain a navy:

> The Bounties and Encouragements offered to Seamen and Marines in the Coarse of the last Session have not had the Effect you expected and I wished; the Officers have engaged a few but, I believe, our Number is not on the Whole at all increased. The Congress having laid aside the Scheme of employing the Baltimore and Independence which indeed we could not Man and the Price of Workmanship Labour and every Article necessary for finishing and fitting out Vessels having greatly advanced the Council and I thought, as things were circumstanced, it would have taken in fitting out the Annapolis and therefore nothing has been done on her in Pursuance of the 15th [section] of the 10th Act of the last Session but if it continues your Desire to have that Galley fitted out, we shall chearfully have it done with all Expedition.

The General Assembly acquiesced, and during the first half of 1779, Maryland sold all of her naval craft except the galleys *Conqueror*, *Chester*, and the schooner *Dolphin*.[5]

It was not possible for Maryland to withdraw completely from naval affairs. An act passed in October 1780 provided for the defense of the Chesapeake Bay by authorizing the construction of four large barges, a galley, and a sloop or schooner. The third paragraph of the act has particular reference to Marines:

> That a company of one hundred men be immediately raised to serve as Marines on board the said galley and sloop or schooner, and occasionally on board the said barges or rowboats; and that the governor and council be authorized and requested to appoint and commission one captain and two lieutenants to command the said company of marines, and to direct such officers to procure by enlistments as soon as possible, the said number of healthy able bodied men, including two sergeants and

> two corporals, to serve in such company for the term of three years, unless sooner discharged.[6]

These Marines were to be paid two pounds, five shillings a month and a bounty of 40 dollars. It was not until 1782 that this new Marine establishment was effected, and the Governor and council proceeded with their efforts to raise and officer 250 able seamen, watermen, landsmen, and Marines who were to serve until 1 January 1783, or longer.

The commission as captain of Marines went to Levin Handy who had served as a lieutenant in the 4th Maryland Battalion of the Flying Camp in 1776 and then as captain, 5th Maryland Battalion, from 10 December 1776 to 1 May 1780. He was appointed to serve on board the barge *Protector* on 3 August 1782, but his service was of short duration as he was captured off Tangier Island in November 1782 in what came to be known as the "Battle of the Barges." As Colonel John Cropper, a volunteer officer on board *Protector*, wrote: "Commodore Whaley was shot down a little before the enemy boarded, acting the part of a cool, intrepid, gallant officer. Captain Joseph Handy fell nigh the same time, nobly fighting with one arm, after the loss of the other. Captain Levin Handy was badly wounded. There went into action in the Protector sixty-five men, twenty-five of them were killed and drowned, twenty-nine were wounded, some of which are since dead, and eleven only escaped being wounded, most of whom leaped into the water to save themselves from the explosion."[7] The two authorized lieutenants, if commissioned, have not been identified.

State Marines generally were stationed on board ships which operated in coastal waterways for defensive purposes. One company of Virginia State Marines, raised in 1782, proved to be the exception.

As the American Revolution drew to a close, George Rogers Clark was faced with the monumental task of maintaining military control over the Ohio Valley. With few men at his disposal, he, nonetheless, devised several schemes which granted him the best possible control with his small force. One method was to establish strong posts at key points, the other was to construct armed galleys for use in controlling the navigation of the Ohio, particularly at the mouths of the Miami River and Licking Creek.

Governor Benjamin Harrison of Virginia heartily approved of such efforts. Of the river galleys, he wrote: "They will certainly render essential service. I wish therefore to have three or four of them built as soon as possible. It you are in Want of cannon to mount on them they shall be sent early in the Spring to Fort Pitt or any other post more convenient that you shall direct."[8]

Even though the Virginia Treasury was about empty, the Governor and Council authorized a warrant of 50 pounds for the building of the boats.[9] Any additional expense was to be borne by Clark himself.

By early May 1782, Clark was able to report that he had two gondolas ready "to mount [their guns]" and a galley on the stocks that would be completed in about 20 days. This last vessel was unusual in several respects. She had a 73-foot long keel, designed for navigation of the Ohio River. As a row galley, she provided for 46 oars with a complement of 110 men. Her armament was to be one 6-pounder, six 4-pounders, and one 2-pounder. At this time the cannon promised by the governor had not yet reached Fort Nelson at the Falls of the Ohio (now Louisville, Kentucky) where the craft was being built.[10]

The two open gondolas built by Clark proved unsuccessful for they were vulnerable to ambushes as they approached shore or traversed narrow parts of the rivers. The unique construction of the row galley avoided this pitfall. Her gunwales were four feet high and thick enough to stop both arrow and bullet. In addition, she had false gunwales mounted on

strong hinges which would, when positioned, raise her sides high enough to be safe from musket fire from shore.[11] This feature seems to have been original, as it is not found in other row galleys of the period.

Only sparse information is to be found on her construction, which was directed by James Asturgas for two dollars per day, Spanish money. Manpower was short, and it was reported that militiamen were employed in building both Fort Nelson and the vessel.[12] While timber was plentiful, commonplace items such as rope was scarce, and a substitute made from papaw bark had to be used. The cannon were brought overland to Redstone (now Brownsville), Pennsylvania and floated down the Monongahela and Ohio Rivers to the construction site.[13]

It was no easy task to assemble the 110-man complement required. Among the first steps taken by Clark was to authorize the recruitment of a company of Virginia State Marines. Selected as captain was Jacob Pyeatt who had been performing commissary duties both as a civilian and as an officer in the Illinois Regiment since September 1778. Pyeatt's company of Marines was to serve for six months and,

George Rogers Clark, by James Barton Longacre.

when finally enlisted, numbered 20 men and one other officer, Lieutenant William Biggs. Many of the enlistees were discharged men who were enticed to re-enter the service on the promise of a suit of clothing and 10 dollars a month pay. The company muster included one captain, one lieutenant, two ship's carpenters, three sergeants, and 15 privates.[14]

Pyeatt's company of Marines was probably enlisted to constitute the permanent military force on board the row galley, and it was probably expected that they would, among other duties, constitute the vessel's gun crews and mount guard over both the magazine and the ship herself. As soldiers had enlisted for service on the water, they were properly designated as "Marines," although still constituting a unit of Clark's loosely bound Illinois Regiment.

Records do not indicate that Clark's row galley was officially given a name, but available evidence indicates that she was referred to commonly as "the Miami galley," after the river of the same name. Private Cornelius Darnel, in a deposition made several years later, noted that he "met with Captain Jacob Pyeatt a captain of the marines, who was out recruiting or enlisting marines for manning a boat or ship on the Ohio river called the Rogalia."[15] The name "Rogalia" is assumed to be no more than a phonetic spelling of the expression "row galley."

The services performed by the *Miami* galley are vague but impressive. Her summer patrol of the Ohio adjacent to Shawnee Territory caused alarm among the Indians, who thought Clark was preparing for an incursion. Two British captains from Detroit, William Caldwell and Alexander McKee, had gathered an Indian army of nearly 1,000 braves to make a strike at Wheeling and were on the march when news of the supposed thrust by Clark caused the Indians to break off the march in order to defend

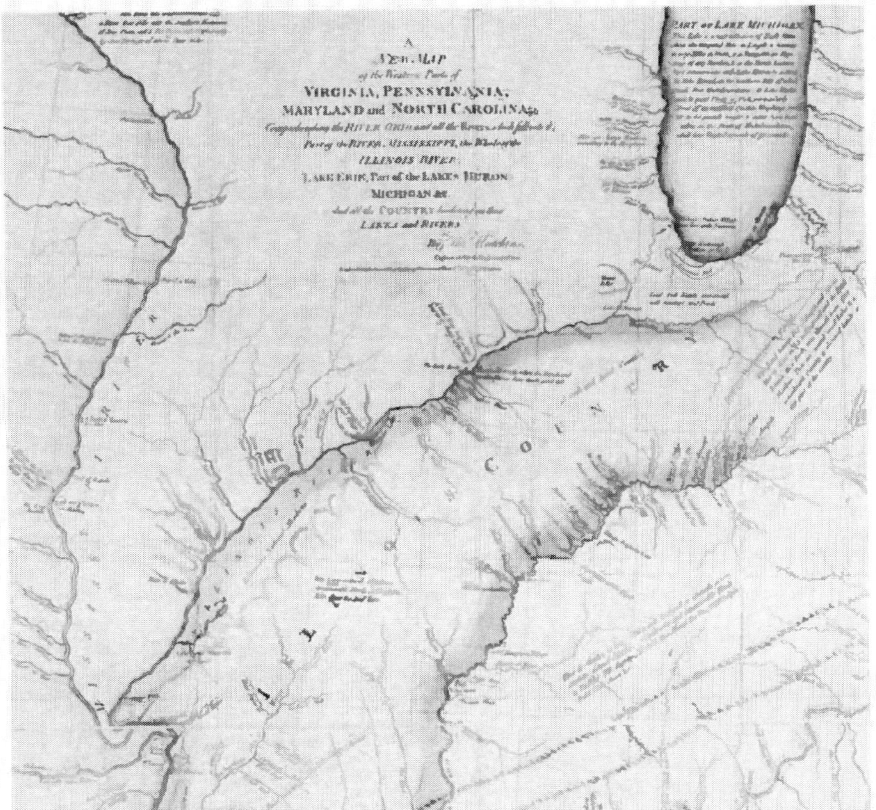
A
VIRGINIA, PENNSYLVANIA,
MARYLAND and NORTH CAROLINA
ILLINOIS RIVER
MICHIGAN
VIRGINIA

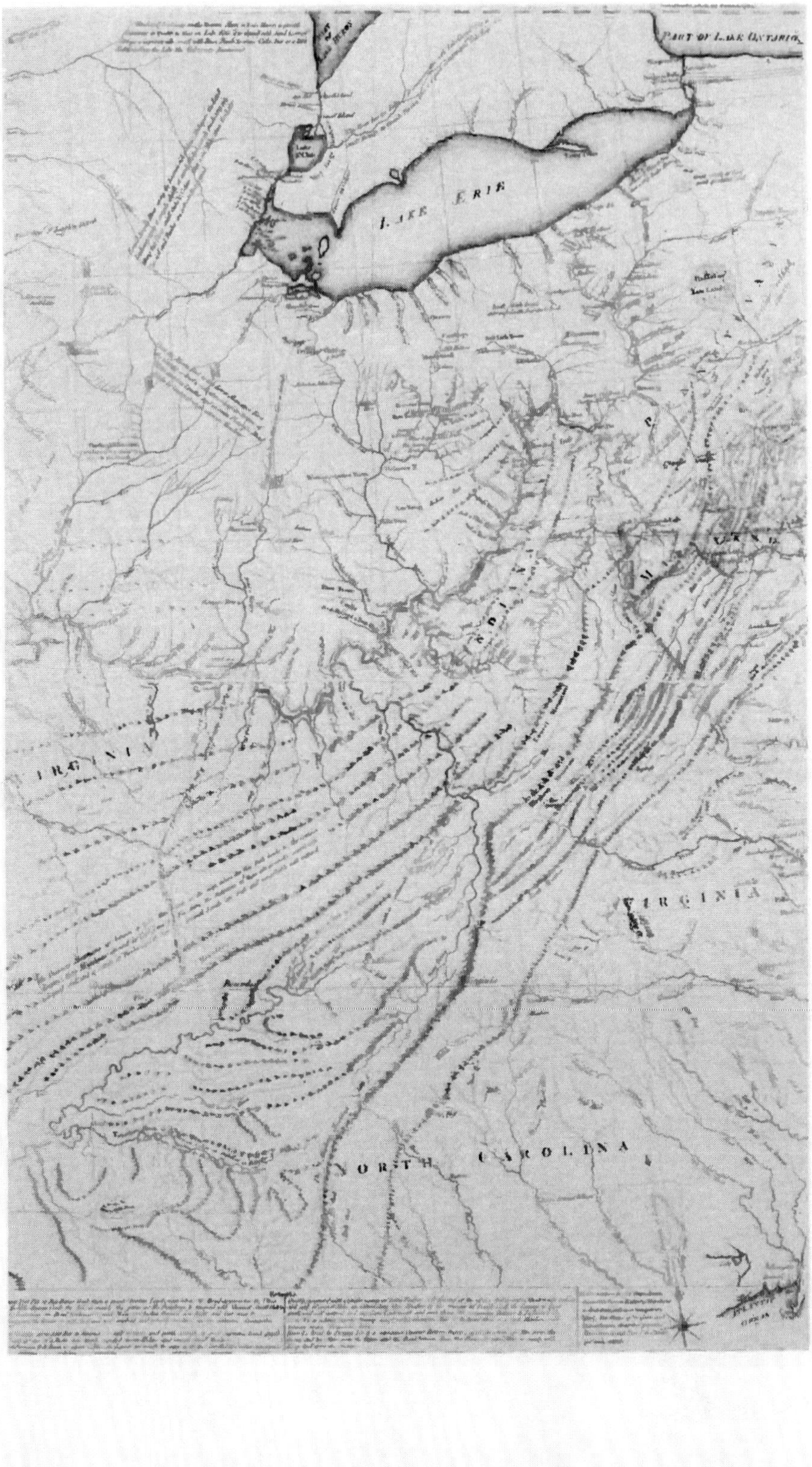
PART OF LAKE ONTARIO
LAKE ERIE
VIRGINIA
VIRGINIA
NORTH CAROLINA

their own country.[16] The postponed attack on Fort Henry (Wheeling) was finally made in mid-September 1782 by a much smaller force estimated at 250 Indians and 40 Tories.

The *Miami* galley made a contribution to the defense of the frontier even though she had but a short life. Private Darnel reported that the vessel sank at Bear Grass at the Falls of the Ohio about 1 September 1782, and that he and the remaining men of Pyeatt's Marine company were then transferred to the Illinois Regiment.[17] Like the Maryland Marines of the ship *Defence*, Jacob Pyeatt and his Marines contributed in their small way to the final defeat of the British.

In Private Pay

In addition to serving on board Continental and state vessels, American Marines served on the thousands of privateers, letters of marque, or private ships of war, as they were often called, commissioned by the Continental Congress or the states. From the very beginning, these private vessels of war roamed the seas, and this implies the existence, at the outbreak of the American Revolution, of an element of naval preparedness in the American colonies which has seldom been appreciated. These privateers at times were the sole reliance of the thirteen states in disrupting the supply system and lines of communication of the enemy, bringing millions of dollars of essential stores and war materiel to American land forces, and capturing or destroying British warships, merchantmen, and privateers, but at a profit to their owners.

On 23 March 1776, the Continental Congress authorized privateering and in less than two weeks approved a form of commission for privateers and resolved to send blank copies, signed by the President of Congress, to the various colonies, there to be issued to bonded privateersmen. A set of instructions was also drafted for privateer officers.[18] Several of the colonies relied solely on these Continental commissions, and did not establish state privateering. Pennsylvania, for example, sent out approximately 500 vessels under Continental commissions. The remaining states actively participating in naval affairs, issued their own commissions, while also using those of Congress. The American commissioners in Paris, and the Continental agent in the

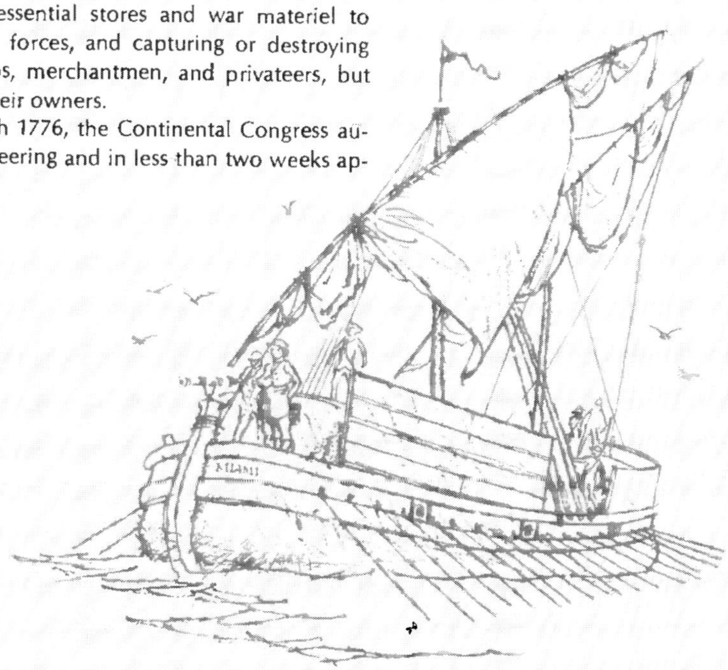

West Indies, also commissioned privateers. The total number of American vessels engaged in privateering is unknown. Making allowances for duplication on the part of the states, Congress, and its agents, private armed vessels numbered approximately 3,000. If this were true, the vessels doubtlessly carried near 100,000 men.[19]

It is equally difficult to ascertain the number of Marines who served on board the privateers, but they probably numbered in the thousands. Naval historians have insufficiently emphasized the fact that all important private vessels carried Marines. Instead, they refer to them as gentlemen sailors, gentlemen volunteers, gentlemen officers, landsmen, and soldiers, whereas the correct term, generally used, was "Marine." This omission has fostered the impression that privateers carried no Marines, or, at most, only a very few, which certainly was not the case.

Recruiting for privateers was always easy, for the inducements offered were far superior to those of the Continental and state navies. Since privateers were devoted to destroying commerce, crews enjoyed fewer restraints, larger profits, and higher pay, for the entire value of their prizes went to the owners and captors. Although there seems little doubt that privateers assisted the Continental cause, particularly because they involved no financial outlay by Congress, they were in many ways detrimental to the regular naval service. "The whole attention of Merchants and Seamen at present seems to be on Privateering through the whole New England Colonies," angrily wrote Commodore Esek Hopkins in September 1776.[20] A month later, he estimated that about one third of the men who signed on board Continental ships had, after drawing a bounty and an advance wage, "been one way or another carried away in the Privateers."[21] William Whipple, writing to Josiah Bartlett in July 1778, while stressing the valuable service of privateers, questioned their impact:

> No kind of Business can so effectually introduce Luxury, Extravagance and every kind of Dissipation, that tend to the destruction of the morals of people. Those who are actually engaged in it soon lose every Idea of right & wrong, & for want of an opportunity of gratifying their insatiable avarice with the property of the Enemies of their Country, will without the least compunction seize that of her Friends . . . There is at this time 5 Privateers fitting out here [Portsmouth, New Hampshire], which I suppose will take 400 men. These must be by far the greater part Countrymen, for the Seamen are chiefly gone, & most of them in Hallifax Gaol. Besides all this, you may depend no public ship will ever be manned while there is a privateer fitting out. The reason is plain: Those people who have the most influence with Seamen think it their interest to discourage the Public service, because by that they promote their own interest, viz., Privateering.[22]

There were times, as in 1779, when an embargo was placed on privateer recruitment. Privateers were required in many instances to give bond not to recruit in any town which had not yet raised its quota for the Continental Army or Navy.

Physicians, lawyers, army officers, politicians, merchants, and even ministers of the Gospel were found serving as Marines on board the privateers. It is reported that when the schooner *Revenge* was captured by the British privateer *Belle Poole,* and the American prisoners ordered to Portsmouth, one of the "Gentlemen Volunteers" was discovered to be a woman.[23] Seemingly, men of all professions, and possibly a few women, were carried away by the craze of privateering.

Unfortunately, many of the cruises and actions of these privateers were not recorded or preserved. Numerous battles were fought, daring raids on enemy coasts were made, and many heroic incidents occurred. However, most of their logs, journals, diaries, and letters have been lost, and the information that exists is, as a rule, meagre and fragmentary. Enough is known, nevertheless, to prove that these private ventures were fraught with thrilling action and were most important in their bearing on the results of the war. The following illustrations of the actions of privateer Marines will demonstrate the validity of their contribution.

Early in September 1781, George Washington's estate on the Potomac River was plundered by the crew of the 16-gun British sloop of war, *Savage*. On the 6th, the Pennsylvania privateer ship *Congress* of 24 guns, Captain George Geddes, came upon the *Savage* and captured her. Captain Stirling of the sloop describes the action:

> Early in the morning of the 6th instant, 10 leagues East of Charles-town, we espied a ship bearing down on us, who when about four miles distant, hauled her wind to the Eastward, shewing by her appearance she was an American cruizer; her force could not be so easily distinguished. I therefore gave way to the pleas-

ing idea that she was a privateer carrying 20 nin-pounders, whom I had intelligence was cruizing off here, and instantly resolved either to bring her to action or oblige her to quit the coast, for which purpose we gave chase, but were prevented continuing it long by her edging down, seemingly determined to engage us. Conscious of her superiority in sailing and force, this manoeuvre coinciding with my wishes, I caused the Savage to lay by till we perceived on her nearer approach she was far superior to what we imagined and that it was necessary to attempt making our escape, without some fortunate shot, in the course of a running fight we saw inevitable, admitted our taking advantages and bring on a more equal conflict. At half past ten she began firing bow chacers and at eleven, being close on our quarter, the action commenced with musquetry, which after a good deal of execution was followed by a heavy cannonade on both sides. In an hour's time I had the mortification to see our braces and bow-lines shot away and not a rope left to trim the sail with, notwithstanding every precaution had been taken; however, our fire was so constant and well-directed that the enemy did not see our situation, but kept alongside of us till accident obliged him to drop astern. The Savage was now almost a wreck, her sails, rigging and yard so much cut that it was with the utmost difficulty we could alter our position time enough to avoid being raked, the enemy lying directly athwart our stern for some minutes. This was the only intermission of great guns, but musquetry and pistols still did execution and continued till they opened again, which was not till both ships were almost on board each other, when the battle became more furious than before. Our quarter-deck and forecastle were soon now nearly cleared, scarce a man belonging to either not being killed or wounded, with three guns on our main-deck rendered useless. In this situation we fought near an hour with only five six-pounders, the fire from each ship's guns scorching the men who opposed them, shot and other implements of war thrown by hand doing execution, when our mizen-mast being shot away by the board, our main-mast tottering with only three shrouds standing, the ship on fire dangerously, only 40 men on duty to oppose the foe who was attempting to board us in three places, no succour in sight or possibility of making further resistance, I was necessitated at a quarter before three P.M. to surrender to the Congress, a private ship of war belonging to Philadelphia, who carried 215 men and mounted 20 twelve pounders on her main-deck and 4 sixes above, fourteen of which were fought on one side. She lost during the action eleven men and had near thirty wounded, several of them mortally; her masts, her sails and rigging were so much damaged that she was obliged to return to port, which partly answered my wishes prior to the action, as a great part of the Carolina trade was daily expected on the coast and this privateer we saw sailed remarkably fast. Three days were employed putting her in a condition to make sail and five for the Savage, who was exceedingly shattered. Indeed it is astonishing more damage was not done, as the weather was fine, the water remarkably smooth, and the ships never 30 yards asunder.[24]

Pennsylvania Militia Major Samuel Allen McLane, who with a part of his command had volunteered to serve as Marines on board the *Congress*, was highly commended for the bravery with which his men responded to the enemy's musket fire. Soon after the *Savage* struck McLane went forward to look for Sergeant Thomas. He found him with both legs broken, lying on his back in the netting, near the foot of the bowsprit, with his musket loaded. He was cheering for victory, and exclaimed: "If they have broken my legs, my hands and heart are still whole."[25] Sergeant Thomas recovered the use of his legs and later served on board another Pennsylvania privateer, the *Hyder Ally*.

In the winter and early spring of 1782, Delaware Bay was infested with privateer barges and other small craft, fitted out by loyalists, which preyed upon Philadelphia's commerce. While awaiting Congressional action, the merchants of the city took it upon themselves to fit out a ship called the *Hyder Ally* as a privateer under Continental commission. She mounted four 9-pounders, 12 sixes, and was manned by a crew of 120, of which a number were Marines. The command was given to Joshua Barney, lieutenant in the Continental Navy, who had recently returned from a long imprisonment in England.

On 7 April, the privateer, with a convoy of merchantmen, stood down Delaware Bay to Cape May Road. The following morning, the British sloop of war *General Monk* and privateer *Fair American*

Joshua Barney, by Eugene L. G. Isabey.

were observed standing in toward the fleet. The Americans immediately got under way, stood up the bay, and dispersed. Later an Englishman on board the *General Monk* wrote:

> We soon came up with the Hyder Ally, nothwithstanding she cut her boat adrift and did everything else to get away. We meant to have run upon her quarter and board her and once, but after firing two of our bow chaces when at 100 yards distance, she put her helm a-port and stood right athwart us, therefore we did the same, to prevent being raked, when the action began and we edged towards her till within close pistol-shot. We with great concern soon found our short guns (carronades) to become totally unmanageable and that two-thirds of the shot we fired did not strike the hull of our antagonist. After having sustained the action for ten minutes with musketry only, the decks full of killed and wounded (among the former the Lieutenant and Master, two brave Officers), our rigging so much shot as to render it impossible to haul off, and lastly, seeing no prospect of assistance from the Fair American, Captain Rogers was under the mortifying necessity of striking his Majesty's colours to the *Hyder Ally*.[26]

Much of the credit was due to a detachment of Buck's County Riflemen under a Captain Scull, who served as Marines and whose accurate fire decimated the enemy. As Mary Barney later wrote:

> Many incidents occurred during the heat of this rapid and vigorous action which are well worthy of notice:—Captain Barney, in order that he might the better see all that was going on and regulate his movements accordingly, remained standing upon the binnacle during the whole action, in the most exposed point of his quarter-deck, particularly to the fire of the musketry from the enemy's tops. On one occasion, a ball passed through his hat, just grazing the crown of his head—another tore off a part of the skirt of his coat: seeing himself thus the aim of the small arms, he called to Mr *Scull*, his marine officer, (whose men were all *Buck's County riflemen*, who had never before been on board a ship—) and ordered him to direct his fire into the top from which he was so much annoyed; the order was promptly executed, and with such good aim that every shot brought down its man.—A few minutes after this, one of these brave fellows, who was much better acquainted with the use of his rifle than with the rules of subordination, called out to Captain Barney, with a coolness of tone and familiarity of manner that evinced anything but intended disrespect:—'Captain! do you see that fellow with the *white hat?*' and firing as he spoke, Captain Barney saw the poor fellow 'with the white hat' make a spring at least three feet from the deck, and fall to rise no more. 'Captain!' continued the marksman, *'that's the third fellow I've made hop!'*—It is a remarkable fact, highly indicative of the deliberate coolness of these Buck's County men, that every man of the enemy who was killed by the small arms, was found to have been shot in the head or breast—so true and deadly was their aim.[27]

This brilliant action caused the Pennsylvania legislature to resolve that it entertained "a just sense of the gallantry and good conduct of Captain Joshua Barney, and the officers, seamen, and marines under his command."[28]

Marines on board privateers visited the West Indies, England, Ireland, Scotland, Spain, France, and numerous other places. They received ample pay, participated in prize money, and were pensioned by the United States Government. Many of them languished in prisons until the end of the war, or until they were exchanged or escaped.

As privateer and state Marines passed from the scene at the end of eight grueling years of war, so too did Continental Marines. By the end of 1781, only two Marine detachments remained.

CHAPTER XIV

Mustering Out, 1782-1783

The Last Two Frigates

With the navy reduced to two frigates, and the nation almost bankrupt, naval administration required simplification and economy. In lieu of an Agent of Marine, the authority over naval and Marine affairs was entrusted to the Superintendent of Finance, Robert Morris, who, five years before, had been the guiding spirit of the original Marine Committee. His first step was to appoint a knowledgeable clerk; he selected John Brown for this important position. As secretary to the Marine Committee and later the Board of Admiralty, Brown was conversant with all past naval policies, commitments, and mistakes. Morris' second step was to concentrate upon getting the *Alliance* and *Deane* outfitted and manned. To accomplish this, and to appoint a Deputy Agent of Marine for the New England area, Brown hurriedly left Philadelphia for Boston on 22 September.[1]

It was largely a matter of a crew with the *Deane*. With the *Alliance*, however, it meant a complete job of refitting the frigate, for she required new fore and mainmasts, a full set of canvas, and a thorough overhaul. With much to be done, and ample funds available at Boston, Brown made excellent progress. By late October the *Alliance* was declared ready for sea. The enlistment of crews for both ships became the next priority.

Captains Barry and Nicholson, with the approval of Brown and the Eastern Navy Board, produced an appeal for seamen and Marines on 30 October. They began by painting a glowing picture to "shew the Advantage of this Service to be superior to any other." Officers and men would be entitled to the whole of any vessel of war captured, to a bounty of 20 dollars for each gun taken, to a bounty of 10 dollars for each man captured, and to half the value of any merchantmen and their cargoes. There was also a bounty for entering—10 dollars to each seamen or Marine, and eight to each landsmen or ordinary sailor. Wages would be eight dollars monthly for seamen and Marines, six and two thirds dollars for landsmen. Both the bounty and wages would be paid "punctually in Silver or Gold," not in devalued Continental paper money.

The allurements did not end there. Clothing and slops would be provided at cost. Men wounded in action were to receive a 200 dollar bonus which was to be paid before the division of prize money. Public hospitals would care for all the wounded, and those disabled in the service would be given half-pay pensions for life. In addition to all that was offered, the two captains held out the prospect of adventure:

> ALL able-bodied Seamen, ordinary Seamen and Landsmen, are hereby informed, That the two fine fast sailing Continental Frigates ALLIANCE and DEANE, the first commanded by JOHN BARRY, Esq; and the other by SAMUEL NICHOLSON, Esq; are bound in Consort on a Cruize against the Enemy, and will sail soon, with every Prospect of making a very adventageous Cruize.[2]

A Crisis and a Cruise

On the day the appeal was published, the *Alliance's* muster roll contained but 102 names, officers included. Of the seven enlisted Marines on board, only the drummer and fifer had served with Barry on a previous cruise.[3] The re-enlistment ratio was about the same among seamen and landsmen. Midshipman John Kessler wrote that "such was the attachment of the crew to Captain Barry that, on their being paid off and the question put whether they would ship again, they cheerfully agreed to enter." Had this been the case, the appeal for new recruits would have been unnecessary. As Kessler went on to point out, Barry "refused to admit such as had conducted themselves grossly amiss."[4] This barred a number of hands, including all those even remotely connected with the incipent mutiny discovered during the last cruise.

Among the three Marine officers who had seen past service with Barry, only Captain Matthew Parke and Lieutenant Thomas Elwood returned; Lieutenant of Marines James Warren still suffered from wounds received in the engagement with the *Atalanta* and *Trepassey*. Warren's post was filled on 1 December by the appointment of William Morris, a young Marine lieutenant who had served on board the *Ranger* until her capture at Charleston in the spring of 1780.

Boston church bells pealed joyously on 10 December, for the illustrious Marquis de Lafayette had arrived. John Barry, however, could not share the city's elation. This lack of enthusiasm was traceable to one reason. With the Marquis had come new orders from Robert Morris which directed the *Alliance* to carry Lafayette and his staff to France. Instead of a cruise, with prospects of rich prizes, he was to avoid all enemy vessels and safely land his distinguished passenger. According to Morris, Barry's sole object was "to make a quiet and safe Passage to some port in France." After depositing the Marquis, Barry was to be permitted a short cruise in European waters as compensation. Even this authorization was circumscribed by instructions to return to a French port by 1 March 1782, receive dispatches, and then sail promptly homeward, putting into the most convenient American port, preferably Philadelphia.

Fearing the *Alliance* was still undermanned, Morris inserted a vague clause in the orders: "should she still want men, and part of a crew are engaged for the Deane, they had best be turned over to the Alliance so as to compleat her complement." Also, Barry was permitted "to take on board such French seamen as the Consul can procure." If these two efforts failed to fill the frigate's roll, Barry was to enlist the aid of Lafayette in a joint appeal to Governor John Hancock for permission to impress.[5]

Still hopeful of recruiting the needed number of crewmen, Barry waited until 18 December before asking Captain Nicholson for men from the *Deane*. Reluctantly, the captain wrote Nicholson, citing the paragraph from Morris' orders and pointing up his failure to procure a full ship's company. "Therefore I am obliged to call upon you for forty men to enable me to proceed to Sea agreable to the orders," he continued, "and I must request you will have the said Number ready to be deliver'd by 12 oClock to morrow to one of my Officers who I shall send for that purpose."[6]

The following morning Nicholson gave his reply. He wished the "orders, for taking my Men were possitive," so he might not become involved in the most disagreeable of circumstances. His officers had heard rumors that men from the *Deane* were to be drafted for the *Alliance*, and immediately protested. From their enclosed petition, Barry could see that the officers had enlisted neighbors and neighbors' children from their respective home towns, engaging them specifically for a cruise in the *Deane*. "The very mention of their being turned over to an other Ship," Nicholson noted, "has caused the greatest uneasiness amongst Officers and Men, which would prevent in my Opinion the future manning of the Ships, on any terms what ever." Nevertheless, so great was his regard for the good of the service, that he offered to "muster every man and join my utmost exertions to yours to induce them to go on board of your Ship." The offer was meaningless since Nicholson had no intention of forcing any man from the *Deane*.[7]

In responding to Nicholson's couched refusal, Barry on 20 December, again demanded he "immediately deliver to one of my Officers . . . forty good men from your Ship."[8] Nicholson again refused, but on the 22d relented and sent 20 men—ten Marines, nine seamen, and one Marine drummer.[9] With the addition of 37 French seamen, the *Alliance's* roll then totalled 255 officers and men, of which approximately 46 were Marines.

With the crew finally made up, the *Alliance* moved down to Nantasket Road. There, on Christmas Day, Barry found opportunity to send back to Boston, an "Account of Stores under the Care of the Several Officers belonging to the Continental Ship Alliance." Included among the lists was one of the most complete inventories of small arms belonging to a Revolutionary War shipboard Marine detachment:

96 Muskets 13 of which without Bayonets
97 good Cartridge Boxes
14 bad ditto
48 Cutlasses very ordinary
40 Pistols ordinary
1 Drum
2 Fifes
1 spare drum head
2 arm Chests with Locks
2 spare Padlocks for same
300 Flints[10]

Late that day, with a strong southerly breeze, the *Alliance* stood to sea.

The passage to France was relatively quiet and uneventful, except for the mutterings of disgruntled crewmen. Barry's orders, as they knew, eliminated any possibility of prize-taking and most of them had shipped for a cruise, not a cautious voyage. All secretly wished that "the Marquis was in France." During the first week of January dissatisfaction was openly expressed when the frigate changed course to avoid "a ship which appeared as if she could give them sport." Ordinarily, Barry would have been quick to punish such insubordination, but on this occasion it was apparent that the captain, torn between duty and desire, also wished to be rid of the Marquis. Whatever his feelings, he concealed them from his passengers and sailed on in "sullen silence."[11] On 17 January, after a passage of 23 days, the *Alliance* came to anchor off L'Orient.

For nearly a month, Barry pleaded with both American and French authorities for men, but to no avail. After releasing the 37 French seamen and enlisting a few volunteers, Barry put to sea in disgust on 9 February for a short cruise in European waters. In less than a month, he was forced to give over his quest for prizes because of poor success, "owing entirely to the greatest part of the Trade of Urope, being Carried on in Neutral Bottoms, of which the Sea is full, add to that I had nothing but gales of Wind ever Since I Sailed, which has been the Occasion of Several of my Men's getting Sick—Not a Sign of an English Vessel to be Seen the whole Cruize."[12]

Upon his return to L'Orient, Barry found himself embroiled in a new conflict. Hoping to find dispatches for America, he instead found orders from Franklin directing the *Alliance* to Brest, there to take on a quantity of public goods and then to sail for America with a French fleet about the middle of March.[13] Annoyed, Barry dashed off a letter to the American Commissioner in which he pointed out that the detour would be counter to the orders he had received from Robert Morris. He would stretch

Robert Morris, by Charles Willson Peale.

his orders further than he wished, and defer sailing, if Franklin would send the dispatches by return post.[14] After much exchange, Franklin's dispatches finally arrived on 15 March with a letter wishing the captain "a Prosperous Voyage," but warning that he would be blamed for refusing to go for the public goods at Brest.[15] With a half-month delay to make up, the *Alliance* headed out to sea on 16 March 1782.

The return voyage was long and tedious. Day after day the winds remained unfavorable, driving the frigate off her course. Disgruntlement flared below decks as the crew thirsted for prizes. Occasionally vessels were stopped, but none proved to be enemy craft. On 10 May Cape Henlopen was finally sighted. Within the bay lay one of His Majesty's 64-gun ships-of-the-line and her tender. Access to Philadelphia barred, Barry put about and fled down the Delaware coast with the enemy vessel not far behind. Keeping to shoal water, he forced his adversary to give over the pursuit and return to her station. The following day, Barry, taking advantage of a strong southerly wind, ran northward and toward sunset on the 13th sailed up the Thames River and anchored off New London, Connecticut.[16]

Mutiny and Resignation

When the *Alliance* reached New London, her crew was on the verge of mutiny. Having been denied prizes, they now demanded liberty and back pay. Their pay was promised, but liberty was out of the question for fear that if permitted ashore half the crew would desert. The situation worsened each day and by the 16th the crew was ready to take action. Shortly after nine in the morning, a shallop came alongside the frigate with a load of fresh meat. As was the custom, the officer of the deck directed the quartermaster to pipe all hands on deck in order to unload. The latter did so, but the crew refused, and again demanded "Liberty and back allowance." Spurred on by a few, they rushed on deck from the main hatchway in an apparent attempt to force the issue. Captain of Marines Matthew Parke quickly armed the other officers who then drove the men below.

Throughout midday the mutineers rioted between decks, smashing everything in sight. Late in the afternoon, Captain Barry returned and the mutiny was brought to a quick end. The conspirators were placed in irons and the rest of the crew ordered to unload the provisions.[17]

That night, Barry penned to Robert Morris an account of his recent voyage and its unpleasant conclusion. Of the trouble which he had just quashed, Barry wrote sparingly: "This Day there has been a Verry Dangerous Mutiny on board. . . . I have Sixteen of the Ringleaders in Irons. Should be glad to Know what is to be done with them, as there is not Officers here to hold a Court Martial." The following morning he gave the letter to Marine Lieutenant William Morris, who was to be his personal courier to Philadelphia, and then turned to more pressing tasks which were to occupy several months of his time.[18]

His saddlebags filled with dispatches, William Morris galloped into the capital at sunset on 22 May.[19] After making his report, Robert Morris suggested that he contact Robert Mullan at Tun Tavern or Samuel Nicholas at the sign of the Conestoga Wagon for a bed. Lieutenant Morris spent the next several days, while he waited for return dispatches, chatting with the other two Marine officers, often far into the night. During one of these long conversations, he apparently mentioned that Matthew Parke intended to resign his commission, for on the 27th Robert Mullan suddenly appeared at Morris' office and offered "his service for going as Captain of Marines on board the Alliance." Morris told the unemployed Marine captain that he had not as yet "received official information," and therefore he could do nothing.[20] On 6 June, Lieutenant Morris filled his bags with "Sundry Letters" and set off for New London.[21]

Shortly after his return from Philadelphia, Lieutenant Morris chose to resign his commission, pleading personal business matters:

> I fear that my leaving of it will in some Measure Occur your Displeasure, but belive me it was not out of Dislike to you or your Officers, or the Ship you Sail in is the Occation, but the Nature of my Bussiness at home which is my whole Dependence is such that I cannot possibly go to sea before I regulate my bussiness at home which demands my immediate Attention, but believe me sir, it is w[i]th regret I leave you from the politeness I have always received from you, as allso from your Good Family in Philadelphia, which my abilities Can never make satisfaction for.[22]

Barry was not to be mollified by such flattery, particularly since the lieutenant had not accounted for money used to defray his expenses to and from Philadelphia. "His Conduct in so doing is so Notorious," the captain wrote Robert Morris, "that I have return'd him a Deserter from the Ship & which I hope will meet your approbation."[23] Lieutenant Morris, however, was one step ahead of the captain, for on the same day he submitted his resignation to Barry, he did likewise to Robert Morris. On 30 July John Brown responded for the Agent of Marine:

> I am directed by the honble. Mr. Morris . . . to inform you that he has accepted your resignation altho he is sorry to see, (what he conceives you to be) a worthy Officer depart from the Service—He wishes you much happiness in private life.[24]

Marine Captain Matthew Parke chose this inopportune moment to recommend Marine Lieutenant Samuel B. Hempsted of New London for the vacancy, advancing the argument that Marine officers seldom left the ship, and hence were "a protection to the ship over Prisoners."[25] Inasmuch as Parke and Elwood had already irritated Barry by going ashore whenever they saw fit, the captain lashed back at

Parke with the statement that "every Good Friend of his Country on board is a Safe Guard, but I dont think the name of a Lt. of Marines adds anything to it, especially Such of them as is Seldom on board." He was determined, he continued, to see that they both stay on board hereafter, and, in his opinion, the *Alliance* already had too many Marine officers, therefore he would appoint no one to fill the vacancy left by the resignation of William Morris, much to the disgust of Parke and Elwood.[26]

To be Sold at Auction

The *Deane's* arrival at Boston on 17 April 1781, was greeted with much enthusiasm for she carried military clothing, and more than 200 thirteen-inch shells.[27] Ardor quickly turned to apathy when Captain Nicholson submitted his report to the Eastern Navy Board. The frigate, he said, would require repair and some alteration to her spars and masts as she had grown old and weak. As usual, the Board's first thoughts were of money. "The Deane has ever been as expensive as any other two Ships in the Service," they wrote to the Board of Admiralty, "where we are to obtain the money Necessary for the Equipment of her we are at a loss," but funds were obtained, and repair work was initiated.[28]

As the frigate was grounded in preparation for scraping and caulking, the 215-man crew was discharged and paid off, their cruise completed.[29] Of the officers, only a few chose to remain with the ship. Since the appointment of Lieutenant Samuel Pritchard to the *Alliance* on 20 October 1780, the frigate's Marine detachment was apparently without a competent officer. To remedy this oversight, the Eastern Navy Board appointed Captain Richard Palmes, a veteran of extensive service on board the *Boston*, to the *Deane* on the first day of May 1781.[30]

Throughout the summer, work continued, even though the Navy Board at times found itself without funds to pay the laborers. By early fall financial problems seemed to have been solved and the frigate was declared ready for service. All she required was a crew. Captain Nicholson and Captain John Barry of the *Alliance* produced an appeal for recruits.[31] Offering much more than those who recruited for privateers, Nicholson soon began to fill the frigate's empty roll. At the same time, the Eastern Navy Board began to appoint the needed officers to fill vacant posts. On 29 October, William Waterman, then at Providence, was selected to assume the post of first lieutenant of Marines. "You are hereby required," wrote the Board, "to repair immediately to Boston to do duty on Board the Deane Frigate. . . . It is our wish that you would Enter all the Marines & Sea-Men in your Power, and send them on to the Ship whose expences will be Paid them, for Traveling."[32] Less than a month later, the post of second lieutenant of Marines was filled with the appointment of Jeremiah Reed.[33]

In December, Captain Palmes unwillingly gave up 11 Marines to the *Alliance*, of the 34 thus far enlisted. In late February, the General Court of Massachusetts allowed Captain Nicholson to enlist not more than 12 men from the garrison at Castle William in Boston harbor. How many of the 12 became Marines is unknown, but by the first of March, the Marine detachment on board the *Deane* numbered 66 men.[34]

The *Deane* sailed from Nantasket Road on 10 March for a two month-cruise to the West Indies. During this eight-week period, according to Robert Morris, the frigate was very active:

> On the ninth of April she captured the Ship Mary bound from Belfast to Barbadoes laden with a Cargo of Provisions and dry Goods to the Amount of five thousand Pounds sterling this Vessel was ordered for and I have the Pleasure to add has arrived at the Island of Hispaniola—On the twelfth of April the Deane captured his britannic Majesty's Schooner Jackall going Express to Barbadoes commanded by Lieutenant Logie mounting fourteen three pounders and having on Board fifty two Men. This Vessel has arrived at Boston. On the fourth of May she captured the Swallow Privateer of Bermuda mounting fourteen carriage Guns and having on Board fifty three Men. I am told this Vessel was driven on Shore by a British Cruizer in Boston Bay. She recaptured also the Brig Elizabeth laden with Tobacco and Staves bound from North Carolina to St. Thomas's this Vessel has also arrived at Boston. On the tenth of May she captured the Privateer Ship Regulator belonging to Bermuda carrying eighteen six Pounders and seventy five Men. This Vessel has also arrived at Boston.[35]

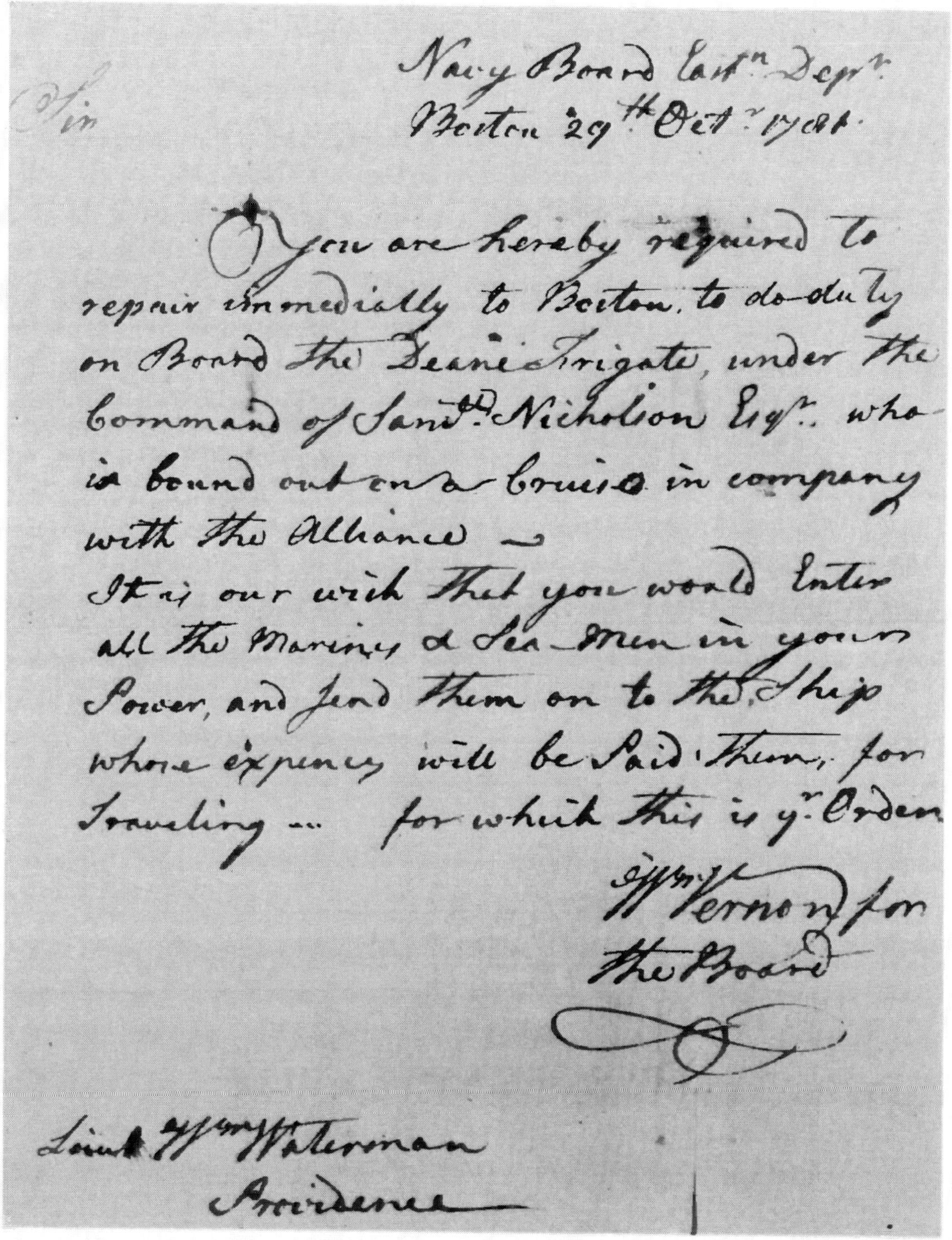

Navy Board Eastn. Depr.
Boston 29th Octr. 1781.

Sir

You are hereby required to repair immediately to Boston, to do duty on Board the Deane Frigate, under the Command of Saml. Nicholson Esqr. who is bound out on a Cruise in company with the Alliance —

It is our wish that you would Enter all the Marines & Sea-Men in your Power, and send them on to the Ship whose expences will be Paid them, for Traveling ... for which this is yr. Order

Wm. Vernon for the Board

Lieut. Wm. Waterman
Providence

William Vernon to William Waterman.

With the frigate filled with prisoners, a violent fever prevailing, most of her officers on board prizes, and her spars damaged, the *Deane* was forced to return to Boston on 17 May.[36]

Shortly after the *Deane's* arrival, Captain Nicholson was relieved of his command and replaced by John Manley who had recently returned from a long imprisonment in England. The circumstances surrounding the appointment and shift of commanders are unknown. It is possible that Nicholson asked to be relieved, or he may have been removed for disciplinary reasons, for the records show he faced a court of inquiry called to examine his conduct during the frigate's spring cruise, and a court martial a year later.[37] Meanwhile, the *Deane* was renamed the *Hague* after Silas Deane had fallen from official favor, and the Netherlands had received John Adams, publicly acknowledging the United States as an independent nation.

Captain Manley reported on board the *Hague* the very day he was appointed and was "welcomed with united acclamations—13 guns were fired in honor of the appointment—the ship beautifully decorated with colors—and every possible demonstration of joy expressed a general satisfaction." Since his ship was ready and manned, it was not likely that he tarried long after assuming command before departing Boston on a cruise to the West Indies.[38]

Making several valuable captures during the southern passage, Manley put into Martinique and then on 26 December, in company with two privateers, sailed for the Leeward Islands.[39] About 300 miles east of Antigua in the early morning hours of 9 January 1783, the 44-gun enemy frigate *Dolphin* was sighted and for the second time in his naval career Manley decided to run. Next morning four 74-gun ships-of-the-line joined the *Dolphin* in the ensuing 36-hour chase. Manley attempted to reach the protection of the French forts covering Baie Mahault on the eastern side of Guadeloupe, but ran aground on a reef laying between the island and Grand Terre. For two days the *Hague* remained captive of the coral while the enemy warships, in Manley's words, "were not very sparing of a heavy and brisk cannonade." The Englishmen tried to warp in on the *Hague,* but each time the shoal water and the frigate's accurate fire kept them at a respectable distance. Furiously working his crew, Manley succeeded in getting the *Hague* into deep water and then under the protection of the French forts, where he remained until late January repairing his damaged hull and rigging.[40]

Early in May, the *Hague* returned to Boston and, after clearing quarantine, was moored alongside Hancock's Wharf. Shortly after the crew was discharged, the *Hague* was decommissioned and then sold at public auction on 2 October 1783.[41]

The Last Cruise

Through a hot, late summer night, the *Alliance* lay becalmed in the mouth of New London harbor. At dawn on 4 August 1782, a fresh breeze sprang up from the northwest and the frigate stood to sea. In less than three hours a brigantine laden with lumber and fish was captured and ordered back to port. "The Cargo is not very Valuable," wrote Captain John Barry to John Brown, "but it is making a Beginning." After a short and unsuccessful chase of an enemy privateer, Barry ordered the frigate's course set for Bermuda.[42]

Six days later, the *Alliance* bagged her next prize, the schooner *Polly,* from Bermuda to Halifax with a cargo of little value—brown sugar, molasses, onions, and limes.[43] On 19 August, with Bermuda off her bow, the *Alliance* hoisted English colors and moved slowly toward St. George's harbor. After luring a pilot on board, the American frigate cruised about the island for several days. A number of vessels were pursued and one of them, the 18-gun privateer *Experiment,* got into the harbor only after a rather stirring chase. Since this procedure was not achieving results, Barry decided to effect an exchange of American prisoners. On the 23d, he put Samuel Tufts, late master of the *Polly,* on board a pilot boat and sent him into St. George with a letter addressed to the Governor of Bermuda. "Unless all the Americans which they held prisoners were sent on board the Alliance," he proposed to "remain three weeks to hinder any vessel

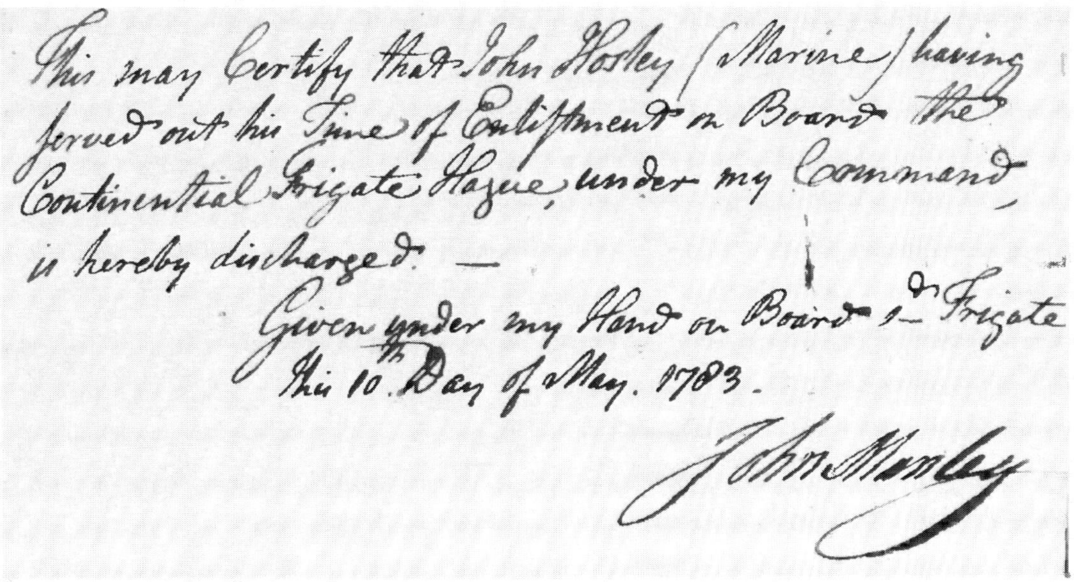

This may Certify that John Hosley (Marine) having
served out his Time of Enlistment on Board the
Continential Frigate Hague, under my Command,
is hereby discharged. —
Given under my Hand on Board sd Frigate
this 10th Day of May. 1783.
John Manley

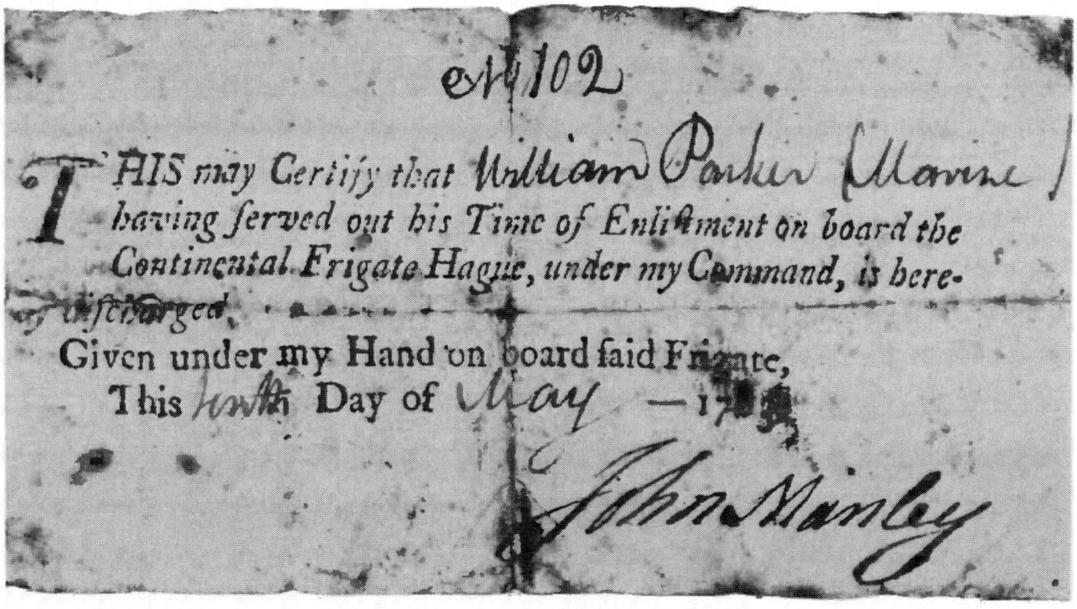

No 102

THIS may Certify that William Parker (Marine)
having served out his Time of Enlistment on board the
Continental Frigate Hague, under my Command, is here-
by discharged.
Given under my Hand on board said Frigate,
This tenth Day of May —17[illegible]
John Manley

Discharge Certificates Issued to Enlisted Marines

from going in or coming out."

Whether the threat succeeded is hard to determine. On the morning of 25 August, a small sloop was observed coming out toward the *Alliance*. She may have been carrying released Americans, but, as she approached, the frigate went off in pursuit of the enemy privateer *Hawk* and her prize, the Connecticut sloop *Fortune*. The sloop was taken, but the privateer escaped to the westward.

The *Alliance* continued the fruitless chase until 30

August when she hailed a brig from Guadaloupe to Rhode Island, which carried intelligence that a large enemy merchant fleet had recently sailed from Jamaica for England.[44] "Finding the prizes I had taken of little value to myself or country, and in all likelihood should be obliged to return into port soon for want of men," Barry determined to alter his cruising ground, and run northwestward toward the Newfoundland Banks.[45]

Again without making Bermuda, the *Alliance* headed for the banks in hope of capturing vessels of the Jamaica fleet. On the way, on 7 September, Barry captured his fourth prize, the whaling brig *Somerset*, which he manned and sent into Boston. Three days later, the *Alliance* entered the Newfoundland Banks, and on 18 September, took a crippled British merchant brig. Barry learned from her master that the Jamaica fleet had been scattered by a terrific storm that had raged over the North Atlantic for two days.[46]

For the next week and a half, the *Alliance* roamed the banks in search of prizes. Four had been taken by dawn on the 28th. Short of water, loaded with prisoners, and with fresh westerly winds, Barry gave up thought of returning to an American port and decided to proceed to France, "with a determined view to get those [prizes] (I had already taken) in safe, and after landing prisoners, to put out immediately."[47] Through the balance of the cruise no more vessels of the fleet were sighted. Instead, the *Alliance* was tossed about by gales and high seas. Barry lost "the sails of the head," and, for a time, he was in danger of losing the headmast itself, but on 17 October the frigate and her four prizes safely moored off Isle de Groix and the following day sailed up past Port Louis to L'Orient.[48]

Parke's Just Deserts

After the prisoners were sent ashore and repairs begun, an advance of prize money was made to the crew. "It was Comput'd," wrote Midshipman Ichabod Perry, "that they [prizes] would amount to three hundred Guineas to a share but they fell short about five sixths. We Receiv'd for them prizes about twenty five Guines in Cash and about the same in Dry Goods, but it set us up finely and we liv'd like good fellows, while our money last'd . . . I brought all my store pay home with me, except a few silk handkerchiefs which was stolen. It was very difficult to keep any thing amongst such fellows as we had."[49] The officers, however, were not to be satisfied by a small advance; they wanted their pay.

In a rather innocuous letter sent to Captain Barry on 17 November, the officers of the *Alliance* formally appointed the captain their agent in the matter of the four Jamaicamen, authorized him to secure their prize money, and to dispose of each officer's share as that individual might direct.[50] Barry agreed to act for them, but, as he replied, he would need more specific instructions, and quickly for he was "Determined to sail this Week if Wind and Weather Permitt."[51] Emboldened by Barry's agreement to act as their prize agent, the officers saw opportunity to enlist him further in an effort to secure back wages, as a number had served on board the *Alliance* as long as four years and had received very little pay—among them Captain Parke and Lieutenant Elwood of the Marines.[52]

The officers' demand, rather than request, provoked the captain. "I certainly have as much Reason to complain as any of you," he wrote, "but I know it to be folly to Grumble." If wages were not paid when they reached America, he concluded, he would join them in a petition, remonstrance, or any other step they might think necessary to secure their just dues.[53] He could do nothing for them in France.

For four days, Barry heard nothing from the officers, but learned that six of them, all except Marine Lieutenant Elwood, had gone ashore and were quartered at an inn in L'Orient. On 24 November, Captain Parke called on the captain and informed him that he and his fellow officers "would not go on board without 2 thirds of their Wages." Barry replied that he had no power to pay them their wages, and warned Parke that if he were not at his post on board the *Alliance* by four that afternoon, he would have to face the consequences.[54]

Since Parke refused to obey Barry's orders, he was placed under arrest the following morning and told "you must look on yourself as having nothing to do with that Ship till you are try'd by a Court Martial in your own country."[55] Barry then had all other officers arrested for insubordination and confined on shore.[56] Only naval Lieutenant Hezekiah Welch and Marine Lieutenant Thomas Elwood were spared.

With inexperienced lieutenants promoted from the lower grades, the *Alliance* sailed out of L'Orient on 9 December. Left behind were the six officers, Captain Parke among them, who were, as Barry wrote Robert Morris, to "get to America as well as they Can, where I hope they will be Try'd by a Court Martial and Meet their deserts."[57] * As he headed out to sea in a cold, bitter northeast wind, Barry set the frigate's course at west by south across the Bay of Biscay, and then southward for a long run down the Portugese and African coasts. By mid-December the *Alliance* was off the island of Madeira where Barry ordered her westward across the Atlantic for Martinique.

The *Alliance* vs. *Sybil*

The passage was uneventful, except for periods of stormy weather, and on 9 January 1783, the American frigate came to in the roadstead of St. Pierre. Scarcely had the larboard anchor slipped away, before Barry was off for shore in the cutter. There he found orders to Havana where he was to take on specie to be carried to Congress then sitting in Philadelphia.

The *Alliance* needed water and repairs to the foretop mast, which had been split during a December gale. After arranging for both, Barry then imparted all the news he had brought from France to the excited editor of the *Martinico Gazette*. Most important was the fact that the peace preliminaries had been signed. Copies of the *Gazette*, and versions of its contents reached America in late February. Garbled accounts appeared in many newspapers, among them the *Boston Evening Post*:

> The Continental frigate Alliance, John Barry, Esquire, Commander, has arrived there [Martinique] from France, with a British passport, and brought Advice, that the ARTICLES OF PEACE were SIGNED on the 22d DAY OF DECEMBER LAST.[58]

Unaware of the incorrect information that was being passed about under his name, Barry completed the repairs, and resumed his voyage on 13 January. His course was northward, past the western shores of Dominica, Guadaloupe, and the southern coasts of Porto Rico and Santo Domingo. After negotiating the Windward Passage, the *Alliance* sailed unmolested along the northern coast of Cuba, and, on the last day of January, passed under the guns of Morro Castle and entered Havana harbor.

The specie was at Havana, but another vessel to receive it was already on hand. She was the 20-gun ship, *Duc de Lauzun*, which Robert Morris had recently purchased for the Continental Navy. Under new orders, Barry was now to convoy this vessel to Philadelphia. After a five-week delay, due to the fact that the port of Havana was closed by Royal embargo, the *Alliance* and her consort got underway shortly after dawn on the morning of 6 March. Amid continuous rain, the two ships tacked northwestward through the Straits of Florida. At mid-morning on the 10th, a warning cry came from the masthead; three large ships were standing directly for the *Alliance*. The strangers turned out to be the British 32-gun frigate *Alarm*, 28-gun frigate *Sybil*, and the 18-gun sloop-of-war *Tobago*. The headmost, which seems to have been the *Alarm*, closed to within gunshot of the *Alliance*, while the other two moved swiftly toward the *Lauzun*. Since the French-built ship was a dull sailer, Barry advised her captain to heave his guns overboard and run off before the wind—"the former he did but still Kept his Course."[59]

For about thirty minutes, Barry refrained from leaving his consort, while he thought of some way to save her. The answer came from an unexpected source. A sail appeared to the southwest which was recognized as the French ship *Triton*, of 64-guns. Encouraged, Barry swung about and bore between the *Duc de Lauzun* and the on-coming frigate, *Sybil*. As the two opponents closed,

* On Matthew Parke's return to the United States he was charged with disobeying orders which caused the *Alliance* to be detained in port for several days, thus incurring additional expenses. He was found guilty on 16 May 1783 of the charge of disobedience, and was sentenced to forfeit his commission, but not any wages or prize money due him on or before 24 November 1782.

John Barry, by Gilbert Stuart.

"Captain Barry went from gun to gun on the main deck, cautioning against too much haste and not to fire until the enemy was right abreast." Barry then ordered the main topsail hove to the mast, the last necessary maneuver to bring the enemy frigate alongside.[60] At ten minutes of twelve, he gave the long-awaited order to fire.

The devastating broadside, delivered at such close range, wrought havoc on board the *Sybil*. An army lieutenant, serving as a Marine officer was killed instantly and several men were wounded.[61] Thereafter, both ships fired as rapidly as guns could be loaded. A few enemy shot found their mark on board the *Alliance*, damaging sails, spars, and rigging, but casualties were kept to a minimum—a total of 10 wounded. Within a half hour, according to Midshipman Kessler, the enemy's "guns were silenced and nothing but Musketry was fired from her. She appeared very much injured in her hull."[62]

After the first smashing broadside, the *Sybil* later had her main and foretop gallant studding sails shot away. Her captain reported a loss of two killed and six wounded, but later accounts listed 37 killed and more than 40 wounded.[63] The actual casualty total lies somewhere between the two differing figures.

When the engagement ended, the *Sybil* sheered off, and stood to the northeast in the wake of the *Alarm* and *Tobago*, neither of which had attempted to assist their comrade in the action. The *Alliance* tacked and gave chase. Her sluggish consort soon fell far behind, forcing Barry to turn back. At dawn the following morning, the northward course was resumed, and for eight days they proceeded without incident. Off Cape Hatteras, shortly after midnight on 18 March, the *Duc de Lauzun* parted company and made her solitary way up the Delaware to Philadelphia.

Less than twenty-four hours later, off the Delaware Capes, the *Alliance* ran into a thick fog. As she neared Five Fathom Bank, east of Cape Henlopen, the fog lifted, revealing an enemy two-decker and frigate. Barry immediately put about and stood toward the shore and fog, but once again the weather lifted to disclose his pursuers slightly to windward. This time he fled in earnest and lost the enemy ships in darkness. "Having Great Reason to suppose the Coast was lined with the Enemys Ships, and no prospect of Getting in till the Weather Cleared up," Barry decided against sailing on to Philadelphia. Instead he made for Rhode Island, and at three in the afternoon on 20 March, he dropped anchor off Newport.[64]

News of the final peace settlement reached the *Alliance* on the last day of March 1783. It found her at anchor off Petuxet, some five miles below the town of Providence, Rhode Island. Since the war and cruise had come to an end, the majority of the crew was discharged; included were 33 out of 41 Marines. By mid-May, with additional reductions, the eight remaining Marines were released, and in compliance with orders from Robert Morris, were given "a certificate that he belonged to the Ship with his Name and Rank so that when he applies here [Philadelphia] he may be known by the paymaster and settled with accordingly."[65] The only Marine not included was Lieutenant Thomas Elwood, who, according to Barry, was to continue at his post so that the *Alliance* could maintain "the appearance of a public Ship."[66]

Throughout the early summer months, Elwood labored to complete a final accounting of the Marine stores. On 20 June he submitted his report:

95 Muskets, 13 of which without Bayonets
29 Pistols.—
111 Cartouch Boxes

41 Cutlasses
1 Drum & Drum Sticks
1 Fife
1 Drum head
2 Arm Chests with Loxks
2 Great Coats.—
300 Flints
Expended.—
1 Musket lost overboard by the Centinal
3 Pistols do. overboard
4 pair Pistols lost in attempting to board a Flumish Ship at L'Orient in February 1782
7 Cutlasses at the same time.—
1 Fife lost by the fifer.—
3 Locks.—

Once accounted for, the stores were locked away in the hold.[67]

The *Alliance*, most of her armament stowed below decks, stood down the river from Petuxet in late June. On orders from Robert Morris, she was to make a short run to Virginia where she was to take on a cargo of tobacco for shipment to Amsterdam. The passage to the Chesapeake proved damaging and, after receiving her cargo, the *Alliance* was forced to put into Philadelphia for repairs. On 5 September, a committee of Congress, appointed to examine her, ended all prospects of continuing the voyage. Morris was directed to unload the tobacco, ship it to Europe in other bottoms, discharge the officers and crew, and have the frigate surveyed. For all intents and purposes, the *Alliance* was finished, even though there was strong sentiment to keep her in national service for the protection of the coast. Others believed that all naval expenditures should cease. The question was broached from time to time until late May 1785, when tightened economic considerations finally prevailed and a committee of Congress recommended the sale of the frigate. She was sold in August of that year.[68]

The *America* and *Bourbon*

Two ships authorized by the Continental Congress on 20 November 1776 were launched during the last two years of the war; the 74-gun ship-of-the-line, *America*, and 36-gun frigate, *Bourbon*. The *America's* keel was laid at Portsmouth, New Hampshire, in May 1777, but work proceeded very slowly. Early in June 1779, Congress instructed Robert Morris to complete the ship as soon as possible. Later in the month it appointed John Paul Jones as her commander. In spite of the efforts of both Jones and Morris, construction of the ship continued at a snail's pace and it was not until 5 November 1782 that she was launched. In the meantime Congress presented her to France to replace the French 74, *Magnifique*, lost in Boston harbor, in order to show their gratitude to the French king for his support against Great Britain.

Laid down at Middletown, Connecticut, early in 1780, the *Bourbon* was launched 31 July 1783, after much delay due to lack of funds. She was advertised as being for sale in September, and nothing further is known about her.

At this time a number of Marine officers petitioned Congress, or were recommended for command of the ships' Marine companies. Among them were George Jerry Osborne, who, at the request of Captain John Barry, was appointed captain of Marines for the *America* on 20 November 1779.[69] But Osborne resigned his commission, having urged "as a reason his private avocations would by no means Admit of his continuing in the Service."[70] Major Samuel Nicholas was next to apply for the position. On 29 September 1781, the Board of War recommended his appointment, but apparently withdrew it later without reason.[71] Lieutenant Zebadiah Farnham seems to have been the only Marine officer to apply for a post on board the *Bourbon*, but Robert Morris, in a letter to Lieutenant Farnham on 21 November 1782, wrote that "no Appointments can be made to that Ship at Present," for the frigate had yet to be fitted out for sea.[72] Apparently, this was Morris' policy until Congress gave the *America* to the King of France and sold the *Bourbon*.

A Few Gallant Memories

With the discharge of Marine Lieutenant Thomas Elwood in September 1783, Continental Marines passed from the scene. For more than seven years, this small force did its part to achieve final victory against the British, but the years took their toll. The original resolution of Congress established a corps of Marines from which battalions could be formed for expeditionary service, and from which detachments could be made for service afloat. Somehow, the intent of the resolve was lost as time passed, not so much from insufficient resources, large distances, and poor communications, as from the lack of centralized control and Congressional neglect. As the war went on, the captains of Continental ships took over the responsibility of raising their Marine detachments in the same way they signed on their naval crews.

Records indicate that approximately 131 officers held Continental Marine commissions. The number of non-commissioned officers and enlisted men is not exactly known, but probably did not exceed 2,000. In comparison with the Army and Navy, the corps of Continental Marines was relatively small, but it contributed measurably to the British defeat. Writing in 1839, James Fenimore Cooper gave the Continental Marine a much deserved, and long overdue recognition:

> At no period of the naval history of the world, is it probable that Marines were more important than during the war of the Revolution. In many instances they preserved the vessels to the country, by suppressing the turbulence of their ill-assorted crews, and the effect of their fire, not only then, but in all the subsequent conflicts, under those circumstances in which it could be resorted to, has usually been singularly creditable to their steadiness and discipline. The history of the navy, even at that early day, as well as in these latter times, abounds with instances of the gallantry and self-devotion of this body of soldiers; and we should be unfaithful to our trust, were we not to add, that it also furnishes too many proofs of the forgetfulness of its merits by the country. The marine incurs the same risks from disease and tempests, undergoes the same privations, suffers the same hardships, and sheds his blood in the same battles as the seamen, and society owes him the same rewards. While on ship-board, necessity renders him in a certain sense, the subordinate, but nations ought never to overlook the important moral and political truth, that the highest lessons they can teach are those of justice; and no servant of the public should pass a youth of toil and danger, without the consciousness of possessing a tenour to a certain and honourable reward, that is dependent only on himself. That this reward has hitherto been as unwisely as it has been unfairly witheld, from all connected with the navy, it is our duty as historians to state, and in no instance has this justice been more signally denied, than in the case of the honourable and gallant corps.[73]

After Robert Morris' retirement in 1784, Congress made no attempt to appoint a successor as Agent of Marine, for there was little need for such an office. The Board of the Treasury (organized early in 1785), aided by the "Commissioner for Settling the Marine Accounts," and by the secretary to the Agent of Marine, wound up the business of the Navy and Marines.

The thought of reorganizing the Navy and Marines under the Articles of Confederation at the conclusion of the war did exist, but the finances of the new government were in no condition to permit such a luxury. This is well illustrated by the offer of Virginia to sell the *Cormorant* to the United States, early in 1783. The Agent of Marine reported on 26 June, which report was agreed to by Congress on 5 August, "that altho' it is an Object highly desirable to establish a respectable Marine Yet the Situation of the Public Treasury renders it not adviseable to purchase Ships for the present, nor until the several States shall grant such Funds for the construction of Ships, Docks, and Naval Arsenals, and for the Support of the Naval Service as shall enable the United States to establish their Marine upon a permanent and respectable footing."[74]

Although records indicate that Marines enlisted for and served in the few armed American vessels of the period, there was no organized corps, and little remained of the Continental Marines but a few gallant memories. This would have to wait until 11 July 1798 when Congress would send to President John Adams, "An Act for Establishing and Organizing a Marine Corps."

APPENDIX A

Notes

Introduction

[1] Colonel Cyril Field, *Britain's Sea-Soldiers: A History of the Royal Marines*, 3 vols. (Liverpool: Lyceum Press, 1924), v. I, pp. 71–73.

[2] *Ibid.*, pp. 75–92.

Chapter I

Birth of the Continental Marines

[1] Worthington Chauncey Ford, et. al., eds. *Journals of the Continental Congress*, 34 vols. (Washington: Government Printing Office, 1904–1927), v. III, p. 316, hereafter *Journals of the Continental Congress*.

[2] "Proposals of the Nova Scotia Committee," in William Bell Clark, ed. *Naval Documents of the American Revolution*, 6 vols. (Washington: Government Printing Office, 1964–1973), v. II, pp. 957–958, hereafter *Naval Documents*.

[3] *Journals of the Continental Congress*, v. III, p. 348.

[4] "Journal of the Rhode Island General Assembly," 26 August 1775, in *Naval Documents*, v. I, p. 1236.

[5] *Journals of the Continental Congress*, v. III, pp. 274–275, 277.

[6] John Adams, *The Works of John Adams*, 4 vols. (Boston: Little, Brown, 1865), v. III, pp. 7–8.

[7] *Journals of the Continental Congress*, v. III, p. 278.

[8] *Ibid.*, pp. 272–312.

[9] *Ibid.*, pp. 313–376.

[10] John Adams to Elbridge Gerry, 5 November 1775; John Adams to James Warren, 5 November 1775, in *Naval Documents*, v. II, pp. 896–897.

[11] *Dunlap's Maryland Gazette*, 19 December 1775; Howard I. Chappelle, *The History of the American Sailing Navy: The Ships and Their Development* (New York: W. W. Norton & Co., 1949), p. 53, hereafter *The History of the American Sailing Navy*; John J. McCusker, Jr., "The Continental Ship Alfred," *Nautical Research Journal*, v. XIII, nos. 2–3 (Autumn 1965), pp. 39–68.

[12] George Washington to John Hancock, 19 November 1775, in John C. Fitzpatrick, Ed., *The Writings of George Washington from the Original Manuscript Sources, 1745–1799*, 39 vols. (Washington: Government Printing Office, 1931–1938), v. IV, pp. 99–100, hereafter *Writings of Washington*.

[13] George Washington to John Hancock, 28 November 1775, *ibid.*, pp. 120–121.

[14] "Return of those Gentlemen's Names who are Recommended to Serve in the Two Battalions of Marrines, as Captains," undated, Revolutionary War Muster Rolls, R.G. 93, Continental Troops, 90, no. 4, National Archives (NA).

[15] George Washington to John Hancock, 14 December 1775, 18 December 1775, 24 January 1776, in *Writings of Washington*, v. IV, pp. 160, 174, 274; *Journals of the Continental Congress*, v. II. p. 393.

[16] John Hancock to George Washington, 8 December 1775, Papers of the Continental Congress, R.G. 360, 12A, v. IV. pp. 23–24, NA, hereafter *PCC*.

[17] George Washington to the Board of Admiralty, 27 May 1780, *PCC*, 37, pp. 423–424.

[18] Commission of Samuel Nicholas, 28 November 1775, Miscellaneous Collections, Historical Society of Pennsylvania.

[19] *A History of the Schuylkill Fishing Company of the State in Schuylkill, 1732–1888* (Philadelphia: Members of the State in Schuylkill, 1889), pp. 367, 407.

[20] Neville B. Craig, *Sketch of the Life and Services of Isaac Craig, Major in the Fourth (Usually Called Proctor's) Regiment of Artillery, During the Revolutionary War* (Pittsburgh: J.S. Davison, 1854), pp. 4–5.

[21] "Assignment of Marines in the Continental Fleet," 4 January 1776, Esek Hopkins Papers, Rhode Island Historical Society.

[22] Miss Anne H. Parke to Captain Dudley W. Knox, 9 March 1936, "Matthew Parke," ZB Files, Operational Archives Branch, Naval Historical Center.

[23] "Letter from Philadelphia," 27 December 1775, in Peter Force, comp. *American Archives: Consisting of a Collection of Authentick Records, State Papers, Debates, and Letters and Other Notices of Publick Affairs, the Whole Forming a Documentary History of the Origin and Progress of the North American Colonies; of the Causes and Accomplishments of the American Revolution; and of the Constitution of Government for the United States, to the Final Ratification thereof.* 2 ser., 9 vols. (Washington: n. p., 1837–1853), 4th ser., v. IV, p. 468, hereafter *American Archives*.

[24] Muster Roll, Isaac Craig's Company of Marines, 19 December 1775, Craig Papers, Historical Society of Western Pennsylvania.

[25] Navy and Marine Corps Military Personnel Statistics (Nav Per 15658), February 1972, Personnel Branch, United States Marine Corps; Men Examined in Armed Forces Induction Centers, August 1969–January 1970, United States Army.

[26] *Journals of the Continental Congress*, v. III, pp. 378–387.

[27] *Ibid.*, v. IV, pp. 35–37.

[28] *Ibid.*, v. III, p. 405.

[29] *Ibid.*, v. III, p. 396.

[30] *Ibid.*, v. III, p. 433; General Knox to his wife as quoted in Edward Field, *Esek Hopkins Commander-in-Chief of the Con-*

tinental Navy During the American Revolution, 1775 to 1778 (Providence: Preston & Rounds Co., 1898), p. 134.

[31] Nicholas Biddle to Lydia McFunn, 15 February 1776, in *Naval Documents*, v. III, p. 1306; William Bell Clark, *Captain Dauntless: The Story of Nicholas Biddle of the Continental Navy* (Baton Rouge: Louisiana State University Press, 1949), hereafter *Nicholas Biddle*.

[32] Samuel Eliot Morison, *John Paul Jones: A Sailor's Biography* (Boston: Little, Brown and Co., 1959).

[33] Samuel Hazard, ed., *Pennsylvania Colonial Records*, 16 vols. (Harrisburg: T. Fenn, 1851–53), v. X, p. 436, hereafter *Pennsylvania Colonial Records*.

[34] Samuel Hazard, et. *al*., eds., *Pennsylvania Archives*, 9 ser., 138 vols. (Philadelphia: J. Stevens, 1874–1919), 1st ser., v. IV, p. 694, hereafter *Pennsylvania Archives*.

[35] *Pennsylvania Colonial Records*, v. X, pp. 441, 443.

[36] *Pennsylvania Archives*, 2d ser., v. X, p. 49; John Hancock to Commanding Officer of the First Pennsylvania Battalion, 2 December 1775, in *American Archives*, 4th ser., v. IV, p. 154.

Chapter II

Outbreak of Hostilities: Early Marine Participation

[1] Benedict Arnold to the Massachusetts Committee of Safety, 30 April 1775, in *Naval Documents*, v. I, p. 250.

[2] Massachusetts Committee of Safety to Benedict Arnold, 3 May 1775, in *ibid*., p. 267.

[3] "Ethan Allen's Account of the Taking of Ticonderoga," in *ibid*., pp. 304–305; Benedict Arnold to the Massachusetts Committee of Safety, 11 May 1775, in *ibid*., pp. 312–313; Ethan Allen to the Committee of the City of Albany, 11 May 1775, in *ibid*., p. 314.

[4] Edward Mott to the Massachusetts Provincial Congress, 11 May 1775, in *ibid*., pp. 315–316; Oscar E. Brendenberg, "The American Champlain Fleet, 1775–1777," *The Bulletin of the Fort Ticonderoga Museum*, v. XII, no. 4 (Sep 1968), p. 249; "Extract of a Letter from Ticonderoga," *New England Chronicle* (Cambridge), 1 June 1775, hereafter *Eleazer Oswald's Journal*.

[5] *Ibid*.; Benedict Arnold to the Massachusetts Committee of Safety, 19 May 1775, in *Naval Documents*, v. I, pp. 364–366; Benedict Arnold to the Massachusetts Committee of Safety, 23 May 1775, in *ibid*., pp. 512–513.

[6] *Eleazer Oswald's Journal*; Benedict Arnold to the Albany Committee of Safety, 22 May 1775, in *Naval Documents*, v. I, pp. 503–504; "A Pay Rool of Captn Samuel Herrick's Company In Colo Arnolds Regment," 25 June 1775, Peter Force Collection, Library of Congress (LC); "Pay Role of the Sloop Interprise &c from 3rd May to 1 July 1775," Massachusetts Archives.

[7] *Journals of the Continental Congress*, v. II, pp. 55–56, 104–105.

[8] Benedict Arnold to the Continental Congress, 13 June 1775, in *Naval Documents*, v. I, pp. 671–673; Massachusetts Committee to Benedict Arnold, 23 June 1775, in *ibid*., p. 743.

[9] Samuel Elmore to Philip Schuyler, 23 July 1775, Schuyler Papers, v. XXIV, no. 609, New York Public Library; Journal of the Continental Schooner *Liberty*, PCC, 153, v. I, pp. 106–107.

[10] *Journals of the Continental Congress, v. II, pp. 108–110.*

[11] Philip Schuyler to Jonathan Trumbull, 3 August 1775, in *Naval Documents*, v. I, p. 1055.

[12] George Washington to Benedict Arnold, 14 September 1775, in *ibid*., v. II, pp. 93–95.

[13] Thomas Hartly to Horatio Gates, 21 July 1776, in *American Archives*, 5th ser., v. I, p. 487.

[14] General Order of Horatio Gates, 23 July 1775, "The Deputy Adjutant General's Orderly Book, Ticonderoga, 1776," *The Bulletin of the Fort Ticonderoga Museum*, v. III, no. 1 (Jan. 1933), p. 37.

[15] Horatio Gates to Thomas Hartly, 23 July 1776, in *American Archives*, 5th ser., v. I, p. 548.

[16] Thomas Hartly to Horatio Gates, 30 July 1776, in *ibid*., p. 683.

[17] *Ibid*.

[18] "The Deputy Adjutant General's Orderly Book, Ticonderoga, 1776," *The Bulletin of the Fort Ticonderoga Museum*, v. III, no. 2 (Jul 1933), p. 87.

[19] *Ibid*., v. III, no. 1 (Jan 1933), p. 48.

[20] Benedict Arnold to Horatio Gates, 18 September 1776, in *Naval Documents*, v. VI, p. 884.

[21] "List of Continental Armed Vessels on Lake Champlain August 18th 1776," Washington Papers, LC; Benedict Arnold to Horatio Gates, 31 August 1776, in *Naval Documents*, v. VI, pp. 371–372; Benedict Arnold to Horatio Gates, 2 September 1776, in *ibid*., p. 654; Benedict Arnold to Horatio Gates, 7 September 1776, in *ibid*., pp. 734–735.

[22] Benedict Arnold to Horatio Gates, 8 September 1776, in *ibid*., p. 747; Benedict Arnold to Horatio Gates, 21 September 1776, in *ibid*., pp. 925–926; Benedict Arnold to Horatio Gates, 28 September 1776, in *ibid*., pp. 1032–1033; Benedict Arnold to Horatio Gates, 10 October 1776, in *ibid*., pp. 1197–1198.

[23] "A List of His Majesty's Naval Force on Lake Champlain," enclosed in Captain Charles Douglas, R.N. to Philip Stephens, 21 October 1776, in *ibid*., pp. 1343–1344.

[24] Benedict Arnold to Horatio Gates, 12 October 1776, in *ibid*., p. 1237.

[25] Jonathan Trumbull to Nicholas Cooke, 3 December 1777, Governor's Letters, v. 11, p. 95, Rhode Island Archives; Heathcote Muirson to Benjamin Franklin, 21 July 1778, Franklin Papers, Historical Society of Pennsylvania.

[26] Benedict Arnold to Horatio Gates, 12 October 1776, in *Naval Documents*, v. VI, p. 1237.

[27] Benedict Arnold to Philip Schuyler, 15 October 1776, in *ibid*., p 1275–1277.

[28] General James Wilkinson, *Memoirs of My Own Time*, 3 vols. (Philadelphia: Abraham Small, 1816), v. I, p. 91, hereafter *Memoirs*.

[29] "The Deputy Adjutant General's Orderly Book, Ticonderoga, 1776," *The Bulletin of the Fort Ticonderoga Museum*, v. III, no. 2 (Jul 1933), p. 112.

[30] George Washington's Instructions to Nicholson Broughton, 2 September 1775, in *Writings of Washington*, v. III, pp. 467–469.

[31] John Hancock to George Washington, 5 October 1775, Washington Papers, LC.

[32] Joseph Reed to Nicholson Broughton, 12 October 1775, *ibid*.; Instructions from George Washington to Nicholson Broughton, 16 October 1775, in *Writings of Washington*, v. IV, pp. 33–34.

[33] Charter Agreement for the Armed Schooner *Harrison*, 22 October 1775, in *Naval Documents*, v. II, p. 572.

[34] Joseph Reed's Report on Washington's Armed Vessels, 29 October 1775, Washington Papers, LC.

[35] William Watson to Stephen Moylan, 23 November 1775,

ibid.; William Watson to George Washington, 29 November 1775, *ibid.*

[36] George Washington to John Hancock, 4 December 1775, in *Writings of Washington*, v. IV, p. 141–145.

[37] Instructions to Samuel Tucker, in John H. Sheppard, *The Life of Samuel Tucker, Commodore in the American Revolution* (Boston: Alfred Mudge and Son, 1868), p. 34, hereafter *Samuel Tucker*.

[38] Thomas Cushing to John Hancock, 17 May 1776, in *Naval Documents*, v. V. p. 135.

Chapter III
The New Providence Raid

[1] Dispatch from New Bern, North Carolina, *Virginia Gazette* (Williamsburg), 2 March 1776.

[2] Intelligence from Philadelphia, Transmitted by Captain Hyde Parker, Jr., R.N., 4 January 1776, in *Naval Documents*, v. III, p. 615.

[3] "A Journal of a Cruise In the Brig Andrew Doria Nicholas Biddle Esqr Commander from ye Port of Philadelphia Begun January 4th 1776," in *ibid.*, v. IV, p. 1489, hereafter Journal of the *Andrew Doria*.

[4] Intelligence on the Continental Navy, 4 January 1776, in *ibid.*, v. III, p. 616.

[5] Assignment of Marines in the Continental Fleet, 4 January 1776, Esek Hopkins Papers, Rhode Island Historical Society.

[6] Continental Naval Committee Order, 4 January 1776, in *Naval Documents*, v. III, p. 613.

[7] Naval Committee to Esek Hopkins, 5 January 1776, in *Naval Documents*, v. III, p. 613.

[8] *Journals of the Continental Congress*, v. III, p. 296.

[9] Naval Committee to Esek Hopkins, 5 January 1776, *PCC*, 58, pp. 239–240.

[10] *Pennsylvania Colonial Records*, v. X, p. 447.

[11] *Ibid.*, p. 451.

[12] "Inventory of Hussars Cloaths Arms &c delivd. Capt. Nicholas on Board the Alfred," undated, copy of unlocated original in History and Museums Division, Headquarters, United States Marine Corps.

[13] *Pennsylvania Evening Post* (Philadelphia), 9 January 1776, repeated 13, 18, 20 January 1776.

[14] "Journal Prepared For the King of France by John Paul Jones," in Robert G. Sands, *Life and Correspondence of John Paul Jones Including His Narrative of the Campaign of the Liman* (New York, n.p., 1830), p. 35, hereafter *Life and Correspondence of John Paul Jones*.

[15] Nicholas Biddle to Lydia McFunn, 15 February 1776, Nicholas Biddle Papers, on deposit at the Historical Society of Pennsylvania.

[16] George Washington to Joseph Hewes, 25 December 1775, in *Writings of Washington*, v. IV, p. 185.

[17] *Journals of the Continental Congress*, v. III, p. 390.

[18] Notes of Esek Hopkins for his Defense Before Congress in August 1776, Esek Hopkins Papers, Rhode Island Historical Society.

[19] Naval Committee to Esek Hopkins, 5 January 1776, *PCC*, 58, p. 240.

[20] Naval Committee to William Stone, 5 January 1776, Esek Hopkins Papers, Rhode Island Historical Society.

[21] Esek Hopkins to Nicholas Biddle, 14 February 1776, Nicholas Biddle Papers, on deposit at the Historical Society of Pennsylvania.

[22] Nicholas Biddle to James Biddle, 15 February 1776, *ibid.*

[23] Esek Hopkins to the President of Congress, 9 April 1776, in *American Archives*, 4th ser., v. V, p. 823.

[24] The table is based on the following sources. Because of the vast contradictions in the several sources, the number of men involved are only approximations. Intelligence on the Continental Navy, 4 January 1776, in *Naval Documents*, v. III, p. 616; Assignment of Marines in the Continental Fleet, 4 January 1776, Esek Hopkins Papers, Rhode Island Historical Society; Quarters Stations for the Continental Schooner *Wasp*, 16 March 1776, in *Naval Documents*, v. IV, p. 375; "List of the People on Board the Andrew Doria From February 1776," Nicholas Biddle Papers, on deposit at Historical Society of Pennsylvania; "A Roll of All the Officers And Men belonging to the Ship Alfred from the time of her being put into Commission until the 5th September 1776," Transcript of Military Papers, Revolutionary War, v. XL, pp. 657–663, Rhode Island Archives; "Muster Roll of the Men and Officers belonging to the Ship Columbus from the time of her being put into Commission to the fourteenth day of November 1776," *ibid.*, pp. 668–675; Journal of the *Andrew Doria*, in *Naval Documents*, v. IV, pp. 1489–1503; "A list of the Seamen and Landsmen that Came out of the Capes of Deliware in the Fly," Esek Hopkins Papers, Rhode Island Historical Society.

[25] Journal of the *Andrew Doria*, in *Naval Documents*, v. IV, pp. 149–1491.

[26] *Journals of the Continental Congress*, v. IV, pp. 352–357.

[27] "Journal Prepared For the King of France by John Paul Jones," in Sands, *Life and Correspondence of John Paul Jones*, p. 35.

[28] Journal of the *Andrew Doria*, in *Naval Documents*, v. IV, p. 1490.

[29] Thomas Gage to Montfort Browne, 29 August 1775, enclosed in John Brown to the Earl of Dartmouth, 7 November 1775, Public Record Office (PRO), Colonial Office Papers (CO) 23/23, London.

[30] John Brown to Thomas Gage, 29 September 1775, *ibid.*

[31] "A Narrative of the Transactions &c on the Invasion of the Island of New Providence by the Rebels in 1776," in John Gambier, et. *al.* to Lord George Germain, 31 March 1779, PRO, CO 23/24, hereafter "A Narrative . . . on the Invasion."

[32] Johann David Schöpf, *Travels in the Confederation*, trans. Alfred J. Morrison, 2 vols. (New York: Bergman Publishers, 1968), v. II, pp. 262–263, hereafter *Travels in the Confederation*.

[33] Disposition of His Majesty's Ships and Vessels in North America Under the Command of Rear [Vice] Admiral Shuldham," 22 March 1776, in *Naval Documents*, v. IV, p. 448; John Brown to Lord George Germain, 2 May 1776, in *The Pennsylvania Magazine of History and Biography*, v. XLIX, (October 1925), p. 349; Montfort Browne to the Earl of Dartmouth, 6 May 1776, PRO, CO 23/23.

[34] Deposition of Captain George Dorsett, 2 April 1779, in John Gambier et. *al.* to Lord George Germain, 31 March 1779, PRO, CO 23/24.

[35] "A Narrative . . . on the Invasion," *ibid.*

[36] John Gambier et. *al.* to Lord George Germain, *ibid.*

[37] Journal of the *Andrew Doria*, in Naval Documents, v. IV, p. 1490.

[38] John Paul Jones to Joseph Hewes, 14 April 1776, in *ibid.*, p. 815.

[39] Schöpf, *Travels in the Confederation*, v. II, p. 262.

[40] Montfort Browne to the Earl of Dartmouth, 6 May 1775, PRO, CO 23/23.

[41] *Bahamian Interlude: Being an Account of Life at Nassau*

in the Bahama Islands in the Eighteenth Century, Reprinted From the "Memoirs of Peter Henry Bruce Esq." Published in London 1782 (London: John Culmer, 1949), p. 27.

[42] [Sir Bede Edmund Clifford], *Historic Forts of Nassau in the Bahamas* (Nassau, n.p., 1952), p. 27.

[43] John Brown to Vice Admiral Clark Gayton, 22 March 1776, in *Naval Documents*, v. IV, p. 461.

[44] "A Narrative . . . on the Invasion," PRO, CO 23/24.

[45] John Brown to Vice Admiral Clark Gayton, 22 March 1776, in *Naval Documents*, v. IV, p. 461.

[46] "Journal Prepared For the King of France by John Paul Jones," in Sands, *Life and Correspondence of John Paul Jones*, pp. 35–36.

[47] Esek Hopkins to the President of Congress, 9 April 1776, in *American Archives*, 4th ser., v. V, p. 823; Clark, *Nicholas Biddle*, pp. 104–105.

[48] New Providence Council Minutes, 3 March 1776, in John Gambier et. *al.* to Lord George Germain, 31 March 1779, PRO, CO 23/23; Schöpf, *Travels in the Confederation*, v. II, p. 264; "Extract of a Letter From the Captain of Marines [Samuel Nicholas] on Board the Ship Alfred," 10 April 1776, in *American Archives*, 4th ser., v. V, p. 846, hereafter Nicholas Letter.

[49] Thomas Atwood to the Earl of Dartmouth, 22 March 1776, in *Naval Documents*, v. IV, pp. 464–465.

[50] John Brown to Vice Admiral Clark Gayton, 22 March 1776, In *ibid.*, p. 463; Nicholas Letter, in *American Archives*, 4th ser., v. V, pp. 846–847.

[51] "A Narrative . . . on the Invasion," PRO, CO 23/24.

[52] *Ibid.*

[53] *Ibid.*

[54] *Ibid.*

[55] Nicholas Letter, in *American Archives*, 4th ser., v. V, pp. 846–847.

[56] William Taylor to Lord George Germain, 20 May 1776, PRO, CO 23/23.

[57] *Ibid.*; John Gambier et. *al.* to Lord George Germain, 31 March 1779, PRO, CO 23/24.

[58] "Narrative . . . on the Invasion," *ibid.*

[59] *Ibid.*

[60] Nicholas Letter, in *American Archives*, 4th ser., v. V, p. 847.

[61] Diary of John Trevett, Trevett Papers, Newport Historical Society, hereafter *Trevett Diary*.

[62] Thomas Atwood to the Earl of Dartmouth, 22 March 1776, in *Naval Documents*, v. IV, p. 465.

[63] Nicholas Letter, in *American Archives*, 4th ser., v. V, p. 847.

[64] Thomas Atwood to the Earl of Dartmouth, 22 March 1776, in *Naval Documents*, v. IV, p. 465.

[65] Nicholas Letter, in *American Archives*, 4th ser., v. V, p. 847.

[66] Thomas Atwood to the Earl of Dartmouth, 22 March 1776, in *Naval Documents*, v. IV, p. 465.

[67] Esek Hopkins to the President of Congress, 9 April 1776, in *American Archives*, 4th ser., v. V, p. 823.

[68] Esek Hopkins Manifesto, 3 March 1776, Esek Hopkins Papers, Rhode Island Historical Society.

[69] John Brown to Vice Admiral Clark Gayton, 22 March 1776, in *Naval Documents*, v. IV, p. 463.

[70] Thomas Atwood to the Earl of Dartmouth, 22 March 1776, in *ibid.*, p. 465.

[71] New Providence Council Minutes, 3 March 1776, in John Gambier et. *al.* to Lord George Germain, 31 March 1779, PRO, CO 23/24.

[72] Montfort Browne to Lord George Germain, 5 November 1776, in *The Pennsylvania Magazine of History and Biography*, v. XLIX (October 1925), p. 353.

[73] Thomas Atwood to the Earl of Dartmouth, 22 March 1776, in *Naval Documents*, v. IV, p. 465.

[74] Montfort Browne to Lord George Germain, 5 November 1776, *The Pennsylvania Magazine of History and Biography*, v. XLIX (October 1925), p. 353.

[75] Lieutenant William Grant, R.N. to Governor Patrick Tonyn, 7 March 1776, in *Naval Documents*, v. IV, p. 225.

[76] Montfort Browne to Lord George Germain, 5 November 1776, in *The Pennsylvania Magazine of History and Biography*, v. XLIX (October 1925), p. 354.

[77] "Journal Prepared For the King of France by John Paul Jones," in Sands, *Life and Correspondence of John Paul Jones*, p. 36.

[78] Thomas Atwood to the Earl of Dartmouth, 22 March 1776, in *Naval Documents*, v. IV, pp. 465–466.

[79] New Providence Council Minutes, 3 March 1776, in John Gambier et. *al.* to Lord George Germain, 31 March 1779, PRO, CO 23/24.

[80] John Brown to Vice Admiral Clark Gayton, 22 March 1776, in *Naval Documents*, v. IV, p. 463.

[81] William Taylor to Lord George Germain, 20 May 1776, PRO, CO 23/23.

[82] Nicholas Letter, in *American Archives*, 4th ser., v. V, p. 847.

[83] John Paul Jones to Joseph Hewes, 14 April 1776, in *Naval Documents*, v. IV, p. 817.

[84] Nicholas Letter, in *American Archives*, 4th ser., v. V, p. 847.

[85] *Trevett Diary*.

[86] Journal of the *Andrew Doria*, in *Naval Documents*, v. IV, p. 1490.

[87] Clark, *Nicholas Biddle p. 107;* Journal of the *Wasp*, photocopy, Naval Records Collections, R.G. 45, NA.

[88] *Trevett Diary*.

[89] John Gambier *et. al.* to Lord George Germain, 31 March 1779, PRO, CO 23/24.

[90] Montfort Browne to Lord George Germain, 5 November 1776, in *The Pennsylvania Magazine of History and Biography*, v. XLIX (October 1925), p. 354.

[91] John Brown to Vice Admiral Clark Gayton, 22 March 1776, in *Naval Documents*, v. IV, p. 463; Esek Hopkins to Governor Nicholas Cooke, 8 April 1776, Esek Hopkins Papers, Rhode Island Historical Society.

[92] Nicholas Letter, in *American Archives*, 4th ser., v. V, p. 847.

[93] Journal of the *Andrew Doria*, in *Naval Documents*, v. IV, p. 1490.

[94] Esek Hopkins to the President of Congress, 9 April 1776, in *American Archives*, 4th ser., v. V, p. 823.

[95] John Brown to Vice Admiral Clark Gayton, 22 March 1776, in *Naval Documents*, v. IV, p. 463.

[96] William Taylor to Lord George Germain, 20 May 1776, PRO, CO 23/23.

[97] Isaac Craig to the Marine Committee, 5 November 1776, Craig Papers, Historical Society of Western Pennsylvania.

[98] Journal of the *Andrew Doria*, in *Naval Documents*, v. IV, p. 1490; Esek Hopkins to Stephen Hopkins, 21 April 1776, Hopkins Letter Book, Rhode Island Historical Society.

[99] William Grant to Governor Patrick Tonyn, 7 March 1776, in *Naval Documents*, v. IV, p. 225.

[100] Governor Patrick Tonyn to Andrew Barkley, 8 March 1776, in *ibid.*, p. 250.

[101] "Minutes of a Council held on Board His Majesty's Ship Scarborough, Savannah River in Georgia This 14th March, 1776", in *ibid.*, pp. 343–344.

[102] Montfort Browne to Lord Dartmouth, 17 March 1776, in *The Pennsylvania Magazine of History and Biography*, v. XLIX (October 1925), p. 366.

[103] Journal of the *Andrew Doria*, in *Naval Documents*, v. IV,

Chapter IV
Uncombined Operations, April-December 1776

[1] Journal of the New York Provincial Council, 11 March 1776, in *Naval Documents*, v. IV, p. 301.

[2] Journal of the New York Provincial Council, 12 March 1776, in *ibid.*, p. 309.

[3] *Journals of the Continental Congress*, v. IV, pp. 199–201.

[4] Diary of Richard Smith, LC.

[5] Chapelle, *The History of the American Sailing Navy*, pp. 53, 78.

[6] Marine Committee to John Barry, 8 April 1776, John Barry Papers, New York Historical Society; Marine Committee to John Barry, 11 April 1776. *ibid.*

[7] *Pennsylvania Archives*, 1st ser., v. IV, pp. 724–725.

[8] *Pennsylvania Colonial Records*, v. X, pp. 525–526.

[9] *Pennsylvania Archives*, 1st ser., v. IV, p. 725.

[10] Vice Admiral Molyneux Shuldham to Captain Andrew Hammond, 17 January 1776, in *Naval Documents*, v. III, p. 834.

[11] *Pennsylvania Colonial Records*, v. X, pp. 527–528.

[12] Diary of Richard Smith, LC.

[13] *Journals of the Continental Congress*, v. IV, pp. 238–240.

[14] Chapelle, *The History of the American Sailing Navy*, p. 78.

[15] Catherine Pennington to Benjamin Franklin, 15 December 1785, Franklin Papers, American Philosophical Society; Samuel Nicholson to Commissioners, 15 September 1777, *ibid.*; John Elliott to James Read, 24 November 1778, John Paul Jones Papers, LC; *Pennsylvania Archives*, 1st ser., v. IV, pp. 745–746.

[16] *Pennsylvania Evening Post* (Philadelphia), 11 April 1776.

[17] *Ibid.*, 13 April 1776.

[18] Marine Committee to John Barry, 8 April 1776, John Barry Papers, New York Historical Society; Marine Committee to John Barry, 11 April 1776, Miscellaneous Collections, New York Historical Society.

[19] *Pennsylvania Journal* (Philadelphia), 8 May 1776.

[20] Log of the *Solebay*, in *Naval Documents*, v. IV, p. 1289.

[21] *Pennsylvania Archives*, 1st ser., v. IV, pp. 743–744.

[22] *Pennsylvania Colonial Records*, v. X, pp. 556–557.

[23] *Ibid.*, pp. 557–558.

[24] *Pennsylvania Archives*, 1st ser., v. IV, pp. 745–746.

[25] *Ibid.*

[26] Will of Captain Miles Pennington, 30 April 1776, no. 92, Will Book R, p. 144, Register of Wills, City and County Archives, Philadelphia.

[27] Catherine Pennington to Benjamin Franklin, 15 December 1785, Franklin Papers, American Philosophical Society.

[28] *Pennsylvania Archives*, 1st ser., v. IV, pp. 747–748.

[29] *Pennsylvania Colonial Records*, v. X, p. 560.

[30] *Pennsylvania Archives*, 1st ser., v. IV, p. 749.

[31] Marine Committee to John Barry, 8 May 1776, Roberts Autograph Collection, Haverford College.

[32] William Duane, ed., *Extracts from the Diary of Christopher Marshall, Kept in Philadelphia and Lancaster, During the American Revolution*, 1774–1781 (Albany: Joel Munsell, 1877), p. 35, hereafter *Diary of Christopher Marshall.*

[33] Journal of HMS *Roebuck*, in *Naval Documents*, v. IV, pp. 1446–1447; *Pennsylvania Archives*, 1st ser., v. IV, p. 746.

[34] Journal of HMS *Roebuck*, in *Naval Documents*, v. IV, p. 1470.

[35] James S. Biddle, ed., *Autobiography of Charles Biddle, Vice-President of the Supreme Executive Council of Pennsylvania, 1745–1821* (Philadelphia: E. Claxton and Co., 1883), p. 85, hereafter *Autobiography of Charles Biddle*; *Pennsylvania Archives*, 1st ser., v. IV, p. 748.

[36] Pension File of John Risdell, S4117, Revolutionary War Pension and Bounty Land Warrant Application Files, NA.

[37] Journal of HMS *Roebuck*, in *Naval Documents*, v. IV, p. 1470.

[38] *Pennsylvania Archives*, 1st ser., v. IV, p. 751; Thomas Read to the Pennsylvania Committee of Safety, 9 May 1776, Navy Papers, Committee of Safety, Pennsylvania Historical and Museum Commission.

[39] *Pennsylvania Archives*, 1st ser., v. IV, p. 750.

[40] *Ibid.*, p. 750

[41] Pension File of John Risdell, S4117, NA.

[42] Narrative of Captain Andrew Hamond, 5–9 May 1776, in *Naval Documents*, v. V, pp. 15–16.

[43] *Pennsylvania Evening Post* (Philadelphia), 29 June 1776.

[44] Journal of HMS *Roebuck*, in *Naval Documents*, v. V, p. 33.

[45] *Pennsylvania Archives*, 1st ser., v. IV, p. 754.

[46] *Pennsylvania Colonial Records*, v. X, pp. 565–566.

[47] Narrative of Captain Andrew Hammond, 15 May 1776, in *Naval Documents*, v. V, pp. 108–109.

[48] *Pennsylvania Colonial Records*, v. X, p. 570; *Pennsylvania Archives*, 1st ser., v. IV, pp. 558–559.

[49] Muster Roll of the Brig *Lexington*, March 1777, Franklin Papers, Historical Society of Pennsylvania.

[50] *Pennsylvania Archives, 1st ser.*, v. IV, pp. 762–763.

[51] Journal of HMS *Liverpool*, 27 May 1776, in *Naval Documents*, v. V, pp. 277–278.

[52] *Journals of the Continental Congress*, v. IV, pp. 365–366.

[53] Marine Committee to Lambert Wickes, 10 June 1776, PCC, 37, pp. 7–8; Committee of Secret Correspondence to William Bingham, 10 June 1776, *ibid.*

[54] Lambert Wickes to the Committee of Secret Correspondence, 16 June 1776, PCC, 78, v. XXIII, p. 289; Lambert Wickes to Samuel Wickes, 2 July 1776, in *Naval Documents*, v. V, pp. 882–884.

[55] *Ibid*; Lambert Wickes to the Committee of Secret Correspondence, 11 July 1776, PCC, 78, v. XXIII, p. 293.

[56] *Ibid.*; Lambert Wickes to the Committee of Secret Correspondence, 13 July 1776, *ibid.*, p. 295.

[57] Statement Made by Captain Lambert Wickes, 28 July 1776, in *Naval Documents*, v. V, p. 1263.

[58] Captain John Chapman, R.N. to Vice Admiral James Young, 29 July 1776, in *ibid.*, p. 1278.

[59] Statement Made by Captain Lambert Wickes, 28 July 1776, in *ibid.*, p. 1264.

[60] Captain John Chapman, R.N. to Vice Admiral James Young, 29 July 1776, *ibid.*, p. 1278.

[61] Baron de Courcy to Count d'Argout, 28 July 1776, in *ibid.*, pp. 1263–1264.

[62] Count d'Argout to Gabriel de Sartine, 31 July 1776, in *ibid.*, pp. 1318–1319.

[63] Josiah Barlett to William Whipple, 27 August 1776, in *American Archives*, 5th ser., v. I, p. 1179.

[64] Marine Committee to Lambert Wickes, 17–18 October 1776, Marine Committee Letter Book, Miscellaneous Papers of the Continental Congress, R.G. 360, NA, hereafter Marine Committee Letter Book.

[65] Marine Committee to John Barry, 2 July 1776, Dreer Collection, Historical Society of Pennsylvania.

[66] *Pennsylvania Evening Post* (Philadelphia), 6 August 1776.

[67] John Hancock to Robert Morris, 2 January 1777, *PCC*, 12A.

[68] Journal of the *Andrew Doria*, in *Naval Documents*, v. IV, p. 1491.

[69] Esek Hopkins' Sailing Orders from New Providence, 18 March 1776, Hopkins Letter Book, Rhode Island Historical Society.

[70] Journal of the *Andrew Doria*, in *Naval Documents*, v. IV, pp. 1491–1492.

[71] Log of the Schooner *Wasp*, photostat, Naval Records Collection, NA.

[72] Journal of the *Andrew Doria*, in *Naval Documents*, v. IV, pp. 1492–1495.

[73] Esek Hopkins to the President of Congress, 9 April 1776, in *American Archives*, 4th ser., v. V, p. 823.

[74] Journal of the *Andrew Doria*, in *Naval Documents*, v. IV, p. 1495.

[75] Nicholas Letter, in *American Archives*, 4th ser., v. V, p. 847.

[76] As quoted in "Remarks on Board His Majesty's Ship *Glasgow* Saturday the 6th Day of April 1776," in *Naval Documents*, v. IV, p. 680.

[77] "A List of the Kill'd & Wounded On Board the Brigantine Cabot," undated, Esek Hopkins Papers, Rhode Island Historical Society.

[78] Journal of the *Andrew Doria*, in Naval Documents, v. IV, p. 1495.

[79] Nicholas Letter, in *American Archives*, 4th Ser., v. V, p. 847.

[80] "Remarks on Board His Majesty's Ship *Glasgow* Saturday the 6th Day of April 1776," in *Naval Documents*, v. IV, p. 680.

[81] *Newport Mercury*, 8 April 1776.

[82] "Remarks on Board His Majesty's Ship *Glasgow* Saturday the 6th Day of April 1776," in *Naval Documents*, v. IV, p. 681; Esek Hopkins to the President of Congress, 9 April 1776, in *American Archives*, 4th ser., v. V, p. 823.

[83] Journal of the *Andrew Doria*, in *Naval Documents*, v. IV, p. 1496.

[84] John Hancock to Esek Hopkins, 17 April 1776, in *ibid.*, pp. 868–869.

[85] *Journals of the Continental Congress*, v. V, p. 662.

[86] *Ibid.*, v. X, p. 13.

[87] Esek Hopkins to John Hancock, 1 May 1776, Hopkins Letter Book, Rhode Island Historical Society.

[88] Journal of the *Andrew Doria*, in *Naval Documents*, v. IV, p. 1497.

[89] Esek Hopkins to John Hancock, 1 May 1776, Hopkins Letter Book, Rhode Island Historical Society.

[90] George Washington to Esek Hopkins, 25 April 1776, in *Writings of Washington*, v. IV, p. 521.

[91] Esek Hopkins to George Washington, 1 May 1776, Hopkins Letter Book, Rhode Island Historical Society.

[92] Marine Committee to Esek Hopkins, 23 April 1776, in *Naval Documents*, v. IV, p. 1217.

[93] Esek Hopkins to John Paul Jones, 13 June 1776, Hopkins Letter Book, Rhode Island Historical Society.

[94] Esek Hopkins to Nicholas Biddle, 12 May 1776, *ibid.*

[95] Esek Hopkins to John Paul Jones ,13 June 1776, *ibid.*

[96] Alpheus Rice to Esek Hopkins, 21 May 1776, Esek Hopkins Papers, Rhode Island Historical Society.

[97] "Muster Roll on board the Sloop Providence Commanded by John Paul Jones Esqr From May 10th [1776]," Transcript of Military Papers, Revolutionary War, v. XL, pp. 663–666, Rhode Island Archives.

[98] John Paul Jones to Daniel Tillinghast, 20 June 1776, in *Naval Documents*, v. V, p. 639.

[99] Esek Hopkins to John Paul Jones, 18 June 1776, Hopkins Letter Book, Rhode Island Historical Society.

[100] John Hancock to Esek Hopkins, 31 May 1776, *ibid.*

[101] "Arrangement of Commissioned and Warrant Officers of the American Fleet," 30 April 1776, Esek Hopkins Papers, Rhode Island Historical Society.

[102] Journal of the *Andrew Doria*, in *Naval Documents*, v. IV, p. 1497; "A List of People on board the Andrew Doria 10th May [1776]," Nicholas Biddle Papers, on deposit at the Historical Society of Pennsylvania .

[103] Nicholas Biddle to Esek Hopkins, 21 May 1776, Esek Hopkins Papers, Rhode Island Historical Society.

[104] Journal of the *Andrew Doria*, in *Naval Documents*, v. IV, pp. 1499–1501.

[105] *Ibid.*, p. 1501.

[106] *Public Advertiser* (London), 16 September 1776.

[107] Journal of the *Andrew Doria*, in *Naval Documents*, v. IV, p. 1501; *Trevett Diary*.

[108] Journal of the *Andrew Doria*, in *Naval Documents*, v. IV, pp. 1501–1503.

[109] Nicholas Biddle to Charles Biddle, 16 June 1776, Nicholas Biddle Papers, on deposit at the Historical Society of Pennsylvania.

[110] Charles Pond to George Washington, 19 June 1776, Washington Papers, LC.

[111] *Trevett Diary*.

[112] "Muster Roll of the Men and Officers belonging to the ship Columbus from the time of her being Put in Commission to the fourteenth day of November 1776," Transcript of Military Papers, Revolutionary War, v. XL, pp. 668–678, Rhode Island Archives.

[113] Deposition of Samuel Nicholas, 20 September 1782, Society Collections, Historical Society of Pennsylvania.

[114] Edward Burke to Esek Hopkins, undated, Esek Hopkins Papers, Rhode Island Historical Society.

[115] *Pennsylvania Ledger* (Philadelphia), 29 June 1776; Journal of HMS *Cerberus*, in *Naval Documents*, v. V, p. 600.

[116] Esek Hopkins to Nicholas Biddle, 16 June 1776, Hopkins Letter Book, Rhode Island Historical Society.

[117] Nicholas Biddle to Esek Hopkins, 22 July 1776, Nicholas Biddle Papers, on deposit at the Historical Society of Pennsylvania.

[118] Accounts of Nathaniel Shaw, Jr., Shaw Collection, Book 39, p. 7, Yale University.

[119] Nicholas Biddle to Esek Hopkins, 22 July 1776, Nicholas Biddle Papers, on deposit at the Historical Society of Pennsylvania; *Newport Mercury*, 22 July 1776.

[120] News Dispatch from Philadelphia, 18 September 1776, in *American Archives*, 5th ser., v. II, p. 378.

[121] *Trevett Diary;* Esek Hopkins to the Marine Committee, 10 September 1776, Hopkins Letter Book, Rhode Island Historical Society.

[122] *Pennsylvania Gazette* (Philadelphia), 18 September 1776.

[123] Commission of Isaac Craig, 22 October 1776, Craig

Papers, Historical Society of Western Pennsylvania.

[124] Alpheus Rice to John Paul Jones, 9 August 1776, John Paul Jones Papers, LC.

[125] "Muster Roll on board the Sloop Providence Commanded by John Paul Jones Esqr From May 10th [1776]," Transcripts of Military Papers, Revolutionary War, v. XL, pp. 663–666, Rhode Island Archives.

[126] *Journals of the Continental Congress*, v. V, p. 715.

[127] John Paul Jones to the Marine Committee, 4 September 1776, John Paul Jones Papers, LC.

[128] John Paul Jones to the Marine Committee, 30 September 1776, *ibid.*

[129] John Paul Jones to Robert Morris, 17 October 1776, *ibid.*

[130] Esek Hopkins to John Paul Jones, 22 October 1776, Hopkins Letter Book, Rhode Island Historical Society.

[131] "A Roll of all the Officers and Men belonging to the Ship Alfred from the time of her being put into Commission untill the 5th September 1776," Transcript of Military Papers, Revolutionary War, v. XL, pp. 657–663, Rhode Island Archives; "List of Officers & Men belonging to the Ship Alfred who are entitled to a Share of the Ship Mellish & Brig Active," undated, Revolutionary War Rolls, v. 52, p. 84, Massachusetts Archives.

[132] *Trevett Diary.*

[133] John Paul Jones to the Marine Committee, 30 October 1776, John Paul Jones Papers, LC.

[134] *Trevett Diary.*

[135] *Ibid.*

[136] John Paul Jones to the Marine Committee, 12 January 1777, John Paul Jones Papers, LC.

[137] *Journals of the Continental Congress*, v. V, p. 601.

[138] Lieutenant Oliver P. Remick (USRCS), *A Record of the Services of the Commissioned Officers and Enlisted Men of Kittery and Eliot, Maine: Who Served Their Country on Land and Sea in the American Revolution, 1775–1783* (Boston: Alfred Mudge & Son, 1901), p. 223; "The Continental Frigate *Raleigh*; Ship *Raleigh's* Book," *New Hampshire Genealogical Record*, v. II, no. 4 (April 1905), p. 181.

[139] William Whipple to John Langdon, 22 July 1776, William Whipple Papers, Peter Force Transcripts, LC.

[140] John Bradford to John Hancock, 15 May 1776, Dreer Collection, Historical Society of Pennsylvania.

[141] *Journals of the Continental Congress*, v. V, pp. 601–606.

[142] John Hancock to Thomas Cushing, 30 July 1776, in *Naval Documents*, v. V, p. 1290.

[143] "List of the Officers of the Navy of the United States," undated, PCC, 37, p. 473.

[144] Journal of William Jennison, LC.

[145] Journal of the Committee Appointed to Build Two Continental Frigates in Rhode Island, Rhode Island Historical Society.

[146] *Ibid.*

[147] *Journals of the Continental Congress*, v. V, p. 696; Gilbert Saltonstall to Silas Deane, 16 August 1783, in *Collections of the Connecticut Historical Society* (Hartford: The Society, 1930), v. XXIII, pp. 188–190.

[148] "List of the Officers of the Navy of the United States," undated, *PCC*, 37, p. 473.

[149] *American Archives*, 5th ser., v. III, p. 355; Resolve of the New York Committee of Safety, 27 December 1776, U.S. Navy Collection, Franklin D. Roosevelt Library.

[150] *New York Journal*, 8 September 1777.

[151] Joseph Hewes to Samuel Purviance, Jr., 25 June 1776, Purviance Papers, Maryland Historical Society.

[152] *Journals of the Continental Congress*, v. V, p. 478; "List of the Officers of the Navy of the United States," undated, PCC, 37, p. 473; Richard Henry Lee to Samuel Purviance, Jr., 24 November 1776, in *American Archives*, 5th ser., v. III, p. 826.

[153] Esek Hopkins to John Hancock, 10 September 1776, Esek Hopkins Papers, Rhode Island Historical Society.

[154] *Journals of the Continental Congress*, v. VI, p. 913.

[155] *Ibid.*, pp. 954–955.

[156] Uniform of Naval and Marine Officers, 5 September 1776, John Paul Jones Papers, LC.

[157] Whitfield J. Bell, Jr., "Addenda to Watson's Annals of Philadelphia: Notes by Jacob Mordecai, 1836," in *The Pennsylvania Magazine of History and Biography*, v. XCVIII (April 1974), pp. 145–146.

Chapter V
The Trenton-Princeton Campaign

[1] Esek Hopkins to John Hancock, 19 June 1776, Hopkins Letter Book, Rhode Island Historical Society.

[2] Samuel Nicholas to the President of Congress, 10 August 1781, *PCC*, 78, v. XVII, pp. 301–304.

[3] *Journals of the Continental Congress*, v. V, p. 478.

[4] *Ibid.*; Accounts of the Continental Fleet, October 1776, Esek Hopkins Papers, Rhode Island Historical Society.

[5] "Pay Roll of Capt. Robert Mullan's Compy Marines to December 1st 1776," Minute Book (AM 3275), Miscellaneous Collections, Historical Society of Pennsylvania.

[6] *Pennsylvania Journal* (Philadelphia), 20 November 1776.

[7] William A. Porter, "A Sketch of the Life of General Andrew Porter," *The Pennsylvania Magazine of History and Biography*, v. IV, no. 3, p. 263.

[8] *Pennsylvania Gazette* (Philadelphia), 14 August, 4 September 1776.

[9] *Journals of the Continental Congress*, v. V, p. 697; *Pennsylvania Archives*, 1st ser., v. V, p. 178.

[10] *Journals of the Continental Congress*, v. V, p. 716.

[11] Minutes of the Pennsylvania Council of Safety, 16 September 1776, *PCC*, 69, p. 277.

[12] *Journals of the Continental Congress*, v. V, p. 811.

[13] Minutes of the Pennsylvania Council of Safety, 8 October 1776, *PCC*, 69, p. 233.

[14] *Pennsylvania Colonial Records*, v. X, p. 757.

[15] *Journals of the Continental Congress*, v. VI, p. 921.

[16] "Regimental Court Martial held at Philada Barracks on the 24th Novr 1776, Minute Book (AM 3275), Miscellaneous Collections, Historical Society of Pennsylvania.

[17] "A Pay Roll of Capn Robert Mullan's Company of Marines from December 1st 1776 to April 1st 1777," *ibid.*

[18] *Journals of the Continental Congress*, v. VI, pp. 950–951.

[19] George Washington to John Hancock, 30 November 1776, in *Writings of Washington*, v. VI, p. 314.

[20] Duane, ed., *Diary of Christopher Marshall*, p. 105.

[21] *Pennsylvania Colonel Records*, v. XI, p. 26; Susanna Reed to James Reed, 3 December 1776, de Coppet Collection, Princeton University.

[22] "Extract from the Minutes of the Council of Safety," 3 December 1776, John Cadwalader Papers, Historical Society of Pennsylvania.

[23] Brigade Return, 17 December 1776, *ibid.*

[24] George Washington to John Cadwalader, 7 December 1776, *ibid.*

[25] George Washington to the President of Congress, 8 December 1776, in *Writings of Washington*, v. VI, p. 336; George Washington to Brigadier William Maxwell, 8 December 1776, in *ibid.*, p. 337.

[26] George Washington to the President of Congress, 9 December 1776, in *ibid.*, p. 339.

[27] George Washington to the President of Congress, 10 December 1776, in *ibid.*, p. 342.

[28] George Washington to the Pennsylvania Council of Safety, 10 December 1776, in *ibid.*, p. 344.

[29] George Washington to John Cadwalader, 11 December 1776, in *ibid.*, p. 347; George Washington to the President of Congress, 12 December 1776, in *ibid.*, pp. 354–355.

[30] Margaret Morris, *Private Journal Kept During a Portion of the Revolutionary War* (Philadelphia: Privately Printed, 1836), pp. 7–13.

[31] George Washington to Philemon Dickinson, 12 December 1776, in *Writings of Washington*, v. VI, p. 358.

[32] George Washington to James Ewing, 12 December 1776, in *ibid.*, pp. 360–361.

[33] George Washington to John Cadwalader, 12 December 1776, in *ibid.*, pp. 362–363.

[34] "Quarters of the several parts of Colo. Cadwalader's Brigade," 14 December 1776, John Cadwalader Papers, Historical Society of Pennsylvania.

[35] *Pennsylvania Archives*, 1st ser., v. V, pp. 99–100.

[36] *Journals of the Continental Congress*, v. VI, pp. 1024–1027.

[37] Robert Morris to the President of Congress, 13 December 1776, *PCC*, 137, Appendix; Robert Morris to Captain Nicholas Biddle, 13 December 1776, Bank of North America Papers, Historical Society of Pennsylvania.

[38] Samuel Chase to James Nicholson, 11 December 1776, Maryland Archives; Robert Morris to the President of Congress, 13 December 1776, *PCC*, 137, Appendix.

[39] Robert Morris to the President of Congress, 17 December 1776, *ibid.*

[40] Robert Morris to George Washington, 13 December 1776, Washington Papers, LC.

[41] George Washington to Robert Morris, 14 December 1776, in *Writings of Washington*, v. VI, p. 375; Brigade Return, 12 December 1776, John Cadwalader Papers, Historical Society of Pennsylvania; Brigade Return, 17 December 1776, *ibid.*

[42] "Return of Capt Robt. Mullans Compy of Marines," 20 December 1776, *ibid.*; "A Weekly Return of Capt Andw Porters Company of Marines to Saml. Nicholas Esqr. Major of Marines," 20 December 1776, *ibid.*; Brigade Return, 17 December 1776, *ibid.*

[43] "Main Guard at Bristol," 17 December 1776, *ibid.*; "Return of Prisoners," 21 December 1776, *ibid.*

[44] George Washington to Horatio Gates, 14 December 1776, in *Writings of Washington*, v. VI, p. 372.

[45] [Thomas Rodney.] *Diary of Captain Thomas Rodney, 1776–1777* (Wilmington: Historical Society of Delaware, 1888), p. 22, hereafter *Rodney Diary*.

[46] *Ibid.*; [Charles Willson Peale.] "Journal of Charles Willson Peale," in *The Pennsylvania Magazine of History and Biography*, v. XXXVIII, p. 276, hereafter "Peale Journal."

[47] Robert Morris to Richard Henry Lee, 29 December 1776, Lee Papers, University of Virginia.

[48] John Cadwalader to George Washington, 25 December 1776, Washington Papers, LC.

[49] George Washington to John Cadwalader, 27 December 1776, in *Writings of Washington*, v. VI, p. 448.

[50] "General Joseph Reed's Narrative of the Movements of the American Army in the Neighborhood of Trenton in the Winter of 1776–1777," in *The Pennsylvania Magazine of History and Biography*, v. VIII, p. 395, hereafter "General Joseph Reed's Narrative."

[51] *Ibid.*, pp. 395–396.

[52] *Ibid.*, pp. 396–397; *Rodney Diary*, pp. 24–26.

[53] *Ibid.*, p. 26.

[54] John Cadwalader to George Washington, 31 December 1776, Washington Papers, LC.

[55] Tench Tilghman to John Cadwalader, 31 December 1776, John Cadwalader Papers, Historical Society of Pennsylvania.

[56] Sergeant R——, "The Battle of Princeton," in *The Pennsylvania Magazine of History and Biography*, v. XX, p. 516, hereafter "The Battle of Princeton."

[57] George Washington to the President of Congress, 1 January 1777, in *Writings of Washington*, v. VI, p. 462; John Cadwalader to George Washington, 31 December 1776, Washington Papers, LC; Spy Map, enclosed in Cadwalader to Washington, 31 December 1776, Map Division, LC.

[58] "Peale Journal," p. 278; *Rodney Diary*, pp. 30–31.

[59] Wilkinson, *Memoirs*, v. I, pp. 137–138; "General Joseph Reed's Narrative," pp. 401–402; *Rodney Diary*, p. 31.

[60] Henry Knox to Lucey Knox, 7 January 1777, Knox Papers, Massachusetts Historical Society; George Washington to the President of Congress, in *Writings of Washington*, v. VI, p. 470.

[61] Wilkinson, *Memoirs*, v. I, p. 140.

[62] Ibid., pp. 140–141; "Peale Journal," pp. 279–280.

[63] *Rodney Diary*, pp. 32–33.

[64] Sergeant R———, "The Battle of Princeton," p. 517.

[65] William S. Stryker, *Documents Relating to the Revolutionary History of the State of New Jersey* (Trenton: John L. Murphy, 1901), 2d ser., v. I, p. 260; "Peale Journal," p. 280; *Rodney Diary*, pp. 34–35.

[66] George Washington to the President of Congress, 5 January 1777, in *Writings of Washington*, v. VI, p. 469.

[67] George Washington to the Pennsylvania Council of Safety, 8 January 1777, in *ibid.*, p. 481; "Peale Journal," p. 280.

[68] *Rodney Diary*, pp. 37–41.

[69] "Journal of Sergeant William Young," in *The Pennsylvania Magazine of History and Biography*, v. VIII, p. 272.

[70] Receipt signed by Samuel Nicholas, 11 January 1777, John Cadwalader Papers, Historical Society of Pennsylvania.

[71] Clothing receipts signed by Hugh Montgomery, 16, 19 January 1777, *ibid.*

[72] *Pennsylvania Eveneing Post* (Philadelphia), 21 January 1777.

[73] Robert Morris to John Hancock, 28 January 1777, *PCC*, 137, Appendix.

[74] Robert Morris to Nicholas Biddle, 30 January 1777, Marine Committee Letter Book.

[75] Memorandum Book of Henry Knox, 16 January–21 March 1777, Knox Papers, Massachusetts Historical Society.

[76] Report of the Commissioners of the Treasury, 7 August 1777, *PCC*, 136, v. I, p. 343.

[77] Isaac Craig to George Washington, 18 March 1779, Washington Papers, LC.

[78] George Weedon to Robert Mullan, 20 February 1777, Roberts Autograph Collection, Haverford College; *Pennsylvania Archives*, 2d ser., v. I, pp. 437–438.

[79] "Copy of a Report made by Mr. Ozeas Depy. Barracks Master," 21 March 1777, Correspondance of the Supreme

Executive Council, Pennsylvania Historical and Museum Commission.

[60] Commission of Isaac Craig, 14 March 1777, Craig Papers, Historical Society of Western Pennsylvania; Statement of Claims by Captains Craig and Porter, 9 March 1782, Washington Papers, LC.

[61] Report of the Commissioners of the Treasury, 7 August 1777, PCC, 136, v. I, p. 343; "Muster Roll of the Frigate Washington," 26 October 1777, Washington Papers, LC.

Chapter VI

New Frigates in Action and the Debacle on the Delaware, 1777

[1] *Trevett Diary*; Esek Hopkins to William Ellery, 13 March 1777, Hopkins Letter Book, Rhode Island Historical Society.

[2] John Grannis, *et al.* to Robert Treat Paine, 11 February 1777, Massachusetts Historical Society.

[3] "The Examination of John Grannis on the Subject Matter of the Petition of several Officers of the Frigate Warren . . .," undated, *PCC*, 58, pp. 235–238.

[4] *Journals of the Continental Congress*, v. VII, p. 204.

[5] *Ibid.*, v. X, p. 13.

[6] Marine Committee to John B. Hopkins, Abraham Whipple, and Dudley Saltonstall, 4 April 1777, Marine Committee Letter Book.

[7] *Trevett Diary.*

[8] Esek Hopkins to Joseph Olney, 15 January 1777, Hopkins Letter Book, Rhode Island Historical Society; John Bradford to Robert Morris, 29 January 1777, John Bradford Letter Book, LC; Esek Hopkins to Robert Morris, 28 February 1777, Hopkins Letter Book, Rhode Island Historical Society.

[9] Isaac J. Greenwood, *Captain John Manley* (Boston: C.E. Goodspeed & Co., 1915), p. 64.

[10] *Pennsylvania Journal* (Philadelphia), 2 July 1777.

[11] Journal of William Jennison, LC.

[12] Hector McNeill to the Massachusetts Council, undated, in Gardner W. Allen, *Captain Hector McNeill of the Continental Navy* (Boston: Massachusetts Historical Society, 1921), p. 33, hereafter *Captain Hector McNeill.*

[13] Hector McNeill to the Marine Committee, 21 May 1777, in *ibid.*, p. 39.

[14] Hector McNeill to the Marine Committee, 16 July 1777, in *ibid.*, pp. 56–57.

[15] Hector McNeill to Thomas Thompson, 21 July 1777, in *ibid.*, pp. 61–62.

[16] *Ibid.*; Journal of William Jennison, LC.

[17] Hector McNeill to the Marine Committee, 16 July 1777, in Allen, *Captain Hector McNeill*, p. 57; Hector McNeill to Thomas Thompson, 21 July 1777, in *ibid.*, pp. 61–62.

[18] Hector McNeill to the Marine Committee, 16 July 1777, in *ibid.*, p. 57.

[19] *Ibid.*, pp. 57–58; *Independent Chronicle* (Boston), 7 August 1777.

[20] List of Prisoners on board the Cartel *Swift*, 9 November 1777, Massachusetts Archives; Sir George Collier to Edward Brooks, 8 November 1777, *ibid.*

[21] Hector McNeill to his wife, 23 July 1777, in Allen, *Captain Hector McNeill*, p. 64; Sir George Collier to Edward Brooks, 8 November 1777, Massachusetts Archives; "A Return of Prisoners that came from Halifax in the Favorite Cartel," *ibid.*; Journal of William Jennison, LC.

[22] Hector McNeill to Richard Palmes, 10 August 1777, in Allen, *Captain Hector McNeill*, p. 69.

[23] Hector McNeill to Richard Palmes, 11 August 1777, in *ibid.*, p. 69.

[24] Journal of Wiliam Jennison, LC.

[25] *Ibid.*

[26] Hector McNeill to the Marine Committee, 25 August 1777, in Allen, *Captain Hector McNeill*, p. 72.

[27] *Ibid.*

[28] Hector McNeill to the Navy Board of the Eastern Department, 9 September 1777, in *ibid.*, p. 75.

[29] Hector McNeill to the Marine Committee, 17 September 1777, in *ibid*, p. 82–83.

[30] Marine Committee to the Navy Board of the Eastern Department, 12 November 1777, Marine Committee Letter Book.

[31] Hector McNeill to the Marine Committe, 17 September 1777, in Allen, *Captain Hector McNeill*, p. 84.

[32] Marine Committee to the Navy Board of the Eastern Department, 12 November 1777, Marine Committee Letter Book.

[33] Colonel Pierce Long to the New Hampshire Committee of Safety, 4 February 1777, in Nathaniel Bouton, et. *al.*, eds., *Documents and Records Relating to the Province of New Hampshire*, 40 vols. (Concord: n.p., 1867–1943), v. VIII, p. 480.

[34] Committee of Safety to George Jerry Osborne, 4 February 1777, in *ibid.*, pp. 480–481.

[35] Thomas Thompson to the New Hampshire Committee of Safety, 6 February 1777, in *ibid.*, pp. 482–483.

[36] George Jerry Osborne to the New Hampshire Committee of Safety, 6 February 1777, in *ibid.*, pp. 484–485.

[37] New Hampshire Committee of Safety to George Jerry Osborne, 8 February 1777, in *ibid.*, p. 489.

[38] Thomas Thompson to the Marine Committee, 28 September 1777, Lee Papers, Harvard University.

[39] Robert Morris to John Paul Jones, 5 February 1777, PCC, 168, pp. 5–8.

[40] John Paul Jones to Matthew Parke, 2 July 1777, John Paul Jones Papers, LC.

[41] *Ibid.*

[42] John Paul Jones to Samuel Wallingford, 15 July 1777, Library of the University of Texas.

[43] John Paul Jones to Matthew Parke, 27 July 1777, John Paul Jones Papers, LC.

[44] John Paul Jones to the New Hampshire Committee of Safety, 29 August 1777, *ibid.*

[45] John Paul Jones to the Speaker and Representatives of New Hampshire, 20 September 1777, *ibid.*

[46] John Paul Jones to John Langdon, 11 October 1777, *ibid.*

[47] Ezra Green to Nathaniel Cooper, 4 December 1777, in Ezra Green, M.D., *Diary of Ezra Green, M.D., Surgeon on Board the Continental Ship-of-War "Ranger," Under John Paul Jones, from November 1, 1777, to September 27, 1778* (Boston: D. Clapp & Son, 1875), pp. 6–7, 17–18, hereafter *Diary of Ezra Green.*

[48] Officers of the *Randolph* to Nicholas Biddle, undated, *PCC*, 78, v. II, pp. 245–246.

[49] Clark, *Nicholas Biddle*, pp. 178–179.

[50] Nicholas Biddle to James Biddle, 11 March 1777, Nicholas Biddle Papers, on deposit at the Historical Society of Pennsylvania.

[51] Nicholas Biddle to Robert Morris, 1 September 1777, *PCC*, 78, v. II, p. 327.

[52] Officers of the *Randolph* to Nicholas Biddle, undated, *ibid.*, pp. 245–246.

[53] Nicholas Biddle to Robert Morris, 1 September 1777, *ibid.*, p. 327.

[54] *South Carolina and American General Gazette* (Charleston), 11 September 1777.

[55] Biddle, ed., *Autobiography of Charles Biddle*, p. 105.

[56] *Ibid.*; Nicholas Biddle to Robert Morris, 12 September 1777, *PCC*, 78, v. II, p. 241; *Gazette of the State of South Carolina* (Charleston), 15 September 1777.

[57] Nicholas Biddle to Robert Morris, 12 September 1777, *PCC*, 78, v. II, p. 241.

[58] Marine Committee to Nicholas Biddle, 24 October 1777, Marine Committee Letter Book; Nicholas Biddle to James Biddle, 22 November 1777, Nicholas Biddle Papers, on deposit at the Historical Society of Pennsylvania.

[59] George Washington to Patrick Henry, 31 May 1777, Washington Papers, LC.

[60] "A Muster Roll of Capt. Andrew Porters Company in Col. Lambs Regiment taken from the time of Enlistment to the 1st of August 1777," Revolutionary War Rolls, Continental Troops, 47–1, NA; Report of the Commissioners of the Treasury, 7 August 1777, *PCC*, 136, v. I, p. 343.

[61] *Pennsylvania Archives*, 2d ser., v. I, p. 117.

[62] Marine Committee to Charles Alexander, 8 April 1777, Marine Committee Letter Book.

[63] Francis Hopkinson, *et. al.* to the Pennsylvania Committee of Safety, 10 April 1777, Simon Gratz Collection, Historical Society of Pennsylvania.

[64] *Pennsylvania Evening Post* (Philadelphia), 3, 8, 10 July 1777.

[65] Muster Roll of the Armed Boat *Vulture*, May 1777, Records of the Comptroller General, Pennsylvania Historical and Museum Commission.

[66] *Pennsylvania Evening Post* (Philadelphia), 8 May 1777.

[67] "Muster Roll of the Frigate Washington's Crew Thomas Read Esquire Commander Abreast of Whitehill 26th October 1777," Washington Papers, LC.

[68] *Pennsylvania Archives*, 1st ser., v. V, p. 465.

[69] *Pennsylvania Gazette* (Philadelphia), 6 August 1777.

[70] *Pennsylvania Archives*, 1st ser., v. V, p. 501.

[71] *Ibid.*, pp. 538–539.

[72] *Ibid.*, 2d ser., v. I, pp. 224–225.

[73] [John Montressor.] *The Journal of Captain John Montressor: Collections of the New York Historical Society for the Year 1881* (New York: New York Historical Society, 1882), p. 458.

[74] *Pennsylvania Evening Post* (Philadelphia), 23 September 1777.

[75] *Journal of Captain John Montressor*, p. 458.

[76] Report of Captain Lieutenant Francis Downman, R.A., to Colonel James Pattison, Commanding the Royal Artillery at New York, undated, in *Journal of the Society for Army Historical Research*, v. 9 (1930), pp. 28–29.

[77] Marine Committee to Charles Alexander, 26 September 1777, Marine Committee Letter Book.

[78] *Pennsylvania Archives*, 1st ser., v. V, p. 637.

[79] Report of Captain Lieutenant Francis Downman, R.A., in *Journal of the Society for Army Historical Research*, v. 9 (1930), pp. 28–29.

[80] *Ibid.*; *Pennsylvania Packet* (Philadelphia), 31 December 1777.

[81] General Sir William Howe to Lord George Germain, 10 October 1777, in J. Almon, ed., *The Remembrancer; or Impartial Repository of Public Events, 1775–1784*, 17 vols. (London: Printed for J. Almon, 1775–1784), v. V, pp. 413–414, hereafter *The Remembrancer*.

[82] *Pennsylvania Archives*, 1st ser., v. V, pp. 664–645.

[83] *Ibid.*

[84] *Ibid.*, v. VI, pp. 27–29.

Chapter VII
Across the Atlantic, 1777–1778

[1] "Considerations on the Affair of the English Colonies in America," 12 March 1776, in *Naval Documents*, v. III, pp. 966–970.

[2] Lambert Wickes to Benjamin Franklin, 14 February 1777, Franklin Papers, American Philosophical Society; Lambert Wickes to the Committee of Secret Correspondence, 28 February 1777, *PCC*, 78, v. XXIII, p. 377.

[3] "Journal of the Proceedings at L'Orient Relative to the Five Prizes Taken by the Reprisal," in Almon, ed., *The Remembrancer*, v. IV, pp. 308–311.

[4] Lambert Wickes to Benjamin Franklin, 14 February 1777, Franklin Papers, American Philosophical Society.

[5] Jonathan William, Jr., to the American Commissioners in France, 19 February 1777, *ibid.*

[6] "Journal of the Proceedings at L'Orient," in Almon, ed., *The Remembrancer*, v. IV, pp. 308–311.

[7] Lambert Wickes to Benjamin Franklin, 14 February 1777, Franklin Papers, American Philosophical Society.

[8] "Journal of the Proceedings at L'Orient," in Almon, ed., *The Remembrancer*, v. IV, pp. 308–311.

[9] Benjamin Franklin and Silas Deane to the Committee of Secret Correspondence, 12 March 1777, in Francis Wharton, ed., *The Revolutionary Diplomatic Correspondence of the United States*, 6 vols. (Washington: Government Printing Office, 1889), v. II, p. 287, hereafter *Diplomatic Correspondence*.

[10] American Commissioners in France to Samuel Nicholson, 26 January 1777, Franklin Papers, LC.

[11] Samuel Nicholson to Silas Deane, 13 April 1777, Deane Papers, Connecticut Historical Society,

[12] "A Narrative respective Luggar surprise & Cutter revenge," in *The Pennsylvania Magazine of History and Biography*, v. XXII, pp. 479–488.

[13] *Ibid.*

[14] Lord Stormont to Lord Weymouth, 8 May 1777, in B.F. Stevens, comp., *Facsimiles of Manuscripts in European Archives Relating to America, 1775–1783; With Descriptions, Editorial Notes, Collations, References and Translations*, 25 vols. (London: Malby & Sons, 1889–1898), no. 1529, hereafter *Facsimiles*.

[15] Montmorency Price de Robecq to the Comte de Vergennes, 9 May 1777, in *ibid.*, no. 1530.

[16] George Lupton to William Eden, 10 April 1777, in *ibid.*, no. 681.

[17] Muster Rolls of the Brig *Lexington*, April, May 1777, Franklin Papers, Historical Society of Pennsylvania.

[18] George Lupton to William Eden, 4 June 1777, in Stevens, comp., *Facsimiles*, no. 168.

[19] Muster Roll of the Brig *Lexington*, May 1777, Franklin Papers, Historical Society of Pennsylvania.

[20] Muster Roll of the Cutter *Dolphin*, May 1777, John Paul Jones Papers, LC.

[21] Lambert Wickes to Samuel Nicholson, 23 May 1777, *PCC*, 41, v. VII, p. 145.

[22] Samuel Nicholson to the American Commissioners in France, 28 June 1777, Franklin Papers, American Philosophical Society; Lambert Wickes to the American Commissioners in France, 28 June 1777, *ibid.*

[23] "A Narrative respective Luggar surprize & Cutter revenge," in *The Pennsylvania Magazine of History and Biography,* v. XXII, pp. 479–488.

[24] Lambert Wickes to the American Commissioners in France, 15 August 177, Franklin Papers, American Philosophical Society.

[25] Samuel Nicholson to Henry Johnson, 25 August 1777, in Stevens, comp., *Facsimiles,* no. 1658.

[26] American Commissioners in France to Lambert Wickes, 28 August 1777, in [Silas Deane.] *The Deane Papers, 1774–1777,* 5 vols. (New York: New York Historical Society, 1887), v. II, p. 117; Lambert Wickes to Henry Johnson, 14 September 1777, in Stevens, comp., *Facsimiles,* no. 1685; Silas Deane to Robert Morris, 1 October 1777, in *The Deane Papers,* v. II, p. 160; Log of the *Lexington,* photostat, Naval Records Collection, NA.

[27] Lieutenant John Bazely to Philip Stephens, 24 September 1777, in Stevens, comp., *Facsimiles,* no. 1695; Charles Herbert, *A Relic of the Revolution* (Boston: Charles H. Peirce, 1847), p. 65.

[28] *Ibid.,* pp. 64–66, 94, 253.

[29] Samuel Nicholson to the American Commissioners in France, 21 September 1777, Franklin Papers, American Philosophical Society; American Commissioners in France to the Committee of Foreign Affairs, 30 November 1777, in Wharton, ed., *Diplomatic Correspondence,* v. II, p. 434.

[30] American Commissioners in France to Captains Thompson and Hinman, 25 November 1777, in *ibid.,* p. 428; M. Gourlade to Benjamin Franklin, 29 December 1777, Franklin Papers, American Philosophical Society.

[31] Log of the Continental Ship *Ranger,* 26 November 1777-14 May 1778, photostat, Naval Records Collection, NA.

[32] "Extrait de Journal des Services principaux De Paul Jones, dans la Révolution des Etats-Unis D'Amerique; écrit par luimême, it presenté avec un profond respect, an Trés illustre Prince Louis XVI," LC, hereafter Memoirs of John Paul Jones.

[33] Benjamin Franklin and Silas Deane to John Paul Jones, 16 January 1778, in Wharton, ed., *Diplomatic Correspondence* v. II, pp. 471–472.

[34] John Paul Jones to the American Commissioners in France, 5 December 1777, *PCC,* 58, pp. 133–135.

[35] Log of the Continental Ship *Ranger,* 26 November 1777–14 May 1778, photostat, Naval Records Collection, NA.

[36] *Ibid.*

[37] Thomas Simpson, Elijah Hall, and David Cullam to John Paul Jones, 14 February 1778, Rosenbach Collection, United States Naval Academy Museum.

[38] Matthew Parke to John Paul Jones, 19 February 1778, John Paul Jones Papers, LC.

[39] John Paul Jones to the American Commissioners in France, 15 August 1778, *ibid.*

[40] Log of the Continental Ship *Ranger,* 26 November 1777–14 May 1778, photostat, Naval Records Collection, NA.

[41] Deposition of Lieutenant Jean Meijer, 14 April 1780, *PCC,* 168, v. II, p. 71.

[42] Memoirs of John Paul Jones, LC.

[43] Green, *Diary of Ezra Green,* p. 20.

[44] Log of the Continental Ship *Ranger,* 26 November 1777–14 May 1778, (photostat), Naval Records Collection, NA.

[45] *Ibid.*

[46] Green, *Diary of Ezra Green,* p. 21.

[47] John Paul Jones to the American Plenipotentiaries, 27 May 1778, *PCC,* 168, v. I, pp. 79–92.

[48] Green, *Diary of Ezra Green,* p. 22.

[49] John Paul Jones to the American Plenipotentiaries, 27 May 1778, *PCC,* 168, v. I, pp. 79–92.

[50] Memoirs of John Paul Jones, LC.

[51] John Paul Jones to the American Plenipotentiaries, 27 May 1778, *PCC,* 168, v. I, p. 79–92.

[52] Log of the Continental Ship *Ranger,* 26 November 1777–14 May 1778, photostat, Naval Records Collection, NA.

[53] *Ibid.*

[54] John Paul Jones to Lady Selkirk, 8 May 1778, John Paul Jones Papers, LC.

[55] Lady Selkirk to Lord Selkirk, 24 April 1778, Collections of Lady Hope-Dunbar, Dundrennan, Scotland.

[56] Lady Selkirk to her Sister the Countess of Morton, 15 May 1778, *ibid.*

[57] Lady Selkirk to Lord Selkirk, 24 April 1778, *ibid.*

[58] Lord Daer to Dr. Blane, 27 May 1778, *ibid.*

[59] Robert Jeffrey to Lord Selkirk, 23 April 1778, *ibid.*

[60] John Schweighauser to Arthur Lee, 30 December 1779, Lee Papers, Harvard University.

[61] *London Chronicle,* 28 April 1778.

[62] Log of the Continental Ship *Ranger,* 26 November 1777–14 May 1778, photostat, Naval Records Collection, NA.

[63] Memoirs of John Paul Jones, LC.

[64] Deposition of Lieutenant Jean Meijer, 14 April 1780, *PCC,* 168, v. II, p. 71.

[65] Memoirs of John Paul Jones, LC.

[66] John Paul Jones to the American Plenipotentiaries, 27 May 1778, *PCC,* 168, v. I, pp. 79–82.

[67] Deposition of John Wentworth, 14 September 1838, in Pension File of James Roberts, W16701, NA.

[68] John Paul Jones to the American Plenipotentiaries, 27 May 1778, *PCC,* 168, v. I, pp. 79–92.

[69] Pension File of John Ricker, S11290, NA.

[70] Green, *Diary of Ezra Green,* p. 25.

[71] Jonathan Williams to John Paul Jones, 26 May 1778, John Paul Jones Papers, LC.

[72] William Morris to John Paul Jones, 26 May 1778, *ibid.*

[73] John Paul Jones to the American Commissioners in France, 15 August 1778, *ibid.*

[74] Sands, *Life and Correspondence of John Paul Jones,* p. 95.

[75] American Commissioners to John Paul Jones, 3 June 1778, in Wharton, ed., *Diplomatic Correspondence,* v. II, pp. 604–605.

[76] John Paul Jones to the American Commissioners in France, 16 June 1778, in *ibid.,* p. 616.

[77] Green, *Diary of Ezra Green,* pp. 26–27.

[78] Journal of William Jennison, LC.

[79] *Ibid.*

[80] Navy Board of the Eastern Department to Samuel Tucker, 10 February 1778, Samuel Tucker Papers, Peter Force Transcripts, LC.

[81] [John Adams.] *The Adams Papers. Diary and Autobiography of John Adams,* 4 vols. (Cambridge: The Belknap Press, 1962), v. II, p. 269, hereafter *Diary and Autobiography of John Adams.*

[82] *Ibid.,* pp. 271–272.

[83] Log of the *Boston,* in Sheppard, *Samuel Tucker,* pp. 262–263.

[84] *Diary and Autobiography of John Adams,* v. II, pp. 275–276.

[85] Journal of William Jennison, LC.

[86] *Diary and Autobiography of John Adams*, v. II, pp. 276–286.
[87] Samuel Tucker to the Navy Board of the Eastern Department, 11 March 1778, Samuel Tucker Papers, Peter Force Transcripts, LC.
[88] Log of the *Boston*, in Sheppard, *Samuel Tucker*, pp. 282–283.
[89] Samuel Tucker to the Navy Board of the Eastern Department, 1 April 1778, Samuel Tucker Papers, Peter Force Transcripts, LC.
[90] Journal of William Jennison, LC.
[91] *Ibid.*
[92] American Commissioners in France to Samuel Tucker, 13 April 1778, Samuel Tucker Papers, Peter Force Transcripts, LC. John Brondfield to Samuel Tucker, 25 May 1778, *ibid.*
[93] Musco Livingston to Samuel Tucker, 29 May 1778, *ibid.*
[94] Petition and Accounts of Sergeant Cazaneuve, 4 June 1778, *ibid.*
[95] Log of the *Boston*, in Sheppard, *Samuel Tucker*, p. 290.
[96] Journal of William Jennison, LC.
[97] Log of the *Boston*, in Sheppard, *Samuel Tucker*, p. 299.
[98] Richard Palmes to Samuel Tucker, 8 July 1778, Samuel Tucker Papers, Peter Force Transcripts, LC.
[99] Samuel Tucker to the American Commissioners in France, 12 July 1778, *ibid.*; Journal of William Jennison, LC.
[100] American Commissioners to Samuel Tucker, 22 July 1778, Samuel Tucker Papers, Peter Force Transcripts, LC.
[101] Log of the *Boston*, in Sheppard, *Samuel Tucker*, p. 304.
[102] John Deshon to Abraham Whipple, Abraham Whipple Letter Book, Clements Library, University of Michigan.
[103] Pension File of Stephen Earl, W21035, NA.
[104] John Deshon to the Navy Board of the Eastern Department, 9 March 1778, Collections, Rhode Island Historical Society.
[105] John Deshon to Abraham Whipple, 1 April 1778, Abraham Whipple Letter Book, Clements Library, University of Michigan.
[106] Commission of William Waterman, 29 April 1778, in Pension File of William Waterman, W25983, NA.
[107] Abraham Whipple to the American Commissioners in France, 31 May 1778, Abraham Whipple Letter Book, Clements Library, University of Michigan; Abraham Whipple to Congress, 10 June 1778, *ibid.*
[108] Abraham Whipple to the American Commissioners in France 31 May 1778, *ibid.*; Abraham Whipple to Benjamin Franklin, 15 June 1778, Franklin Papers, American Philosophical Society.
[109] William Jones to Abraham Whipple, 1 July 1778, Abraham Whipple Papers, Clements Library, University of Michigan.
[110] Abraham Whipple to Congress, 10 June 1786, *ibid.*
[111] Log of the *Boston*, in Sheppard, *Samuel Tucker*, pp. 304–306.
[112] *Ibid.*, pp. 309–327; Journal of William Jennison, LC; Abraham Whipple to the Navy Board of the Eastern Department, 16 October 1778, Abraham Whipple Letter Book, Clements Library, University of Michigan.
[113] *London Chronicle*, 7 May 1778.
[114] "Account of Prizes taken in the *Revenge* by Capt. Gustavus Conyngham," in [Gustavus Conyngham,] *Letters and Papers Relating to the Cruises of Gustavus Conyngham* (New York: Naval Historical Society, 1915), table facing p. 152.
[115] "A Narrative respective Luggar surprize & Cutter revenge," in *The Pennsylvania Magazine of History and Biography*, v. XXII, pp. 479–488.
[116] *Ibid.*
[117] *Journals of the Continental Congress*, v. XIII, p. 236; Marine Committee to Joseph Reed, 12 March 1779, Marine Committee Letter Book.

Chapter VIII
A Year of Landings and Leg Irons, 1778

[1] *Trevett Diary.*
[2] John Gambier to Lord George Germain, 25 February 1778, PRO, CO 23/24.
[3] John Brown to Lord George Germain, 22 February 1778, *ibid.*
[4] *Trevett Diary.*
[5] *Pennsylvania Packet* (Philadelphia), 25 March 1778; Deposition of Cornelius Anabil, *et. al.*, 21 February 1778, *PCC*, 41, v. X, p. 24.
[6] *Pennsylvania Packet* (Philadelphia), 25 March 1778; Deposition of Cornelius Anabil, *et. al.*, 21 February 1778, *PCC*, 41, v. X, p. 24.
[7] *Trevett Diary.*
[8] Minutes of the Council, 27 January 1778, in John Gambier to the Lords Commissioners of Trade, 16 June 1778, PRO, CO 23/9.
[9] *Trevett Diary.*
[10] *Pennsylvania Packet* (Philadelphia), 25 March 1778; *London Chronicle*, 28 April 1778.
[11] *Trevett Diary.*
[12] John Bradford to the Marine Committee, 17 June 1778, John Bradford Letter Book, LC.
[13] William Ellery to William Vernon, 6 May 1778, "Papers of William Vernon and the Navy Board, 1776–1794," *Publications of the Rhode Island Historical Society*, v. VIII, no. 4, p. 241, hereafter *Vernon Papers*.
[14] John Bradford to the Marine Committee, 17 June 1778, John Bradford Letter Book, LC.
[15] James Warren and John Deshon to William Vernon, 10 June 1778, *Vernon Papers*, p. 247.
[16] Manuscript, undated, *PCC*, 157, p. 204.
[17] John Bradford to the Marine Committee, 2 September 1778, John Bradford Letter Book, LC.
[18] William Vernon and James Warren to the Marine Committee, 9 December 1778, Letter Book of the Navy Board of the Eastern Department, 23 October 1778–29 October 1779, New York Public Library, hereafter Eastern Navy Board Letter Book.
[19] Nicholas Biddle to Robert Morris, 23 November 1777, U.S. Navy Collection, Franklin D. Roosevelt Library.
[20] William Moultrie, *Memoirs of the American Revolution, so Far as it Related to the States of North and South Carolina, and Georgia*, 2 vols. (New York: David Longworth, 1802), v. I, p. 198, hereafter *Memoirs*.
[21] Alexander S. Salley, Jr., ed., *Journal of the Commissioners of the Navy of South Carolina, October 9, 1776–March 1, 1779* (Columbia: The State Company, 1912), p. 102, hereafter *Journal of the Commissioners*.
[22] John Rutledge to General Robert Howe, 12 December 1777, in Moultrie, *Memoirs*, v. I, pp. 192–193.

[23] Salley, ed., *Journal of the Commissioners*, pp. 113–114.

[24] *Ibid*., p. 114.

[25] John Rutledge to General Robert Howe, 12 December 1777, in Moultrie, *Memoirs*, v. I, pp. 192–193.

[26] "Minutes of the Council of War," 13 December 1777, in *ibid*., p. 195.

[27] *Ibid*., p. 196.

[28] William Moultrie to Robert Howe, 23 December 1777, in *ibid*., p. 197.

[29] Salley, ed., *Journal of the Commissioners*, pp. 114–132.

[30] "An Order Book of the 1st Regt., S.C. Line, Continental Establishment," in *The South Carolina Historical and Genealogical Magazine*, v. VII, pp. 139–140.

[31] "Biographical Memoirs of the Late Captain Nicholas Biddle," in *The Port Folio*, v. II, no. 4 (October 1809).

[32] Salley, ed., *Journal of the Commissioners*, pp. 133–134.

[33] Moultrie, *Memoirs*, v. I, pp. 199–201.

[34] Salley, ed., *Journal of the Commissioners*, p. 135.

[35] Moultrie, *Memoirs*, v. I, p. 201.

[36] Salley, ed., *Journal of the Commissioners*, p. 135.

[37] "An Order Book of the 1st Regt., S.C. Line, Continental Establishment," in *The South Carolina Historical and Genealogical Magazine*, v. VII, pp. 197–199.

[38] John Blake to Thomas Hall, 7 October 1804, in Biddle, *Autobiography of Charles Biddle*, pp. 393–394.

[39] Robert Fanshawe to Viscount Richard Howe, 13 February 1778, PRO, Admiralty Records, Admirals' Dispatches, 1/488.

[40] John Blake to Thomas Hall, 7 October 1804, in Biddle, *Autobiography of Charles Biddle*, p. 394.

[41] Ibid.; Rawlins Lowndes to Henry Laurens, 30 March 1778, in *The South Carolina Historical and Geneological Magazine*, v. X, pp. 171–173.

[42] John Blake to Thomas Hall, 7 October 1804, in Biddle, *Autobiography of Charles Biddle*, p. 394.

[43] Certificate of John Davis, 21 August 1801, in *ibid*., pp. 108–109.

[44] William Moultrie to Robert Howe, 18 April 1778, in Moultrie, *Memoirs*, v. II, p. 372.

[45] Log of the *Yarmouth*, PRO, Admiralty Records, Captains' Logs, 51/1091.

[46] William Moultrie to Robert Howe, 18 April 1778, in Moultrie, *Memoirs*, v. II, p. 372.

[47] Biddle, *Autobiography of Charles Biddle*, pp. 107–108.

[48] Captain Nicholas Vincent to Admiral James Young, 17 March 1778, PRO, Admiralty Records, Admirals' Dispatches, 1/309.

[49] *Independent Chronicle* (Boston), 9 April 1778.

[50] *Continental Journal* (Boston), 30 April 1778.

[51] *Ibid*.

[52] Log of HMS *Ariadne*, PRO, Admiralty Records, Captains' Logs, 51/68.

[53] William R. Cutter, ed., "A Yankee Privateersman in Prison in England, 1777–1779," in *The New England Historical and Genealogical Register*, v. XXXI, pp. 286–287, v. XXXIII, p. 38.

[54] Elisha Hinman to the Council of Massachusetts, 12 February 1779, Revolutionary War Letters, v. 200, p. 312½, Massachusetts Archives.

[55] Pension File of Nathaniel Richards, S35624, NA.

[56] William Story to William Vernon, 29 April 1778, *Vernon Papers*, p. 240.

[57] Marine Committee to the Navy Board of the Eastern Department, 8 May 1778, Marine Committee Letter Book.

[58] Elihu Trowbridge to Congress, 8 February 1779, *PCC*, 41, v. X, pp. 57–58, NA.

[59] John Deshon to the Navy Board, 9 March 1778, *Vernon Papers*, p. 214.

[60] William Vernon to the Navy Board, 25 March 1778, *ibid*., p. 229.

[61] John Deshon to the Navy Board, 9 March 1778, *ibid*., p. 214; William Vernon to the Navy Board, 25 March 1778, *ibid*., p. 229.

[62] *Ibid*.

[63] James Warren to William Vernon, 20 April 1778, *ibid*., p. 236.

[64] Elihu Trowbridge to Congress, 8 February 1779, *PCC*, 41, v. X, pp. 57–58.

[65] Marine Committee to the Navy Board of the Eastern Department, 24 July 1778, Marine Committee Letter Book.

[66] James Warren to William Vernon, 23 July 1778, *Vernon Papers*, p. 254.

[67] John Bradford to the Marine Committee, 16 September 1778, John Bradford Letter Book, LC; *Independent Chronicle* (Boston), 17 September, 15 October 1778.

[68] William Vernon to the Navy Board, 25 March 1778, *Vernon Papers*, p. 230.

[69] *Independent Chronicle* (Boston), 16 April 1778.

[70] *Ibid*.

[71] Hoysteed Hacker to William Vernon, 3 April 1778, Vernon Papers, Newport Historical Society.

[72] W. Griffith to Viscount Howe, 28 March 1778, PRO, Admiralty Records, Captains' Letters, 1/1838.

[73] Hoysteed Hacker to William Vernon, 3 April 1778, Vernon Papers, Newport Historical Society.

[74] John Deshon to the Navy Board, 9 March 1778, *Vernon Papers*, pp. 214–215.

[75] William Vernon to the Navy Board, 25 March 1778, *ibid*., pp. 229–230.

[76] Mary Barney, *A Biographical Memoir of the Late Commodore Joshua Barney* (Boston: Gray and Bowen, 1832), pp. 63–67, hereafter *Memoir of Joshua Barney*; Marine Committee to James Nicholson, 4 March 1778, Marine Committee Letter Book.

[77] William H. Browne, ed., *Archives of Maryland: Journal and Correspondence of the Committee of Safety of Maryland, March 20, 1777–March 28, 1778* (Baltimore: Maryland Historical Society, 1897), pp. 534–535.

[78] *Pennsylvania Evening Post (Philadelphia)*, 20 May 1778; Log of HMS *Emerald*, PRO, Admiralty Records, Captains' Logs, 51/311.

[79] Henry Laurens to John Nixon, Francis Hopkinson, and John Wharton, 27 April 1778, *PCC*, 13, v. 1.

[80] Barney, *Memoir of Joshua Barney*, pp. 66–75.

[81] "Return of Prisoners onboard the St. Albans," undated, Papers of the New Jersey Department of Defense, New Jersey State Library; List of Prisoners on board HMS *Ardent*, undated; *ibid*.

[82] Thomas Pownal to Congress, 8 May 1787, *PCC*, 41, v. VIII, p. 222.

[83] Marine Committee to John Barry, 29 January 1778, Marine Committee Letter Book.

[84] "Narrative of Lieut. Luke Matthewman," in *The Magazine of American History*, v. II, p. 178; Marine Committee to John Barry, 11 March 1778, Marine Committee Letter Book.

[85] "Narrative of Lieut. Luke Matthewman," in *The Magazine of American History*, v. II, p. 178; William Ellery to William Vernon, 16 March 1778, *Vernon Papers*, p. 223.

[86] John Barry to George Washington, 9 March 1778, *PCC*, 152, v. V, pp. 367–368; William Ellery to William Vernon, 16 March 1778, *Vernon Papers*, pp. 223–224.

[87] Almon, ed., *The Remembrancer*. v. VI, pp. 148–149.

[88] William Ellery to William Vernon, 25 April 1778, *Vernon Papers*, pp. 237–238.

[89] John Brown to John Barry, 21 May 1778, Hepburn Collection, William B. Clark Transcripts, Naval Historical Center.

[90] John Barry to Robert Morris, 3 September 1778, U.S. Navy Collection, Franklin D. Roosevelt Library.

[91] Deposition of David Phipps, 16 October 1778, PRO, High Court of Admiralty, Prize Papers, 32/441.

[92] A List of officers and crew of the Continental brigantine *Resistance*, 21 August 1777, Nathaniel Shaw Collection, Yale University.

[93] Marine Committee to John Barry, 24 August 1778, Marine Committee Letter Book.

[94] John Barry to the Navy Board of the Eastern Department, undated, Hepburn Collection, William B. Clark Transcripts, Naval Historical Center; *Pennsylvania Evening Post* (Philadelphia), 19 October 1778.

[95] *Ibid.*

[96] Log of HMS *Unicorn* PRO, Admiralty, Captains' Logs, 51/1017.

[97] *Pennsylvania Evening Post* (Philadelphia), 16 October 1778.

[98] *Ibid.*; John Barry to the Navy Board of the Eastern Department, undated, Hepburn Collection, William B. Clark Transcripts, Naval Historical Center.

[99] *Ibid.*

[100] *Ibid.*; *Pennsylvania Evening Post* (Philadelphia), 16 October 1778; Log of HMS *Unicorn*, PRO, Admiralty Captains' Logs, 51/1017.

[101] Libel Against the Continental Ship *Raleigh*, PRO, High Court of Admiralty, Prize Papers, 32/441; William Vernon and John Deshon to the Massachusetts Council, 13 January 1779, Massachusetts Archives.

[102] Marine Committee to the Navy Board of the Eastern Department, 6 March 1778, Marine Committee Letter Book; John Bradford to the Marine Committee, 19 December 1777, John Bradford Letter Book, LC.

[103] "Copy of Prize List of the U.S. Brig General Gates John Skimmer, Comr.," Revolutionary War Rolls, v. 52, p. 79, Massachusetts Archives.

[104] John Bradford to the Marine Committee, 2 July, 30 July, 2 September 1778, John Bradford Letter Book, LC; James Warren to the Marine Committee, 9 December 1778, Eastern Navy Board Letter Book.

[105] James Warren and William Vernon to Joseph Hardy, 3 November 1778, "Joseph Hardy," ZB Files, Naval Historical Center.

[106] William Vernon and James Warren to Joshua Huntington, 6 November 1778, Eastern Navy Board Letter Book.

[107] Jonathan Trumbull to the Marine Committee, 2 April 1778, in James L. Howard, *Seth Harding Mariner: A Naval Picture of the Revolution* (New Haven: Yale University Press, 1930), p. 61.

[108] Pierre Landais, *Memorial, to Justify Peter Landai's Conduct During the Late War* (Boston: Peter Edes, 1784), pp. 3–5, hereafter *Memorial*.

[109] "A List of Officers, Seamen and Marines belonging to the Continental Ship of War the Alliance under the Command of Peter Landais Esqr.," Franklin Papers, University of Pennsylvania.

[110] Commission of James Warren, Jr., 2 October 1778, Private Collection, copy in History and Museums Division, Headquarters United States Marine Corps.

[111] Pension File of Thomas Elwood, S37907, NA.

[112] *Independent Chronicle* (Boston), 7 May 1778.

[113] Pension File of Samuel Pritchard, W22066, NA; Pension File of Samuel Gamage, W16266, NA.

Chapter IX
Marines on the Mississippi, 1777–1779

[1] J. J. Blomart to Oliver Pollock, 29 January 1775, Pollock Papers, LC.

[2] Edmund C. Burnett, ed., *Letters of Members of the Continental Congress*, 8 vols. (Washington: The Carnegie Institution, 1921–1936), v. II, pp. 421–423, 445–447, v. III, p. 96, hereafter *Letters*.

[3] *Ibid.*, v. II, p. 565.

[4] Commerce Committee to Oliver Pollock, 21 November 1777, PCC, 50, pp. 40–41.

[5] James Willing to Edward Hand, 7 January 1778, Miscellaneous Collections, Historical Society of Pennsylvania.

[6] James Willing to Congress, *PCC*, 42, v. VIII, pp. 237–238.

[7] James Willing to Oliver Pollock, 30 May 1778, Archivo General de Indias, Papeles procedentes de la isla de Cuba, Legajo 2370, hereafter A.G.I., Cuba, 2370.

[8] "List of sundrys for Expedition to New Orleans," 21 December 1777, Doc. 17443, Miscellaneous Numbered Records, War Department Collection of Revolutionary War Records, R.G. 93, NA.

[9] "Return of Men who Voluntarily entered to Serve on the Expedition to New Orleans," 22 December 1777, Revolutionary War Rolls, 13th Virginia Regiment, NA; "Pay Abstract of the following Men sent on Command with Capt. James Willing on board the Continental arm'd Boat Rattle Trap destined for New Orleans—from the Date of their respective Enlistments untill the 8th Day of January—1778," Records of the Comptroller General, Pennsylvania Historical and Museum Commission; "Pay Abstract of a Company of Marines Commanded by Captain James Willing of the United American States Commencing the 10th January and Ending the 3d June 1779 inclusive," 16 May 1782, *ibid.*; James Willing to the Board of War, 2 July 1781, *PCC*, 19, v. II, p. 223.

[10] Edward Hand to Richard Peters, 24 December 1777, *PCC*, 159, v. II, p. 223.

[11] Edward Hand to the Commercial Committee of Congress, 17 January 1779, *PCC*, 159, p. 431, NA.

[12] Reuben Thwaites and Louise P. Kellogg, eds., *Frontier Defense of the Upper Ohio, 1777-1778* (Madison: Wisconsin Historical Society, 1912), pp. 287–288, 302–303; Justin Winsor, *The Westward Movement: The Colonies and Republic West of the Alleganies, 1763–1798* (Boston: Houghton, Mifflin, 1897), p. 129.

[13] Burnett, ed., *Letters*, v. III, p. 144n.

[14] "Pay Abstract of a Company of Marines Commanded by Captain James Willing of the United American States Commencing the 10th January and Ending the 3d June 1779 inclusive," 16 May 1782, Records of the Comptroller General, Pennsylvania Historical and Museum Commission.

[15] Consul W. Butterfield, *History of the Girtys: Being a Concise Account of the Girty Brothers—Thomas, Simon, James and George, and of their Half-Brother, John Turner—Also of the Part Taken by Them in Lord Dunmore's War, in the Western Border War of the Revolution, and in the Indian War of 1790–95* (Cincinnati: Robert Clarke & Co., 1890), p. 105.

[16] Petition of Fourteen Americans, 2 February 1778, Archivo General de Indias, Papeles procedentes de la isla de Cuba, Legajo 191, hereafter A.G.I., Cuba, 191.

[17] Anthony Hutchins to Bernardó de Galvez, 14 March 1778, *ibid.*

[18] Governor Peter Chester to Peter Parker, 26 April 1778, PRO, Admiralty, Admirals' Dispatches 1/241.

[19] Capitulation of Natchez, 21 February 1778, A.G.I., Cuba, 2351.

[20] Governor Peter Chester to Peter Parker, 26 April 1778, PRO, Admiralty, Admirals' Dispatches 1/241.

[21] Mrs. Dunbar Rowland, *Life, Letters and Papers of William Dunbar* (Jackson: Mississippi Historical Society, 1930), pp. 60–62.

[22] Affidavit of Frederick Spell, 25 July 1778, A.G.I., Cuba, 191.

[23] Peter Chester to Peter Parker, 26 April 1778, PRO, Admiralty, Admirals' Dispatches 1/241; Oliver Pollock to Congress, 18 September 1782, *PCC*, 50, pp. 1–8.

[24] Affidavit of Peter Sellier, 4 May 1778, A.G.I., Cuba, 191; Affidavit of Benjamin Hodgdon, 3 May 1778, *ibid.*; John Campbell to Bernardo de Galvez, 12 March 1778, *ibid.*

[25] Galvez Proclamation, 3 March 1778, Archivo General de Indias, Audiencia de Santo Domingo, Legajo 2596.

[26] Governor Peter Chester to Peter Parker, 26 April 1778, PRO, Admiralty, Admirals' Dispatches 1/241.

[27] Oliver Pollock to the Commerce Committee, 1 April, 6 July 1778, *PCC*, 50, pp. 59–66, 77–81.

[28] John Fergusson to Bernardo de Galvez, 14 March 1778, A.G.I., Cuba, 191.

[29] Bernardo de Galvez to John Fergusson, 14 March 1778, PRO, Admiralty, Admirals' Dispatches 1/241.

[30] Bernardo de Galvez to John Fergusson, 14, 18 March 1778, *ibid.*

[31] James Willing to Bernardo de Galvez, 24 March 1778, A.G.I., Cuba, 2370.

[32] Bernardo de Galvez to James Willing, 6 April 1778, A.G.I., Cuba, 191.

[33] Commissions of Robert Elliott and Daniel Longstreet, April 1778, *PCC*, 41, v. III, pp. 33–34.

[34] Oliver Pollock to the Commerce Committee, 6 July 1778, *ibid.*, 50, pp. 77–81.

[35] Joseph Nunn to Governor Peter Chester, 14 April 1778, PRO, Admiralty, Admirals' Dispatches 1/241.

[36] Bernardo de Galvez to Congress, 5 May 1778, in Burnett, ed., *Letters*, v. III, p. 494n.

[37] Joseph Nunn to Bernardo de Galvez, 16 April 1778, A.G.I., Cuba, 191.

[38] Andrew Ellicott, *The Journal of Andrew Ellicott* (Philadelphia, n.p., 1814), pp. 130–132.

[39] J.F.H. Claiborne, *Mississippi, as a Province, Territory, and State* 2 vols. (Jackson: Power & Barksdale, 1880), v. I, p. 121.

[40] James Willing to Bernardo de Galvez, 24 May 1778, A.G.I., Cuba, 2370.

[41] Governor Peter Chester to "all his Britannick Majesty's loyal Subjects in Louisiana," 28 May 1778, A.G.I., Cuba, 191.

[42] Oliver Pollock to the Commerce Committee, 6 July 1778, *PCC*, 50, pp. 80–81.

[43] James Willing to George Rogers Clark, 1 September 1778, in James A. James, ed., *George Rogers Clark Papers, 1771–1781*, 2 vols. (Springfield, Ill., 1912–1926), v. II, pp. 67–68, hereafter *George Rogers Clark Papers*.

[44] Bernardo de Galvez to James Willing, 27 March 1778, A.G.I., Cuba, 191.

[45] James Willing to Bernardo de Galvez, 1 April 1778, A.G.I., Cuba, 2370.

[46] James Willing to Oliver Pollock, 30 May 1778, *ibid.*

[47] Oliver Pollock to James Willing, 31 May 1778, *ibid.*

[48] Oliver Pollock to Commerce Committee, 6 July 1778, *PCC*, 50, pp. 77–81.

[49] Oliver Pollock to the Commerce Committee, 11 August 1778, *ibid.*, pp. 84–87.

[50] Oliver Pollock to Bernardo de Galvez, 16, 18 June 1778, A.G.I., Cuba, 2370; Oliver Pollock to the Commerce Committee, 11 August 1778, *PCC*, 50, pp. 84–87.

[51] Bernardo de Galvez to "todos los Comandantes de la Costa del Rio Misisipi hta. la Punta Cortada," 14 July 1778, A.G.I., Cuba, 2351.

[52] Robert George to Bernardo de Galvez, 14 August 1778, A.G.I., Cuba, 2370.

[53] Bernardo de Galvez to Robert George, 18, 19 August 1778, *ibid.*

[54] Robert George and Richard Harrison to Bernardo de Galvez, 19 August 1778, *ibid.*

[55] "Pay Abstract of a Company of Marines Commanded by Captain James Willing of the United American States Commencing the 10th January and Ending 3d June 1779 inclusive," Records of the Comptroller General, Pennsylvania Historical and Museum Commission; "Pay Roll of Capt Robert George Company of Artillery in the service of the Commonwealth of Virginia and Illinois Department commencing the 4th June 1779 and ending the 8 December 1781," Virginia Half-Pay File of Robert George, NA.

[56] James Willing to Congress, *PCC*, 42, v. VIII, pp. 237–238; *Pennsylvania Gazette (Philadelphia)*, 5 January 1779.

[57] Marine Committee to Colonel John Beatty, 5 July 1779, Marine Committee Letter Book; James, ed., *George Rogers Clark Papers*, v. II, p. 108.

[58] Oliver Pollock to the Commerce Committee, 25 August 1778, *PCC*, 50, unnumbered.

[59] Oliver Pollock to the Commerce Committee, 27 July 1779, *ibid.*, p. 110.

[60] *Ibid.*, p. 111.

[61] Oliver Pollock to the Commerce Committee, 25 August 1779, *ibid.*, p. 112–113.

[62] Robert Elliott to Congress, 17 January 1781, *PCC*, 41, v. III, pp. 61–64.

[63] Oliver Pollock to the Commerce Committee, 25 August 1779, *PCC*, 50, p. 113.

[64] J. Henderson to the Commerce Committee, 17 September 1779, *ibid.*, pp. 114–115.

[65] "Copy of Capitulation of the Inhabitants on the Settlements on Lake Pontchartrain," 16 October 1779, *ibid.*, p. 120; Oliver Pollock to Congress, 20 January 1780, *ibid.*, pp. 125–126.

[66] *Ibid.*

[67] Oliver Pollock to William Pickles, 20 January 1780, *ibid.*, pp. 123–124.

[68] Joseph Calvert to Congress, 24 April 1780, *ibid.*, pp. 357–358.

[69] John Henderson to the Commerce Committee, 17 April 1780, *ibid.*, p. 132; Joseph Calvert to Congress, 24 April 1780, *ibid.*, p. 358.

[70] Report of the Board of Admiralty, 7 July 1780, *PCC*, 37, p 251; Report of the Board of Admiralty, 5 December 1780, *ibid.*, p. 535.

[71] James Willing to Congress, *PCC*, 42, v. VIII, pp. 237–238.

Chapter X

Success at Sea and Disaster at Penobscot, 1779

[1] Navy Board of the Eastern Department to the Marine Committee, 9 December 1778, Eastern Navy Board Letter Book.

[2] Navy Board of the Eastern Department to the Marine Committee, 6 January 1779, *ibid.*

[3] Navy Board of the Eastern Department to the Marine Committee, 9 December 1778, *ibid.*

[4] Navy Board of the Eastern Department to the Commissioners of the Navy Board at Philadelphia, 13 March 1779, *ibid.*

[5] Elihu Trowbridge to Congress, 8 February 1779, *PCC*, 41, X, pp. 57–58.

[6] Navy Board of the Eastern Department to the Commissioners of the Navy Board at Philadelphia, 13 March 1779, Eastern Navy Board Letter Book.

[7] *Continental Journal* (Boston), 18 February 1779.

[8] Navy Board of the Eastern Department to the Marine Committee, 5 February 1779, Eastern Navy Board Letter Book.

[9] Navy Board of the Eastern Department to the Marine Committee, 15 January 1779, *ibid.*; *Independent Chronicle* (Boston), 21 January 1779.

[10] Navy Board of the Eastern Department to the Marine Committee, 22 February 1779, Eastern Navy Board Letter Book; William Ellery to William Vernon, 23 March 1779, *Vernon Papers*, p. 257.

[11] *Independent Ledger* (Boston), 8 February 1779.

[12] *Ibid.*, 10 May 1779.

[13] Log of the Continental Ship *Ranger*, 24 August 1778–10 May 1780, photostat, Naval Records Collection, NA.

[14] Navy Board of the Eastern Department to the Marine Committee, 13 March 1779, Eastern Navy Board Letter Book.

[15] Marine Committee to Joseph Olney, 10 February 1779, Marine Committee Letter Book.

[16] Log of the Continental Ship *Ranger*, photostat, Naval Records Collection, NA.

[17] *Pennsylvania Gazette* (Philadelphia), 28 April 1779.

[18] *Ibid.*

[19] *Independent Ledger* (Boston), 19, 26 April 1779.

[20] Navy Board of the Eastern Department to the Marine Committee, 28 April 1779, Eastern Navy Board Letter Book.

[21] *Ibid.*

[22] *Ibid.*

[23] Marine Committee to the Navy Board of the Eastern Department, 20, May 1779, Marine Committee Letter Book.

[24] Navy Board of the Eastern Department to the Marine Committee, 28 April 1779, Eastern Navy Board Letter Book.

[25] Navy Board of the Eastern Department to the Marine Committee, 18 May 1779, *ibid.*

[26] Navy Board of the Eastern Department to the Marine Committee, 17 April 1779, *ibid*; *Continental Journal* (Boston), 8 April 1779.

[27] Marine Committee to the Navy Board of the Eastern Department, 7 June 1779, Marine Committee Letter Book.

[28] "A Muster Roll of Officers, Seamen and Marines belonging to the Frigate Boston, Saml Tucker, Commander," undated, Revolutionary War Rolls, v. 52, unnumbered, Massachusetts Archives.

[29] Navy Board of the Eastern Department to the Marine Committee, 3 April 1779, Eastern Navy Board Letter Book.

[30] Navy Board of the Eastern Department to the Marine Committee, 17 May 1779, *ibid.*

[31] Navy Board of the Eastern Department to the Marine Committee, 6 May 1779, *ibid.*

[32] Navy Board of the Eastern Department to the Marine Committee, 22 April 1779, *ibid.*

[33] *Independent Ledger* (Boston), 17 May 1779; See also *Pennsylvania Packet* (Philadelphia), 25 May 1779.

[34] Navy Board of the Eastern Department to Captain Hoysteed Hacker, 13 May 1779, Eastern Board Letter Book.

[35] Navy Board of the Eastern Department to the Marine Committee, 19 May 1779, *ibid.*

[36] "List of men on Sloop Providence entitled to Prize Money as of 7 July 1779," Newport Historical Society.

[37] Navy Board of the Eastern Department to the Marine Committee, 10 June 1779, Eastern Navy Board Letter Book.

[38] Navy Board of the Eastern Department to the Marine Committee, 16 June 1779, *ibid.*

[39] Joseph Hardy to Joshua Huntington, 5 March 1779, *Confederacy Papers*, Historical Society of Pennsylvania.

[40] Joseph Hardy to Joshua Huntington, 2 April 1779, *ibid.*

[41] Marine Committee to Seth Harding, 17 April 1779, Marine Committee Letter Book.

[42] Marine Committee to Samuel Tucker, 2 June 1779, *ibid.*

[43] "A Muster Roll of Officers, Seamen and Marines belonging to the Frigate Boston, Saml Tucker, Commander," undated, Revolutionary War Rolls, v. 52, unnumbered, Massachusetts *Archives*; *Independent Ledger* (Boston), 12 July 1779.

[44] William Blodget to Nathanael Greene, 24 May 1779, Nathaniel Greene Papers, American Philosophical Society.

[45] Marine Committee to Seth Harding, 25 June 1779, Marine Committee Letter Book.

[46] Marine Committee to Samuel Nicholson, 25 June 1779, *ibid.*

[47] *Pennsylvania Packet* (Philadelphia), 21 September 1779; *Independent Chronicle* (Boston), 9 September 1779.

[48] Navy Board of the Eastern Department to Abraham Whipple, 12 June 1779, Abraham Whipple Papers, Clements Library, University of Michigan.

[49] Navy Board of the Eastern Department to the Marine Committee, 24 June 1779, Eastern Navy Board Letter Book.

[50] Andrew Sherburne, *Memoirs of Andrew Sherburne, A Pensioner of the Navy of the Revolution* (Providence, n.p., 1831), pp. 21–24.

[51] *Ibid.*, p. 24.

[52] General Sir Henry Clinton to Brigadier General Francis McLean, 11 February 1779, copy in Papers Relating to Penobscot, Maine Historical Society.

[53] General Sir Henry Clinton to Brigadier General Francis McLean, 13 April 1779, *ibid.*

[54] Brigadier General Francis McLean to General Sir Henry Clinton, 16 May 1779, *ibid.*

[55] Brigadier General Francis McLean to General Sir Henry Clinton, 26 June 1779, *ibid.*

[56] John Murray to Jeremiah Powell, 18 June 1779, in James P. Baxter, ed., *Documentary History of the State of Maine* (Portland: Lefavor-Tower, 1910), v. XVI, pp. 289–292, hereafter *Documentary History of Maine.*

[57] Charles Cushing to the Council of Massachusetts Bay, 19 June 1779, *ibid.*, pp. 295–296.

[58] Navy Board of the Eastern Department to the Marine Committee, 30 June 1779, Eastern Navy Board Letter Book.

[59] Navy Board of the Eastern Department to Jeremiah Powell, 30 June 1779, *ibid.*

[60] Resolves of the Council of Massachusetts Bay, 29 June 1779, v. 201, p. 126a, Massachusetts Archives; Jeremiah Powell to Mesech Weare, 29 June 1779, v. 201, p. 126, *ibid.*

[61] Deposition of Jeremiah Hill, 29 September 1779, v. 145, pp. 284–300, *ibid.*

[62] Deposition of Captain Philip Brown, v. 145, pp. 260–265, *ibid.*

[63] General Orders, 24 July 1779, General Solomon Lovell's Orderly Book, in Baxter, ed., *Documentary History of Maine*, v. XVI, pp. 394–395.

[64] Narrative of Thomas Philbrook, in Benjamin Cowell, *Spirit of '76 in Rhode Island: or, Sketches of the Efforts of the Government and People in the War of the Revolution* (Boston: A. J. Wright, 1850), p. 316.

[65] "At a Council of War on Board the Warren," 26 July 1779, v. 145, p. 118, Massachusetts Archives.

[66] Narrative of Thomas Philbrook, in Benjamin Cowell, *Spirit of '76 in Rhode Island*, p. 317.

[67] "At a Council of War on Board the Warren," 27 July 1779, v. 145, p. 119, Massachusetts Archives.

[68] Lieutenant and Masters to Dudley Saltonstall, 27 July 1779, v. 145, pp. 50–51, *ibid.*

[69] Gardner W. Allen, *A Naval History of the American Revolution* (Boston: Riverside Press, 1913), v. II, pp. 425–426.

[70] *Continental Journal* (Boston), 2 September 1779.

[71] General Orders, 30 July 1779, General Solomon Lovell's Orderly Book, in Baxter, *Documentary History of the State of Maine*, v. XVI, pp. 410–411.

[72] "At a Council of War holden on board the Warren," 6 August 1779, v. 145, p. 68, Massachusetts Archives.

[73] "Proceedings of Council of War held at Head Quarters on the heights," 6 August 1779, v. 145, p. 69, *ibid.*

[74] Navy Board of the Eastern Department to Dudley Saltonstall, 12 August 1779, in Baxter, ed., *Documentary History of Maine*, v. XVI, p. 456.

[75] Kenneth Scott, "New Hampshire's Part in the Penobscot Expedition," *The American Neptune*, v. VII, no. 3 (July 1947), p. 211.

[76] *Journals of the Continental Congress*, v. XIV, p. 837.

[77] Report of the Marine Committee, 15 September 1779, Reports of the Marine Committee, Miscellaneous Papers of the Continental Congress, R.G. 360, NA.

[78] Joseph Pennell to Samuel Nicholas, 7 August 1779, copy in Manuscript Collections, History and Museums Division, Headquarters, United States Marine Corps.

[79] "An Account of Cash Laid out in Recruiting And to whom Given Nov 1st 1779," Minute Book (AM 3275), Miscellaneous Collections, Historical Society of Pennsylvania; "Return of Cloathing issued from the Magazine at Philadelphia from 30th August to 1st March 1780," Doc. 27411, Miscellaneous Numbered Records, War Department Collection of Revolutionary War Records, R.G. 93, NA.

[80] "A Copy of Return Given In to Major Samuel Nichlas of Men Enlisted by Capt. Robert Mullan, Since Augst 9th 1779 to Januy 1st 1780," Minute Book (AM 3275), Miscellaneous Collections, Historical Society of Pennsylvania; *Pennsylvania Gazette* (Philadelphia), 6 October 1779, 10 November 1779.

[81] Report of the Marine Committee, 15 September 1779, Reports of the Marine Committee, Miscellaneous Papers of the Continental Congress, NA.

[82] *Journals of the Continental Congress*, v. XV, p. 1262.

[83] Report of the Board of Admiralty, 8 January 1780. PCC, 37, p. 173; *Journals of the Continental Congress*, v. XVI, p. 28.

[84] Marine Committee to Seth Harding, 17 September 1779, Marine Committee Letter Book.

[85] Marine Committee to Seth Harding, 17 October 1779, *ibid.*

[86] *Pennsylvania Archives*, 1st ser., v. III, pp. 761–762; *Pennsylvania Colonial Records*, v. XII, pp. 147–148.

[87] John Jay to the President of Congress, 24 December 1779, in Wharton, ed., *Diplomatic Correspondence*, v. III, pp. 436–445.

[88] *Ibid.*

[89] Seth Harding to Samuel Huntington, 30 December 1779, PCC, 78, v. II, pp. 487–488; John Jay to the President of Congress, 26 December 1779, in Wharton, ed., *Diplomatic Correspondence*, v. III, pp. 448–449.

[90] Navy Board of the Eastern Department to the Marine Committee, 29 September 1779, Eastern Navy Board Letter Book.

[91] Navy Board of the Eastern Department to Abraham Whipple, 19 November 1779, Abraham Whipple Letter Book, Clements Library, University of Michigan.

[92] Pension File of Samuel Gamage, W16266, NA.

Chapter XI
Sailing Against England, 1779–1781

[1] Landais, *Memorial*, pp. 6–10.

[2] *Independent Chronicle* (Boston), 29 April 1779.

[3] *Ibid.*

[4] Landais, *Memorial*, pp. 11–14.

[5] *Diary and Autobiography of John Adams*, v. II, p. 358.

[6] Landais, *Memorial*, pp. 14–15.

[7] *Diary and Autobiography of John Adams*, v. II, pp. 370–371.

[8] *Ibid.*, v. IV, p. 125.

[9] Wharton, ed., *Diplomatic Correspondence*, v. II, pp. 599, 610.

[10] Memoirs of John Paul Jones, LC.

[11] *Ibid.*

[12] *Ibid.*

[13] Chavalier de Fitz-Maurice to John Paul Jones, 5 February 1779, John Paul Jones Papers, LC; Eugene Robert MacCarthy to John Paul Jones, 3 March 1779, *ibid.*

[14] Comte de Walsh-Serrant to John Paul Jones, 14 June 1779, *ibid.*

[15] *Diary and Autobiography of John Adams*, v. II, p. 371.

[16] Alexander Dick to Arthur Lee, 26 May 1779, Lee Papers, University of Virginia.

[17] Benjamin Franklin to John Paul Jones, 14 July 1779, Delaware Historical Society.

[18] Memoirs of John Paul Jones, LC.

[19] John Paul Jones to de Sartine, 11 August 1779, John Paul Jones Papers, LC.

[20] Landais, *Memorial*, pp. 18–23.

[21] John Paul Jones to Benjamin Franklin, 1 July 1779, John Paul Jones Papers, LC.

[22] Memoirs of John Paul Jones, LC.

[23] Benjamin Franklin to John Paul Jones, 8 July 1779, John Paul Jones Papers, LC.

[24] Henry Lunt to John Paul Jones, undated, Doc. 7541, *ibid.*

[25] "Proceedings of Court Martial in Trial of Robert Towers," 28 July–2 August 1779, *ibid.*

[26] John Paul Jones to Gourlade and Moylan, 28 July 1779,

ibid.; Benjamin Franklin to John Paul Jones, 28 July 1779, *ibid.*

[27] John Paul Jones to "every Commander in the Squadron," 10 August 1779, *ibid.*

[28] Landais, *Memorial*, pp. 25–26; Memoirs of John Paul Jones, LC.

[29] Louis F. Middlebrook, ed., *The Log of the Bon Homme Richard* (Mystic: The Marine Historical Association, 1936), pp. 23–24, hereafter *Log of the Bon Homme Richard.*

[30] John Paul Jones to Benjamin Franklin, 3 October 1779, John Paul Jones Papers, LC.

[31] *Ibid.*; Middlebrook, ed., *Log of the Bon Homme Richard*, pp. 27–29.

[32] John Paul Jones to Benjamin Franklin, 3 October 1779, John Paul Jones Papers, LC.

[33] *London Chronicle*, 9 October 1779.

[34] *Public Advertiser* (London), 18 November 1779.

[35] John Paul Jones to Benjamin Franklin, 3 October 1779 John Paul Jones Papers, LC.

[36] *Ibid.*

[37] "Plan of Attack upon Leith," 14 September 1779, John Paul Jones Papers, LC.

[38] Memoirs of John Paul Jones, LC.

[39] John S. Barnes, ed., *Fanning's Narrative: Being the Memoirs of Nathaniel Fanning an Officer of the Revolutionary Navy 1778–1783* (New York: Naval Historical Society, 1912), p. 29, hereafter *Fanning's Narrative.*

[40] John Paul Jones to de Chamillard, 14 September 1779, John Paul Jones Papers, LC.

[41] John Paul Jones to the Provost of Leith, 14 September 1779, *ibid.*

[42] John Paul Jones to Benjamin Franklin, 3 October 1779, *ibid.*

[43] *Ibid.*

[44] Barnes, ed., *Fanning's Narrative*, pp. 35–38.

[45] *Ibid.*, p. 39.

[46] Memoirs of John Paul Jones, LC.

[47] John Paul Jones to Benjamin Franklin, 3 October 1779, John Paul Jones Papers, LC.

[48] *Ibid.*

[49] *Ibid.*

[50] Memoirs of John Paul Jones, LC.

[51] Barnes, ed., *Franning's Narrative*, p. 47.

[52] Memoirs of John Paul Jones, LC.

[53] Barnes, ed., *Fanning's Narrative*, pp. 48–49.

[54] Captain Thomas Piercy to Captain Richard Pearson, in Almon, ed., *The Remembrancer*, v. IX, p. 48.

[55] Memoirs of John Paul Jones, LC.

[56] John Paul Jones to Benjamin Franklin, 3 October 1779, John Paul Jones Papers, LC.

[57] Narrative of John Kilby, September 1810, Collections, United States Naval Academy Museum.

[58] John S. Barnes, ed., *The Logs of the Serapis-Alliance-Ariel under the Command of John Paul Jones, 1779–1780* (New York: Naval Historical Society, 1911), pp. 25–27, hereafter *Logs of the Serapis-Alliance-Ariel.*

[59] *Ibid.*; pp. 28–29.

[60] John Paul Jones to Lieutenant Colonel Antoine-Felix Wuibert, undated, Franklin Papers, University of Pennsylvania.

[61] Deposition of Edward Stack, 22 October 1779, John Paul Jones Papers, LC; Barnes, ed., *The Logs of the Serapis-Alliance-Ariel*, p. 32.

[62] Andrew Paton to Sir Anstruther of Elie, 6 January 1780, in *The Navy Miscellany* (London: Naval Records Society, 1902), v. I, p. 403.

[63] *Pennsylvania Packet* (Philadelphia), 24 February 1780.

[64] John Paul Jones to Lieutenant Edward Stack, 4 November 1779, John Paul Jones Papers, LC.

[65] Comte de Walsh-Serrant to Edward Stack, 29 November 1779, in Memoirs of John Paul Jones, LC.

[66] Prince de Montharey to Edward Stack, 12 December 1779, in *ibid.*

[67] Certificate as to the service of Lieutenant Colonel Wuibert, 7 December 1779, John Paul Jones Papers, LC.

[68] Wharton, ed., *Diplomatic Correspondence*, v. III, p. 379.

[69] Memoirs of John Paul Jones, LC.

[70] Wharton, ed., *Diplomatic Correspondence*, v. III, p. 425.

[71] Barnes, ed., *The Logs of the Serapis-Alliance-Ariel*, p. 48.

[72] *Ibid.*, pp. 48–51.

[73] Barnes, ed., *Fanning's Narrative*, p. 80.

[74] "Regulations to be strictly Attended to in Working the Ship," 3 January 1780, John Paul Jones Papers, LC.

[75] Memoirs of John Paul Jones, LC; Barnes, ed., *The Logs of the Serapis-Alliance-Ariel*, pp. 51–55.

[76] Barnes, ed., *Fanning's Narrative*, pp. 77–78.

[77] Barnes, ed., *The Logs of the Serapis-Alliance-Ariel*, p. 59.

[78] *Ibid.*, pp. 61–69.

[79] Memoirs of John Paul Jones, LC.

[80] Benjamin Franklin to John Paul Jones, 19 February 1780, John Paul Jones Papers, LC.

[81] Memoirs of John Paul Jones, LC.

[82] Benjamin Franklin to John Paul Jones, 19 February 1780, John Paul Jones Papers, LC.

[83] John Paul Jones to Benjamin Franklin, 25 February 1780, *ibid.*

[84] Benjamin Franklin to John Paul Jones, 31 March 1780, *ibid.*

[85] Barnes, ed., *The Logs of the Serapis—Alliance—Ariel*, p. 81.

[86] Benjamin Franklin to John Paul Jones, 31 March 1780, John Paul Jones Papers, LC.

[87] Landais, *Memorial*, pp. 97–99; Officers of the *Alliance* to Benjamin Franklin, 7 June 1780, PCC, 193, p. 431.

[88] Testimony of Matthew Parke, 29 November 1780, PCC, 193, pp. 679–702.

[89] Barnes, ed., *The Logs of the Serapis—Alliance—Ariel*, p. 89.

[90] Pierre Landais to Matthew Parke, 13 June 1780, PCC, 193, p. 617.

[91] Matthew Parke to John Farnam, 13 June 1780, *ibid.*, p. 619.

[92] Barnes, ed., *The Logs of the Serapis—Alliance—Ariel*, p. 89.

[93] Court Martial of Pierre Landais, 20 November 1780, *PCC*, 193, pp. 451–595.

[94] John Paul Jones to Matthew Parke, 20 June 1780, John Paul Jones Papers, LC.

[95] Memoirs of John Paul Jones, LC.

[96] Narrative of John Kilby, September 1810, Collections, United States Naval Academy Museum.

[97] Court Martial of Pierre Landais, 20 November 1780, PCC, 193, pp. 451–595.

[98] Le Marquis de Puysegur to John Paul Jones, 2 June 1780, John Paul Jones Papers, LC; "A List of Officers. and Men belonging to the Ship of War Ariel, Commanded by the Honorable J. P. Jones Esqr.," undated, Franklin Papers, University of Pennsylvania.

[99] Barnes, ed., *The Logs of the Serapis—Alliance—Ariel*, p. 97–110.

[100] *Ibid.*, p. 109.
[101] Memoirs of John Paul Jones, LC.
[102] Barnes, ed., *Fanning's Narrative*, p. 87.
[103] Memoirs of John Paul Jones, LC.
[104] *Ibid.*
[105] Board of Admiralty to John Paul Jones, 19 May 1781, John Paul Jones Papers, LC.
[106] Board of Admiralty to John Paul Jones, 4 June 1781, *ibid.*
[107] Report of the Board of Admiralty, 6 July 1781, *PCC*, 37, pp. 495–496.
[108] Report of a Committee of Congress, 16 July 1781, *PCC*, 28, p. 141.
[109] Report of Committee on the report of the Board of War, 25 August 1781, *PCC*, pp. 143–144.

Chapter XII
Charleston, 1780

[1] Benjamin Lincoln to William Moultrie, 29 February 1780, in Moultrie, *Memoirs*, v. II, pp. 55–56.
[2] Abraham Whipple to the Marine Committee, 8 January 1780, Abraham Whipple Letter Book, Rhode Island Historical Society.
[3] Samuel Tucker to Richard Palmes, 8 December 1779, Samuel Tucker Papers, Peter Force Transcripts, LC.
[4] Abraham Whipple to the Marine Committee, 8 January 1780, Abraham Whipple Letter Book, Rhode Island Historical Society.
[5] Journal of William Jennison, LC.
[6] Abraham Whipple to the Marine Committee, 8 January 1780, Abraham Whipple Letter Book, Rhode Island Historical Society.
[7] Thomas Simpson to Abraham Whipple, 11 January 1780, *ibid.*
[8] Abraham Whipple to Thomas Simpson, 12 January 1780, *ibid.*
[9] Benjamin Lincoln to Abraham Whipple, 17 January 1780, *ibid.*
[10] Journal of William Jennison, LC.
[11] Abraham Whipple to the Marine Board, Philadelphia, 13 February 1780, Abraham Whipple Letter Book, Rhode Island Historical Society.
[12] Journal of Captain Peter Russell, in *American Historical Review*, v. IV, no. 1 (October 1898), p. 749.
[13] Vice Admiral Arbuthnot to Admiralty, 14 May 1780, in Almon, ed., *The Remembrancer*, v. X, p. 45.
[14] A. S. Salley, Jr., ed., *Journal of the Commissioners of the Navy of South Carolina, July 22, 1779—March 23, 1780* (Columbia: The State Company, 1913), pp. 71–72.
[15] Journal of William Jennison, LC.
[16] Governor Rutledge to Abraham Whipple, 18 February 1780, Abraham Whipple Letter Book, Rhode Island Historical Society.
[17] Benjamin Lincoln to Abraham Whipple, 30 January 1780, *ibid.*
[18] Naval Officers to Abraham Whipple, 1 February 1780, Samuel Tucker Papers, Peter Force Transcripts, LC.
[19] Benjamin Lincoln to Abraham Whipple, 13 February 1780, Abraham Whipple Letter Book, Rhode Island Historical Society.
[20] Governor Rutledge to Abraham Whipple, 13 February 1780, *ibid.*
[21] Log of the Continental Ship *Ranger*, photostat, Naval Records Collection, NA.
[22] Governor Rutledge to Benjamin Lincoln, 15 February 1780, Samuel Tucker Papers, Peter Force Transcripts, LC.
[23] Abraham Whipple to Samuel Tucker and Thomas Simpson, 15 February 1780, Abraham Whipple Letter Book, Rhode Island Historical Society.
[24] Log of the Continental Ship *Ranger*, photostat, Naval Records Callection, NA.
[25] Abraham Whipple to John Rathbun, 13, 22 February 1780, Abraham Whipple Letter Book, Rhode Island Historical Society.
[26] Abraham Whipple to Samuel Tucker, 20 February 1780, Samuel Tucker Papers, Peter Force Transcripts, LC.
[27] Abraham Whipple *et. al.* to Benjamin Lincoln, 27 February 1780, Abraham Whipple Letter Book, Rhode Island Historical Society.
[28] Benjamin Lincoln to Abraham Whipple, 28 February 1780, *ibid.*
[29] Council of War, 20 March 1780, *ibid.*; Benjamin Lincoln to Abraham Whipple, 20 March 1780, *ibid.*
[30] Admiral Arbuthnot to the Admiralty, 14 May 1780, in Almon, ed., *The Remembrancer*, v. X, pp. 45–46.
[31] Journal of William Jennison, LC.
[32] "List of the Batteries and the names of the Commanders in Charlestown 25th March 1780," Historical Society of South Carolina.
[33] Regimental Book of Captain James Bentham, 28 March 1780, copy in Area File, Naval Records Collection, NA.
[34] "Articles of Capitulation," 12 May 1780, in Moultrie, *Memoirs*, v. II, pp. 99–103.
[35] Abraham Whipple *et. al.* to Admiral Arbuthnot, 15 May 1780, Abraham Whipple Letter Book, Rhode Island Historical Society.
[36] Journal of William Jennison, LC.
[37] Navy Board of the Eastern Department to the Marine Committee, 18 August 1779, Eastern Navy Board Letter Book.
[38] Board of Admiralty to James Nicholson, 22 May 1780, Marine Committee Letter Book.
[39] Libel of James Nicholson against the Schooner *Queen Charlotte*, 27 May 1780, Court of Admiralty Papers, Historical Society of Pennsylvania.
[40] Gilbert Saltonstall to Gurdon Saltonstall, 14 June 1780, in *Records and Papers of the New London County Historical Society* (New London: The Society, 1893), p. IV, v. I, pp. 51–53.
[41] Gilbert Saltonstall to Gurdon Saltonstall, 19 June 1780, in *ibid.*, pp. 53–56.
[42] Gilbert Saltonstall to Gurdon Saltonstall, 14 June 1780, in *ibid.*, pp. 51–53.
[43] Gilbert Saltonstall to Gurdon Saltonstall, 19 June 1780, in *ibid.*, pp. 53–56.
[44] *Independent Chronicle* (Boston), 7 September 1780.
[45] Journal of Joseph Hardy, Mystic Seaport, Inc.
[46] Joseph Hardy to William Bingham, 29 January 1780, Simon Gratz Collection, Historical Society of Pennsylvania.
[47] Seth Harding to the President of Congress, 4 February 1780, *PCC*, 78, v. XIII, p. 5.
[48] Journal of Joseph Hardy, Mystic Seaport, Inc.
[49] *Ibid.*
[50] "Return of Arms belonging to the Ship Confederacy wanting repair," 23 May 1780, Area Files, Naval Records Collection, NA.
[51] Board of Admiralty to the Navy Board of the Eastern Department, 7 April 1780, Marine Committee Letter Book.

[53] George Washington to the Board of Admiralty, 27 May 1780, *PCC*, 37, p. 424.

[54] *Ibid.*

[54] *American Archives*, 5th ser., v. I, p. 918; v. II, p. 522.

[55] Elias Boudinot, *Journal, or Historical Recollections of American Events During the Revolutionary War* (Philadelphia, 1894), pp. 15–16.

[56] George Washington to the Board of Admiralty, 27 May 1780, *PCC*, 37, pp. 423–424.

[57] Report of the Board of Admiralty, 15 July 1780, *PCC*, 37, p. 415.

[58] Report of the Board of Admiralty, 23 July 1780, *PCC*, 37, p. 275.

[59] Board of Admiralty to the Navy Board of the Eastern Department, 16 June 1780, Marine Committee Letter Book.

[60] James L. Howard, *Seth Harding Mariner: A Naval Picture of the Revolution* (New Haven: Yale University Press, 1930), p. 129.

[61] Report of the Board of Admiralty, 10 July 1780, *PCC*, 37, p. 261; *Journals of the Continental Congress*, v. XVII, p. 603.

[62] *Pennsylvania Colonial Records*, v. XIII, pp. 427, 430, 436, 447, 487, 522, 535–537.

[63] Navy Board of the Eastern Department to the Board of Admiralty, 18 April 1780, Papers of the Eastern Navy Board, LC; *Independent Chronicle* (Boston), 20 April 1780.

[64] *Ibid.*, 25 May 1780.

[65] Navy Board of the Eastern Department to the Board of Admiralty, 19 June 1780, Papers of the Eastern Navy Board, LC.

[66] Navy Board of the Eastern Department to the Board of Admiralty, 3 August 1780, *ibid.*

[67] Board of Admiralty to John Young, 11 August 1780, Marine Committee Letter Book.

[68] Board of Admiralty to the Navy Board of the Eastern Department, 12 September 1780, Marine Committee Letter Book; *Pennsylvania Journal* (Philadelphia), 20 September 1780.

[69] Richard Langdon to Nathaniel Thwing, 5 October 1780, in *London Chronicle*, 2–5 December 1780.

[70] Report of the Board of Admiralty, 6 November 1780, *PCC*, 37, p. 517.

[71] Board of Admiralty to the President and Council of Pennsylvania, 7 September 1780, Marine Committee Letter Book.

[72] James Wilson to John Holker, 6 September 1780, in *Pennsylvania Archives*, 1st ser., v. VIII, p. 542.

[73] Board of Admiralty to the President and Council of Pennsylvania, 7 September 1780, Marine Committee Letter Book.

[74] Dennis Leary to William Moore, 1 September 1780, in *Pennsylvania Archives*, 1st ser., v. VIII, pp. 531–532.

[75] Board of Admiralty to the President and Council of Pennsylvania, 7 September 1780, Marine Committee Letter Book.

[76] Joseph Reed to John Holker, 14 March 1781, in *Pennsylvania Archives*, 1st ser., v. IX, p. 10.

[77] *Journals of the Continental Congress*, v. XVIII, p. 1097.

[78] Joseph Reed to the Pennsylvania Delegates to Congress, 30 November 1780, in *Pennsylvania Archives*, 1st ser., v. VIII, p. 643.

[79] Journal of Joseph Hardy, Mystic Seaport, Inc.

[80] *Pennsylvania Gazette* (Philadelphia), 10 January 1781.

[81] *Pennsylvania Journal* (Philadelphia), 28 February 1781.

[82] Journal of Joseph Hardy, Mystic Seaport, Inc., Harding Memorandum on Activities of the *Confederacy*, undated, Seth Harding Papers, *ibid.*

[83] *Ibid.*; *Pennsylvania Gazette* (Philadelphia), 2 May 1781.

[84] *New York Gazette*, 21, 30 April 1781.

[85] Stephen Gregory to Benjamin Franklin, 3 September 1781, Franklin Papers, American Philosophical Society.

[86] Joseph Hardy to Benjamin Franklin, 19 August 1782, Franklin Papers, Historical Society of Pennsylvania; Joseph Hardy to Charles Thompson, 13 September 1785, *PCC*, 55, pp. 289–291.

[87] Navy Board of the Eastern Department to the Board of Admiralty, 21 December 1780, Papers of the Eastern Navy Board, LC.

[88] *Acts and Laws of the Commonwealth of Massachusetts*, 1780–1781 (Boston, 1890), pp. 240–260.

[89] "Proceedings of Court-Martial for Trial of James Degge," 19 January 1781, *PCC*, 193, p. 773.

[90] Navy Board of the Eastern Department to Board of Admiralty, 15 February 1781, Papers of the Eastern Navy Board, LC.

[91] Ichabod Perry, *Reminiscences of the Revolution* (Lima, New York: Daughters of the American Revolution, 1915), p. 51, hereafter *Reminiscences*.

[92] John Barry to the Board of Admiralty, 6 June 1781, *PCC*, 168, v. I, p. 709.

[93] Perry, *Reminiscences*, pp. 51–52.

[94] John Barry to Benjamin Franklin, 10 March 1781, Franklin Papers, American Philosophical Society.

[95] John Barry to Benjamin Franklin, 23 March 1781, *ibid.*

[96] Narrative of John Kessler, John Barry Papers, LC.

[97] *Ibid.*; John Barry to the Board of Admiralty, 6 June 1781, *PCC*, 168, v. I, p. 709.

[98] *Ibid.*

[99] Narrative of John Kessler, John Barry Papers, LC.

[100] John Barry to the Navy Board of the Eastern Department, 6 June 1781, *PCC*, 168, v. I, p. 708.

[101] Mercy Warren to William Warren, 30 September 1781, Warren Papers, Massachusetts Historical Society.

[102] *Continental Journal* (Boston), 14 June 1781.

[103] *Ibid*; Narrative of John Kessler, John Barry Papers, LC.

[104] John Barry to the Board of Admiralty, 6 June 1781, *PCC*, 168, v. I, p. 709.

[105] "Records of the Vice Admiralty Court at Halifax, Nova Scotia," *Essex Institute Historical Collections*, v. XLV, pp. 37–38.

[106] John Barry to the Board of Admiralty, 6 June 1781, *PCC*, 168, v. I, p. 709.

[107] *Continental Journal* (Boston), 13 September 1781.

[108] *Independent Chronicle* (Boston), 30 August 1781.

[109] Diary of Robert Morris, Superintendent of Finance, v. I, pp. 56–57, LC.

[110] Robert Morris to Tench Francis, 11 September 1781, Robert Morris Letter Books, A, pp. 357–361, LC.

[111] Diary of Robert Morris, Superintendent of Finance, v. I, pp. 56–57, LC.

[112] Robert Morris to Tench Francis, 11 September 1781, Robert Morris Letter Book, A, pp. 357–361, LC.

[113] Diary of Robert Morris, Superintendent of Finance, v. I, pp. 93–94, LC.

[114] *Ibid.*, p. 105.

Chapter XIII
State and Privateer Marines

[1] Journal of the Maryland Convention, 14 January 1776, in *Naval Documents*, v. III, pp. 786–787.

[2] Joseph Smith to Elnathan Smith, 20 March 1776, in *ibid*., v. IV, p. 422.

[3] Samuel Purviance, Jr., to Daniel of St. Thomas Jenifer, 8 March 1776, in *ibid*., v. IV, p. 241.

[4] "A List of the Officers & Men on Board the Ship Defence the 19th of September Anno Dom. 1776 Geo. Cook Commander," in *ibid*., v. VI, pp. 905–907.

[5] William H. Browne, ed., *Archives of Maryland: Journal and Correspondence of the Council of Maryland, April 1, 1778—October 26, 1779* (Baltimore: Maryland Historical Society, 1901), pp. 322–323.

[6] ———, *Archives of Maryland: Muster Rolls and Other Records of Service of Maryland Troops in the American Revolution, 1775–1783* (Baltimore: Maryland Historical Society, 1900), p. 609.

[7] Charles O. Paullin, *The Navy of the American Revolution* (Cleveland: Burrows Brothers, 1905), p. 450.

[8] Benjamin Harrison to George Rogers Clark, 20 December 1781, in H.R. McIlwaine, ed., *Official Letters, Governors of Virginia, Letters of Thomas Nelson and Benjamin Harrison, June 7, 1781—February 27, 1783* (Richmond: Virginia State Library, 1929), v. III, p. 114.

[9] *Ibid*., p. 198.

[10] George Rogers Clark to Benjamin Harrison, 2 May 1782, in *Calendar of Virginia State Papers and other Manuscripts*, 11 vols. (Richmond: 1875–1893), v. III, p. 150.

[11] Stephen Clark to Benjamin Harrison, 11 October 1782, in *ibid*., pp. 346–347.

[12] A. McKee to Major De Peyster, 28 August 1782, in Theodore Roosevelt, *The Winning of the West* (New York: C.P. Putnam's Sons, 1900), p. 306.

[13] Leland D. Baldwin, *The Keelboat Age on Western Waters* (Pittsburgh: University of Pittsburgh Press, 1941), pp. 18–19.

[14] "Pay Roll of Capt Jacob Pyeatts Company of Marines under the Command of B. Genl Clark," undated, Virginia State Archives.

[15] Pension File of Cornelius Darnel, S1802, NA.

[16] Roosevelt, *The Winning of the West* (New York: C. P. Putnam's Son's, 1900), pp. 58, 308.

[17] Pension File of Cornelius Darnel, S1802, NA.

[18] *Journals of the Continental Congress*, v. IV, pp. 229–233, 249–254.

[19] Gardner W. Allen, *A Naval History of the American Revolution* (Boston: Riverside Press, 1913), v. I, pp. 46–47.

[20] Esek Hopkins to John Hancock, 10 September 1776, Hopkins Letter Book, Rhode Island Historical Society.

[21] Esek Hopkins to John Hancock, 2 November 1776, *ibid*.

[22] William Whipple to Josiah Bartlett, 12 July 1778, in *Historical Magazine*, v. VI, no. 3 (March 1862), pp. 73–78.

[23] Edgar S. Maclay, *A History of American Privateers* (New York: Library Press, 1970 Reprint), p. 8.

[24] Almon, ed., *The Remembrancer*, v. XIII, pp. 48–49.

[25] *Daily National Intelligencer* (Washington), 28 May 1829.

[26] *London Chronicle*, 10 September 1782.

[27] Barney, *Memoir of Joshua Barney*, p. 116.

[28] *Ibid*., p. 311.

Chapter XIV
Mustering Out, 1782–1783

[1] Robert Morris to John Brown, 19 September 1781, Dreer Collection, Historical Society of Pennsylvania.

[2] *Continental Journal* (Boston), 1 November 1781.

[3] "List of Officers and Men on board the Continental Frigate Alliance Jno. Barry Esqr. Commander," 17 May 1782, John S. Barnes Collection, New York Historical Society.

[4] Narrative of John Kessler, John Barry Papers, LC.

[5] Martin I.J. Griffin, *Commodore John Barry* (Philadelphia: Author, 1902), pp. 154–158.

[6] John Barry to Samuel Nicholson, 18 December 1781, John Barry Letter Book, Naval Historical Society Collections, New York Historical Society.

[7] Samuel Nicholson to John Barry, 19 December 1781, *ibid*.

[8] John Barry to Samuel Nicholson, 20 December 1781, *ibid*.

[9] "Muster Book and Pay Abstract of the Continental Frigate Deane Commencing May 1, 1781. Ending May 31, 1782," Revolutionary War Rolls, v. 66, pp. 115–128, Massachusetts Archives.

[10] "Account of Stores under the Care of the Several Officers belonging to the Continental Ship Alliance," 25 December 1781, John Barry Papers, Hepburn Collection, William B. Clark Transcripts, Naval Historical Center.

[11] Narrative of John Kessler, John Barry Papers, LC.

[12] John Barry to Benjamin Franklin, 27 February 1782, Franklin Papers, American Philosophical Society.

[13] Benjamin Franklin to John Barry, 10 February 1782, John Barry Letter Book, Hepburn Collection, William B. Clark Transcripts, Naval Historical Center.

[14] John Barry to Benjamin Franklin, 29 February 1782, Franklin Papers, American Philosophical Society.

[15] Benjamin Franklin to John Barry, 9 March 1782, John Barry Letter Book, Naval Historical Society Collections, New York Historical Society.

[16] John Barry to Robert Morris, 16 May 1782, *ibid*.; Narrative of John Kessler, John Barry Papers, LC.

[17] Court Martial held on board the Alliance, 27–29 July 1782, Simon Gratz Collection, Historical Society of Pennsylvania.

[18] John Barry to Robert Morris, 16 May 1782, John Barry Letter Book, Naval Historical Society Collections, New York Historical Society.

[19] Diary of Robert Morris, Superintendent of Finance, v. II, p. 80, LC.

[20] *Ibid*., p. 82.

[21] *Ibid*., pp. 91–92.

[22] William Morris to John Barry, 29 June 1782, John Barry Letter Book, Naval Historical Society Collections, New York Historical Society.

[23] John Barry to Robert Morris, 2 August 1782, *ibid*.

[24] John Brown to William Morris, 30 July 1782, Robert Morris Letter Book, v. II, pp. 102–103, LC.

[25] Matthew Parke to John Barry, 30 July 1782, John Barry Letter Book, Naval Historical Society Collections, New York Historical Society.

[26] John Barry to Matthew Parke, 30 July 1782, *ibid*.

[27] Navy Board of the Eastern Department to the Board of Admiralty, 18 April 1781, Papers of the Eastern Navy Board, LC.

28 Navy Board of the Eastern Department to the Board of Admiralty, 26 April 1781, *ibid.*

29 Navy Board of the Eastern Department to the Board of Admiralty, 18 April 1781, *ibid.*

30 "Muster Book and Pay Abstract of the Continental Frigate Deane Commencing May 1, 1781. Ending May 31, 1782," Revolutionary War Rolls, v. 66, pp. 115–128, Massachusetts Archives.

31 *Continental Journal* (Boston), 1 November 1781.

32 Navy Board of the Eastern Department to William Waterman, 29 October 1781, Pension File of William Waterman, W25893, NA.

33 "Muster Book and Pay Abstract of the Continental Frigate Deane Commencing May 1, 1781. Ending May 31, 1782," Revolutionary War Rolls, b. 66, pp. 115–128, Massachusetts Archives.

34 *Ibid.*

35 Robert Morris to the President of Congress, 20 June 1782, Robert Morris Letter Book, v. II, pp. 93–94, LC.

36 *Independent Chronicle* (Boston), 23 May 1782.

37 Diary of Robert Morris, Superintendent of Finance, v. II, p. 218, LC; "Order for a Court of Enquiry on Capt. Samuel Nicholson and an Order for a Court Martial for his Trial," 18 October 1782, Robert Morris Letter Book, v. III, pp. 122–125, LC.

38 *Independent Chronicle* (Boston), 26 September 1782.

39 *Ibid.*, 30 January 1783.

40 *Ibid.*, 27 February 1783.

41 *Continental Journal* (Boston), 16 September 1783.

42 John Barry to John Brown, 4 August 1782, John Barry Letter Book, Hepburn Collection, William B. Clark Transcripts, Naval Historical Center.

43 John Barry to John Brown, 10 August 1782, *ibid.*

44 Narrative of John Kessler, John Barry Papers, LC.

45 *Independent Chronicle* (Boston), 2 January 1783.

46 Narrative of John Kessler, John Barry Papers, LC.

47 *Independent Chronicle* (Boston), 2 January 1783.

48 *Ibid.*

49 Perry, *Reminiscences* p. 60.

50 John Buckley *et. al.* to John Barry, 17 November 1782, John Barry Letter Book, John Barry Papers, LC.

51 John Barry to the Officers on board the *Alliance*, 18 November 1782, *ibid.*

52 Samuel Cooper *et. al.* to John Barry, 18 November 1782, *ibid.*

53 John Barry to Patrick Fletcher, *et. al.*, 19 November 1782, *ibid.*

54 John Barry to Matthew Parke, 24 November 1782, *ibid.*

55 John Barry to Matthew Parke, 25 November 1782, *ibid.*

56 John Barry to Thomas Barclay, 27 November 1782, *ibid.*

57 John Barry to Robert Morris, 7 December 1782, *ibid.*

58 *Boston Evening Post*, 22 February 1783.

59 John Barry to Robert Morris, 20 March 1783, John Barry Letter Book, John Barry Papers, LC.

60 Narrative of John Kessler, *ibid.*

61 "A Journal of the Proceedings of His Majys. Ship Sibyl James Vashon Esq. Comr. From the 12th day of July 1782 to the 15th day of July 1783," PRO, Admiralty, Captains' Logs, 51/875.

62 Narrative of John Kessler, John Barry Papers, LC.

63 "A Journal of the Proceedings of His Majys. Ship Sibyl . . .," PRO, Admiralty, Captain's Logs, 51/875.

64 John Barry to Robert Morris, 20 March 1783, John Barry Letter Book, John Barry Papers, LC.

65 Robert Morris to John Barry, 29 March 1783, Robert Morris Letter Book, v. III, pp. 160–161, LC.

66 John Barry to Robert Morris, 31 March 1783, John Barry Letter Book, John Barry Papers, LC.

67 "Account of Small arms &c. belonging to the Marine department left on board the Continental Frigate Alliance . . .," 20 June 1783, John S. Barnes Collection, New York Historical Society.

68 Samuel Osgood and Walter Livingston to the President of Congress, 5 August 1785, *PCC*, 140, v. II, p. 45.

69 Marine Committee to John Barry, 20 November 1779, Marine Committee Letter Book.

70 Navy Board of the Eastern Department to the Marine Committee, 4 December 1779, Papers of the Eastern Navy Board, LC.

71 "Report of a comee on Major Nicholas' Memorial," 29 September 1781, *PCC*, 19, v. IV, pp. 533–534.

72 Robert Morris to Zebadiah Farnham, 21 November 1782, Robert Morris Letter Book, v. III, p. 138, LC.

73 James Fenimore Cooper, *The History of the Navy of the United States of America*, 2 vols. (London: Richard Bentley, 1839), v. I, pp. 295–296.

74 Report of the Agent of Marine, 31 July 1783, *PCC*, 137, v. II, p. 725.

APPENDIX B

Bibliography

Manuscript Sources

Archivo General de Indias, Savilla, Spain
- Papeles Procedentes de la isla de Cuba; Legajos 191, 2351, 2370
- Audiencia de Santo Domingo; Legajo 2596

American Philosophical Society, Philadelphia, Pennsylvania
- Sol Feinstone Collection
- Benjamin Franklin Papers
- Nathanael Greene Papers

Boston Public Library, Boston, Massachusetts
- Manuscript and Special Collections

Carnegie Library, Pittsburgh, Pennsylvania
- Isaac Craig Papers
- George Morgan Letter Books

City and County Archives, Philadelphia, Pennsylvania
- Register of Deeds
- Register of Wills
- Sheriffs' Dockets

William L. Clements Library, University of Michigan, Ann Arbor, Michigan
- Josiah Harmer Papers
- Abraham Whipple Papers

Connecticut Historical Society, Hartford, Connecticut
- Silas Deane Papers

Darlington Library, University of Pittsburgh, Pittsburgh, Pennsylvania
- Darlington Manuscript Collection

Essex Institute, Salem, Massachusetts
- Institute Collections

Harvard University Library, Cambridge, Massachusetts
- Arthur Lee Papers

Haverford College Library, Haverford, Pennsylvania
- Charles Roberts Autograph Collection

Historical Society of Delaware, Wilmington, Delaware
- Caesar Rodney Papers
- Unpublished Revolutionary Archives

Historical Society of Pennsylvania, Philadelphia, Pennsylvania
- Account Book at Morristown (AM 642)
- Bank of North America Papers
- Nicholas Biddle Papers (On Deposit)
- William Bradford Manuscripts
- John Cadwalader Papers
- Confederacy Papers
- Ferdinand Dreer Autograph Collection
- Frank Etting Autograph Collection
- Benjamin Franklin Papers
- Simon Gratz Autograph Collection
- Joshua Humphreys Papers
- John Langdon Papers
- Maps of the Revolution (AM 602)
- Minute Book (AM 3275)
- Robert Morris Papers
- Orderly Book on the German Battalion (AM 632)
- Society Collections
- David Stauffer Collection
- Anthony Wayne Papers
- Wharton Papers
- William Whipple Papers
- Samuel H. Woodhouse Collection

Historical Society of Western Pennsylvania, Pittsburgh, Pennsylvania
- Isaac Craig Papers
- James O'Hara Papers

Library of Congress, Washington, D.C.
- John Barry Papers
- John Bradford Letter Book
- 'Extrait du Journal des Services principaux De Paul-Jones, dans la Révolution des Etats-Unis D'Amerique; écrit par lui-même, it presenté avec un profond respect, an Trés illustre Prince Louis XVI."
- Peter Force Manuscript Collection
- Peter Force Transcripts of the William Whipple and Samuel Tucker Papers
- Benjamin Franklin Papers
- John Holker Papers
- Journal of William Jennison
- John Paul Jones Papers
- Robert Morris Diary, Letter Books, and Papers
- Papers of the Navy Board of the Eastern Department, 1779–1782
- Diary of Richard Smith
- George Washington Papers

Maine Historical Society, Portland, Maine
- Papers Relating to the Penobscot Expedition, 1779

Mystic Seaport, Inc., Mystic, Connecticut
- Seth Harding Papers
- Journal of Joseph Hardy

Mariners Museum, Newport News, Virginia
- Journal of Captain Hector McNeill in the Frigate *Boston*, May–July 1777

Maryland Historical Society, Baltimore, Maryland
- Purviance Family Papers

Massachusetts Historical Society, Boston, Massachusetts
- Adams Family Papers
- William Heath Papers
- Joshua Huntington Papers

Henry Knox Papers
Solomon Lovell Papers
Timothy Pickering Papers
Mercy Otis Warren Papers
Massachusetts State Archives, Boston, Massachusetts
American Revolution, Maritime (v. 157)
American Revolution, Military (v. 146)
Minutes of the Massachusetts Board of War (vs. 150–151)
Papers of the Massachusetts Council (vs. 170, 176)
The Penobscot Expedition (v. 145)
Revolutionary War, Letters (vs. 200–201)
Revolutionary War, Miscellaneous (vs. 138–143)
Revolutionary War, Rolls (vs. 52, 66)
National Archives, Washington, D.C.
Records of the Veterans Administration (Record Group 15)
Revolutionary War Pension and Bounty-Land Warrant Application Files, 1800–1900
Naval Records Collection of the Office of Naval Records and Library (Record Group 45)
War Department Collection of Revolutionary War Records (Record Group 93)
Miscellaneous Numbered Records (Manuscript File)
Record Books of the Revolutionary War
Revolutionary War Rolls, 1775–1783
Records of the United States Marine Corps (Record Group 127)
Records of the United States General Accounting Office (Record Group 217)
Miscellaneous Treasury Accounts of the First Auditor (Formerly the Auditor) of the Treasury Department, September 6, 1790–July 15, 1840
Records of the Supreme Court of the United States (Record Group 267)
The Revolutionary War Prize Cases: Records of the Court of Appeal in Cases of Capture, 1776–1787
Records of the Continental Congress and the Continental Convention (Record Group 360)
Papers of the Continental Congress, 1774–1789
Miscellaneous Papers of the Continental Congress, 1774–1789
Naval Historical Center, Washington, D.C.
Hepburn Collection, William B. Clark Transcripts
New England Historic and Genealogical Society, Boston, Massachusetts
Society Collections
New Hampshire Historical Society, Concord, New Hampshire
John Langdon Papers
New Jersey State Library, Trenton, New Jersey
Papers of the New Jersey Department of Defense
Newport Historical Society, Newport, Rhode Island
Revolutionary War Papers
John Trevett Diary and Papers
John Vernon Papers
New York Historical Society, New York, New York
John S. Barnes Collection
John Barry Papers
Cooper Manuscript
Silas Deane Papers
Naval Historical Society Collection
New York Public Library, New York, New York
Emmet Autograph Collection
Letter Book of the Navy Board of the Eastern Department, 1778–1779
Myers Autograph Collection
Philip Schuyler Papers
U.S. Navy Manuscript Collection
Peabody Museum, Salem, Massachusetts
Josiah Fox Papers
Pennsylvania Historical and Museum Commission, Harrisburg, Pennsylvania
Committee of Safety, Navy Papers
Correspondence of the Supreme Executive Council
Muster and Pay Rolls, Pennsylvania State Navy
Records of the Comptroller General
Princeton University Library, Princeton, New Jersey
de Coppet Collection
Private Collection of Lady Hope-Dynbar, Dundrennan, Scotland
Lord and Lady Selkirk Correspondence
Public Records Office, London, England
Admiralty, Admirals' Dispatches
———, Captains' Letters
———, Captains' Log; HMS *Ariadne*, *Emerald*, *Sybil*, *Unicorn*, and *Yarmouth*
Colonial Office Papers, Bahamas
High Court of Admiralty, Prize Papers
Rhode Island Historical Society, Providence, Rhode Island
Esek Hopkins Letter Book and Papers
Journal of the Committee Appointed to Build Two Continental Frigates in Rhode Island
Abraham Whipple Correspondence and Letter Book
Rhode Island State Archives, Providence, Rhode Island
Governors' Letters
Transcripts of Revolutionary War Military Papers
Franklin D. Roosevelt Library, Hyde Park, New York
United States Navy Collection
South Carolina Historical Society, Charleston, South Carolina
Society Collections
United States Naval Academy Museum, Annapolis, Maryland
Museum Manuscript Collection
University of Pennsylvania Library, Philadelphia, Pennsylvania
Benjamin Franklin Papers
University of Virginia Library, Charlottesville, Virginia
Lee Family Papers
Yale University Library, New Haven, Connecticut
Nathaniel Shaw, Jr., Papers
Abraham Whipple Letter Book, 1778
Jonathan Williams, Jr., Letter Books

Printed Diaries, Journals, Reminiscences, and Document Collections

"An Account of the Battle of Princeton." *The Pennsylvania Magazine of History and Biography*, v. VIII (1884), pp. 310–312.

"Account of the Loss of the Randolph as Given in a Letter from Rawlins Lowndes to Henry Laurens." *The South Carolina Historical and Genealogical Magazine*, v. X (1909), pp. 171–173

Acts and Laws of the Commonwealth of Massachusetts, 1780–1781. Boston, 1890.

[John Adams.] *The Adams Papers: Diary and Autobiography of John Adams*. Ed. by L. H. Butterfield. 4 vols. Cambridge: Belknap Press, 1961.

———. *The Works of John Adams*. Notes by Charles F. Adams. 4 vols. Boston: Little, Brown, 1865.

[James Allen.] "Diary of James Allen, Esq., of Philadelphia, Counsellor-at-Law, 1770–1778." *The Pennsylvania Magazine of History and Biography*, v. IX (1885), pp. 176–196, 278–296, 424–441.

J. Almon, ed. *The Remembrancer; or, Impartial Repository of Public Events, 1775–1784*. 17 vols. London: Printed for J. Almon, 1775–1784.

[Benedict Arnold.] "Benedict Arnold's Regimental Memorandum Book." *The Pennsylvania Magazine of History and Biography*, v. VIII (1884), pp. 363–376.

[Jeduthan Baldwin.]"Baldwin's Diary." *The Bulletin of the Fort Ticonderoga Museum*, v. IV, no. 6 (January 1938), pp. 10–40.

John S. Barnes, ed. *The Logs of the Serapis-Alliance-Ariel, under the Command of John Paul Jones, 1779–1780, with Extracts from Public Documents, Unpublished Letters and Narratives, and Illustrated with Reproductions of Scarce Prints*. New York: Naval Historical Society 1911.

[Charles Biddle.] *Autobiography of Charles Biddle, Vice-President of the Supreme Executive Council of Pennsylvania, 1745–1821*. Ed. by James S. Biddle. Philadelphia: E. Claxton and Co., 1883.

Elias Boudinot. *Journal or Historical Recollections of American Events During the Revolutionary War*. Philadelphia: Frederick Bourquin, 1894.

A Brief Narrative of the Ravages of the British and Hessians at Princeton in 1776–77: A Contemporary Account of the Battles of Trenton and Princeton. Ed. by Varnum L. Collins. Princeton: The University Library, 1916.

Edmund C. Burnett, ed. *Letters of Members of the Continental Congress*. 8 vols. Washington: The Carnegie Institution, 1921–1936.

[John Calef.] *The Siege of Penobscot*. New York: William Abbatt, 1910.

Calendar of Virginia State Papers and Other Mansucripts. 11 vols. Richmond: Virginia State Library, 1875–1973.

[George Roger Clark.] *George Rogers Clark Papers, 1771–1781*. Ed. by James A. James. 2 vols. Springfield: Trustees of the Illinois State Historical Library, 1912–1926.

Connecticut. *The Public Records of the State of Connecticut, From October, 1776, to February, 1778, inclusive, With the Journal of the Council of Safety From October 11, 1776, to May 6, 1778, inclusive, With an Appendix*. Comp. by Charles J. Hoadly. Hartford: Lockwood & Brainard Co., 1894.

[Timothy Connor.] "A Yankee Privateersman in Prison in England, 1777–1779." Ed. by William R. Cutter. *The New England Historical and Genealogical Register*, v. XXX (1876), pp. 174–177, 343–352; v. XXXI (1877), pp. 18–20, 212–213, 284–288; v. XXXII (1878), pp. 70–73, 165–168, 280–286; v. XXXIII (1879), pp. 36–40.

[Gustavus Conyngham.] *Letters and Papers Relating to the Cruises of Gustavus Conyngham, a Captain of the Continental Navy, 1777–1779*. Ed. by Robert W. Neeser. New York: Naval Historical Society, 1915.

[———.] "A Narrative respective Lugger surprisé & Cutter revenge." *The Pennsylvania Magazine of History and Biography*, v. XXIII (1898), pp. 479–488.

Silas Deane. *The Deane Papers: Correspondence Between Silas Deane, His Brothers, and Their Business and Political Associates 1771–1795. Collections of the Connecticut Historical Society*, v. 23. Hartford: The Society, 1930.

[———.] *The Deane Papers, 1774–1777*. 5 vols. New York: New York Historical Society, 1887.

[Mrs. Henry Drinker.] "Extracts from the Journal of Mrs. Henry Drinker, of Philadelphia, From September 25, 1777, to July 4, 1778." *The Pennsylvania Magazine of History and Biography*, v. XIII (1889), pp. 298–308.

[William Dunbar.] *Life, Letters and Papers of William Dunbar*. Ed. by Mrs. Dunbar Rowland. Jackson: Mississippi Historical Society, 1930.

Andrew Ellicott. *The Journal of Andrew Ellicott*. Philadelphia, n.p., 1814.

"Engagements at Trenton and Princton, January 2 and 3, 1777." *The Pennsylvania Magazine of History and Biography*, v. X (1886), pp. 263–265.

[Nathaniel Fanning.] *Fanning's Narrative: Memoirs of Nathaniel Fanning*. Ed. by John S. Barnes. New York: Naval Historical Society, 1912.

Peter Force, comp. *American Archives: Consisting of a Collection of Authentic Records, State Papers, Debates, and Letters and Other Notices of Publick Affairs, the Whole Forming a Documentary History of the Origin and Progress of the North American Colonies; of the Causes and Accomplishments of the American Revolution; and of the Constitution of Government for the United States, to the Final Ratification thereof*. 2 ser., 9 vols. Washington: n. p., 1837–1853.

[Ebenezer Fox.] *The Revolutionary Adventures of Ebenezer Fox, of Roxbury, Massachusetts*. Boston: Munroe & Francis, 1838.

[Ezra Green.] *Diary of Ezra Green, M. D. Surgeon on Board the Continental Ship-of-War "Ranger," Under John Paul Jones, from November 1, 1777, to September 27, 1778*. Boston: D. Clapp & Son, 1875.

Charles Herbert. *A Relic of the Revolution*. Boston: Charles H. Peirce, 1847.

[Esek Hopkins.] *The Correspondence of Esek Hopkins, Commander-in-Chief of the United States Navy*. Ed by Alverda S. Beck. Providence: The Rhode Island Historical Society, 1932.

[———.] *The Letter Book of Esek Hopkins, Commander-in-Chief of the United States Navy, 1775–1777*. Ed by Alverda S. Beck. Providence: The Rhode Island Historical Society, 1932.

[John Kilby.] "Narrative of John Kilby, Quarter-Gunner of the United States of the U.S. Ship *Bonhomme Richard*, Under John Paul Jones." With an Introduction and Notes by Augustus C. Buell. *Scribners Monthly*, v. XXXVIII (July–December 1905), pp. 23–41.

Pierre Landais. *Charges and Proofs Respecting the Conduct of Peter Landais*. New York: Francis Childs, 1787.

———. *Memoiral, to Justify Landai's Conduct During the Late War*. Boston: Peter Edes, 1784.

———. *The Second Part of the Memorial to Justify Peter Landai's Conduct During the Late War*. New York: Samuel Loudon, 1784.

"A List of the Officers and Crew, of the Continental Frigate Raleigh." *New Hampshire Genealogical Record*, v. III, no. 2. October 1905), pp. 70–76.

[Solomon Lovell.] *The Original Journal of General Solomon Lovell Kept during the Penobscot Expedition, 1779; with a Sketch of His Life by Gilbert Nash; Together with the Proceedings of the Society for 1879–80*. Ed. by Gilbert Nash. Boston: Weymouth Historical Society, 1881.

Maine. *Documentary History of the State of Maine*. Ed. by James P. Baxter. vs. XVI, XVII. Portland: Lefavor & Tower, 1910.

[Christopher Marshall.] *Extracts from the Dairy of Christopher Marshall, Kept in Philadelphia and Lancaster, During the American Revolution, 1774–1781.* Ed. by William Duane. Albany: Joel Munsell, 1877.

Maryland. Committee of Safety. *Archives of Maryland: Journal and Correspondence of the Committee of Safety of Maryland, March 20, 1777–March 28, 1778.* Ed. by William H. Browne. Baltimore: Maryland Historical Society, 1897.

———. Council. *Archives of Maryland: Journal and Correspondence of the Council of Maryland, April 1, 1778—October 26, 1779.* Ed. by William H. Browne. Baltimore: Maryland Historical Society, 1901.

———. Troops. *Archives of Maryland: Muster Rolls and Other Records of Service of Maryland Troops in the American Revolution, 1775–1783.* Baltimore: Maryland Historical Society, 1900.

Massachusetts Soldiers and Sailors of the Revolutionary War: A Compilation from the Archives Prepared by the Secretary of the Commonwealth. 17 vols. Boston: Wright and Potter, 1896.

[Luke Matthewman.] "Narrative of Lieut. Luke Matthewman of the Revolutionary Navy." *The Magazine of American History,* v. II, pp. 175–185.

[Hector McNeill.] *Captain Hector McNeill of the Continental Navy.* Ed. by Gardner W. Allen. Boston: Massachusetts Historical Society, 1922.

Louis F. Middlebrook, ed. *The Log of the Bonhomme Richard.* Mystic: Marine Historical Association, 1936.

[John Montressor.] *The Journal of Captain John Montressor: Collections of the New York Historical Society for the Year 1881.* Ed. by G. D. Scull. New York: New York Historical Society, 1882.

[Margaret Morris.] *Private Journal Kept During a Portion of the Revolutionary War.* Philadelphia: Privately Printed, 1836.

[Robert Morton.] "The Diary of Robert Morton; Kept in Philadelphia While that City was Occupied by the British Army in 1777. *The Pennsylvania Magazine of History and Biography,* v. I (1877), pp. 1–39.

William Moultrie. *Memoirs of the American Revolution, so Far as it Related to the States of North and South Carolina, and Georgia.* 2 vols. New York: David Longworth, 1802.

[Henry Mowat.] "Captain Henry Mowatt's Account." *Magazine of History,* v. 3, extra no. 11 (1910), pp. 46–55.

The Navy Miscellany. v. I. London: Naval Records Society, 1902.

New Hampshire. *Documents and Records Relating to the Province of New Hampshire.* Ed. by Nathaniel Bouton, *et. al.* 40 vols. Concord: n. p., 1867–1943.

New Jersey. *Documents Relating to the Revolutionary History of the State of New Jersey.* Comp. by William S. Stryker. 2d ser., 5 vols. Trenton: John L. Murphy, 1901.

"An Open Letter to Captain Pringle." *The Bulletin of the Fort Ticonderoga Museum,* v. I, no. 4 (July 1928), pp. 14–20.

"An Order Book of the 1st. Regt., S.C. Line, Continental Establishment." *The South Carolina Historical and Genealogical Magazine,* v. VII (1906), pp. 75–80, 130–142, 194–203; v. VIII (1907), pp. 19–28, 69–87.

[Charles Willson Peale.] "Journal of Charles Willson Peale." *The Pennsylvania Magazine of History and Biography,* v. XXXVIII (1914) pp. 271–286.

[Joshua Pell, Jr.] "Diary of Joshua Pell, Junior: An Officer of the British Army in America 1776–1777." *The Bulletin of the Fort Ticonderoga Museum,* v. I, no. 6 (July 1929), pp. 2–14.

Pennsylvania. *Pennsylvania Archives.* Ed. by Samuel Hazard, *et. al.* 9 ser., 138 vols. Philadelphia: J. Stevens, 1874–1949.

———. *Pennsylvania Colonial Records.* Ed. by Samuel Hazard, 16 vols. Harrisburg: T. Fenn, 1851–1853.

Ichabod Perry, *Reminiscences of the Revolution.* Lima: Ska-Hase-Ga-O Chapter, Daughters of the American Revolution, 1915.

"Records of the Vice-Admiralty Court at Halifax, Nova Scotia: The Condemnation of Prizes and Recaptures of the Revolution and the War of 1812." *Historical Collections of the Essex Institute,* v. XLV (1909), pp. 28–48, 161–184, 221–244, 309–332; v. XLVI (1910), pp. 68–80, 150–160, 257–272; 317–324; v. XLVII (1911) pp. 20–24, 189–196, 236–249.

[Joseph Reed.] "General Joseph Reed's Narrative of the Movements of the American Army in the Neighborhood of Trenton in the Winter of 1776–1777." *The Pennsylvania Magazine of History and Biography,* v. VIII (1884), pp. 391–402.

"Report of Captain Lieutenant Francis Downman, R.A., to Colonel James Pattison, Commanding the Royal Artillery at New York." *Journal of the Society for Army Historical Research,* v. 9 (1930), pp. 28–30.

[Thomas Rodney.] *Diary of Captain Thomas Rodney, 1776–1777. Introduction by Caesar A. Rodney. Papers of the Historical Society of Delaware,* v. VIII. Wilmington: The Historical Society of Delaware, 1888.

Rolls and Lists of Connecticut Men in the Revolution, 1775–1783. Collections of the Connecticut Historical Society, v. VIII. Hartford: Connecticut Historical Society, 1901.

[Peter Russell.] "The Siege of Charleston; Journal of Captain Peter Russell, December 25, 1779, to May 2, 1780." *The American Historical Review,* v. IV, no. 1 (October 1898), pp. 478–501.

Johann David Schöpf. *Travels in the Confederation.* Trans. by Alfred J. Morrison. 2 vols. New York: Bergman Publishers, 1968.

Sergeant R———. "The Battle of Princeton." *The Pennsylvania Magazine of History and Biography,* v. XX (1896), pp. 515–519.

Andrew Sherburne. *Memoirs of Andrew Sherburne, a Pensioner of the Revolution.* Providence: H. H. Brown, 1831.

[Molyneux Shuldham.] *The Dispatches of Molyneux Shuldham Vice-Admiral of the Blue and Commander-in-Chief of His Britannic Majesty's Ships in North America, January–July 1776.* Ed. by Robert W. Neeser. New York: Naval Historical Society, 1913.

South Carolina. Commissioners of the Navy Board. *Journal of the Commissioners of the Navy of South Carolina, October 9, 1776–March 1, 1779.* Ed. by A. S. Salley, Jr. Columbia: The State Company, 1912.

———. ———. *Journal of the Commissioners of the Navy of South Carolina, July 22, 1779–March 23, 1780.* Ed. By A. S. Salley, Jr. Columbia: The State Company, 1913.

Benjamin F. Stevens, comp. *Facsimiles of Manuscripts in European Archives Relating to America, 1775–1783, With Descriptions, Editorial Notes, Collations, References and Translations.* 25 vols. London: Malby & Sons, 1889–1898. 2,107 facsism.

[John Trevett.] "Journal of John Trevett, 1775–1780." *Rhode Island Historical Magazine,* v. VI (1886), pp. 72–74; v. VII (1887), pp. 194–199, 271–278.

[John Trumbull.] "The Deputy Adjutant General's Orderly Book, Ticonderoga, 1776." *The Bulletin of the Fort Ticonderoga Museum*, v. III, no. 1 (January 1933), pp. 30–57; v. III, no. 2. (July 1933), pp. 84–123.

[John Trumbull.] "John Trumbull at Ticonderoga From his Autobiography." *The Bulletin of the Fort Ticonderoga Museum*, v. III, no. 1 (January 1933), pp. 3–12.

Bernard A. Uhlendorf, ed. *The Siege of Charleston: With an Account of the Providence of South Carolina: Diaries and Letters of Hessian Officers From the von Jungkenn Papers in the William L. Clements Library*. Ann Arbor: University of Michigan Press, 1938.

United States. Continental Congress. *Journals of the Continental Congress, 1774–1789*. Ed. by Worthington C. Ford, et. al. 34 vols. Washington: Government Printing Office, 1904–1912.

———. ———. *Secret Journals of the Acts and Proceedings of Congress, from the First Meeting Thereof to the Dissolution of the Confederation, by the Adoption of the Constitution of the United States*. 4 vols. Boston: Thomas B. Wait, 1820–1821.

———. Department of the Navy. Naval Historical Center. *Naval Documents of the American Revolution*. Ed. by William B. Clark and Dr. William J. Morgan. 6 vols. Washington: Government Printing Office, 1964—

———. Department of State. *The Revolutionary Diplomatic Correspondence of the United States*. Ed. by Francis Wharton. 6 vols. Washington: Government Printing Office, 1889.

———. Marine Committee and Board of Admiralty. *Outletters of the Continental Marine Committee and Board of Admiralty, August 1776–September 1780*. Ed. by Charles O. Paullin. 2 vols. New York: Naval Historical Society, 1914.

[William Vernon.] "Papers of William Vernon and the Navy Board, 1776–1794." *Publications of the Rhode Island Historical Society*, v. VIII, no. 4 (January 1901), pp. 197–277.

Virginia. Governors. *Official Letters, Governors of Virginia, Letters of Thomas and Benjamin Harrison, June 7, 1781–February 27, 1783*. v. III. Richmond: Virginia State Library, 1929.

[George Washington.] *The Writings of George Washington, from the Original Sources, 1745–1799*. Ed. by John C. Fitzpatrick. 39 vols. Washington: Government Printing Office, 1931–1933.

[William Whipple.] "Stray Leaves from an Autograph Collection: Correspondence of Joshia Bartlett, of N.H., during the American Revolution." *Historical Magazine*, v. VI, no. 3 (March 1862), pp. 73–78.

General James Wilkinson. *Memoirs of My Own Time*. 3 vols. Philadelphia: Abraham Small, 1816.

[William Young.] "Journal of Sergeant William Young, Written During the Jersey Campaign in the Winter of 1776–7." *The Pennsylvania Magazine of History and Biography*, v. VIII (1884), pp. 255 278.

Newspapers

Boston Evening Post
Continental Journal (Boston, Massachusetts)
Gazette of the State of South Carolina (Charleston, South Carolina)
Independent Chronicle (Boston, Massachusetts)
Independent Ledger (Boston, Massachusetts)
London Chronicle
Maryland Gazette (Annapolis, Maryland)
Daily National Intelligencer (Washington, D.C.)
New Hampshire Gazette (Portsmouth, New Hampshire)
Newport Mercury
New York Gazette
New York Journal and Weekly Advertiser
Pennsylvania Evening Post (Philadelphia, Pennsylvania)
Pennsylvania Gazette (Philadelphia, Pennsylvania)
Pennsylvania Journal and Weekly Advertiser (Philadelphia Pennsylvania)
Pennsylvania Ledger (Philadelphia, Pennsylvania)
Pennsylvania Packet (Philadelphia, Pennsylvania)
Public Advertiser (London, England)
Providence Gazette
South Carolina and American General Gazette (Charleston, South Carolina)
Virginia Gazette (Williamsburg, Virginia)

Bibliographies, Reference Works, and Finding Aids

Charles M. Andrews, comp. *Guide to the Materials for American History, to 1783 in the Public Record Office of Great Britain*. 2 vols. Washington: Carnegie Institution, 1912–1914.

Charles M. Andrews and Frances G. Davenport, comps. *Guide to the Manuscript Materials of the United States to 1783, in the British Museum, in Minor London Archives, and in the Libraries of Oxford and Cambridge*. Washington: Carnegie Institution, 1908.

Calendar of the Papers of Benjamin Franklin in the Library of the American Philosophical Society. 5 vols. Philadelphia: American Philosophical Society, 1908.

Samuel Hazard, ed. *General Index to the Colonial Records, in 16 Volumes, and to the Pennsylvania Archives, in 12 Volumes Prepared and Arranged by Samuel Hazard, under an Act of the General Assembly of Pennsylvania*. Philadelphia: J. Stevens, 1860.

Francis B. Heitman. *Historical Register of the Continental Army During the War of the Revolution*. Washington: Rare Book Shop Publishing Co., 1914.

Marion Dexter Learned, comp. *Guide to the Manuscript Materials Relating to American History in the German States Archives*. Washington: Carnegie Institution, 1912.

Waldo G. Leland, comp. *Guide to Materials for American History in the Libraries and Archives of Paris*. Washington: Carnegie Institution, 1932–1943.

William Matthews, comp. *American Diaries: An Annotated Bibliography of American Diaries Written prior to the Year 1861*. Boston: J. S. Canner & Co., 1959.

Myron J. Smith. *Navies in the American Revolution: A Bibliography*. Metuchen: Scarecrow Press, 1973.

United States. Department of the Navy. Naval Academy Museum. *Catalogue of Manuscripts*. Annapolis: United States Naval Academy Museum, 1957.

———. ———. Naval Historical Center. *The American Revolution, 1775–1783; An Atlas of 18th Century Maps and Charts: Theatres of Operations*. Comp. by W. Bart Greenwood. Washington: Government Printing Office, 1972.

———. ———. ———. *United States Naval History: A*

Bibliography. Washington: Government Printing Office, 1972, 6th ed.

———. Library of Congress. *The American Revolution: A Selected Reading List*. Washington: Government Printing Office. 1968.

———. ———. *Periodical Literature on the American Revolution: Historical Research and Changing Interpretations, 1895–1970*. Comp. by Ronald M. Gephart. Washington: Government Printing Office, 1971.

———. ———. American Revolution Bicentennial Committee. "Guide to Manuscripts in the Library of Congress from the Period of the Revolution, 1763–1789." Forthcoming.

———. ———. Division of Manuscripts. *A Calendar of John Paul Jones Manuscripts in the Library of Congress*. Comp. by Charles H. Lincoln. Washington: Government Printing Office, 1903.

———. ———. ———. *A Guide to Manuscripts Relating to American History in British Depositories Reproduced for the Division of Manuscripts of the Library of Congress*. Comp. by Grace G. Griffin. Washington: Library of Congress. 1946.

———. ———. ———. *Handbook of Manuscripts in the Library of Congress*. Washington: Government Printing Office. 1918.

———. ———. ———. *Index to the George Washington Papers, Presidents' Papers Index Series*. Washington: Government Printing Office, 1964.

———. ———. ———. *Naval Records of the American Revolution, 1775–1788. Prepared from the Originals in the Library of Congress by Charles Henry Lincoln*. Washington: Government Printing Office, 1906.

———. National Archives. *Guide to the Records in the National Archives*. Washington; Government Printing Office, 1948.

———. ———. *Miscellaneous Papers of the Continental Congress, 1774–89: National Archives Microfilm Publications Pamphlet Accompanying Microcopy No. 332*. Washington: National Archives, 1962.

———. ———. *Papers of the Continental Congress, 1774–1789: National Archives Microfilm Publication Pamphlet Describing M247*. Washington: National Archives 1971.

———. ———. *Pre-Federal Maps in the National Archives: An Annotated List*. Washington: National Archives, 1971.

———. ———. *Preliminary Checklist of the Naval Records Collection of the Office of Naval Records and Library, 1775–1910*. Comp. by James R. Masterson. Washington: National Archives, 1945.

———. ———. *The Revolutionary War Prize Cases: Records of the Court of Appeals in Cases of Capture, 1776–1787: National Archives Microfilm Publications Pamphlet Describing M162*. Washington: National Archives, 1954.

———. ———. *War Department Collection of Revolutionary War Records: Preliminary Inventories, Number 144 (Revised)*. Comp. by Mabel Deutrich. Washington: National Archives, 1970.

Articles

Gardner W. Allen. "State Navies and Privateers in the Revolution." *Massachusetts Historical Society Proceedings*, v. XLVI (1913), pp. 171–191.

Whitfield J. Bell, Jr. "Addenda to Watson's Annals of Philadelphia: Notes by Jacob Mordecai, 1836." *The Pennsylvania Magazine of History and Biography*, v. XCVIII, no. 2 (April 1974), pp. 131–170.

"Biographical Memoirs of the Late Captain Nicholas Biddle." *The Port Folio*, v. II, no. 4 (October 1809).

Oscar E. Brendenberg. "The American Champlain Fleet, 1775–1777." *The Bulletin of the Fort Ticonderoga Museum*, v. XII, no. 4 (September 1968), pp. 249–263.

John Caughey. "Willing's Expedition Down the Mississippi, 1778." *The Louisiana Historical Quarterly*, v. 15, no. 1 (January 1932), pp. 5–36.

"The Continental Frigate *Raleigh*; Ship *Raleigh's* Book." *New Hampshire Genealogical Record*, v. 2, no. 4 (April 1905), pp. 177–187; v. 3, no. 1, (July 1905), pp. 19–30; v. 3, no. 2 (October 1905), pp. 65–76.

Louis S. Fagan. "Samuel Nicholas, First Officer of American Marines." *Marine Corps Gazette*, v. XVIII, no. 11 (November 1933), pp. 5–15.

Joseph Foster. "The Continental Frigate *Raleigh*." *The Granite Monthly*, v. LX (November 1928), pp. 558–566.

Malcolm Harcourt, Jr., "The Taking of the Bahamas by the Continental Navy in 1776." *The Pennsylvania Magazine of History and Biography*, v. XLIX (October 1925), pp. 349–366.

Edgar Howell. "Officer's Coat of the Revolution." *Military Collector and Historian*, v. XIII, no. 2 (Summer 1961), pp. 44–45.

E. P. Jewell. "A Curious Relic [Log of the *Ranger*]." *The Granite Monthly*, v. 5, no. 2 (November 1881), pp. 64–68; v. 5, no. 3 (December 1881), p. 100; v. 5, no. 4 (January 1882), pp. 129–132; v. 5, no. 7 (April 1882), p. 244; v. 5, no. 10 (July 1882), p. 340; v. 5, no. 12 (September 1882), pp. 407–408.

Herbert Knoetel and Frederick P. Todd. "Hesse-Cassel Field Jaeger Corps, 1776–1783." *Military Collector and Historian*, v. VII, no. 2 (Summer 1955), p. 45.

Richard Koke. "The Britons Who Fought on the Canadian Frontier: Uniforms of the War of 1812." *New York Historical Society Quarterly*, v. XLV, no. 2 (April 1961), pp. 150–164.

Major Edwin M. McClellan, USMC. "Marine Officers of the Revolution." *Daughters of the American Revolution Magazine*, v. LV, no. 6 (June 1921), pp. 303–312; v. LVI, no. 1 (January 1922), pp. 23–33; v. LVI, no. 9 (September 1923), pp. 560–568.

John J. McCusker, Jr. "The American Invasion of Nassau in in the Bahamas." *The American Neptune*, v. XXV, no. 3 (July 1965), pp. 189–217.

———. "The Continental Ship *Alfred*." *Nautical Research Journal*. v. XIII, nos. 2–3 (Autumn 1965), pp. 39–68.

Lynn Montross. "Money by the Cartload." *Leatherneck*, v. 39, no. 8 (August 1956), pp. 48–51.

Charles O. Paullin. "The Administration of the Continental Navy of the American Revolution." *United States Naval Institute Proceedings*, v. XXXI, no. 9 (September 1905), pp. 625–675.

Robert E. Peabody. "The Naval Career of Captain John Manley of Marblehead." *Historical Collections of the Essex Institute*, v. XLV, no. 1 (January 1909), pp. 1–27.

William A. Porter. "A Sketch of the Life of General Andrew Porter." *The Pennsylvania Magazine of History and Biography*, v. IV, no. 3 (1880), pp. 261–301.

Frank H. Rathbun. "[John Peck] Rathbun's Raid on Nassau." *United States Naval Institute Proceedings*, v. XCVI (November 1970), pp. 40–47.

Kenneth Scott. "New Hampshire's Part in the Penobscot Expedition." *The American Neptune*, v. VII, no. 3 (July 1947), pp. 200–212.

Henry I. Shaw, Jr. "Penobscot Assault—1779." *Military Affairs*, v. XVII (Summer 1953), pp. 83–94.

"The Story of the Vessels Built in Connecticut for the Continental Navy." *Records and Papers of the New London County Historical Society*, pt. IV, v. I (1893), pp. 46–64. London County Historical Society, 1893.

"William Bingham, Agent of the Continental Congress in Martinique." *The Pennsylvania Magazine of History and Biography*, v. LXI, no. 1 (January 1937), pp. 54–87.

Secondary Works

M. Almy Aldrich. *History of the United States Marine Corps*. Boston: Henry L. Shepard, 1875.

Gardner W. Allen. *A Naval History of the American Revolution*. 2 vols. Boston: Riverside Press, 1913.

———. *Massachusetts Privateers of the Revolution*. Boston: Massachusetts Historical Society, 1927.

Helen Augar. *The Secret War of Independence*. New York: Duell, Sloan and Pearce, 1955.

Bahamian Interlude: Being An Account of Life at Nassau in the Bahama Islands in the Eighteenth Century, Reprinted From the "Memoirs of Peter Henry Bruce Esq.," Published in London 1782. London; John Culmer, 1949.

Leland D. Baldwin. *The Keelboat Age on Western Waters*. Pittsburgh: University of Pittsburgh Press, 1941.

Mary Barney. *A Biographical Memoir of the Late Commodore Joshua Barney*. Boston: Gary and Bowen, 1832.

Donald W. Beattie and J. Richard Collins. *Washington's New England Fleet: Beverly's Role in Its Origins, 1775–77*. Salem: Newcomb & Gauss, 1969.

Alfred Hoyt Bill. *The Campaign of Princeton, 1776–1777*. Princeton: Princeton University Press. 1948.

Ledyard Bill, ed. *History of the Bill Family*. New York: n. p., 1867.

George A. Billias. *General John Glover and His Marblehead Mariners*. New York: Holt, 1960.

Harrison Bird. *Navies in the Mountains: The Battles on the Waters of Lake Champlain and Lake George, 1609–1814*. New York: Oxford University Press, 1962.

Charles K. Bolton. *The Private Soldier Under Washington*. New York: Charles Scribner's Sons, 1902.

Consul W. Butterfield. *History of the Girtys: Being a Concise Account of the Girty Brothers—Thomas, Simon, James and George, and of their Half-Brother, John Turner—Also of the Part taken by Them in Lord Dunmore's War, in the Western Border War of the Revolution, and in the Indian War of 1790–95*. Cincinnati: Robert Clarke & Co. 1890.

Howard I. Chappelle. *The History of the American Sailing Navy: The Ships and Their Development*. New York: W.W. Norton & Co., 1949.

J. F. H. Claiborne. *Mississippi, as a Province, Territory and State, With Biographical Notice of Eminent Citizens*. 2 vols. Jackson: Power & Barksdale, 1880.

William B. Clark. *Ben Franklin's Privateers: A Naval Epic of the American Revolution*. Baton Rouge: Louisiana State University Press, 1956.

———. *Captain Dauntless: The Story of Nicholas Biddle of the Continental Navy*. Baton Rouge: Louisiana State University Press, 1949.

———. *The First Saratoga, Being the Saga of John Young and His Sloop-of-War*. Baton Rouge: Louisiana State University Press, 1953.

———. *Gallant John Barry 1745–1803: The Story of a Naval Hero of Two Wars*. New York: Macmillan, 1938.

———. *George Washington's Navy: Being an Account of his Excellency's Fleet in New England Waters*. Baton Rouge: Louisiana State University Press, 1960.

———. *Lambert Wickes, Sea Raider and Diplomat. The Story of a Naval Captain of the Revolution*. New Haven: Yale University Press, 1932.

William L. Clowes. *The Royal Navy, A History from the Earliest Times to the Present*. 7 vols. Boston: Little, Brown, 1899.

Jack Coggins. *Ships and Seaman of the American Revolution*. Harrisburg: Stackpole Books, 1969.

[Sir Bede Edmund Clifford.] *Historic Forts of Nassau in the Bahamas*. Nassau, 1952.

James F. Cooper. *The History of the Navy of the United States of America*. 2 vols. London: Richard Bently, 1839.

Benjamin Cowell. *Spirit of '76 in Rhode Island; or, Sketches of the Efforts of the Government and People in the War of the Revolution*. Boston: A. J. Wright, Printer, 1850.

Neville B. Craig. *Sketch of the Life and Services of Isaac Craig, Major in the Fourth (Usually Called Proctor's) Regiment of Artillery, During the Revolutionary War*. Pittsburgh: J. S. Davison, 1854.

Major Richard S. Cullum, USMC. *History of the United States Marine Corps*. New York: Hamersly, 1903, 2d ed.

Mrs. Reginald de Koven. *The Life and Letters of John Paul Jones*. 2 vols. New York: Charles Scribner's Sons. 1913.

Robert A. East. *Business Enterprise in the American Revolutionary Era*. Gloucester: Peter Smith, 1964.

William H. English. *Conquest of the Country Northwest of the River Ohio, 1778–1783; and Life of Gen. George Rogers Clark*. 2 vols. Indianapolis: Bowen-Merrill, 1896.

Colonel Cyril Field. *Britain's Sea-Soldiers: A History of the Royal Marines*. 2 vols. Liverpool: Lyceum Press, 1924.

Edward Field. *Esek Hopkins, Commander-in-Chief of the Continental Navy During the American Revolution, 1775–1778*. Providence: Preston and Rounds, 1898.

John C. Fitzpatrick. *The Spirit of the Revolution: New Light from some of the Original Sources of American History*. Port Washington: Kennikat Press, 1970.

Esther Forbes. *Paul Revere & the World He Lived In*. Boston: Houghton Mifflin Co., 1942.

Worthington C. Ford, ed. *Defences of Philadelphia in 1777*. Brooklyn: Historical Printing Club, 1897.

John W. Fortescue. *A History of the British Army*. 13 vols. New York: Macmillan, 1899–1930.

Charles W. Goldsborough. *United States Naval Chronicle*. Washington: J. Wilson, 1824.

Elbridge H. Goss. *Life of Colonel Paul Revere*. 2 vols. Boston: Joseph G. Cupples, 1891.

Isaac J. Greenwood. *Captain John Manley: Second in Rank in the United States Navy, 1776–1783*. Boston: C.E. Goodspeed & Co., 1915.

Martin I. Griffin. *Commodore John Barry: A Record of His Services for Our Country*. Philadelphia: The Author, 1902.

Colonel Robert D. Heinl, Jr., USMC. *Soldiers of the Sea: The United States Marine Corps, 1775–1962*. Annapolis: United States Naval Institute, 1962.

A History of the Schuylkill Fishing Company of the State in

Schuylkill. Philadelphia: Members of the State in Schuylkill, 1889.

James L. Howard. *Seth Harding Mariner: A Naval Picture of the Revolution.* New Haven: Yale University Press, 1930.

James A. James. *The Life of George Rogers Clark.* Chicago: University of Chicago Press, 1928.

———. *Oliver Pollock: The Life and Times of an Unknown Patriot.* New York: D. Appleton-Century Co., 1937.

Dudley W. Knox. *The Naval Genius of George Washington.* Boston: Houghton Mifflin, 1932.

Cecil C. P. Lawson. *A History of the Uniforms of the British Army.* 4 vols. London: Norman Military Publications, 1940–1961.

Charles M. Lefferts. *Uniforms of the American, British, French, and German Armies in the War of the American Revolution.* New York: New York Historical Society, 1926.

Philip K. Lundeburg. *The Continental Gunboat Philadelphia and the Northern Campaign of 1776.* Washington: Smithsonian Institution, 1966.

Major Edwin N. McClellan, USMC. *History of the United States Marine Corps.* 2 vols. Washington: n:p., 1925–1932.

John J. McCusker. *Alfred: The First Continental Flagship, 1775–1778.* Washington: Smithsonian Institution, 1973.

Edgar S. Macley. *A History of American Privateers.* New York: Appleton, 1899.

Alfred T. Mahan. *The Influence of Sea Power on History, 1660–1793.* Boston: Little, Brown, 1890.

———. *The Major Operations of the Navies in the War of American Independence.* Boston: Little, Brown, 1913.

Lawrence S. Mayo. *John Langdon of New Hampshire.* Concord: Rumford Press, 1937.

Lieutenant Colonel Clyde H. Metcalf, USMC. *A History of the United States Marine Corps.* New York: Putnam's Sons, 1939.

Louis F. Middlebrook. *History of Maritime Connecticut During the American Revolution, 1775–1782.* 2 vols. Salem: Essex Institute, 1925.

William J. Morgan. *Captains to the Northward: The New England Captains in the Continental Navy.* Barre, Mass: Barre Gazette, 1959.

Samuel Eliot Morison. *John Paul Jones: A Sailor's Biography.* Boston: Little, Brown, 1959.

Robert W. Neeser. *Statistical and Chronological History of the United States Navy, 1775–1907.* 2 vols. New York: Macmillian, 1909.

Ralph D. Paine. *The Ships and Sailors of Old Salem: The Record of a Brilliant Era of American Achievement.* Boston: Charles E. Lauriat Co., 1923.

Captain William D. Parker. *A Concise History of the United States Marine Corps, 1775–1969.* Washington: Historical Division, Headquarters U. S. Marine Corps, 1971.

Charles O. Paullin. *The Navy of the American Revolution: Its Administration, its Policy, and its Achievements.* Cleveland: Burrows Bros., 1906.

Howard H. Peckham. *The War for Independence, a Military History.* Chicago: University of Chicago Press, 1958.

Harold L. Peterson. *The Book of the Continental Soldier, Being a Compleat Account of the Uniforms, Weapons, and Equipment With Which He Lived and Fought.* Harrisburg: Stackpole Co., 1968.

Colonel Robert H. Rankin, USMC. *Uniforms of the Sea Services: A Pictorial History.* Annapolis: United States Naval Institute, 1962.

Lieutenant Oliver P. Remick (USRCS). *A Record of the Services of the Commissioned Officers and Enlisted Men of Kittery and Eliot, Maine: Who Served their Country on Land and Sea in the American Revolution 1775–1783.* Boston: Alfred Mudge and Son, 1901.

Ernest R. Rogers. *Connecticut's Naval Office at New London During the War of the American Revolution, Including the Mercantile Letter Book of Nathaniel Shaw, Jr.* New London: New London County Historical Society, 1933.

Theodore Roosevelt. *The Winning of the West.* New York: C.P. Putman's Sons, 1900.

[Robert G. Sands.] *Life and Correspondence of John Paul Jones Including His Narrative of the Campaign of the Liman.* New York: n.p., 1830.

J. Thomas Scharf and Thompson Westcott. *History of Philadelphia: 1609–1884.* 3 vols. Philadelphia: L. H. Everts & Co., 1884.

John H. Sheppard. *The Life of Samuel Tucker, Commodore in the American Revolution.* Boston: Alfred Mudge and Son, 1868.

Samuel S. Smith. *The Battle of Princeton.* Monmouth Beach: Philip Frenau Press, 1967.

———. *The Battle of Trenton.* Monmouth Beach: Philip Frenau Press, 1965.

———. *Fight for the Delaware, 1777.* Monmouth Beach: Philip Frenau Press, 1970.

Robert A. Stewart. *The History of Virginia's Navy of the Revolution.* Richmond: Mitchell & Hotchkiss, 1933.

William C. Stinchcombe. *The American Revolution and the French Alliance.* Syracuse: Syracuse University Press, 1969.

William S. Stryker. *The Battles of Trenton and Princeton.* Boston: Houghton, Mifflin and Co., 1898.

———. *The Forts on the Delaware in the Revolutionary War.* Trenton: John L. Murphy, 1901.

Ray Thompson. *Washington Along the Delaware. The Battles of Trenton and Princeton—As Told by the Men Who Were There and Through Washington's Own Official Dispatches.* Fort Washington: The Bicentennial Press, 1970.

Reuben G. Thwaites, and Louise P. Kellogg, eds. *Frontier Defense of the Upper Ohio 1777–1778.* Madison: Wisconsin Historical Society, 1912.

Christopher Ward. *The War of the Revolution.* Ed. by John R. Alden. 2 vols. New York: Macmillan, 1952.

Justin Winsor. *The Westward Movement: The Colonies and the Republic West of the Alleganies, 1763–1798.* Boston: Houghton Mifflin, 1897.

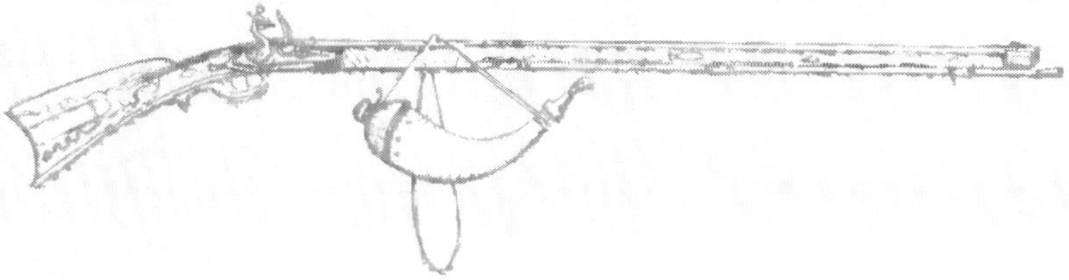

APPENDIX C

Diary of John Trevett, Captain of Marines

John Trevett was born at Newport, Rhode Island, in 1747, the son of Mary (Church) and Eleazar Trevett. In early life he was in the merchant service, and made a number of voyages from Newport to such places as Lisbon. On the breaking out of the Revolution he entered the navy. In November 1775, he took passage in the sloop *Katy* (later renamed the *Providence*) to Philadelphia, where the first Continental Fleet was fitting out, and accepted the position of midshipman on board the ship *Columbus*. Early in February 1776 he was promoted to lieutenant of Marines, and the following month participated in the taking of New Providence.

When the fleet returned in April, Trevett left the *Columbus* and was attached to the Continental brig *Andrew Doria*. Later in the year he joined the sloop *Providence*, under the command of Hoysteed Hacker, then Jonathan Pitcher, and later Captain John Peck Rathbun. In late January 1778, Captain Trevett, with 28 Marines, landed on the island of New Providence and took possession of the island's two forts for a second time.

Throughout the remaining years of the war, Trevett served in various capacities on board both Continental and private ships of war. While a volunteer on board the *Trumbull* in June 1780, he lost an eye and was wounded in the foot during the frigate's engagement with the British ship *Watt*.

In the years following the Revolution, Trevett worked as a joiner (carpenter) "until infirmities disease & blindness" rendered him incapable of any further labor. On 5 September 1823, he died at Newport, aged 76 years.

The diary of John Trevett, which spans almost eight years of service from November 1775 to June 1783, is now in the Newport Historical Society, Newport, Rhode Island. The words, spelling, and punctuation are Trevett's own, but because the first several pages of the original diary no longer exist, the missing material has been taken from a copy made by Eleazor Trevett following his father's death.

This is to Whom it May Concern that I, John Trevett, sailed from Providence in a Sloop called the Catea [*Katy*], Commanded by Abra[ha]m Whipple, Esq., of Providence, with a number of passengers, to sail with a fleet of armed vessels fixing at Philadelphia, in the month of Nov. 1775. Arrived there the same month [actually 5 December] and found ship called the Alfred 1 ship called the Columbus 1 brig called the Calbot [*Cabot*] 1 brig called the Andrew Doria, and then our Sloops name was altered and she called the Providence. I went on board the Ship Columbus as first Lieutenant of Marines, and when we had got to sea we stood to the Southward until we made the Island of Abbaco one of the Bahama Islands. There we came to anchor until we took two small craft belonging to New Providence and got pilots, and went in the night over to New Providence and landed all the men we could spare. I took command of one of the companies and marched to the first fort. They fired a few pound Shot, but did no damage we saw an officer coming, I went up to him to know what he wanted he informed me that Gov. Brown would wish to know who we were and what our business was we gave him his answer and the first fort stopped firing and that night we lodged in the fort. The next morning by break of day we marched for the town where the Governor lived which was 4 miles from the fort. We arrived within a mile we halted about one hour, and had a parley, when the gate of the Capital fort, Nassau, was opened, and the British colours hauled down, and we took possession. The Commodore and our vessels lay back of Hog Island, and in about 2 hours after, they came into the harbour, and as the Commodore was landing we saw Governor Brown and his council walking his Piazza and his servants waiting below with horses. Capt. [Henry] Dayton, and myself Asked Major [Samuel] Nichol[a]s liberty to go and take him. The Major informed us he had no orders from the Commodore to take him, but we may do as we pleased. Capt. Henry Dayton, and a young officer from Philadelphia and myself went to the Governor's house and informed him he must go with us to the fort. He made reply it was beneath his dignity as Governor to go to the fort. We made him this reply he must go, then he says it must be by the force of arms, then we told him it was by the force of arms and he walked down with us to the fort and in a short time Com. Hopkins and the Captains landed and came to the fort and after some time spent the Commodore sent for me, and gave me orders to take as many men as would be sufficient, and keep the Governor prisoner in his own house; and I kept him safe until our Fleet was ready for sea. Then I had orders to wait on him down to the barge, and we brought him to New London with 2 more gentlemen passengers which were prisoners, nothing material happening until we made Long Island, when we took what was called the Bum Brig [bomb brig] and an armed Schooner and let the Ship Glasgow run away from us and after 2 or 3 years we were ordered to be paid in Continental money, When we wished to spend it, it would about pay for 1 pair of Shoes, a grand cruise, and I am glad it ended so well. May 1776. I am now about to begin a new cruise in the Continental Brig Andrew Doria, Nicholas Biddle Esq. Commander. [Here a break occurs in the narrative] we took 3 prizes 2 of them large transport Ships bound from Glasgow to Boston we captured them of the banks of Newfoundland, after we took on board as many of the principal officers as we had room for and all their warlike stores we took out all the Soldiers which amounted to 220 besides Sailors Women and Children, how many there were, I never knew, Lieut. James Josiah went [as] Prize Master of one of them [*Crawford*] and Lieut [John] Mac Dougal of New York the other [*Oxford*]. I went on board Mac Dougal as Mate &c and we had about 300 on board; we kept company with each other 17 days, when we got near Nantucket we fell in with Lord Howes Fleet in the fog, we steered different courses, and before we got out of sight of Capt. Biddle the prisoners rose on the ship and took her from us, and they put one Canada (formerly Mate of the Ship) in Commander. I could not blame them for I would do the same, he stood to the southward for Hampton Roads in Virginia where we arrived in 15 days after they took possession, we got into Hampton roads about sunset, we immediately came along side 2 small pilot boats, and they informed us the Foy [Fowey] ship of war lay 40 miles up James river and they must immediately get under way, after giving 3 cheers we weighed anchor and stood up the River with a light air of wind, about 12 o clock at night after the Capt. of the Pilot boats had found out that we were officers on board the Andrew Doria they called on me to know how they should retake her. I informed them, it was exactly done, and at day light I informed our new Capt. Canada that he had no more command, and that he must go forward and all but the women, and children, they might stay on the quarter deck; which was done, we stood up James River until we arrived at Jamestown, and there we landed 220 highlanders, which was escorted by a part of a regiment of riflemen with their rifle frocks, which I think the finest sight I ever saw, they encamped near Williamsburg, at the same time there was a convention setting there, Lieut. Mac Dougal and myself applied to the Speaker of the House for money to pay our expenses for ourselves and seamen, 11 in number, we were supplyed with what cash we wanted, with pleasure, and were treated very politely by one and all at Williamsburg, and we drew a bill on Hon. John Hancock then President for the cash we received from the State of Virginia, we carried with us one of the bills, which the Hon. President received with pleasure. We tarried one day longer than we needed, on account of seeing Independence declared, which was on the 4th day of July 1776. The next day we set out for New York on our way to Rhode Island when we arrived at Newport, we went immediately on board the Privateer [*Andrew Doria*] which was lying in the harbour at New Port, we brought dispatches for Capt. Biddle whom we had the pleasure of finding on board, we found they had made a short cruise since we were taken, and had captured one Jamaica Ship, but being chased hard by a British Frigate, they ran her on shore on Fishers Island reef, and lost the Ship, and all her sugars, but saved 100 puncheons of rum, sails & stores, which were sent to Mr. Nathaniel Shaw, of New London, who was appointed Agent, and Capt. Biddle gave me my share of prize money, which I received from Mr. Shaw so ends this Cruise!

July, 1776. I was at Newport but a few days before I sailed again with Capt. Biddle, we took several prizes, some of them from Lord Dunmores, 1 brig from Barbadoes bound to Newfoundland, this prize I went on board of and arrived safe at Providence. When our cruise was out, Capt. Biddle went into Philadelphia, and by that means, I went on board the Sloop Providence Captain Histed [Hoysteed] Hacker; and I soon found we were to sail under command of John Paul Jones Esq. we sailed in a short time, we stopped at Tarpolin cove, where we found a small privateer belonging to Providence commanded I think by Capt. Rhodes. Com. Jones sent for me on board of his Ship late in the day, and gave me orders to arm and man our barge on board the Sloop, and he would have his barge manned and armed and sent along side our sloop, and then we were to go along side [the privateer], and while I was examining the Ships Articles, I was to give the barges crews order to press all we could, I did so, and we pressed 25 men out of 35 and carried them out on a cruise on board the Ship Alfred Com. Jones. We sailed to the eastward of Halifax, the first prize we took was a snow from England bound to Halifax her cargo dry goods, the next prize was a Ship called the Malech [*Mellish*] her cargo 10,000 suits of Soldiers Clothing ready made 1 set light horse accoutrements with carbines and a valuable invoice of Medicine chests! the Ship the most valuable out of 45 sail! the rest of her cargo trunks of Silk gowns and dry goods suitable for Gen. Burgoynes army at Quebec, the Ship haled down her colours to the Sloop Providence; she mounted 12 carriage guns and had between 60 & 70 men, the Alfred and the Snow coming down on us we then manned the Malech and ordered both for New Bedford, where they arrived safe. As soon as they arrived, without trial, for the Malech, she was onloaded and all the clothing taken out and waggons prepared to send them on to Gen. Washingtons army, at that time his army being in a distressed situation for clothing, and in this Ship was every article for a Soldier from the hat, to the shoes and at that time I can say with pleasure I had rather taken her, than a Spanish Gallion with hard money, although we took Continental money for our parts of all the prizes. We cruised off Halifax until we took 3 more Ships their cargoes seacoal &c when we had a violent snow storm, it being in the month of Nov. we [Providence] parted with Com. Jones and then we put away for Rhode Island and arrived the last of Nov. and the Alfred arrived safe at Boston. Shortly after (the 6th of Dec.) the British took possession of Newport. The ship Warren, Ship Providence & Sloop Providence, lay near Gould Island but we made the best of way to Providence, while the British fleet was running into Narriganset bay, the Sloop Providence had some men on shore on Gould Island cutting wood. I perceived a large quantity of hay stacked up there. I ordered one of the men to give me a brand of fire. I stepped into the barge and our sloop hove too, until I set fire to all the hay on the Island, as I well knew it would fall into the hands of the British, and all I received for this was the loss of silver knee buckle, and a Waistcoat, but had great contentment of mind which money cannot purchase. then we hauled our wind for the north end of Jamestown Island, the wind being S.W. as soon as we opened Narriganset bay, there was nothing to be seen but Ships, we under easy sail wishing some of them to give chase, we lay in the way until we gave them 3 shot, when immediately 3 of their ships with all sail they could pack gave us chase, which we wanted, we under easy sail stood up for Warwick neck; they finding we intended to get them a ground, signal was given from the Com. of the British Fleet and they gave up the chase, & then we went up for Providence. A short time after, one of our prize Ships running in for New Port was taken after receiving a number of shot, being cut off by one of their Ships near Prudence Island, and carried back to Newport, so ends their cruise.—Dec. 6th 1776. This day my Father and Mother and a kinswomen and a young son of my Brothers went off for East Greenwich, they had but a few hours notice, they took with them some beds, and bedding, and a few trunks with clothing, and left there home with all the remainder of the furniture behind, with their wood, provisions, and every thing necessary for the Winter, and fortunate for them, the day they arrived at East Greenwich they fell in with Mr. Peleg Olden, who took them into his house, and treated them with every kindness that a good man could do, but to end this affair, all that he left behind, was lost partly by the British, but mostly by our own people. I will stop here, and now our Sloop Providence is at Providence fitting for another cruise and now it is Jan. 1777. we heard at Providence, that a British frigate called the Diamond had get a ground near Warwick neck, we went down with the Sloop, and one armed Galley, and we had 2. 18 pounders, which were [placed] on the point, at Warwick neck, which were played well on her during the night, but they started the water, and lighted her, that she floated before day, and went down for Newport, so that next day we returned to Providence, and soon after another ship took her station, not so near Warwick neck, at the same time we had a fire Brig, and Sloop fitting at Providence and soon was ready we went down the river with them, in the night waiting for a favorable time to chain them together. The time shortly came when we under took to chain them, but a sudden breeze of wind sprung up before we could chain, and the sloop-fire-vessel, got so near the ship she was obliged to run ashore, near East Greenwich, and we sat her on fire rather than she should fall into the hands of the enemy. The Brig and Sloop Providence returned to Providence. In a day or two after, I went over on a party to Capt. Nicholas Websters, to Rehobeth, while I was on this party Governor Hopkins & Com. Hopkins sent over Capt. Henry Dayton to me, for to come to Providence immediately, after they informed me that they contemplated fixing out a Cartel to send to Newport and for me to take out of the Sloop Providence 2 men to go with me, as there was some prisoners coming from Boston, and one Capt. Ayeres of Boston was to go Captain of the Cartel, as he was not to know who we were, that was a going his men, from that time I began to let my beard grow so as to disguise me, I took 2 of our midshipman from the Sloop Providence. The time soon arrived when the British prisoners came with Capt. Ayeres from Boston, we one and all disguised ourselves in the sailors dress, and made sail for Newport, and came to anchor near the long wharf, when Shortly—a barge come from the ship renown of 50 guns, laying outside—the fort. I informed Capt. Ayers and the British officer that our cable was so poor, we should be [here John Trevett's original diary begins] Apt to go ashore If we was to go out side of the fort Island, by that Meains the Barge went on Board and Brought A Midshipman and Some Men with him to take Charge of the Cartele and the British ALowed Capt. Ares to Gon on Shore When the Prisoners was Landed and he went to Mr. Thomas Townsends. Very Cold Wether and the Cove Froze over I had Plesher of a Veu of the Diment [*Diamond*] Friget laing on a Crene beLow the Long Wharfe stoping up her Bruzes we Gave her the Week before att Warrick Neck—and now I was Contriving how I Should Gett on Shore I Says nothing on My Tew Ship mates but noing we had but one Gang Cask of Warter on Board we had Plenty of Good Rum and Sugar noing that Sailors Liked a Sling or Can Hook in the Morning Such Bitter Could Wether; in the Night I stepe Down the Hole and Turnd the Gang Cask Bung Down So thatt in the Morning we had no Water on Board. You must think I went by the Name of Jack by My Ship Mates So as to Deseve Capt. Ares. Att Day Light as I Expected our British officer Wanted A Sling as Jack was Very Attentive he Says to Me have You Got any Good Spirits Aboard I informd as Good As Jamaca Could Afford then he Says Make a Good Sling well to to the Northward I Emeadely Told My Ship Mate Tom as that was the Name he went by then; to Gett the Warter While I would Gett the Rum and Sugar Redy Tom went in to the Hole and Sune Returnd swearing; and Said by Some Axsident the Cask had Gott Bung Down and thare was no Warter on Board this was a Short Time After Day Light in the Morning the Officer Turns to Me and Says Jack Do You Know Whare You Can Gett Warter handy. I informed him that I had Sailed from this Place Some time before and told him there was some Good Warter Neair the Long Wharfe Jack Says this officer Step in the Boat with tew hands and that was our Tew Midship Man Tom & Bill that was the Names we Sailed by then Now the officer Gave us a Strick Charge Not be Gone More than Twenty Minets—This was Jest What I wanted I went into Mr. Philip Wantons Dock took out the Gang Cask and My Tew Mas Mates Carred itt up I went with them into Mr. Wantons Wash Room Whare tha had a Pump with Good Warter in the well [who] Should I See thare must Mr. George Lawton A Washing His Hands I Asked him to Len Me a funel to fill the Gang Cask he inform Me he had none but told Me Mrs Batte on the Long Wharfe had one I See Mr Lawton Cast is Eye on Me hard butt made My Self Scase I gos to Mrs. Battes on the Long Wharfe finding She was A-Lone and Making a fire I new her well I asked Mrs. Battey to Le[nd] Me a Funel She Ansered Very Short no as no Soul was [near] I Goes up to her and Told her She Shou[ld] Lend Me one S[he] New My Mode of Speking and Says for God Sake Whare are You from I informed her from Providence She Says How Did You Leve My Son Mening Capt. Henry Dayton I informed her well now Says Mrs. Batte Speake Low for I Gott Over head Severel of the officers of the British Transports Boards with Me and I Expect Mr. Batte a Shore this Morning as he is a pilot on Board of one of the Ships of War and iff he Sees You he May nowe You heair is the Funel will You Eat or Drink Any thing I told her no . . I must Remark fore or 5 Months before I took up this same Batte for a Torri. I had the Warter filled in a short time and Carred the Funel Back she had a Good Eole of Conversation and before I Left her she Said she was Affraid her Son and I should be Hanked for the British surtenly would Beait the Americans.

I Larfed att her and said that Can never be; & I Never see no More of her I than went to the Longboat & Gott in the Gang Cask of Water and than the Tide was About Half Down I Says To Tom & Wm. them was My Shore Mates Lett the boat now Ground for I want to take A Cruse Around the Town Which thay Did I then went to Mr. Peleg Barkers Whare the Commander of the Hashens Quarterd I found Sentres att the Frunt Doors and Likewise Att the Wharf ware Sentres; I went Down Across Tew Wharfes Below and went into the Back Doors I found the Ketchen full of Hushens As I was well Acquainted with the House I shaped My Corse for the S West Rum thare I found Mr. Barkers Familey and Likewise Deacon Peckham from MedelTown Which I was Very Glad to see Alltho tha ware Very much Frited to see Me Thare I sune Got them Reaconsiled & told tthem to Make them Selves Ezea for I believed I new Whate I was Aboute we sune Got into Conversation I made All Inquira What and Whare and What Name of Trupes ware on the Island and Whare tha ware statined I was Much Plesed to Gett this Account I had no Expectation of; After I Got All the Information I could Gett thare I went to Mr. Waldrons thare tha had no Trupes Quarted thare I found a small family Dog My Farther had Left behind Moving away in Such hast. I Lett him stay until I went to Capt. Lillebreges on the Parade then It was About Eleven A Clock AM he Kept A Tavern I went in & Called for a sling the Room was Crowded with Some British & some Heshan officers. I Emeadily went into the Kitchen Whare the Family ware Noing that Capt. Lillibrige had bene ill Treated by the British and had no Regard for them In a short time followed him out to the Barn and no one [being near] I made My self none to him he Emeadly Left the Barn . . . & went into the East Roome by our Selves he Gave Me what Refreshment I wanted and thare I Could see All the British officers & solgers & old Refege Torres Walking About the Parade but he Could not help Sheding Tears for My safty for feair of one of our Towns Men that visited his house all times in the Day of noing Me I was A Veuing the Parade When This one of the worst of Villians His Name Was William Crosen Came Runing up the Steps Came Right to the East Door ware Capt. Lillibrige and I had been for some time he was not sune Anuf for I steped to the Door and I had My finger on the Lache and he suposed itt was fast went Emeadily thrue the Bar Room into the Ketchen I never bid My friend Capt. Lillibrige Good by steped out on the Parade and Direct before Me Was Mr. John Wanton he spoke to Me and Called Me by My Name but I Did not Anser him and went Emeadetly Down the Long Wharf Quck Step I went Round the point and Came to Mr. Waldrons Whare My family Dog was thare I spent some time Got some Refreshment then itt was About 3 Aclock I thort by this time our boat would be Afloat I than bed Mrs. Waldron and the Family Good by took my faverite Dog and Went for the boat Jest as I Gott Neair Whare Mr. Jacob Richardson Leves on the Long Wharfe I mett the Midshipman Jest Goot on shore he handed Me out a Few Cuses & Dams I informed him that our boat had Gott Aground he told Me to go on Board Emeaditly for says he You Got some Dam Good frinds heair I said but Littel Made the Best of My way to the boat found her Afloat and My Comrades Waiting for Me we soone Roaed of but I must not forgett I saw Mr. Peleg Barker and his son Peleg Agoing Down the Long Wharfe And Eyeing our Boat and I well new the Antsiety tha had to see Me Gott on Board the Cartelle as she Lay of the head of the Long Wharfe we Very sune Goot Aboard and had not bene Long on Board Before Capt. Ares was orderd on board as itt had bene Reported that I was att Newport but Very Few would Beleve itt, Capt Ares said if itt had not bene for Me he Could have staid on shore I informed him I thort itt Must be A mistake but the Next Day we took in some Pasingers & some American Pisners and orderd for Providence the wind being Partly a head we ware obliged to beat some before we Goot by a British Ship of War that Lay off from Prudence but as sune as we Goot by the uper Ship I Left Tending the Jib Sheat and went Arft I says to Wm Come now take Your Turn forward so I took the Helm I see Capt. Areas Eyed Me; as I thort itt was not Long before we Got Abrest of Pertucksett Whare our sloop Providence Lay the Barge Emeadely Came ALong side I took one Man in My Rome to Work the Cartile up to Comadore Whipoles ship and I went on Board the sloop Providence I had then Anserd All and More than All My Expectations for I had in My head All thare Ships of War and Ware tha ware stained I Emeadely Shaved & Dresed My Self in My Sunday Dress I proseded Emeadely on Board Commadore Whipels Ship Whare I found Capt Ares; Arfter Capt Ares had Gone threu his Conversation the Comadore Turns to Me and asked Me What I had Descoverd Capt. Ares Looked Att Me and Did not now me; Arfter we had Gone Thru Capt Ares Turns to Me I Would not Run the Resk for the Cartele Loaded with Dollars I told him I had Anserd All My Expectations and More than I Expected—

And now I am once More on Board the Sloop Providence and I find Johnathon Pitcher Esqr. is to Take the Command and we Only waiting for a Good Chance and now itt is Febuar and we Made sail Att sunset A Light Wind to the NE standing Down the River past one ship of Prudence and Went thrue Narraganset Bay And Run so Neair a 50 gun ship About 2 A M as to heair them A Talking on Board And Att sun Rize we ware becarmd About Tew Miles from the Light House and Could See Newport and the ships in the Harbor Abought Ten A M A small Breze sprunk from the S. West we standing to the East Ward We Arrived Att Bedford that Night we had but few Men on Boad as tha Could not Expect that we should Gett out of Providence River we Got Her Mandd and sailed on a Cruse to the Eastward Nothing Material Hapening untel we Arrived of Cape Bretton About 5 Legs. East Cape Breton in Sight; Saw several sail and herd a Number of Heve Cannon the Night before one Brigg Bore Down on us and began A fire Att Long Shot we Run from her About one Hour untell we Got in Good order for Action then we took in sail and

Lett her Come up Close A Long Side and the Sea Very Smuth in forty Menets we Cut Away All her Cullers and tha began to be slak in a few Menets the began the fire as Brisk as Ever and Cut our sails and Riging Badly one 6 lb short we I I Lettel below the Hounds of the Mast itt Lasted About 40 Menets Longer When we Cut away her Main Top Mast and we Hailed them without A Trumpet being Close in on her Starboard Quarter and the Stoped Firing to no Wether tha give up or not And the Answer was Yes Capt. Petcher was badly Hert but Kept the Deck ontill she Gave up but I can tell You itt was Dimon Cut Dimon Capt. Petcher sent me in the Barge first on board I found them a Vere Bloody Deck and Quarter Deck and her Spar's sailes & Riging Very Much Damaged I staied on Board untel I sent the Capt. on board our Sloop but a Going Down in the Cabin the Flouers was Spread as full of Wounded Men, as You Could scasley find Room to put You foot and I foun tha ware some of them Irish as tha Cry out for Jeses Sake to Spar's thare Lives tha ware Very Badly Wounded we found her to be A Brigg Directly from England she had Twenty fife Soldyers and Tew British officers and thare own Crew she was Loaded With Kings Stores, and Bound for Quebeck As sune as the Action was Over we found our Sloop so Wounded in the Marst & Spairs that we ware obliged to send Down her Top sail Yard & Top Mast and we Maned the Brigg and that Night the Wind Blew Hard with Squarls & Dark the Next Morning we found no More of Her Capt. Pitcher orderd her for Bedford and we ware Obliged to Give up our Cruse and put Away for Nantucket and in a short time we Arrived att Bedford Carring with us all our New Bergoine Soldyers and Officers and some sea Men we are now fixing our Vesel this is beginin of May 1777 and I find that Capt. John Peck Rathbun is to take Capt. Pitchers plase as he Went on to Congress and Got His Appointment one Werd More A Better officer then Capt. Petcher I think Can not be our Prisners ware sent to Boston only a few of them Run Away as tha Chose to stay in a frea Cuntry So Ends this short Cruse—

May 1777 We are Now fixing our Vesel for Capt. John P. Rathbun And I have Jest Recevd an Account of My Brother Constant Church Trevett being Taken. He Commanded a Mercht. Vesel Bound from the West Endeas to Carolina and was Taken by the British and sent for New York and put on Board A British Prison ship Called the Old Jersey I Emeadely sent on to Boston and Procurd a British Capt to send to New York and itt was Done in a short time but Tew Late; for before the British Capt. Arrived he was no More He Died with hared Treatment from the British Piruts as I May say & say the Truth, Black Bard the Notoris Pirut was A Christian to them Bilingate Villins that had the Command Att New York I shortly Arfter saw some Americans that was on Board the same Prison ship My Brother ware tha ware Exchanged and had Gott Home Butt tha Most of them lucked as if tha ware in a Deap Consumtion I herd a Nuff from My Poor Americans to Convince Me that iff I had My Choise I had Rather be Taken by Turks; but I must stop short Heair and say but Lettel Mark well Revenge Is swiet; we shall Sail in a short Time on A Cruse—Now is June 1777 we are under sail bound to the Vinyard to Gett a Few Men and I took a step on shore att Bedford and presed one John Scranton one of My Towns Men and thre More and then Proseded and Arrived att Old Town the Next Day Got What Men we could; Lay there Tew Days and then Proseded and went threw a Very Narrow Pasege Called Muskekek betwikt old Town & Nantucket and stood to the Westward intending to Make a Short Cruse we Run for Sandy Hoock and Made itt saw some Large ships Lay thare and Att the same time Discovered A ship Brigg scuner and sloop Gett under way & Come out standing to the south East we Doged then untill the Next Day when we Thout we had Gott them A good Distance from the Hoock When we stood for them About 3 P M we came up with the ship the other Vesels Near to her on her wether Bow we Hailed the ship as she had her Pendent Jack and Ensine Fixing at her Mizen Peake she Gave us no Anser we Gave her a Bow Gun intending to Brake her Cabin Winders Drew Very Neair her the Wind so scant we found we Could not Gett to Wind Ward we bore Away and went under her Lee as Neair As we Could and gave her a good Broad side Emeadily she Gave us As good a one When she Run us Aboard on our starboard Quarter and Hung thare About 5 Menets untill she Brook All of sweps that ware Lashed thare att the same time the Brigg of 10 Guns and the scuner of 8 Lost no time All three of them firing into us att one time as the ship fell off us she Gave us the starboard Broad side we shot a head of them with our sailes and Riging much Cut to Peces and then we bore Away and All Hands Employed in fixing our Rigging we had put a poor Crew att this time on Board our Loss att this time was our sailing Master Capt. George Sinkens of Newport he was killd and only Tew or 3 slightly Wounded we Hove him Over board Gott our Rigging As sune As Posabel Redey Made sail for the ship Come up with Her Jest arfter sun sett with a Determenation to Board her for we well new if we Carred the ship the Rest of the Vesels would fall into our Hands we Ran within Half of Pistel short Gave her a full Broard side but All thre of them plaied thar part so well we Gave itt up as then itt was Neair Dark att this time the scuner was a head of All these Vesels we Run for her and Run a Long side and took her we found her to be A scuner Neair 140 Tons and found the ship to be a ship of sixteen Guns Henry Johnson Commander All of them Bound to Iamaca in Ballest the scuner had a spand of Horses some Carriages & Triffering Was her Cargo we ordered her for Bedford Whare she Arrived safe we had 12 Guns Only we stered the same Cours as tha steard Intending Att Day Light to take the 3d heat att them, but in the Morning saw but one sail A Long way to the South ward, we stood for her & she for us, we saw she was a Privateer butt she Ran and sune Goot from us, we stell sailing south ward and Eastward, Nothing Hapening for some Days, saw & spoke a Brigg from Staseka [Eustatia]

Bound to Holland Lett Her Pass, saw severel Others, All Deains and Duch, being in Gulf Strene About Merideon saw a sail As far as we Could Discover, stood for her, att sunsett we Discovered her to be a ship and tha on board Appeared to us to Acket strangly as she was Decoing us for some times before the Wind some times a shakeing in the Wind and Top gal sails and All sail out About 2 Aclock att Night itt being star Light we Got Neair to her, some said she had Lights but the Officers Could not see Any, we Gott Within Pistel Shot and, we Hailed Her, no Anser, We Gave her thre Guns att once Made a Craking into her and still no Lights we thout itt Was Strange, Capt. Rathbon orderd the boat out and Armed her and he told Me for My Conserlation if tha Killed Me, he would not spair a Man; About one Aclock A M I sett out, I order the Coxen to steair under the starn of the ship, which he Did, I had a Lanthon in the Barge, I Discovered her Ruder Gone, I Hayled, but no Anser, I orderd the Coxen to steair Round her Larboard Quarter & Go A Long side, I sent one Man up with the Lanthon and I followed him, I discoverd no Boats on Deck, I Discoverd on the Quarter Dack A Depsea Lead and Line, Then we went for the Cabin, I Discoverd All the Beds, & Trunks All full of Clothing, and the Keys in the Trunks & Chest, Emeadetly one of our Men Cries out A man, a Man, by God, I asked him ware, itt Proved to be A small Dog, that opened All the Eyes he had, and Could not Speake our Yanke Tong, I Emeadely went into Ships Hole, & found she was Ballest, What we Call shingel Bables, we could find no Cargo, nor any Provision, Except very Lettel Bread, found forty Casks of Nails, a Few Kases of French Cordels, I Emeadely sent Trunks & Chests &c, on Board, and Took out All small Affairs we could find, that would Anser our Purpose, so Lately out, we could not stow Away so maney things as we Could Wish, moving without Any Ruder itt would Make itt Difficelt to have Gott Her Into Port, we Capt the Barge All Day untill Late in the Evening Pling from the ship, Last of all we took out the Dog, and for Feair she should fall into the hands of the English, we sett her on fire, and Burnt her Down to the Warter Edge, by What Judgment we had of her, she Appeard to be a French ship About Thre Hundred Tons, as the Trunks had Wemens silk shues & Gowns, & aparrel of Finirea, and the Mens shirts all Rufel, & small shirts, French Pocket Handerchefs &c. She was a tite ship, & we Except that she must Have Got in with Cape Hatteras Shoals. & and the Crew, & Pasengers if tha hand Any; Took to the Boat, and Tride to Gett to the Land, but the ship being under full sail, Top Gallen sails Histed a Tanto, & sheated Hom; we Expect tha Left her in Hast; When we Arrived we Examined the Southern Papers, but never Could Gett any Information of such a ship, we supposed the Westerly Winds Gott her into the Gulf Stream, thare itt was we Came Across her, Nothing Material Hapening, we Crused our Cruse out, & Returnd into Bedford, We find she was Considerabel out of Repairing, August 1777, The Carpenters and Corkers to Work, and I sett out to spend a Few Days att Providence, and East Grenwitch, so Ends this Cruse

Now beginen to Recrute for a New Cruse with the same Commander John Peck Rathbon Esqr. Now itt is Novm. 1777. All Redy for sailing, intending for a short Cruse, we Anker a Way standing for Block Island, A Light Wind At North East, Att Night began to snow, the Next Day Blowing A Gale sprung our Bolsplit, Hove Tew, one Night Bloing Hard, the Next Day put Away standing to the S. West, intending for South Carolina, Nothing Material Hapening, untill we Arrived off Charles town Bar, we Run in for the Bar, & Goot neair to itt in fife Farthum Warter, Night Coming on saw no Pilot boat of, we staning of and on, Waiting for Day Light, our Guns All in, and Ports All secured, a Very Plesent Moon Light Night, About Tew Aclock in the Morning, saw a sail in shore, we took her to be a Droger Going from Charlestown to Georgetown, but shortly Arfter, one of our Officers Came Down and Informd Capt. Rathbon that she was standing for us, and Very Neair, Capt. Rathbun & Myself went on Deck, and she shortly Arfter Hailed us, and orderd the Dam Yanke Burgers to Hall Down the Cullers, You must now We had a foule Wether Jack att Mast head, she in a few Menets Run under our Lee Quarter, and Gave us a Broad side, and Run a head of us, Capt. Rathbun informd the Botswen to Call and Gett All Hands to Quarters, as Still as he Could, and Not Make any use of His Calle, as this Privatear Bore Away, and Coming up Again, As He informed us Arfter wards, he was sune A Long side, we ware All Redey as sune as he Made the first flash we Gave him A handsome Brard side he up helm And Run to the Eastward we never had a singel Man hert of Any Conequence we Made sail Arfter her our rigging sails was so Much Cutt Away itt was sume time before we could Gett under full sail she Emeadetly Histead a Lanthon att Mast head but such a buteful Mone Light Moring itt Could not be seen any Considerabel Distance but by thare Heving out the signel we new thare Must be an Enemy not a Grate Distance of we fired no More of our Cannon att Her and found we Ganed A Lettel Every Hour Day Made her Apperence the Man Crys out a Large ship under the Land standing Close on a Wind to the Southward & East Ward the Wind Being S. West. About sun Rise we Neard the Privatear so as thare Lieut. Goot on thare Round House and fired severul times, the Balls went a Long Way bee End us I informd Capt. Rathbun we had I Did not dout as Good Muskets as tha had I spoke to Mr. Mickel Molton and one Mr. Bayle of Boston and Dezired them to go forward with My self and take a shot Att him as he was a fine Mark to shute att he standing on thare Round House You Must note this Vesel steard With a Wheal we had not fired More then thre shot before we saw him fall & Emeadetly the Privatear Gott in the Wind an we ware A Long side in a few Minets When we Boarded her we found itt was the Lieut. we had shot and he fell on the Man stering with the Wheail. This Lieut. was a Man belonged in the State of Virginia and he Expectd to be punished If he was taken by the Americans and he was Determind to fight as Long as he could he had a pair of Hansom pistels to his side As he Lay Dead on Deck tha found 5 Hands bad Wounded on board of her 3 of them ware Blacks our shot went into one Quarter & threw the other and we shaterd her well we found this ship we saw to Wind ward was a Freget that tha was on Board the Day before and tha ware to Meait her the Next Day of Charlestown Bar we got so far to the Eastward that we stered for Georgetown we had taken 6 Negros that tha had Taken the Day before that ware out of Charleston a fishing Come to Examen the Blacks we found that ware Pilots belonging to some of the Citersens of Charlestown. The Capt. officers and Crew All as one Consented to Delever them up to Mrsrs. Harcott and Tucker Mercht. att Georgetown to be sent to thare oners Cleair of Expence a Day or Tew Arfter we Arrived att George Town we put All the Priseners in Gaile but the Capt. I Escorted him to Charlestown an Deleverd him to the Commisare of Prisners . . . . and thare I saw My Old frind Commadore Beddel he had Thre ships under his Command All Redey for sea Att that time thare was an Embarg threw the State of South Carolina Commadore Beddel Persuaded Me Very hard to

Go with him I informed him that I had Made an agrement with Capt. Rathbun to go and Take New Providence as one Capt. Newton had Lately arrived from New Providence and Gave us an Acct of the ship Marry from Jamaca had put into New Providence in Distres and Capt. Newton said we could Gett thar before she would be Redey for sea and as this ship had kiled our sailing Master About six Month before of New York Mr. George Sinkens and Gave us a god Buning we ware now Determind to Take Fort Nassau and then we Could Have Command of the Town and Harber and take What we Pleased Capt. Rathbun was Present att the Corfea House att Charlestown Commadore & My Self went from Mr. Dorskes [Dorsius] than Continentel Agent to the Corfea House as I said If Capt. Rathbun would Consent I would go with him and All this Time telling Me that itt Was Presumiton to Make such an Atemt And Aluded to the scoth ship that was taken from Lieut. Macdougel & My self but finirly I told him tha ware all well Landed Att Virginia he Pled Likewise with Capt. Rathbun but Ansered no Perperse for saze Capt. Rathbun I have Made the Agrement with Capt. Trevett I will not Give itt up. So this Ended All this Conversation we shuk Hands and the Comadore says I am Very sorre for I never shall see you more, Capt. Rathbun & My self sett out Emeaditly for Georgetown in South Carolina so End this short Cruze.

Now we are Agoing on a New Cruse and a New Yeair Jan 1778 I have had A Long time to think of What I am A Going to undertake but I am Very well satisfied that we Are in a Good Cause & we are a fiting the Lords Battel And we are Getting under sail Runing Down from George town that Next Day put to sea standing to the Southward the Next Day att day Light saw a sail to the Eastward then saw tew More tha Pruved to Biritish one ship Brigg and sloop tha Gave Chase the ship Ganed farst on us by 2 PM the ship we Could Discaper her Teair of Guns Night Come on And Dark we Haled Down All sail & put our Lights out of sight and in a Few hours we Could see her and she pased and When she Gott out of our sight we Alterd our Corse and the Next Morning we Could see no sail Att all now we had hove over a Board so Much of Our Wood and started so Much Warter that we Concluded to Make All sail for Abaco we had A short Pasage to Abaco we Come to Anker and went to fixing A scailing Lader in Tew Days Arfter we stood over for New Providence we sent Down our Top Mast and Top sail yard and Howsed our Guns and New Providence in sight Kept all our Men out of sight and About 12 Aclock att Night we Got a Brest of The Harber and A Light are of Wind of the Land I had now picked out My Lams better I could never Wish All smart but one was Lame he says to Me I can not Run I Made him this Replia You are the Man I would Chuse we sune Embarked on Board fifteen Men first as our Barge would Take no More and them twelve besides My selve thus Made 28 All; we Carred Nothing with us to Eat or Drink but filled our Pockets with Carteredges we Landed About A Mile from the Fort we Got our scailing Lader and All in order and I Recolected When I was with Comadore Hopkins I Recolected one of the Peckets was out and I thout itt would be faverabel I Left our Men and Went My self I found the Pecket had never bene put in a New I went thrue Near the Ambezears I heard talking in the fort Emeadetly one of the sentres Came to the Corner of the Fort and Hallowed All is well tha Anserd att the Other End of the Fort All in well. The ship Lay Neair the foort all is well I Lay a few Minets as I Expected tha ware Walking Round I went

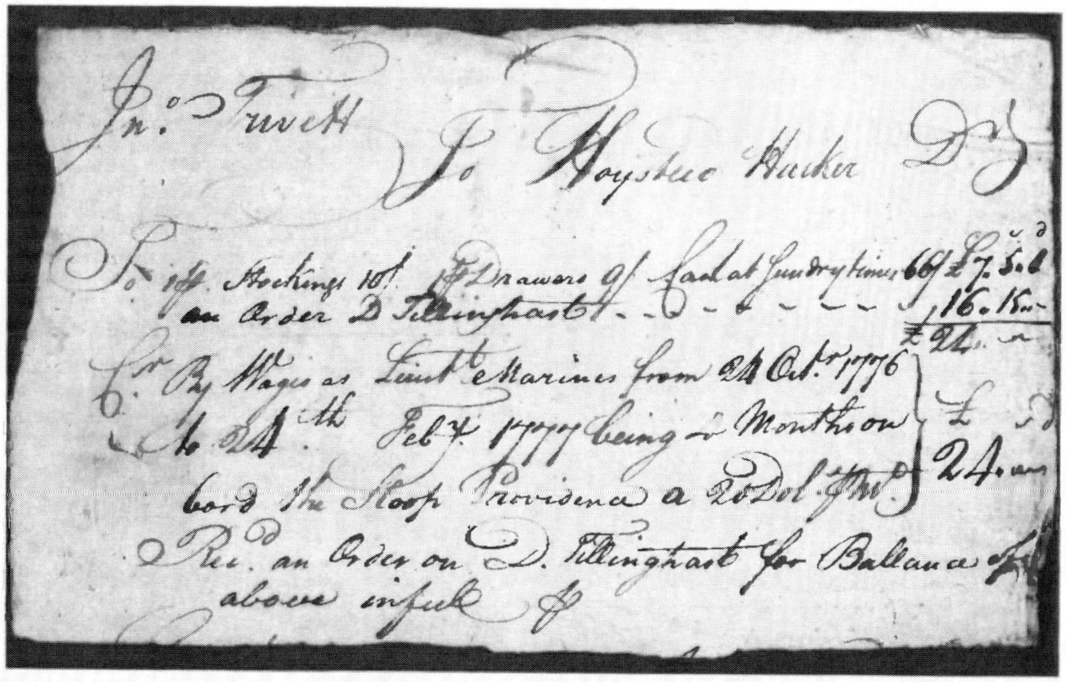
Jno. Trevett To Hoysteed Hacker Dr

To 1 pr Stockings 10/ 1 pr Drawers of Cash at Sundry times 66/ £7.5.6
an Order D Tillinghast 16.15.–
£24.–.–

Cr By Wages as Lieut Marines from 24 Octr 1776 to 24th Feby 1777 being 4 Months on bord the Sloop Providence a 20 Dol pr Mo } £24.–.–

Recd an Order on D. Tillinghast for Ballance of above in full

John Trevett's Account With Hoysteed Hacker, February 1777

back and we Came on with the scailing Lader and Lay Down Neair the fort untill tha should Come Round Agan for I Expected Every Half our tha would Cry out which we had bene but a short Time before tha Came Round and Cryd. All is well I Waited a Few Minets and then we Plased the scailing Lader neair one of the Ambazears I Went over Every Man followed I Gave Positive orders for no Man to fire a pistel as I new itt would ALarm the town As I was a Turning Round the Corner of the Barrucks I met one of the sentres full Butt I Took Hold of His Coler and ordered him Rite into the first Barruck Door; he was much frighted and said for God sake What have I Dune the Next Man to Me fired a Pistel over My shoulders att the Man I had Hold of butt Did no Damage I spoke a sorft Werd and put him into the Barruck and Examend him I found thare was only one more sentinel att the Other End of the foort I sune put him in a Nother Barruck and Examend them apart When I Gott into the fort I found severel of the Eighteen pounders Loaded and Matches Litt by them I Examend the Men we took in the foort and tha Each of them Told one storry I found About thre Tons of Pooder and tha informed Me that Governer Gambeair had sent into the Fort Every Arterkel Necesery About thre Weeks before we paid them these Veset; I Asked these priseners the Realson of only Tew sentres tha informed Me If tha had only time to fire a one or Tew of them Guns in Less than Ten Minets tha would have fife Hundred Men in the forte we Emeadetly orderd one of the priseners to see What Cartriges ware filled Took A Lanton and Proseded to the Magazine of Pooder and finding Not so Many as I thout Necessary sett Tew Men a filling Cartriges now itt was Tew Aclock in the Morning and we Kept up there Custom Cring out All is well Every Half our and the Jamaca Ship Anserd and a Nother Vesel up the Harber we Emploied the Remainder of the Night in Plasing some of the Cannon on the Different Streets and on Ship Day Light now Appeairs we sett our Therteen Stripes A Fliing; and you Must Remember we ware Landed Without A Singel Bisket of any thing to Eatt or Dring that Gave Me Now Consern as we Had Tules to Work with in the fort. The first Move I made was to send a Flag not to the Governer or Commandig officer but I sent itt to Mr. James Gould A Mercht. and a Gentlemen formerly a Native of Newport Rhodeisland he Emeadely Came Down to the Fort I had the scalen Leder Lanched out of one of the Ambezeairs he New Me Emeadely he Asked Me What our fleet Consisted of I made yuse of Commadore Beddels fleet I informed him the Fleet was of the Island Abbaco and I was sent into see some of My Old Firinds In a Tender and that we ware bound the North side of Jamaca only Hearing Capt. Henry Johnson was in Your haber & I Came to Waite on her and was Determend to take her of and My Orders from the Commadore was not to Molest or Disturb any Private Property onless war Like stores; Then I informed My frind Gould I had Tew Hundred Men and thirty officers and I had Plenty of Provision for the Men but I mus have Breckfast Gott for My thirty officers he Emeadely said itt should be Gott As sune as Posable he Emeadely Called on the Bakers and Stoped a Very Large Alowence of Bread & Down Came Butter Corfea and Every Arterkel we wanted and A Very Good Breckfast & Dubel Allowence Before our Breckfast Came I sent Mr. Mickel Molton fore Miles thru the Town to take the Other Fort with Only Tew Men I Gave Pertickeler Orders and I new thare ware only 2 sentres and Go give them no time to parle and to inform them that we had Posesion of Fort Nassau with Tew Hundred Men and thirty officers and Keep Posesion untill you herd from Me he succeded with out Any Difficulty the Next Afare or bezness was to Gett Posesion of the ship she Mounted sixteen Carrege Guns & was All most Redy for See. She Lay withen Pistel Shot of the fort I sent A Long side Aboat I Took from one of the Wharfs With fore Men and a Midship Man to Bring to the foort All the boarding pikes and Muskets Pistels &c the Capt. Was sick on shore and the Second in Command Refused to Lett them Come on board I was All Redey to Settel the Affair with them I Haled her with hard Langege & some hard Names and tha Admited them on board the Midship man staid on Board & sent thare Crew and Arms into the fort than Our Breckfast being Redy we sett Down and Made A Very Harte Breckfast the Town in Grate Confusion and we Could see fife or 6 Legs att see and Could not see nothing of our Vesell Arfter Breckfast was Over Mr. Gould Came Down and asked Me Wither we had sufficent I informed him we had I then Asked Mr. Gould If thare was Any Turtel in the Crouls he said thare was I said I should like A Turtel Cucked he said itt should be Don Accordingly tha Lades Orderd thar servents to Work and we had a Grand Turtel Dinner that Day About 2 PM we saw a sail standing for the Harber and a Nother sail A Runing Down from Harber Island one was the sloop Providence and the other was Capt. Chambers in a Privateer of sixteen Guns fitted out att Jamaca the sloop Providence Gott safe into the Harber About Tew Hours before Capt. Chambers Gott Neair the Barr I Had information of Capt. Chambers being Looked for I informed Capt. Rathbun New the Privater was we Concluded to Hall Down All our Cullers I Likewise Halled ours flagg Down att the fort but the Devel was in the Men Wemen and Children for the Hills being Covered Men a waven Hatts wemen the Aprons and she Runnin in for the Bar and att Last tha Gott About out side of the Harber and she Roing hard for her and Waving thare Hatts she bore A Way and Going out side I Run our flagg up and Gave her thre Eighteen Pound shot one of the 18 pound shot went into her Main Beame And never hert a Man

She went Round the Island an Came in A Nother way and Came to Anker in sight of us—I Fore saw I Could Give Mr. Molton Now Assistance att the other fort as itt was fore Miles from us I sent a Mesage to him to spike all the Guns & brake All the Ramers & Spunges and heve All the Powder in the sea so as to Discomode them All he Could and Return Emeadetly to our fort Which he did Compleat; and Arived with All His forse A Lettel before Dark his fors was Tew Men and himself; You Must Remember before our sloop Got to our Asistance All this was Dun And A Good Diner of Turtel Reserved for them and then our sloops Crew ware Quality that Day for the Dined att 7 Aclock in the Evening Arfter Dining I may say super; Capt. Rathbun had the sloop Morred Abrest of the Town & springs on her Cabel and Likewise the ship so that we ware All in good Order; Arfter Mr. Molton had Distroid All the Amanition and spiked the Guns att the Other fort Capt. Chambers Came by that fort and Ankerd About Gun Short from us and thare he Now Lays Now itt 9 Aclock in the Evening and as for our parts we are Very Eazea but the inhabitens in Grate Consternation Moving thare Efects out of the Town and the Culerd people Very Riatus Are I May say Culerd people for tha are of All Cullers from Cole Black to White. So Ends this first 24 Hours—

This 24 Hours begans with Pleasent wether All Hands on board The Vesels bezea in Getting the ship Marry formely Commanded by Capt. Henry Johnson Redey for sea and Lixwise A Brigg with A Large Quanity of Indego and other Valueble Goods on board this Brigg Mr. Danel Beairs or therd Lieut is to take the Command; The Town for New Providence in Grate Confusion Many of them Moving back thare Valueble Goods Grate Numers of All Cullers under Arms Round the Goveners House & All the Hills Round is a Live with thare Guns &c Glesining itt Gives Me but A Very Lettel unezeness for I have now in the foort Twenty fife Men with My self and All in Good spirits as Many of them is Very Ugent for Me to fire on What tha Cal Black Berds; but I think I know What Ground I stand on and If I Can do without I am Determend in My one Mind not to shed Blude—this Morning Came within Half a Pistel shot of the fort some of thare Culerd Trupes one of them Hollowing to some Neair att Hand seing me on the Walls of the fort says thare is that Dam Buckerer Come Again that Carred away Governer Brown I paid no Attention to no such small Affairs—Sune Affter the Colecter Mr. Hunt and Tew Gentlemen Came Down Neair the fort I purseved tha would wich to speak to Me I orderd our scailing Lader to be Lanched over the Walls of the fort and went out to Meait them I found that Governer [John] Gambeair [Gambier] had sent Down Mr. Hunt Colecter of the Custum House and these Tew Gentlemen to Enquir of Me What our Intenshons Ware I informed him that Commadore Beddel had Gave orders to Capt. Rathburn to take fort Nassau and All Armed Vesels and all Amerian Properte we Could find in the Harber of New Providence and Likewise to hold sacred All Privet Property Which I meain to Comply witt and the Governer May Rest Ashured that He may give himself and his friends My Werd an oner as an American that His person & All the inhabitens I shall Protect but that I understood that Capt. Chambers Agent had the sailes and some small Regin in his store neair his House and I wanted them sent Down under the fort—and Likewise I hear the Agent Commands a parte of these Men that are under Arms on the hiles—I wated About one Hour and herd Nothing more from the Governer—then I informed Mr. Michel Molton that I would have the scailing Lader Lanshed of the Walls of the fort and Go up Amongst the Rabel as I have informd you thare was All Cullers and tha ware All in sight of the fort and if he saw any Confusion Amungs them to not mind Me but to give them Tew or thre Eighteen pounders as we had them Redy Loaded with Grave and Langrege shot I Gave My self one hour to go to them we Lanched the Lader and I took My Hanger in My Hand as a Walking starf and then seet out I had Walked but a Few steps before I came Across Collecter Hunt he Asked Me Whare Bound I informed him he says If itt is a Greabel to You I will go with you I anserd itt was Very Agreabel he says itt is All in Your Way to stop Att the Agents House first I informed him What orders I had Left att the fort and I had but a short time to spend he Anserd itt would take but a Few Minets Accordingly we sune Went to the Agents House and the Lady sune Made her Apperence and semed not very well Pleased I Asked her Whare her Husband was she sune Informed Me he was on the Hills and had not bene att Home since we had taken the fort Mr. Hunt Asked Me to sett Down and handed Me a Chair I said I had no time to sett the Lady Did not seme to be Pleased I informed her I was Agoing to see him and would or May Come home When he Pleased for I would Protect him and his Property Mr. Hunt says to her Whare is Your Bottels of Wine A servt sune Brought them she Did not Take any Mr. Hunt & My self Took a Glass Each and then Proseded up the Heall I sune saw this Gentlemen I informed him What I came for he Redely Complied and Dezired Mr Hunt to Walk with Me and Dezire His Wife to send All the sailes & Reging Down by her servents which was Dun Now Every Arterkel that was Nesessery for the Tew American Scuners was Gott them I then informed the Capt. tha had but a Few hours to gett Redy as our ship & Brigg would be Redy by the Next Night and tha most Luse no time and if that Wanted any Arterkel that was to be had att New Providence My Cr. was Good as Long as I had the Command of the fort. I should not have Made them that offer If itt had not bene for a Refege Vesel Lay thare from George with a Valuabel Cargo of Goods & Provishens and we had put A Man on Board to Take Charge of Her and the Man or Men that oned her Asked Me that day What Capt. Rathbun was A Going to Do with Her I anserd him I Could not say he then informed Me and shew Me that the Onerabel Congress had Given Liberte to All that ware A Mind to Leve the united States that tha may Go I new All this; he Emeadely offered me Any servis that Lay in his Poar att the same time Informd Me he had Plenty of Provison If I wanted I should be Wellcome I Gave for Anser we had Riche oners and we Wanted for Noting we Did not All tho we Kept A Man on board we Did not Entend bringing her of. This 24 Hours Ends Pesabel—and Pesabel—

Now Day Light Appears and A Plesent Morning; Capt. Rathbun Came Ashore and Came to the fort and Informs Me tha well be All Redey for sea the next Morning and that we should Want thre Pilots one for the ship one for the Brigg and one for the sloop and the Scuners would be in Redeness to follow the ship all as itt is a Bard Harber we concluded on one Way to Procure the Pilots and Expected itt would Make a Bone of Contenshon betwext the Whig & Torre parte that was I have About 25 Casks of Rice beloning to Capt Chambers & his Crew in the fort have them Turned out under our Muskets and have A Vandue I have a man in My Eye I will Gett to be thare and to Give Me a hint Hew the Pilots are; You have Your Barge Mand and Armd on the operset side from the shore so as the Inhabitents should not Discover Any of our Proseding And I will have All but Tew Casks Rold out them I want to Give to the Tew Men that we have Priseners in the fort as tha have famelys and are Poor; Capt. Rathbun Agreed to Al this; the Next thing was When I Had My Vendue itt was to be Erly Next Morning and When I had Gott My Pilots Marked out I was to take My Hatt of & screch My head that was the signel, tha Next Morning Twenty thre Casks ware Rold out before Sun Rise A short time After sun Rise I had our Drumer with his Drum I orderd him to beat with Me as I was A Going to Market Square Whare most of the Gentlemen Resorted and A grate Number Gott Round Me to heair What I had to say I informd them I was well Convensed of Breed Kind being Very scase and I had a number of Casks of Rice Rold out from the fort that I would sell them very Cheap and if tha would not Purshles them I would have them started in the sea & I should have My Vandue att 8 Aclock I then Returned with my Drumer and sent him into the fort itt was not Long before A number of Men and Boys Came Round I had thre Casks headed up and one head of Each knoked in them I Adresed My self to the Children to Run and Gett some Baggs or Baskets and I would Give them What Rice tha wanted I told them to tell thare Mothers itt was Banyon Day and tha must have Rise Puden and Butter; I Kept Laughing and Talking with them the same as if tha had bene Pertickeler Aquainted thare was Boys & Black Girls Came Plenty of them tha sune went of with the thre Casks Rise; All this time my frind was a fixing on the Best Pilots I sune Gott the Marks & Numbers of them I gos up to one of them In some Movement I says What Price will itt best to fix I intend tha shall go sune and Vire Cheap I Pules of My Hatt and began to scrash as if I had some Crepers thare I Luked a one side now our Barge was Making for the shore as sune as she struck the officer steped out I Pecked out My thre Pilots and informed them and Gave them My Wird an oner that as sune as tha Carred the Vesels over the Bar I would have them Landed; one and All Declared tha ware not Pilots I order them Emeadelly in the Barge and Opened a Book of Hard Names upon one and All the Pilots tha Carred All over safe I must Remind You the Arfternune before we sailed I Dined with My Frind Gould I had not sett Down att Tabel Long before one of the Lades A-Loocking out of the Winder saw a Man Runing for thare House I steped up and Mett him in the Peazer He Informed Me Capt Chamber was a Landing His Men I sent him Back to the Fort and sune followed him I had spys All Round I came athort one sune Arfter he informed Me tha had Gott some Peses of Cannon on the Goveners Hill and that Night tha ware Determened to Attack us I told My Good Frinds tha would have a hard time of itt If tha Did for I should not mind fore Times thare force I new All this stoff I said; would sune be herd by them this Capt. Chambers I was Entemetly Acquainted with Him in the Bay of Hondorus not Long Arfter this the sun being About 2 Hours Hie frind Gould Came Neair the fort and his Family Going over to Hog island I went out of the fort And Smilingly said I Hope You are not Afraid No I am not Affraid but I Do not Expect to see You any More for tha have All the Privaters Crew att the Goveners House and Making Every Preparation to Attack you to Night; for tha have Discoverd Your Strength; and If I thout as he Expresed himself that You had nothing Butt that Damd Bumboat I would fight Against You My self I smiled and told him itt Gave Me no Concern but if he Came over to Morrow Morning he would find Me in the fort we parted and he went to Hogisland but You Must Remember I never invited no one to Come into the fort but I Kept them out untill I was Redey to Leve the Place Arfter My Frind Gould Left Me I Returned and went into the fort I Asked one Jack Scranton a Rhodisland Young Man that was the Spryes Mand I had into the fort wither he Could shin itt up to the head of the Top Mast &c spoke A hard Werd and said Yes then I told him to Gett a hamer and Nailes Redey untill I prepared A Flagg and sent one Capt Tower of New London with itt I Did not Right but a few Lines to the Governer informing him that the Men ware Very Numerous on the Hills and Made An apperence Horstile and unfrindly and If tha should not Dispurse in 15 Minets arfter the Flag Returns to the fort I shall begin to Cananade the Town and he May Rely on itt that I will Give no Quarter nor take none att this time of his Reciving this Bellit Jack Scranton had Gott up the Top Mast head and was Nailing our flagg All in sight and With in hearing the Flag sun Returnd from the Governer and in Less then Twenty Minets scarsly a Man to be seen on the hills Att sunsett and the Evening Vere Still and Quait this Night All Hands Emploied Getting All the Poder and Cartreges out of the fort onboard our sloop and then began to spike our Guns & Breeck all the Ladels spunges and Distroing Every Warlicke utensil by Day Light we ware All Redy to Embark I must inform You that this is the scond time of My being att the Taken of New Providence and that thare never was a Dollar Taken from Any of the Inhabitens but this one Arterkel that is to say Mr. Dungcan Mercht att New Providence had fore Hundred Waite of Powder As I was Informd by one of My Frinds I sent a Bellet to him and Demand itt and that was All the Properte I Ever herd of Being Taken from them—About ten Aclock this Day Al our Peopl was out of the fort only My self and that was no more then the Boat Could Take of I had tha Tew Prisoners in the fort Confind so as tha should Give no Information untill we Got of to the sloop Providence we took our Scailling Leader with us & Toed itt part of the way and then Lett itt Go A Drift All our Vesels over the Bar but our sloop and she had her Mansel up and our Anker short a peke for Tew hours before I Could Gett al in Redeness About ten Minets before I Gott into the Barge A Gentlemen Came with A Mesage to Me from Capt. Chambers that Comanded the British Privatear as he was Neair Att Hand Att Mrss Bunches Tarvern with an invitaiton to Drink one Glass of Punch before I went onboard Att the same time swaring that no one should Hert Me and Profesed A Great *Del of Frendship;* not old Anuf or a Yanke Yett I sent by this Gentlmen Like Man an Anser to Capt. Chambers If he would Come over the Bar and take the sloop Providence then I would take some Punch with Him—but I new I had left a bone of Contention betexet the Whigs & Tore so Caled for the 20 ord Casks Rice I Turnd out to Trap My Pilots arfter I had them that was All the Vandue I wantd; then I Made out Bills some one Cask and some Tew and sold itt without Money or Price to the Whig Parte and made them take them away before I Left the fort I anserd Every Point I Could wish for the Very Time we Got under way to Come over the Bar the Privatear Did the same and Ancored Whare we Left and went on shore tha Goot All by the Ears a fiting with Each other and Confined Mr. Gould and some others we are over the Bar now and have Anserd Every Purpose I could wish no Blood being Split the ship Brigg & Tew scuners a Legea head standing for Abaco we Released About thirty Americans that was thare Priseners Abaco or the Hole in the Rock in sight, Arfter I had Dined this Day Capt. Rathburn and Lieut Vezea Dezired me to Lay Down to Gett a Littel Rest as not one of our Men nor My self Had any sleep from the time we Landed untill now for When we Relived gards we took them from ond part of the fort and put them on duty att A Nother so as to Deceve the inhabitens and Make them think we had a grate Number of Men in the fort tha Most men I Ever had in the fort from the first to the Last was Twenty Eight besides My self—Now itt is Allmost sun sett Capt. Rathbun and tha Liut. thinking I was a sleep was in the Caben & talking with Each other I all this time was a Lising to hear what tha had to say I found in a short time tha ware ordering a Signull to be Hove out for our ship to Bare Down to us and to take Liut. George House out of Her and put Me in his Plase I then

Turn out of My Caben and Interferd in this and said I had bene Acquainted with Lieut. since the first of the War and A better officer tha was not in the service Lieut House ware sune on board the sloop And sided in with Capt. Rathbun and our other Liut. so I finelrly went on board and I had one half of our sloops Crew Amounting to thirty Men I had not Got on Board the ship for we saw a ship Bearing Down on us I Got on Board the ship and then we parted the ship standing for the sloop we Allterd our Course in the ship and the Next Morning we Gott by Abaco and see no More of ship Nor sloop and now we Making the Best of our way for Block Island I had the Log book Kept in the same Maner as If Capt. Henry Johnson had the Comand bound for New York this I had Dune in Case I had Come Acrors any British Cruser and I Did not intend to Run out of My Way from Any ship! We saw a Number of sails on our Pasage but never spoke with Any the first Land we Made was the Vinyard we are now Runing Down for Nantucket itt is now 18 Days since we Parted with the sloop Providence and we have had Very Cold Wither and hard Gailes out of thirty Men thare is not more then ten but Whos has thare Hands & feet froze one Man froze to Death his Name was James Dark. He informed Me When I took him att New Providence that he had bene taken by a British Privatear and that He belonged to Vergenea this Day we Ankerd under Nantucket the Wind still blowing Very heard so that we Can not Gett on shore we firing segnul guns the Next Day the Wind Continus Blowing so no boat Can Come on Board I thout itt Nesesere to bure James Dark and we Did itt in a Desent Maner the therd Day severel Boats from the shore Came on board being Very short of Provison sent on shore and Purchesed fife sheep so we ware well Provided with fresh Peck; the Wind Abates now A sail in sight Runing Down the south side of Nantucket; the Nantucket Men Not Less than Twenty on board our ship she stood Down for our ship these Men Agred one and All that itt was the Harlem Privatear from New York Now we Got under Way I Made the Nantucket Men an offer to sheair All Equal A Like for I shoud take her; out of the 20 I thing thare was Tew stood by us we stood for Each Other and before we Got up with her we Discoverd her to be the sloop Providence that we had parted with of Abaco As we Pased her we ware Rejoced to see them we had not time to Draw our shot and we Gave them A salute and tha Returnd in the same Maner we hove About and followed the sloop in; the Wind Modorrates Run and under Nantucket and we Came to Anker as the Wind would not Admit of Going over the Shoules this is the 20 of Feba. 1778 A Light are of Wind att N. Eeast and Lukes Like A storm Runing by the Round Shoule the snow begins to fly Quick & the blow Encreses; now a snow storm and so thik we can not see but a short Distanc Runing for Cape Poge About 12 Aclock our ship struck very hard on a shole Neair the Hors shue we had Very hard time A Considrabel of a sea we Kept. All sail on her until we Got Acrost the shole and Depened our warter and then we Lett go our Anker but Before we Goot of this shoul we Lost our Ruder and stove our Boat on Deck the Next Morning we found the sloop Providence Gott on shore Going into Old Town but sune Got of Again the Next Day by the Asistance of Boats we Got safe into Old Town thare we Discharged our Cargo Consisting of Rum Molases Sugers Corfea some Indego & Cotton a Valuabel Cargo some part of our Cargo sent Down to the Hi anners [Hyannis] some to Boston but the Most sent to Bedford so Ends this Cruse

Now is March 1778 we are Arrived with the sloop Providence at Bedford the ship we Left att Old Town Reparing and Getting a New Ruder then we are A going to Bring her to Bedford—Capt. Rathbun and My self sett out for Boston to Call on the Board of Warr [Navy Board of the Eastern Department] setting thare we spent part of Tew Days and Returnd to Tarnton [Taunton] Capt. Rathbun sett out for his Home and Left Me Att Tarnton to Go to Plimouth with Lawyer Paine of Tarnton to the Triall of the ship we had the Triall and itt Did not sute the Capters as she was a Commishon ship we Appeailed to a Hier Pope so that I Returnd to Tarnton with Lawyer Paine; Letters Rote to one Mr. Lewis Att Congress from Mr. Paine and a Number of Papers Lawyer Paine furnished Me with to send on to Congress then sett out for York Town [York, Pennsylvania] and in June I went on My Jurnea and I never Experenced so Hott Wether in America in My Life I Cripeled one Hors by the time I Arrived att A plase Called Crumpond Whare Col. Green and Majr. Flagg Lost thare Lives by the Refeges from New York Nothing Material Hapining till I Gott wel on to a Town Called Monmouth the Day before I Arrived thare fell in with A Grand Trupe of Light Hors Going Threw a Large an Loong Wood I was a Lone; I Drew My self a Lettel one side to see them parse Me and I was of a Pinyon that thare ware British Light Horse until I Rode fore Miles on My way and stoped to an inn and thare tha Informed Me that tha Ware Generul Washingtons Light Horse In thing I never wished to see Abutefule sight and so Grate a Number I had on a uniform Coat and a Coxade in My Hatt I thout tha ware no End to them tha pased Me on Acanter a Compliment by Hand not one Werd was said by officers nor Men I then Proseded on My Jurne the Next Morning Early I herd a Numer of smal Arms Discarged I sune Lernt that the British Trupes had Left Philadelphia on thare Way to New York and the Americans had Come up with A part of them this was as Hot Wether I think as Ever I felt in the West indes the Next Day I parsed thru Whare I Came a thart some Dead Bodeys that was not barred I was informed the Day before that a Number Died with the Heait A Dringing Warter out of the Broocks Nothing Material Hapening untill I Arrived att York Town, I saw Governer [John] Collins & Wm. Ellery Esqr. was our Tew Dillegates att the Time from the state of Rhodisland in Tew or thre Days Arfter I Arived att York Town Congress Adjirned to Philadulphia, I then I sett out for Philadulphia Arfter I Arrived there I Called on Mr. Wm. Lewis An Attorney and Employed him to Do My Buzness I Left Henry Johnsons Comisson & the Log Book of the ship Mary with Mr. Lewis and so much Money What he Agread to Do the Buzness for, Then I Made the Best of My Way for Providence in the state of Rhodisland I Arrived att Providence in July When I Arrived att Providence I herd of Bedford being Burnt I sett out for Tarnton & Bedford I found A Grate part of our Cargo Bunt in the stores att Bedford and our ship Mary had got Feted att Old Town Gott as far As Woodes Hole When the British burnt Bedford and Farehaven tha Burnt our ship att that time att Woodsis Hole so this Finished My Jurne to York Town—

This is now July 1778 [1779] Now thare is an Expedition fiting out att Boston for PerNobscut and our sloop [*Providence*] Praparing to Join the Fleet now I have some buzness to Settel and I have no Enclenaton to go to Pernobscot as I think the British well Gett information Either att New York or Newport before tha Can Gett Redy to sail and If tha Due I now thre or fore Large British ships can Block them in and that will be the Lors of All our shiping; and as I have since the first of the War had no time to settel My own besness I will take this Oppertunity and I sett out for East Grenwitch

Whare My Farther then Lived I had not bene att My farthers but A short time before itt so hapened Early in A Morning Very fogge Morning I herd the firing of Musketry I Emmeadtly Looked out of the Winder and saw severel of Col. Greens Black Regment Making the Best of thare Way for Portewume Directly Came by Lieut. Snow with A small fild pese I went back of the House My Farther Lived in Whare was a small fort and put some Musket Cartreges in My Pocket and then sett out for the shore the fogg Breaking Away I saw the Remainder of Capt. John Alens House Burning att the same time saw the Refege Fleet Comanded by one Leonard; one of his Tenders Aground Neair the shore Lieut Snow Gott His field Pese Abrest of her and with the Help of some Muskets; we Kept the Men from Getting her of and firing att the Remainder of the Fleet untill tha bore Away and Run for Newport we Goot the Tender of and carried her into Grenwitch and found the Refeges had not forgott Thare theving buzness for thare was Plunder that tha had stole from Capt. Allen some Life Calves and Other Afaires—All that I Did was Priming the field Pese with the Carteriges I had in My Pocket and I had no Expectaton neither Did I want any thing When the Prize was sold Col Green orderd Twenty fife Dollars for My sheair—itt was Plesing [to me] for itt was not more then fore or fife Hours that I was gone from Hone; that I Made Mone in fore or fife Hours att this time then I Did With Comadore Hopkins in our Long Cruse of Take in New Providence the first time. So Ends this frakus—

Now is August. 1778 [1779]

Now thare is a plan Laied with Mr. Johathun Waldron and Cap Isack Freborn Capt. James Phillups and Others to Build a Brigg of 14 Guns as we have Got one a Mafiridite Brigg Called the Black snake of 10 Guns Agred to and I was to go to Benington to Procure floor & Tallow for the New Brigg in a short time I took My Departure the first Place was to be Lanesberrow to James Barker Esq. I sent on Tew Waggons with Goods to Williamstown Whare I was to have the Goods Left. I sune sett out and Arfter a short Jernea I Goot to Esqr. Barkers I had not bene Long thare before a Gentlemen Arrived from Boston and Informed us that all our Fleet was Laiing Att Nantasket Road Redy for see; this Gentlemen saies we shall sune have Penobscutt. I was Allways unezea from the first fixing out this fleet for feair of information should be Gott by the British and tha would Block our Fleet in so this Conversation Ended the Next Day I sett out for Benington in the state of Vermont I put up Att Generull Fays I sune fell in with Mr. Tetchener att that time he was Comesarre General of the American Arme of this Gentlemen I Gott My supply of Whatt I was in Want of; now My Tew Waggons had Arrived.

I Goot My Tew Waggons Loaded and sent them on for Providence with such Goods as I wanted—now I have some spair Time I am now Going a small jarne towards What is Caled Bumers [Bemis] Higths Whare Generul Bergoine was taken and I was att the Redouts Whare Generul Benadick Arnule was Wounded I was Likewise on the spot and threw All the Plases from benington Whare All the Scermishash took Plase as I had in Company Tew officers that was in the Benington fight. I had An Agreabel time I then Returnd to James Barkers Esqr. now itt is Jana 1779 [1780]—

Now is Febuare I am a setteling All My Affairs and preparing My self to go for Stonningtown in the state of Coneticut as I have Recevd a Letter from Capt. Isaack Frebon Whare in he Rites Me the New Brigg Ratelsnake will be Redey for see in the Month of March—Aboutt the 10 of March I Left Lansberow for Stoningtown Arrived in a short time sent My Hors on Fishes Island and then I went by Water to Newport. when I Arrived att Newport I found the Brigg Black Snake Commanded by James Phillups Esqr. & Brigg Rattel Snake Commanded by Isack Freborn Esqr. the New Brigg Mounting forteen Carege Guns All most Redey we sune was Redey sailed the beginin of April in Compane Bound of Sandy Hoock we had not bene out but fore Days had Very fogge Wether in the Morning of the 5 the fogg Clearing we saw Tew British ships to Windard of us Not More then Musket shot the Wind being att South East blowing in for the Hook We bore Away before the Wind Every shott tha fired over us a Long Distance we on Board the Rattel snake Run to the southward of the Hoock Capt. Phillups Halled to the Northward. the ship Gullite [*Galatea*] in Chase of us and the ship Delight in Chase of Phillups. Capt. Phillups was Run on shore Near the West End of Long island and he and all his Crew ware taken Priseners by the English our Brigg was Run on the Jerzey shore of Shrwsbery [Shrewsbury] About fore Legs to the Southward of the Hook we as sune as we struck the shore we cut Away our fore Mast and itt sune went over our bow and we Hove over Aboard All our stores that we Could Come att the ship Galatea Runing in Neair us and Kept up A Contenual Fire as she Ankerd within Musket shot of us we Gott our boat out and put in some Muskets & Pistels Capt. Freborn and Doct John Turner and as Many as Could Convenently Go in her went but as sune as she Drew neair the shore she upset and All the Muskets & Pistels ware Lost a Number of our Men Endevering to swim on shore our Guner Mr. Hathaway of Fretown was Drowned severel other was Halled on shore supposed to be Drowned but all Came to them selves now tha sent Tew of thare boats to Come on Board The wind Blowing Very Hevea Att South East and a Hevea Sea hove in and Brock All over us I went on the Round House and tha Lay on thare ours I spoke with them, tha ware not more then 30 yds. from us I informd the officer that if tha Attemted to Come ALong side thar Boats would fill; and I put them in away that suted them I informed the officer that if tha Gott out thare Large boat and Drop a Kedge of by that mens tha may Gett a Long side tha Emeadely went on board the ship to gett out thare Large boat the Tide Going out faverd us Every Minet More & More to Make our Escape for att this time we had upwards of Sixty Men on board in About one hour I saw them a Coming with thre boats I then Called on Mr. John Rider our sailing master and informed him I was Determined to get on shore If Posabel and Advised one and All to Do the same I Emeadetly Called on Capt. Freborn and the Doct and the Men on shore and Dezired them to give us Whatt Asistance tha Could as I found there was A Dredfull underto—I Emeadely went forward and Took Hold the Jeb Down Hall and Give My self such A swing I fell on My feet to neair the shore I never Wett My self but very Triffeling tha one and All followed Me and att the same time the ship made a Constant fire for some time but never Wounded nor hert A Man itt was but a short time before all our men was on shore but the Steward he being Lame Could not Make his Escape we Could Gett no Asistance from the Amerians; itt was so neair the British Line sune Arfter the boats wont on board made but a short stay before we saw Defferent smokes and saw that tha had her on fire in Defferent Plases and in a short time Arfter the fire Gott to the Magazene of Powder and A Very small part of the Brigg Could be seen arfter this happened; and As for the Young Man our Steward we never herd no More from—one Consolaition we had tha had not A Gradel of Plunder for we Hove Every thing we Could Come att of Any Consequence over a board then itt being Late in the Day we went back into the Jersey Cuntre the Most of our officers and Crew went on to Philadulphia Doct John Turner and His Kinsman and My self sett out for Newport whare we Arrived in one Day short of Tew Weeks from the Time we sailed—I think may properly be said we ware Turned on shore with an ouer on our Backs—A Poor Mizerabel Cruse—

Now I am All Redey for a Nother Cruse This is now the beganing of May 1780. Now I sett out for New London and as I Arrived thare I found the ship Trumbl [*Trumbull*] under sail the wind Very Light thare was in company with Me thre Young Men we Agreed to Go out as Voluntears the Gratest Indusement I had was in Hopes to fall in with the ship Galatea as to be Revinged on them The Wind Brezed and we went to see and Run out by Block Island and steard to the Southward we Run of Bermuda tha first Eneme we saw was A small Privatear fitted att New York we Histed English Cullers and she Run Close under our Lee we sune manned her and Orderd her to Philadulphia Nothing Material Hapening untill the 1 Day of June we saw a Large ship Barring Down on us we sune Discoverd her to be an Armed ship we Run from her About Tew Hours untill we Got All Redey for Action she firing her Bow Guns we toock no Notes of her all this time but Kept English Cullers att our Mizen Peak and Arfter our Commadore Gott All Redy we Hove our Main Top sail to the Mast And she sune Came ALong side we then Halled Down our English Cullers and Histed the American flagg and then we began the Contest itt Lasted About Tew Hours in that time we Lost forty thre Men Killed and Wounded with them was Tew Lieut. Mr. Bill was shot Dead and Mr. Star Lieut. Died with His Wounds in a Few Days Mr. Star Midshipman went over Aboard with Tew More Men When our Mizenmast Was shot Away we had not one boat but was shot to Peses in such A maner that we Could Give them No Asistance Altho the sea was Very smuth we herd them some time arfter Dark beging to us for Relive but itt was not in our Power as we had only our fore Mast and About Half of our fore top Left and that in A shaterd situation and When the ship went of and Left us and att that time we Gave them thre shot to thare Tew and we Drove thre ports in one for them we new we Must Have Killed Tew to one we Gott our Wounded Dressed that Night The Next Morning we began to Cleair our Deck and fixing Jure Mast and the 17 June we Arrived Att Boston we saw A Number of Vesels but none of them would Come Neair us our ship was hardly wors Repairing she was shot to peses so much Arfter I Arrived att Boston A wound I had in My foot made me so Lame that I was Obliged to Hire a Hors to Carre Me to East Grewitch and All I Gott by this Cruse was to pay My one Expences for the Privateer we took was of no Grate Consequence now have a Lettel time to Rest, but My face and Eyes Which was blown sune Gott the Better but My foot Very Lame—I Arrived att Grenwitch 23. of June so Ends this unfortunitt Cruse

Commadore James Nicholson Commanded and I Called him a Brave officer—

Now is towards the Last of June I have bene home but A short time My face and Eyes is well in a Maner; but My foot is Very Lame I have Jest Recevd an Envitaiton to go with an old Acquaintence Capt. Elisha Hinman of New London he has a Good ship Called the Deain [Deane] Mounting thirty Tew Guns and I think I can Gett well on board as well as on shore—and I am a fitting My self in My own Mind to go—well I have made Tew Bad Cruses I will Try Again I have salted Down thirty thousand Continentell Dollors and I have some Contintell and some Rhodisland state securites besides Continentell A nuf More then My farthers family will want. The most of Money I have salted is Dated the 10. of May 1775 and I can not think but itt will be Made Good—

Now the ship is Arrived att Newport and to sail Emeadely for the Vinyard for to Gett More Men the Second of July sailed Arrived att Old Town the same Day oppened a Randevose Att Old Town Lay thare fife Day and Gott Twenty fife Men the Most of them Copper Culerd and then Capt. Thomas Tripp of Newport Piloted us over the shouls and he Left us and we sailed to the Eastward Nothing Material Happening untill we Gott off the Banks of Newfound Land the first Prize we took was a Brigg from Ganzea bound to Hallefax Her Cargo Jenevea & Brany Carstel soap and other Goods we orderd her for Boston the Next was a Letter of Mark ship Called the Elizabeth from the strates she Mounted sixteen Guns & When we took her she had sixty Men, we had taken All her Men out butt 16 When Tew ships Bore Down before we had time to take out any More of thare Men so that Capt. Hinman told the prize Master Mr. Ebberns and My self incase tha should be Armed ships for us to Make the Best of our way for Boston itt was All most Dark and we Lay under the Lee of our ship When the head Most Vesel Gave Abroad side and had a Return we Emeadetly Made the Best of our way for Boston the Next Morning had a Man att Mast head but Nothing to be seen tha Wind Blowing Heve att N. N. West A Large ship passing us we shew No Cullers Neither Did the ship so she pased of Nothing Happening untill the therd when att sun Rise saw A Large ship under our Lee seeing her Mast was a Grate Distance Apart she Appeard to sail farst And in one hour we Discoverd her to be an Armed ship and we ware senseasabel itt was not our ship we under Close Refe Top sailes we sune out Reefs and Carred All sail we Could pack; About 8 A M we Concluded she was an Eneme we then Concluded to have A Large stew Poie for All Hands as we had Plenty of fouls on board we sune Got itt underway before itt was Dune she Histed English Cullers and Gave us a shot and Repeated and Kept . . . [Illegible] Every short time Kept Firing but she to Leward About one Aclock P M she Neaird us so that Every shot went A Long way over us she Went to Heve in stays And Mist stays a Very Heave see Agoing that Hapend in our faver as we had more time to Dine in we sune satt Downe and Dined Very Harte we had jest time to Cleair all away When she Got Neair to us and Hailed us an orderd us to Heave our Main Top sail to the Mast and orderd our Cullers Down we complied and thare Barge was sune on board and we as sune had a New ship she was Called the New Marcary of thirty six Guns Commanded by Isaack Prescot he was Brother to Generel Prescut as sune as we ware on board Capt. Prescut sune began to Examen us and was not Very Pleasent oing Partely to his firing upwards of fifty shot before we Haled our Cullers Down he was determend as he said to Me If we had fired one shot att His ship he would have sunk us and would not take up a Man I Made him for Anser he would not Gained Much by itt for we had 15 of thare Men to 13 of our Men he Examened several of our Men and tha said tha ware Obliged to go for tha had to Ether Besness to Due and most of our Men told this same story all these Men he put Emeadetly to Due Deuty on Board the ship and than he began to Examen Me I Asked him if He Ment I should tell him the Truth his anser to Me Yes Young Man I Expect you will tell me the Truth The first Question to Me was How Long have You bene fitting Aganst the King; If Posabel sir before the War Commensed Well Young Man says Capt. Prescot suppose I was to Come A Long side of a French friget would you not Assest in fiting Her; No sir for I Luck on the french to be our Good Allies and Gratetude would Oblige me to Assist the French as Much as If tha ware Americans I was Asked some few more questions and he Left Me and Examened Tew More and tha Geve him Much the same Anser he Capt. Prescut Appeared to be Very Angre; Then Came the British Priseners one of them was a Doctor and the Other Was a Prize Master Capt. Prescut Asked them Turning to us Thre Americans How says Capt. Prescut is these fellows Treated You While A Prisener the Doct Replied tha ware Treated well Capt. Prescut Turns Round to us & says Very well for you Young Men I Could not Help Ansering Capt. Prescut and said sir since the war has Commenced I have bene taken severel Hundred Prisener and If tha behaved well I Allways Treated them well If other ways I treated them Accordingly; Att this Time I Expected Nothing More then to be put in Irons—att this time thare was but Lettel More said Capt. Prescut now Goes Down to His Caben; itt is now All Most sun sett he Came on the Quarter Deck and orderd one of the Officers to have one of the Midshipman Apartment Cleard for thre of us; sune Arfter we Tuck Posseson of our Berth Thare was no Chest Carred on board the friget butt Mine and that well stood with small stores as I had Time to Do itt as we had A Long Chase I fixed My self before the Lieut. Came on board I had some Punch and Wine on the Tabel as I new the Custom was Allways to be in a Grate Hurry for such officers Emeadetly Arfter he had Taken a Glass of Wine I Emeadetly Present him with My Hanger the Reaison I Gave him this was to Endever to Gett My Chest on board the Friget itt Anserd a Good Purpos for he Emeadetly orderd My Chest taken in the Barge and itt was Carred on Board the Fregitt the Methord I took on Board to save My small stores was to hand the Key to Capt Prescut and informed him I had no Warlike Wepens in itt he Emeadetly orderd Me to put itt in My Pocket; If I had a handed itt to Any under officer I should not only Lost My small stores but I Had A Large New french Ensine in My Chest that I Expected to Have yuse for When I was put on board a Prisen ship Which was Very useful to Me When I Gott thare Nothing Materel Hapening untill we Arrived att Sent Johns [St. Johns] Newfound Land which was seventeen Days Arfter we ware Capterd All this time Mr. Erben Mr. Channing and My self ware Treated well by the Capt and officers on Board now we Are Arrived Capt. Prescut said to Me have You Gott any Money If you have not I will Lett You have some for I have A Master heair Admarel Edwards I said I had a Lettel Money with Me for says Capt. Prescut the Capt. of the Prisen ship is a Scottsman and You will be half starved without You have Money Emeadetly the boat was Ready to Carre us on board Capt. Prescuts Servant Brout A Large Basket of Caben Bread and a Doz Bottels of spirits that I took and sent My Complements to the Capt. Wheu sent Me Werd back Again before I Left His ship; That His barge would Parse the Prison ship Every Day and If I was in Want of Any thing to send and I should have itt—

The Next Day Mr. Eberns & Channing and My self was sent for before a Notre publick on shore to be Examened Conserning the Time of our Takeing the ship Called the Elizabeth Mounting sixteen Carrege Guns Carring About sixty Men I happened to be the first Examend Arfter some saremoney Conserning the ship then he began to Questen Me Asked me My Age and Whare I was born I anserd him then How Long I had A fiting Aganst the King I anserd him in the same Manner I had Capt. Prescut that If itt was Posabel before the war Commensed he Discarged me Emeadetly Arfter he Examened My Tew ship Mates we ware sent Emeadely on board the Porteas Prison ship as that was the Name of the ship and the Capt. Name Was Brocannon and A Grater Villen Could not be Left un hung The Next Day Appeard on Board the Prison ship A Young Gentlemen the Name of Warren Belonging to Plimouth son to Generul Warren that Was one of the Board of War [Navy Board] att Boston he was a Young Gentlemen that was Left as a Hostage for Capt. MacNeail [Mc Neill] and Crew so that he was a Prisener att Large I had butt a Very slighlea Acquaintence with him Arfter some Conversaiton with him he went on shore and said he would Come and see Me the Next Day Accordingly he went to Admerrel Edwards and Goot Leberte for Me to go on shore and Dine with him Accordingly I went in the boat with him on shore to the House he Boarded att whare fore or fife Scotch Merchts. Boarded While we ware Dining one of these Scotch Merchts. had Consederabel to say to Me and in the Conversaiton Asked Me how long itt was since I had left Boston I anserd him then he says how Does the Rebels Carre on in Boston I was hard of Heairing to this he Repea itt thre times and then he sais Mon are You hard of Heairing I sais No Why Mon I Asked you how the Rebels Carred on in Boston I Gave him for Anser I never had herd of Rebels sence the Yealr forty fife this Ended all our Conversaiton as sune as we Ware Dune Dining tha All Left the House and Proseded to the fort Whare the Admerel was in a short time Mr. Warren and Myself setting att winder saw fife High Landers with thare Guns & Bayenets Making for this House I Emeadely spoke to Mr. Warren and Informed him that I Expected tha ware My Waters and he had better Gett the Money he had Promised Me he Gott Me Ten Geneaz and by the time I had put itt in My Pocket tha Nock att the Dore I sais to Mr. Warren I will go to the Dore I Accordingly went without My Hatt I suppose one that was a Corporul says is thare a Mon heair belongs on Board the Prison ship I Ansers I am the Mon he says we have orders to see You on Board Emeadetly I says I suppose You will Allow Me to Gett My Hatt Yes Mon I sune Goot My Hatt and Mr. Warren went Down to the shore with Me and I was safe Abord in a hurre Who Ever has an Ideair of a Furter Punishment Can not Make a worse Floating Hell then a British Prison ship—Nothing Material Hapening only once in some Time tha Bring in a Privatear tha Brought in one ship Called the Fanne from Baltamore and a Packet bound to Holland with the Honbe. Henry Lawrence [Laurens] on board the British was so Clost on board that When the Americans Hove thare Dispaches over a board for want of More Wate tha British took them Mr. Lawrence had not Arrived thirty Hours before he was sent of in a British ship to England and then the war Brock out with the Dutch and English itt was supposed oying to some information tha got by Taken them Dispatches—Nothing Material Hapening untill Octobr. When thare was a Fleet of Mercht. ships Came from Deffirent parts of Newfound Land to say some for Lisban and sume for the West indeas to sail under a Convoy The first I new of the Fleet sailing was a signul Given to All Mercht. ships from the Admerul ship and in a short time thare ware Hanging all Round the Prison ship boats from Most of the Merch. ships; and with them thare ware A barge from a fifty Gun ship with A Lieut. on board, I sune found What those Besness was that is Every fife Men the ships had to take one American I spoke to this Lieut. as he Appeared to Me to be as Tirancul As a Negro Driver in Jamaca is this Your mode of Exchangeing Priseners as for My part If I had none of this I would have Goot some Parm Oil and that woulf Made them

Look Much better for tha ware Turnd out in a Scandilus Maner. The Fleet all und sail now; itt was but a short time before this More then a Common savege Brocannon Came with My Name and fife More, to be sent on board a Brigg Letter of Mark Bound to Sonkets [St. Kitt]. I spoke to the Lieut. and to Capt. Brocan and said I had None a number of Americans Exchanged, but none in this way, and I Loocked on My self a Prisener of War, and I Ment to be Exchanged in a Deferent maner. All that I Recvd. for Anser Was Dam Your Blud Mon, go and Peck up Your Duds or I will have You hove in the boat; I Emeadely went Down in the Lower Reguns As tha Expected to peck up My Duds, but I was Determend not to go as I new the Fleet Must All sail that Day, this Enfernul Scoundrel Brocanon Repeatedly with a Hand Trumpet Called Me to now weather I was Common I anserd Him Kindly with as hard Werds as he Could Give I would not, then he says Dam his Blud Gett Another Man in His Rume tha Did so; Capt. Brocannon was More of A nave then a fool, but I Beleve he was a coward, for I was in the Lower Reguns and he New If we put out our Candels we should be All of A Culler, and I new My self safe Down thare and I Remaind below untill the Evening, and All that Fleet sailed—Nothing Pertecker you must now this Brocannon went on shore Every Night Gaming As I was informed by some of his officers; and some Mornings he would Appeair Like A sevell Man but nott offen in Tew or thre Days he semed to be Very Good Tamerd I then spoke to him About Taken out Priseners in such a Maner; I says to him I have none ships of War to Pres Men out of Marct. ships but I never heard of taking Men out of A Prisen ship to Man Mercht. ships All that he said itt was Orders from the Admerul I all way thout tha had taken so Many Men out of thare Merct. ships to Man thare ships of Warr and the Admerul took our Poor Yanke and french Man to pay the Deat for tha ware french Man Taken out the same as Americans—itt was but a few Days Arfter a Carteail Arrived from Boston past by our Prison ship and went up the Harber a Long side of one of the ships of War I was Very unezea untill I Could heair What news I new What Capt. Brocannon said would be a Lie; and I Could Gett on shore any Night I was Amnd to go You All no that money speaks all Lanweges the Next Night a Young Man beloning to the Eastward and My self Went on Shore we was Att severul Tavvirns and we sune found that severul Quebeck ships had bene taken by our Privateairs and Carred to Salem & the Eastward and Number of thair people had Come in the Carteail I Goot on board the Prison ship before 12 Aclock this hapened Very well for Me Capt. Brocannon would have some thing to say Every Morning the Next Morning I Asked him What news from Boston he says to Me You well All be Exchanged sune for the Rebels is all Acoming in and is Come in butt 5 states so that You have Goot butt fife stripes so Your Bunch of Bugers as he steled the American stripes I can not Beleve itt and in a Smiling Way I says I should think the Americans had Taken Bunton [bunting] Anuf out of the Quebeck fleet to Add thirty stripes More to our Cullors that Made this Scoth vilen Angre and he Cald Me a Damd Rebell and orderd Me Down in the Lower Reguns for that Day: that finished My Days Work with him—

Now for severel Days not but Lettel spoke; thare was About ten to fifteen Eastward Men and Tew or thre french Man that had Brout forward a Plan to Run Away from the prison ship and tha Lett Me into the Affair tha had some Matters well Provided but as I thout not as I would Wish I told them I Liked a part of the sceme Very well I told the head Leders If tha would Kepe All A scrett and not Lett no More no any thing About the Affair I would Assist them All in My Poor [Power] Accordingly the Night Come You must now thare was some times Tew some times thre Tenders Lay Betwixt us & the Enterance of the Harber but Every Night thare ware More or Less of these fishing Vesels that Vesels a Coming in I took this Oppurtunety to Right to a frind of Mine att Boston and that was the first News My farther and frinds had from Me in this Letter I Rote to My farther to pay Generul Warren ten Geneas that I had him att Newfound Land and Long Before I Got to America that was paid threw the Hands of Mr. John Manly [Manley] Deperte Contenentel Agent the Poor fellows that went Away tha Took a Dark and Dismal Night tha Bought that Los ther way that Night of the harber had as was Reported Att Sent Johns upwards of one Hundred quentels of Cod fish and was said to be a good Boat Every Day for a Week Alltho the Wind Blew hard itt was fair that I Entertain A Grate Hopes My Frinds would heair from Me Arfter this a Dubel Look out was Kept on us; Nothing of Any Consequence Happening untill Novm. When the Last Fleet was to sail for this season Gott together att St. Johns some for Lisbon some for the West indes—att this Time Capt. Isaak Prescut had Made Tew Cruses since I was taken by him he Came on board the Prison ship I put My self in His way he sais I see You are heair I Anserd him Yes sir and All that was taken with Me and Hundreds of others had bene Released he says how Hapens all this I Emeadetly informd him Cap Brocann standing by the Reason is this becors I have told Capt. Brocanon the Truth he smiled and Walked Round and then he spoke Again to Me and said I am a Going now to Convoy the Lisbon fleet If You are amine to go with Me to Lisbon I shall from thare Go to England and as You Call them I well Gett You a Pasage to Your Good Allays and You shall be Treated as well as You had bene on board My ship; I said to Capt. Prescutt I took itt very Kind of him but I was not Prepared for a Winter Voige and should Like to Go to the West Endeas well says he When I Leve the Prison ship I will Leve a paper for You with Capt. Brocannon to Go on shore I sais If You Pleais to put Down Capt. Parker of Epsage [Ipswich] Which he Did we both went on shore he sune found a Mercht. he new we Gott Every see store we Wanted and went on board a Brigg Bound to Barbados the Next Night Blowing Very hard we sprunk A Leak that one Pump Would scarsly Keep us frea Arfter the Gale Abated we found our Leake to be in the Transome and so Ruttan that we Could do Nothing but put Oackham and old Canvis to stop itt that Helped but Very Lettel for If we Lett the Pump stand fifteen Minets itt would take as Long to free her this was untill the 10 Day att Night When we fell in with a French friget Called the Amazune [*Amazon*] Comanded by Capt. Latoch the first Lieut Name was Costabele he spoke Good English tha boarded our Brigg About 12 Aclock att Night the officer that boarded us treated us one and all well; Allowed us to bring All our Cloth and What we Ware Amind but still thare was a Nother Misforton Redey for us the sea Runing Hie and our boat Rether Crowded the first stroke A Long side of the friget stove our boat and Left us a padeling in the sea but fortenet for us not one Life was Lost so Expedeshus tha ware with Ropes hove over that tha saved Every Man; but Nothing of the boat or any thing we had in the boat ware saved What was the Most of My Misforten I never untill then Lost My Commison and some other papers I sett by this was in Novm. 1780 Thatt Night was Very Cold we had some old sails on the Gun Deck to Cover us but wett and Cold untill the Next Day I sune had an opertunety to speak to the first Lieut. Arfter I informed him I was an American and thare was one More on Board he Asked Me a Number of Questions About Newport that Gave him All the satisfaction he Warnted he Emeadely had some Cloth for Me to shift and Plased Me with the Docters of the ship and Provided Me with Every thing I Could Wish to Make me Comfortabel I found this ship had bene from Newport butt ten Days tha Maned the Prize and orderd her for france we Made all said and went on to france as this ship had Dispaches and Col. Rochambo [Rochambeau] Was Pasenger on Board Nothing of Any Consequence hapeng and in fifteen Days from the Time we ware taken we Arrived safe att Leorient. When I Arrived thare I found My Old Commadore Paul Jones he was on board A sloop of Warr Called the Ariel and All Redey to sail for America the Next Day I went on board his ship I thare found Mr. Nathanel fanning and Mr. Thomas Potter that I had Long bene Aquainted with tha Came to the side of the ship itt was now 12 Aclock tha Emeadely Asked Me to Dine with them as I was a Going Down to Dine with them I saw Commadore Jones Walkeing the Quarter Deack And as he Did not speake to Me I was not A Mind to Speak first I was talking with fannin & Potter and said I was surprised that the Commadore Did not speake to Me Mr. Fannin & Potter mad this Replia he seldom spoke to Any of the Officers I found by the Conversaiton tha Did not Like him I said he would Certenly send for Me If he new I was A board tha said no but in a few Minets a Midshipman Came Down to the Berth and said the Commadore Dezird Me to Come up thare I went and Arfter Asken of Me some Questens Whair I had bene &c he informed Me he should sail in a few Days for America and said he would Geve Me a Good Chance If I would Go with him and Provide Any thing I Wanted I informed him thare Was one American on board the Amazune I would Wish him to go Likewise he was to Apersunce well Pleased Arfter a short time I went below and Dined with Mr. Potter and fannin I Emeadely found our Commadore had Treated the most of His officers with disrespect and Had Grone More Naugty then When I sailed with him I Told Mr. Potter and fanning I was a Going on board the French friget and from thare on shore and I should think more About Going; tha both of them Larfed Att Me and said You have not Gott out of our ship Yett And You will find hard Work to Gett out. I said but Lettel on this head I new Commadore Jones in My Openyon better then tha Did for I Had sailed with him since he went As Lieut. I sune went on the Quarter Dack and spoke to the Capt. of the Arieal for the Comadore had Gone on shore I informd him that I had A mind to go on board the French friget. he Made Me this Reply I have no orders from the Commadore and I Can not Lett any one Go out of the ship I sais no More att 12 Aclock att Night the Commadore found his way on board All this time Potter and fanning Loughing and said to Me You will not go out of this ship untill You Gett into America for says Potter we have Goot Governer Wards son of Rhodeisland on Board and he was Cheaf Mate of A Large french ship sailed out of Loreon and Every American that is att that Place If the Commadore says the Werd tha are sent on Board and says fanning thare is now on shore Confined in Gail severul of our under officers tha will be Kept thare untill we Gett all Redey for sea I says but Very Lettel More; and Made My Self Ezea untill the Next Morning Arfter we had Breckfasted I went on Deack and sune Arfter the Commadore Made his Apperence on the Quarter Deck I found orders Given to the Boatswen to have the Barge Manned to put the Commadore on shore; I Emeadely Goes On the Quarter Deck and Informed the Commadore in the Presence of the Capt that in the Run of the Day I should Like to go on Board of the french ship to see My Companyon Capt. Parker of Epesage and to see whether I Could not Get Him to go with us to America and I should go on shore to take a Veu of the Town he Made Me this Reply and Turnd to the Capt. of the ship and said that When I was Ready I should be put on Board or a shore When I Pleased to go; I Did not make My Self in a hure About this time Waited for him to go on shore first; but I was Determind Neither Parker nor My self should go with him to America, I sune went Down and see My frinds Potter and fanning and informed them What I Had Done tha said tha Ware supprized att his Letting me out of the ship I Did nott Tell them My intention I went on board the french ship and Capt. Parker and My self went on shore I went on shore I went to the jail I saw the officers that was Confined thar tha Gave Me to understand What Treatment tha had from the Comadore I then took a Rambel All over the to[wn] of Leorient and then went on board the french frigett; the Next Morning Capt. of the friget informed us that he was bound to Brest to Convoy some ships Laden with Provishens and then he was agoing into the Dook to New Copper his ship and Expected to go shortly for America; This Was What I wanted for I wanted to sea some part of Yeurope before I Returned Home; Accordingly the Second Day After I had Left My Old Commadore but I must not Omit one Affair Arfter My Rambel was out on shore I Did not intend the Commadore should see Me on shore Again for I new he had All Influence with the Coverner and we should sune find his ship Which I Carded Against well now our french ship is under sail Bound for Brest and we are Obliged to Run so neair Comadore Jones that I Could have jurked a Besket on board I Gott in full sight of the Commadore & My frinds Potter and fanning and Wished them A good Time to America and Informed them I was A Going to take the Touer of Urope; I new then How the Commadore felt for itt was his intetion to Trapt Parker & My self but Old fowle is hard to be Kept in such snairs; now we have Run out our frind Jones and the Victelen [victualling] ships and Tew More french ships Runing out the Harber bound to Brest Nothing Happening Worth Remarks, we Are All safe Arrived att Brest All our Guns out and stores and we halled into a Dry Dock and as Many Men to Work as Can stand by Each Other; I have Leberty to Go in an out of all the forts or Whar I Pleas I have bene in a Very Large Hospatel and All Round the Citey but itt is a Durty Hole of a place I meain the Citey but the Hospetel is Kept Very Cleain and Every part was Desent, I Have bene in some of thare Bateres [batteries] whare I could take a fare Veu of the Harber and itt is a Very Large and Commodus Harber as I Ever saw; our ship was in the Dock butt thre Days tha took All the Old Copper of and Calked Every sem in her and went out of the Dock in thre Days; Now I have bene as Long in Brest as I can see no More then I have seen and thare is a French Privatear ship of thirty Guns Jest Arrived and Brought in a Large English ship a Prize one Luke Rion An Irish Man Commands her she is fitted out from Dunkirk And I understand thare is a Number of Americans on board I sune Gott an Oppertunety to Go of in a French Barge tha put Me on board I Enquired to Know wether thare ware Any Americans on board I found but Tew only one beloning to Philadulphia and one to Vergenea both of them under officers; I sune saw the Topsails Luse and Anker short apeak I spoke to this Philadulphia Young Man an Asked him wether the ship was bound out he says Yes and the Pilot is a board; I says how am I to Go on shore he says our boats are All in and we shall be A Considerabel Distance before the Pilot will Leve us And itt is More then Likely we shall put him on board some Other Vesel; I sune found out the Capt. and I goes on the Quarter Deck and spoke to him as If I Had bene Acquainted with him before I informed him I was an American; I came on board to see What Americans tha ware on board And I had Brought no Cloth with Me he smild and said we shall be out Young Man but a few Days and More then Likely we shall be att Brest Again All now in Confushion Making All sail and a fresh Leading Breze; now I thout to My self of a bad barging Make the best, I must Confess old fouls is some times intrapt I sune had time to think how I Left My old ship Mate Capt. Parker and some others in brest and Left att My Lodgings What few Cloth I had and My Capt. of the french frigett and none to no What became of Me and from that time I never saw no More of Any of them for we had Left Brest butt a short time before we [came] in the English Chan[nel] vesels . . . [Illegible] in sight The first sail was an English Cutter Run Down for us and Arfter Getting neair Did not Like us Halled His Wind for the Land we sune Gave Chase with All sail we Could pack Neard him farst but he Neard the Land faster all this time a french Boye was att the Main Top Gal Mast head and by some Accident fell of and struck in the Warter a stern a Consedeable Distance we sailed Very fast threw the Water and Endevering to overtake the Cutter before she Gott a shore but All invain we Lost the Boy & the Cutter Run into the Brakers so we Left her an Haled our Wind of and Never see no More of the Boy Arfter; now a Number of Vesels in sight. Night Coming on under Ezea sail towards Daylight Rain and Hazey Wether About Merideon saw Tew sail we take them to be British sloops of War as tha are standing had on us tha Neard us fast. and we found our Mistake When itt was to Late to Run but Capt. Luke Rion was att no Lorss About this Affair he orderd About Twenty of His Musket Men To fire in the first ships &c fire Arrows as he was Determind to Run Neair Aboard of them so as to sett the first ship sails on fire Which I Beleve saved us. the first ship that Came up we Gave them thre Chers tha Returnd itt we under full sail we Gave them a full Broad side and out of our Tops Likewise; she Returnd itt as well by the time we Could Load which was not to Exceed fife Minets. The Other ship was Redey Gave her thre Chairs and att itt . . . [Illegible] tha Returnd the same we found the fire Arrers had a Good Esfext for the first ship [was seen] to be a fire and the Last ship Laid by her Above an hour Gave us a Good start a head itt was not Long before both ships ware in Chase of us and before Darck tha fired shot over as but a Dark and Squarle Night saved us this Time, the Next Morning saw Nothing of these ships our fore Yard Was much shatterd and a Number of our spars & sails, no one Killed and only 3 slightly Wounded. The Next Day fell in with A Large ship of Twenty Tew Guns Cananades, and took her, she Pruved to be a ship from New York Called the Lord Dunmore bound to England we took a Number of British officers and I all ways thout one of them was the Trater Benedick Arnull, we took out her Guns and Warlike stores and Ransomed her, we Kept Her with us thre or fore Days untill we Made King sail in Ierland, and we sent her a head of us for King sail Harber, Each under English Cullers we saw Tew shipps and a Brigg Coming out of King sail we Run [in] and Cutt them of from the Land and [took] All thre of them. Tew of them Letters of Mark we orderd [all of] them for france, we then Made sail for franch to Land our Priseners, we Arrived att Morlox [Morlaix] in franch and Landed our Priseners and Laid thare six Days, and then we sail on A nother Cruse in the Channel When we toock An English ship A Privatear of Eighteen Guns. Att the same time we took Tew English Cutters, and Retook a Dutch Brigg, the ship Privatear we Carread her in to Austend in Flanders and Lay thare fife Days, and then sailed for Dunkirk in France, Whare we Arrived safe and then the ship went into Dock to Repair; I must say I have sailed with Commadore Paul Jones but this Capt. Luke Rion beait him out; this Ends Tew short Cruses.

A New Cruse I went from Dunkirk to Calas, And then to Bullen [Boulogne] Whare the same ouners Ware A fixing out a Cutter Privatear of Twenty Tew Guns, Carring Tew Hundred Men Mr. John Erven and My self partely Agread to sail in the Cutter, but finding A Number of Americans that had bene a sailing in the same Employ, and had Carred in A Grate Number of Prizes, and never Could Gett no settlement. with the Oners, we John Erven and My self, Concluded to go threw Flanders into Holland, so as to Gett to Amsterdam As we had herd of the ship South Carolina Lay thare, we sune satt out and Arrived att Amsterdam in March 1781. The Next Day Arfter we Arrived Att Amsterdam I fell in with an old Acquantance Capt. Wm. Haden of N. Bedford, he had a ship of sixteen Guns A Letter of Mark He offered Generosly a Pasage to Boston, I Gave him for Anser I was Obliged to him for his Generous offer, but his ship was not of forse for Me, and Att the same time I informd him I would Wish to see Commadore Gillon; Capt. Hadden informed Me he was Ackquainted with the Commadore and Emeadetley Capt. Haden Mr. John Ervan and My self went the same Day and saw the Commadore and Engaid to sail in the South Carolina frigett, she then Lay Att the Tuxel sixty Miles from Amsterdam, Arfter we tarred in Amsterdam some time, we went Down on board the ship att the Taxel and thare we Lay untill Septmer and then we sailed for America, and we went North About and in the North See we Took an English Privatear of sixteen Guns she was a Brigg, we ordered her for france Nothing wath Remearking Arfter some time Crusing we put into Corona in Spain, thare we Goot supplys for the ship and went on shore and Took a Vew of the Town and I . . . [Illegible] itt to say the Bildings are built of stone and Appeared to be in a Very Old Stile if itt had bene bult A short time Arfter the Floäd—we Lay thare some time and then we putt to see; Att Crona Doct Benjm. Warterhouse and some other Gentlemen Pasengers Left our ship and went for Billboa to Gett A Pasage to America, the Next Prize we fell in with was a Brigg from St. Johns Newfound Land Bound to Lisbun, she was Called the Venus Mr. Jere Peirce was Prize Master, and I went his Mate we had Al french Man on board, Nothing Material Hapenining only by the Neglect of our stewards Mate one Morning About 8 A M the steward Deling out the Crew thare Allowance of Brandy his Candels warnted snuffing, he had severul pailes of Brandy Drew, he took the Candel from the Lanthon to snufe Hove itt by Accident into a pale of Brandy, in an instent All the pails took fire and beait the Hole out of the Room and sett the ship on fire as to blaze six out of 8 feet Above our Gun Deck, but the Culiness of the Commadore & the Americans tha put itt out. When we put itt out & Went to the Stewards Room A number of Casks of Brandy was All most burnt threw and under the Brandy was ten Tons of Gun Powder, Loockey for us itt was Gott under so sune!—Now, we are sailing in the Venus Brigg, Teneveffe in sight, and the South Carolina frigett in Company, we Are now Arrived att Santa Cruze in Tennevefe, Arfter Discarging part of our Cargo, the Commadore sent orders to Delever her to a Gentlemen from the shore and then Mr. Perce and our Crew went on board and we sailed Again on Another Cruse, we Run to the

Southward and Run Down the Trade Winds till we Gott well to the Westward, then we Halled to the Northward and Arrived of Charlestown Barr and saw English Cullers Afling and a Number of Large ships, we then Bore Away for Abaco, one of the Bahama Islands Arfter Making Abaco, we Ran Down between Abaco and New Providence, and as we Goot into the Gulf we fell in with thre Large ships from Jamaca Letters of Marks, and Tew Briggs without Guns, we Made Prizes of the Hole fife sail, Mr. James Dick was put in Prize Master, and I went As his Mate on board of one of the Prizes, and we all Arrived safe Att the Havanah, and Arfter Laing thare a few Days the Commadore took Possesion, and we then went on board ship, shortly Arfter the Commadore had our ship Halled into the Custom House Wharf to Repair, we sune found that the Spanish Government Had Hired her and her Crew, to Go up with a Fleet fitting att that time by the Spanish Govn. Ment to take New Providence, Accordingly Arfter About thre Month we sailed, the Hole fleet Consting of fifty seven sail with a spanish Govener, and About Twenty fife Hundred spanish Troops on board, we Capterd and Took Possesion on the 11 Day of May 1782 And that Day we Doused the British Cullers and Histed spanish Cullers in thare Place—

NB. one small Cercumstance happened While we ware A fixing our ship Commadore Gillon sett Very Much by Mr. Samuel foot of Salem and Mr. Wm. Coal of Marvelhead Each of them ware Masters Mates on board, and tha belonged to the same Mess that I did; and While I was on board A Prize, the Commadore fearfull tha would Leve our ship, and Go to Salem, he had them both put on Board A Spanish Gardship, until we Gott our ship Redey for see, I found Meins to send them A Line or Tew, and sune Gott An Anser, I Provided some Mederson for the Gards When tha went to pray, as tha Did nit and Morning, I Accordingly Hired A Bum Boat, and Tew spanish Negros And I went with them and took them out and putt them on board A Brigg belonging to Mr. Grarfton of Salem, the Next Morning the Brigg sailed and tha Arrived safe att Salem—the spanish Capt. of the Gard ship Came on board our ship and informd Our officers that he was fearfull that tha had Atemted to swim on shore and the Sherkers had Destroid them—and Commadore and All our officers sett a Gratdel by them, and Wore Very Long faces, & I never Lett None of them into the Affair untill I Arrived Att Philadulphia—then I Gave them A Gradel of satesfaction to think tha ware still ALive—

I Enterd on board the South Carolina friget in March 1781 and Mr. John Erven of Philadeulphia the same Day and we Arrived att Philadulphia the 28 of May 1782—and we Agread with the Commadore att fore pounds. Sterling p Month—

Now Comes A List of the Volenteairs that was on Board ship While I was on board—

Capt. Thomas Smith Capt. John Tiler Capt. Abraham Freland Mr. Caleb Foot Mr. Jacob Tucker Mr. Samuel Williams of Salem Masacusets	Capt. Joseph Dizamore Mr. Johnathon Glovyer Mr. Thomas Williams Marvehead
Capt. Jacob Higens Doctor Wing Cape Cod	Capt. Robert Burvel Mr. Edward Mc Agroth of Boston
Capt. Eber. Warters of Gilford in Conitecut	Capt. John Erven Philadulphia
Mr. Joshaway Bowen of Swanzey Neair Tarnton	Capt. Nathanel Tibbets and Mr. Samuel Rise of East Grenwikh
	Capt. Josiah Arnol of Providence in the State of Rhodeisland
	Capt. Elias Ellwell of Cape Ann

NB Now You May say Wheiu is this John Trevett—I will inform You My Grandfarther on My Farthers side Came from Wailes in England and Marred att Marvelhead My farther was born att Marvelhead and att the Age of Twenty thre Yeairs he Came to Newport in the Coloney of Rhodeisland and in A Few Yeairs Arfter he Marred to Marry Church of Fretown Daughter to Capt. Constant Church hew Was Second son to Col Benjm. Church then Leving Att Fretown in the Colany of Masachusets And this John Trevett was born att Newport Now in the state of Rhodeisland—

[End of Diary]

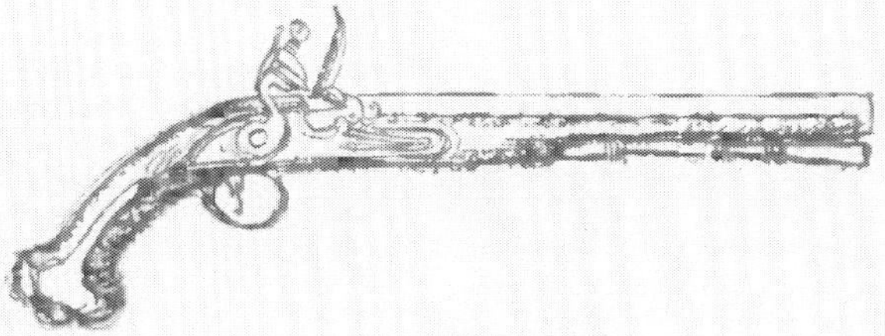

APPENDIX D

Journal of William Jennison, Lieutenant of Marines

William Jennison was born in Mendon, Worchester County, Massachusetts, 4 August 1757. As a wealthy and influential man in the county, Doctor William Jennison was able to provide his son with a private tutor, the Reverend Amariah Frost, who prepared young William for his entrance to Harvard in 1770. Graduating four years later, with highest honors, he traveled to Providence, where, for six months, he studied law in the office of Daniel Hitchcock, a celebrated attorney in Rhode Island. Following the custom of the day he also enrolled himself in the fashionable Cadet Company of Providence, commanded by Captain Joseph Nightingale.

In the spring of 1775, he returned home and enlisted in his father's company of Minute Men as a sergeant. When news of Lexington arrived, Captain William Jennison, Sr., marched his company for Cambridge. Twelve days later, the Mendon Minute Men started back for their farms. Sergeant Jennison, however, remained behind and enlisted in Colonel Joseph Read's regiment of the Massachusetts Line, in which he was returned as its quartermaster on 18 May 1775.

During his nine month service with Read's regiment, Jennison began his journal. Unfortunately only a fragment remains of his service in 1775. Instead of reenlisting when the army was reorganized before Boston on 1 January 1776, he returned to Mendon where his journal was continued.

Having served in the Continental Army before Boston in 1775, in the Continental Marines for a short time at Providence during the spring of 1776, and again with the arrmy around New York in the summer and fall, William Jennison entered as a volunteer on board the frigate *Boston*, in January 1777. He participated in the joint cruise of the two Continental frigates *Boston* and *Hancock*, begun in May of that year. When the British frigate *Fox* was captured and manned with officers and men from the *Boston*, Jennison was made acting lieutenant of Marines on the frigate. On 26 December 1777, he was commissioned a full lieutenant of Marines, and sailed in the *Boston* for France with John Adams, newly appointed American Commissioner, as passenger.

Upon the frigate's return to Boston in late October 1778, Jennison was appointed purser. While the *Boston* lay outfitting in the harbor, he secured a leave of absence and sailed in the private schooner *Resolution*. The privateer was captured on 8 May 1779, and the crew sent to prison in *Halifax*. After his exchange in September, Jennison rejoined the *Boston* as a midshipman. The *Boston* with the *Providence, Ranger*, and *Queen of France*, arrived at Charleston, South Carolina in December 1779, and on 16 January 1780, Jennison was authorized to act as a volunteer. As a result of the American defeat, he was captured but allowed to return home. With his arrival at Boston on 16 July 1780, the journal concludes.

There being very little call for lawyers after the war, William Jennison devoted himself to classical teaching and established a number of academies at Boston, St. Johns, Nova Scotia, Vicksburg, Mississippi, and Baton Rouge, Louisiana. In 1828, while at Vicksburg, he published a small pamphlet on political economy. Returning to Boston, he died at the venerable age of 86, on 24 December 1843.

The journal, which now reposes in the Library of Congress, is printed just as William Jennison wrote it, with certain annotations to help clarify the narrative for the reader. Additions to the text or corrections in spelling have been bracketed, and all places where it has been difficult to transcribe the journal due to its condition have been noted.

[1775]

the 1st. Division boat
1 bottle of London Porter was distributed—1 bottle to 2 Men—The 2d. & 3d. Divisions drove 30 swine—10 head of horned cattle and an elegant mare belonging to a Brigadier Genl. whose name I believe was FOX Piercy [Brigadier General Hugh Percy] perhaps. a son of the Duke of Northumberland (scott) supposed to be the *present* Duke of Northumberland—All of which were delivered safely on the opposite shore

Augt. 9 At 11. oClock the party returned safe to Roxbury Camp.

1776

Apr. In April I went to Providence & offered myself to engage in the [Marine] service A board then sitting to appoint Officers for the Warren Frigate As I had been in the Cadet Company under Capt. [Joseph] Nightingale, they gave me Instructions & money to recruit Marines for the Warren Capt [John Burroughs] Hopkins

May & part of June was employed in recruiting men for the service

June In June Genl. Washington called for men, and Volunteers turned out under Capt. [James] Foster at Mendon called the 5 month men—[1]
I relinquished Recruiting, and joined Capt. Foster—[2] march'd to New York—The Regiment was commanded by Colo. [Jonathan] Smith—[3] The Adjutant being taken sick on his way to join his Regt. I was selected by Col. Smith to do the Adjutants Duty—

[July] We were [lying] at the time in Greenwick—There we were very frequintly annoyed by a Sloop of War, which afterwards carried Genl. [Benedict] Arnold from West Point to New York.

[Aug] When the British landed on Long Island, our Regiment was ordered and marched to a part of N.Y. which was then called the "Holy Ground"—Genl. Washington narrowly escaped from the enemy this afternoon on Brooklyn— At sunset we were orderd to march to mile square NE of Kingsbridge, where we halted for a few days—Then marched again towards the White Plains—where the American Army Encamped

[Oct] 27 Genl. [Israel] Putnam with 1500 men, was ordered to proceed down 7 or 8 miles to the Encampment of the British Army and reconnoitre their position, which was performed without any Interruption
28 by 2. oClock A.M.—At early dawn we began to move towards the White Plains, falling trees into the Roads, and casting stones and whatever might obstruct the Ordnance of the Enemy—the Party arrived at 9 a m—at 11 a m the Enemy's Light Horse said to have been 800 strong hove in sight on a Ridge of Hills South of the River Bronx (but fordable) which r[an] between the American Camp and the Ridge—They maneuvered [a] little while when Genl. [Henry] Knox complimented them with a shot which caused some commotion among them—This Light Horse had filed off to the Left which brough them near opposite the Artillery of Genl. Knox—Had the Enemy contin[inued] to keep the main Road for One mile, they must have taken all the Provisions collected for the American Army, but by filing off to the left, a chance was left for Us to get most of them to a safer place.
At noon the British Forces came up and formed on the Ridge when the Action was commenced and continued till sunset—"A remarkable instance of Providence happened to Genl. Washington when riding on another Ridge N. of the Brook a shot came from the Enemy & struck directly under his Horse's belly when on full Gallop,—but did him no Injury".—
As soon as it was dark, Orders were given to retreat farther into the Interior, Every One carrying what he could—
For a Week or two we lay undisturbed by the British—but soon retreated to Croton Heights where Winter Quarters were intended—

Nov. At the close of this month the 5 months service ended—when I left the Army at Fish Kill with a full Intent not to be in the Land Service again—after a fortnight walk I reached my fathers house—where I rem[ain]ed until

1777 Jany 8, I set out for Mendon

Jany 11 I passed on to Boston put up at Colo. D[avid] Brewers—

14 I entered as a Volunteer on board the Frigate Boston Hector McNeil[l] Captain—[4]

20 The ship was hauled into the Stream in order to keep her Crew on board—
Note [By Jennison] The Ship rode in Boston Harbour until May—Proceeded to Sea in Company with the Hancock Frigate Commodore [John] Manl[e]y—

29 Captd. a Brig from Ireland to New York & sent her for Boston, where she arrived safe consigned to Colo. [Gabriel] Johonnot our Ship's Company's Agent—[5]

June 5 Discovered a Sail at Dawn of Day and gave Chase—had a running fight until Noon when we in the Boston got along side of her and Exchanged a Broadside or two with her, and she hailed down her Colours & struck to the Boston—by 3 oClock p m All the prisoners were brought on board the Commodore and Boston—The Capt. of the Fox first came on board the Boston, but was ordered on board the Commodore's Ship much against Capt. [Patrick] Frothingham's [Fotheringham] Wishes, as Capt. McNeil was a more Gentlemanly Character than Commodore Manly—[6]
At 6. pm upwards of 30 vessels had collected under our Flags which had all English Ensigns kept up as a Decoy—among the number of Vessels was a Brig into which the Commodore ordered all the prisoners to be put on board under a [comis]sioned Officer to proceed to Saint Johns in Newfoundland to release an equal

number of Prisoners—Capt. McNeil remonstrated against this Measure, which the Commodore persisted in—

(1777) We supposed our Ships not farther from St Johns than 200 Miles, and on the Arrival of the Brig (as a Cartel) a Sail Boat impelled by Wind & Oars could pass to Halifax in 12 hours—which would give the Admiral on that Station a Chance of intercepting our Squadron in its Return to Boston—The Brig sailed at 9. oClock pm—named Polly—

July On our return from the Banks it was very Foggy on the Coast of Nova Scotia and a Rocket was frequently fired to discover the proximity of our Ships—A little to the Eastward of Halifax the Fog being very dense Capt. McNeil spoke the Commodore & enquired whether he would ring the Bells as usual, who replied Yes, if all the British Navy was near him—But (note. [By Jennison] The Fox when captured was manned with Mr. [Stephen] Hill as Capt. from the Hancock, and Simon Groce [Gross] 2d. Lieut. of the Boston & 67 Men & under officers from the Boston—they were ordered to keep in Company during our Cruize—The Fox ahead, the Commodore in the Center, & the Boston in the Rear—)

6 Discovered a Sloop & gave Chace, she proved from Spanish River bound to Halifax with coal & Dry Goods—The Commodore took her in Tow—

This day made Cape Sables—Cast off the Sloop & scuttled her—when

7 At 6 am discovered a Vessel standing athought our Fore Foot—
At 8 she hawled round & stood in our water—giving each of us a Broad side without trailing up even the Courses (that is the Fore & Main sails) to get along side of the Fox which before belonged to the British Navy and was their determined purpose to recapture The Flora being graved and pressing all sails passed us who had been 4 months at Sea—and overhaled the Fox which continued her Course to the Southward till 2 pm when we saw her Colours struck, & Boats passing & repassing from Ship to Ship—during the time of the Flora's Chase, Guns from the Boston called Stern Chases were continued—At 10 oClock the Flora passed us and fired a Broadside on the Commodore,—we received three shots between Wind & Water, several were killed and wounded on board of the two Ships—After the Flora had passed the Commodore, still pushing in for our prize, The Commodore hove about to plug up the Shot holes and ordered the Boston to follow him—The Captn. told him that his Ship was injured and could not wear Ship until he had the injury remedied—By the time the Shot holes were stopped, a large Ship we saw to leeward some while before had got in reach of our Guns shot, which caused Captn. McNeil to consult his Officers whether to follow the commodore after the strange Ship which was afterwards known to be the Rainbow of 44 guns, the only ship in the Halifax station with a Poop, which gave reason to suspect she was a Ship of the Line—This consultation terminated in our Standing to the Northward—The Commodore made all sail to the Southward the Rainbow pursuing after her—The Fox now a prize to the Flora was at a considerable distance to Leeward—we lost sight of them all by Sunset—

1777 July 11—We arrived & put into Sheepsgut [Sheepscott] River and anchored. .

12 Weighed & sailed up the River to Pownalboro'—

18 Unmoored Ship & sailed down the River

23 Came to sail from the mouth of the River 10 Leagues, when we arrived at Falmouth in Casco Bay—

30 Illuminated the Ship with 53 Lanthorns—

Augt. 7 Mr. [Ebenezer] Hogg a midshipman was put into Irons for selling Tickets for a Lottery on board—

8 The Ship's company were put on Allowance by deducting 1/3—

10 Weighed anchor & came to Sail at 11 oClock A.M.—At 3 p.m. a Brig gave us chase till 4 pm when we put about and drove two Coasters ashore & fired 9 Guns on the Brig. which escaped—
This Day Capt. [Richard] Palmes of the Marines was arrested, and confined to his Cabin—I the lieutenant of Marines was ordered to confine him to his cabin at 40 minutes past 6 pm—
Mr. Hogg had his Irons taken off—*

11 Came to Anchor in Piscatagua River at 1 pm—The Sailing Master Lawrence Furlong was put under arrest When we passed the lower fort 8 Guns were fired, and 13 more at Anchoring—

15 A two mast Brig arrived with a few Prisoners who had escaped from Halifax—

16 Unsuccessfully employed in trying to get out of the Harbour

17 The Raleigh Frigate & the Alfred sailed on a Cruize

18 Weighed Anchor & arrived at Marblehead at ¾ after 7. pm—

21 Weighed & came to Boston on the 23 am—
Thus terminated the 1st Cruize—

Sept 15 Recd. a Furlough for 10 days—Rode to Mendon & was sollicited to accompany Otis Whipple & ——— Taft Esq. to the Secret Expedition to Rhode Island—but some negligence in procuring Boats prevented a Landing on the Island, the plan of the Expedition was relinquished—

Dec 4 I received a summons to attend a Court of Enquiry on Capt. McNeil respecting his Cruize & Treatment of his Officers—

25 Capt. Samuel Tucker was appointed to the Command of the Boston, & ordered to refit & man her out for Sea in 60 Days.—

26 Recd. an Order from the Honourable Navy Board for the Eastern Department to repair on board the Ship Boston as Lieutt. of Marines—

27 Recd. a Furlough with 200 Dollars to recruit Marines with—

28 Rode to my Father's in Douglas & recruited 3 privates & 1 Corporal pd their Advance Wages— Corpl. £6—private £ 2.00—

[1778] Jany 12 Returned to the Ship

Settled with Honble. Board for assisting the Gunner ([Thomas] Foster) £ 3.9—

[Copies of two orders were inserted into the Journal at this point]

Copy of Orders given to William Jennison Leut. of Marines—June 9. 1778.—

As it is essentially necessary We guard against all possible apprehensions of Danger, I in future shall expect that the Serjeants & Corporal of Marines be very particular in keeping a vigilant & cautious Watch whenever they have any Centinels under their more immediate Eye, and whenever All hands are called to Quarters, they are to muster all the Marines on the Quarter Deck, there to await further Orders.—

As soon as they are paraded, the Serjeants of each Division, are to apply for the Keys of the Arm chest—in order that each man may be supplied with a Gun & Cartridge Box filled with Cartridges, or any other proper Weapon for Defence—

After they are supplied with their armour, they are to be paraded & marched to their particular Quarters, there to be kept as closely covered from the sight of the Enemy as possible—

No one is allowed to speak, but the officers Commanding, or the Serjeants, or in their Absence, the Corporal, are most particularly to see that every man's Gun & other accoutrements are in the best order possible—

If any thing should be wanted, they are immediately to inform the commanding Officer therewith—The Serjeants & Corporals are to parade with the Main Body as soon as in Gun Shot of the Enemy—After an Engagement, or otherwise— Every man is to return his Gun in as good Order as when delivered to him—during the time from being called to Quarters & engaging, every man is to pick Oakam sufficient to clean his Gun—

No man whether Officer or private, is suffered to desert the post assigned him, but in case he attempts to run, or fly from his Colours he is to be shot upon the Spot, wherefore I hope, as the Caution is now given, None will attempt to be in the least disconcerted or disheartened, but act worthy of the Glorious Cause in which *they are* Engaged.

Given under my hand on board
of the Boston June 9, 1778

Richard Palmes Capt.

Copy of Capt. Tucker's Orders to the Capt. of Marines on his appointment to the Command of the Boston Frigate

Jany. 1778—

Sir You being Commanding Officer of Marines on board the Ship Boston under my Command—My Orders are that the Commissioned & Non commissioned Officers are to go on board with the Men under your Command & reside there constantly at their Duty—

You are to be particularly careful that a commissioned Officer does constantly lie on board every Night—

You are to obey such Orders as you shall receive from me, or the Commanding Officer of the Ship in my absence—

The Commanding Officer of the Ship is not to give any Marine Leave to go on shore upon their own Occasions, if you have any reasonable Objection thereto

The Marines are to be exercised in the Use of their Arms as often as you shall think proper and are to be employed as Centinels, and upon any other Duty & Service on board the Ship which they shall be capable of—Such as not on Duty that 24 hours as Centinels, therein to be subject to the Officers of the Ship—but they are not to be obliged to go aloft, or be beat or punished for not shewing an Inclination to do so—being assured the ambitious will do it without driving—Those Marines not on immediate duty are to be turned on Deck by a Serjeant to do the Duty as above—

The Marine Arms, Cartooch Boxes, Drums, Fifes & Accoutrements are under the immediate Inspection of you or your Officers—

You are to be careful that the Arms be kept in good Order, & keep a sufficient number of cartridges & Balls in readiness for Action—You are welcome to go on shore when you please, leaving such Orders with your Officers, as for the Safety & good Order of the Ship—

Relying wholly on your good Judgement for the Honour and Glory of your Country in the

Defence of which, make no Doubt of your doing your duty as a faithful Officer—

Given under my hand on Board the Continental Ship Boston this 20th. day of January A.D. 1778.—

[Journal continues]

1778 Feb. 1. The Ship fell down to Nantasket Road—

2 Took our Powder on board—(3) Wrote to President [Samuel] Langdon for a Diploma—[9]

6 A heavy Snow Storm from N by E which drove ashore several vessels—

9 Sold 1 Share of my Cruizing Prize money to Willm. Spooner for £ 33.6.8

13 Capt. Tucker went to Braintree in his Barge & brought the Honble John Adams and suite on Board—[10]

14 Unmoored Ship—

15 Sailed with a W S W wind, & put in to Marblehead at 2 pm—

16 We had great Difficulty in getting under way—At 7 pm in passing halfway Rock, a distance of 5 or 6 miles from the Harbour, Mr. [William] Barron the 1st. Ship's Lieutenant fell overboard, & by catching hold of the Flukes of the Anchor, which he was trying to fish—was safely caught & got on board—Course E S E—

19 Discovered a Sail to the Eastward at 5 am—
At 10 am discovered 2 others supposed to be cruizers—Stood for them till noon At 3 pm One of the Ships in Chase of Us—Courses S E by S—

20 The Ship still in Chace of Us—In the latter part of the day it became very Cloudy with Rain & exceedingly dark—
About 1 at night it began to blow strongly with Squalls—
A Clap of Thunder with sharp Lightning broke upon the main mast just at the upper Woulding—which burnt several of the Men on Deck—a most terrible night—The Capt. of the Main mast was struck with the Lightning which burnt a place in the Top of his head about the bigness of a Quarter Dollar—he lived three Days & Died raving mad—
It was supposed by Officers and Men, that the Lightning striking the Mast had descended down to the Hold of the Ship & forced its way along the Keelson & found its passage thro' the Counter of the Ship—The Horror of Darkness, our imminent Danger, & it blowing almost a Hurricane, filled all with apprehension not communicable to every One The apprehension was expected as following the Grain of Wood—Such a Crisis may be conceived of, but Relative Ideas can not equal—In such a conjunction the Mind, the Soul is absorbed in the Abyss of Reflection—
At 2 pm or perhaps later—The Ship in Chase of us—we making from her as fast as we could—Capt. Tucker having placed Me at the Main Hatchway to communicate the Results of the Soundings in the Well Room—the first report was 4 feet Water in the Hold, the second 5 feet—the 3d. more, which confirmed Us in the Idea as above mentioned, (that is) That the Lightning had passed from the step of the Main Mast to the Keelson, & continuing along the Keelson thro' the Transum and made its Exit thence but Providence ruled, In 10 or 15 minutes the Carpenter told me that there was but three feet

1778

Feb 20 This great Impuls to the force of self Preservation, viz Captn. Tucker went on the Quarter Deck and ordered the Quarter master at the Helm to alter the Ship's Course two points to the Eastward—The Officer on Deck was ordered to set the Fore Stay sail—In the interim the Ship was in pursuit—but on this Alteration of our Course and no Lightning for some Minutes, We passed from the usual View of the Ship pursuing Us so that she lost sight of us, and thus we escaped being captured, as we surely should have been in the morning following—as she was a heavier Ship than the Boston & directly from Halifax, a large Frigate of 36 Guns, well mann'd & by some information from Boston knew of the Ship's Destination with Mr. Adams on board—Capt. Tucker had Instructions not to risque the Ship in any way that might endanger Mr. Adams, & was ordered to land him safe in France or Spain—

22 Discovered a Vessel to Windward—At 7. am. At 12 she fired a Gun to Leeward & hoisted Danish Colours, upon which we fired a Gun to Leeward & hoisted American Colours, & in making sail, had a heavy squall which carried overboard our Main Top Mast—which gave her an opportunity to escape—Course N.

25 Got up a New Main Top Mast, Main Topsail Yard & Top Gallant Mast & yards and bent their Sails—Lat. by Obs. 38.35 N.

Mar. 10 Prize At 11 am Discoverd a Vessel to Windward—gave Chace & came along side at Noon—she fired 3 Guns at Us, one of which carried away our Mizen yard—we returned a few Guns shot & hoisted American Colours, upon which she struck her Colours.—name Martha Our Boats were got out immediately, there was a heavy swell, which prevented the boats getting to the Ship before they had thrown overboard the mail which sunk not more than boat's hook length before the Boat reached the Ship—
This was a Ship of 16 nine pounders Commanded by Capt. Peter Mc. Intosh bound from the Thames to New York

Her invoice (which we found stuck up in the Carlings [cowlings] over his Cabin) was footed £ 97,000 Stirling—

Hezekiah Welsh our 3d Lieut. was put on board the Prize as Master & left us at 7 pm—Our Course S S E—
At 10 pm Capt. Tucker presented me with a small Sword &c

13 At 10 am discovered a Vessel standing for us—At 11 am we fired at 4 pounder to Leeward—At 12 fired a 12 pounder & hoisted American Colours—came along side of her—she proved from Bordeaux to the Grand Banks for Fish—At 2. pm descried a Polacre [11] at a Distance from us—We gave Chase, but not overhauling her—At midnight Mr. Barron, Capt.

1778 Mar 13—Palmes & myself were sitting on the Main Gratings, the Capt. called for the Gunner to fire a 9 pounder Mr. Barron proffered his Service, and went forward to the 2nd Starboard 9 pounder which burst into many large and small pieces—One of which almost cut off a Beam that crossed the Forecastle—he had his right Leg broken, his face blacken'd with exploded powder—& 3 others badly wounded—

He was carried to the Cockpit, to Doctr. [Nicholas] Noel (who had been the Principal Surgeon in the American Army under Genl. Washington) who amputated his Leg & several flesh wounds, & was laid into his Cabin in 10 minutes time—

The Brig proved to be French property bound to the Grand Banks—

14 At 7 discovered 2 large Ships on our Weather Beam—

15 At Day dawn saw 2 large Ships—At 3 pm the headmost hoisted French Colours—We then put about S W by S—

16 Very rough weather—At 2 pm discovered 2 Vessels in our Weather Quarter—made Sail from them—they were Men of War

22 A violent Hurricane overtook us and I lost my hat overboard—

23 Discovered LAND distant about 12 Leagues—We supposed it to be the Asturius or Cantabrian Mountains covered with Snow—[12]

24 Stood in for the Land S.S.E. & at 11 am discovered a Ship lying off Bilboa Bar—
At 2 pm a Spanish Pilot came on board in order to pilot the Boston into the Harbour of St Antonio—but seeing a Brig standing down the Bay of Biscay—made sail for her—spoke her—proved to be a Spanish Vessel bound to St Sebastins—but after vain Attempts to get in, the Wind blew directly out of the Harbour & the Shore being ragged, we put to Sea—

25 Rough boisterous Weather and the Ship very wet—

26 It was impracticable to land on the Spanish Coast, we steered away for the Garonne a large River which runs from the Grand Canal of Languedoc passing by the City of Bordeaux to the Bay of Biscay

Mr. Barron the Ship's first Lieut. died—

27 The Funeral Service was read & Mr. Barron was cast into his watery Grave—
At 1 pm discovered a Sail to windward—

Recd. 1 pair of Ladies Slippers taken in the Martha captured Mar. 10—also 1 yard of Silver Fringe—taken in the Martha—

28 At 5 am saw several Vessels—a large Ship passed us—then a number of other Vessels one of which we spoke—they were all French—
At 4 pm. Mr. [Benjamin] Reed 1st Lieut. was ordered in the pinnace to board a smuggler, but did not succeed—made the Isle De. Rhe

1778 Mar 29—At sunrise I counted 30 Sail of Vessels—At 8 made a Signal for a Pilot, who came on board and told us that we were 15 Leagues from the Corduan [Cordaise] a light house at the mouth of the Garonne 27 League from Bordeaux—

30 At 1 pm we passed the Corduan & sailed 13 Leagues to Blaye—then took another Pilot (because the Kings Pilots were not allowed to go beyond 6 League at one time) & came to until flood tide—for it is a vary rapid & runs at the rate of 7½ knots or miles per hour then sailed up the River as far as Larmoon Stern foremost—This day Mr. Adams, Capt. Tucker, Capt. Palmes & Doctr. Noel were invited to dine on board a French Frigate lying in the River—

31 Landed Capt. Mc. Intosh & the prisoners taken in the Martha Mar. 10—

Apr. 1 Arrived & moored Ship opposite the Chatroons or Suburbs—

For 10 days following the Ship was crowded with Men Women and Children for the Boston was the first American Ship of War at Bordeaux

2 Mr. Adams & the other Passengers disembarked, who were treated with every Mark of Politeness & Respect—

4 Mr. Adams set off for Paris accompanied by Capt. Palmes, Mr. Lee & J[ohn] Q[uincy] Adams—

7 The Capts. Clerk, Peter Cavey a midshipman & myself walked about the City—

9 Fell down to Larmoon where the Ship was graved—

10 At 10 the Ship was moved up to the Pontoons & made fast opposite the City & the beautiful Exchange of Bordeaux—

13 Wrote a Billet to Mesr. John Bonfield the American Agent for 10 Guineas which I received—by Mr. Cooper—

15 Procured a Brig along side to put the movable Contents of the Ship in whilst she was hove down to be graved—

16 The Ship was hove down & the Larboard side graved—

24 Mr. [Benjamin] Bates the Sailing Master & myself were ordered on Shore after 6 men who had deserted the Ship—but could not discover them—

25 Capt. Tucker got Information of them & took them on board & put them in Irons—

27 Capt. Palmes returned from Paris—where he had gone for Orders w[it]h. Dr. Noel—

29 Michael Mc Laughlin a Marine deserted his Post while Centinel and took off 2 others with him—

31 Was ordered on Shore in search of them & patrolled the City & Chatterooms until the next morning but unsuccessfully—

May 5 At 5 am a Ship from Virginia a Letter of Marque was seen coming up the River with American Colours & passed the Boston discharging 21 Cannons—

Peter Cavee [Cavey] the midshipman on duty informed the Capt. who ordered the Ship cleared and 13 Guns to be discharged—

Cavee jumped from the Gang way on to one of the Gun ports which gave way & he fell overboard and was drowned—

The Ship's Colours were displayed halfmast high & at 12 discharged 5 Minute Guns—

[A clothing requisition was inserted into the Journal at this point]

1778. Apr. 1. On board Ship Boston.

Wanted for the Use & Service of Marines belonging to this Ship

40 Green Coats faced with white—
40 White Waistcoats &
40 White Breeches—

The Buttons for the whole to be a plain White Coats to be opened sleeved, & a Belt to every Waistcoat.

In behalf of the Capt. of Marines
William Jennison Lieut. of Marines
[Journal continues]

1778

May 11 Went ashore in search of Deserters & at 12 caught 2 seamen who had been absent three weeks—sent them on board & dined ashore with Mr. Bates & Mr [Benjamin] Tucker the Capts. Brother—In the Evening went to the Playhouse with them and saw the Deserter acted—

12 The Body of Peter Cavee was driven ashore 6 miles above Bordeaux & an Order was sent by the 1st Lieut. to bury, but the Priests would not allow him to do that Office, but performed it themselves with 16 dollars cost—

13 Regimentals for the Marines an Uniform ordered by Congress was brought on board and distributed to all Officers & men—[18]

21 At 9 am the Pilot came on board with Orders to unmoor Ship—

22 At 5 am fell down to Larmoon—At 4 pm came to Anchor 3 miles above Blaye—& discharged 13 Cannon—The Fort returned the Salute—

24 Sunday a Billet from Mr. Bonfield the Agent invited All the Commissioned Officers to dine with him—a large Number of both Sexes were present—The Crew of the pinnace under M. [John] Vickery a midshipman being sent to Bordeaux rose upon him—as going up after the Captain—

25 Capt. Palmes & Mr. Bates were ordered to Bordeaux in search of them

At 8 pm [Musco] Livingston who acted as a Lieut. and was Capt. of a Ship from Virginia with Tobacco on Congress Account, was ordered to go up to Bordeaux to take the Depositions of Serjt. [Jerome] Cazneau & Azema, two of whom had been importuned to enter into a Conspiracy & rise upon the Officers when at Sea & carry her into England—By the Desire of Mr. Livingston I was ordered to accompany him taking one Martin a Swede into Custody as principal in the Conspiracy with Us—

26 We arrived at Bordeaux at 9 am. & it was discovered that the Major part of the Ship's Seamen were engaged in the Black project of taking the Captain alive if possible & kill every other Officer who should resist—

27 A holy day—we went to the Rope dance

28 At 10 am. Embarked in a passage boat with 5 Recruits—At 11 pm arrived at the Ship which had fallen down to Polyac [Pauillac] 5 miles from Blaye—

June 2 Passed down 6 or 7 miles—

4 At 3 pm fell down to the mouth of the River—

6 At 4 am. Came to Sail in Compy with a French Frigate, a Lugger and 40 Merchantmen, & Capt. Jones bound to Baltimore—

7 At 3 pm. made the Isle De. Dieu—

8 Capt. Jones, his Doctr. & Capt. Ward spent the Day on board the Boston—

9 At 11 pm a Lugger came along side, when discovering our Guns, they plied their Oars and got off—

10 Mr. [William] Snowden a Midshipman was confined to his Cabin—

11 Made the Isle De. Groa & barely cleared it—At 1 pm came to Anchor off Le'Orient—

12 This Afternoon Mr. Snowden was ordered to his Duty—

13 At 10 am Sailed, Wind N E.y—spoke a snow from Marseilles to St. Malois—

1778 June 14. At Dawn Discovered a Brig from Bordeaux to Nantz—

At 7 am. 2 other Vessells—Our Course W N W. At 8—2 Vessels—one of which chased the Boston 40 minutes, then hawled the Wind for the other—We then made sail & chased her till half after 7. when she took in Studding sails, Top Gallant Sails, and hawled up her Courses—we followed her Example & both prepared for Action—she proved a 36 Gun Frigate—she hoisted French Colours & we Hoisted American—Saw two other Vessels—

15 Saw the Frigate & Brig again—

16 At 9 am discovered a fleet or Squadron of 9 Ships of the Line as a Fleet of Observation which we had heard of when in France—

At 5 pm we had run them down that is we got out of sight of them—
At 6 pm we discovered a sail which pursued us—it blowing very fresh—but we had the good luck to escape—

17 Saw several Vessels & spoke one—we ascertained them to be Friends—

18 At 10 am All the Commissioned Officers were summoned to the Cabin & held a Consultation concerning our Cruze, the Capt. having discretionary Orders, & they unanimously advised the Destruction of the Fishery on the Banks of Newfoundland.—

19 Discovered 2 Vessels—Gave chase & spoke a Swede from Venice—the other a Dutchman—

20 At Dawn of Day discovered 2 Vessels both Dutch—At 4 am saw an other a head of us—We gave Chace & spoke her—

Prize

At 6 am—she was from Cephalonia in Greece, bound to Glasgow which we made a Prize her name was John & Rebecca—
Capt. Finlay Mr. [Joshua] Goss was put on board as Prize master, who was ordered to L'Orient—Capt. Finley was master of the Brig—which was cargoed with Cream of Tartar, Currants &c
At 10 am saw a sail—At 2 pm saw a White Flag at her Fore top Gallant Masthead—soon she fired a Gun to Leeward—Capt. Tucker ordered a Continental Jack at his Main Top Gallant Mast's head & fired a Gun to Windward.
At 4 pm she came under our Stern, she proved to be Capt. Jones who left us at Isle de. Dieu—He & Capt. Ward dined with us—

21 Capt. Tucker at 1 pm told Capt. Jones that he could not keep Company with him farther South than N W by W. upon which Capt. Jones gave us 3 Cheers & discharged 13 Guns—
Capt Tucker returned 3 Cheers & 4 Quarter Deck Guns—

22 We had a great Battle with the Engine [Fire Engine]—

23 At 9 am discovered a Sail to Windward, gave chace & spoke her at 1 pm—she proved a Jersey man from the Banks with Fish—her name Brittannia Capt. Baker—a prize—put Mr. Gaseard [Giscard] on board as prize master & ordered him to L'Orient—

Prize

24 At 5 pm. Capt. Tucker asked me when on the Quarter Deck if I wished [go] to Boston where possibly we should be lying unemployed, or tack about and make for L'Orient whither All our prizes had been ordered—I answered that I should prefer to go to L'Orient—he immediately went down to the Cabin and called all his Officers & proposed to alter his Course—for L'Orient—they all declared in the Affirmative except Mr. Livingston who was anxious to get to his Family in Virginia—but the American Decision "A majority" overruled him, and we proceeded for L'Orient, where we arrived safe July 2nd

25 At 9 am. discovered a Sail—gave Chase, spoke her at 7 pm. she was bound from the Grand Banks to the Isle of Jersey with Fish—Captured her & put Mr. Vickery as prize master on board her & ordered her to L'Orient—her name Nontom-pa Capt. Anguiteel—At 3 pm. a sail chaced us for some time but seeing us make a prize of the Vessel above mentioned—she altered her Course and left us—We pursued after her—

Prize

26 At 3½ am. discovered the Vessel that chaced Us yesterday, gave Chace & spoke him at 4 pm—he heaving to, & hoisting American Colours & firing a Gun to Leeward—She was a Brig of 12 Guns from Baltimore to Nantz with Despatches from Congress 23 Days out—Capt. [Alexander] Murray—

27 At 11 am. Gave Chace to a Ship, but she outsailed Us and escaped—

28 Fresh Breeze & rough Sea—At 9½ am saw a Schooner, gave Chace & carried away the Main Topsail Clue [Clew], soon after she carried away her Topmast—but she escaped Us—

At 6 pm Very foggy—when it cleared away, saw 2 Brigs—Gave Chace, & spoke One with English Colours, bound from Glasgow up the Streights—Captured her & put Monsr. Tanais prize master, ordered her to L'Orient—her name—Capt. Allen—

Prize

29 Spoke a Petatch, Dutchman—

30 Spoke a Dutchman to Port a Porte—
At 6 pm Saw a Cutter, made Sail after her & came so nigh as to fire a shot over her, she eat us out of our wind, & escaped—

July 1 At 9 am Gave Chace to a Sail which proved a Dutchman—At 2 pm saw 5 Sail—which were French—
At 3 pm made the West Penmarks, & at 5 pm the Isle D'Groa—
At 8 pm came to Anchor off the Island—

2 At 2 pm hove up & moored Ship along side of the East India Pontoons in Port Louis—4 miles below L'Orient—there discharged 13 Guns. Visited by many Officers of the Irish Brigade said to be One of the best in the French Army—

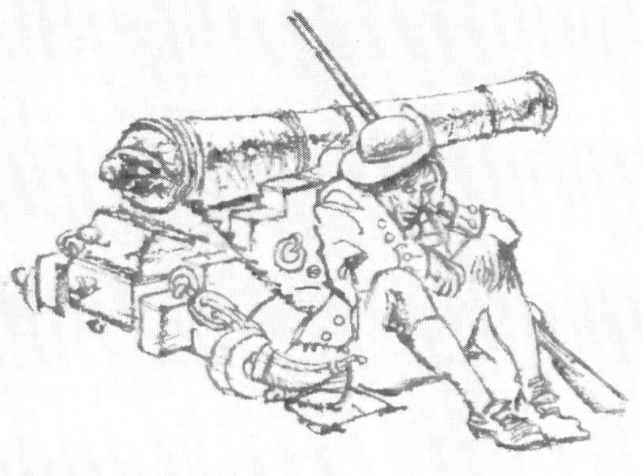

1778 July 4—By the Desire of the Prisoner Capts. I was ordered by Capt. Tucker to attend them on Shore whenever they wished to go—

5 Several of the Marines had permission to go ashore—in the Afternoon sent for their Things—At 2 pm attended the Prize Capts. to the Broker Mr. Field at Port Louis & were all invited to dine & then were waited on to a public Ball—
At 7 pm the Brig Non-tom-pa arrived—upon which we discharged 3 Guns—
At 8 pm the John & Rebecca arrived—both of which came in with American Jacks at their Fore Top Gallant Masts head—[14]

7 At 8 am I was ordered to attend Capt. Tucker to L'Orient on Account of the Marines—At 1 pm went to the Broker's & dined, & received 6 Crowns of him—At 8 pm the Brittania arrived, captd. June 23—

8 Went to L'Orient & spent the Day with Mr. [Titus] Ogden & Capt. Palmes.—
Purchased of Solomon a Jew a few necessaries Amounting to Livs. 256.10 sous

11 At 3 pm the French Genl. of Marine came on board and took away 24 French Marines whom we had recruited while at Bordeaux—

13 Accompanied the prize Capts. to Port Louis—

14 At 3 pm Mr. Reed drove a Centinel from his post & disarmed him—

15 William Roberts & William Smith seamen brot. to the Gang way & flogged 12 lashes each

16 A black man Cuff Jennings for Abusing the Officers was flogged with 12 lashes—

17 At 4 pm Monsr. De Gascard complained of William Atkins' having taken the Command from him & attempting to stab him—The Officers adjudged him worthy of 12 lashes which he received—
At 5 pm went ashore at Port Louis to see a Genl. Review of the Garrison—

19 There was some Disagreement between Mr. Reed the Lieut. & Mr. [Samuel] Cooper the Capts Clerk on Account of an Irish Officer—

20 This morning Capt. Tucker pd the 3d. Class their prize money 13 in the Class amounting to 47 Crowns Each—for Prizes taken June 23, 25 & 28th.

23 Presented my Account of Expenses on Ship's duty while at Bordeaux, and

24 Received the Sum of Livs. 93—

25 Recd. 9 Crowns for my proportion of 1/20th part which was ordered by Congress to be reserved, but was relinquished by them in favour of Captors—
Went to L'Orient & purchased some Clothing Amo[untin]g to Livs. 119 . . 16 su.

Augt. 1 At 5 am the Boston convoyed the 3 prizes to Nantz where they had been bought by the Merchants—At 1 pm passed Bell[e] Isle

2 Came to Anchor at noon off the Crozick [Le-Croisie]—At 2 pm sailed up the River till 8 pm—

3 Sailed up the River as far as St Nazarie & discharged 13 Cannon—

4 At 7 am—the 4 Capts. with their men were sent to Nantz—6 marines were ordered to guard them—

6 Discharged 13 Guns because the Providence Frigate, Capt. Murray[15] & several Continental Vessels, had come down River to our Moorings—

8 At 3 in the morning, Weighed Anchor, & sailed in Company with the Providence & 11 other Vessels for Brest—Saw a sail to Windward—

10 Commodore [Abraham] Whipple dined on board the Boston—All the Convoy left us—

11 Discovered several large Vessels—At 5 pm we hoisted English Colours to a Vessel & she displayed Swedish Colours—

12 We were chaced by a Gaff Topsail Schooner, saw 8 Ships ahead—wore Ship—

1778 Augt. 14. Bad Weather, the Commodore fired several Guns for a Pilot who came off at noon & piloted Us up to Brest by Night—

15 At sunrise the Commodore & Boston fired their Salutes of 13 Guns each which was answered by the Le Bretange[16] with 21 Guns—

16 The French Admiral Count De. Arvilliers came on board the Providence with his Retinue, & on his departure was saluted with the Yards mann'd, Huzzas & 13 Guns—

17 At 11 am 20 Ships of the Line sailed down the River—

18 15 others followed—We were ordered not to unmoor our Ships until the French Fleet had proceeded 6 Leagues from the Shores—

20 We unmoored & came to sail down the River—

21 Sailed out of the River in Company with the Providence & Ranger for the United States—with a Brig & a Cutter in Co.

22 Thomas Shaw & John Churchill flogged 12 lashes for bad conduct—

23 At 10 am Discovd. a Sail & gave Chace

Prize 24 Spoke the Vessel at 6 am—she proved a Brig from London to Pensacola[17] We captured her & the Commodore put a prize master on board—At 7 am discovd. a Sail—proved from N. Carolina to France—
Saw an other sail, gave Chace & she proved from Bordeaux to St Nichola—saw several Vessels—

Sept. 1 At 5 am gave Chace, but a large Ship heaving in Sight, the Commodore hove out a Signal to pursue her—but we could not catch her—Gray & Jones fell overboard—

Prize 8 At 8 am a Signal from the Commodore for a Sail—gave Chace & spoke her—bound from the

Grenadas [18]—which we captured—a prize master from the Commodore—

11 Daniel Sword & several seamen were put on board the Commodore for mutiny—Articles of War read on Board—

14 Discovd. a sail & gave Chace & she proved to be our prize Brig—

Prize 15 Discovd a sail, gave Chace—At 4 pm hoisted English Colour & fired a Gun to Leeward, she bore away from Us, & at 6 pm, Mr. Bates was sent on board her, & struck the Colours—She was from Newfoundland with 2500 Quintals Fish [19]—

16 Gave Chace to several Vessels, but it being very foggy, we lost sight of them—

25 Got soundings in 90 Fathoms Water—

26 " " " 42 " "— Caught some Codfish—

28 Hogged the Ship [20]—Saw a Number of Islands of Ice.

Oct. 6 At 3 pm Andrew Berry died—buried him 7th.

12 At 11 am De Lombe died—

13 Got Soundings in 45 Fathoms Water—Chased a Vessel which proved to be our prize—William Atkins—he had lost his Mate washed overboard by the Sea—

15 Got Soundings in 70 fath'—suppose we were on St Ledger's Bank—
At 2 pm Espied Land—At 5 Bore away for Piscataque River—Made Cape Ann & Thatcher's Island at 3½ pm—Caugh Mackerell—
Arrived within the Light house a 9 pm & anchored—in Kittery—

16 Capt. [Elisha] Hinman a Passenger in the Providence, set off for Boston with our Dispatches—delivered all our blank Cartridges to the Gunner—

20 At 4 pm Lieut. [Hezekiah] Welsh arrived from Boston with Orders to proceed to Boston—

22 At 1 am Unmoored Ship, & a head wind obliged Us to put into Marblehead—In entering the harbour, the Providence & Boston saluted the Forts with 13 Guns—which was returned—Lieut. Welsh informed us of the Loss of the prize Martha—

23 Guns were fired to collect the Crew on board & got under way at 4 pm—passed the Lighthouse at 7 pm, took a Pilot, & anchored under the Stern of the French Admiral the Languedoc at 11—a part of Count D'Estaing's Fleet.

1778 Oct. 24—In passing Castle William now called Fort Independence, we saluted it with 7 Guns, which returned 11—

28 Called on our Agent Johonnot, & recd. 40 Dollars as Share of Prize money for the Brig taken Augt. 24—

Nov. 3 Got our Ships Guns on Shore—

14 I accompanied Capt. Tucker to the Navy Board, who appointed me Purser of the Boston Frigate—from Nov. 14. 1778 to Apr. 29, 1779—

1779

Apr. 29 Our Ship lying in Boston Harbour, & expecting to continue some time, by permission of Capt. Tucker, I went on board a Privateer called the Resolution to proceed to the Gut of Canso—The Vessel belonged to Messrs. Adams, Martin & Blake who had News that 2 Brigs, loaded & waiting for a Convoy to England were there—We did not expect to be absent more than 10 or 14 days—she was commanded by Abel Gore a Master's Mate of the Boston—
mounting 6 Guns & mann'd with 35 men—
At noon we sailed from Hancock's Wharf—got aground on the Middle Ground where we lay till the Afternoon flood Tide—

30 At 9. am Came to sail thro' the Sound—

May 2 Made Persew where we anchored among the Tusket Islands—Foggy Weather

6 Made sail from the Tuskets & passed Cape Negro [21]—

7 Fresh Breeze—At 9 am saw a Vessel ahead of us—chaced & spoke her—she was from Halifax to Annapolis, but had permission from our House of Assembly to trade in both sides—

8 Spoke a Privateer from Marblehead—At 10 saw a Vessel to Leeward, bore away for her on finding her a ship of War, we tried to escape her—but the Wind blew strong and drove us almost under Water—we were constrained to heave to—she fired several shot over our Schooner—
she proved to be the Blonde Frigate of 32 Guns Andrew Barcklay Esq. Commander from the Halifax Station—James Bruce a seamen was recognized by them as a Deserter, & put in Irons—The rest of our Crew were put under the Starboard Quarter Deck under a Guard of 2 Centinels—

9 She cruized to the Eastward, chaced 2 Vessels & spoke them, they were British—

10 She ran into the Gut of Canso after American Privateers, & got aground which afforded them an Opportunity to escape—

14 Arrived at Halifax at 11 am—during our stay with the Blonde we were served with ⅔ Allowance—

In Prison At Halifax 17 Eighteen of Us were carried on Shore, & put into a Goal called the Stone Jug—where we found 23 others taken in the Fly by the Sloop Howe—

25 Five of the remaining were sent from the Ship—Peter Clarke, Francis Tree, Thomas Harrington, James Bruce & John Murray were detained as Deserters—

28 Robert Gordon & a boy entered from the Prison on board the Annapolis of 20 Guns

29 Mr. Brookman & an other prisoner brought in, taken in a Privateer from Baltimore—

June 1 The Blonde, 7 Transports & 2 Sloops of War, sailed from Halifax—

6 Eighteen prisoners brought in, & put on George's Island, taken in the Rambler from New Hampshire—

10 21 of us were sent from the Stone Jug to George's Island—

14 Capt [David] Ropes in the Wild Cat from Salem chaced a Brig in & drove her ashore.—
Thomas Avery went on board a 20 Gun Ship—
11 Men were brought in, taken in a Whale Boat from Marblehead—

1779 June 19. The Raven from Marblehead taken & brot. in—

20 Ship Eagle of 20 guns sailed with a Number other Vessels ('twas said) after Coal from Spanish River—

24 Mr. Dunham from the Vineyard entered on board a Transport for Quebeck—

28 Three Ships, 2 Brigs & 2 Schooners arrived from St John's & the West Indies—

July 5 The Blonde arrived at 1 pm.

7 Seven men arrived, who had broke Confinement at Boston—

23 A mate of the Dutchess of Gordon Ship came to the Goal, & procured 7 men—

29 The Treatment We received at Halifax induced Us to meditate an Escape, and at muster in the Afternoon, it was somewhat favourable for the Effect—as we had sawed off 3 of the pickets previous thereto—At Dusk it became very foggy, so that the Centinels retired into their Boxes—thereby giving an Opportunity which was embraced by
Mr. Lewis senr.—Mr. Lewis—Mr. Ryan—Mr. Hinds—Mr. Thomas—One Cummings who had deserted from Colo. Gorham's Regt. at Windsor & who (thro' a Difference between the Colo. & the Major who knew him) was not put into Close Confinement Mr. Hampton—Mr. Griggs—Mr. Cole—& John Hildreth—but Hildreth had not got through before there was an Alarm—& in half an hour, 20 different parties were dispatched in Quest & pursuit—At 11 pm Hildreth returned to Goal & surrenderd himself up—

30 At 4 pm All the prisoners in the Stone Jugg—were sent to George's Island—

31 The Rachael of 10 Guns arrived from Penobscott, & several others fr. NYork—

Augt. 6 Mr. William's belonging to the Hibernia taken & brought in—

12 The Falcon Sloop of War with a Vessel she had retaken from the Providence, Queen of France & Ranger arrived, Powers prize master—

13 8 prisoners brought into Goal—

15 The Robuste of 64 Guns, the Lecorne Frigate & several other Vessels arrived from Newfoundland with 200 prisoners taken in the Sullivan & Wild Cat and carried into that place—

19 Capt. Ropes was brought to the Island & put into Irons hands & feet—

22 The Robuste, Lecorne, Falcon & sevl. Transports sailed for Penobscot—

24 The Romulus of 44 Guns arrived & brought in the Glasgow of 20 Guns which the Boston & Deane had captured—The prize master was put in Irons for the Audacity of firing upon the Romulus—

30 The Fleet whh. sailed the 22d. All returned, except the Falcon & Sloop Howe whh. they suppose to have been foundered, having been overtaken by a Hurricane in the Bay of Funday whh. dismasted the greater part of them—

Septr 11 A Cartel arrived from Boston with 340 prisoners—Capt. Scott—

15 At 6 pm we were carried on board the Cartel.—

16 At 10 am got under way but they brought Us to by firing a Shot over us—

17 At 8 am Orders came for Us to take our departure—

22 At Sunrising we made Cape Ann—landed at 9 pm—had a pleasant passage—

29 Went on board ship Boston at 6 pm

Oct. 21 At 9 am Mr. Welsh came on board with Orders for me to repair to the Honble. Navy Board & to act as Purser until the one appointed should come from Philadelphia [He was] left there to receive the Ship's prize money for the last Cruize—
At 4 pm procured Slops for the Boston as follows—viz

50 pair Shoes	Bread for 200 men for a 4 month's Cruize 24,000 lbs
100 Blankets	Beef ....10,285¾ lbs
100 pair Stockings	Pork ....10,285¾ do.
28 pair Breeches	Potatoes .....10,285¾ do. say 150 bushels
23 outside Jackets	Peas ....643 Galls.
100 Shirts	Cheese ....2,571½ lbs
Butter ....428 lbs	Vinegar ....106¼ Galls.
Rice 214¼ Galls.	

Rum 1,500 Galls. Candles, Soap & Sundries—
Which I receipted for as midshipman—

24 Went on board at Sunrise & made the Ward Room Indent as follows
1 Doz. Plates—2 Dishes—1 doz Knives & Forks—2 quart Mugs—2 Table Cloths—1 doz spoons—2 barrals Sugar—150 lbs. Coffee—2 Coffee Pots—½ doz tin Coffee cups—

Nov. 6 Went down to the Ship which lay in Nantasket Roads—

11 At 4 am unmoored Ship—

12 At 8 am arrived in Piscataqua River & came to—Mr. Grant was sent to Capt. [John Foster] Willams of the State Ship Protector of 20 Guns—

13 At 11 am Sailed to Newbury Bar to convoy the Protector to Boston, but he could not get over Bar under 8 or 10 days—then returned to Nantasket Roads—
The Thames Letter of Marque, a french Frigate & Lugger sailed—

18 15 men were put in Irons for mutinizing—

23 At 9 am upon a Signal being given—At 11 am The Ship Providence as Commodore, Queen of France & Ranger, The Irish Tartar of 20 Guns & several Merchantmen, accompanied by the Boston sailed, Our Destination not to be known until we had passed Bermudas—
A man belonging to the Providence (while passing the Light House) jumped overboard & was Drowned—passed Cape Cod at 10 pm distt. 3 Leagues—from whence we took our Departure. Course S.S.E—

24 Discovd. a Sail, gave Chace—she was bound from Guadeloupe to New London—

27 A heavy Gale for several Days—Reefed the Main Sail, which was torn from the yard—

Dec. 1 At 11 am discovd. Land, bearing S.W.—distant about 7 Leagues—which we judged to be Bermudas—

2 Cloudy with Rain—Several Vessels in sight—At 8 am chased a Frigate & a Brig—til 9 pm when thro the darkness of the night, they escaped—
At 5 pm the S.W. part of the Island bore N.E. about 5 Leagues, lying in Lat. 32.40 N & Longte. 63.50 W. from which We take a new Departure—

6 At 6 am discovd. a Sail to Windward—gave
Prize Chase—The Providence spoke her at 9. am She was a Cruiser of 14 Guns from [St.] Augustine—he put a prize master on board—

13 Samuel May the Capt. of the Fore top fell overboard at 8 pm—

16 Got Soundings in 20 Fathoms Water—

17 At 6 got Soundings in 17 Fathoms—at 11 am in 14—At 4 pm discovd. Land, which we judge to be the Shore between St Johns & Beaufort—sounded in 7 Fathoms at 1 League from the Shore—bore away till Midnight & sounded in 10 fathoms Water—

18 Stood down the Shore to the Northward—

19 Came to Anchor 4 miles from Fort Moultrie now called Sullivan's Island—
At 9 pm sent Mr. [Isaac] Collins to Town for a Pilot—

20 At 4 pm Mr. Collins returned with 4 Pilots—

21 Attempted to get over the Bar, but did not effect it—

22 At 3 am George Bell a seamen died—

1779 Dec 23—At 4 am came to sail, & got over the Bar, & anchored in 5 fathoms hold-till Floodtide when we sailed up to Charleston—got aground at 11—

24 Having floated, sailed up & moore Ship opposite the Town.—

31 Francis Lateff tried by a Court Martial & punished with 260 Lashes—& a man of the Providence with 39 lashes—

1780

Jany. 4 A jury of Carpenters sat on the Queen of France & condemned her as unfit for Cruising—

5 The Ranger ordered to sea—

7 Francis Mehany died—

10 Painted Ship, Guns, Carriages, Pikes, Spunge Staffs, & every Article on Deck—

11 At Noon the Lively of 22 Guns which had been taken by the French, fell down to Rebellion Road

12 At Sunrise a Flag was hoisted at Fort Moultrie for a Ship off the Bar, which proved to be the Ranger—In her Cruise she chaced a Ship with 200 Recruits bound from N. York to Georgia—but did not capture her—

16 At 11 pm Capt. Tucker sent for me into his Cabin, & being informed that I was disagreeably confined to the Company of his petty Officers—allowed me to act as a Volunteer which by Advice of Capt. Palmes & Lieut. [Jeremiah] Reed of the Marines who was appointed in my stead by my detention in Halifax prison—I accepted with Gratitude—

17 Some Altercations between the Officers of the different Ships concerning an Equal Share of all prizes taken, by reason that no more than two could be permitted to cruise at a Time—

19 At 11 am Genl. [Benjamin] Lincoln with his Aids Colo. [John] Laurens with several Others, came on board—At 2 pm they disembarked accompanied

by Capt. Tucker for Town, when the Yards were mann'd, & 13 Guns discharged with 3 Huzzas—

20 At 10 am Unmoored Ship & fell down to the Commodore, having Orders to fit for a short Cruise—

21 Mr. Ash let me 100 Dolls. Continental Money— The Commodore & Ranger fell down to the Road, in order to put to Sea, by the first fair Wind— confining the Boston in Port, which was considered as a precedent yet unknown in the British Navy—

23 At 10 am Matthew Watson a Seamen fell from the Starboard Main yard Arm athot. the Muzzle of a 9 pounder, which caused his Death in 10 minutes—

Prize At 4 pm a Brig hove in sight, which being decoyed by the Commodore's boats & a pilot boat, was made a prize of.—she was bound from Newfoundland to Savannah in Georgia—
In her were taken several British Land Officers & 19 Soldiers mounting 12 Guns, named the Lady Crosbie.—

24 The Commodore & Ranger, the French Ships Lively of 22 Guns, Earle of 20 & LeSensible of 32 Guns fell down to 5 Fathom Hold—
At—pm. They sailed over the Bar with a S. W wind—

27 At 9 am The Flag was hoisted at Fort Moultrie, & 2 Sloops came in, prizes to the Tender Capt. Page—they were part of a Fleet of 90 Sail from NYork to Savannah—At 11 am the Commodore & Ranger hove in sight, & brought Information of 13 large. English Men of War being seen—

28 40 men were sent on board the Brecole [25] to fit her for Defence—& I was desired to act as purser on her—

1780

Jan. 29 At 9 am Discovered 2 Men of War, when the Commodore & Ranger crossed the Bar.

30 The Boston fell down to the Road—

Feby. 1 The Commodore with the several Capts. went down to sound & buoy the Bar

8 Several of the St. Eustatia fleet arrived—

This Afternoon the Lady Crosbie fell down along side the Providence to be fitted as a Privateer—

5 1000 of the North Carolina Line arrived under Brigr. Genl. [William] Woodford.—

6 A large Ship appeared off the Bar—

7 She came up, having been chased upon the Bar—when she was forced to throw over 17 hogsheads Rum & a large Quantity of Salt— She brought Information that Capt. Manly in the Jason of 22 Guns was taken by the Surprise Frigate of 28 Guns, & carried to England.—

10 Invited to a party of Pleasure to the Quarter House—

18 I wrote to my father by a Vessel bound to Salem—Mass—

26 At 7 am discovd. a party of the Enemy at Fort Johnson which the Americans had demolished the 20th.[26] The Providence & Ranger, who lay within reach fired upon them—
This Fore noon a Flag was hoisted on the S.E. Battery of Charleston—

27 A Flag was sent to the Enemy from the Queen of France who lay in Ashley River—8 miles from Fort Moultrie, & opposite the S.W. part of the Town—

28 The Commodore came on board the Boston, & ordered the Boston & Ranger to Fort Johnson in order to disposses the Enemy if possible & they brought 2 field pieces to play upon Us at 2 pm—

Mar. 2 The Providence, Boston & Ranger fell down opposite the Fort, & moored in a Line—
At 11 am—The Brocole, Notre Dame & Galleys fell down to our Line—
At 3 pm—The Galleys lay off Fort Johnson, & kept up a scattering Fire upon the Enemy—
This Night was taken up a Canoe with 2 Negroes bound to James' Island—

4 2 Ships & 2 Schooners appeared off the Bar—

7 At Midnight Our Cutter being on Row Guard, took up a Canoe with 1 Negro man 1 Woman—

8 A number of ships appeared off the Bar— The Commodore & the Capts. went in the Commodore's pinnace to reconnoitre them—At 3 pm the English Commodore fired a Gun, when they all came to Anchor [87]—

9 The Enemy were busy in Buoying the Channel, till Capt. [Hoysteed] Hacker in the Lady Crosby, Notre Dame of 16 Guns, & a Galley were sent to drive them on board—

12 This forenoon the Enemy appeared in Wappoo Creek with 6 square rigg'd Vessels-under cover of a 6 Gun Battery, which they opened & began to play upon the Galleys which lay in Ashley's River, which caused them to remove down the River—

1780 Mar. 12. This afternoon 4 Stone Jarrs were procured for the Boston to be used as Stink pots.

14 A plan was laid to moor Boats in a line, althoht the Entrance of the Harbour capable of bearing a Cable to bring their Ships up, as they attempted to pass—14 Boats were procured & moored directly opposite Fort Moultrie—

16 This day One Balentine was hanged for secret Correspondence with & attempting to sacrifice several Posts, to the Enemy— The Enemy by Sea busily employed in carrying the Marines ashore being entirely useless for passing the Fort—

20 This morning at Daylight, the Fleet came to Sail & passed the Bar at 6 am, where they came to anchor in 5 fathom Hold.— At 11 am Genl. Lincoln came down, & in the Evening all the Captains held a Council of War on board the Commodore, at midnight they returned to their Ships—Whereupon Every ship unmoored.—

21 This morning all the Vessels sailed up & came to opposite Gadson's [Gadsden's] Wharf E N E of Charleston—

24 Got our Guns, Powder & provisions out & put our provision under the Exchange—

25 Lieut [Silas] Deval was ordered to command a Fort at Governor's Bridge, to which I was attached—[88]

31 The Enemy's fleet weighed & passed Fort Moultrie under a severe Fire from the fort—At 6 an Ordnance Store Ship in passing had her Rudder shot away & ran aground—her Crew took to their boats having first set her on fire, making a terrible Explosion at blowing up.—She mounted 20 Guns— The Fleet came to anchor under Fort Johnson at 5 pm—Not any Vessel was stopped by the Fort—except one had her Fore Top Mast shot away & fell overboard—

Apr. 8 The Enemy during last night hove up 3 Redoubts in front of our lines.

10 A summons was sent to Genl. Lincoln demanding a Surrender of the Garrison, Stores &c

15 The 3 Batteries above mentioned. were opened upon our Lines—

21 A Flag was sent to Genl. [Henry] Clinton for a truce of 6 hours to digest Articles of Capitulation—

27 The Fort at Lamprie's point was Evacuated, & the Troops with the Major part of the Garrison of Fort Moultrie were ordered up to defend the Town—

28 Colo. [Francis Marquis De] Malmode [Malmedy] was surprized & taken with his party at Hadarel's point—

May 6 Fort Moultrie surrendered—& Genl. [Isaac] Eugee [Huger] surprized & Colo. [William] Washington obliged to swim across Santee River—his Cavalry cut all to pieces—

8 A Flag from the Enemy with a second summons—

1780

May 9 A flag was sent to the Enemy with Alterations of the Articles of Capitulation—but they were refused—

11 A Flag to the Enemy accepting the Terms offered—

12 At 2 pm the British Grenadiers marched in, & took possession of the Horn Work—Flags at all the Forts were struck & the Forts taken possession of by different parties of Soldiers—

June 10 The Commission Officers by Sea had permission to charter a Schooner for Rhode Island & they sailed at 6 pm—Those of Us who had been taken belonging to the Sea Service whether Publick or Private were sent on board 3 Vessels as Cartels—

13 Sailed & crossed the Bar at 4 pm—Course E. S. S.y

18 At 9 am discovd. a Sail in Chace of Us— At 3 pm she fired a shot after us At 3½ she fired another Shot—in which we hove to—She proved to be the Sir George Collier of 22 Guns from Bermudas on a Cruise—

23 We arrived at Chester 30 miles below Philadelphia—and were transported in a Schooner to the City Took Lodgings in Spruce Street until the Eightienth day of July—

July 19 Left Philadelphia & sailed to Trenton, where I travelled

30 to New Haven on Foot—

31 Embarked in a Perryawger [89] for New London—

Augt. 2 Set off from Groton & arrived at New Port—on Foot—

5 At 3 pm Took passage for Providence—

14 Set off for Boston & arrived on the 16th. at 11 oClock—

[End of Journal]

[1] On 3 June 1776, Congress ordered that 13,800 militia be raised in New England to reinforce the army at New York, and serve until 1 December. Massachusetts was requested to furnish 2,000 of this number.

[2] Captain James Foster, with his company of 86 men, marched from Mendon, Massachusetts, on 22 July 1776.

[3] Colonel Jonathan Smith commanded a regiment of militia raised in Berkshire and Worcester counties, Massachusetts.

[4] The Marine officers on the frigate *Boston* were Captain Richard Palmes, First Lieutenant Robert McNeill, a nephew of the captain, and Second Lieutenant John Harris. Jennison at this time had no other status than that of a volunteer.

[5] The prize was the 80-ton brigantine *Littleton*, William Johnson master, laden with duck and cordage.

[6] When the *Fox* was captured, Marine lieutenants McNeill and Harris of the *Boston* were detached to her and Jennison became acting lieutenant of Marines on board the *Boston*.

[7] Pownalboro, now Wiscasset, on the Sheepscott River, Maine.

[8] Captain Hector McNeill preferred charges against Captain Palmes and Midshipman Hogg. They were subsequently tried by a court martial and the latter was dismissed from the service.

[9] President of Harvard College from 1774 to 1780.

[10] John Adams had been appointed by Congress to succeed Silas Deane as a Commissioner to France.

[11] A two or three masted merchant vessel used chiefly in the Mediterranean.

[12] Mountain ranges on the north coast of Spain.

[13] It has been a popular misconception that immediately after the Marine Committee specified the Marine uniform on 5 September 1776, all officers and men were immediately supplied with them. By Jennison's requisition of 11 April, and his entry of 13 May, it is evident that insofar as the *Boston's* Marines were concerned, they were not wearing the prescribed uniforms until almost two years after it had been specified.

[14] This was not the *John & Rebecca*, but an unnamed brig taken 28 June 1778.

[15] Captain Alexander Murray commanded the Maryland privateer brig *Saratoga*, of 20 guns.

[16] The *Bretagne*, of 110 guns, was the flagship of the French Fleet.

[17] The brig *Sally*, Captain Ward.

[18] The brig *Friends*, Captain McFarland, bound to Glasgow.

[19] The snow *Adventure*, Captain Symes, bound to Port-au-Port.

[20] Cleaned the ship's bottom with brooms called hogs.

[21] Northeast of Cape Sable, at the lower tip of Nova Scotia.

[22] The recaptured vessel was the brig *Three Sisters*.

[23] Not the *Glasgow*, but the ship *Glencairn*, taken 9 August and ordered to Boston.

[24] William Ash had been appointed purser of the *Boston* after Jennison had secured a leave of absence the previous April.

[25] The *Bricole* was a French armed merchantmen taken over by the South Carolina Navy for harbor defense.

[26] Fort Johnson had been erected on a point of James Island below the city and commanded the south side of Charleston Harbor.

[27] The British naval commander was Rear Admiral Marriott Arbuthnot, and his flagship was the ship *Europe*.

[28] Governor's bridge crossed an inlet of the Cooper River directly above the anchorage of the frigate *Boston*.

[29] Piroque, a two-masted, flat-bottomed boat having a lee-board, and decked at the end or ends but open in the middle.

APPENDIX E

Journal of Joseph Hardy, Captain of Marines

Without his personal record, any impression of Joseph Hardy would have remained dim. Little more is known of him than that his service in the Marines began on 25 June 1776. Later, when a captain, he joined the 36-gun frigate *Confederacy* as the Marine commander, then sailed with her to West Indies in the distinguished company of the French minister and his family, as well as the American ambassador to Spain, his family, and secretary.

Hardy begins his journal on 20 December 1779 at Martinique where they were forced to harbor by "late misfortunes" that dismasted the *Confederacy* and left her adrift in mid-Atlantic. A decision to head for calmer seas and port in the Caribbean took the party to Martinique. There the ship repaired and refitted while Captain Hardy attended to his naval duties, interspersed with private business and entertainment ashore. Like the other officers, Hardy felt embarrassed in Martinique's bustling society by his lack of funds. He alludes to it often in the journal.

After 4 February, the *Confederacy's* captain gave up hope of properly refitting her and informed Congress that he would return to America—provided they had no further orders for him—after receiving the essentials to sail. The ship warped out of the St. Royal harbor, under the protection of the French fleet, and made sail on 13 March. After a call at St. Pierre to complete repairs and taken on provisions, she finally got away on 30 March 1780. A month later she was back in America. The rest of the journal is devoted to the technicalities of sailing home, the continuation of repairs after arriving, and a second voyage to the Caribbean.

Very few similar journals kept during the period have survived. This one details the daily affairs of life on board a continental ship, including some evidence of what the Marines did. Captain Hardy frequently mentions exercising them, even stating in one entry that "for the first time [they] performed beyond expectations." This provides an indication of an attempt to train and discipline shipboard Marines. Little enough is cited elsewhere about them to make their duties known.

This transcription is intended as a letter-by-letter reproduction of the original. There seemed little need for tampering with Captain Hardy's style. The log form of the journal makes it easy to follow and his sentences are generally punctuated, although periods are omitted sometimes and the sentences run together. The occasions will cause the reader little disruption, if any. The discrepancies that exist then result from the problems of reproducing handwritten work in printed form rather than the intrusion of the editor. Additions to the text or corrections in the spelling are bracketed; all places left blank by Captain Hardy are noted. Footnotes elaborate upon the text or define important place, names, and terms.

It is assumed that the reader has a knowledge of common sailing terminology, so words and phrases describing a ship's sails and rigging are not defined. Some others, dealing particularly with the ship's movements and repairs, are. The original journal is in the possession of Mystic Seaport Incorporated, Mystic, Connecticut, by whose courtesy this journal is reproduced.

Remark's &c. on Board the Confederacy Frigate. now at St. Pierr's.

Monday 20th Decemr. 1779

I returned on Board this Morning from Town.—The Governor of this Island, several Gentlemen of distinction, and a few Officers of the Army and Navy came on Board this morning to Breakfast and view the Ship.—Ships Crew employed about sundries.—went ashore again in the afternoon & returned in the Evening without anything particular happening.

Tuesday 21st. Employed myself all this day on Board writing home.—the Weather very warm notwithstanding it's very agreeable and cool on board the Ship having the advantage of the Sea breezes. Cleaned Ship fore and Aft &c.

Wednesday 22d. At 4 this Morning Mr. Gerrard went in our Barge accompani'd with Mr. Jay and Bingham to Fort Royal.—[1] as he passed the Ship he was saluted with 21 Guns as a farewell One of the French Frigates in this Harbour sailed the forenoon, soon after a Sloop of War and several Merchant Ships arrived, from Fort Royal.— At Meredian 4 large Ships supposed to be English in Chace of three french Frigates from Granada they were some distance to Leeward. This afternoon being in Town had good information from several Gentlemen of Character that D . . . from whom I bought my Bills had formerly lived here and run away to America greatly in debt after defrauding a number of Gentlemen and otherwise bears a most villianous Character. From other circumstances and all I can collect I too much fear my Bills is not worth a Shilling. a disagreeable piece of News to find myself here a Stranger, Moneyless, Friendless and Bills not even Saleable. too severe a stroke on me after escaping the late misfortunes.—

Thursday 23d. This Morning met Mr. P.—in town who acted the part of a polite Friend in respect to my private business.—Dined with Capt. Daniels on board his Brig lately from Boston. in company with the Doctor & Purser. after dinner went ashore and took a little Exercise— at 6 PM. went in company with all our Officers to the playhouse, The Piece that was performed is an English Farce called the Deserter and much esteemed by Europeans but in the French is much altered from the English Mode and ruined with a multiplicity of Singing it was uncommonly short beginning and concluding the Evenings entertainment without any addition. The Sloop Hancock Capa. Leister, Schooner Dean Capa. P Richards and another Sloop arrived this Evening from Statia and lately from N London in Connecticut.—

Friday 24th. A Dispute having happened a few Nights ago in a Bagnio, between Mr. P. of our Ship and an Acting Lieutt. of one of the French Frigates in this Harbor a Challenge ensued. in consequence thereof the Parties with their proper attendants was this morning at 6 O'Clock conducted in our Barge to the Northward of the Town where they Landed and exchanged two Shot without either party's receiving injury afterwhich the Enmity subsided and the Partys became Friends. It may be easily judged what was the cause of this Dispute from the place where it happened, but it is too frequently the cause of Strife and promoter of Bloodshed.—Mr. Gerrard and Jay returned this morning from Fort Royal and ensured our being fitted there with all expedition, but as it will take a considerable time to compleat us the French Admiral has favored them with a Frigate to carry them to France, so that we shall return to America from hence except orders to the contrary from Congress.—The Iphigena Frigate of 32 Guns and the Sloop of War Series of 18.—Sailed this forenoon on a Cruize.— Unmoored Ship this Evening in order to go to Fort Royal tomorrow.—

Saturday 25th. Hove up this morning and dropt outside of the French Frigates in order to come to Sail.—Dined on board with the Officers tho' it was Christmas Day after which we went to Town.—This Day is highly celebrated here by the Inhabitants in their Churches. some amusing themselves at various Entertainments. whilst others of the World are carrying on their usual business and grasping at every glittering Bauble they can see.—Spent the Evening and Supp'd at Mr. Binghams with Mr. Jay and Lady, of whom I took a final leave and returned on board.—Light Winds and Shore business prevented our Sailing today.

Sunday 26. At Daybreak this morning hove up and got under way with a light breeze off Shore. kept along shore to the Southward with Winds very variable. passed several neat Village's and small Batteries.—at 12 O'Clock got into the mouth of Fort Royal Bay. at 1 the Town and Shipping in sight about four Leagues distance.—The Wind being down the Bay obliged us to Work the Ship up Tack & Tack which she performs beyond expectation. The Land about Fort Royal appears more even than at St. Pierrs[2] and some well cultivated Plantations— about half way this Bay on a small Island called Pigeon Island near the South shore is a very formidable Fort with a number of heavy Cannon and Mortars. On the North side of the Bay another small Fort situated on Point Negro about four Miles below the Town another opposite Fort Royal on the S. shore with Fort Borboun the back of Fort Royal appearing very large and spacious— at 5 PM a Pilot or Harbor-master came on-board and at 8 came to Anchor above the Ships.—

Monday 27th. The Pilot came onboard this morning got the Ship underway and run into the Carnage [careenage] it's a small Bason but deep where all the Kings Ships heaves down, situated on the East side of the Fort. Adml. LaMot Piquet lays off in the Harbor with 6 Sail of the Line, whom we saluted as we passed which was returned with Eleven Guns. Also Saluted the Fort Royal which was returned in the same manner. at 12. Moored Ship head and Stern close under the Walls of the Fort all hands employed in the afternoon striking Yards & Topmasts.— went ashore in the Evening to view the Town of Fort Royal found it to be a very uniform Built place situated on a perfect Level it's about half as large as St. Pierrs but the Buildings not so good in general there are several good Walks and Malls about the Town. The greatest part of the Inhabitants here are Mustees—[3] very little Trade here tho' the seat of Government.—

Tuesday 28th. Having private business at St. Pierrs went ashore this morning with Mr. V. at 10 took passage in a Canoe from Town.—These Canoes will carry 6 or 7 hh'ds [hogheads] Sugar and ply with Freight or Passengers from here to St. Pierrs. they are rowed by five or six Negroes whose Lives appears to be as wretched as any part of the Human race. some of them are chained by one leg to the Boat and others shews the stripes of cruelty on their Body's in this manner these unhappy Mortals row in these Boats for Weeks without 10 hours intermission and as Naked as the moment of their Birth not even the Galley Slaves in Barbary is more misirable.—It is not only these that feels the stripes of inhumanity but many on shore are to be seen with a heavy Iron ring round his Neck from which leads a heavy chain to another ring around his Waist and from that to another Ring round his Anckle. and others dragging by one foot 50 or 60 lb. of Chain and in this situation are obliged to go thro' their usual services. But to return to our passage we arrived at St. 3 PM. but contrary to expectation found that Mr. Jay and Lady with Mr. Gerrard, Carmichel Penet and Livingston[4] had that Morning Sailed for Europe in the Aurora French Frigate, which frustrated our intentions.— spent there at sundry places.—

Wednesday 29th December. several Vessels arrived to day at St. Pierr's from Statia. Dined & spent the afternoon with Mr. Bingham the Continental Agent. in the Evening at the Dancing School where the Young Ladies and Gentlemen of the Town practice twice or thrice a Week they Dance extraordinary well particularly Cotillions, afterwards Supt with Mr. Byrne at his Lodgings and retired to our own.—

Thursday 30th. The forenoon spent in transacting a little trifling Business.—after Dinner procured a passage in a small wherry Boat at 4 PM set off with the addition of Mr. W.—in company—on our passage saw a French Frigate beating up Fort Royal Bay. at 9 O'Clock arrived safe at the Ship and found all well but in low Spirits having but an indifferent prospect of getting Spars to fit us. since my absence the Powder and Guns sent ashore. and the Admiral has demanded and taken on board his Fleet all our Frenchmen to the amount of 40 or 50 a good riddence of Lubbers. The french Fleet is going out on a short Cruize and being in want of Men the Admiral has borrowed 30 Men from us for 20 days amongst which is a Serjt. & 15 Marines[5]

Friday 31st. Intillegence received here yesterday from St. Lucié [St. Lucia] that 15 Sail of the British fleet had Sailed from thence their destination unknown. a Crew employed rigging the Cutter sent a number of water Casks ashore.—Admiral La Mot Piquet aboard this afternoon viewing the Ship—no spars procured yet. at sundown the Amphitrite french Frigate came to Anchor in the Fleet having carried away her foretopmast and Maintop Gallant mast in a flow of Wind at the mouth of this Bay in the forenoon. These flows of wind comes off the Land unexpected and are very common.—Working the Chain pumps this Evening one of the Men had his Arm caught between the Crank and Stantion which tore great part of the Flesh off his Arm is feared it will cause Amputation. he sent ashore immediately to the Hospital.—

Saturday 1st. January 1780. The Gunners Crew employed carrying Shot ashore &c Mr. Bingham came to Town. spent the afternoon ashore being Newyears day—in the Evening returned on board.—at 7 in the Evening the Centry on the Larboard Gangway fired on a Man swmming away from the Ship without injuring him. the Jolly Boat pursued him alongside the 74 laying along side of us; he proved to be one of her Men which had Swam on board our ship in the evening and was returning back. but not answering the Centry when hailed he was obligated to Fire on him.—

Sunday 2d. Several Gentlemen on board this morning viewing the Ship, also Mr. B. [Bingham] our Agent, who informs us he cannot procure Spars to fit us immediately, therefore must wait the arrival of a Spar Ship which is expected here, so that from appearances at present there is no probability of our leaving here this several Months. Rigged the Cutter this afternoon Sloop fashion. nothing more particular.—

Monday 3d. January Rainy Foggy Weather this morning. the French fleet 6 Sail of the Line & 2 Frigates hove up came to Sail and run out to Sea their destination unknown. at 9 O'Clock Capa. 2d. Lieutt. & Purser Sailed in the Cutter for St. Pierrs got two Spars from shore to take out the Mizenmast. employed myself all day writing home.—The Cutter returned in the Night with the 2d. Lieutt.—

Tuesday 4th. Early this morning appeared in the Mouth of this Bay an English Fleet of 14 sail having intillegence I suppose that Monsieur Piquet was going to Sail, but they were a Day too late the Bird they seek'd had flown. at 11 they made Sail and stood to the Southward. as soon as this Fleet was discovered all the Ships in the Bay was ordered into the Carnage. In the morning went with Mr. V. in the Cutter to the other side of the Bay. hauled the Seine several Times but without success, which obliged us to purchase a few fish at a Neighbouring House which we had cooked and served for a very good repast. at 1 made Sail again into another Bay where we made another attempt attended with the usual success. We then throwed by our Nets and went up to an adjacent Planters house where we were received and treated with great politeness we found the Master of the House to be a Man of good property.—after regaleing us with a Bottle of Wine and fruit he conducted us thro' his Gardens, were I found all kinds of

Sallads in great perfection with a variety of West-India fruits such as Cocoa Nuts, Bennania's, Plantain, Oranges and the Garden surrounded with a hedge of Lime trees growing about as high and thick as the Thorn hedge Trees are full of small thorns and the leaves like the American Laurel always Green. at 5 we parted with him after receiving many pressing invitations to renew the Visit.—got to the Ship at sundown and found the Mizenmast and stump of the Bowsprit taken out.—

Wednesday 5. Got up all our Sails to dry. the Cutter sent to water this morning early to Casnapiere Bay, on their return the Wind blowed so fresh they cou'd not get here before 2 P.M. myself ashore on Ship business in the Morning. Carpenters at work ashore on the Mizenmast, the Cutter with the 2d. Lieutt. went to our former Fishing Grounds. went to Town in the Eveng.—

Thursday 6th. Very warm Weather. All Hands employed about sundries. The cutter went to St. Pierrs this morning for the Captain. spent the Evening ashore nothing further particular.

Friday 7th. The Carpenters ashore Tongueing the Mizenmast, others preparing to Caulk the Ship in the Evening the Capa. & Purser returned in the Cutter from St. Pierrs and brought with them a quantity of Sauce and Coffee for the Ships use. came round in the Cutter a french Lieutt. taken in one of the 3 Frigates (the 3 being taken) that saw chaced by 4 74's whilst we lay at St. P.—The Officers had everything plundered from them when taken except the Cloaths they had on, and in many other respects treated Cruelly.—

Saturday 8th. Carpenters caulking the Ships sides. Rigg'd the Foreyard fore and aft on the fore Mast in order to get in the Bowsprit (when we get one). Capta. Harding ashore all day endeavoring to procure Spars for the Ship but can get no satisfaction from the Intendant.—here we have been three Weeks and no nearer being fitted than the Hour we arrived.—the inattention that we have been treated with here as a Vessel in distress and an Ally calls loudly to his Christian Majesty for Redress as it is most undoubtedly a breach of the Treaty of Alliance—After Meredian a French frigate came into the Carnage to heave down.—

Sunday 9th. Spent the day on board a Sermon Preached on board in the forenoon. Dined with the Captain.—the 3d Lieutt went to St. Pierrs on Ships business. Mr. Minot a Mercht. from Boston spent the day on board.—a Flagg from St. Lucee but last from St. Piers came up the Bay and Landed several Prisoners. came on board in the forenoon 3 Americans just made their escape from St. Lucee, they bring no intillegence of certainty.

Monday 10th. This morning Capa. Harding with the first Lieutt. went in the Cutter to St. P. in order if possible to expidite our refitting.—at 11 I went ashore to the Fort and Town on Ships business amongst other things waited on the Intendent for a few Articles for the Ship but to no effect. Dined on board after dinner went in company with some of the Officers up to Fort Borboun situated about a Mile to the Nd. & Ed. of the Town on a high Mountain and very Commanding piece of Ground its the largest and most extensive piece of work of the kind I ever saw the Walls is upwards of 50 feet high surrounded with a very wide Ditch which we were informed can be filled in a few minutes on occasion from Water works in the Fort. but the Commanding Officer not being in the Fort we cou'd not view the Water works or many other curosities. I imagine the whole Works comprehending Redoubts &c extends a Mile and half the whole looks down into Fort Royal.—It was this hill the English availed themselves of last War unknown to the French and beat down the Walls of Fort Royal and obliged the Island to Capitulate. since, the French has taken the hint and fortified it themselves.—returned on board in the Evening.—

Tuesday 11th. This Morning took a view of Fort Royal. it's situated on the South side of the Town on a Penninsula the gratest part of it surrounded by the Bay. it's large and very strong composed of a great number of Batteries mounting in the whole to upwards of 130 pieces of heavy Connon & 6 large Mortars with very large Barracks sufficient to contain 6 or 7000 Troops—also several large Magazines well fitted. in this Fort there is a large Cav'ty cut out of a Rock sufficient to contain all the Inhabitants of the Town secure in case of a Seige.—Latter part of the day waited on the Marquis de Bouille—the General, to demand two Marines who had lately deserted and Listed into their service upon which he immediately released them.—Tho' seemingly against the inclination of the Officer that Listed them who amongst the rest endeavoured to add to our distress by wishing to detain them

Wednesday 12th. Rainy thick weather all day. the Capta. returned in the Cutter from St. P. Mr. Byrne came on board with him. the three Men which I retrieved yesterday was brought on board under Guard and confined in Irons.—Several of the Seamen having Liberty to go ashore in the afternoon returned on board much intoxicated with Liquor and in consequence of it like all Sailors began to grumble at their wage and insisted on being sent to St. Lucee to be exchanged these were joined by several who was taken out of English Privateers and in a few minutes the whole ship appeared in a flame of Mutiny. the Ringleaders were immediately confined in Irons hands and feet and some of the most obstropolous [obstreperous] were Gagged, which soon cooled them down. however they were kept in confinement all Night.—

Thursday 13th. Released all the Prisoners from confinement under promise of good behaviour Prevented all Bomb boats from coming alongside which was a great means of our Crews getting so much Liquor. The Captain has agreed with the Officer of the Flag to exchange these men for Americans. imployed myself all day writing home.—The Flag sailed after dark for St. Lucee she is to return in a few days with Americans for us. The Officer of this Flag had the most Liberty of any Officer of the kind I ever saw, being under no restrictions going thro' the Ships Forts &c where ever he pleases he'll be able to give a good representation of the state of affairs there. the Ifant States of America wou'd not allow themselves to be duped in this manner.—

Friday 14 January Nothing particular this morning. all Hands employed picking Oakum making Sinnet and fitting Mizen rigging &c.

Saturday 15th. Took down the Sheers which is wanted on board the French Ships.—Our Friends here have at last granted us what Spars they have to spare being only two or three pieces to Tongue the Mizenmast and Bowsprit and two Spars for lower Yards having none for a Mainmast or foremast, but they have granted us as much Rigging as we want.

Sunday 16th. This morning Signal displayed for a great number of Ships off the South part of the Island. arrived here a Brig from Dartmouth in America Loaded with Lumber Our Chaplain delivered an extraordinary good Discourse. after Dinner went a Sailing with the Officers spent the afternoon aboard of an American Ship.—

Monday 17th. The Capa. and first Lieutt. went to St. Pierrs

this morning in the Cutter the Master ashore picking suitable Rigging for the Ship.—got a Spar on board to Tongue the Mizenmast which we are under the necessity of making it serve for a Mainmast.—arrived here this afternoon a French flag of Truce which had been sent from here to Barbadoes some time ago for a number of French Prisoners which is due the French, but they wou'd not permit the Officer to go ashore or pay any attention to his demands and during the time he lay there a strong Guard was kept on Board the Ship. Treatment far different to what the English Flag received when here the other day.—

Tuesday 18th. I was taken this Morning very ill with a Fever which confined me to my Bed all day.—

Wednesday 19th. Found myself much better this morning so as to be able to go upon deck the Cutter returned this afternoon with the Captain they took up the Armorers Mate whilst at St. Pierrs.—got part of the Rigging on board yesterday. Carpenters caulking[6] the Ship and working ashore on the Mizenmast.

Thursday 20th. several people sent to the Hospital sick and others taking sick daily—got the Rigging on board.— Our polite Planter on the South side of the Bay with whom we spent the afternoon when last a Fishing spent the forenoon and dined with us. the Carpenters Tongueing the Mizenmast, all hands employed working up Junk &c.—

Friday 21st. Seamen employed fitting the Mainrigging.— Cutter taking the Shingle ballast ashore, got off from Shore two Spars to make lower Yards and two Spars for Sheers to get in the Mainmast.—

Saturday 22d. Raised the Sheers and got out the Jury—mainmast. The first Lieutenant went in the Barge to St. Pierrs to send round a number of Spars for the Ship—carrying Ballast ashore all day. The 74 alongside of us this Evening hauled off to heave down. a heavy Firing was heard the day before yesterday from the South part of the Island at St. Lucee. The reason of it is not yet known.—

Sunday 23d. The Cutter returned this morning from St. Pierrs with a Sloop's mainmast to make us a Foremast.— very pleasant day, squally with a good deal of Rain in the afternoon nothing further particular. it being the Sabbath day staid on board all day.—

Monday 24th. This morning the Barge sent to St. Pierrs for more Spars. our Carpenters employed making the lower Yards. one hand Caulking the Ship. Employed myself the forenoon writing home. in the Evening in Town with some of our Officers. amongst many passing Scenes one presented to my view this afternoon which I was before totally ignorant of a Lady was passing along Street dressed in a brown riding Habit and petticoat, it was cried out she was Booted. I looked more narrowly and greatly to my surprize found it so as her Petticoats was short they were esily perceived with high Cork heels. I don't know how high they drew up but I suppost to her I looked under her Umbrello to satisfy myself whether it was Man or Woman & there I beheld a face all bedaubed with paint and head as big as a bushel. she was an Original she had been riding with a whip in her hand.—

Tuesday 25th January. Heavy Rains with smart Sqaulls of Wind. employed four French Carpenters at work on the lower Yards—procured a spar this afternoon to Splice the Foremast. The Barge returned in the Evening from St. Pierr's with two large Spars, having left our Purser there, very Sick with a fever. we are informed a Philadelphia Brig arrived at St. P. yesterday in distress, having met with severe Weather on the Coast of America from where she was drove off and lost her Topmasts she was last from Cailais.—two of our People taken with severe fitts this Evening the fifer tumbled over Board and had near being Drowned A Man tumbled into the Mainhold without receiving much injury, and one of the Centrys tumbled off the Gangway in upon the Deck. I believe the D—l got aboard to Night the whole Ship seemed to be in a Tumult all the Evening.—

Wednesday 26th The Boat returned to St. Pierrs this morning.—Squally weather the forenoon—the Carpenters at work ashore on the Spars.—fixing the Main rigging aboard, sending the Shingle Ballast ashore.—very heavy Squalls of wind with Rain all the Evening.—Signals displayed about Sundown up at Fort Borboun for a Fleet being seen off. not more particular.—

Thursday 27th January. Fresh Breezes this Day from the S.E. piece of Timber which we got to splice the Foremast we have concluded to make a Bowsprit of it by joining the heel of the old one to it. got on shore this afternoon for that purpose. Signals displayed at different times to day for a Fleet being seen off—Yesterday application was made formally to the Intendant for (only) Canvas sufficient to make our Courses but he preremptorily refused it at the same time we were certainly assured of their being a large Quantity of it in the Public Stores. Thus is our refitting wantonly retarded by this Man, until we can be redressed by the General, who is at present out of Town, and owing to nothing more than a difference in Public Party affairs subsisting between the General and Intendant.—[7]

Friday 28th Late last Night the 1st Lieutent. returned in the Barge from St. Pierrs but without any Spars the Continental Agent there having not Money to purchase anymore.—this morning 13 or 14 Sail of the English Fleet in sight.[8] Several Seamen entered on board to day having left some American Vessels at St. Pierrs.—Ebenezer Tanner one of the Cooks Mates Sleeping on deck last Night was accidently trod on the Belly by some Person in the dark from which he died this afternoon it's supposed some of his Entrails was broke.—heavy squalls with rain this afternoon a Ship arrd. from St. P. with public Stores.—6 Sail of the English fleet in the mouth of this bay this Evening.— several of our Officers taken sick also the People.—

Saturday 29th January. This morning the Barge and Cutter went to St. Pierrs for Provisions &c. the Capta. & 2 Lieutts. went in them.—Seven Sail of Admiral Parkers Fleet off the mouth of the Bay this forenoon.—Carpenters at work ashore on the Masts. Seamen fixing rigging &c.—

Sunday 30th. Last Night the Series Sloop of War came up the Bay. the English Fleet in sight this morning.—no business Transacting to day it being the Sabbath. after Dinner took a walk about a Mile in the Country with my 2d Lieutenant. saw several good Country Seats and fine Plantations covered mostly with the Sugar Cane. but for want of Linguister returnd early in the afternoon.—

Monday 31st At 3 O'Clock this morning the Barge returned from St. Pierrs with four small Spars. by which we learn that the British Fleet being off St. P. yesterday under French Colours one of them took an American Brig bound into St. P. close under one of the French Batteries which fired on the Ship and was returned from her but without any injury on either side. at 9 AM. the Barge went to St. Pierrs again. two of my Marines died to day at the Hospital Vizt. John Wainright—fifer, & Enoch Bretts.—Carpenters at work on sundrys sawing Paunches for the Masts and Tonguing the Bowsprit.—

Tuesday 1st February. At 11 AM. set off from the Ship in

the Cutter for St. Pierrs in company with several of our Officers on our way round saw eight sail of the English to the Southward of this Island. being calm made our Passage very tedious at 6 PM got aboard of Capa. Thompsons Brig from Philadelphia saw an old acquaintance Cap. Mullet who I delivered Letters for Phila. went to Town about Dark and met Cap. H.—spent the Evening at the American Coffe-House with several Gentlemen from America.—

Wednesday 2d. Set off this morning at 6 O'Clock in the Cutter, also the Barge with Capa. Harding and Mr. Bingham. I breakfasted with Captain Thompson on board his Brig—afterwards made the best of our way to the Ship and arrived about three O'Clock.—The Carpenters, at work on the Spars. &c.

Thursday 3d. Pleasant agreeable Weather all this Day. Carpenters finishing off the Mainmast.—Cutter employed all Day carrying Shingle Ballast ashore.—a Crew fixing the Rigging &c. nothing more worthy of Remark.—

Friday 4th. Went ashore this forenoon on business with Mr. Bingham. He has made a formal demand of Spars &c to fit us which at last is complied with thro' the assistance of the General.—at 2 P.M. the 2d Lieutt. went to St. Pierrs in the Barge to employ a Number of Carpenters.—[9] Mr. B. returned in the Barge. This Evening got from the Kings Yard a number of Spars for Topmasts, Top-gallant Masts, Steering Sail Booms &c. Carpenters finished the Mainmast this Evening all ready to take in tomorrow Morning so that we shall soon begin to appear a little formidable. it really Animates us the sudden change and prospect we have to day of getting Refitted, tho' we might now been much forwarder in fitting had it not been for the Intendant's ungenerous and Villianous conduct in denying us those Articles which he had in his power to serve us with [10]

Saturday 5th February. Launched the Mainmast this morning at 11 got it in its place and before Night got the Main-rigging overhead and seized in the Dead Eyes. finished the Bowsprit also ready to take in, and two Topmsts. Purchased two Caps ashore for the Mainmast and Foremast. Our Sail-makers employed altering the Sails. procured Canvas from the Kings Stores for our Courses and sixteen Barrels of Provisions Sailmakers employed ashore to make the Courses the rest of our Sails making in St. Pierrs the weather very pleasant and favorable for our Work.—

Sunday 6th The Lieutenant returned this morning from St. Pierrs without effecting the business he went on. Yesterday ten Sail of the English Fleet being to the Northward of the Island chaced in a Brig and Schooner under a three Gun Battery which the Ships engaged for some time and Landed upwards of 200 Men upon which the few Troops that was in the Battery abandoned it after Spiking up the Guns. the Enemy burnt the two Vessels in this grand acquisition One of the Ships lost her Bowsprit another a Top G. Mast & Crossjack Yard and two Boats Sunk in landing the Troops.—The great necesity of getting the Ship fitted with all Expedition obliged us to keep the Ship Crew at work to Day tho' it is the Sabbath. Notwithstanding our Chaplain delivered us abt. 11 OClk an excellent well adapted Sermon on the abominable Practice of Swearing.—afterwards turned all Hands to work raising Sheers and getting ready to take the Bowsprit in.—after Dinner launched the Bowsprit brot it alongside and fixed it in its proper place.—Carpenters at work ashore on the Topmasts finished taking out the Shingle Ballast.

Monday 7th Raised Sheers this morning on the Forecastle and got the Jury foremast we are obliged to make a Mizen-mast of our two old Topmasts by tonguing and fishing them together.[11] Carpenters at work on them. got out the Stump of the foremast.—Seamen at work on the Bowsprit fixing the Bobstays Gammoning &c.—Went in the Cutter this afternoon over the other side of the Bay to Bath in Salt Water returned on board early.—The Magnifique 74 compleated Graving this Evening having in days hove both sides keel out Corked and and Graved them.

Tuesday 8th The Cutter went this morning to St. P. with the 1st Lieutt.—set up the Main rigging and maintop fixed. Carpenters making Spars &c.—This being Shrove Tuesday and a great Holidays or what is called Carnival time amongst the French just before the Commencement of Lent is kept and Celebrated with all kinds of diversions amongst which is a celebrated one. the Inhabitants of the town dressed in odd figures and masked parading the Streets with 100 negroes all spattered with paints following them others carrying Images of the Betrayer, Judas Iscariot, as it said it was on this day he was betrayed, and on board the Ships they have the same Effagies hung up to their Bowsprits which duck pelt and Shute at all day & finally drown it—in this ignorant stupid manner this day is spent and all the holidays as ridiculous which is not equalled by the natural Savage Indian—

Wednesday 9th February.—Very dry warm weather blowing fresh Trade winds. Rattled down the Main Shrouds. got the Main-cap up in his place. Rigged the Main-yard and got him atho't.—Carpenters at work on board making large Wedges for the mainmast out of the Stamps of the old Masts. got a quantity of Oak Plank and Pine Boards from the Kings Yard.—

Thursday 10th Nothing very remarkable the forenoon all Hands busily employed about sundries.—at 3 P.M. the Cutter returned from St. Pierrs with the Lieutt. and Mr. P. Richards, Brother to our Purser who spent the Evening with us the Cutter bro't round some Slops &c.

Friday 11th This forenoon the Cutter returned with the Captain and Mr. Richards the Purser and Doctor to St. Pierrs.—after Meridian 8 Sail of the British Fleet passed this Bay standing to the Southward. Carpenters finishing the Mizenmast &c.—

Saturday 12th The Ceris [12] sloop of War got under way and went down the Bay. the British fleet in sight—Ships Crew fixing the Rigging—Carpenters finished the Mizenmast ready to get in tomorrow.—Gunners Crew painting the Gun Carriages &c.—Dry warm weather.—

Sunday 13th Notwithstanding it being the Sabbath necessities of the Ship obligated the Officers to keep all Hands on board & the french Carpenters ashore to Work—at 8 A.M. got the Mizen mast in it's place and the Rigging up took the small Sheers down off the Forecastle and the Sheers from the Quarter Deck and moved them on the forecastle.—a Sloops Mast taken to make the foremast Carpenters at work on it fixing the Cheeks &c. its to be tongued with one of our old Topmasts.—In the Evening at 8 OClk. the Cutter returned from St. Pierrs with a few Quoils of Rigging.—

Monday 14th This Morning got the Maintopmast thro' the Cap and rigging on it went in the Pinnace to Casnavier River to water there a small Town there where a number of Fishermen lives I saw them hauling Seines in the Bay but no fish caught the Seine is upwards of 400 Fathom long and depth proportionable they take above thirty Negroes to Work them. On our way up the Bay saw a small Whale or Grampus assaulted by the Sword fish and Thrasher the conflict was so sharp it obliged the Grampus to forsake his Element several times jumping out of water his whole length this continued about fifteen Minutes being some considerable distance cou'd [not] discover in whose favor it terminated.—very squally with rain this afternoon.—Swayed the Mizentopmast thro' the Cap before dark. the Cutter went to St. P. for the Captain.—

Tuesday 15th Seamen employed fitting Maintop Gallant rigging. the afternoon swayed the Maintop Gallant mast up and rigged it. Mizentopmast rigged. Carpenters at work on the Foremast and making light Spars. Calm and very Sultry warm weather the Evening pleasant Breeze and rain.—

Wednesday 16th Early this morning the Fandange 74 arrived from Chesapeake Bay in America she was closely pursued and fired on by the English Fleet at the Mouth of this Bay.—at 4 O'Clock this morning Mr. R. sent in the Pinnace watering to Casnavier as usual with a Crew of six Men. after filling the water and on their return its supposed the Crew mutined and took the Boat as she was afterwards seen standing for the English Fleet. the Enemy will now have a particular detail of our situation. Rattled down the Maintopmasts Shrouds. got the Crossjacks up in its place and rigged. Heavy Rains this afternoon.—Cutter returned from St. P. with a load of Rigging—

Thursday 17th Cutting sent to Casnavier for water the English Fleet off the mouth of the Bay.—Very warm weather Trade wind fresh at N.E. In the afternoon arrived here safe a Letter Marque Ship of 18 Guns in 23 days from Boston Loaded with Provision for the French Government.—Late last Night the Captain returned from St. Pierrs. Carpenters at work on the Foremast.—set up the Main and Maintopmast Rigging.—Crew employed at Sundry other necessary Duties.—

Friday 18th February. Received thirty Barrels of Provisions on board from the King's Stores. Rigged the Spritsail Yard athwart. several employed Reeving Topsail Ties and Hallyards and other running rigging. Swayed up the Mainyard in its proper place. Cutter made two turns to Casnavier watering filling water for a Cruize. Carpenters busily employed. &c.

Saturday 19th At 8 O'Clock this morning the Captain and Purser went to St. Pierrs in a Passage Boat. The Cutter on her return from Casnavier this forenoon over set in a sudden Squall of wind she lost her rudder by the assistance of a Schooner she was righted without the loss of a Man.—The British fleet in sight all day.—Reeved the running rigging and otherwise employed about the Rigging.—

Sunday 20th All Hands kept at work to day as usual. The English off this Harbor in the Morning and having intillegence of the French Fleets being near the Island the made the best of their way into St. Lucee.—Carpenters at work ashore got 4 Stages alongside in order to finish Caulking the Ship. Blacked all the Yards Mastheads, &c—

Monday 21st February The french Fleet early this morning beat up the Bay having been joined by three of Mossr. De Grasse's Fleet who now commands this Fleet during their Cruize they took six Prizes, two English Frigates off this Bay reconnitreing. The 50 Gun Ship which lately hove down hauled out of the Carnage and joined the Fleet which makes a Fleet of Eleven Sail of double Deckers and two Frigates ready for Sea. got a large piece of a Timber for a rudder Post.—20 Caulkers employed this morning Caulking the Ship Carpenters at work on the foremast and Foretop &c. Seamen fixing the forerigging &c. Made a demand of our Men which was lent Mossr. Piquets fleet.—

Tuesday 22d. The Corkers at work heeled the Ship in order to Caulk under the Water. Sixteen Sail of British Ships amongst which is four or five Frigates Cruizing to day in the Channel between St. Lucee and this Island it's reported the French fleet here, is going out to Tomorrow to meet them.—very Warm dry weather wind S.E. Carpenters at work ashore on the foremast. Seamen fixing fore rigging. Cutter filling water Took up the Main Shrouds seized in the dead Eyes and set them up. Sailmakers employed on board altering our Former Topsails. &c.—

Wednesday 24th (13) Employed all Hands at Sundry duties such as filling Water Rigging the Fore and foretopsail Yard and compleating the forerigging.—about 500 Troops Embarked on board the French fleet in the afternoon with Intrenching Tools &c. which seemed to indicate a Land interprize.—but a strong report is propagated that they are determined to meet the British Fleet—more especially as to day they went into St. Lucee—but if the Britons has enqual force I make no doubt but they will meet the french and by the superior sway should it be the cast I shall be very uneasy for our dear Allies.—

Thursday 25th The General of the Island this morning sent on board of us three Americans taken up by the Guards, two of which had made their escape from a British Man of War at St. Lucee one of them Mr. R. belonging to Philada. taken a Prisoner about Eight Months ago. he brings certain intillegence of Admiral Rodneys arriving at Barabadoes with eight Sail of the Line.—Capa. Harding arrived last Night from St. Pierrs—this morning he waited on Admiral D'Grasse and demanded our Men which was complied with and sent on board in the afternoon.—Corkers Caulking the Waterways of the Main deck Carpenters finishing the foremast &c.—

Friday 26th February A Midshipman sent in the Cutter this morning to St. Pierrs for Provisions, Slops, &c. Signals displayed for the English Fleet being in the South Channel upon which the French Admiral sent out their Tender and the Series Sloop of War to Reconnitre. Carpenters compleated the foremast at last.—Ships Crew employed about sundry Duties.—at 9 in the Evening one of the Centinals on the Gangways fired on a Man attempting to Swim away from the Ship and narrowly escaped being killed he was brought on board and confined with five others who had consulted with him to make their escape.—

Saturday 27th The Ma[g]nifique 74 hauled from the Head of the Carnage alongside of us in her usual Birth.— The Cutter late last Night returned from St. Pierrs with Provisions Slops &c.—After Dinner Launched the Foremast and got it in its proper place got all the rigging over the Mast head before Night.—Signals displayed this afternoon for the English Fleet being in the South Channel. at Meridian the Man who last Night attempted to runaway and being the Ringleader of this Scheme was publickly Whiped at the Gangway by order of the Capa. this first regular Punishment that has been inflicted on any this Ships Crew for this common crime which will have a good effect provided there is continuance of it— the rest are confined yet in Irons.—

Sunday 28th February This morning got the fore top up (having kept all hands to work) seized in the Dead Eyes and set the fore Shrouds up before Night got the foretop mast up and rigg'd. Carpenters at work on the Rudder took in a number of Casks of water to bring the Ship by the head in order to find out the Leak. went ashore in the afternoon for a few hours and visited Mr. G. who is Sick in the Hospital. at 9 AM a Signal was made in the fleet upon which one of the Frigates got under way and run down the Bay.—

Monday the 29 Feby. Two or three french Ships arrived from Chesapeak Bay Loaded wth Tobacco. the Prisoners discharged out of Irons.—All Hands employed rigging the Foremast. got the fore and foretopsail Yard athwart. took in and water Shot &c to bring the Ship by the Head, very fresh Trade winds and warm weather.—

Tuesday 1st March.—The English fleet consisting of twenty two laying off and on the Mouth of this Bay it is suppos'd they are reinforced. The french Fleet has declined going out to meet their Enemies now because their force is inferior, but it's what was expected as they might have a few days ago met their Antagonists on equal Terms.—the Ship brought by the Head got a Hauser out a Stern and hauled her Stern ashore at high water. in the afternoon at low water examined for the Leak but cou'd not reach it. secured the Rudder braces and hauled the Ship off at Night. Caulkers employed Caulking the water ways of the Main deck. Carpenters at work on the Rudder which is ordered to made two feet narrower than the former one being now only five Feet.— Swayed up the foretop Gallant Mast and rigged it. ashore this Evening on business nothing very material Transpired.

Wednesday 2d March. Employed this day reeving running rigging preparing to Careen tommorrow, taking in water &c.—The British Fleet in sight from this Island.—at Meridian Mr. W—C— of Providence in Rhode Island came on board having about four days ago made his Escape from the Island of Antigua he has been a Prisoner Eight Months. he brings Letters of intillegence to the Governor here that there is a very Secret expedition now on foot at Antigua it is thought to be agt. some of the French Islands abt. 5 or 6000 Troops had embarked the day he came away and was to proceed to St. Kitts there to be joined by more Ships &c and all to Rendezvous at St. Lucee—Capa. C. was assisted to make his escape to bring this interesting intillegence this News seems confirmed by the appearance of the British Fleet being strongly augmented. Shou'd the Britains make an attempt upon one of the French Islands they have a competant force here to instantly retalliate by making a descent on Antigua which Mr. C. assures us is now almost defenceless not having two hundred troops on the Island.—

Thursday 3d March.—Careened the Ship down to her Ballast port. Scraped her bottom to the Keel with Hogs and Bruches in the afternoon payed about five feet under her Wale with white Lead and Tallow. in the Evening righted the Ship. Signals this afternoon displayed for the English Fleet being in the South Channel. The Magnifique alongside of us fitting with the greatest expedition to join Count D'Grass fleet now lying in the Bay.—

Friday 4th—Several Gentlemen this morning visited us from St. Pierrs amongst whom is Mr. B. Capa. Fosdyke who lately lost his Brig on this Island by the Enemy and Mr. C—y a Merchant lately from Boston. they dined with us afterwards went ashore & embarked for St. Pierr's.—compleated Careening and hogging the Ship. Ryan died very suddenly to day he was carried out into the middle of the Bay it being a custom here when dies so sudden tho' by the Laws of this Island there is a fine of Five hundred pounds if a Man is permitted to Die on board a Vessel in this harbor, except in the above mentd. case.—

Saturday 5th March All Hands employed to day taking in water and stowing it away. took in a Canoe Load of Wood and stowed it away. Fresh Trade winds weather very dry and warm.—Carpenters at Work on the Rudder &c. nothing more remarkable.

Sunday 6th This morning borrowed the Admirals Launch to get our Guns off. he politely sent thirty Men with his Gunner in her to assist us. his Boat wou'd have been of much more service to us without the men than with them for during the whole Day they got but Eight Guns off— Kept all hands at work notwithstanding it being the Sabbath. continued stowing Water in the Hold. One of the French Frigates in the Harbour went out this Morning. Carpenters and Caulkers at work as usual.—

Monday 7th But little work going on this morning having discharged the Launch being unfit for the Service and waiting for the Magnifique Stage to get our Guns off took on board all the Water we had filled. likewise all the Gun Carriages.— The 3d Lieutt and Purser returned from St. Pierrs with a number of Sails and a few Slops. Signals displayed this afternoon for the English Fleet being in the South Channel. it is a something extraordinary that the Bold Britains has not ventured as far as this Bay since Admiral D'Grass arrived here notwithstanding their superior force.

Tuesday 8th March.—Employed the forenoon fitting Sails ready to bend. Bent the Jibb fore and Main Staysail. after Dinner got the Stage and took off the remainder of our Guns.—Launched the Rudder and hung it in it's place. One of the French Ships Launches and Crew assisting us brought on Board one Load of Shingle Ballast Several Sail of Merchantmen came round here this afternoon from St. Pierrs. Cutter watering to Casnavier. Capenters making Partners to the Mainmast on the Quarter Deck &c.—Fresh Trade wind to day very pleasant.—

Wednesday 9th. All Hands employed bending Sails. bent the Courses Topsails Top G. Sails &c. a Canoe employed taking off Shingle Ballast. Washed and Scraped the Maindeck this morning.—afternoon several of the Ships Launches taking from Fort Royal a variety of Military Stores amongst which is Travelling Artillery Carriages Scalling Ladders Shot &c. from which appears an Expedition to be in agitation against some of the Islands. In the Evening a Negro Man was Executed on the Parade in Town for Robbery. Last Monday Morning arrived here three Americans who made their escape from the British Fleet at St. Lucee in a Yawl which they took from a Transport on their way over they were drove to Leward of the Island & fortunatly taken up a by a Statia Vessel they inform us of Mr. R. being carried in there in our Pinnace.

Thursday 10th March.—This morning Mr. D. an intimate acquaintance in Philadelphia came on board to see me, he came here Passenger in the Fondonge 74.—at 10 AM John Griffin departed this Life of a very short illness.—finished taking in Shingle Ballast this Morning after Meredian took on Board part of our Powder—set up all the Shrouds and Stays. Sailmakers employed altering the Mizen.—Finished the Partners of the Main mast and all the Spars. discharged the french Carpenters.

Friday 11th Rain and Squally Weather. three Sail of Merchantmen arrived here from Granada. employed taking on board all the empty Water Casks—Light Spars &c. fixed in the Tiller and putting the running Rigging in order for Sailing.—Admiral D'Grass honored our Captain with a Visit this afternoon he is endeavouring to Negotiate an exchange of American Prisoners before we leave this Island.—

Saturday 12th Very heavy Rain this Morning received on board a Boat Load of Wood. English Fleet in sight all day.—took everything on board this forenoon—Afternoon employed clearing and getting the Ship in order to Sail tomorrow morning for St. Pierrs.—took two Men, Americans, out of the French Fleet.

Sunday 13th March.—Break of Day this Morning hove the Ship ahead to her Anchor. took a Pilot on board at 8 AM Warped out of the Carnage and made Sail with a light Breeze of wind. passed under the Admirals stern and Saluted him with Thirteen Guns and returned from him with Nine, afterwards made Sail weather growing Squally, with Rain, obliged us to take in the Top G. Sails and double Reef the Topsails the Rigging being new and unstretched.—Beat into St. Pierrs and at 3 PM. came to Anchor off the South part of the Town. hoisted out the Yawl Capta. went ashore.—employed the afternoon Mooring Ship in 33 Fathom Water.—At 4 PM. Simon Ashbow a Marine died suddenly having but a few days ago been discharged from the Hospital at Fort Royal after a severe illness.—In the Evening sent three Sick Men to the Hospital at St. Pierrs.

Monday 14th This morning Unship'd the Round house and sent it ashore unbent the foresail and sent it to the Sailmakers to be altered. Cutter watering at the Fountain Received on board One hundred and Eleven Barrels of Beef, a Quantity of Cheese Rice Butter &c. a Crew employed Scraping the Quarter deck. at Sundown a Fleet of 20 Sail of Merchantmen with 4 Ships of the Line came out of Fort Royal Bay and stood the Northward.—

Tuesday 15th Went ashore this Morning on Publick business with the Agent. returnd. on board before Dinner. received on Board a Quantity of Provisions of different kinds. continued taking on board Water. sent on board this afternoon by an Officer of the Ship Mr. G. twenty Hhds. of Rum in a Clandestine manner supposed to be on freight this extraordinary Proceeding has caused no small contentions in the Ship. soon after the Capa. in a Letter to G. prohibated anything being taken in the Ship on Freight but tomorrow must produce the Consequences arising from this unwarrantable transaction.

Wednesday 16th The Rum mentioned Yesterday to be taken on board was ordered out this Morning by the Captain it was accordingly done but before it went from alongside thirteen Hhds of it was purchased for the use of the Ship and hoisted in again.—The Resolution Frigate arrived here this afternoon. took on board sundry Provisions.—Crew employed Stowing the Hold.—Others fixing Rigging—altering Sails—and Marines making Musket Cartridges.—ashore at St. Pierrs a short time on business in the Evening.

Thursday 16th The Capta. and Agent came on board this fornoon. bent the foresail fitting Steering Sail Booms.—Mates stowing the Hold.—took on board a Quantity of Pees. arrived a brig this Evening from Salem 24 Days out—No News.

Friday 17th March. All Hands engaged about Sundry Ship Duties. Mates Stowing the Hold. at 11 AM went ashore and Dined with Mr. B.—spent the remainder of the Day in Town, returned on board in the Evening. This being St. Patricks day the Ships Crew endeavored to Celebrate it in their way among others one of the Centinels was found drunk laying down on the Gangway he was immediately confined in Irons.—

Saturday 18th Early this morning two large English Ships and a Snow to Leward Standing to the Northward.—took on board a scow load of Bread this forenoon. at 11 AM went ashore and dined with Mr. T—spent the afternoon in Town.—The Embargo being taken off to day a number of Vessels sailed this Evening for different Ports.—received on board this Evening 21 Hhds. of Sugar on Account of the Continent to Reimburse a Sum of Money belonging to a Private Gentlemen in America which has been expended by Capa. Harding for the use of the Ship—

Sunday 19th Capta. Billings in a Pilot Boat sailed for Statia, by whom I forwarded Letters to Philadelphia. went ashore about 11 AM. Dined and spent the afternoon with Mr. N—in St. Pierrs, very rainy Squally Weather all this day several Vessels sailed out of this Port in the Evening—nothing more worthy Remark.

Monday 20th—This forenoon continual Rain and very Sultry—at 11 AM went ashore with some of the Officers on private business—Dined with Capta. H—at Mr. Binghams who is gone to Fort Royal to make a final Settlement of his Publick business as he designs to leave the Island and go Passenger with us to America.—returned on board in the Evening.—All Hands employed about sundry necessary Dutys fitting the Ship. took a quantity of Bread on board about Noon.—filling water Stowing the Hold &c.—

Tuesday 21st At 1 O'Clock this morning an Officer came onboard from the Resolution French Frigate with a request from the Governor to take on board Troops and proceed to Dominique in order to join the long expected Fleet. the Request was refered to the Captain who, in a proper Manner Answered it in the Negative.—at Daylight Count D'Grass Fleet in sight standing to the Northward in order to join the Grand fleet.—Almost all the Vessels in the Harbour this forenoon is Pressed as Transports in the Kings Service to carry Troops to join the Fleets.—from which it appears there is already an Expedition going forward against some of the English Islands.—It is reported that the English Fleet went into St. Lucee to Day.—the Revolution at day light Joined the other Ships.—All Hands employed about sundry Duties such as scraping the Ships sides in order for Painting. Gunners Crew getting Steering Sail Booms aloft fixing Rigging. Marines filling Musket and Pistol Cartridges and Sailmakers altering Sails &c.

Wednesday 22d at 9 O'Clock this Morning the long expected Fleet appeared coming from the Northward having being joined by Admiral D'Grass making in the whole upwards of 25 Sail of the Line exclusive of 8 or 10 Frigates having under their Convoy about 50 Sail of Transports, Merchantment &c amounting in the whole to upwards of 100 Sail making a very grand appearance—at 12 when abreast of this place about 30 Sail left them and beat up into this Harbour several others stood for Fort Royal—immediately afterwards the Ships of the Line made all Sail and stood over for St. Lucee in order to Besiege that Island its said they have about 13,000 Troops in this fleet besides the Troops of this Island a Regiment of which embarked this forenoon on board Transports and followed the Fleet. Ashore this forenoon on business our Commander has concluded to join this Fleet therefore the Ship is getting ready with all expidition Paying the Ships sides with Lampblack and Tar.—took on hhds. of Sugar being the remainder of the Sugar taken on account of the Continent.—

Thursday 23d March—Very wet Weather winds variable all the fore Part of the Day—One of the french Ships which arrived Yesterday in Warping in like *Frenchmen* and Men unacquainted with the Management of a Ship came on board of us & carryed away our Driver Boom—Last Night a Schooner arrived Express from Gl. Lincoln in S. Carolina requesting a Naval reinforcement being lately Invaded by a Body of Sea and Land Forces.—Two Boats with a Lieutt. in each Pressing thro' all the Vessels in the Harbor for Americans got but 5 Men.

Friday 24th March This Morning the Captain sent off orders to Unmoor Ship and get ready for Sailing in order to join the French Fleet at Royal going on an Expedition Against the Island of St. Lucee. at 11 AM the Captain came on board soon after the Ship was under way. standing under her Topsails alongshore at half past twelve boarded by a Boat with an Express from Mr. Bingham to return to St. Pierrs upon which the Capta. went to Town. at 1 O'Clock wore Ship and worked into our Usual Birth. before we put about saw the french Fleet coming out of the Bay and standing over the St. Lucee. we are now to return to America as soon as possible.—

Saturday 25th March.—Shifted all the Guns on one side the Deck and heeled the Ship and Boutoped[34] her both sides at the same time scraped the Gun deck. at 10 AM. being the Anniversary of the Resurrection all the Ships colours for two days past being hoisted half Mast on discharge of a Gun from one the Ships were hoisted in their proper places and every Vessel in the Harbor fired a Salute at the same instant the shore being crowded with Negroes on the firing of the first Gun they all Leaped into the Sea and Bathed this strange Custom prevails I'm informed thro' all the Roman Catholick Country's. spent the Day in Town dined with Messrs. B-t-n & M-t—No certain accts. from the Fleet.

Sunday 26th March.—The Ships Crew on board employed all the forenoon cleaning Ship between Decks fore and Aft—very warm sultry Weather with frequent Showers of Rain.—a strong Press ashore today for Men for the French fleet an instance not common and I believe unprecedented before in the French Navy. reports prevails of a long and heavy Fire being heard yesterday at St. Lucee, but no particulars Authenticated.—Ashore after Dinner Walked a few Miles out of Town with an acquaintance to a Very fine Sugar Plantation upon which is very good Improvements and well Stockd. on our return took a turn thro' one of the Holy Fathers Gardens in the invirons of the Town, found it a large plain Garden will planted with Sallads & some West-India fruits in the Evening appeared a Ship to leward having lost some of her principal Spars.—

Monday 27th Two Merchantmen arrived this Morning from Fort Royal Stationed and Quartered our Men this fore noon. ashore all day on business Dined at M. Binghams.—Signals displayed ashore for a fleet of 20 Sail being in the South Channel its reported an English Frigate is taken at St. Lucee.—Mr. Pr— & his Lady with three others in company came on board in the afternoon to view an American Frigate—

Tuesday 28 March In the forenoon received on board a large quantity of Cocoa Ship'd by Mr. Bingham.—ashore the greatest part of the Day on private business. several sail of Transport Sloops with Troops which Sailed from here on an Expedition under Convoy of the Resolution Frigate. Hoisted the Cutter on board in the Evening.—cleared Ship in order for Sailing.—

Wednesday 29th At Sunrise this Morning unmoored Ship and hove short to one Anchor in order for Sailing—in Town the fore Part of the Day on Private business. The Captain came on board in the forenoon. the Wind being ahead all Day prevented our Sailing. The Captas. of four American Vessels on board receing Signals and Orders who is going under our Convoy. In the Evening four or five Sail of Small Ships with a 74 beating into the Harbor from Fort Royal. Mr. Bingham sent on board the greatest part of his Baggage this Evening.—at 9 O'Clock the above mentd. 74 coming in hove in Stays alongside of us and run stern on against our Mizen Chains carried away her Gibb Boom we cleared her with any other loss or damage than one of our Mizen shrouds.

Remark's &c. on board the Confederacy Frigate at Sea.—

Thursday 30th March 1780. Signal for Sailing displayed early this Morning, got under way at St. Pierrs at 10 AM. with 3 Briggs 1 Schooner and Sloop under our Convoy bound to America.—stretched under easy Sail along the Town, past Eleven Mr. Bingham come on board who is going Passenger with us to America.—at 12. Saluted the

Ships and Town with 13 Guns the Commodore returned it with Eleven. bore away and set Top G. Sails.—two of the Brigs Sailing very dull obliged us to short Sail at sundown under double reifted Topsails Brigs astern with all Sail set—at sunset the North end of Martinique bore E.—distance 7 Leagues. Dominica NE by N distance 8 Leagues moderate weather Course NW.—

Friday 31st.—All the Convoy in company this day.—Two of the Ships company taken this Morning with small Pocks built an Awning over the Cutter on the Booms for them to prevent it spreading in the Ship as there is a number of the Crew that have not had it, who is preparing by a course of Diet and Medicine in case they shou'd catch it.—Light Trade Winds Steering this last 12 hours W.N.W. under the three Topsails and foresail, the Brigs astern with all Sail set Steering sails &c.—Exercised the Great Guns and small Arms after Dinner.—at 3 PM. a small Sail in sight on the Weather Quarter.—Lattitude by Observation at Meredian 15° 27" N. at sunset the Island of Dominica in sight bearing N by E. distance 15 or 16 Leagues. Nevis

Saturday 1st April.—Pleasant moderate weather Wind S.E. Course NW by W. set the foresail and Main T.G. Sail. Fleet close up with us in the forenoon.—some of the Islands in to the N. Westward.—in the afternoon dropt the Fleet astern took in the Foresal & Main TG. Sail.—in the forepart of the Night small Showers of Rain.—Latt. by Observ. at Meridian 16° 04" N.—Nevis bearing NE. by N. distance 16 or 17 Leagues.

Sunday 2d This morning light Squalls of Rain thick and cloudy all the forenoon at 12 saw a Sail on our Weather Bow standing for us supposed to be a Brig or Schooner almost Calm made all Sail set steering Sails &c. and gave Chace very little headway at half past twelve the weathering clearing away discovered the Island St. Croix right a head bearing NNW continued in Chace course NW sent the 2d. Lieutt. with 20 Men on board the fast Sailing Brig in order to give Chace but the wind being light & the Sail a great distance gave over Chace, took in Sail and kept away for the West end of the Island Course NW. Lattitude at Meredian by Observation 17° 12" North. at 12 at Night ran close up with the Land saw several Lights ashore. kept away West running for the West end of the Island.

Monday 3d At 5 O'Clock this morning passing the West end of St. Croix with a light Breeze of wind hauled up N. the Convoy in company. This Island is much more low and Level then Martinique extremely well Cultivated with the common produce a small Town in sight at the W. end where 7 or 8 Sail of large Ships lay. at 6 O'Clk a Schooner some distance astern supposed to be a Cruizer at 9 the Island of St. Thomas's insight ahead and other of the Virgin Islands—the same time saw a large Brig to windward between the two Islands appearing to be a Cruizer by her Action. at 5 PM past the Sail Rock it appears very much like a large Sail when you first make it as its very white, run between it and the Island of St. Thomas on our right with a great number of small Islands uninhabited a very heavy swell heaving from the Northward. Course NNW. Wind East, after Sunset the Island of St. Thomas bore E by S. distance 6 Leagues. The Sail Rock . . . . . distance . . . . . all the Fleet in company. Lat. by Obs. at Meredian 17° 57" North.—

Tuesday 4th April.—Light Winds and very Variable all day with a heavy head swell from the Northward. Course N.N.W. The Ship being out of Trim and light Winds &c. the Fleet run ahead of us. in the Evening the fleet joined us. Latt. by Observation at Meredian 19° 07" N.—

Wednesday 5th.—This morning before Sunrise a sail discovered from the Masthead three points on the Lee bow, light wind made all Sail and gave Chace at 8 OClk. lost sight of the Chace in the haize. two of our Convoy left us and kept their Course during the Chace. the others following on. at ½ past nine cou'd not perceive the Sail gave up the Chace and took in light Sails kept our former Course again N.N.W. with a pleasant Breeze but a Tumbling Sea under foresail two Topsails and Top G. Sails. After Meredian the wind very light and Ship rolling very heavy.—at 4 PM. the Maintopmast discovered to be badly sprung in two differenc places occasioned its supposed by the immoderate Rolling of the Ship. this may be called a very unfortunate accident for Us in our present situation as we have but a very indifferent and only spare one to replace it with, but when we recollect the Distress and and disagreeable events of our late Passage we can or ought to look upon this as a trifling circumstance. sent the Main T.G. mast and Yard down.—at 5 PM all our Convoy joined us again. hauled the Ship up NE. to ease her.—in the forenoon Crews employed shifting Ballast out of the after hold, forward, to bring the Ship in better Trim.—The Commander has concluded to fish the Topmast. Carpenters preparing a Fish for it—stood our course again in the Evening N.N.W. almost Calm. double riefed the MT sail.—Lattitude by Observation at Meredian 20° N.—

Thursday 6th April. Light Airs almost Calm the forepart of the Day—afternoon entire Calm. Ships head alround the Compass no head way. Fished the Topmast in the Evening Latt. by Observation 21° 19" N.

Friday 7th Entire Calm all the forenoon Ships head all round the Compass. the Convoy all in sight. got up the Main T.G. mast and Yard. set up the Rigging fore and aft. After Meredian took a fresh Breeze from the Northward made Sail Course NW. by W.—breeze freshens in the middle of the Afternoon at sundown two of the Brigs a great distance a Stern. the french Brig Sloop and Schooner in Company all Night. very Squally in the Night. Lattitude by Observation at Meridian 21° 33" N. Longitude 65° 19" W.—St. Thomas's bearing S. 28° E. distance 231 Miles.—Cape Hatteras N 36° W. distance 1013 Miles.—

Saturday 8th April This morning Capta. Coolidge and the Salem Brig out of sight. the French Brig, Schooner and Sloop in company, Fresh Breezes under easy Sail. Wind hauled to the Eastward last Night Course NNW.—in the after part of the Day breeze lighter. saw a small Sail from the Masthead to windward, being late in the day did not Chace.—Latt. at Meredian

Sunday 9th.—Light moderate Breezes from the [15] with a tumbling Sea. under all Sail with the Convoy in company in the forenoon our Chaplain delivered a short discourse to the Ships Comapny. in the Evening the Wind hauled to blowing a fresh Breeze all Night.—double Riefed the Topsails under the three Topsails and foresail Course NW by N.—Latt. by Observation.

Monday 10th April.—Fresh Breezes and Cloudy the forepart of the day. at Meredian a heavy Squall of Rain Thunder & Lightening took us from the Westward weathered it very well having taken in all Sail but Foresail and forestaysail. at 2 PM cleared away Light winds hauling to the Northward made Sail Course In the Evening the Brig and Sloop in company the Schooner almost out of sight astern. the Wind continues hauling to the Eastward. No observation to Day.—

Tuesday 11th Last Night George Reynolds one of the Man who had the small Pox died of it, the other is perfectly recovered. very pleasant moderate Weather Wind Course under all Sails. D. O'Bryan received Corporal Punishment this Morning at the Gangway for Stealing Cloaths.—Fresh Breezes in the afternoon at 3 PM. saw a Sail from the Masthead to windward on the Beam standing before the Wind at ½ past four saw her off Deck when she hauled her Wind to the Southward. she was a large Schooner she was judged too far to Windward to Chace at this time of day the Brig Schooner and Sloop in company—Lattitude at Meredian by Observation.

Wednesday 12th April.—All this day Calm very little head way Course NNW.—no particular Transactions this 24 hours—dead Calm All Night all Sail set with the Brig Schooner and Sloop in Company.—Latt. by Obs. 28°—05" North—

Thursday 13th.—This morning the Sailmakers Mate died of a Dissentry. Calm all the forepart of the day. Exercised the Men at the Great Guns, and the Marines with Powder.—Light Air of wind in the afternoon from the Nd. Course W by S. at 4 PM saw a Sail right ahead nearing us fast made all Sail displayed Signal for the Brig to Chace. the Chace at ½ past four put about and stood from us and perceived her to be Cutter—continued the Chace untill Sundown when we lost sight of her. gave over Chace, almost Calm throwed out Signals for the Brig to give over Chace.—Lattitude by Observation at Meredian 28°—05 N. Long.

Friday 14th.—At break of day this morning discovered the abovementioned Sloop or Cutter under our Lee bearing SW distance[16] Leagues. bore away made all Sail & gave Chace with a light wind at NE The Chace bore away and made All sail—the French Brig gave Chace continued it some time without gaining at 10 finding she gained from us fast gave over Chace and hauled our Wind, Rieft the Topsails. Breezes freshens Course NNW. Last Night Southerly Grinnell, one of the Marines departed this Life.—in the afternoon blows a fresh Breeze double Riefed the Topsails very thick Cloudy weather. Lat at Meredian 28° 09" N Long. 71° 13" W.—Cape Henlopen N 16° W. distance 670 Miles.—

Saturday 15th.—One of this Ships company taken this Morning with the small Pox.—Fresh Breezes at East in the forenoon under all Sail with the Brig and Sloop in company. after Meredian the Wind freshens and very variable took in Sail. in the Evening blows fresh at SE and by S. with Squalls. close Rieft. the Topsails sent down the Top G. Yards. Lat by Obs. 30° 05" N Long. 72° 13" W.—Cape Henlopen bearing N. 14° W. distance 538 Miles.

Sunday 16th Fresh Breezes this Morning but variable under Snug Sail Wind Course[17] at 10 AM. wore Ship with our head[18] the Sloop Capa. Munroe out of sight astern. afternoon the Winds moderates veers and hauls—split the Mizen Topsail got it down and mended it by 12 at Night and bent it again.—at 6 PM wore Ship with the Brig in company Lat. by Obs. at M. 31° 41" North.

Monday 17th.—Fresh Breezes all the forepart of the Day Course under easy Sail with the Brig in company. after Meredian almost Calm. in the Evening frequent Squalls of wind and heavy Rains attended with Thunder and Lightening No Observation to Day.—

Tuesday 18th.—Constant heavy Squalls of wind and Rain. Landed the Topsails under snug Sail Winds SSW Course variable at Meredian weather more moderate made Sail Wind SW—Course in general NNW.—No Observation to day very thick and Cloudy.

Wednesday 19th.—At six O'Clock this morning heavy Squall of Wind carried away our Mizen Gnaff got him down. Carpenters employed fishing him.—at Meredian fresh Gales of wind at NW. Course N by E. with a rough Sea being in the Gulf Stream. at 2 PM continues blowing fresh in foresail Lay the Ship too under her fore-Main and Mizenstaysails head up N by E off NE. the Brig in Company. at 7 weather moderates made Sail and got up the Mizen Gaft. at 11 brot. too again under Staysails and lay all night Lat by poor Obs. 35° 09. Long. 72° 53" W. Hatteras N. 54° W. dist 166 miles. Cape Hentopen N 24° W dist 242 Miles.—

Thursday 20th April.—Early this morning made Sail Wind WNW. Course N. at 7 perceived the Water coloured Sounded with 138 fathom of Line no bottom found—At Meredian fresh Gales and clear weather took in Sail under the foresail and Staysails Blows fresh all the afternoon Wind Course[19] towards sunset the Wind moderates made Sail again with the Brig in company.—Lattitude by Observation at Meredian 36° 38" N.—Longt. 72° 22" W.— Cape Hatteras S. 69° 15" W. distance 206 Miles. Cape Henlopen N. 40° 58 Wt. distance 175 Miles.

Friday 21st.—Pleasant Breezes at NW with smooth Sea course NNE. at 4 AM wore Ship Course W.S.W. under the Topsails and Courses. at 3 P.M. winds variable and ahead Tacked Ship course N. by W. out Rief in the Topsails.—the weather pleasant but wind continues variable.—Lat by Obs. 37° 19" N. Long. 72° 04" W. C. Henlopen N 56° W. dist. 163 Miles.—

Saturday 22d.—At 7 O'Clock this morning Sounded no bottom found fresh Breezes at W. by N. course N. and clear weather at 11 AM saw a sail on the Lee bow made Sail and gave Chace the Chace was large Sloop at 1 PM. the Chace leaving us fast gave over Chace the Ship Sails exceeding bad. hauled our Wind Course N.N.W. at sundown a fresh breeze very thick and Cloudy close Rif'd the Topsls. sent down T.G. Yds. 12 at Night wore Ship in compy with the Brog course S. by W. Lat at Meredian by Obs. 38° 15" W. Long. 71° 40 Wt. Cap H. dist 156 Miles.—

Sunday 23d April.—Cool, clear and Pleasant weather Wind NW. Course WSW.—fresh Breezes and flying Clouds at

Meredian took in Top G. Sails.—several of our People broke out with the small Pox. afternoon weather moderate and very smooth Sea. Winds variable but continues ahead. At 6 AM sounded with 130 fathom of Line and no bottom at 7. the wind more favorable at WNW. wore Ship Course NNW. Lattitude at Meredian by Obs. 38° 15" N. Long 72–04 W. Cape H bearing N. 75° W. dist 138 Miles.

Monday 24th The weather moderate this forenoon—Winds WNW. After Meridian a Schooner discovered to windward standing down for us—hove out English Colours and fired a Lee Gun she soon answered by first hauling her wind with an English Jack at her Mainmast head—she soon afterwards hove about and stood the Southward. kept on our Course thinking it ineffectual to Chace in Light Winds The Brig hoisted English Colours over American to appear as a Prize to us but it had no effect. very moderate Weather in the Evening very little Wind Course WNW. under all Sail at 9 AM. saw two Lights on the Land sounded in 16 fathom and Wore Ship stood off to the Eastd. untill 11. deepend. the Water to 24 and wore Ship again Course NW under easy sail. Lat at Mered. 38–11 N. Cap. H N 63° 24 W. dis 72 Miles

Tuesday 25th April.—Light Airs of wind this Morning at 7. Wore Ship Course W by N.—at 11. AM Calm sounded in 23 Fathom Water at 12 very thick and Cloudy with Rain sounded in 21 Fathom water.—at 1 PM took a pleasant Breeze from the Southward stood our Course in for the Land. at 3. the Light house and Cape Henlopen discovered on the Lee bow immediately made all Sail set Steering Sails &c. and stood in for the Land at 8 O'Clock came to Anchor in company with the Brig off the Light House, very dark and Squally in the Evening.—

Wednesday 26th.—At Daylight this morning found ourselves within the Light-house it bearin E.S.E.—blowing very fresh at N.E.—made Signals for a Pilot by firing several Guns &c. at 6. the Ship drifted bent a shroud hauser to the small Bower Cable as it is but indifferent and let go the small bower. at 10. hove in the best bower Cable and found the Anchor gone bent the stream Cable to the Stream Anchor in order to let him go. At 12 the Gale increasing got down T.G. Masts. Continued making Signals for a Pilot but no appearance of any coming off and the wind increasing with a heavy Sea accompany'd with Rain about 1 AM lowered down the lower Yards, got the Topsail Yards in the Tops and Stript the Mizen. blow so heavy that we expected to go ashore on the Cape. about 3 PM. the wind and sea very heavy and the Ship thought in eminent danger of going ashore, having no Pilot on board in case the Ship shou'd part her other Anchor the 2d Lieutt. at his own request went with 6 Men in the Yawl to Lewistown by the way of the Creek for a Pilot.—continues blowing heavy all the afternoon let go all the Ground tackling we had.—at Dark no Pilot off or appearance of our Boat returning. the Brig in company riding at Anchor.—

Thursday 27th April. The Gale moderating this morning still safe at Anchor the Weather very thick cold and Cloudy no Boat off. Fired signal Guns. very moderate at Meredian. much surprised at the delay of our boat. sent a Midshipman ashore to enquire about our Boat the barge returned on board the Brig. about 2 AM. discovered Mr. Fisher launching a Boat on the Beach without any of our People.—a bad Omen. my forbodeing heart, tells me some unhappy unforeseen disaster has taken place

From the aforegoing Date [Thursday 27th April] to the present Day the Ship lay at Philada. and Chester.—Careening, Rigging, Manning &c. being near seven Months

Tuesday December 5th 1780. Having a Pilot on board, got under way with light Wind at WSW. and dropt down below Marcus-hook where we came to Anchor with our best Bower.

Wednesday 6th December. pleasant moderate weather light winds, but still contrary at S.W.—the Cutter returned from Philadelphia with a number of Men.—at 11 AM got underway and dropt down to the Highlands of Wilmington and came too.

Thursday 7th.—Calm thick, cloudy weather. Officers employed all day stationing and Quartering our Men. at 5 O'Clk. PM got under way with light winds from the Northward and Westward run down below Wilmington came too. in the Night the third Lieutt. came on board from Philadela accompanied by two french Gentlemen one the late Counsel the other Vice-Counsel at Charles-town in S. Carolina, which indicates that we are bound on another Merchant-Voyage to the West-Indies.—

Friday 8th Sent up T.G. Yards—got under way at 10 AM with light winds ahead dropt down about a mile below N. Castle, brought a Shallop alongside took out of her 5 Cords of Wood and 8 Quarters of Beef—took three Deserters out of some Merchtmen. blows very fresh at SSE.—prest 3 Men (being the first) out of a Schooner

Saturday 9th Took a Cutter load of Ballast on board, thick cloudy weather Winds ahead—went ashore in afternoon in company with the Surgeon to Newcastle, spent the Eevening very agreeably with a number of female acquaintances. continued ashore all Night.—

Sunday 10th The Commander of the Shallop from whom we got the Wood informed us of three Seamen being at Port Penn in Cog. jud[g]ing them to be some of our Deserters last Night the 3rd Lieutt. was dispatched in the Barge to take them which effected by surprising in them Bed, and returned with them this morning about 10 O'Clock. at the same time returned on board from N Castle with my Companion. immediatly after Dinner accompanied the Captain ashore to N Castle in order to Purchase his Stock in the Evening engaged Mr. Fallow to Purchase them.—staid ashore all Night, spent the Evening with the Ladies—wind continues ahead.—

Monday 11th Turned out early this morning with the Captain and walked several miles along shore gathering the Stock at the farm-houses—returned in Town about Eleven having compleated our business returned on board about 4 OClock PM. in the Evening bending the Mainsail a Man fell off the Main yard on the Gunwale of a Boat alongside which killed him by his immediatly sinking.—Winds ahead—preparing the Ship for Sea.—

Tuesday 12th Went in the Barge this morning to Newcastle on business for the Captain returned on board about Eleven. Impressed seven Men out of a Schooner from Sea a light Breeze of Wind from the North. got under way and run down and Anchored at Reedy Island.—

Wednesday 13th.—Pleasant Weather this morning with the Wind to the Southward which brought up three Vessels from Sea the Brig Trooper, Randolph and Schooner Neptune, fired several Shot and brought them to Anchor. Mann'd and Armed the Barge and Cutter, impressed seven Men of the Brig Randolph and three out of the Schooner. a Lieutt. in the Barge boarded the Brig Trooper mountg. 16 Guns between Decks, close Quarters, where her Crew was Mustered amountg. to near 100 well Armed with a resolute determination not to be Pressed but at the expence of Blood—after using menaces and threats of hauling the Brig alongside &c. finding it had no impression and that our Commander had no Authority to impress Men the design was given up.—but

had he been invested with power similar to the British Navy in such Cases, we cou'd have found methods to have taken the refusal of her Men, without endagering lives. finally she was permitted to get under way and proceed on her Passage to Pha,—the Barge went on board the Count D'Grass to search for Men but found none in the afternoon exercised the Great Guns and Marines.—

Thursday 14th At 6 O'Clock AM hove up, got under way run 6 or seven Miles below Reedy Isld. the Wind failing came to Anchor the Wind to the Northd. and Westward, the Ships Company in the forenoon employed cleaning Ship the remainder of the Day employed in Exercising the great Guns and small Arms.

Friday 15th.—Thick cloudy disagreeable Weather, winds variable and ahead. Cutter went to Port Penn Watering.—Mates employed birthing the Volunteers and sick.

Saturday 16th At 5 AM the Cutter returned with a load of water. all hands employed clearing Ship. at one PM. the wind being ahead and likely to blow fresh got the Ship under way and run back to the Island—a Man died sent him ashore to be Buried.—the Gunners Mate sent to Philada. to the Hospital.—Exercised at the Guns and small Arms.—

Sunday 17th Very foggy thick weather all day. the Ship Genl. Greene just Built at Morris-River, went up the River to be fitted.—Winds very light and variable cleaned Ship, with other Dutys took up the Day.—

Monday 18th at 10 A.M. the Sloop of War Saratoga, came down and Anchored at Reedy Island. foggy weather wind to the Southward, Mates employed in Birthing the People—Capa. Young of the Saratoga paid us a visit this afternoon he is bound on a Cruize and designs to go out with us. in the Evening exercised the Cannon and Marines.—

Tuesday 19th December. Cloudy weather warm pleasant Winds from Westward at 10 O'Clock A.M. the Schooner Fame from St. Croix, came up, brought her too & pressed 6 Men out of her.—the Barge went to Fort penn & returned in the Evening.—loosed sails to dry. in the afternoon Officers employed, Mustering the Men correcting the Watch and Quarter bills &c.—Exercised the Marines with Powder for the first time performed beyond expectations.—We now Muster near 260 Men.—

Wednesday 20th Light winds from the Northd. & Westward, got under way in company with the Saratoga and 13 sail of Merchantmen run down about 2 leagues being Calm the fleet came too.—at 10 AM. sprung up a fresh Breeze at NW the fleet got under way and stood down the Bay as low as Bombo-hook when the Weather growing very squally and tide most spent the Pilot thinking it dangerous to run in such weather came to Anchor again. the remainder of the fleet with the Saratoga stood down the Bay.—continued squally all day.

Thursday 21st This morning clear weather with a fresh breeze at NW. by W. got under way, stood down the Bay, spoke with the Minerva Capa. Earle from the Havannah at Anchor in the Bay. at 1 PM. abreast the lighthouse hove the Barge out, set the Pilot ashore with two Women and a sick Man at 2 the Barge returned hoisted her in, and made sail Course S.E.—at 4 PM. the light house bore NW by W. distance by estimation 4 leagues.—at sundown two Brigs in Company that joined us at Cape Henlopen the weather moderate—

Friday 22d pleasant moderate weather this forenoon under easy Sail Winds light at S.E. steering E by N. in the afternoon let a Reef out of the Topsails & set stay sails light Winds from S. to SW. steering S.E. the Brigs in Company. quite Calm at dark No observation to day.—

Saturday 23d Light Winds at NNW. Course S.E. in the morning.—at ½ past six saw a sail on the Larboard bow. gowing entirely Calm cou'd not Chace, at 9 a very light breeze from the Southward made Sail and gave Chace—out Riefs in. the Topsails set the Mainsail and staysails fore and aft at 11 set all sail, breeze at 1 AM freshens under all sail in Chace—at 2 PM finding we did not gain much on the Chace gave it up and took in all light Sails—during the Chace Winds SW. Chaced E. NE—and NE by E.—Lattitude by Observation at Meredian 39° –30″ N. Longitude in 72° –29″ W.—

Sunday 24th December Moderate pleasant weather under easy Sail. Winds at North. Steering S.E. in the Morning saw a Sail to the Southward take her to be the Brig that came out with us.—the other Brig parted from us during the Chace yesterday. got up T.G. Yards this morning. Lat by Observation at Meredian 37.° 40.″ N.—

Monday 25th.—at 7 AM made a Sail distance three four Leagues ahead gave Chace, at 8 made sail by the wind Steering from ESE. to SSE. with a fresh breeze at SW. at 10 AM came up with the Chace, proved to be the brig our Consort, shortened Sail. at 1 PM the wind freshens handd. the fore & Mizen topsails.—a Crew employed taking in the slack of the Lee rigging. at 4 close reifed the fortopsail, at 6 blowing very fresh handed the Maintopsail and set storm Mizen staysail.—blowed fresh all Night with a heavy Sea from WSW. Ship Labours excessively. Main rigging being very bad stretches greatly.—Latt. by Observ. at Meredian 36° 41 N.—

Tuesday 26th The Main rigging during last Night stretched amingly [amazingly] being my morning Watch on Deck, at 6 O'Clock blowing heavy and ship roling immoderately, sprung the Mainmast about 15 foot above Deck, lowered down the Main Yard immediately to ease the Mast.—all hands imployed taking in the slack of the Main rigging, and securing the Mast by preventer Shrouds, Runners & Tackles &c. at 1 PM fresh Gails at W.N.W. Course S.E. Ship under her foresail and storm Staysails, labours exceedingly at PM all hands employed securing the Mainmast.—the Mainrigging continues stretching, took in the slack of it again as well as the Weather wou'd admit—set up the fore-rigging also, and topmast back stays. continues blowing With a heavy Sea. Lat by Ods. 35°—09" N.—

Wednesday 27th Weather continues bad blowing at NNW, very heavy, at 9 set the Mainstaysail and close rief't foretop-sail, at 1 AM set MTM. staysail and Mizen t. sail.—at 2 Wore Ship Course ENE. took in the topsails. at 4 employed getting up other preventer Shrouds under the Cheeks of the Main-mast.—ships labouring so much with the slackness of the Main rigging drawed three of the Main chain plate bolts of the weather shrouds. Carpenters employed refitting them. in the Evening very squally with rain.—Lat by Obs. 33° 00" at Meridian Bermudas bearing S 71° E. distance 133 Miles.—

Thursday 28th at 5 O'Clock this morning wore ship S.S.E. our Course—Winds WNW—Weather moderates but squally, set fore & Mizen topsail, MT.M. staysail & Mizen at 12 saw the Island of Bermudas bearing SW distance 6 or 7 leagues. Wore [20] ship to the Eastward, fresh Breezes and Cloudy. at 2 PM saw a Brig standing down for us on our Weather bow bearing N.N.E. got up fore top G. Mast.—the day being far advanced, the Mainyard down and Mainmast not properly secured or fished it was thought most proper not to Chace her.—at 6 PM Course S.E. the weather moderating, some of our People broke out with the small Pox. Lat at Meridian by Observation 32° 37" N.

Friday 29th.—Pleasant moderate weather Winds at East Course at Meridian South.—all the Carpenters employed fishing the Mainmast.—under moderate sail—pleasatnt Breezes in the afternoon with squalls of Rain.—Lattitude by Observation at Meridian 31°—16" N.—Bermudas bearing NW by W ½ W distance 124 Miles.—Turks Island [21] SW by S ¼ W. distance 749 miles.

Saturday 30th.—at 5 this morning Samuel Collins a Marine departed this Life. Winds very variable general course S by W. with squalls of Rain.—a Crew employed woulding [22] the Mainmast.—in the afternoon squally & hazy handed the Top-sails, and at 4 finished the Mainmast.—at 6 wore Ship Course WNW. Wind SE by E.—Lat. at Meridian 29° 43" N.—

Sunday 31st December 1780.—Blowing squally weather with heavy Rain. Wind S.W. laying too under the foresail and Mizen staysail. Ships head up ESE. off East.—at 5 in the Morning Asa Durfey a Boy Departed this Life.—the weather more moderate and pleasant in the afternoon made a little Sail Course S. Wind W.—Lattitude at Meredian by Obs. 29° 37" N.

Monday 1st of January 1781.—at 7 A.M. made a sail bear-ing West hauled our Wind and gave Chace.—the Main yard being on deck swayed it up.—swayed up T.G.M. and yards, set the Mainsail M.T. sail and T.G. sails a very rough tum-bling Sea, but weather pleasant fresh breezes at at NW steer-ing S.W. continued the Chace untill 3 PM when finding we did not gain much on her took in sail gave over Chace.—the Chace a great advantage of us before we cou'd make sail our Mainyard being upon deck &c. or we shou'd have come up with her. during the Chace carried away the Jaws of the derrick lowerd. it down Carpenters employed Night mending it.—in the Evening Rief't Topsails—Weather very pleasant with a Trade Wind at E. Course S by E. [23] died of the small Pox.—Lat by Ob. 28° 22" N. Bermudas bearing N 21 W. dist. 251 miles. Turks Island S 47° W dist 611 Miles.—

Tuesday January 2d 1781.—Pleasant Weather out riefs in the Topsails, set T.G. Sails, Staysails, & fore T.M. steering sail Wind E. Course S by W.—got up the Mizen derrick.—all hands on Deck employed cleaning Ship—in the Evening took in light Sails.—Lattitude by Observation at Meridian 26° 20" N—

Wednesday 3d.—This morning pleasant weather and fresh Breezes at E.N.E. Ship under fresh way, at 8 OClock a Man dropt overboard, hove the Ship too throwed out the Jolly boat,[24] the Man considerable distance astern swimming well, Jolly got to him took him, he was expiring—took the boat in and made Sail. Course S.W.—Ships Crew employed clean-ing hammocks, & between decks.—set up the Topmasts Rig-ging.—Lattitude at Meridian 23° 35" N. Longitude 62° 51" W. Turks Island bearing S. 73° W distance 432 Miles.—

Thursday 4th pleasant weather and fresh Breezes at ENE Steering S.W. lay too all Night under the Topsails—11 hands cleaning Ship. Lat by Obs. 21-39" N.—Yesterday having passed the Tropick line this day was celebrated in the usual manner.—

Friday 5th January.—At 5 this morning made Sail Course W. Winds E.N.E.—at 12 O'Clock a sail discovered from the fore T G mast head distance 8 or 9 Leagues. made all sail and gave Chace. by sundown gained more than half the dis-tance. kept sight of her after Night—suppose her to be a Ship—continued the Chace—got Ship clear for Action.—at 10 PM came up alongside of the Chace—hailed her—she answered from Liverpool in Nova-scotia bound to Turks Island —concluded her a prize. sent the Barge on board of her—brought the Capta. on board who finding us to be Americans said he was from Salem bound to the aforementd. Island and produced two sets of Papers, upon which the Capta. thought proper to take her into Cape francay for Tryal—she is loaded with Balast—lay too all Night under the Topsails.—Lat at Meredian 21°—31"—N. Turks Isld S. 88 W dist 170 Miles.

Saturday 6th at 7 O'Clock this Morning saw a Sail in the S.E.—made sail and gave Chace Course S.S.E.—at 12—the Chace kept away—our Course WSW. gaining on her.—at 4 P.M. having gained within two Miles of the Chas—and making the Turks Islands ahead—to bring her too—fired a shot but had no effect we continued the Chace gaining but little—at sundown the Chace discovering the Land Wore Ship and stood for it—we fired eight or ten bow Chaces at her without the desired effect—and the Land being close aboard bearing W by N. distance six Leagues thought proper to give over Chace—she appeared to be an American Brig.—took in sail reifed topsails and hauled by the Wind Course[25] the Prize Brig in Company. Lat by Obs. at Meridian 21°—21" N Long. 68"—58°—Wt.—

Sunday 7th Stood off and on last Night under the Top-sails to the Northward and Southward.—at daylight made sail Course SSW. at 9 saw the Grand Turk bearing S by W.—saw a brig to windward running along the Land—at 10 she anchored in Turks Island road—at 12 run into the road astern of her under English Colours. hove too and sent the Barge on board of her as belonging to the British Venus Frigate. she proved to be from Nantucket, therefore quitted her. the Barge taken in made sail. at this harbor a Town composed of a number of small huts inhabited by Bermudians who come there to make salt—of which we saw sufficient to load a hundred sail of Ships—this Island is about 5 or 6 Miles long entirely barren being covered with Rocks and shrubs, pro-ducing nothing but salt which is formed by nature in large Ponds on the Island, there is six or seven more of these Islands all in sight but very small without Inhabitants some of them has salt Ponds. at 6 ½ OClock passed the sand key the Southermost & Westermost of these Islands &at sun-down it bore N.E. Courses S by Y.—at 9 AM SW by S.—

Monday Janüy. 8th.—At 7 this morning saw the Island of Hispanolia [26] and the Cape Francis bearing S. 7 or 8 Leagues

Wind ESE. Course SW by S. found ourselves to Leeward of the Cape. at 8 Tacked Ship stood to the ENE. Wind to the S.E. Tacked again W—and stood in for the Cape. met a number of small Vessels coming out at 2 PM up with the Cape and took on board a Pilot—on the this Cape Land stands a very strong fort at ½ past 2 came to Anchor opposite the Town—it appears near as large as St. Pierrs and very regularly Built situated under two very high mountains.—oppisite the Town over the harbor is a small Village called Petit Anse which some years ago was laid out and intended for the Town of Cape Francway (the present one at that time being Destroyed by fire) but they rebuilt it as at present therefore the other tho' situated on a fine and extensive Plain did not succed in their attempts, tho' equally as advantagous as to the Harbour but the Inhabitants at the time of the Conflagration being possessed of Ground Estates in the Town thought it less expensive to Rebuild on their Lots.—The Town Village aforementd. being 2 Miles distance & oppisite forms the harbour in breadth its about 5 Miles in length, capable of containing 300 Sail of Ships and very safe in all Weather being defended at the Mouth by several riefs of Rocks which break off the Sea in the greatest hurricanes—at 3 O'Clock AM hoisted out the Barge the Capa. & Passengers went ashore Moored Ship—a great number of Merchantmen in the harbor from differt. Ports—the Ship Flora arrived here about half an hour before we did—who sailed in Company with the Saratoga the day before we did—

Tuesday 9th.—Exceeding Warm here untill 10 or 11 O'Clock when the Sea Breeze generally comes in and blows fresh—Saluted the Town at sun rise this Morning with 21 Guns, it was returned with the same number from a Battery in Town, several Vessels arrived amongst whom a Brig from N London. unbent some of the Sails, &c.—

Wednesday 10th.—Unbent all our Sails—Crew employed mending them.—hauled a french Brig alongside—got the Yard takle on her foremast and hoisted it out.—sent Twenty two Sick Men ashore to the Hospital. others taking Sick hour—the [crew] grows very sickly.—at 10 Men swam away from the Ships Bows the french Brig laying at a small distance took them up—brought them on board and confined them in Irons.—

Thursday 11th.—At 9 AM Went ashore to Town with our Commissioned Officers, who in company with our Capta. waited on the General, Intendant—Commanding Officer—and several other Officers and principle Inhabitants of the Town. this is a Custom made use of and expected by our Allies.—Dined with Mr. P—d one of our Passengers—who has treated us with every mark of politeness—in the Evening accompanied him to the Play—their playhouse is large and very neat—superior much in every respect to the Theatre in Martinique in which is two or three good performers with an exceeding grand Orchestra of Musick this Town is much larger than I expected being superior to St. Pierrs it is most regularly built in small Squares.—the Inhabitants dress very elegant—the Ladies ware most enormous large heads. bolsters on their hips by way of hoops.—

Friday 12th Went ashore this morning—Dined with a large Company at Mr. Calabash, a principle Merchant in this Town.—in the afternoon took a Walk thro' the Kings Stores and Yards. Visited several elegant seats along shore towards the fort.—in the Evening accompanied Mr. P.—d to a Ball at the Playhouse which is converted into an Assembly Room by laying a Stage over the Pit.—the Company small but elegant.—

Saturday 13th Accompanied the Captain and Officers this morning waiting on several Gentlemen, amongst whom was Mr. Choisel—Governor of the Southern part of the Island.—Waited on Mr. Vevie and Mr. Ravy—Merchants in this Town to whom I had Letters and was received with the utmost Politeness.—Dined at Mr. P.—spent the afternoon viewing the Town.—

Sunday 14th Dined in Town to day with a large company at Mr. Denny's a Mercht. here from Boston, spent the afternoon there. in the Evening went to the Comedy being too late our seat was filled did not stay. afterwards was conducted to a Mustee Ball—[37] they dress in same taste as the white Ladies and dance tolerable well—slept ashore at Mr. D.—ys

Monday 15th Returned on board to dress this Morning—at 10 AM accompd. several of the Officers out to the Hospital where we spent the Day most agreeably with the chief Surgeons and several other Gentlemen their Hospital is very spacious about a mile and half from Town, situated on a rising plain at the foot of a Mountain, the Surgeons Hall is elegant from which you have an agreeable and extensive prospect of a large Plain for many Leagues distance at the back of the house there is an elegant Garden with a variety of Fruits and a very fine Fountain which throws the Water perpendicular upwards of 20 feet—here we have 20 odd sick all recovering.—in the Evening returned on board the Ship.—arrived this afternoon a small Schooner from Phila. and a Brig from Salem.—

Tuesday 16th.—Spent the day on board fine cool Sea Breezes and pleasant Weather.—arrived this Evening to large American Ships from France belonging to Philadelphia.—

Wednesday 17th.—Went ashore this Morning with several of the Officers to the Admiralty on the Condemnation of our Prize.—dined aboard—afternoon ashore again—in the Evening went to the Ball—where we found a very large and brilliant Company—at 9 O'Clock waited on Mr. R. and his Lady home.—Supped with them and returned aboard.—a Brig arrived from Connecticut.—

Thursday 18th.—this Morning a small Schooner arrived from Philadelphia.—several Caulker employed Caulking the Ship.—all Hands employed clearing Hold in order to take in our Cargo.—Dined ashore—spent the Evening at the Comedy.—at 9 returned on board.—

Friday 19th Went ashore in the Morning on business.—Dined with the Marquis [38] General and 2d in Command in this Island—after dinner the Card Tables introduced spent an hour at them.—General appears to be a very polite agreeable Genta.—but as he did not understand English or we french rendered our Company less familiar and agreeable than if we cou'd have supported Conversation.—paid a few friendly visits in the Evening and returned on board.—Caulkers at Work. hold unstoe'd &c.

Saturday 20th.—Early this Morning the Capa. went out of Town with Mr. Lorey on a Visit in the Country for a few days.—at 9 O'Clock a Messenger arrived from the Governor desiring the Commanding Officer to wait on him—at 11 AM the first Lieutt. sent off orders on board to bend sails and get the Ship in order for Sailing on a short Cruize to Convoy in an expected French fleet from Martinique, at 1 AM the Lieutt. returned on board and informed us the Cruize was given up—as the Governor just received advice of the Fleets safe arrival at Port Louis on the S. part of the Island—spent the afternoon.—

Sunday 21st.—Spent the forenoon aboard. several Americans Captains dined with us.—several Seamen belonging to the Phila. Ships lately arrived here offered to Enter with us on account of the bad treatment they received from their

Captains consequently the Officers of the Ship were in duty bound to receive & enter them in the service—upon the Officers going on shore met with the Capas. of the Ships who insulted the Officers and the service despised our Authority & rejected our orders—our barge in the afternoon went aboard one of the Ships for the People Cloaths the Commandg officer confined the Barge Crew for some time—these Audicious [insults] Offered to the American flag by the Subjects of America, highly required us to punish the Offenders—for which purpose we stated the matter to the General who ordered the Commander of the Ships before him the next morning.—In the Evening went to the Comedy—Supt with a friend—at 10 went to a Masquerade Ball—but the Evening being rainy the did not assemble—slept ashore.—

Monday 22d.—Early this morning the 2d Lieutt. went on board one of the Ships Capt.[29] to demand the Mens Wages which we had entered—but the Capa. treated our Lieutt. with the utmost disrespect and scuralty [scurrility] but it served no other purpose than an addition to his Crimes yesterday and determined us to bring him to a sense of his erroneous conduct—at 8 O'Clock our Officers with the Masters of the two Merchantmen went before the General who after a long Examination offered to put them in Goal and punish them according to the Laws of the Country if we required it—but the delinquents acknowledged their errors begged every Officers pardon belonging to the Ship that they had insulted and promised for the future to pay due respect to the Officers and flag of the United States, upon which they were discharged.—a frequent Conduct of this kind where insult is offered wou'd bring our Flag in more respect.—at 11 AM. arrived here the Privateer Brig fair American belonging to Phila. 26 Days out—during which time she captd. two Brigs off Carolina one Loaded with dry Goods which she sent into Phila. the other loaded with fish which she brought here with her—also arrived a sloop from Connecticut—and the Brig Neptune Capa. Burroughs from Phila.—afternoon went ashore with Mr. P——d— paid a visit to Mrs. V—— the finest Woman I have seen in the Island—yet I cannot converse with her—a most unhappy circumstance—returned aboard in the Evening.—

Tuesday 23d On board the greatest part of the day. the Cutter and Barges Crew ashore getting 5 Anchors down on the Wharf ready to take on board—took a Deserter out of the Fair American.—ashore in the Evening went to the Play sup't ashore and returned on board.—

Wednesday 24th This morning a fleet of Deans & Dutchmen sailed from here bound to Windward.—aboard all the forenoon—blows fresh—a Schooner one of the fleet that sailed this morning returned having sprung a Leak Sailing in the Cutter this afternon.—

Thursday 25th Went ashore this afternoon on business—Capa. Harding returned from the Country. Dined aboard—at 2 PM arrived a Brig—Capa. Earl fom Rhode Island—ashore in the afternoon went to the Comedy and supt with a friend.—returned on board about 10 O'Clock

Friday 26th Two flats employed filling water Crew clearing the hold to receive the Anchors Dined with Mr. R—y—spent the afternoon ashore—in the Evening the Anchors came alongside—arrived a Schooner from Boston—returned on board in the Evening.—

Saturday 27th At 12 O'Clock the Saratoga Sloop of War Capa. Young arrived here with a Prize Ship Mounting 20 Guns Loaded with Turpentine the Saratoga sailed the day before us and had made three Prizes.—ashore after dinner.—Went to the Play—and supt with a number of American Gentlemen and returned aboard.—

Sunday 28th Two Gentlemen from Town breakfasted with us. at 8 O'Clock this morning the Saratoga saluted the Town—the first Gun fired unfortunately proved to be Loaded with a round shot and Grape supposed to be put in by the Prisoners which killed a negro Woman ashore—and in loading one of the Guns without Springing killed a Man and wounded another—Dined in Town with Mr. V—— Waited on the General at 4 O'Clock respecting the misfortune attending the Sarata. Salute. at the Comedy in the Evening—supt with a friend at 11 went to a Masquerade Ball—but only two or three Masks—and the Assembly of Ladies but few it appeared more like a Country frolick than a Masquerade at 12 the Company broke up.—Slept ashore

Monday 29th Warm weather People growing Sickly 60 at the Hospital.—filling Water and Caulking Ship.—With the Capta. & Officers dined with Mr. Lavo—the Genta. who Transacts the Loading of the Ship.—The Goods not being ready which are to carry home—the Governor had requested to Cruize for 12 or 13 Days.—accordingly Orders sent on board to bend sails and prepare the Ship for Sea.—at 1 PM arrived the Kings Brig Cat—who is going with us as our Tender.—

Tuesday 30th Hauled the Prize Brig alongside and took out of her 30 Ton of stone Ballast—all Hands employed stowing hold—bending Sails &c. Dined with the Officers on board of Capa. French—at 3 went ashore—Spent the Evening at the Comedy and returned aboard.—

Wednesday 31st.—Preparing for Sea—taking in stone Ballast—Wrote home to Philadelphia by Capa. Burrows Brig Neptune.—Ashore afternoon at 4 sent on board 31 french Troops for Marines[30] Unmoored Ship—Lodged ashore.—

Thursday the 1st February Went aboard this Morning at sunrise—Calm weather.—about 8 got under way with the Land breeze soon out of the Harbour with the Brig Cat in Company having four American Vessels under our Convoy at 9 under moderate sail standing out from the Land—employed Quartering the Men—a Ship to Windward standing down for us—made Signals and fired two Guns to Windward—it proved to be the Dean Frigate having been Cruizing on the Coast 8 or 9 Weeks without success—she parted with us after giving them directions how to run into the Cape.—a very fresh Trade Course N.—at sundown the fleet Rief'd Topsails. the Land of Hispa. in sight hearing[31]

Friday 2d February.—Weather moderate and Calm this Morning—the fleet in Company—in the forenoon frequent squalls of Wind and Rain Course N. & N by E. at 3 let the Rief out of the Topsails and set the Mainsail—at ½ past three discovered the West Cocas ahead—run down to it by Sundown hove the Barge out and sent Signals on board of the french Brig—and Letters on board of the other Vessels for America.—at 6 hove too and lay under the Topsails untill 12 O'Clock then made Sail Wind E by N. Course N by E.—

Saturday 3d.—pleasant Weather and fresh Breezes—at 7 AM saw the West Cocas bearing S.E. 5 or 6 Leagues—at 9 saw Maguana on the Lee bow bearing WNW distance 6 Leagues.—Steering N by E.— at 9 PM lost sight of all our Convoy the steered more the Northward.—the Brig Cat in Company.—Latde. at Meredian by Observation 22° 20" N.—

Sunday 4th.—fresh Breezes at E by S.—the Brig in Company—the Ship thought to be out of trim run the four bow Guns aft with all the People sails somthing better but the Ship wants more balast—blows fresh at [illegible] the Topsails—at 1 Wore Ship to set up the Starboard rigging—at 4 wore Ship again Course NE by E.—Lat. by Observ. at Meredian 23-25" N. Maguana bearing S 24° W. dist. 103

Miles

Monday 5th February.—Cloudy Squally Weather in the forenoon Winds ESE. Course N.E. the Cat in company—employed trimming Ship took the Provision & Ballast out of the fore hold into the after hold for that purpose the Cat in company—Exercised at the Cannon and small Arms.—Lat. at Meredian by Oms. 25°—1" N—Long. 71° 17" W. Maguana bearing S 29° W. 185 miles

Tuesday 6th Pleasant moderate Weather the fore part of the day Course NE by E. Wind SE by E.—at 12 wore ship—Course SW by S. Wind SE by S.—at 2 PM squally with Rain— at sundown the Cat astern lay by for her—at 8 PM. Tacked Ship Course E. Wind SSE. and pleasant weather. Lat. by Obs. 26°—07" N. Long. 70°—18" W Maguana bearing at noon S. 31° W 264 miles

Wednesday 7th.—Light Winds and smooth Sea Course the forepart of the Day E by N.—afternoon Winds variable and light in the afternoon Exercised the Cannon and Marines at small Arms with Powder. the Cat in company—Calm in the Evening—Lat by Obs. 26°—02" N. Long. 69°—36 W. Turks Island bearing S. 6° W dist 279 miles

Thursday 8th.—Light Winds almost Calm at S.—Course ESE.—Washed Ship fore and aft—Caught a Dolphin. after Meredian light Airs very variable at 5 squalls of rain took in T.G. Sails Wind SW Course SSE.—Lat by Ob. 26—12" N Long. 68. 45" W.—

Friday 9th.—Light Airs of Wind At South this morning—at 9 AM discovered a small sail from the T.G. mast head in the NE.—made signals to the Cat and both gave Chace—continued the Chace Calm at Meridian finding we did not gain much on the Chace gave it over and took in the steering Sails at two OClock—soon after a Brig discovered astern in Chace of us.—almost Calm. laid the sprit sail Topsail a back, and trimmed the other sails as not to make much headway in hopes to decoy her alongside—she gaining on us—at half pas three discovered another sail on our Weather Quarter standing for us supposed to be the others Consort hauled all the Guns but 5 of a side & Lowered down the Ports—at sundown she was within about 2 Legues of us but not liking our appearance she hove about & stood for the Northd & Westd. & for the other Vessel—being Calm thought it unnecessary to Chace—after dark lay the Main Topsail to the Mast—the Cat in Company. Lat by Obs. at Meredian 26° 12 N. Long. 67°—51" W. Turks Isld. S. 24° W. dist 244 Miles.—

Saturday 10th At 12 last Night Wore Ship head WSW.—saw the two sail bearing W by N. at 5 this morning made all sail and gave Chace with the Cat to the two Brigs ahead—very light Winds at SSW. at 9 one of the Brigs hove about and stood to Windward of us—We hoisted our Coulors and fired a Gun to Leeward the Cat being to Windward the Brig bore down to her and spoke with her then came down to us. proved to be the fair American the other Brig of 14 Guns and 70 Men belonging to N. Providence—the Fair American made sail and stood after her Prize ahead.—almost Calm all day Course WSW. Exercised the Ships Marines with Powder in the aftenoon. Philip Carney a Marine died last Night about sundown the fair American got up with her Prize put about & stood for us in Company with them all Night—Lat. by Obs. 25—59" N. Long. 68°—11" Wt.—

Sunday 11 February.—Pleasant Breezes at S.E.—at 5 this morning discovered the Fair American ahead in Chace, made sail and gave Chace Course SSW. at 7 the Fair American brought the Chace too—Tacked and stood for us, at ½ past seven sent our barge on board of her she was from St. Croise bound to N. Carolina. during the Chace she throwed her Boat overboard therefore our Capta. thought proper to spare him our Jolly Boat. at 10 took in the Barge and made sail in Company with three Brigs—Light Winds all the afternoon at SE. by S. Course SSW. Lat. by Obs. at Meredian 25°—09" N. Long. 69°—16" W. Turks Islands bearing S. 12° W dist 228 Miles.—

Monday 12th Light Airs of Wind at SE by E. and very smooth sea. the Cat—Fair American and her Prize in Company—the Afternoon Wind E.—Course SSE. took in the middle & T.G. staysails at 6 Olock PM—Lat by Obs. at Mered. 24°—00 N Long. 70°—02" W. Turks Isld. S. 2° W dist 155 Miles.—

[End of Journal]

---

[1] Conrad Alexandre Gerard, John Jay, and William Bingham. Bingham as commercial agent of the Continental Congress in Martinique attempted to supply the needs of the *Confederacy*. The other two were passengers; Gerard had completed service as French minister to the United States and was returning to France while Jay, former President of Congress, was enroute to Spain as American Ambassador.

[2] St. Pierre, the largest town on Martinique and metropolis of the French West Indies.

[3] Offspring of white and black; a contraction of Mestizo.

[4] The last three are other members of the sailing party.

[5] The cruise lasted longer than Admiral Piquet expected. A superior English fleet compelled him to take shelter at Guadeloupe for a month. Not until 25 February were the men returned to the *Confederacy*.

[6] Caulking: driving oakham into the seams of the sides, decks, etc.

[7] The "General" or governor, Marquis de Bouille, assured them that no effort would be spared to recondition the ship.

[8] This is the same fleet that appeared off the bay on 4 January, the day after the French fleet sailed. See Hardy's entry of the day.

[9] Hardy skipped two pages of his journal before continuing the 4 February entry that appears above the one of the 7th.

[10] Captain Harding wrote this day to Congress about his refitting difficulties:

I have been endeavoring to procure the necessary articles for the ship's outfit, which I find it very difficult to obtain, and some impossible to be had at this place, especially suitable masts . . . am apprehensive that my crew will suffer very much with sickness by a further detention.

Therefore shall Fit her out with such necessary's as are to be Procured immediately . . . If I do not receive them . . . shall proceed as I informed you in my last, that is to Boston.

[11] To strengthen the mast by fastening on it a long piece of oak, convex on one side, and concave on the other.

[12] HMS *Ceres*. Hardy seems unsure of its spelling; here *Series* is corrected to *Ceris*.

[13] This should be Wednesday 23d; the discrepancy continues until Thursday, March 16.

[14] Scraping off the grass, slime, shell, etc., which adhere to the bottom near the surface of the water, and daubing on a mixture of tallow, sulfur, and resin, as a temporary protection against worms.

[15] Left blank by Hardy in the journal.

[16] Left blank in journal.

[17] Left blank in journal.

[18] Left blank in Journal.

[19] Left blank in Journal.

[20] Wear: to turn a ship around from the wind.

[21] Island near Haiti.

[22] Woolding: winding a piece of rope about a mast or yard to support it.

[23] Left blank in Journal.

[24] A smaller boat than a cutter, but likewise clinker-built.

[25] Left blank in journal.

[26] Hispaniola: it now conprises Haiti and the Dominican Republic.

[27] See note 2.

[28] Left blank in journal.

[29] Left blank in journal.

[30] To strengthen Captain Hardy's Marine company.

[31] Left blank in journal.

APPENDIX F

The Narrative of Thomas Philbrook

Thomas Philbrook was born 9 May 1760 in Boston, Massachusetts, but moved to Somerset, Connecticut, as a youth. In May 1776, he joined Captain Vine Elderkin's company in Colonel Samuel Mott's Connecticut State Regiment and served six months as a private at Fort Ticonderoga. Philbrook entered on board the Continental sloop *Providence* under Captain Hoysteed Hacker in the spring of 1779, serving first as a sergeant and then as an acting lieutenant of Marines. During his four-month tour on board the sloop, he took part in the "reduction of the British post on Penobscot River where our vessel with the rest of the fleet were blown up & destroyed." After the debacle, Philbrook joined Colonel Jonathan Mitchell's Massachusetts Militia Regiment and served four and a half months as its "Quarter Master Sergeant." At the conclusion of his service he made his home in Providence, Rhode Island, where he achieved a reputation in later life as a "gentleman well known for his urbanity of manners, and as a teacher in several useful branches of learning."

The following is an extract from a more detailed account of his military activities during the war, and is one of the best historical accounts of the Penobscot expedition in existence. The account was prepared some time after 1832 at the request of Benjamin Cowel for inclusion in his volume *Spirit of '76 in Rhode Island: or, Sketches of the Efforts of the Government and people in the War of the Revolution*, published in 1850. The original narrative is now lost.

In the spring of this year, 1779, I engaged with Capt. [Hoysteed] Hacker, to go a cruise in the continental sloop Providence, but before we could get ready for sea, the Penobscot expedition was fitting out, and we were ordered to join that fleet. All the movements were slow and tardy, and it was not until near the last of June that the fleet could be got together. The armed force consisted of the Warren frigate, 36 guns, and Providence sloop, 14 guns; these were the only continental vessels, the rest were all private property belonging to Boston and Salem, and hired by the State of Massachusetts. They consisted of nine ships of from twenty-two to eighteen guns: six brigs of sixteen and eighteen guns, and forty coasting sloops of about one hundred tons each, employed as transports. The fleet was commanded by Commodore [Dudley] Saltonstall, of New London, and the land forces by Gen. [Solomon] Lovell, of Hingham. We sailed from Boston, (I think the last of June), the next day we anchored in Portland harbor to receive on board a part of the troops. We tarried here several days, and then sailed for Broad bay, where we were to receive the remainder of the militia. Hence we wasted several more days, seemingly, for no other purpose, but to give the enemy sufficient time to prepare for us. We, however, got to Penobscot at last, having been twelve or fourteen days in making the passage, which we certainly might have done in four or five. We sailed up the bay, until abreast of the British garrison, where we anchored; taking special care to keep out of the reach of their guns. The fort is situated on a point of land formed by the junction of the Bagaduce and Penobscot Rivers, on a gently elevated piece of land which appears to have been cleared and cultivated for a number of years. The breast-work or fort was rather a rough looking concern, built with logs and dirt, and not more than three or three and a half feet high, which our long-legged militiamen would have straddled over without much difficulty. We expected that the troops would have been landed the next morning after our arrival, but the morning came and the day passed without any movement; some hard speeches were handed round at the expense of the General. The next passed as the former, but not without much grumbling. The British had two ships of twenty-eight and twenty-four guns, laying in the mouth of Bagaduce river, (this was all the naval force they had.) Our captain and the captain of one of the armed brigs, thought to have a little diversion and drive them from their moorings. This they got liberty to do. At sunset, I landed with thirty marines from the Povidence, with as many more from the brig, all under the command of Capt. R.[obert] Davis, of the continental army. We landed on a small island of two or three acres lying in the mouth of

the Bagaduce, about a mile below the ships. The island was thickly covered with wood and underbrush, we soon found our way to the summit (which was not very much elevated,) and commenced erecting a breast-work. We were very industrious through the night, making as little noise as possible, that we might not be heard on board the ships. Before the break of day, we had our works completed and received from one of the ordinance transports, three long 18 pounders, which we soon had mounted; as it began to grow light, we trimmed up a tall spruce tree on which we hoisted our flag and saluted our neighbors with the well loaded guns. This I believe, was the first they knew of our being so near them, as they appeared to be perfectly still and quiet; and by the time it took them to get ready to return our morning call, a brisk fire was kept up from us and from the ships; we could frequently see our shot hull them, so that we must have done them considerable damage; they generally over-shot us, but unfortunately, one of their shot struck the top of our breast-work and killed two men and wounded three others. After about three hours' firing the ships slipped their cables and moved up the river, out of the reach of our shot. We next went to work, cleared a piece of ground, and built us some comfortable huts to lodge in until further orders. Our little settlement we called Hacker's Island. We had now been here five or six days and the troops were not landed. A general uneasiness pervaded all ranks, both among the sailors and soldiers, something must be wrong. Our general was said to be a very *good sort of man*, but these good sort of men seldom make good Generals. I recollect that I thought then, and I still think that Mr. Lovell would have done more good, and made a much more respectable appearance in the deacon's seat of a country church, than at the head of an American army. It was, however, at last agreed upon to land the troops. (I think it was the morning of the seventh day after our arrival.) The place selected for landing was very injudiciously chosen, being a high bank covered with small trees and shrubs, with an ascent of at least forty-five degrees, whereas, about half a mile distant was a fine level cleared spot, sufficiently large to hold the whole army, where we might have landed under the cover of the guns of one or two of our ships without the loss of a man. This appeared very strange to us all, at the time, but I believe that all were of the opinion afterwards that the old General had agreed to go snacks with the Commodore in whatever they were to have for defeating the expedition. We were however, landed at the place appointed. The marines were first set on shore on the beach, some musket shot were fired at us from the brow of the hill, but we were at two great a distance from them to receive any damage. When the marines were all landed and about half the militia, we began our ascent, which was indeed a very difficult one; had it not been for the shrubs growing on the sides of the hill, we might have lost half our men before we gained the height. Though I was not encumbered with a musket, I found it difficult to keep my footing. When we had ascended about one third of the distance, the British from the brow, began a brisk fire upon us, which they kept up till we were within a few rods of the top; they then courageously fled and left the ground to us. In this ascent we had forty men killed and Capt. [John] Welch, of the marines on board the Warren frigate, a very amiable young gentleman, and a brave officer. Our brave General did not lead the van in their ascent, neither did he bring up the rear, probably he and the Commodore were walking the Warren's quarter deck with their spyglasses to see the fun. I saw him two hours after on the hill, giving orders about building huts, for which he probably was well calculated. With the marines belonging to the Providence and brig Tryall, we returned to our little isle, where we found good cheer and comfortable quarters. In three or four days the militia were comfortably housed as if we had come to spend the summer with our English neighbors, when everybody knew, that knew any thing, that an express had been sent to Halifax, and that they would be prompt in sending a reinforcement which might be expected in a very short time. Councils of war were held every day on board the Commodore's ship; the result commonly was, the Commodore and General could not agree, probably they had agreed in one point, and engaged not to agree in any other. Thus, day after day passed away without any thing being done. Some of the captains of the fleet frequently landed on the Bagaduce shore, either to amuse themselves by rambling in the woods, or to reconnoiter the enemy. In one of these excursions, the Commodore being with them, they spent the most of an afternoon; towards evening they were discovered from one of the English ships, and immediately a boat with a company of soldiers was sent on shore. On seeing the boat put off from the ship, they each made the best of their way for their boat which they had left waiting for them. They all reached the boat nearly at the same time, but the Commodore was missing. They waited, but he came not; it began to grow dark, the British boat had returned to the ship; finally, at 9 o'clock, they concluded to leave him to his fate and take care of themselves. The next morning at sunrise, he was seen on the beach, and a boat immediately sent to take him off. He said he was closely pursued, and in making his escape he had got so far into the woods, that he could not find the way out in the dark, but it was generally believed that he found a very good berth on board one of the British ships. Tumults now ran high, the General was hissed and hooted at wherever he made his appearance, and the Commodore cursed and execrated by all hands. Capt. Hacker offered, with the Providence and the six brigs that if the General would attack the fort, all the men from the ships would gladly join him; he, (Capt. Hacker) would enter the river with the small vessels and engage to bring out the English ships. But no, it was not feasible;—according to the best accounts we could get, the British had only five hundred effective men in their garrison. Their naval force we knew was only two small ships. We had between three and four thousand militia, with sixteen ships and brigs well armed and manned. It is strange that these spirited fellows were kept peacable so long; the sailors indeed, talked hard of leaving their ships, and under a commander of their own choice, pushing forward and storming the fort. A single word of encouragement from any of the captains in the fleet, would have set them in rapid motion, and I have no doubt but they would have succeeded, but we had dallied away our time too long: more than thirty days had been spent in idleness since our first anchoring, and not a single movement made to annoy the enemy. On a fine summer's morning, five lofty ships were discovered in the offing. All knew who they were, for all had expected them for some time; it was now all hilter skilter. The men from the shore were ordered to embark as soon as possible, and the fleet to weigh anchor or slip their cables and proceed up the river. We were soon under sail with a fair wind. The English ships were not more than three miles from us, but the wind so favored us, that we kept clear from their shot. As we came towards the head of navigation, the Warren frigate grounded, was immediately cleared of her men and blown up. The other ships soon followed her example, and as fast as they could land their men and some stores, set fire to their vessels and left them. Our retreat was as badly managed as the whole expedition had been. Here we were, landed in a wilderness, under no command; those belonging to the ships, unacquainted with the woods, and only knew that a west course would carry us across to Kennebec; whereas, there were hundreds of the militia that were old hunters, and knew the country. Some of these ought to have been detained as pilots, and we might have got through in three days; but we had no one to direct; so every one shifted for himself. Some got to their homes in two days, while the most of us were six or seven days before we came to an inhabited country. I got through on the seventh day, after keeping a fast of three days. From Portland, I took passage in the frigate Boston, Capt. [Samuel] Tucker, was treated with much politeness by him and his officers.

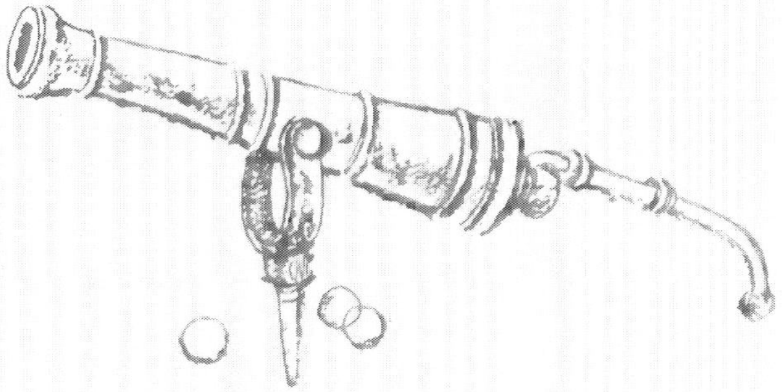

APPENDIX G

Extract of a Letter From Captain Samuel Nicholas, April 10, 1776

"Extract of a Latter From the Captain of Marines [Samuel Nicholas], on board the Ship Alfred, dated at New-London, April 10, 1776." [1]

Before this comes to hand, I make no doubt you will have heard of our arrival in this port, and of our engagement with the Glasgow man-of-war; but as I intend giving you an account of our cruise, must beg your patience for a while respecting that matter. The 17th of February, left Cape-Henlopen, and after a very pleasant passage of fifteen days, came to anchor off the Island of Abaca, about seventeen leagues distance from New-Providence, where we brought to several small vessels belonging to Providence, which gave the Commodore an opportunity of inquiring into the state of the Island, as to its defence; and found it very well supplied with warlike stores, and an object worthy of our attention, as it was not sufficiently manned to give us opposition. The vessels we then had in our possession were detained as transports, to carry the Marines over to Providence. We embarked, and made sail on Saturday evening, March 2, and on Sunday, at two o'clock, landed all our men (two hundred and seventy in number) under my command at the east end of the Island, at a place called New-Guinea. The inhabitants were very much alarmed at our appearance, and supposed us to be Spaniards, but were soon undeceived after our landing. Just as I had formed the men, I received a message from the Governour, desiring to know what our intentions were. I sent him for answer, to take possession of all the warlike stores on the Island belonging to the Crown, but had no design of touching the property or hurting the persons of any of the inhabitants, unless in our defence. As soon as the messenger was gone, I marched forward to take possession of Fort Montague, a fortification built of stone, about half way between our landing place and the town. As we approached the fort, (within about a mile, having a deep cover to go round, with a prodigious thicket on one side and the water on the other, entirely open to their view,) they fired three twelve-pound shot, which made us halt, and consult what was best to be done; we then thought it more prudent to send a flag to let them know what our designs were in coming there. We soon received an answer, letting us know that it was by the Governour's orders that they fired. They spiked up the cannon and abandoned the fort, and retired to the fort within the town. I then marched and took possession of it, in which were found seventeen pieces of cannon, (thirty-two, eighteen, and twelve pounders,) and not much damaged; they were spiked with nails and spikes, which are easily taken out. I thought it necessary to stay all night, and refresh my men, who were fatigued, being on board the small vessels, not having a convenience either to sleep or cook in. The next morning by daylight we marched forward to the town, to take possession of the Governour's house, which stands on an eminence, with two four-pounders, which commands the garrison and town. On our march I met an express from the Governour, to the same purport as the first, I sent him the same answer as before. The messenger then told me I might march into the town, and if I thought proper, into the fort, without interruption; on which I marched into town. I then drafted a guard, and went up to the Governour's, and demanded the keys of the fort, which were given to me immediately, and then took possession of Fort Nassau. In it there were forty cannon mounted, and well located for our reception, with round, langridge, and canister shot. All this was accomplished without firing a single shot from our side. We found in this fort a great quantity of shot and shells, with fifteen brass mortars; but the grand article, powder, the Governour sent off the night before, viz: one hundred and fifty casks. Immediately after we were in the fort, I sent for the Governour, and made him prisoner until the Commodore arrived, which was soon after. We remained at Providence till we got all the stores on board the fleet, and then took our departure, the 17th of March. We have brought with us from Providence the Governour, his Secretary, and one Mr. Irving, Receiver-General of his Majesty's Customs, who belongs to South-Carolina.

On the 4th instant, we made the east end of Long-Island, and discovered the Columbus (who had parted with us the night before) to windward, with a schooner of six guns, one of Captain Wallace's tenders, which she had taken that morning. We made Block-Island in the afternoon, when the Commodore ordered the brig to stand in for Rhode-Island, to see if any more of the fleet were out, and join us the next morning; which was accordingly done, but without seeing any vessel except a New-York sloop, which Captain Biddle brought to the fleet, and after her papers were examined, she was released. At daylight we discovered a brigantine to leeward: we made sail, and soon came up with her, and after a few shots, took her. She proved to be a bomb brigantine belonging to Wallace's fleet, mounting eight guns and two howitzers, commanded by one Snead [Edward Sneyd], a Lieutenant in the Navy. We continued to cruise all day within sight of Block-Island, and in the evening took a brigantine and sloop from New-York, and have brought them into port with us, not being satisfied as to their clearances. At sunset we were twelve sail in all, and had a very pleasant evening. At twelve o'clock went to bed, and at half past one was awakened by the cry of "all hands to quarters." We were soon ready for action; the main body of my company, with my First Lieutenant, was placed in the barge on the main-deck, and remaining part, with my Second Lieutenant and myself, on the quarter-deck. We soon discovered a large ship standing directly for us. The Cabot was foremost of the fleet, our ship close after, not more than one hundred yards behind, but to windward withal. When the brigantine came close up, she was hailed by the ship, which we then learned was the Glasgow man-of-war; the brigantine immediately fired a broadside, and instantly received a return of twofold, which, owing to the weight of metal, damaged her so much in her hull and rigging, as obliged her to retire for a while to refit. Our ship then came up, (not having it in our power to fire a shot before without hurting the brigantine,) and engaged her side by side for three glasses, as hot as possibly could be on both sides. The first broadside she fired, my second Lieutenant fell dead close by my side; he was shot by a musket-ball through the head. In him I have lost a worthy officer, sincere friend and companion, that was beloved by all the ship's company.[2] Unfortunately for us, our tiller-rope and main-brace were shot away soon after the firing began, which caused the ship to broach to, and gave the enemy an opportunity of raking fore and aft. The battle continued till daylight, at which time the Glasgow made all the sail she could crowd, and stood in for Newport; and our rigging was so much hurt, that we could not make sail in time to come up with her again. At sunrise, the Commodore made the signal to give over the chase, he not thinking it prudent to risk our prizes near the land, lest the whole fleet should come out of the harbour. The Glasgow continued firing signal guns the whole day after.

In the action I lost three of my people out of twelve that were on the quarter-deck, and two others who were in the barge, were slightly wounded. Captain [John Burroughs] Hopkins, of the Cabot, is wounded, his Master killed and the Second Lieutenant of Marines wounded, and since dead.[3] Upon the whole, it was a very hot engagement, in which our ship and the brig were much damaged; but we have this consolation, that the enemy is full as badly off; for by several expresses from Rhode-Island, we are assured that it was with much difficulty she got into port, both pumps going. We are now, thank God, in harbour, and shall stay some time to refit.

[1] Force, comp., *American Archives*, 4th ser., v. V, pp. 846–847.
[2] Second Lieutenant John Fitzpatrick.
[3] Second Lieutenant James Hood Wilson.

APPENDIX H

Advertisements for Marine Deserters

The succeeding is a selection of Marine deserter descriptions extracted from newspapers published during the Revolutionary War. Their value as a source for military dress has been demonstrated in numerous studies, but they also provide information as to the individual Marine's personal appearance, age, height, and in some cases, previous occupation.

Deserter From Captain Welsh's Company[1]

TWO DOLLARS Reward.

DESERTED the third inst. from Captain John Welsh's company of marines, PETER M'TEGART, born in Ireland, about thirty years of age, five feet seven inches high, smooth faced, brownish complexion, short dark brown hair. Had on a light brown coat, white cloth jacket and breeches, blue stockings and new shoes. Whoever delivers said deserter to James Guthrie, at the corner of South and Water streets, shall have the above reward.

Philadelphia, January 9, 1776.

[1] *Pennsylvania Evening Post*, 9 January 1776; repeated 13, 18, 20 January 1776.

Deserter From Captain Shaw's Company[1]

FOUR DOLLARS Reward.

DESERTED from Capt. Samuel Shaw's company of marines, belonging to one of the Continental frigates, ANGUS CAMERON, born in Ireland, about five feet eight or nine inches high, dark complexion, short black hair curled behind, a little pock marked, and about thirty years of age. It is said he inlisted in Captain Craig's company of Lancaster militia. Whoever apprehends said deserter, and secures him in this city, shall have the above reward and reasonable charges,

FRANKLIN READ, 1st Lieut.

Philadelphia August 13, 1776.

[1] *Pennsylvania Evening Post*, 13 August 1776; repeated 24, 29 August 1776.

Deserters From Captain Mullan's Company[1]

SIXTY DOLLARS Reward.

DESERTED from Captain Robert Mullan's Company of Marines, lying in the Barracks at Philadelphia, the following persons, viz.

THOMAS MEWKINNY, labourer, about 5 feet 9 or 10 inches high, short fair hair, ruddy complexion, strong built, and lived near Jacob Miller's tavern, on Lancaster road.

JOHN M'LOSKY, a shoemaker by trade, about 5 feet 5 or 6 inches high, well set, short brown hair, one of his thumbs stiff at the joint, and withered so as to appear less than the other; he lived in the same neighbourhood with the former.

JOHN FRITZINGER, labourer, about 5 feet 9 inches high, fair hair, tied, slender built, German extraction, but speaks English well.

JOSEPH LOWRY, labourer, about 5 feet 6 or 7 inches high, short fair hair, full faced, well set, and resided in the same neighbourhood with the former, and they are all four great cronies together.

THOMAS CALWELL, labourer, about 5 feet 9 or 10 inches high, a little stoop shouldered, strong built, and born in Ireland.

THOMAS SAPPINGTON, labourer, aged 21 years, 5 feet 10½ inches high, sandy hair, slender built, born in Kent county, Maryland, and is supposed to have gone towards Warwick.

JESSY REDDING, by trade a blacksmith, about 22 years old, 5 feet 7¾ inches high, and born in Maryland.

JOHN M'CASHON, labourer, 5 feet 4 inches high, sandy complexion, stoops forward, and has the appearance of a hard drinker.

NEIL FARRON, labourer, born in Ireland, aged 26 years, 5 feet 4¾ inches high, short black hair, down look, and dark complexion.

JOSEPH BOYS, labourer, born in Ireland, about 27 years of age, 5 feet 9 inches high, dark complexion, he cannot write, and is supposed to have gone towards Cumberland county. It is probable he may visit his friends in Chester County.

HUGH CONOLEY, labourer, born in Ireland, about 27 years of age, 5 feet 9 inches high, short black hair curled; was a servant before he enlisted, with one James Drum, West-Fallowfield township, Chester county, and is supposed to have gone that way.

WILLIAM TAYLOR, a carpenter, born in England, 36 years old, 5 feet 4 inches high, lost his fore finger of his right hand.

JOHN HILL, born in Ireland, about 25 years of age, 5 feet 2 inches high, and is a sailor.

JOHN CAMPBLE BROWN, about 5 feet 6 or 7 inches high, has lost his left eye, and is of a sandy complexion.

SAMUEL M'CORMICK, labourer, enlisted at Port Penn, is about 37 years of age, 5 feet 5 or 6 inches high, is well set.

Whoever secures said deserters, and sends them to their respective company at the barracks in Philadelphia, shall have the above reward, or FOUR DOLLARS for each, and all reasonable charges, paid by

ROBERT MULLAN, Captain of Marines.

[1] *Pennsylvania Journal*, 20 November 1776.

Deserters From Captain Shaw's Company

SIXTY-EIGHT DOLLARS Reward.

DESERTED from Captain Samuel Shaw's company of marines, lying in the barracks at Philadelphia, the following persons, viz.

BENJAMIN WRIGHT, born in Jersey, by trade a carpenter, 26 years of age, 5 feet 11 inches high, light coloured hair, fair complexion, stoop shouldered, thin faced, long nose, had on a brown coat, other clothes forgot.

DANIEL GORISON, born in Jersey, by trade a blacksmith, 19 years of age, 5 feet 6½ inches high, black curled hair, smooth face, stoop shouldered, his fore teeth black.

PHILIP ASTON, a miller by trade, 22 years of age, 5 feet 6¼ inches high, light coloured curled hair, ruddy complexion, full face, stout made, served his time with William Craig, miller.

ROBERT CRUTHERS, by trade a weaver, born in Ireland, 24 years of age, 5 feet 8¼ inches high, well-made, dark coloured hair, stoop shouldered, pitted with the small-pox.

JONATHAN TAYLOR, a weaver by trade, 25 years of age, 5 feet 5¼ inches high, full face, short red hair; had on, when he went away, a dark coloured hunting shirt.

WILLIAM NEWEL, 20 years of age, 5 feet 5½ inches high, born in Ireland, black hair, blue eyes; had on, when he went away, a light coloured blue coat, faced with white; served part of his time near Newport.

BENJAMIN ADAMS, born in Old-England, 51 years of age, 5 feet 3½ inches high, very talkative, and much given to liquor.

JOSEPH LANCE, born in Jersey, 28 years of age, 5 feet 2¾ inches high, sandy coloured hair, pitted with the small-pox, swarthy complexion, walks wide at the knees; had on a brown hunting shirt, with red cuffs and collar.

PETER ELWELL, born in Jersey, 18 years of age, 5 feet 4 inches high, swarthy complexion, black hair, blue eyes.

JOHN KETCHAM, by trade a cooper, 37 years of age, 5 feet 9 inches high, pale complexion, grey hair, blue eyes; formerly lived at Oakford's mill; born in Jersey.

CHRISTOPHER WHITE, born in Ireland, by trade a weaver, 26 years of age, 5 feet 3½ inches high, black hair, thin faced and slim made.

EDWARD RILEY, born in Old-England, cordwainer, 42 years of age, 5 feet 2½ inches high.

DAVID GORMAN, born in Ireland, labourer, 38 years of age, 5 feet 2 inches high, black hair, pitted with the small-pox.

PHILIP M'LAUGHLIN, 19 years of age, 5 feet 5¼ inches high, born in Ireland, light coloured hair, which he wore tied, smooth and full faced.

JOHN ALLEN, 22 years of age, 5 feet 7 inches high, born in Ireland, brown hair, grey eyes, marked with the small-pox.

JAMES BROWN, about 28 years of age, 5 feet 8 or 9 inches high, dark hair, brown eyes; deserted from the English army last October, and had a pass, signed by General Ewing; had on when he went away, a brown coat, black stock, white cloth breeches, grey yarn stockings, and new shoes.

EDWARD M'CARTY, born in Ireland, much pitted with the small-pox, 40 years of age, about 5 feet 5 or 6 inches high, his hair a little grey; had on a grey coloured jacket, the sleeves of blue stockings.

Whoever secures the above deserters, and sends them to the barracks in Philadelphia, shall have the above reward, or FOUR DOLLARS for either of them, and reasonable charges paid by

SAMUEL SHAW, Captain of Marines.

[1] *Pennsylvania Gazette*, 20 November 1776.

Deserters From Captain Shaw's Company[1]

DESERTED, the 15th of January, four marines belonging to Captain Samuel Shaw's company, from on board the Randolph frigate, lying at Fort-island, viz.

PHILIP MULHOLAND, five feet five inches high, dark brown complexion, a weaver by trade, and about twenty-seven years of age.

NEIL FARAN, five feet five inches and three quarters high, brown complexion, black hair, full faced, and well made.

JOHN CLEMENTS, five feet six or seven inches high, brown complexion, black hair, twenty-two years of age, and a taylor by trade.

THOMAS M'NAMIE, five feet six or seven inches high, brown short hair, brown complexion, down look, and twenty-seven years of age.

January 18th. EDWARD ROWIN, a landsman, about five feet eight inches high, pale complexion, and twenty-five years of age.

HENRY SPEAR, a marine, about five feet eight inches high, remarkably squint eyed, sandy hair, fair complexion, lusty built, about twenty-eight years of age, and plays the fife tolerably well.

EDWARD HIGGINS, a landsman, about five feet nine inches high, stout built, pitted with the small pox, brown complexion, and American born. All the rest are Irishmen.

The above men were well clothed. FIVE POUNDS reward will be given for each of them, or THIRTY FIVE POUNDS for the whole of them, on their appearance on board the Randolph frigate, or if secured in any of the jails of Philadelphia.

NICHOLAS BIDDLE

[1] *Pennsylvania Evening Post*, 21 January 1777; repeated 4 February 1777.

Deserters From the Brig Andrew Doria[1]

DESERTED last night from the Andrew Doria, lying at Fort-island, two marines, viz. THOMAS ROBINSON, an Irishman, between thirty and forty years of age, about five feet seven inches high. THOMAS FURMAN, an Irishman, about thirty years of age, near five feet six inches high. They took away with them a small whaleboat, two or three great coats, and two watches, one of which is numbered 515, and maker's name Edward Vaughan, London. Whoever secures said deserters, so that I get them again, shall have Forty Dollars reward, and reasonable charges, or Twenty Dollars for either of them.

THOMAS VAUGHAN, Lieut.

N.B. If they should offer the watches for sale or pawn, it is hoped they will be stopped. May 21.

[1] *Pennsylvania Evening Post*, 24 May 1777.

Deserter From Captain Matthewman's Company[1]

DESERTER from on board the Continental frigate Montgomery, John Hodge, Esq; commander, on Tuesday the 2d instant, DAVID HARDY, about twenty-seven years of age, five feet six inches high, born in Ireland; had on when he went away, a rifle frock much tarred, a pair of woollen over-alls, a round hat, the crown of which painted red, short brown hair, fore shins, very talkative when in liquor. He was draughted about two months since from Capt. STRONG's company, in Col. HENRY B. LIVINGSTON'S regiment. It is imagined he has steered his course for *Pennsylvania*, as he formerly lived there. Whoever apprehends said deserter, and delivers him to me on board said ship, lying at Fort Montgomery, shall receive FIVE DOLLARS reward and all reasonable charges.

WILLIAM MATTHEWMAN, Captain of Marines.

Ship Montgomery, Sept. 4, 1777.

[1] *New York Journal and the General Advertiser*, 8 September 1777; repeated 15, 22, 29 September, and 6 October 1777.

Deserters From Captain Mullan's Company[1]

DESERTED from Captain Robert Mullan's Company of Marines, in Philadelphia, the following men, viz. THOMAS PECKWORTH, about 37 years of age, 5 feet 9 inches high, dark complexion; had on when he went off a light coloured cloth coat, his other clothing not remembered, a shoemaker by trade. Thomas Mathews, about 25 years of age, 5 feet 10 inches high, a stout well built fellow; had on when he went away, a blue cloth jacket, with sleeves, a stripe linsey under waistcoat, brown strait hair, full face, born in England. Richard Coats, about 30 years of age, 5 feet 11 inches high, much pitted with the small-pox, had on when he went off a straw-hat, linen coat, a pair of tow trowsers and old shoes. James Nicholas, about 20 years of age, about 5 feet 4 inches high, has lately been employed in a shallop up and down this river; had on when he went off a round laced hat, a swanskin jacket with sleeves, striped trousers, stoops much as he walks. John Smith, about 27 years of age, five feet five inches, had on when he went off, a brown sailors jacket, says he serv'd his time to one Mr. Dorsey, on Elk Ridge, Maryland, he is a down ill looking fellow. Richard Babington, about 35 years of age, about 5 feet 7 inches high, had on when he went off, a short green coat, leather breeches and shoes; he has lately come from New York, and has a pass with him to go to Annapolis, where he says he has a wife and family. James Deacon, about 24 years of age, 5 feet 9 inches high, he says he served his time to farming in Bucks county, born in England, a fair complexion, short brown hair, slender built, and very talkative; had on when he went off, a pair of stripe cotton stockings, nankeen breeches much worn, a woollen bound hat with a white lining. Nathan Marchal, about 27 years of age, 5 feet 7 inches high, swarthy complexion, black hair tied behind, remarkable grey eyes, by trade a stocking weaver, says he served his time in Bonam town, Jersey, had on when he went off, a brown woolsey short jacket with sleeves, a pair of stripe trowsers, a swan skin covered cap with an old silver loop. Thomas Sappington, about 24 years of age, 5 feet 10½ inches high, a labourer, born in Maryland, slender built, had on when he went off, an old short brown coat, a pair of long tow trowsers, old shoes and an old felt hat, he formerly lived within seven or eight miles of Port Penn, New-Castle county. Whoever takes up said deserters and secures them in any goal of this State, or delivers them to me at my quarters the corner of Pine-street in Front-street, shall have the reward of One Hundred and Eighty Dollars, or Twenty Dollars for each; and all reasonable expences paid by me

ROBERT MULLAN, Captain of Marines

Philadelphia, Sept. 29, 1779.

[1] *Pennsylvania Gazette*, 6 October 1779; repeated 13, 20 October 1779.

Deserters From Captain Mullan's Company[1]

SIXTY DOLLARS Reward.

DESERTED from Captain Robert Mullan's company of Marines, lying in Philadelphia, the following men, viz. WILLIAM WARNER, house-carpenter, born in Philadelphia, 5 feet 8 inches high, went off in his regimentals, a green coat with red facings, white woollen jacket, a pair of light coloured cloth breeches, woollen stockings, a round hat, with white binding, and new shoes. JOHN M'CASHION, born in Ireland, a labourer, 5 feet 3 inches high, was seen going up Lancaster road with an old woman, whom he has taken with him as a companion; he is a down-looking fellow, with sandy hair; had on a long old blue coat, new white woollen jacket, light coloured cloth breeches, woollen stockings, a round hat with white binding, and new shoes. JOHN MACKEANY, about 5 feet 8 inches high, fair complexion, has lately worked at ditching for Col. Potter, in New-Castle county, and has a wife near the Red Lyon; went off in his regimentals, a green coat, with red facing, white woollen jacket, light coloured cloth breeches, woollen stockings, a round hat, with white binding, and new shoes. Whoever takes up and secures said Deserters in any goal in this State, shall have the above reward, or Twenty Dollars for each, and reasonable charges paid by

ROBERT MULLAN, Captain

[1] *Pennsylvania Gazette*, 10 November 1779; repeated 17, 24 November 1779.

APPENDIX I

Continental Marine Muster, Pay, and Prize Rolls

The following selected muster, pay, and prize rolls of Continental Marines are as complete as it has been possible to make them. Since the names of Marines were often included within the naval rolls, they have been extracted for the purpose of reproduction in this volume. However, in rolls where a distinction was not made between Navy and Marine personnel, the complete roll has been reproduced. In a few cases the format of the original roll has been altered to facilitate its publication.

The Continental Ship *Alfred*

"A Roll of All the Officers And Men belonging to the Ship Alfred from the time of her being put into Commission untill the 5th September 1776.—"[1]

[Extract] [20 November 1775–5 September 1776]

Time of Entry	Place Shipped at	Station in which they have entered	Dead, discharged Run, turned over, &c.	Time when	Number	Men's Names
1775	Rhode Island	Comr in Chief			1	Esek Hopkins Esqr.
Dec. 7th	Connecticut	Captain			2	Dudley Saltonstall
Dec. 7th	Philada.	1st Lieut	Promoted	May 10th 1776	3	John Paul Jones
November 20	Rho. Island	Lieutenant	Run	May 1st 1776	4	Benjamin Seabury
" 20	ditto	Lieutent.			5	Jonathan Pitcher
December 18	Connecticut	ditto			6	Jonathan Maltbie
			[Marine Roll]			
1775						
Nov. 20	Philada.	Captain	Promoted		1	Samuel Nicholas Esq.
" 28	"	1st Lieut.	do		2	Matthew Parke
" 28	"	2nd Lieut	Killed	Apl. 6, 1776	3	John Fitzpatrick
Dec. 14	"	Sergeant	Turned over to Columbus	Aug. 7	4	William Hamilton
" 8	"	do			5	Alexander Nilson
" 14	"	Corporal	Turned over to Columbus	Aug 7	6	James Marshall
" 19	"	"	Run	Feby. 8	7	John Creely
" 8	"	Drummer			8	James Edgar
" 13	"	Private	Run	May 19	9	John Dingham
" 17	"	"	Run	Septr 3	10	Richard Evitt
" 13	"	"	Turned over to Columbus	June 9	11	Thomas Lew Allen
" 19	"	"	discharged	Nov. 7	12	Israel Vanluden

1 Transcript of Military Papers, Revolutionary War, v. XL, pp. 657–663, Rhode Island Archives.

"A Roll of All the Officers And Men belonging to the Ship Alfred from the time of her being put into Commission untill the 5th September 1776.—"[1] (Cont.)

[Extract]

[20 November 1775–5 September 1776]

Time of Entry	Place Shipped at	Station in which they have entered	Dead, discharged Run, turned over, &c.	Time when	Number	Men's Names
Dec. 16	"	"	Run	May 1	13	Samuel Mickery
" 16	"	"	Turned over to Columbus	May 30	14	Thomas Vanluden
Dec. 12	"	"	Turned over to Columbus	May 30	15	Daniel Lafferty
Dec. 13	"	"			16	Thomas Pendall
" 6	"	"	Turned over to Columbus	Augt. 7	17	John Nick
" 7	"	"	Run	May 17th	18	Leonard Cooker
" 16	"	"	Died	May 11th	19	John Hannah
" 8	"	"	Turned over to Columbus	June 9th	20	John Curtin
" 11	"	"	ditto	Augt. 7th	21	Robt McAuley
" 12	"	"	Run	May 23d	22	John Tork
" 21	"	"	discharged	Oct 22d	23	John Long
" 24	"	"	Turned over to Columbus	May 30th	24	John McLocklin
" 13	"	"	Run	April 9	25	William Howell
" 13	"	"	died	May 10th	26	John Coombs
" 15	"	"			27	Isaac Cogan
" 14	"	"	Run	Aug 6	28	John Steward
" 14	"	"			29	Jeremiah Connel
" 10	"	"	Turned over to Columbus	May 30	30	Robert Robinson
" 10	"	"	ditto	ditto	31	John Stolcoss
" 8	"	"	ditto	ditto	32	Robert Richey
1776 Jany 14	"	"	discharged	April 16th	33	James Kirkland
1775 Dec. 15	"	"	Turned over to Columbus	Aug 7	34	Thomas Burns
" 19	"	"	Run	May 11	35	Joseph Makood
" 16	"	"			36	Matthew Green
Dec 7	"	"	discharged	May 12	37	William Leonhard
" 13	"	"	Run	June 7	38	William Williams
" 15	"	"	Turned over to Columbus	Aug 7	39	Thomas Donahoe
" 21	"	"	killed	April 6	40	Thomas Owens
" 11	"	"	Turned over to Columbus	Aug 7	41	Patrick O'Brian
" 21	"	"	"	ditto	42	Edward Hagan
" 17	"	"			43	Charles Bryant
" 11	"	"	died	April 15	44	George Shrack
" 15	"	"	Turned over to Columbus	Aug 7	45	Ellis Rowland
" 13	"	"	discharged	April 16	46	Andrew Reily
" 15	"	"	Turned over to A. Doria	May 13	47	John Glacken
" 15	"	"	Run	March 13	48	John Fox
" 14	"	"	Turned over to Columbus	Aug 7	49	John Bell
" 8	"	"	ditto	ditto	50	James Wall
" 19	"	"	Turned over to Columbus	ditto	51	Patrick McNamara
" 14	"	"			52	Stephen Bell

[1] Transcript of Military Papers, Revolutionary War, v. XL, pp. 657–663, Rhode Island Archives.

"A Roll of All the Officers And Men belonging to the Ship Alfred from the time of her being put into Commission untill the 5th September 1776.—"[1] (Cont.)

[Extract] [20 November 1775–5 September 1776]

Time of Entry	Place Shipped at	Station in which they have entered	Dead, discharged Run, turned over, &c.	Time when	Number	Men's Names
" 7	"	"	Run { Taken at New London May 30 & turned over to Columbus	May 25	53	Isaac Hart
" 8	"	"	Run	June 7th	54	Timothy Lynch
1776 Jan. 5	"	"	discharged	Novemr 6th	55	William Chester
1776[5] Dec. 16	"	"	killed	April 6	56	John Dougherty
" 15	"	"	Run (say Sick)	Sept 28	57	Michael Cooley
1776 Jany 2d	"	"	Turned over to Columbus	Aug 7th	58	James Connell
Jan. 5th	"	"	Run	May 25	59	Philip Fitzpatrick
1775 Dec. 14	"	"			60	Michael Moore
" 18	"	"	Left at Philada		61	Henry Stalcoss
" 18	"	"	ditto		62	Edward Harwisky
" 19	"	"	Run	Jany 29	63	Christ. Cazner
" 19	"	"	Dead		64	Saml Cornelius
" "	"	Corporal	Turned over to Columbus	7 Augt	65	John Hamilton

1 Transcript of Military Papers, Revolutionary War, v. XL, pp. 657–663, Rhode Island Archives.

"List of Officers & Men belonging to the Ship Alfred who are entitled to a Share of the Ship Mellish & Brig Active."[1]

[Extract] [22 October–15 December 1776]

Mens Names	Station	Agents Names	Sums	paid
John P Jones	Captain		846..13..4	
Jonathn Pitcher	1st Lieut	Nathl Greene		
John P. Rathburn	1st Lieut	Jos: Greenleaf		
Robert Sanders	2d Lieut		180..	
Piere Deville	3d Lieut	Jos: Greenleaf		
	[Marine Roll]			
Ezra Winslow	Marine	Jos Greenleaf		
Daniel Bennet	"	N Appleton		
Gideon Grinman	"	J. Greenleaf		
Joseph Cundall	"	L Jarvis	30...	
Lionard Evans	"	N Appleton		
Franklin Tennant	"	Colo Angell		
Edward Clanning	Marine	Jo: Greenleaf		
Durfy Springer	"	Colo Angell		
Stephen Northup	"	Jo: Greenleaf		
Edwd Washburn	"	Jo Greenleaf		
Thomas Potter	"			
Patrick McMullen	"			
David Frank	"	Jo: Greenleaf		
		Phila. £9...		
Archibd Edmunson	"	Jo Greenleaf		
		Phila. £12..10		
William Griffiths	Marine	Jos Greenleaf		
James Sinnex	ditto	Wm Duncan		
Jeremiah Motrany	ditto			
John Long	ditto			
John Nick	ditto	Nat Appleton	30...	
		Jacob Cooper		
Thomas Burns	ditto			
John McLaughlin	ditto			
Philip Fitzpatrick	ditto	Nat Appleton		
John McIndo	ditto			
John Howell	ditto	Phila £5		
		Nat Appleton		
John Gillen	ditto	Jo Greenleaf		
Michael Garret	ditto			
Peter Franklin	ditto	Nat Appleton		
Edwd Thurston	ditto	Jos Greenleaf		
Prince Williams	ditto	Perez Morton		
James Toney	ditto			
Joshua Moses	ditto	John Wye		
John Monely	ditto	Jos Greenleaf		
Peter Conya	ditto	Perez Morton		
William Anderson	ditto	John Winslow		
James Connel	ditto			
Thos Donrahu	ditto	James Boserts	30...	
Danl Pillegas	ditto	J.P. Rathburn	30...	
		£15 Phila		
Nathl Cook	ditto	Elisha Morey Jr.		
Nathan Tucker	Marine	D Tillinghast		
Joab Mann	ditto	E. Morey junr	30...	
Joseph Tuck	ditto	J Greenleaf		
Joseph Rider	ditto	D. Tillinghast		
John Richards	ditto	Danl Aborn	30...	
Isaac Carpenter	ditto	D. Tillinghast	30...	
James Merritreau	ditto	J. Greenleaf		
James King	ditto	J. Greenleaf		
James Morey	ditto	E Morey junr	30...	

[1] Revolutionary War Rolls, v. 52, p. 84, Massachusetts Archives.

"List of Officers & Men belonging to the Ship Alfred who are entitled to a Share of the Ship Mellish & Brig Active."[1] (Cont.)

[Extract] [22 October–15 December 1776]

Mens Names	Station	Agents Names	Sums	paid
John Fiske	ditto	Squire Fiske	30...	
John Deruse	ditto			
Barnabas Rider	ditto	Nat. Appleton		
Ephraim Dimick	ditto	S. Dimick	30...	
John Sales	ditto	John Jenkins	31...	
William Steward	ditto	Nat. Greene		
Daniel Duffy	ditto	J. Greenleaf		
Saml Tiler	ditto	Nat. Greene		
William Burns	ditto	Wm Reed	30...	
Thomas Clark	ditto		5..5.	Cap Hacker
Edmd Arrowsmith	Capn Mars	J. Greenleaf		
William Hamilton	1st Lieut	J. Greenleaf		
Alexandr Nelson	2d Lieut	D. Tillinghast		
Gilbert Tomkins	Serjt			
John Caswell	ditto			
Matthew Green	Corporal	J. Winslow		
James Edgar	Drumer			
Job Woodworthy	Marine	J Greenleaf		
Wm Langworthy	ditto	N. Appleton		
Benjn Allen	ditto	N. Appleton		
William Congell	ditto	J Greenleaf		
Patrick Doran	ditto			
Charles Bryant	ditto	Nat Appleton		
Thomas Pendall	ditto	Nat. Greene		
Stephen Bell	ditto			
Robt Robinson	ditto	John Slade	25...	
Jeremh Connal	ditto			
Isaac Cogan	ditto			
Stephn Dexter	ditto	E Morey junr	30...	
Danl Bartlett	ditto	E Morey junr	30...	

[1] Revolutionary War Rolls, v. 52, p. 84, Massachusetts Archives.

The Continental Frigate *Alliance*

"A List of Officers, Seamen and Marines belonging to the Continental Ship of War the Alliance under the Command of Peter Landais Esqr.—"[1]

[Extract] [17 March 1779]

Stephen Hills	1st Lieutenant	Danl. Cumstock	do.	William Riley	Marine	John Sadler	do.
Joseph Adams	2d do.	Stephen Turner	do.	John Kennady	do.	William Scott	do.
James Degge	3rd do.	Jeremiah Perry	do.	Oliver Arnold	do.	Michl Lyons	do.
[Marine Roll]		George Skipper	do.	John Weatherly	do.	John Fitz Gerrald	do.
Mathew Parke	Captain	James ONeill	do.	Benja. Taylor	do.	James Dickinson	do.
James Warren	1st Lieut	James Haslum	do.	Abra: Bradley	do.	✓John Morand	do.
Thomas Ellwood	2d do.	Samuel Dale	do.	John Forester	do.	✓Patrick Martin	do.
John Farnum	Sergeant	Nichl. Woodbury	do.	John Doile	do.	Them with this mark	
Edmund Ogden	do.			Samuel Platt	do.	✓ are still on board	
George Cox	Marine			John Theaf	do.	the Prize Ship—	
Moses Stocking	do.			Edward Fling	do.		

[1] Franklin Papers, University of Pennsylvania.

"List of Officers and Men &c. on board the Continental Frigate Alliance John Barry Esqr. Commander, when We left the Harbour L'Orient 29th March 1781, including those who have entered since also of those who have been put on board of Prizes."[1]

[Extract] [29 March 1781]

Hoysteed Hacker	1st Lt.	
Hez. Welch	2d do.	
Patk. Fletcher	3d do.	
[Marine Roll]		
Mathw. Parke	Capt. Mar	
Saml. Pritchard	Lt. do.	(dead 30th May 1781)
James Warren	do.	[Wounded in engagement with *Atalanta* and *Trespassy*]
Thomas Elwood	[do.]	
Benoni Simmonds	Sargt. Ma.	
Wm. Moody	Corpl. Ma.	
John Fuller	Marine	
David Brewer	Sargt. Ma.	dead [Killed in engagement with *Atalanta* and *Trespassy*]
Robert Green	Marine	[Put on board prize captured 2 May 1781]
Chr. Perry	Marine	
Saml. Orr	Corpl. Marines	
John Chewing	Marine	
Jnoa. Snow	Marine	
David Sloper	Marine	[Put on board prize captured 2 May 1781]
Eastus Hatch	do.	
Simn. Crossman	Drummer	
Josiah Perkins	Marine	
John Campbell	Marine	[Put on board prize captured 2 May 1781]
Danl. Waddens	Marine	[Put on board prize captured 28 May 1781]
Abner Hotchkiss	do.	
Saml. Hotchkiss	do.	[Put on board prize captured 28 May 1781]
Jona. Bemiss	do.	
John Finigin	do.	[Put on board prize captured 2 April 1781]
Benoni Condall	do.	
John Burtan	Marine	
Benja. Gould	Marine	[Put on board prize captured 2 May 1781]
Wm. Ryan	Marine	
Joseph Ransford	fifer	
James Pratt	Landsm.	Dead [Promoted from Landsman to Corporal of Marines]
Edward Harper	Marine	
Benja. Burnham	do.	

[1] John Barry Papers, LC.

"List of Officers and Men on board the Continental Frigate Alliance Jn.° Barry Esqr. Commander May 17. 1782—"[1]

[Extract]

Names	Station	Names	Station
John Barry	Captain	Robert Bell	Marine
Hezekiah Welch	Lieut.	Morris Fitz Jerald	Marine
Patrick Fletcher	do.	Jno. Wakeland	Marine
Nicholas Gardner	do.	Alpheus Morgan	Marine
[Marine Roll]		Benoni Evens	do.
Mathew Park	Captn. Marines	Jno. Brown	Marine
Thomas Elwood	Lieut. do.	Jno. Cartwright	do.
Willm. Morris	do. do.	Natl. Abbott	Marine
Thos. Joice	Sergt. Marines	Thos. Osborne	Marine
Jno. Keese	Sergt. Marines	Peleg Osborne	do.
Joseph Ransford	Fifer	Wm. Osborne	do.
Simion Crossman	Drummer	Elisha Turner	Marine
Isaac Royal	Marine	Aaron Abbot	ditto
Mich. Carter	Marine	Charles Moon	Marine
Jas. Hotchkins	Marine	Charles Adams	Marine
Christo. Mosier	Marine	Jno. Sears	Marine
Luke Durphy	Marine	Jno. Turril	Marine
Jno. Taylor	Marine	Stephen Curtis	ditto
Thadeus Parling	ditto	Hugh Osborne	Marine
Wm. Erwin	ditto	Jno. Fletcher	do.
Peter Clark	Marine	Nat. Veal	do.
Edmund Holden	ditto	Jerry Doty	do.
Abraham Bump	Marine	Timo. Dwire	do.
Job. Clary	Marine	David Fairbanks	Marine

[1] John Barry Papers, John S. Barnes Collection, New York Historical Society.

"List of Officers and Men On board the Continental Frigate Alliance December 8th, 1782"[1]

[Extract]

No.	Time of Entry	Names	Qualities	Remarks
	1781			
1	Novemr 1	John Barry Esqr	Captain	
2	do	Hezekiah Welsh	Lt 1st	
	1782			
3	July 5th	Hugh Smith	Lieut	protempy from 8 Decemr
4	12th	Wm Wroth	do	do from do do
		[Marine Roll]		
	[1781]			
6	Nov 1	Thos Elwood	Lut Mar	do from do do
25	Nov 1	John Keese	Sergt. Marines	
	[1782]			
28	June 24	John Wright	Sergt Marines	
	1781			
39	Decr 4	Nathl Abbot	Marine	
42	Nov 15	Mathias Martin	Marine	sent on board French Ship at Havannah 16 Feby 1783
	1782			
43	Augt 4	Josiah Sawyer	ditto	
44	Sept 18	Antonio Larego	ditto	Run Jany 11, 1783
	1781			
45	Nov 17	John Scott	ditto	R [Run]
46	Dec 20	Thomas Curtis	ditto	R
47	Sept 24	Robert Taylor	ditto	
48	Nov 21	Antonio Francisco	ditto	Sent on board Admiral at Havannah 13 Feby 1783
49	do 17	Evan Stevenson	ditto	
	1782			

[1] John Barry Papers, LC.

"List of Officers and Men On board the Continental Frigate Alliance December 8th, 1782"[1] (Cont.)

[Extract]

No.	Time of Entry	Names	Qualities	Remarks
50	June 29	Thomas William Blk	ditto	
51	May 30	James Knowles	ditto	
52	Jany 27	Robt Stout	ditto	D Apl. 10 1783
53	March 14	Peter Larry	ditto	Dichargd. Jany. 22d 1783
	[1781]			
68	Nov 23	Alpheus Morgan	Marine	
69	d 28	Timothy Dwire	do	
70	d 6	Morris FitzGerald	ditto	
	1782			
71	Dec. 6	Fridrick Woodsight	ditto	D-D [Discharged]
	1782			
72	Dec 18	Elisha Turner	ditto	
	[1782]			
193	Nov 1	Luke Durphy	Marine	
194		Francis Ricoro		Sent on board a French Ship at Havannah 16 Feby 1783
195	Nov 24	Peter Cuday	Marine	
196		Antonio Moore		
197		Jno Carthwt		
198		James Halfpenny		
199		Joseph Trippet		
200		Simeon Crossman		
201		Joseph Ransford		
202		Isaac Royall		
203		Hugh Osborne		
204		Thos Osborne		
205		Benoni Evans		
206		Robt Bell		
207		Abm Bump		
208		Robt Frazer		
209		Stephen Curtis		
210		John Sears		
211		John Turril		
212		Jerry Doty		
213		James McGuire		
214		Nat. Veal		

[1] John Barry Papers, LC.

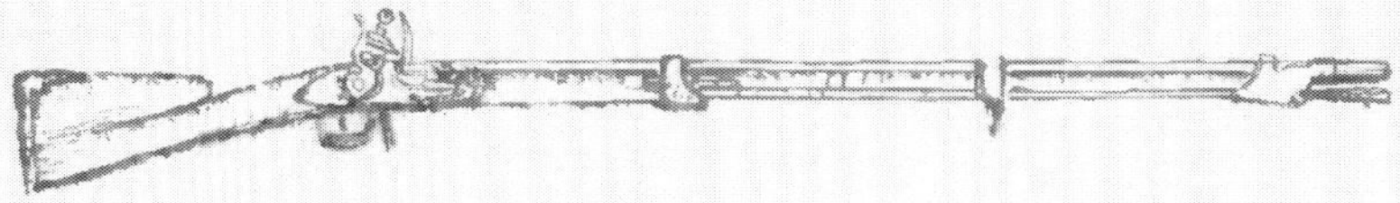

"Pay Role Contl. Frigate Alliance—"[1]

[Extract] [1 August 1781–1 May 1783]

Time of Entry	Names	Qualities	D[ea]D D[ischarged] R[un]	Time When	Time Serv'd	
	John Barry Esq	Captain				
Nov 1. 1781	Hezekiah Welsh	Lieutenant		Apr 1. 1783	17 Months	0 D
Nov 1. 1781	Patrick Fletcher	do	Arrested	Nov 26. 1782	12 Mo	26 D
do	Nichs E Gardner	do	ditto	ditto	12 Mo	26 D
		[Marine Roll]				
do	Mathew Parke	Cap Mar	ditto	Nov 25. 1782	12 Mo	25 D
do	Thomas Elwood	Lieut do		Apr 1. 1783	17 Mo	—
Dec. 1. 1781	William Morris	do		July 3. 82	7 Mo	5 D
[Nov 1. 1781]	John Keese	Sergt Marines		Ap 1. 1783	17 Mo	
[1781] Nov: 12	Thom: Joice	Sergt Marines }	D	June 25. 82	7 Mo	16 D }
1782 June 25	ditto	Marine }		Ap 1. 1783	9 Mo	10 }
[1781] Decemr 18	John Roper	Marine	D-D	Mar 22. 1782	3 Mo	5 D
Nov 1	Joseph Raniford	Fifer		Apr 1. 1783	17 Months	
Nov 1	Simon Crossman	Drummer		Apr 1. 1783	17 Mo	
Decr 17	Isaac Royall	Marine		Apr 1. 1783	15 Mo	15 D
August 29	Thomas Ellis	Marine	D	Apr 1. 1783	19 Mo	5
Decr 18	Michl Carter	Marine	D	Ap 1. 1783	15 Mo	14
Decemr 18	William Eaton	Marine	D-D	March 19. 1782	3 Mo	2 D
Decr 18	Christr Mosier	Marine	D-D	June 3. 1782	5 Mo	18 D
[Novem 1]	Luke Dursey	Marine		Apr 1 1783	17 Mo	
Decemr 11	John Taylor	Marine	R	May 17. 1782	5 Mo	8 D
18	Thadeus Pailing	Do	D	June 14. 1782	5 Mo	29 D
Novemr 2	William Erwin	Do	R	June 7. 1782	7 Mo	8 D
Sep. 26	Peter Clark	Marine		Ap 1. 1783	18 Mo	7
Decr 18	Abraham Bump	Marine	D	Ap 1. 1783	15 Mo	14
Decr 18	Job Clary	Marine	D	Ap 1. 1783	15 Mo	14
Decem 12	Robert Bell	Marine		Ap 1. 1783	15 Mo	20
[Novem] 6	Morris Fitzgerald	Marine	D	Ap 1. 1783	16 Mo	26
Decr 8	Charles Deagan	ditto	D	May 22. 1782	5 Mo	6
[Novem] 20	John Wakeland	Marine	D	Nov 20. 1782	12 Mo	
Novr 1	Jno Brown	Marine	D	Nov 1. 1782	12 Months	
Novem 23	Alpheus Morgan	Marine		Apl 1. 1783	16 Mo	9 D
Decr 20	Warren Sorrey	do	D-D	Ap 28. 1782	4 Mo	10
21	Benony Evans	do		Ap 1. 1783	15 Mo	11
Novem 8	Jno Cartwright	do		Ap 1. 1783	16 Mo	24
Decr 4	Nath Abbot	do		Ap 1. 1783	15 Mo	28
Decem 20	Thos Osborne	do		Ap 1. 1783	15 Mo	12
Decr 20	Pely Osborne	Marine	D-D	June 6. 1782	5 Mo	19
ditto	William Osborne	do	D-D	June 17. 82	6 Mo	
Decr 18	Elisha Turner	Marine	D	Apl 1. 1783	15 Mo	14 D
Novem 29	Ephraim Keith	ditto	D-D	March 12. 82	3 Mo	14 D
Decr 4	Aaron Abbot	ditto		Ap 1. 1783	15 Mo	28 D
Decr 20	Charles Noone	Marine }		May 24. 1782	5 Mo	6 D }
May 24. 82	ditto	Corpl do }	D	Ap 1. 1783	10 Mo	12 }
1781 Dec 20	John Sears	Corpl do }		May 24. 1782	5 Mo	6 D }
May 24. 82	ditto	Marine }		Ap 1. 1783	10 Mo	12 }
Oct. 9. 1781	Charles Adams	do	R	June 16. 1782	8 Mo	11 D
Decr 6	John Adams	do	R	June 16. 1782	6 Mo	13 D
Decr 18	John Turril	Marine	D	Apr 1. 1783	15 Mo	14 D
"	Stephen Curtis	ditto	D	Ap 1. 1783	15 Mo	14 D
Decr 20	Hugh Osborne	Marine		Ap 1. 1783	15 Mo	12 D
Decr 10	John Fletchner	ditto		Ap 1. 1783	15 Mo	2 D
Decr 19	Jarathmial Dotty	ditto	D	Ap 1. 1783	15 Mo	13

1 Hepburn Collection, William B. Clark Transcripts, Naval Historical Center.

"Pay Role Contl. Frigate Alliance—"[1] (Cont.)

[Extract] [1 August 1781–1 May 1783]

Time of Entry	Names	Qualities	D[ea]D D[ischarged] R[un]	Time When	Time Serv'd	
Sep 24	Nathl Neal	ditto		Ap 1. 1783	18 Mo	9
Nov 28	Timothy Dwire	ditto		Ap 1. 1783	16 Mo	4
Decr 18	David Fairwell	ditto	D-D	Ap 1. 1783[2]	3 Mo	17
Octr 2	David Fairbanks	Marine		Ap 1. 1783	18 Mo	1 D
1782						
[March] 3	James McColon	Marine	D	May 14. 1782	2 Mo	13
[June 14]	Robert Greene	Marine	D	Aug 2d [1782]	1 Mo	20 D
July 5	Ebenezer Martindale	Marine	R	July 17. 82	0 Mo	12 D
7	Rufus Reynolds	Marine	D	July 11	0 Mo	4 day
" 3d	Jas Halfpenny	Marine		May 1. 83	10 Mo	2
July 3. 82	James McGuire	Marine		May 1st 1783	10 Mo	2 D
[June] 24	Jno Wright—	Sergt. Marine		do	10 Mo	11 D
July 6	Joseph Frippet	Marine	D	do	9 Mo	29 D
June 29	Jno Trepon	ditto	D	Nov 28. 1782	5 Mo	3 D
July 8	Jno Blanchard	Marine	D	ditto	4 Mo	24 D
July 5	Charles Duboir	Marine	D	Nov 28th 82	4 Mo	27 D
[July] 28	Robert Frazer	Marine	D	May 1st [1783]	9 Mo	7 D
[Nov.] 15	Mathias Martin	Marine	D	Feb. 16 1783	3 Mo	4
"	Antonio Moore	do		May 1. 1783	5 Mo	17
Nov 15	Francis Ricow	Marine	D	do	5 Mo	17
Decr 6	Fredk Woodsight	Marine	D-D	Apr 1. 1783	3 Mo	26
Nov 24th	Peter Cuddy	Marine	D	[Feb 16, 1783]	2 Mo	25
Nov 17.	Baltes Matlow	do	R	Feb 13. 1783	2 Mo	29 D
[1783]						
[Feb] 10	Edw Miller	Marine	D-D	Ap 21. 83	2 Mo	11 D
11	Simon Dinman	do	R	Ap 14. 83	2 Mo	3 D

1 Hepburn Collection, William B. Clark Transcripts, Naval Historical Center.

The Continental Brig *Andrew Doria*

"List of the People on Board the Andrew Doria From February 1776"[1]

[Extract]

Mens Names	Turn'd over or Dead	Sick in the Hospital
Nicholas Biddle Comr.	1	
James Josiah 1 Lieutt.	2	
Elijah Warner 2d Do.	3	
John McDougle 3d Do.	4	
[Marine Roll]		
[Isaac] Kraigg 1 Lieut	1	
——— Turner Serjant	Dead	
Robert Hunter Do		Sick
John Garriagues		Do
Willm Steward		Do
James Moulsworth	2	Do
Charls Parker	Runaway	
Simon Harwood		Do
Isac Dewes	3	
Patrick Kenny	4	
Andw Scott	5	
Henry Jarett	6	
Saml Johnston		Sick
David Clark	7	
John Irazie	8	
Archibald Neilson	9	
Richd Owens		Sick
Patrick Crafford		Do
Andw Campbell		Do
John Campbell		Do
John Broomfield	Dead	
Thoms Wattson	11	
Robt Anderson		Sick
Willm Haselip	12	Do
James Campbell		Do
Fracis Dowie	13	
John McNeal		Do
Robt Kearns	14	
James Kite		Do
Benjn Tate	15	
Timothy Ohara		Do
Willm Skilling		Do
Henry Frazer		Sick
James Williams	16	
Edmund Lee	17	Do
Saml Harvey		Do
John Ponsett	18	
Cornelius Griner		Do
Willm Lock	19	

[1] Nicholas Biddle Papers, on deposit at the Historical Society of Pennsylvania.

"A List of People on board the Andrew Doria 10th May [1776]"[1]

[Extract]

Number	Names	Station
1	Nicholas Biddle	Commander
2	James Josiah	1st Lieutenant
3	Elishar Warner	2 Ditto
4	John McDougal	3rd Ditto
	[Marine Roll]	
1	Isaac Craig	Captain
2	Patrick Kenney	Private
3	Andw Scott	Do
4	Henery Savett	Do
5	Samuel Johnston	Armourer
6	Davd Clark	Private
7	John Treasey	Ditto
8	Andw Campbell	Ditto
9	Willm Haselip	Ditto
10	Francis Dowie	Ditto
11	Robert Kearns	Sergant
12	John Ponsett	Private
13	Cornelius Grimes	Ditto

New Marines

Lieut Wadsworth	James Hall	James Lepthorn
Joshuah Whyley	Charles Liebreth	Samuel Eldred
James Cook	Ekilat Cumstock	Edward Benett
Jonathan Barrett	Francis Lyon	

[1] Nicholas Biddle Papers, on deposit at the Historical Society of Pennsylvania.

"A List of Officers Seamen and Marines on board the Andrew Doria the 25th June 1776"[1]

[Extract]

Number	Names	Station
1	Nicholas Biddle Esq	Commander
2	Elishar Warner	
3	Benjamin Dunn	
	[Marine Roll]	
1	Isaac Craig	Captain
2	Jonathn Seaton	
3	Robert Kearns	
4	Samuel Johnston	
5	John Treasey	Private
6	David Clark	Ditto
7	Willm Haselip	Ditto
8	Henery Gavett	Ditto
9	John Ponsett	Ditto
10	Patrick Kenney	Ditto
11	Andrew Scott	Ditto
12	John McNeal	Ditto
13	Isaac Deweere	Ditto
14	Cornelius Grimes	Ditto
15	Andrew Campbell	Ditto
16	Edmond Leo	Ditto
17	Henry Fraser	Ditto

[1] Nicholas Biddle Papers, on deposit at the Historical Society of Pennsylvania.

The Continental Ship *Ariel*

"A List of Officers, and Men belonging to the Ship of War Ariel. Commanded by the Honorable J. P. Jones Esqr."[4]

[Extract] [December 1780]

Richd Dale	Lieutns	
Henry Lunt		
Saml Stacy	Master	
Mattw Maze	Purser	
Amos Winship	Surgeon	5
John Franksford	Mrs Mates	
Thomas Potter		
Beaut Groube	Midshipn	
Arthur Robinson		
John Peacock	Surgeons Mate	
Abisha Perkins		
Chas Glover	Captns Clark	
De' Lavollette		
Johathan Wheeler	Gunner	
John Gunneson	Master Carpenter	15
John Bourbanks	Do at Arms	
John Hatton	Stewards Mate	
Edwd Garrett	Boatswains Do	
John Lawrence	Do Do	
Wm Roberts	Cooper	20
John Gates	Gunners	
Preserved Syssell	Mates	
John Woulton	Quarter Masters	
Vale Strong		
Elijah Johnston		25
John Down	Quarter	
Thomas Austin	Gunners	
Thos Knight		
John Handerham	Carpenters	
Wm Priest	Crew	30
Edwd Cooney		
Wm Pool		
Benjm Stubbs		
Lawrence Furlong		
John Brown		35
Andrew Ryan		
Wm Lee		
John Wilson		
John Mosey		
Antonio Sponza		
Antonio Masingay		
Thos Golligan		40
Richd Stephens		
Joseph Antoine		
John McDonald		45
John McKenzie		
John Robertson		
Dudley Wright		
John Garner		
Joseph Moria		50
Andrew Roach		
Robert Thomas		
Thomas Burns		
Frederick Wagenor		
John Seaburg		55
Wm Whittham		
David O'Hara		
John McDanald		
James Pinner		
Saml Cavender		60
Micheal Hosman		
Chas Ward		
John Walker		
John Hawkins		
Ezekiel Hibbs		65
Daniel Wingate		
John Murphy		
Matthew M. Caftney		
John Marshall		
John Carroll		70
George Johnston		
Saml. Matthews		
Daniel Swain		
John Rutherford		
Peter Nuddle		75
Richd Wilson		
Lewis Lennard		
Lewis Martin		
John Obrey		
Vincent Marc		80
Gulliam Langlois		
Jacob Henry		
John Thomas		
Amos Wait		
Johanna Chusers		85
Jean Romaino		
John Thompson		
Wm McCullock		
Chas Riley		
John Warren		
Alexr Mayson		
Nichs Caldwell		
Antony Jeremy		
Abraham Martell		
John Wier		95
John Dupee		
Saml Gray		
James Mahanny		
John Duffy		
John Hackett		100
Thos Waterson		
Wm Ears		
John Hall		
Elijah Myddleton		
John Mai		105
Jaque Blorge		
Joseph Curdevia		
Jusa Lucas		
Charles Priley	Captns Stewd	
James Mackenzy		110
John Jordon	Cripples	
Joseph Brusson		

[4] Franklin Papers, University of Pennsylvania.

"A List of Officers, and Men belonging to the Ship of War Ariel. Commanded by the Honorable J. P. Jones Esqr."[1] (Cont.)

[Extract]

Volunteers

Duclos
George
Jolicaeur 115
Caron
Janique
Pontcarner
Daire
Rousseau 120
Baret
Muzeau
Poul
Maudru
Vallé 125
Chaillot
Dumont
Clement
Darnvire
Le Tret 130
Maruis
Le Trem
La Tulive
Joseph Audrain
Vieuville 140
Fattebert
Guillebert
Plaisant
Quentin
Augereau 145
Renould
Desnant
Silvain
Roland
Jouvaute 150
Daunery
Audrain Tambour
Cudrio Cuisinier du Ce.
Le Maitre, Voulanger
Capitaine de Marine
M. Nicholson
Lieutenant de Marine
M. Joriss

1 Franklin Papers, University of Pennsylvania

The Continental Frigate *Boston*

"A List of Officers and Men on board the Ship Boston"[1]

[14 November 1777].

Hector Mc Neill	Captain
Hezekiah Welch	Lieutenant
William Faris	Acting do
Patrick Connor	Acting Master
Samuel Carlton	Mates
Daniel Webber	Mates
Hector McNeill Junr.	Mates
Thomas Ward	Mates
John Hewes	Mates
Alexander Robertson	Mates
Samuel Gragg	Pilott for the Coast
Benj Crowningshield	Midshipmen
Peter Cavey	Midshipmen
William O Brien	Midshipmen
Peter Clarke	Midshipmen
William Lamb	Capts. Clerk
Joseph Lewis	Boatswain
Thomas Waite Foster	Gunner
Gideon Woodwell	Carpenter
John Linn	Surgeon
Moses Sawyer	do Mate
Benja. Bangs	Steward
Chace Rogers	Cooper
William Jennison	Serjeant [Marines]
James Sims	do [Marines]
Josiah Waite	
Benja. Foster	Sailmaker
Caleb Parker	do Mate

Benja. Tapping
John Smith
Joseph Henderson
Benja. Henderson
Daniel Harraden
Arthur Lloy'd
John Willson
Thomas Morgan
Lewis Leblanch
James Fitzgerald
Joseph Pettit
Richard Lubey
John Fullerton
Michael Burns
Winter Calef
Hants Pederson
Thomas Balch
John Elliott
Daniel Horne
Francis Tree
Thomas Shaw
John Calderwood
Thomas Perkins
Christopher Crow
Arthur Lloy'd Junr.
Joseph Harraden
Arbraham Buzzell
John Keefe
Joel Philbrook
John Murray
John Crossley
Richard Webber
James Elmes
William Liscomb
Elkanah Gray
George Smith
David Willson
Robert Dorne
John Mc Leod
Thomas Gilliam
John McLane
Prince Gilbert
Nero Freeman
Cato Austin
Cato Wood
Jack Faris
London Terry
Francis Heck
Charles Hathaway
Quashey Hazard
Hampshire Dodge
Cornelius Cowart
Benja. Peirce
William Newman
Ephraim Pettingale
Abel Wetherell
George Lamb
John Keeper

[1] Boston Public Library

"A Muster Roll of Officers, Seamen and Marines belonging to the Frigate Boston, Saml Tucker Commander."[1]

[Extract] [Undated]

No	Mens Names	Time of Entry	Stations	Wages	Remarks
		1778			
1	Samuel Tucker	Octr 28	Command	£ 18	
2	David Phips	Feby 20, 1779	1 Lieut	£ 9	
3	Hezekiah Welch	Octr 28	2d Do	£ 9	
4	Benjamin Bates	Octr 28	3d D	£ 9	
		[Marine Roll]			
6	Seth Baxter	Feby 17, 1779	Capt Mars.	£ 9	
7	Jeremiah Reed	Decr 3d	1 Lt Do		
8	William Cooper	March 28, 1779	2d Lt Do		In the Pole
36	Benjn. Newhall	Decr. 31, [1778]	Sergt. Marns.	48s.	
37	Daniel Mears	Jany 5,	Sergt. Marns.	48s.	Discharged Sick
73	Chevier Cadozo	Jany 4	Marine	40s.	at Boston Feby 16
79	Samuel Hazelton	Jany 14	Marine	40s.	Dischd March 1
80	John Brown	Jany 16	Corpl Mars.	40s.	Run Feby 9
82	Thomas Yates	Jany 16	Marine	40s.	
101	James Harrison	Feb 16	Marine	40s.	[On board *Tyranncide*]
102	Willm Day	Feby 22	Marine	40s.	
104	Willm Harris	Jany 23	Marine	40s.	
105	Jacob Harvey	Feby 23	Marine	40s.	Run 20 March
120	Christian Rodiker	Feby 25	Marine	40s.	
127	Moses Arnold	March 2	Marine	40s.	
128	Robert Hunt	March 2	Marine	40s.	
134	John Handerfozin	Jany 3	Marine	40s.	
136	Francis Gibson	Octr 28	Marine	40s.	Drowned Apl 12
147	Benjn Luce	March 4	Marine	40s.	
151	Silas Johnson	Feby 17	Marine	40s.	In the prize Boyd
157	John Cashaw	March 11	Marine	40s.	Run June 12
158	James Lemerick	March 8	Do	40s.	On board the Warren
150	Francis Reiley	March 8	Do	40s.	
160	Alexander Bebe	March 10	Do	40s.	
161	James Whalan	March 10	Do	40s.	
164	Oliver Tillson	Novemr 11	Marine	40s.	
172	Jabez Post	[March] 19	Marine	40s.	
174	Willm Kimball Jr	19	Fifer	48s.	
175	Josiah Higgins	19	Marine	40s.	
176	Oliver Drake	19	Do	40s.	
177	Elkanah Gage	23	Do	40s.	
179	Timothy Hicks	23	Marine	40s.	
182	Daniel Learned	24	Marine	40s.	
186	Gideon Nye	31	Marine	40s.	
187	Obediah Nye	31	Do	40s.	Discharged Apl 2d.
188	Thomas West	31	Do	40s.	Sent on shore sick to Philadelphia
189	Eliphalet Perrin	31	Do	40s.	In the Mermaid
195	Hezehk Breman	April 3	Marine	40s.	Run April 6
196	John Langlarde	March 8	Do	40s.	In the Prize Boyd
197	John White	8	Do	40s.	
198	Pierre Fico	8	Do	40s.	Run 20 March
203	Abraham Tempel	6	Marine	40s.	Sick at Philadela
204	Morris Wade	April 9	Do	40s.	
205	James Gordon	9	Do	40s.	
206	John Bacon	9	Do	40s.	Run April 12
208	James Doughton	9	Marine	40s.	
209	Willm Mollineaux	April 8	Marine	40s.	
210	William Wheeler	Feby 23	Do	40s.	Died 27 April
212	Samuel Thatcher	18	Marine	40s.	Sent on shore sick in the Confedcy to Philadelphia

[1] Revolutionary War Rolls, v. 52, unnumbered, Massachusetts Archives.

"A Muster Roll of Officers, Seamen and Marines belonging to the Frigate Boston, Saml Tucker Commander."[1] (Cont.)

[Extract] [Undated]

No	Mens Names	Time of Entry	Stations	Wages	Remarks
222	Josiah Pierce	Feby 18	Drumr	40s.	
224	Morgan McNamarough	[March] 12	Marine	40s.	
225	Thomas Swindell	April 3	Do	40s.	on shore sick at
226	Peter Liscough	4	Do	40s.	Philadelphia
241	Samuel Daily	April 24	Marine	40s.	
242	Antony Routeaux	Feby 16	Marine	40s.	
252	John Hebo	Feby 6	Marine	48s.	
255	George Bird	6	Marine	40s.	Run April 1

1. Revolutionary War Rolls, v. 52, unnumbered, Massachusetts Archives.

The Continental Ship *Columbus*

"Muster Roll of the Men and Officers belonging to the Ship Columbus from the time of her being Put in Commission to the fourteenth day of November 1776."[1]

[Extract] [20 November 1775–14 November 1776]

Time of Entry	Men's Names	Number	Station in which they entered	Promotions Decease & Deserted	Discharged Turned over &c.	Time When
1775 Nov. 20	Abraham Whipple	1	Captain			
"	Rhodes Arnold	2	1st Lieut.		Discahrged at Prov.	1st June 1776
	Ezekiel Burroughs	3	3d ditto	promod. to 1st Lieut.		
1776 Dec. 25	Joseph Olney	4	2d ditto	promted to Captain		
1776 Jan. 1st	William Grinnell	5	Lieutenant		turned to S. Providence	16th April
			[Marine Roll]			
1775	Joseph Shoemaker	1	Captain	Deserted		
	John Trivett	2	Lieutenant	Deserted		
	Robert Cummings	3	Lieutenant	Deserted		
1776 May 25	Matthew Parke	4	Captain			
"	Christopher Hopkins	5	Lieutenant			
1775 Dec 29	Edward Burke	6	Lieutenant		Discharged at Newport	1776 19th Oct.
Dec 11	Ernest Grace	7	Serjeant			15 March
" 14	John Gould	8	ditto	promoted to Cook		
" "	John Segeson	9	ditto			
" 16	Thomas Anderson	10	Corporal	promd to Serjt Aug 16		
1775 Dec 11	Frederick Kerker	11	Corporal			
" 14	John Ragan	12	Private			
" 16	Nicholas Tobin	13	Private	died at Providence		16 May
" 14	John Hart	14	"	Deserted		
" 18	Thomas Murray	15	"	Deserted		

1. Transcript of Military Papers, Revolutionary War, v. XL, pp. 668–675, Rhode Island Archives.

"Muster Roll of the Men and Officers belonging to the Ship Columbus from the time of her being Put in Commission to the fourteenth day of November 1776."[1] (Cont.)

[Extract] [20 November 1775–14 November 1776]

Time of Entry	Men's Names	Number	Station in which they entered	Promotions Decease & Deserted	Discharged Turned over &c.	Time When
" "	Joseph London	16	"			
" 18	Robert Hawthorn	17	"	promoted to Armourrs Mate		
" 10	John Smith	18	"	Deserted		
" 14	John Reily	20	"	ditto		
	James Barr	21	"	ditto		
	Francis Higgins	22	"	ditto		
	John Davis	23	"	ditto		
" 16	Martin Ingle	24	"		dischard Newport	19 Oct
" "	Robert Jackson	25	"	Promoted to Corporal 16th May		
" 16	John Jones	26	"	Deserted		
" "	Jacob Brand	27	"		dischd at New London	21 April
" 13	Henry Kess	28	"			19th Oct
" 14	Geo Longcomer	29	"		dischd at Newport	
" 14	John McDole	30	"			
" 15	Michael Garret	31			do at Newport	19th Oct
1776 Jany 2	John Auskill	32	"	Deserted		
1775	John Hussey	33	"		dischd at Newport	19 Oct
Dec 10	John Miller	34	"	Deserted		
" "	Jacob Crouse	35	"	ditto		
" 16	George McDonald	36	"	died at Newport		14th April
	Arthur Nagle	37	"		dischd at New London	
	John Somers	38	"	Deserted		
" 27	Andrew Uler	39	"		dischd at Newport	17th Oct
" 14	William Tate	40	"	Deserted		
" "	Geo A Mauffit	41	"	ditto		
" 15	Frederick Sinquit	42	"	ditto		
	David Coleman	43	"	ditto		
" 16	John Parker	44	"	ditto		
" 26	John Huston	45	"			
20	Michl Dougherty	46	"			
"	Johnson Spear	47	"			
"	John Nowell	48	"			
16	John Clark Jr.	49	"			
1775 Dec 16	John Gillin	50	Private			
10	Emanuel Furnance	51	"			
14	Peter Nagle	52	"			
	Christr Green	53	"	Deserted		
	Matthias Graffe	54	"		dischd at Newport	19th October
1775 Dec 11	Andrew Allen	55	"		dischd at Providence	25th May
16	William Loughridge	56	"		" at Newport	19th October

1 Transcript of Military Papers, Revolutionary War, v. XL, pp. 668–675, Rhode Island Archives.

"Muster Roll of the Men and Officers belonging to the Ship Columbus from the time of her being Put in Commission to the fourteenth day of November 1776."[1] (Cont.)

[Extract] [20 November 1775–14 November 1776]

Time of Entry	Men's Names	Number	Station in which they entered	Promotions Decease & Deserted	Discharged Turned over &c.	Time When
"	Frederick Hackle	57	"	Deserted		
14	John Clark senr	58	"		dischd at Newport	
15	George Marks	59	Drummer			
14	David Congleton	60	Private	Deserted Reedy Island		13th Feby
"	Thomas Milan	61	"	ditto		ditto
14	John Forgey	62	"	ditto		do
"	James Ward	63	"	ditto		"
15	Nicholas Grossman	64	"	do		"
16	Thomas Scott	65	"	do		"
	Connell McNeilus	66	"	do		"
	John McIndow	67	"	do	turned from sloop Prov. & returned back	11th Oct
	James Ogg	68	"		turned from Alfred	30th May
	Peter Morris	69	"		turned from Alfred dischd at Newport	30th May &
	John Curtain	70	"		turned from Alfred	30th May
	Thomas Vanlewan	71	"		ditto	"
	Philip Fitzpatrick	72	"			
	Robert Piker	73	"			
June 5	Ashley Brown	74	"			
	Edward Hagan	75	"		ditto	7th August
	Robert Robinson	76	"		turned from Providence 9th June & put on board the Alfred	
	John McLaughlin	77	"		turned from Alfred	30th May
	Daniel Lafferty	78	"		ditto	do do
	Robert Richey	79	"		ditto	" "
	John Stalcuss	80	"		ditto	" "
	Able Burns	81	"	Deserted		
	Richard Owen	82	"			
	Hugh McCowan	83	"			
	Thomas Lewellan	84	"		turned from Alfred	9th June
1776 May 18	Michael Ryan	85	"			
June 17	John Garrison	86	Fifer			
	John Robinson	87	Private		turned from Alfred	30th May
	William Hamilton	88	Serjeant		ditto	7th August
	James Marshall	89	Corporal	Pressed in Liverpool frigate	do.	"
	John Nick	90	Private		turned from Alfred	"
	Robert McCauley	91	"		do	"
	Thomas Burns	92	"		do	"
	Thomas Donahoe	93	"		do	"
	Patrick OBryan	94	Private	Pressed in the Liv. Frigate		
	Edward Hagan	95	"			
	Ellis Rowland	96	"			
	John Bell	97	"	Pressed in the Liverpool		
	James Wall	98	"			
	Patrick McNamara	99	"			
	James Connell	100	"			

[1] Transcript of Military Papers, Revolutionary War, v. XL, pp. 668–675, Rhode Island Archives.

The Continental Frigate *Deane*

"Muster Book & Pay Abstract of the Continental Frigate Deane S. Nicholson Esq Commr. commencing [May 1] 1781; ending May 31, 1782."[1]

[Extract]

No Entered	Time of Entry Month	Day	Year	Names of the Men	Qualities Rated	Remarks D, D, R, Time when
	October	10	1781	Sam Nicholson	Commdr	
	Septbr	13	1781	Benjamin Page	1st Lieutenant	
	May	1	1781	Arthur Dillaway	Second Lieutenant	
	May	1	1781	Michael Knies	Third Lieutenant	
					[Marine Roll]	
4	May	1	1781	Richard Palmes	Captn Marines	
21	October	10	1781	Tildon Crooker	Sergt Marines	
28	October	29	1781	Wm Waterman	Lt of Marines	
43	Novbr	15	1781	Jeremiah Reed	Lt of Marines	
50	Novbr	19	1781	Alexis Bremont	Marine	Turnd over to Alliance Decbr 21 1781
51	Novbr	19	1781	John Arcant	Marine	
57	Novbr	20	1781	George Wall	Marine	
60	Novbr	20	1781	John Borgatt	Marine	Run
61	Novbr	20	1781	William Richards	Marine	Run
64	Novbr	21	1781	John Scears	Drummer	Turnd over to Alliance Decbr 21 1781
66	Novbr	22	1781	Rufus Davis	Marine	
67	Novbr	22	1781	Benjamin Bates	Marine	
72	Novbr	26	1781	Warren Tony	Marine	Turnd over to Alliance Decbr 21 1781
79	Novbr	28	1781	William Coomer	Fife	Returned to the Army
80	Novbr	29	1781	Moses Saunders	Marine	
84	Decbr	1	1781	John McNeil	Corp Marines	
85	Decbr	1	1781	William Havens	Marine	
86	Decbr	1	1781	Albermarle Stone	Marine	
87	Decbr	1	1781	Isaac Bennet	Marine	
88	Decbr	1	1781	Benjamin Coats	Marine	Died May 31. 1782
91	Decbr	1	1781	Edward Simmonds	Corpl Marines	
92	Decbr	1	1781	Josiah Chandler	Serj Marines	
95	Decbr	4	1781	Charles Moon	Marine	Turnd over to Alliance Decbr 21 1781
97	Decbr	4	1781	David Oherin	Marine	ditto
98	Decbr	4	1781	William Scurry	Marine	ditto
99	Decbr	4	1781	John Divertis	Marine	
101	Decbr	6	1781	Elijah Smith	Corp Marines	
102	Decbr	6	1781	Benjamin Barber	Marine	
119	Decbr	8	1781	Peter Brimmer	Marine	
120	Decbr	12	1781	Hugh Osbourne	Marine	Alliance Decbr 21 1781
121	Decbr	12	1781	Peleg Osbourne	Marine	Alliance Decbr 21 1781
122	Decbr	12	1781	Thomas Osbourne	Marine	Alliance Decbr 21 1781
123	Decbr	12	1781	William Osbourne	Marine	Alliance Decbr 21 1781
131	Decbr	17	1781	Elias Cox	Marine	Discharged Mar 1. 1782
132	Decbr	17	1781	Jean Sarvain Gosse	Marine	Alliance Decbr 21 1781
134	Decbr	20	1781	Eli Green	Marine	
135	Decbr	20	1781	Mial Camp	Marine	
143	Decbr	24	1781	Mical Osbourne	Marine	
147	Decbr	27	1781	John Baptista Martini	Marine	
148	Decbr	27	1781	Francis Shubarn	Marine	
149	Decbr	29	1781	Jesse Richards	Marine	Run
150	Decbr	29	1781	Thomas Lewis	Marine	
152	Decbr	29	1781	John Jackson	Drummer	Run
154	Jany	1	1782	Levi Frances	Marine	Run
164	Januy	10	1782	Luther Turner	Marine	
168	Jany	10	1782	James Holden	Marine	
170	Jany	13	1782	Samuel Willys	Marine	

[1] Revolutionary War Rolls, v. 66, pp. 115-128, Massachusetts Archives.

"Muster Book & Pay Abstract of the Continental Frigate Deane S. Nicholson Esq Commr. commencing [May 1] 1781; ending May 31, 1782."[1] (Cont.)

[Extract]

No Entered	Time of Entry Month	Day	Year	Names of the Men	Qualities Rated	Remarks D, D, R, Time when
171	Jany	20	1782	Prince McClanning	Marine	
172	Jany	21	1782	William Bennet	Drummer	
175	Jany	22	1782	Chace Pillsbury	Marine	
185	Febry	4	1782	Noah Threshear	Marine	
190	Feby	6	1782	Winslow Turner	Marine	
191	Feby	6	1782	Ephraim Luce	Marine	
193	Feby	11	1782	John Fisk	Marine	
194	Feby	11	1782	Nathan Lock	Marine	
195	Feby	11	1782	David Fisk	Marine	
197	Feby	12	1782	Matthew Stutson	Marine	
199	Feby	14	1782	Joseph Willcut	Marine	
201	Feby	15	1782	Philip Proctor	Marine	
202	Feby	15	1782	Josiah Proctor	Marine	
203	Feby	15	1782	Job Spaulding	Marine	
224	Feby	22	1782	Joseph Hawkins	Marine	
226	Feby	22	1782	Veline Hoisington	Marine	
229	Feby	23	1782	Timothy Hibbard	Marine	
230	Febry	23	1782	David Hutchins	Marine	
231	Febry	23	1782	Rogers Lourence	Marine	
232	Febry	23	1782	Moses Ames	Corpl Marine	
233	Febry	23	1782	Thomas Farrington	Marine	
234	Febry	23	1782	Ezra Porter	Marine	
235	Febry	23	1782	Eli Ames	Marine	
241	Febry	25	1782	Benjamin Laurence	Marine	
242	Febry	25	1782	William Parker	Marine	
243	Febry	25	1782	Royal Blood	Marine	
244	Febry	25	1782	Solomon Whitney	Marine	
245	Febry	25	1782	Jonathan Frost	Marine	
246	Febry	25	1782	William Conn	Marine	
247	Febry	26	1782	William Smith	Marine	
248	Febry	26	1782	Imla Parker	Marine	
259	Febry	28	1782	Charles Cutler	Marine	
261	Marh	1	1782	John Munn	Marine	
265	Mar	1	1782	Richard Holden	Marine	

1 Revolutionary War Rolls, v. 66, pp. 115–128, Massachusetts Archives.

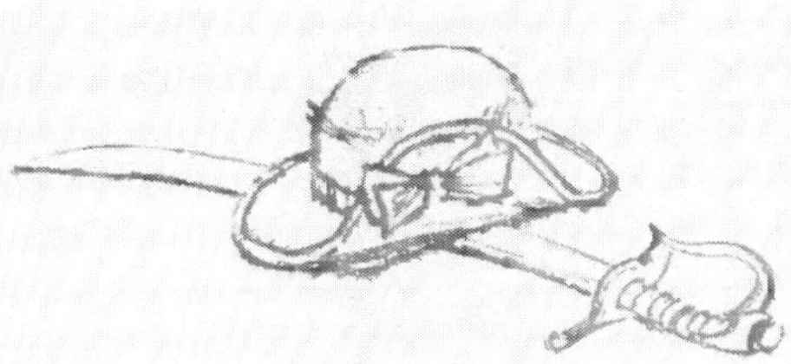

The Continental Sloop *Dolphin*

"Muster Roll for the Continental Sloop Dolphin
Samuel Nicholson Esq Commandr on A Cruize from France
May 1777"[1]

[Extract]

Time of Entry	Names	Stations
Arpil 16th	Sam Nicholson	Captain
	Seth Clark	1st Lieut
	James Degge	2d Do
	[Marine Roll]	
[May] 29th	Cape Pravo	Marine
	Rainey Pravo	Do
	John Lemance	Do
30th	Wm. Cannow	Marine
15th	Pettre Pravo	Marine
	Charles Dedo	Do
	James Cann	Do
[May 8th]	Champaine	Marine

[1] John Paul Jones Papers, LC.

The Continental Frigate *Effingham*

"A Return of the People on Board the Frigate Effingham John Barry Esqr. Commanding . . . Octr. 27th. 1777."[1]

[Extract]

			Amt. Slops
Nicholas Hayes (belongs to Capt. Ashmiad		@ 5/ per day	
Robert Taylor	Do.	@ 0	
Nicholas Deane	Midshipman		
Matthew Clarkson	Do.	2 Jackets (inside & outside)	4.0.11
Edward O'Hara	Clerk	1 Jacket 20/ & Shirt	
John Little	Armourer	1 pr. Breeches 48/9 & Shirt	
William Barry	@ 10/ per day		
Capt. Ryan	of Militia		
Lieut. Lishen	of Do.		
John Tanny	of Do.	1/p Trouses 26/9 & 1 pr Shoes 25/	3.1.9.
Conrad Houseman	of Do.		
Jacob Shaffer	of Do.		
John Campbell	of Do.	1 Jacket 18/10 & 1 pr Shoes 25/	2.3.10
James Smart	of Do.	1 Shirt	
James Long	of Do.		
John Capelle	of Do.	1 pr Shoes	1.5.0
Henry Wannell	of Do.	1 Shirt	
Zachariah Davis	of Do.	1 pr Trousers 36/9 & 1 Shirt	
John Pindel	8th Pennsylva Regt.	Colo. Broadhead & Capt Montgomerys Compy 1 pr Breeches	2.8.9
William Jump			
Samuel Jones	12th Virg Rejt	Colo Woods, Capt Wagner 1 green Jacket	
James Curtis	15th Do.	Colo Mason, Capt Gray—1 pr Shoes	1.5.0
Francis Litchfield	9th Do.	Colo. Matthews. Capt Snead 1 pr Do.	1.5.0
Joseph Guie	Carptr.		
Joshua Phillips	Negro	1 Jacket	3.4.11
James Bryan	Do.		
Jeffrey Matthews	Do.		
Isaac W. Laughlin			
David Edminston		1 green Jacket & 1 Shirt	

[1] Washington Papers, LC.

The Continental Brig *General Gates*

"Copy of Prize List of the U.S. Brig General Gates John Skimmer, Comr."[1]

[Extract] [24 May–3 August 1778]

Number	Names of Men	Rank	Shares	Adv. Wages
1	John Skimmer	Captain		
2	William Dennis	1st Lieut.		
3	John Lewis	2d Lieut.		
		[Marine Roll]		
7	Richard McClure Jr	Lt. of Marines		
13	John Edwards	Marine	1	£6
17	David Williams	Marine	1	£6
18	Solomon Shute	Do	1	£6
19	John Shute	Do	1	£6
26	Penn Sale	Marine	1	£6
27	John Williams	Do	1	£6
39	James Buckman	Marine	1	£6
66	Joseph Swain	Marine	1	£6
71	Thomas Brintnall	Marine	1	£6
74	John Kenny	Marine	1	£6
82	John Bootman	Sergt of Marines		£6
87	Nathl Bootman	Marine	1	
92	William McIntire	Marine	1	£12
93	Elias Hammond	Do	1	£12
106	William Webber	Marine	1	£6

1 Revolutionary War Rolls, v. 52, p. 79, Massachusetts Archives.

The Sloop *Interprise*

"Pay Role of the Sloop Interprise &c from 3rd May to the 1 July 1775"[1]

[Extract]

Names	Rank	Time Entd.	Dischard	Days to Return	Whole Time	Wages pr Month	Whole Wages
John Proat Sloan	Captn.	May 3rd	July 1	15	2 Months 18 Ds.	£6,,–,,–,,	£15,,17,,2
			[Marine Roll]				
James Watson	Lieutenant	[May] 3	July 1		2 Mont. 3	2,,10,,–	8,,8,,7
Ephraim Bill	Sergant	7	1	8	2 Mont. 7	2,,8	5,,8,,–
Josiah Sanbum	Sergant	3	1		2 Mont. 3	2,,8	5,,1,,1¾
Amos Gelucia	Marine	7	1	8	2 Mont. 7	2,,–,,–	4,,10,,–
Abner Rowe		8	1	8	2 Mont. 6	2,,–,,–	4,,8,,7
Thomas Fitch		8	1	8	2 Mont. 6	2,,–,,–	4,,8,,7
William Draper		8	1	8	2 Mont. 6	2,,–,,–	4,,8,,7
James Brakenage		8	1	8	2 Mont. 6	2,,–,,–	4,,8,,7
Alijah Beardsley		7	1	8	2 Mont. 7	2,,–,,–	4,,10,,–
Uriah Cross		6	1		2 Mt. 0	2,,–,,–	4,,–,,–
Samuel Allen		8	1		2 Mt. 6	2,,–,,–	4,,8,,7
Ephraim Masters		8	June 15		1 Mt. 10	2,,–,,–	2,,14,3¼
Ichabod Parker		3	July 1		1 Mt. 10	2,,–,,–	4,,17,1¼
Jonas Galusha		8	1	8	2 Mt. 6	2,,–,,–	4,,8,,7
Ichabod Hawley		3	June 21		1 Mt. 21	2,,–,,–	3,,9,,11¼
John Hart		8	25		1 Mt. 20	2,,–,,–	3,,8,,7
David Crowfort		7	18		1 Mt. 14	2,,–,,–	3,,–,,–
John Lochrain		8	25		1 Mt. 20	2,,–,,–	3,,8,,7

1 Collections, Massachusetts Archives.

The Continental Sloop *Providence*

"Muster Roll onboard the Sloop Providence Commanded by John Paul Jones Esqr From May 10th [1776]"[1]

[Extract] [10 May 1776–14 June 1776]

Men's Names	Stations in which they have entered	time when	Numbers	Dead, discharged, Run, turned over Promoted &c.—	Time When	For what reason
John Paul Jones	Captain	May 10 1776	1	T. from ship Alfred		Order
William Grinnell	1st Lieut	April 16	2	T. from Columbus		ditto
John Rathbun	2d "	Feby. 10	3			
			[Marine Roll]			
Alpheus Rice	Lt. Marines	June 12	40	T. from Andw Doria June 10th & discharged	Aug. 12	
Patrick Russell	Marine	Dec 29 1775	41	T. from Hospital	June 9	Inability
Cornelius Dax	ditto	Jan 4th 1776	42	Taken from Hospital & exchd for Wm Bryant	May 22	
Archibald Nielson	ditto	ditto	43	Taken from Hospital	June 4th	Order
Samuel Allen	ditto	June 23	44	Td to Columbus	June 9th	do
Cornelius Connelly	do	June 4th	45	Run from Hospital at Providence	June 3d	
Matthew McCaffray	do	14th	46	Left sick at Providence	June 6th	Fever
John Robinson	do	Feby. 8th	47	Td to Columbus	June 10th	Order
Daniel Pillager	do	Jany. 3d	48	Td to Columbus	June 9th	do
Bryan Ryan	do	" 5th	49	T. from Cabot Jan 5 & Run	Aug. 14	
Achibald Edmonston	ditto	" 5th	50	T. from Cabot	Jany. 5	Order
Patrick McMullan	do	" 5th	51	ditto	ditto	do
Samuel Wright	do	" 5th	52	do at the Hospital Rho Isd	ditto	do
James Sinney	Marine	Jany. 6th 1776	53	T. from Cabot	Jany. 5th	Order
David Franks	ditto	" 5th	54	ditto	ditto	do
William Griffiths	ditto	" 5th	55	ditto		do
John McCundoe	ditto	" 5th	56	Td from Cabot Jan. 5th to the Colum	June 4th	do
William Ricketts	ditto	Jany. 23	57	Left at Narrowshock	Sept. 25	
Frederick Ruffman	Marine	Dec. 20 [1775]	59	T. from Hospital at N. London	June 17	Order
William Babcock	Marine	[June] 14 [1776]	65	T. from R. Island Brigade	June 28	
Elias Thomas	Serjeant	" 14	66	ditto	" 28	
Joseph Nocke	Marine	" 14	67	ditto	dischd Oct. 8	
August Sanders	ditto	" 14	68	ditto	dischd " 15	
Richard Griffiths	ditto	" 14	69	ditto	Run	
Elias Miller	ditto	" 14	70	ditto		
Thomas Potter	ditto	" 14	71	ditto		
Jonathan Jenks	Marine	June 14	72	T. from R. Island Brigade		
Nathan Munro	ditto	" 14	73	ditto		
James Jacques	ditto	" 14	74	ditto	dischd	Oct 8th
James Searles	ditto	" 14	75	ditto	Run at Boston	June 28th
Constant Whitford	ditto	" 14	76	ditto	Run at R. Isld	June 22d
Richard Pierce	ditto	" 14	77	ditto	Run	Oct. 22
Thomas Harris	ditto	" 14	78	ditto	Run. Boston	June 28
John Robinson	ditto	" 14	79	ditto	Run	Oct. 22
John Hicks	ditto	" 14	80	ditto		
Isaac Steams	ditto	" 14	81	ditto		
James Merryhew	ditto	" 14	82	ditto	Run. Boston	June 29
Nathan Newman	Fifer	" 14	83	ditto	discharged	Oct. 8
Samuel Peckham	Drummer	" 14	84	ditto		
Adin Trask		" 14	85	ditto	dischd	Oct. 15

1 Transcript of Military Papers, Revolutionary War, v. XL, pp. 663–666, Rhode Island Archives.

"List of Officers, Seamen & Marines belonging to the Providence Sloop of War who are entitled to Shares in the Ship Alexander Captur'd Sepr. 20th 1776—"[1]

[Extract]

Mens Names	Stations	Agents	Sums	paid
Jno. Paul Jones	Captain			
Wm. Grinnell	1st. Lieut.		30,,–	13,,2,,10
Jno P. Rathburn	do. ditto	J Greenleaf		
Wm. Hopkins	2d. Lieut.	J Greenleaf	22,,10	
		[Marine Roll]		
Edmd Arrowsmith	Lt. Marines	J Greenleaf		
Elias Thomas	Serjt	J Greenleaf		
Archd. Nelson	Private	John Winslow		
Archd. Edmunson	ditto	J Greenleaf		
Patk McMullen	ditto			
Saml Wright	ditto		4,,10–	,,12–
James Sinnex	ditto			
Davd Franks	Marine	J Greenleaf		
Willm Griffies	ditto	J Greenleaf		
Wm Rickarts	ditto			
Fredk Ruffman	ditto			
Joseph Macater	ditto			
Adin Trask	ditto	Ebenr Trask		4,,5,,11
Auguste Sanders	ditto	Jno Newman	4,,10	
Richd Griffies	ditto			
Elias Miller	ditto			
Thos Potter	ditto			
Nathn. Munro	ditto			
Richd. Pearce	ditto			
John Hicks	ditto	D Tillinghast		
Isaac Stearns	ditto		6,,–	
Jons Jenks	Marine			
Jos Jaqueys	ditto	Jas Clarke	4,,10	
Jno Robinson	ditto	J Greenleaf		
Ebnr Newman	Fifer	Jno Newman	4,,10	
Sam Peckman	Drummer			

[1] Revolutionary War Rolls, v. 52, pp. 8–11, Massachusetts Archives.

"Muster Roll of the Providence Sloop of War"[1]

[Extract] [20 October 1776]

Number	Time of Entry Month	Year	Names	Station	Run Dead Discharg'd	Turn'd over from	Turn'd over to	Time When Month	Year
1	May 10	1776	John Paul Jones	Captain		Ship Alfred	Ship Alfred	Octr. 20	1776
2	April 16	1776	William Grinnell	1st Lieutenant		" Columbus	Prize Brigt. Brittannia	Augt. 27	"
3	Feby 10	do.	John P. Rathbun	do. do.			Ship Alfred	Octr. 20th	"
				[Marine Roll]					
	June 12	do.	Alpheus Rice	Lt. Marines	Run Augt. 16. 1776	Andrew Doria		June 10	"
	Decr. 29	1775	Patrick Russell	Marine	Run Augt. 10. 1776			Augt. 10	1776
	Jany 4	1776	Cornelius Dax	Do.	Discharg'd May 24th			May 24	"
	June 4	1776	Archibd. Neilson	Do.	June 4th Taken from Hospital		Ship Alfred	Octr. 20	"
	Jany 23	"	Samuel Allen[2]	Do.			Columbus	June 9	"
	Jany 4	"	Cornels. Conelly	Do.	Run June 3d. 1776			June 3	"

[1] Massachusetts Historical Society

[2] Used as Marines on board the *Providence*

"Muster Roll of the Providence Sloop of War"[1] (Cont.)

[Extract] [20 October 1776]

Number	Time of Entry Month	Year	Names	Station	Run Dead Discharg'd	Turn'd over from	Turn'd over to	Time When Month	Year
	Jany 14	"	Mathew McCaffry	Do.	Sent Sick Quarters			" 6	"
	Feby 8	"	John Robinson	Do.			Columbus	" 10	"
	Jany 3	"	Danl. Pillegar	Do.			do. do.	" 9	"
	" 5	"	Bryan Ryan	Do.	Run Augt. 27. 1776	Brigte. Cabot		Jany 5	"
	" "	"	Archibald Edmunson	Do.		do. do.	Ship Alfred	Octr. 20	1776
	" "	"	Patk. McMullen	Do.		do. do.	do. do.	" 20	"
	" "	"	Saml. Wright	Do.		do. do.	Sick Quarters	" 20	"
	" "	"	James Sinnex	Do.		do. do.	S. Alfred	Octr. 20	"
	" "	"	David Franks	Do.		do. do.	do.	" 20	"
	" "	"	William Griffies	Do.		do. do.	do.	" 20	"
	" "	"	John McIdoe	Do.		do. do.	Columbus	June 14	"
	Jan 23	"	William Rickets	Do.	Run Sept. 24. 1777			Sept. 24	"
	Decr. 20th	1775	Fredk. Wm. Ruffman	Do.		Hospital	Prize Brigt. Favourite	Sept. 8	"
			[Members of the Rhode Island Brigade][2]						
	June 14[3]	1776	Edmund Arrowsmith	Ensign		Rhode Isd Brigade	Ship Alfred	Octr. 20	1776
	" "	"	Elias Thomas	Serjt.		do do	Prize Brigt Brittana	Augt 27	"
	" "	"	William Babcock	Soldier	Run June 28				
	" "	"	Joseph Nocake	ditto		do do	Columbus's Prize	Sept 10	"
	" "	"	Adin Trask	do	Discharg'd Octr 10	do do			
	" "	"	Augustus Saunders	do		do do			
	" "	"	Richard Griffies	do			Prize Brigt. Britannia	Augt 27	"
	" "	"	Elias Millar	do		do do	do do Favourite	Sept 8	"
	" "	"	Thomas Potter	do		do do	Ship Alfred	Octr 20	1776
	" "	"	Jonathan Jenks	do		do do	Prize Brigt. Britannia	Augt 27	"
	" "	"	Nathan Munroe	do	Discharg'd Feby 1st	do do	do do Defiance	Sept. 23	"
	" "	"	Joseph Jaqueys	do	Discharg'd Novr 1	do do	Columbus Prize	Sept 10	"
	" "	"	James Searles	do	Run June 19. 1776	do do			
	" "	"	Constant Whitford	do	Run June 29	do do			
	" "	"	Richard Pearce	do		do do	Prize Ship Alexander	Sept 25	"
	" "	"	John Robinson	do		do do	Ship Alfred	Octr 20	1776
	" "	"	Thomas Harris	do	Run June 29th 1776	do do			
	" "	"	John Hicks	do		do do	Prize Brigt Defiance	Sept 23	"
	" "	"	Isaac Stearns	do	Discharg'd Feby 10	do do	Prize Brigt. Sea Nymph	Sept	"
	" "	"	James Merrihew	do	Run July 6th	do do			
	" "	"	Elnathan Newman	do Fifer	Discharg'd Octr 10th	do do			
	" "	"	Saml Peckham	Drummer		do do	Prize Brigt Defiance	Sept 23	"

1 Massachusetts Historical Society
2 Used as Marines on board the *Providence*

"List of Officers & Seamen on board the Sloop Providence who have a Share in the Mellish & Active"[1]

[Extract] [30 October 1776–27 November 1776]

Mens Names	Stations	Agent	Sums paid
Hoysteed Hacker Esq.	Captain		
Philip Brown	1st Lieut	Nat Green	10 . . 8 . . .
Adm Thaxter	2d Lieut	Nat Green	
	[Marines]		
Jno Trevett	Lt Marines	Nat Green	
Jno Abhoms	Serjt Marines	---------	100 . . .
Jonl Lillibridge	Drummer	Nat Green	
Sampson George	Fifer	Saml Smith	30 . . . .

1 Revolutionary War Rolls, v. 52, pp. 12–14, Massachusetts Archives.

"List of Officers & Seamen on board the Sloop Providence who have a Share in Schr. Loyalty &c."[1]

[Extract] [1 August 1777–27 September 1777]

Mens Names	Stations	Dead Shares	Sums paid
John Peck Rathbun	Captain		109 . . 18 . . 10
Joseph Casey	1st Lieut		54 . . 19 . . 58
Daniel Bears	2d Lieut		54 . . 19 . . 58
		[Marine Roll]	
John Trivett	Cap. Marines		54 . . 19 . . 58
James Clarke	Serjt. Marines		16 . . . 9 . . . 8
Nicholas Stoddard	Marine		8 . . 11 . . . 2
Joseph Allen	ditto		8 . . 11 . . . 2
Joseph Weedon	ditto		8 . . 11 . . . 2
James Ciall	ditto		8 . . 11 . . . 2
Joseph Stewart	ditto		8 . . 11 . . . 2
Bazillai Luce	ditto		8 . . 11 . . . 2
Thomas Allen	ditto		8 . . 11 . . . 2
Thomas Collins	ditto		8 . . 11 . . . 2
John Tinkam	Marine		8 . . 11 . . . 2
Ezek Whipple	ditto		8 . . 11 . . . 2
Joseph Shaw	ditto		
Saml Browning	ditto		8 . . 11 . . . 2
Seth Baker	ditto		8 . . 11 . . . 2
Saml Wood	ditto		8 . . 11 . . . 2
Benj. Harding	ditto		
Stephn Read	ditto		8 . . 11 . . . 2
Elnathn. Lake	ditto		
Henry Stoddard	ditto		8 . . 11 . . . 2
Thos. Hawes	ditto		8 . . 11 . . . 2
James Blohorn	ditto		8 . . 11 . . . 2
Andrew Burnet	ditto		8 . . 11 . . . 2
Dennis Marator	ditto		

1 Revolutionary War Rolls, v. 52, pp. 15–17, Massachusetts Archives.

"A Muster Roll of all Officers Seamen & Marines belonging to the Continental armed sloop Providence Commanded by John Peck Rathbun Esqr. from June 19.. 1777 to"[1]

[Extract] [Bedford, July 31—1777]

Number	Names	Stations	Time of Entry Year Month	Wages per month	Run, Dead, Sick or Discharged	Promotions	Cash advanc'd	slops advanc'd
1	John Peck Rathbun	Captain	1777	48				
2	Joseph Nesey	1st Lieutenant	"	24				36/
3	Daniel Bears	2d ditto	"	24				
	[Marines]							
5	John Trevett	Capt Marines	"	30				
32	James Clarke	Serjt Marines	1777 May 3	8			48/	78/
39	Edward Clanning	Marine	do June 19	6 2/3		Promoted to Seamen	78/	36/
40	Joseph Weeden	do	do " 26	6 2/3			36/	15/
41	James Vial	Marine	do July 24	6 2/3				
42	Barazillai Luce	ditto	do June 24	6 2/3	Sick Quarter July			
45	Thomas Allen	Marine	do [July] 1				40/	51/
46	Thomas Collens	ditto	do " 8				40/	15/
47	John Tinckom	ditto	do " 1				48/	
48	Esek Whipple	ditto	do " 1				48/	
49	Joseph Shaw	ditto	do June 19					20/
50	Saml Browning	ditto	do July 14				42/	27/
51	Seth Baker	ditto	do " [Torn]				40/	
60	Bery Harding	Marine	" [July] 29	6 2/3			42/	15/
61	Nathl Arnold	do	" " 26	6 2/3		[Promoted to Seamen]	48/	36/
63	Elnathan Lake	Marine	" " 24	6 2/3	Run July 25 1777			
64	Stephen Read	do	" " 30	6 2/3			48/	
67	Henry Stoddard	Marine	" " 31	6 2/3				
70	William Howell	ditto	" Augst 1					
71	Solomon Hallett	ditto	" " 1					
79	Joseph Allen	Marine	" [Torn]					

[1] Miscellaneous Collections, Rhode Island Historical Society.

"List of men on Sloop Providence entitled to Prize Money as of 10 April 1779"[1]

[Extract]

Hoysteed Hacker	
Philip Brown	1 Liut
Nicholas Gardner	2d Do
[Marines]	
John Chilton	Lieut Marines
Bailey Grinnell	Sergt Marines
Gedeon Wrightington	Marine
James Bennet	Marine
Nathanal Saider	Do
Nathanel Dring	Marine
Elnathan Desated	Marine
Taiol Arges	Marine
Elnathan Pope	ditto
Joseph Crosman	Marine
Obed Drew	ditto
Nathel Ingraham	ditto
Benjamin Tompkins	ditto
Elisha Randall	ditto
Thomas Hillet	ditto
James Mitchell	ditto
John Hill	ditto

[1] Newport Historical Society.

"List of men on Sloop Providence entitled to Prize Money as of 7 July 1779"[1]

[Extract]

Hysteed Hacker	Capt.
Abraham Hawkins	1st Lieut
Nichs Gardner	2d Do
[Marines]	
Robt. Davis	Lieut. Mar
Thos. Philbrook	Sergt. Marines
Robert Love	Marine
Roger Haskel	Marine
Bervilla Butlar	Ditto
John Riddell	Ditto
Lephaniah Stdron	Ditto
Elijah Jones	Ditto
William Hearsey	Ditto
Gideon Wrightington	Ditto
Thomas [Illegible]	Ditto
Samuel Smith	Marine
John Parker	Ditto
Cesar Wyman	Ditto
Salithiah Hammon	Marine
John Hill	Ditto
Elias Mallisson	Ditto

[1] Newport Historical Society.

The Continental Frigate *Raleigh*

"A List of all Officers, Volunteers, Seamen, Marines &tc on Board the Raleigh Frigate, Thomas Thompson Esqr Commander in the Service of the United States of No America January 22d, 1778."[1]

[Extract]

No.	Names	Quality	No.	Names	Quality
2	Peter Shores	1st Lieut	167	George Loud	Do
3	Josiah Shackford	2nd Do	168	John Libbey	Do
4	Hopley Yeaton	3rd Do	169	Simeon Gray	Do
	[Marine Roll]		170	John Wedgwood	Marine
6	George J. Osborne	Captn Marines	171	Ebenezer Whitehouse	Do
7	Stephen Meeds	1st Lt Do	172	Wiggin Evans	Do
8	Nathl Twing	2nd Lt Do	173	William Mendum	Do
145	William Hart	Sergt Marines	174	Mark Nelson	Do
146	John Mills 1st	Corpo Do	175	David Tina	Do
147	Geo: Rollins	Do Do	176	Saml Clay	Do
148	Joseph Jackson	Do Do	177	Benjamin Worster	Do
149	John Scates	Do Do	178	John McCoy	Do
150	James McIntyer	Drummer		List of Frenchmen Entered at L'Orient for three years.	
151	Theodore Fernald	Fifer		[Marine Roll]	
152	Simeon Tibbits	Marine	[206]	Nicholas Naquant	Marine
153	John Mills 2nd	Do	[207]	Francois Duval	Do
154	Daniel Smith	Do	[208]	Guillaume Morin	Do
155	Ichabod Tibbits	Do	[209]	Oliver Huit	Do
156	John Hill	Do	210	Augustin DeLorme	Marine
157	George Hohn	Do	[211]	Jean Quornon	Do
158	Satchewel Rundlet	Do	[212]	Louis Canvel	Do
159	Alexa Goold	Do	[213]	Jean Francois	Do
160	James Allen	Do	[214]	Claude Salonette [Lalonette]	Do
161	Wm Knight	Do	215	Mathurin Gautie	Do
162	G. Andrew Rankin	Do	[216]	Henry Quelee	Do
163	Jontham Booker	Do	[217]	Pierre Menguy	Do
164	Isaac Worthen	Do	[218]	Jean Fady	Do
165	John Davis	Do	[219]	Jacque Lamour [Samour]	Do
166	George Parcher	Do	220	Louis Alain	Do

[1] *New Hampshire Genealogical Record*, v. III, n. 2 (October 1905), pp. 70–76.

The Continental Sloop *Ranger*

"Roll of the Officers and men belonging to the Continental Ship of War Ranger Thomas Simpson/Commander, at the Capture of the Brig Lydia Crosby and Sloops Henry and Swift."[1]

[Extract] [December 1779]

Number	Names	Quality	Bounty	Dead run or Discharged
1	Thomas Simpson	Commander		
2	Elijah Hall	1st Lieutenant		
3	David Cullum	2nd "		

[Marine Roll]

Place & Time of Entry	Number	Names	Quality	Bounty	Dead run or Discharged
Portsmouth August 25	1	William Morris	Lieutenant		
" "	2	Richd. Lr. Nelson	Sergeant		Disch'd Nov. 10
" "	3	John Grouard	"		Disch'd Nov. 10
Sept 28	4	Theopelus Wiggen	Corporal	6	
" "	5	Benjamin Shute	"	6	
	6	[Blank]	Drummer		
	7	[Blank]	Fifer		
Sept. 28	8	Joseph Wardwell	Marine	6	
" 29	9	Thomas Allard	"	6	
" 28	10	Meturen Ricker	"	6	Put on board the Prize Brig Dolphin
" "	11	Abraham Hodgsdon	"	6	Put on board the Brig Dolphin
Oct 6	12	John Leavey	"	6	
" "	13	Henry Shillings	"	6	
" "	14	Daniel Rowell	"	6	
" "	15	John Gowen	"	6	
" "	16	James Hill in the Seamans Roll			
" "	17	Noah Ricker	"	6	
" "	18	Stephen Jones	"	6	
Portsmouth Octr 11	19	Amos Pike	Marine	6	
" "	20	David Woodsun	"	6	
" 13	21	Samuel Staples	"	6	
" 16	22	James Neal	"	6	
" 27	23	Joseph Roberts	"	6	
" "	24	Nathaniel Watson	"	6	Put on board the Brig Dolphin
" 28	25	James Boyce	"	6	
14	26	Daniel Nason	"	6	
" 20	27	Samuel Hill	"	6	
" 22	28	Benja. Fishley	"	6	
" 23	29	Thomas Badington	"	6	
" 25	30	David Maxwell	"	6	
" 26	31	Charles Ricker	"	6	
" 8	32	Elias Lord	"	6	
" 27	33	Simeon Applebee	"	6	
" 16	34	Nathl Cobbet	"	6	
" 22	35	Winthrop Langley	"	6	
Octo 26	36	John Fitfield	Sergeant	6	

1 Miscellaneous Collections, Historical Society of Pennsylvania.

The Continental Frigate *Trumbull*

"List of Officers and men on board the Continental frigate Trumbull James Nicholson Esq. Commander. at the Capture of the Brigg Little William Captain Lang. in Compy with the Deane frigate. Septemr. 1780"[1]

[Extract]

James Nicholson Esqr.	Captain	John Knott
Jonathan Maltbee	Lieut.	Quibee Hotchlziss
Jacob White	do.	Daniel Bostwick
John Fanning	do.	Godfrey Harper
[Marine Roll]		Lewis Shaller
Gilbert Saltonstall	Captn.	William Gates
William Fielding	Lieut.	Isaac Armstrong
Abraham Smith Hayden	Sergt.	Andrew Walters
Elias Swan	do.	John King
Manassah Cadey	Corporl.	Lemuel Fitch
Jacob Reynolds	ditto	John Paterson
Timothy Lockwood		William Jackson
Jonathan Meecham		John Hmstead
Joseph Horton		Lemuel Mead
David Lockwood		
William Howlett		
William Copely		
Ebenezr. Downs		
Francis Goodsale		
Benjamin Willson		
Daniel Howard		

[1] Manuscript Collections, History and Museums Division, Headquarters, United States Marine Corps.

The Continental Frigate *Washington*

"Muster Roll of the Frigate Washington's Crew Thomas Read Esquire Commander Abreast of Whitehill 26th October 1777."[1]

Names	Stations	To what Regiment they belong	Colonels Names	Captains Name to whose Compy they belong
John Angus	Lieutenant			
Abel Morgan	Lt: Marines			
John Feares	Carpt. Mate			
James Feares	do. yeoman			
Robert Dougherty	Serjt. Marines			
David Finny	Seamen			
William Midcalf	Landsman			
Patrick Hayes	do.			
William Ewing	do.			
William Galloway	do.			
John Clivery (Negro)	Cook			
Benjamin Stewart	Marine			
Soldiers Vizt.				
Jacob Smock	Sergeant	8th Virginia		Robt. Higg[ ]
Jeremiah Smith	Private	Pennsylv	Patten	Allen M[ ]
Silas Dollahide	do.	1st Nt. Carolina	Clark	Thompson
Thomas Ridder	do.	15th Virginia	Masons	Jno. Griggery
John Farrel	do.	4th Jersey	Martin	Holmes
John Smith	do.	ditto	ditto	Martins

[1] Washington Papers, LC.

Miscellaneous Rolls

"Muster Roll of Captain Craigs—Company of Mariens Philadelphia 19th December 1775"[1]

Mens Names	Age	Size feet	Size Inches	When Inlisted	Where Inlisted	What Country Born	By Trade
Patrick Craford	30	6		9th Decr 75	Philadelphia	Irland	Labourer on Furlough
William Steward	19	5	9	10th Decr 1775	Philadelphia	Philadelphia	Breeches Maker
Henry Javet	28	5	8¾	18th Do Do	Philadelphia	Switzer Land	Servant
Willm Wood	24	5	8¾	10th Do Do	Philadelphia	Irland	Labourer
John Nortan	28	5	8	18th Do Do	Do	New England	Carpenter
Thoms Byrnes	19	5	8	9th Do Do	Philadelphia	Irland	Doctor
Saml Johnson	24	5	7¼	9th Do Do	Philadelphia	Chester County	Brass founder
Patk Kenney	23	5	7	18th Do Do	Do	Irland	Brick Layer
Henry Frazier	40	5	6½	10th Do Do	Do	Holland	House Carpenter
Saml Harvy	21	5	6¼	12th Do Do	Do	Christien	Cooper
John McNail	19	5	6¼	9th Do Do	Do	Irland	Cooper
John Porcet	35	5	6¼	9th Do Do	Do	Britain	Jewiler
Simn Harwood	30	5	6¼	13th Do Do	Do	Philadelphia	Labourer
Jams Stevenson	33	5	6	11th Do Do	Do	North Britain	Mason—Discharged 19th Decr. 75
John Collins Junr	21	5	5¾	15th Do Do	Willmention	Brandy Wine	Cooper
Jams Williams	18	5	5¾	13th Do Do	Philadelphia	Brittain	Labourer
Jams Campbell	30	5	7½	18th Do Do	Do	Irland	Do
Isaac Dewees	25	5	5½	15th Do Do	Do	Kent County Delevar	Cabinet maker
Willm Hopkins	27	5	5½	9th Do Do	Do	New England	Baker
Andw Scott	25	5	5¼	11th Do Do	Do	Irland	Barber
John Collins Senr	25	5	5¼	14th Do Do	Do	Do	Labourer
Richd Owens	22	5	5	13th Do Do	Do	Chester County	Do
Fredk Roughman	30	5	5	11th Do Do	Do	Garmany	Hosier
Edwd Leo	35	5	5	15th Do Do	Do	Irland	Butcher
Benjn Tate	19	5	5	10th Do Do	Do	Dover	Taylor
Christr Warren	26	5	4½	18th Do Do	Do	Irland	Weaver
Thoms Mitchell	19	5	4½	9th Do Do	Do	Brittain	Miller
Willm Haislop	23	5	4½	13th Do Do	Do	Do	Painter
Jams Kite	18	5	4½	13th Do Do	Do	Do	Do
David Clarke	24	5	4½	9th Do Do	Do	Irland	Wool Comber
Willm Lock	20	5	3¼	12th Do Do	Do	Britain	Taylor
Fredk Bowman	32	5	5¼	11th Do Do	Do	Germany	Carpenter
Willm Skinell	30	5	3½	9th Do Do	Do	Irland	Brick Leyer
John Bohonan	25	5	3¾	19th Do Do	Do	Do	Barker
John Traycie							
Michl Goodman							
John Thomson							
Jame Magrady						33 Effective	
Archld Nielson						11 Ditto	22 Decr
Thms Macanaly						—	
Thoms Saltter						44	

[1] Craig Papers, Historical Society of Western Pennsylvania.

"Pay Roll of Capt. Robert Mullan's Compy Marines to Decembr. 1st 1776"[1]

No	Men's Names		Date of Enlist.	Mo.	Days	When desertd or dead	
1	Robt. Mullan	Capt.	June 25th	5	6		
2	David Love	1st. Lieut.	do. do.	5	6		
3	Hugh Montgomery	2nd Lieut.	do. do.	5	6		
4	James Coakley	Serjt.	July 1st	5			
5	Andrew Read	do.	Augt. 22d	3	9		£3..18..0
6	John McKinley	do.	do. 2d	3	29		6.. 8...
7	Warwick Flamabough	do.	Sept. 13th	2	18		2.. 6...
8	George Murray	Corpl.	Augt. 27th	3	4		3.. 2..4
9	Adam McFerson	do.	Octr. 22d	1	9		1.. 1..6
10	John Cribs	do.	do. 13th	1	18		1..13...
11	Joseph Grumley	do.	Sept. 17th	2	14		6..15..8
12	Collin York	Drumr.	June 25th	5	6		8..16...
13	Peter York	Fifer	do. do.	5	6		8..16...
14	John Hogg	Private	Augt. 21st	3	10		3.. 6..8
15	William Barnett	do.	Sept. 1st	3	—		2..10...
16	Lawrence Lessee	do.	do. 3d	2	28		2.. 6..8
17	Benjamin Woodin	do.	Augt. 12th	3	19		4..11..8
18	Robt. Gilmore	do.	do. 28th	3	3		2..15...
19	William Allison	do.	Sept. 2d	2	29		2.. 8..4
20	John Stone	do.	do. do.	2	29		2.. 8..4
21	Danl. Forrman	do.	do. do.	2	29		2.. 8..4
22	William Carcill	do.	Augt. 19th	3	12		3..10...
23	Henry Sharp	do.	Sept. 1st	3	—		2..10...
24	George Campbell	do.	Augt. 4th	3	27		4..15...
25	James McTllear	do.	do. 8th	3	23		4.. 8..4
26	Stephen Rutledge	do.	do. 22d	3	9		3.. 5...
27	James Stevenson	do.	do. do.	3	9		3.. 5...
28	Votier Gawden	do.	Sept. 9th	2	22		1..16..8
29	Thomas Murphy	do.	do. 2d	2	29		2.. 8..4
30	Robt. Work	do.	Augt. 16th	3	15		3..15...
31	Patrick Quigley	do.	July 16th	4	15		7.. 5...
32	Mark Sullivan	do.	Sept. 10th	2	21		1..15...
33	John McFall	do.	Augt. 5th	3	26		4..13..4
34	William Stone	do.	Sept. 5th	2	26		2.. 3..4
35	Stephen Archer	do.	Augt. 13th	3	18		4.. 0...
36	James Cane	do.	Sept. 9th	2	22		1..16..8
37	Danl. McCarty	turn'd over to A Doria	Augt. 10th	3	21		
38	Micl. Kelly	do.	Sept. 12th	2	19		1..11..8
39	Neil Farron	do.	Augt. 16th	3	15		3..15...
40	William Beauchamand	do.	Sept. 4th	2	27		2.. 5...
41	Henry Dehart	do.	do. 2	2	29		2.. 8..4
42	Wm Campin	do.	do. 11th	2	20		1..13..4
43	John Speer	do.	Augt. 16th	3	15		3..15...
44	George Lasberry	do.	do. 5th	3	26		4..13..4
45	Jacob Guy	do.	do. 19th	3	12		3..10...
46	Francis Quinn	do.	do. 15th	3	16		3..16..8
47	Owen Ward	turn'd over to A Doria	do. 4th	3	27		
48	Robert Douglass	do.	Sept. 2d	2	29		2.. 8..4
49	John McClure	do.	Augt. 16th	3	15		3..15...
50	John Gilmore	do.	do. 28th	3	3		2..15...
51	Thomas Gough	do.	do. do.	3	3		2..15...
52	Richard Keys	do.	Octr. 3d	1	28		2.. 6..8
53	Micl. Millar	do.	do. do.	1	28		2.. 6..8
54	William Rivelly	do.	do. 10th	1	21		1..15...
55	Edward Smith	do.	do. 2d	1	29		2.. 8..4
56	William Rich	do.	Sept. 18th	2	13		1.. 1..8
57	Robt. Elder	do.	do. 7th	2	24		1.. 0...

[1] Minute Book (AM 3275), Miscellaneous Collections, Historical Society of Pennsylvania.

"Pay Roll of Capt. Robert Mullan's Compy Marines to Decembr. 1st 1776"[1] (Cont.)

No	Men's Names		Date of Enlist.	Mo.	Days	When desertd or dead	
58	Edward Asberry	do.	Augt. 29th	3	2		2 . . 13 . . 4
59	Barney Maloy	do.	Sept. 12th	2	19		1 . . 11 . . 8
60	Thomas McKey	do.	Augt. 27th	3	4		2 . . 16 . . 8
61	Allan McKey	do.	do. do	3	4		2 . . 16 . . 8
62	John Getty	do.	Sept. 11th	2	20		1 . . 13 . . 4
63	Enoch Jenkins	do.	do. 13th	2	18		1 . . 10 . . .
64	Henry Hassan	do.	Sept. 10th	2	21		1 . . 15 . . .
65	John Lewis	do.	do. 25th	2	6		. . 10 . . .
66	Henry Ripshon	do.	Octr. 21st	1	10		. . 16 . . 8
67	Patrick Harvy	do.	Sept. 17th	2	14		6 . . 3 . . 4
68	William Dougherty	do.	Novr. 12th	0	19		1 . . 11 . . 8
69	Isaac Walker (Negro)		Augt. 27th	3	4		5 . . 6 . . 8
70	Orange . . . (Negro)		Octr. 1st	2	—		5 . . 0 . . .
71	Thomas Caldwell (Desertd)		Augt. 20th	3	11	Octr 1st	
72	Jesse Redding (Deserted)		Sept. 2d	2	29	do. 21st	Returned May 10th 1777
73	Patrick Russell (Desertd)		Augt. 11th	3	20	do. 15th	
74	Alexr. Cummins (Desertd)		Sept. 1st	3	—	Novr. 1st	
75	John McCashon (Desertd)		Augt. 21st	3	10	Octr. 25	Returned April 13th 1777
76	Hugh Conolly (Desertd)		Sept. 8th	2	23	do. 21st	
77	John McClosky (Deserted)		Augt. 29th	3	2	do. 7th	
78	Thos. Meishinney (Desertd)		do. 31st	3	1	do. 7th	
79	Jno. Fritzinger (Desertd)		do. do.	3	1	do. do.	
80	Joseph Lowrey (Desertd)		do. do	3	1	do. do	
81	John Hill (Desertd)		do. 16th	3	15	do. 25th	
82	Thos. Sappington (Desertd)		Sept. 7th	2	24	do. 17th	Returned May 1st 1777
83	Joseph Boyce (Desertd)		Augt. 29th	3	2	do. 25	Returnd Decemr 3d 1776
84	William Taylor (Desertd)		Octr. 10th	1	21	do. 15th	
85	Danl. Cloud (Dead)		Augt. 21st	3	10	do. 20th	
86	Thos. Atkinson (Dead)		do. 23d	3	8	do. 27th	

1 Minute Book (AM 3275), Miscellaneous Collections, Historical Society of Pennsylvania.

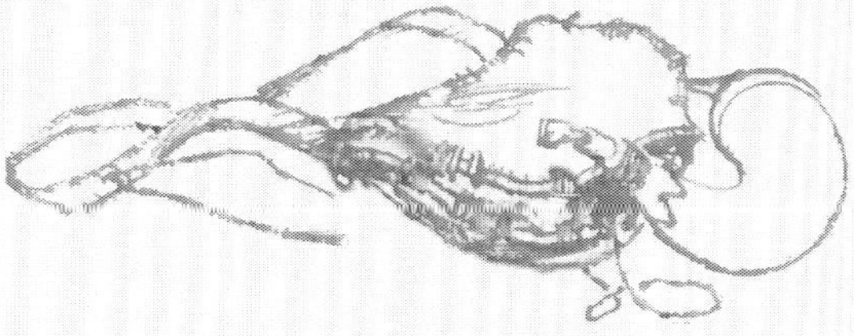

"A Pay Roll of Capn. Robert Mullan's Company of Marines from December 1st 1776 to April 1st 1777"[1]

No	Mens Names	Quality	Months & Days		Total due	Amount of Cloaths at Mr. Tods	Amount of Cloathing at Morristown	Ballance due
			Mo.	Dy.	£sd	£sd	£	£
1	Robert Mullen	Capn	4.	–	45, –	–15,10	2,8,9	41,15,5
2	David Love	Lieut	4.	–	22,10	7,16,5		14,14,7
3	Hugh Montgomery	Do	4.	–	22,10	2,16,4		19,13,8
4	Thomas Hart	Serjt	4.	–	12, –	6,1,5	2,17,6	3,1,1
5	Andrew Read	Do	4.	–	12, –	11,5,–		–15,–
6	John McKinley	Do	4.	–	12, –	13,3,2		
7	Barny Moloy	Corpl	4.	–	11, –	2,3,11		8,16,1
8	Adam McFarson	Do	4.	–	11, –	9,17,1	0,17,6	–5,5
9	James Butler	Do	1.		2,15			2,15,–
10	Collin York	Drumr	4.	–	11, –	7,13,–	1,18,1	1,8,11
11	Peter York	Fifer	4.		11, –	7,13,–	1,18,1	1,8,11
12	Wm Allisson	. . .	4.	–	10, –	9,17,8	1,7,6	
13	James Cane		4.	–	10, –	7,11,7		2,8,5
14	Jacob Guy		4.	–	10, –	8,18,7		1,1,5
15	Wm Williams		4.	–	10, –	7,12,11	1,11,3	–,15,10
16	Isaac Negro		4.	–	10, –	6,6,5		3,13,7
17	Benjn Woodin		4.	–	10, –	11,9,4	0,17,6	
18	John Hogg		4.	–	10, –	8,11,11	2,6,3	
19	John Stone		4.	–	10, –	9,15,11	1,18,9	
20	Wm Stone		4.	–	10, –	10,10,1	1,1,3	
21	Allen Mc. Key		4.	–	10, –	8,5,6	0,3,9	1,10,9
22	Thos. McKey		4.	–	10, –	3,15,5		6,4,7
23	Geo: Campbell		4.	–	10, –	8,14,4		1,5,8
24	Stephn. Rutledge		4.	–	10, –	11,8,4	2,8,9	
25	Jas. Stephenson		4.	–	10, –	10,14,2		
26	Robt. Work		4.	–	10, –	9,8,9	6,11,3	
27	Stephen Archer		4.	–	10, –	9,14,3	3,–,–	
28	Henry De Hart		4.	–	10, –	8,18,3		1,1,9
29	John Spear		4.	–	10, –	10,4,8	3,11,3	
30	Fras. Quin		4.	–	10, –	18,10,3		
31	Mickl. Kelly		4.	–	10, –	9,17,5		0,2,7
32	Robt. Douglass		4.	–	10, –	8,15,8		1,4,4
33	Richd. Keys		4.	–	10, –	8,13,5	2,5,–	
34	Wm. Rivelly		4.	–	10, –	11,–,2	1,7,6	
35	Edwd. Smith		4.	–	10, –	11,6,2	1,2,6	
36	Robt. Elder		4.	–	10, –	12,5,–		
37	Heny. Ripshin		4.	–	10, –	12,7,9	5,11,9	
38	Wm Dougherty		4.	–	10, –	10,13,7		
39	George Price		4	–	10, –	13,6,2		
40	James Wilson		4.		10, –	11,18,1		
41	Patrk: Clinton		4.	–	10, –	12,1,3		
42	John Brown		4.	–	10, –	11,5,10		
43	Orrang Negro		4.		10, –	5,10,2	4,18,6	
44	Wm. Caysy		4.	–	10, –		1,11,3	8,8,9
45	Joseph Boyce		4.	–	10, –			
46	Danl. McCarthy							
47	Jno. McCashon							
48	James Lee							
49	John Connollin							
50	Barnabs. Allen							
51	Thos. Lesley							
52	Phil. Kennedy							
53	Nichl. Miller							
54	Jacob Murry							
55	Patk: Preston							
56	Patk. Brannon							

[1] Minute Book (AM 3275), Miscellaneous Collections, Historical Society of Pennsylvania.

"Weekly Return of Capt Benjamin Dean's Company Marines to Major Nicholas"[1]

December 20th 1776

Mens Names	Capt	Lieutt	Sergeants	Corporals	Privates fit for Duty	Sick
Benjamin Dean	1					
Peregrine Brown		1				
Abel Morgan		2				
Robt Dougherty			1			
James Buchanan			2			
John Ryan			3			
James Morgan				1		
Nicholas Henry					1	
Thomas Pratt					2	
James Savery					3	
Benje. Stewart					4	
William Smyth					5	
Peter Fitzpatrick				2		
Thomas Gruson					6	
Joseph Frame					7	
Jerimiah Driscall					8	
John Kist					9	
Edwd. Edwards					10	
Edwd. Lanagon					11	
Patrick Murray					12	
John Anderson					13	
John Kelley					14	
James Smyth					15	
Patrick Donelly					16	
Robt. Derrumple					17	
Alexander Jorden					18	
William Bird					19	
Neal Cook						
William McKindley					20	
Edward Kelley					21	
John Wilson					22	
Thomas Coxwell					23	
Thomas Hail					24	
William Johnston					25	
Thomas Cooper					26	
Abraham Avenell					27	
Morgan Holliner					28	
John Douglas					29	
Garritt Hornett					30	
Robert Kilpatrick					31	
Andrew Ross					32	
Samuel Gother					33	
Total fit for Duty					42	
John Nicholson						1
Samuel Seegar						2
David Clemmens						3

[1] John Cadwalader Papers, Historical Society of Pennsylvania.

"A Weekly Return of Capt Andw Porters Company of Marines to Saml. Nicholas Esqr. Major of Marines Bristol Decembr. 20th. 1776"[1]

	Captain	Lieutenants	Sergeants	Corporals	Privates	Sick
Andrew Porter	1					
James McClure		1				
James McNeir			1			
Ezekiel Howell			2			
Patk. Porter			3			
John McGregor				1		
James Stevenson Senr.				2		
James Stevenson Junr.				3		
David Stewart					1	
George Gadsby					2	
John Dunn					3	
Garrit Crow					4	
Michael Linch					5	
William Wattson					6	
George Knox					7	
Barry Allen					8	
Nicholas Murphy					9	
David Davis					10	
George Gillass					11	
Alexr. Chrisham					12	
John Powers					13	
Paul Frider					14	
John Fisher					15	
Thomas Keys					16	
Hugh McBride					17	
John Caskey					18	
Reubin Ellis					19	
Robert Parker					20	
William Crow					21	
John McKinley					22	
Thomas Ford					23	
James Allen					24	
Samuel Assop					25	
James Davidson					26	
William Dunkan					27	
James Smith						28

[1] John Cadwalader Papers, Historical Society of Pennsylvania.

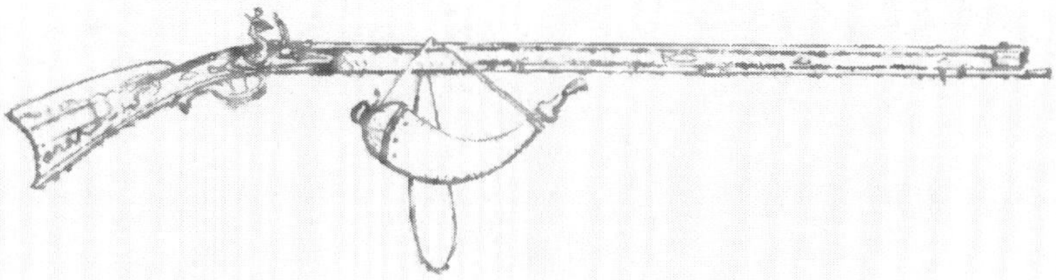

"Return of Capt Robt Mullans Compy of Marines Decr 20th 1776"[1]

No	Mens Names		Capt	Lieutt	Serjts	Corpls	Drum & Fife	Rank & File	Sick
	Robt Mullan	Capt	1						
	James Coakley	1st Lieut	2	1					
	Peter Bedford	2nd do	3	2					
	Thos Hart	Serjt		3	1				
	Leo: Green				2				
	Adam McFarson					1			
	Jn Burnet					2			
	Jn Cribs					3			
	Bary Maloy					4			
	Jno Hogg							1	
	Benjm Woodin							2	
	Robt Gilmore							3	
	Jno Stone							4	
	Allen McKey							5	
	Thos McKey							6	
	Edwd Smyth							7	
	Wm Stone							8	
	Robt Work							9	
	Patk Quigly							10	
	James Stephinson							11	
	Michl Miller							12	
	Heny Ripshin							13	
	Wm Revely							14	
	Jno Spear							15	
	Robt Douglass							16	
	James Casy							17	
	Thos Jinnings							18	
	James Ligate							19	
	Jerh Murry							20	
	Patk Clinton							21	
	Geo: Price							22	
	Lawr Leacy							23	
	Orange/Negro							24	
	Collin York Dr.						1		
	Peter York fife						2		
	James Cane							25	
	Thos Murphy							26	
	Stephn Rutledge							27	
	Wm Campin							28	
	Heny Dehart							29	
	Wm Dougherty							30	
	Robt Elder							31	
	James Wilson								1
	Wm Carsill								2
	Richd Keys								3
	Wm Williams							32	
	Wm Beaukanan							33	
	Wm Allison								4
	Jno Lewis								5
			1	2	2	4	2	33	5

[1] John Cadwalader Papers, Historical Society of Pennsylvania.

"A Muster Roll of Capn. Robert Mullan's Compy. of Marines to April 1st 1777."[1]

No.	Mens Names	Time of Enlistment	Months & Days	Total due	Amount of Cloaths at Mr. Tods	Amount of Cloathing at Morristown	Blankets revd. from the States	Ballance due	D Dead DD Discharge R Run
		1776		£	£	£	£ s	£ s d	
1	Robert Mullan Capn	June 25	4	45,..		2,8,9		41,15,5	
2	David Love 1Lt.	June 25	4	22,10	7,16,5		1 1:4:--	13,9,7	
3	Hugh Montgomery 2d	June 25	4	22,10	2,16,4		1 1:4:--	18,9,8	
4	Thos Hart	Nov. 25	4	12--		2,17,6	1 1:4:--	7,18,6	
5	Andrew Read	Aug 22	4	12	8,12,1	0,17,6	1 1:4:--	1,6,5	
6	John McKinley	Aug 2	4	12	13,3,2		1 1:4:--		
7	Barny Moloy	Sepr. 12	4	11	2,3,11		1 1:4:--	7,12,1	
8	Adm. McPherson	Oct: 22	4	11	9,9,11		1 1:4:--	0,6,1	
		1777							
9	James Butler	Mar: 1	1	2,15				2,15,-	
10	Collin York	1776 June 25	4	11	8,-,11	3,2,6	1 1:4:--		
11	Peter York	June 25	4	11	7,5,1	0,13,9	1 1:4:--	1,17,2	
12	Wm Allison	Sep: 2	4	10	8,14,8	1,7,6	1 1:4:--		
13	James Cane	Sepr. 9	4	10	6,11,7		1 1:4:--	2,4,5	
14	Jacob Guy	Aug. 19	4	10	8,6,11		1 1:4:--	0,9,1	
15	Wm Williams	Nov. 25	4	10		1,11,3	1 1:4:--	7,4,9	
16	Benjn. Woodin	Aug. 12	4	10	10,3,4	0,17,6	1 1:4:--		
17	John Hogg	Aug. 21	4	10	8,-,3	2,6,3	1 1:4:--		
18	John Stone	Sep. 2	4	10	8,9,11	1,18,9	1 1:4:--		
19	Wm Stone	Sep: 5	4	10	9,6,1	1,1,3	1 1:4:--		
20	Allen McKey	Aug: 27	4	10	7,1,6	0,3,9	1 1:4:--		
21	George Campbell	Aug. 4	4	10	7,8,4		1 1:4:--	1,7,8	
22	Steph: Rutledge	Aug 22	4	10	10,4,4	2,8,9	1 1:4:--		
23	Jams. Stephensen	Aug 22	4	10	9,10,2		1 1:4:--		
24	Robt. Work	Aug. 16	4	10	8,4,9	6,11,3	1 1:4:--		
25	Steph: Archer	Aug. 13	4	10	6,17,1	3,-,-	1 1:4:--		
26	Henry Dehart	Sep: 2	4	10	7,14,3		1 1:4:--	1,1,9	
27	John Spear	Aug. 16	4	10	9,1,8	3,11,3	1 1:4:--		
28	Fras. Quin	Aug. 15	4	10	17,6,3		1 1:4:--		
29	Michl. Kelly	Sepr. 12	4	10	9,17,5		1 1:4:--		
30	Robt. Douglass	Sep: 2	4	10	6,11,8	1,1,-	1 1:4:--	1,3,4	
31	Richd. Keys	Octr. 3	4	10	8,2,9	2,5,-	1 1:4:--		
32	Wm Rivelly	Oct: 10	4	10	9,16,2	1,7,6	1 1:4:--		
33	Edwd. Smith	Oct. 2	4	10	10,2,2	1,2,6	1 1:4:--		
34	Robt. Elder	Sep: 7	4	10	12,-,10		1 1:4:--		
35	Hy. Ripshon	Oct: 21	4	10	11,3,9	5,11,9	1 1:4:--		
36	Wm Dougherty	Nov. 12	4	10	8,5,7		1 1:4:--		
37	Thos. McKey	Aug. 27	4	10	3,3,9		1 1:4:--	5,12,3	
38	Joseph Boyce	Aug. 29	4	10		4,9,9	1 1:4:--	4,6,3	
39	Danl. McCarthy	Aug. 10	4	10			1 1:4:--		
40	John McCashon	Aug. 21	4	10			1 1:4:--		
41	John Conolly	Sep: 8	4	10			1 1:4:--		
42	Phil: Kennedy								
43	Nichs. Miller	1777 Mar: 1	4	2,10					
44	Jacob Murry	Mar: 1	4	2,10					
45	Geo: Rice	1776 Nov. 22	4	10.--		2,1,3	1 1:4:--	6,14,9	
46	James Wilson	Nov 25	4	10.--		4,-,-	1 1:4:--	4,18,8	
47	Patk. Clinton	Nov 22	4	10.--		3,16,3	1 1:4:--	4,19,9	
48	John Brown	Dec: 1	4	10.--		1,11,3			
49	Wm Casey		4	10.--				8,8,9	
50	Thos Lesley								
51	Pat: Preston								
52	Patk Brannon								
53	Isaac Negro	Aug. 27	4	10	8,2,11		1 1:4:--	0.13,1	
54	Orang Negro	Oct. 1	4	10	4,6,2	4,18,6	1 1:4:--		
55	James Cakley	1776 July 11			4,16,6	1,5,-	1 1:4:--		1776
56	Warwick Hallabough	Sepr. 13					1 1:4:--		R Nov. 15

[1] Minute Book (AM 3275), Miscellaneous Collections, Historical Society of Pennsylvania; arrearages have been deleted.

"A Muster Roll of Capn. Robert Mullan's Compy. of Marines to April 1st 1777."[1] (Cont.)

No.	Mens Names	Time of Enlistment	Months & Days	Total due	Amount of Cloaths at Mr. Tods	Amount of Cloathing at Morristown	Blankets revd. from the States	Ballance due	D Dead DD Discharge R Run
57	George Murry	Aug: 27		11	9,19,5	2,8,9			D 1777 Apr. 1
58	John Cribs	Octr. 13		11	7,5,1		1 1:4:__		R Apr. 6
59	Joseph Guimly	Sepr: 17		1,7,6	4,6,_		1 1:4:__		R 1776 Dec 16
60	Wm. Barnet	Sepr. 1		0,6,8	11,15,3		1 1:4:__		R Dec 5
61	Lawce. Lesey	Sep: 3	3,12	8,10,_	10,15,7		1 1:4:__		R 1777 Mar 13
62	Robt. Gilmore	Aug: 28	4	10	7,11,9	4,12,6	1 1:4:__		R Apr: 1
63	Danl. Forsman	Sep: 2	2	0,3,4	8,_,11				R 1776 Dec: 3
64	Wm Carcill	Aug: 19	4	10	3,4,1		1 1:4:__	5,14,7	DD 1777 Apr. 1
65	Henry Sharp	Sepr. 1	_5	8,4	13,8,4		2 2:8:__		R 1776 Dec: 6
66	Jas. McIllear	Aug. 8	_4	6,8	13,7,10		1 1:4:__		R Dec 5
67	Votier Gawdon	Sepr. 9	_4	6,8	5,2,9				R Dec: 5
68	Thos. Murphy	Sepr. 2	_15	1,5	9,8,4		1 1:4:__		R Dec: 6
69	Patk. Quigley	July 16	1	2,10	13,4,_		1 1:4:__		R 1777 Jany. 1
70	Mark Sullivan	Sep: 10	_4	0,6,8	13,_,4		1 1:4:__		R 1776 Dec: 5
71	Jno. McFall	Aug. 5	1,14	3,13,4	9,_,6	0,17,6	1 1:4:__		R 1777 Jany. 15
72	Neil Farron	Aug. 16	_4	0,6,8	2,16,2		1 1:4:__		R 1776 Dec: 5
73	Wm Buchanan	Sepr. 4	4	10	6,12,9	1,7,6		1,19,9	DD 1777 Apr: 10
74	Wm Campin	Sepr. 11	_4	0,6,8	9,18,11		1 1:4:__		R 1776 Dec: 5
75	Geo: Lasbery	Aug: 5	1,15	3,15,_	6,12,9		1 1:4:__		D 1777 Jan 16
76	Jno. McClure	Aug: 16	4	10					DD 1776 Dec: 1
77	Jno. Gilmore	Aug: 28	_11	_,18,4	7,11,9	4,12,6	1 1:4:__		DD Nov. 20
78	Thos. Gough	Aug: 28	_4	_,6,8	5,13,4		1 1:4:__		R Dec: 5
79	Owen Ward	Aug: 4					1 1:4:__		
80	Michl Millar	Octr. 3	4	10	7,13,11	0,10,_	1 1:4:__	0,12,1	R 1777 Apr: 10
81	Wm Rich	Sep: 18	3	7,10,_	4,1,3		1 1:4:__	2,4,9	D Mar 1
82	Edd. Asberry	Aug: 29	_14	1,3,4	6,16,11		1 1:4:__		D 1776 Dec 15
83	Jno Getty	Sepr. 11			2,15,5		1 1:4:__		
84	Enoch Jenkins	Sep. 13	2,18	6,10,_			1 1:4:__		R 1776 Nov:15
85	Hy. Hassan	Sepr. 10	1,6	3,_,_			1 1:4:__	1,16,_	R 1777 Jany 7
86	Jno Lewis	Sepr. 25	4	10	4,8,1		1 1:4:__	4,7,11	R Apr 16
87	Patk Harvy	Sepr. 17			1,7,10	4,1,3	1 1:4:__		DD
88	Pat Harvey								
89	Thos Livingston	Augt 25							

[1] Minute Book (AM 3275), Miscellaneous Collections, Historical Society of Pennsylvania; arrearages have been deleted.

"Pay-Abstract of a Company of Marines Commanded by Captain James Willing of the United States, Commencing the 10th. January 1778 and Ending the 3d. June 1779 inclusive."[1]

Names	Rank	Commencing	Ending	Pay per Month Dollars
Robert George	1st Lieutenant	Jany 10th 1778	June 3d 1779	33⅓
Richard Harrison	2d Lieutent.	Feby 12th	do.	33⅓
George Girty	Do	6th	Deserted May 4.	33⅓
John Hodgson	Carpenter	Jany 10th	Died Aug 30th	10
Solomon Burny	Coxwain	10th	Deserted 26th	10
Thomas Beard	Serjeant	10th	June 3d	10
Nathaniel Downs	Corporal	10th	Do.	9
John Marny	Serjeant	10th	April 20th	10
Thomas Love	Corporal	10th	Taken Pris. Apl. 12th 1778	9
John Ash	Private	10th	June 3d	8⅓
Samuel Fury	do	10th	Died 3d Augt.	
John Walker	do	do	June 3d	
Daniel Whitaker	do	do	do	
Philip Hupp	do	do	do	
Henry Haut	do	do	do	
Mark Foley	do	do	do	
Henry Hawk	do	do	Died 31st Oct.	
John Kilpatrick	do	do	Died 1st Sept.	
Nathaniel Kennison	do	do	Taken Pris. Apl. 12-1778	
Richard Murray	do	do	Desert. Augt. 15th	
Leven Spriggs	do	do	March 10th	
James Taylor	do	do	June 3d	
John Henwood	do	do	Died Octob. 3d	
Lazarus Ryan	do	do	June 3d	
William White	do	do	do	
Richard Roddy	do	do	Desert. Augt. 15th	
Laurence Keenan	do	do	June 3d	
Jacob Wheat	do	do	do	
Solomon Walker	do	do	do	
Nicholas Walker	do	do	do	
William Poston	Gunner	do	do	8⅓
Ephraim Carey	Gun. Mate	do	Died Augt. 10th	
John Burroughs	Private	do	June 3d	
William Johnston	Private	Jany 10th 1778	June 3d 1779	8⅓
Patrick Doyle	do	do	Died March 28th	
Jacob Oadham	do	do	June 3d	
William Dove	do	do	do	
Edward Matthews	Serjeant	16th	do	10
David Wallis	Private	March 1st	do	
William Brown	do	April 1st	do	
John Stampley	do	Feby 16th	Desert. 28th May	
Henry Repard	do	do	June 3d	
John Bryan	do	28th	do	
Valentine Bolsinger	do	do	do	
Andrew Conoro	do	June 4th	do	
Thomas McDonald	do	March 10th	do	
George McKnight	do	Feby 16th	do	
John Reiley	do	28th	do	
Thomas Mitchell	do	16th	do	
Joseph Anderson	do	March 1st	do	
James McDonald	do	16th	do	
Cornelius Cornoble	do	June 4th	Deserted 20th Septem 1778	
Henry Sholes	do	do	Deserted do	
John Hinman	do	do	do	

[1] Military Accounts, Office of the Comptroller General, Pennsylvania Historical and Museum Commission.

"Pay-Abstract of a Company of Marines Commanded by Captain James Willing of the United States Commencing the 10th. January 1778 and Ending the 3d. June 1779 inclusive."[1] (Cont.)

Names	Rank	Commencing	Ending	Pay per Month Dollars
John Seeder	do	1st	June 3d	
Peter Collins	do	Feby 16th	do	
John Roberts	do	May 8th	do	
John Mann	do	April 1st	Died 19 Augt.	
Sufret Gion	do	March 9	June 3d	
John Bush	do	Augt 17	do	
Gideon Thomas	do	April 1st	Died May 17	
Nicholas Smith	do	Feby 9	Died March 23	
Charles Suffroy	do	Sept 4	June 3d	
Adam Lainhart	do	Feby 9	do	
John Ceasar	do	Augt 12	do	

I do hereby Certify that I received the above Named Men & Officers from Capt. James Willing with Orders to proceed to the Illinois & Fort Pitt, which Men are disposed of (Deaths and Desertions excepted) as specified within, that is to say such as end on the 3d June 1779 having joined the Service of the State of Virginia in the Illinois Department under my Command, and for whom I am accountable, signed in presence of Brigadier Genl. Clark at Fort Nelson this sixteenth day of May One thousand seven hundred and Eighty two.—

G. Clark B.G. signed—Robt. George Lt.

1. Military Accounts, Office of the Comptroller General, Pennsylvania Historical and Museum Commission

"A Coppy of Return Given In to Major Nichlas of Men Enlisted by Capt Robert Mullan, Since Augst 9th 1779 to Jany 1st 1780"[1]

	Mens Names	On Duty	Confind	Sick in Hospital	Deserted	Dead		Mens Names	Deserted	Dead
1	Abraham Lewis Serjt	1					37	William Warner	1	
2	Jeremiah Mahon		1				38	Andrew Anderson	1	
3	Richard Kisby	1					39	James Carney	1	
4	Francis Deuprey	1					40	Thos Paul	1	
5	Edward Butler	1					41	James Murdock	1	
6	John Smith		1				42	John Kean	1	
7	John Henry	1					43	John Coil	1	
8	Andrew Elbrew	1					44	John Woodford	1	
9	Samuel Everfelt	1					45	John Mackearney	1	
10	John Fife	1					46	John Miller	1	
11	John Egan	1					47	Patrick Higgins	1	
12	Joseph Hardunck	1					48	John Radue	1	
13	Thos Conner	1					49	Nichos Ultzman	1	
14	James Plowmer	1					50	John Barbro	1	
15	Peter Gurney	1					51	Thos Smith	1	
16	Joseph Desaign	1					52	William Baker		1
17	James Deacon		1							
18	James Beyan		1							
19	Joseph Craft	1								
20	Theodora David	1								
21	James Crampton			1						
22	Robert McGinnis					}		on Board Confedrysy with Captn Boyce		
23	Charles Stephenson					}				
24	Thos Harden									
25	Nathan Marchal				1					
26	Thos Mathews				1					
27	Thos Sappington				1					
28	James Nicholas				1					
29	Thos Peckworth				1					
30	Richard Coats				1					
31	John Smith				1					
32	John Dougherty				1					
33	Richard Babington				1					
34	John McCashion				1					
35	Thos Sellords				1					
36	George McCray				1					

[1] Minute Book (A 3275), Miscellaneous Collections, Historical Society of Pennsylvania.

APPENDIX J

Biographies of Continental Marine Officers

Introduction

The majority of the names of Continental Marine officers in the following biographies were taken from "A List of Commissioned Officers who served in the Navy of the United States in the late War [Revolution.]" It was compiled by Mr. Doyle Sweeny, Auditor's Office Treasury Department, Philadelphia, Pennsylvania, and was dated 18 March 1794.

Mr. Sweeny had this to say regarding the names:

> No regular record appears to have been kept of the appointments made in the Marine Department and it is not to be wondered at, when it is considered how many persons and Boards were vested with Authority to make appointments.
>
> The foregoing list is formed from the Minutes of the Marine Committee and Navy Boards and from the rolls of several vessels.—Many of the Officers served only for a Cruize.

In several instances, names in Sweeny's list researched by this compiler have proved to have been naval officers. In many other instances, subsequent research of both primary and secondary sources supported Mr. Sweeny's wondering observation, that little or nothing exists concerning the lives and careers of certain officers. In fact, some have been found to have been enlisted men, who for reasons currently unknown, may have been elevated to a temporary rank or to that of a volunteer officer for a short period of time.

One might also suspect that Mr. Sweeny may have included state and privateer Marine officers who later received federal payment for their Revolutionary services.

A complete search of Mr. Sweeny's original sources is nearly impossible today. While portions survive in the National Archives and Records Service, Washington, D. C., the majority are believed to have been destroyed in a fire in one of Washington's "The Six Buildings."

Three authors of later periods also published lists of Continental Marine officers, but each, for the most part, appears to have relied on Sweeny's original list. One or another may have given the date of the officer's commission, but, in some cases, gave no date at all. These were:

Charles Oscar Paullin. *The Navy of the American Revolution: Its Administration, Its Policy, and Its Achievements*. Cleveland: Burrows Brothers Co., 1906. 549 pp. notes, bibliog.

Gardner W. Allen. *A Naval History of the American Revolution*. 2 vols. Boston: Riverside Press, 1913. illus., notes, bibliog.

Major Edwin N. McClellan, USMC. *History of the United States Marine Corps*. 2 vols. Washington: Historical Section, Headquarters, U. S. Marine Corps, 1925–, 1932. notes, bibliog.

The latter, although unpublished, is, in the view of the compiler, the most informative genealogically of the three works. Major McClellan's copious footnotes led to sources that otherwise might not have been searched. McClellan, however, also wrote summaries of many state and privateer Marine officers, which are not included in this appendix.

In several instances, the compiler has listed a first choice of surname that has not been used in the body of the text. That choice has been influenced first, according to the usage of a proven genealogy or the use by a majority of families of a particular surname; second, by its consistency of use in records of the Continental Navy and Marines; third, as stated by Major McClellan; and fourth, as stated by Mr. Sweeny.

Extensive work was done by the compiler in the primary and secondary sources available in Washington,D. C. As time and monetary restrictions permitted, the records of several primarily genealogical repositories of the east coast, especially those of the New England states, were consulted.

No attempt has been made to overly exalt or detract from the career or life of any officer. Many of the summaries are detailed and others are quite short, but all information presented is based solely upon documentation researched to date. When the proper identity of a person was in doubt the compiler used the best evidence available. In other instances, lack of primary and substantial evidence resulted in the necessity of citing vital statistics of several persons of the same given name or surname.

Some of the subsequent biographies, printed in alphabetical order, will contain the fact that the name was recorded in Mr. Sweeny's list of 1794, and perhaps in those of Allen, Paullin or McClellan, followed by any information considered pertinent.

The Marine Corps will be most grateful to all readers of these biographies, whether they be descendants or authorities of a particular person or family represented herein, for any additions or corrections they may see fit to contribute. Inasmuch as 1975 is the bicentennial of the founding of the Corps, donations of artifacts and documents pertaining to these officers are also very desirable for permanent exhibit purposes. All donated material will be accessioned in the collections of the History and Museums Division, Headquarters, United States Marine Corps, Washington, D. C. 20380.

ARROWSMITH, EDMUND [EDMOND, EDWARD]. The origins of Edmund Arrowsmith are unknown, but he may have been a native of Rhode Island or New York.

On 14 June 1776, 'Edd. Arrowsmith" entered the sloop *Providence* as an ensign, transferring with him from a Rhode Island brigade 20 enlisted men. He was apparently promoted to lieutenant of Marines shortly after, for he was entitled to prize shares for that rank in the ship *Alexander*, captured by *Providence* on 20 September 1776.

Captain John Paul Jones, commanding *Providence* on that day, wrote a letter to the Marine Committee of Continental Congress, describing a humorous incident which occurred off the Isle of Sable. *Providence* was being chased by a British frigate, which she easily outsailed for eight hours. The frigate continuously expended useless powder and shot. At length, to show his contempt for a particularly long range broadside, Jones ordered his Marine officer—presumably Arrowsmith—"to return the Salute with only a Single Musquit."

Arrowsmith probably served on *Providence* in late October 1776, for he was on the court martial of Gunner James Bryant, brig *Hampden*, held on board *Alfred* on the 23d.

A series of promotions and transfers may have prompted a new assignment for Lieutenant Arrowsmith, perhaps in late October or early November. Captain Samuel Nicholas was promoted to major and began land duty in June. Lieutenant Matthew Parke, of *Alfred*, was promoted to captain and transferred to *Columbus* when Captain Joseph Shoemaker left the Marine service in April. Arrowsmith then transferred from *Providence* to *Alfred*.

With Captain Jones and Navy Lieutenant John Rathbun of *Providence*, Arrowsmith boarded the privateer schooner *Eagle*, Captain Issac Field, at Tarpaulin Cove, Massachusetts on the night of 1 November to search for deserters from the two Continental ships. He also shared prize money of the brig *Active* and ship *Mellish*, which were captured 12 and 13 November, respectively.

On 2 July 1777, Jones was appointed to command the sloop of war *Ranger*, 18 guns, built at Portsmouth, New Hampshire. He was extremely short of men, and ordered Captain of Marines Matthew Parke, at Providence, Rhode Island, to recruiting duty. He also instructed Parke "As I obtained for Capt. Arrowsmith his present commission and introduced him into the service . . . he will now have an opportunity of recommending himself to further notice. . . ." Arrowsmith's orders were to "go around with a Drum, Fife and Colours as often as may be proper . . ." to enlist men for service with Jones.

Papers of the Continental Congress, however, do not show that Arrowsmith was promoted to captain until 20 October 1777, nor is information forthcoming of his activities for the following two years. He did not serve on board *Ranger*.

On 15 June 1779, Arrowsmith was captain of Marines in the ship *Queen of France*, Captain John Rathbun. Both he, and his lieutenant of Marines, Peter Green, were captured at the fall of Charleston, South Carolina on 13 May 1780.

Nothing further is known of the career or life of Captain Edmund Arrowsmith.

BARNEY, WILLIAM STEVENSON. William Stevenson Barney was born 28 December 1754, one of the 14 children of William and Frances Holland (Watts) Barney, of Patapsco Neck, Maryland. His brother, Joshua, later Commodore, U. S. Navy, was born 6 July 1759.

The exact date of William Barney's commission is unknown, but he is reputed to have been a lieutenant of Marines in the brig *Andrew Doria*, 14 guns, when the crew of that vessel evacuated the fort at Billingsport, New Jersey on the Delaware River on 2 October 1777.

Following the fall of Philadelphia in late 1777, Navy lieutenant Joshua Barney took the crew of the *Andrew Doria* from Bordentown to Baltimore to man the frigate *Virginia*. William Barney was her second lieutenant of Marines.

On 30 March 1778, *Virginia* attempted to get to sea during the night, but her pilot ran her aground between the Virginia Capes. Abandoned by her captain the following morning, she was forced to surrender to HMS *Emerald*. Although Joshua Barney was imprisoned on board the enemy vessel,

William was sent to Baltimore for exchange on 11 April 1778.

Further Marine service of William Barney in the Revolution is not recorded. However, he was appointed a first lieutenant of the 30th Infantry on 30 April 1813 and served through 1814. He is reputed to have kept a hotel in Georgetown, District of Columbia.

Doyle Sweeny, in his list of Revolutionary naval officers, compiled by the Auditor's Office, Treasury Department, and dated 18 March 1794, however, listed William Barney as "dead."

BAXTER, SETH. Seth Baxter was born at Braintree, Massachusetts in 1732, one of 11 children of John and Mehitable (Willard) Baxter. He was married first, 26 October 1758, to Elenor, daughter of Jonathan Allen, of Martha's Vineyard, by the Reverend Anthony Wibird, of Braintree. He married second, Mary Saunders, 20 December 1767. He had three children by his first wife, and four by his second.

He was appointed a captain of Marines 10 October 1776, but was unemployed in Boston for a time. Meanwhile, frigate *Hancock*, Captain John Manley, had been built at Newburyport, Massachusetts, launched 10 July 1776, and was commissioned 23 June 1777. She was outfitted in Boston harbor.

Captain Baxter became *Hancock*'s senior Marine officer, probably reporting on board in late 1776 or in January 1777. In the latter period, Jacob Spear enlisted as corporal of Marines under Baxter in the *Hancock*. William Bubier, of Marblehead, Massachusetts, was her lieutenant of Marines.

Hancock sailed with frigate *Boston* from Boston mid-May 1777 searching for the Jamaica fleet. On the 30th, *Hancock* had a brush with the *Somerset*, 64 guns, but escaped in the night.

On the morning of 7 June, off the Banks of Newfoundland, they encountered the British frigate *Fox*, 28 guns. *Hancock* captured her, manned her with a prize crew, and sailed in company with her. First they sailed toward Charleston, South Carolina, but then changed course for Massachusetts.

On 7 July, they were chased by the British ships *Rainbow*, 44 guns, *Flora*, 32, and *Victor*, 18. They separated, each going a different direction, *Rainbow* pursuing *Hancock*. Late the following morning, Captain Manley was forced to strike his colors. *Fox* was also captured, by *Flora*, but *Boston* escaped into the Sheepscot River, Maine.

Captain Baxter and Lieutenant Bubier were sent on board *Rainbow*, and were later imprisoned at Halifax. Baxter was exchanged on 9 November 1777 for Captain John Aire.

By early 1779, Captain Samuel Tucker had taken command of the frigate *Boston*, and Seth Baxter joined the ship as her captain of Marines on 17 February, replacing Captain Richard Palmes. First Lieutenant of Marines Jeremiah Reed had joined in December, and William Cooper became the frigate's second lieutenant of Marines in March 1779.

In April, *Boston* was ordered to Philadelphia for a supply of bread, and thereafter to cruise off the Delaware Capes. There, cruising with frigate *Confederacy*, three prizes were taken, among them the British privateer *Pole*, 26 guns. Lieutenant Cooper was a part of her prize crew and took her into Philadelphia.

The Marine career of Captain Seth Baxter thereafter is unknown. Captain Baxter died 8 August 1805, aged 74.

BECKER, HENRY. Doyle Sweeny, in his "A List of Commissioned Officers who served in the Navy of the United States in the late War," compiled for the Treasury Department in 1794, states that Henry Becker was a lieutenant of Marines in the Revolution.

Apparently, Sweeny's source of information is now lost. Current research has failed to provide a clue to his life or career.

BEDFORD, PETER. Doyle Sweeny, in his Treasury report of 18 March 1794, credits Peter Bedford as having been a Continental lieutenant of Marines, and further stated that Bedford resigned his commission on 5 July 1779.

In December 1776, Major Samuel Nicholas was confronted with the necessity of forming a Marine detachment for the frigate *Delaware*, then fitting at Philadelphia. This occasioned a reassignment of officers appointed in June of that year. Lieutenants David Love and Daniel Henderson took command of *Delaware's* Marines. Second Lieutenant Hugh Montgomery was assigned the duty of quartermaster. To replace the latter, Peter Bedford, reputedly a friend of Captain Robert Mullan, was then commissioned a second lieutenant of Marines. His duties thereafter are not known.

A study of the Bedford families of Pennsylvania, New Jersey, and Delaware is a genealogist's nightmare. Two Gunning Bedfords, cousins, had similar careers, one calling himself "Jr." and the other "Sr." The best available evidence seems to indicate that "Gunning Bedford, Sr.," from either New Jersey or Pennsylvania, was the father of Peter Bedford. He was born in 1742 and in 1775 was a major and deputy quartermaster-general of the New York Department. As a lieutenant colonel in a Delaware regiment, he was wounded at White Plains on 28 October 1776. He was muster-master general of the Army in 1776 and 1777.

He became a prothonotary, or clerk—probably in Philadelphia—in 1779 and, after a succession of higher state and federal offices, was elected governor of Delaware in January 1796. His wife, Mary, was a sister to George Reed, a signer of the Declaration of Independence.

He settled in New Castle Hundred, Delaware and, according to the *Reconstructed 1790 Census of Delaware*, his estate extended to several other localities in the New Castle area. He died in September 1797. His cousin, Gunning Bedford, "Jr.," on the other hand, did not die until 30 March 1812, and further computations in this biography are based on the date of death of the former.

Possibly following his resignation from the Marine service in July 1779, Peter Bedford may have married Mary Vanzoling at Philadelphia's Second Presbyterian Church on 10 August 1779. A person of the same name signed the petition of one Thomas Moore in 1784 which the latter was submitting for the appointment to health officer of the Commonwealth of Pennsylvania.

Francis White's *The Philadelphia Directory* for 1785 listed Peter Bedford as a grocer on Second Street, between Market and Arch Streets, and a rival directory of the same year gave his address as 544 Second Street. One Gunning Bedford, a house carpenter—and quite possibly a brother of Peter—was also in both directories. One of them also listed "Bedford and Barclay, grocers," on Spruce Street, between Second and Front Streets.

Evidence that Gunning Bedford, "Sr.," was the father of Peter lies in the fact that on 7 May 1803, Gunning Bedford and Peter Bedford, two of the sons and executors of Gunning Bedford, Philadelphia, deceased, disposed of land on the west side of Second Street and on the north side of Plumb Street, in the Southern Liberties [Southwark], Philadelphia. The will of Gunning Bedford, dated 9 October 1795, has not been found.

Peter Bedford, the grocer, apparently died before 15 December 1818, for the Bond of Administration of his estate was filed on that date. One of his administrators was Joseph Bedford, printer. Nothing material in the inventory indicated a previous occupation as a Marine officer.

Further complicating the history of this family is the fact that in March 1780, Joshua Barney, a Continental navy officer, married Anne, the daughter of Alderman Gunning Bedford of Philadelphia. They then spent a short holiday at the residence of her brother [unnamed] in Delaware.

In April 1782, Barney commanded the privateer *Hyder Ally* and engaged the 20-gun British ship *General Monk* off Cape May. Joseph Bedford, a brother of Barney's wife, was a volunteer in his crew, stationed in the main-top, and was severely wounded in the groin.

The relationship of these persons to Peter Bedford remains uncertain.

BICKER, VICTOR, JR. Victor Bicker was from the state of New York, but his place of nativity and parents are unknown.

From 28 June 1775 to 31 March 1776, he was a second lieutenant of the 1st New York Regiment.

Frigate *Congress* was built by Lancaster Burling of Poughkeepsie, New York and was launched 29 October 1776. However, she never received her armament and was never commissioned. She was burned at Kingston, New York on 6 October 1777 to avoid capture by the British.

On 4 December 1776, the New York Committee of Safety named Victor Bicker, Jr., a lieutenant of Marines for *Congress*, directing him to recruit 30 Marines on the same terms as for other Continental troops.

The same committee, meeting at Fishkill on 27 December 1776, voted Lieutenant Bicker the sum of $200.00 for the purpose of raising his Marine detachment. He was assigned the duty of guarding both the *Congress* and the frigate *Montgomery*, which was laid down at the same time. Other ships and stores "belonging to the Continent" also came under his surveillance.

Francis Lewis, in the name of Continental Congress, then meeting in Baltimore, notified the New York Committee on 27 December that a commission would be sent for Bicker as soon as it was received from Philadelphia.

Nothing further is known of Victor Bicker, Jr.

BILL, DAVID. Ledyard Bill, in his 1867 *History of the Bill Family*, states that David Bill was born in 2 October 1751, the son of Samuel and Martha (Wheeler) Bill, of New London, Connecticut. He was a cousin of Ephraim Bill, Jr. and Gurdon Bill, both Continental Marine officers. He was married in May 1776 to Temperance Harris, by whom he had a daughter in 1777 and a son, David, in 1779. He moved to Boston, where he was purportedly living on 25 June 1776.

One David Bill, of Connecticut, was an ensign of Colonel Samuel Selden's Connecticut State Regiment from June to December 1776.

U. S. Treasury records assert that David Bill was an acting lieutenant on board frigate *Confederacy* from 12 February to 2 June 1780, when he was supposed to have been killed. However, *Confederacy* was in port at Chester and Philadelphia at that time.

Captain of Marines Gilbert Saltonstall, writing to his father, Gurdon Saltonstall, from Nantasket Roads on 14 June 1780, included a list of killed and wounded in the engagement between *Trumbull* and *Watt* on 1 June 1780, with the following notation: "David Bill, volunteer, killed."

BILL, EPHRAIM. Born 31 May 1759, Ephraim Bill was the son of Ephraim and Lydia (Huntington) Bill, Norwich, Connecticut. He was a younger brother of lieutenant of Marines Gurdon Bill.

One "Ephraim Bills" was a sergeant of Marines on the sloop *Enterprise*, Captain John Prout Sloan, on Lake Champlain in 1775. He entered on her roll 7 May and was discharged 1 July 1775. His wages per month were two pounds and eight shillings. His whole wages, approved by the Provincial Congress of Massachusetts Bay Colony and sworn to at New Haven, Connecticut, 24 July 1775, were five pounds and eight shillings.

The date that Ephraim Bill was commissioned a second lieutenant of Marines is unknown. On either 22 or 27 February 1779, he wrote to Major Joshua Huntington, Norwich, from Philadelphia, stating that he had obtained an order on the Continental Treasury for a sum of money, but that it would not be available until the following week. Out-letters of the Marine Committee appear to confirm this transaction, for in February or March 1779, Lieutenant Bill carried the sum of $500,000.00 to the Navy Board, Eastern Department, in Boston.

At the same time prior to April 1779, Ephraim Bill is believed to have entered the frigate *Confederacy*, Captain Seth Harding, outfitting at New London, Connecticut. His seniors were Captain Joseph Hardy and First Lieutenant Gurdon Bill. He was probably on the ill-fated cruise to France with John Jay and the French Minister, Alexandre Gerard, when the vessel was dismasted and put into the West Indies for repairs. *Confederacy* reached Philadelphia for more permanent repairs in the following spring.

For most of the summer of 1780, she awaited repairs. Lieutenant Ephraim Bill occupied some of his time on messenger duty. One such trip took him to New London, carrying a letter to Nathaniel Shaw, the Continental agent there, directing him to sell properties he was holding and to collect all money possible for fitting *Confederacy*.

The length of his duty on board the frigate is unknown. His biographer, Ledyard Bill, asserted that he died at sea in November 1780. If so, he must have been on duty in a different ship, for *Confederacy* was still in Philadelphia. Ephraim Bill was not—as was his brother, Gurdon—mentioned in the will of his father, dated 7 October 1800.

BILL, GURDON. Gurdon Bill, one of the sons of Ephraim and Lydia (Huntington) Bill, was born in Norwich, Connecticut 26 August 1757. His father and his maternal grandfather—Major Joshua Huntington—were instrumental in construction of the Connecticut ships *Defence* and *Oliver Cromwell* and of the Continental frigate, *Confederacy*.

In December 1778, Gurdon Bill applied to Major Huntington for the position of lieutenant of Marines of *Confederacy* or, in case it was already filled, that of steward. Despite the fact that some of his early letters were signed as a lieutenant of Marines, he apparently acted in the capacity of steward or purser from 18 January to 4 February 1779, while the vessel was fitting out at New London.

On 5 January 1779, he refuted in a letter to a newspaper a rumor that the commissioned officers of the ship were discontented with their captain, Seth Harding. He signed it as lieutenant of Marines.

In a muster roll, "Officers and Crew in 1779 . . ." Gurdon Bill and Samuel Holt headed the list of "Men's Names," but the rank of neither was entered.

Gurdon Bill was on board *Confederacy* when she departed New London on her maiden voyage on 1 May 1779 and captured the British frigate *Pole*, 24 guns, and two smaller vessels on 6 June.

He was still on board during the unfortunate cruise beginning 26 October 1779. Enroute to Europe with John Jay and French Minister Alexandre Gerard, she was dismasted and had to put into the West Indies for a jury rig. After considerable difficulty, she again reached Philadelphia on 27 April 1780, where she remained for the better part of the summer.

Lieutenant Gurdon Bill was apparently on board when *Confederacy* began her last cruise on 5 December 1780, in company with *Saratoga*. Their destination was the West Indies, to convoy merchant vessels to the United States. With *Saratoga* and frigate *Deane*, she departed Cape François on 15 March 1781.

On the morning of 14 April, *Confederacy* found herself alone with the merchant ships off the Delaware Capes. She was soon confronted by two large British ships, the *Roebuck*, 44 guns, and *Orpheus*, 32 guns, whose overwhelming firepower induced Captain Harding to surrender without unnecessary bloodshed.

One secondary source relates that all the officers were paroled from New York and that the enlisted men were confined in the *Jersey* prison ship. A prominent primary source, however, asserts that at least some of the officers were sent from New York to prison in England.

Apparently Lieutenant Gurdon Bill was one of the latter. He probably escaped from prison in England, or was among those fortunate few who were paroled and exchanged to France. There, on 6 July 1782, he received from the U. S. Commissioners at Paris the sum of $120.00 to defray his expenses to America. No further service of Gurdon Bill has been found.

Bill married Betsey Tracy, also of Norwich, on 23 December 1797. She was born 20 March 1781 and died 31 July 1847. They had eight children between 1799 and 1814.

Following the Revolution, he continued to go to sea, primarily in merchant vessels to London and the Continent. One of his ships, the schooner *Hannah*, was seized by the French in the Naval War with France. In 1801, at the solicitation of his wife, he abandoned seafaring life.

The family genealogist, Ledyard Bill, asserted that Gurdon Bill was a lieutenant of Marines in 1798, but no proof of this allegation has been found. Further he was supposed to have been a favorite of Captain Thomas Truxton. Bill's eldest son's middle name was "Truxtun."

He was among 15 Norwich townsmen who organized the Chelsea Grammar School in 1806, which was incorporated in 1821. He was admitted to membership in the Society of the Cincinnati on 7 July 1790. He died in Norwich on either 4 or 6 March 1815.

BOYCE [BOYS], ABRAHAM [ABRA'M]. Although Abraham Boyce was reported in Doyle Sweeny's 1794 Treasury List as both a lieutenant and captain of Marines, some of his service is a matter of conjecture. He may have been from New Jersey.

In the spring of 1776, the schooners *Wasp* and *Hornet* were ordered by Continental Congress to harass the British frigates *Roebuck* and *Liverpool* in the lower Delaware Bay. In addition, the 16-gun brig *Lexington* was purchased in March in Philadelphia, put under the command of Captain John Barry and ordered to the same duty.

Whether *Lexington* had an officer of Marines when she engaged the sloop *Edward*, tender to *Liverpool*, in Delaware Bay on 7 April 1776, is not certain. It was not until 8 April that Congress notified Captain Barry that a vessel enroute to Cape May would "put on board your Lieutt of Marines and some men." Also, Congress informed him on 11 April that "Capt. Boys with four or five Marines went down to Cape May in the Schooner *Wasp*, where they were to land if you did not fall in with them consequently you will take them off." The "Lieutt" and "Capt. Boys" may have been the same person. On the other hand, Abel Morgan was identified as *Lexington's* lieutenant of Marines about this same time. Apparently, *Lexington* continued to cruise in the vicinity of the Delaware Capes for the remainder of the year.

In October 1776, *Lexington*, then under the command of Captain William Hallock, was sent to the West Indies on secret business of Congress. On her return in December she was captured by the British frigate *Pearl*. Robert Morris informed John Hancock on 2 January 1777 that Captain Boyce reported the incident to him at Baltimore. Only the Navy officers of *Lexington* were removed from her, and *Pearl* put aboard only seven men as a prize crew. *Lexington's* crew rose under the command of Captain Boyce, retook the brig and sailed her into Baltimore.

Further details of the career and life of Abraham Boyce are scarce. A return of men enlisted by Captain Robert Mullan given to Major Samuel Nicholas, dated 9 August 1779 to 1 January 1780, named two Marines "on Board *Confederacy* with Captn. Boyce." This entry may possibly have been in error, for *Confederacy* was in the later portion of this period refitting in the West Indies. At the time, her captain of Marines was Joseph Hardy, and Gurdon Bill and Samuel Holt were her lieutenants of Marines.

A descendant who was a member of the Daughters of the American Revolution identified Abraham Boyce simply as a "soldier in the Revolution."

BROWN, PEREGRINE. The vital statistics of Peregrine Brown are unknown, but he may have been from Philadelphia.

On 25 June 1776, Brown was one of four men commissioned a first lieutenant of Marines by the Continental Congress. He was one of two lieutenants designated for the frigate *Washington*. Captain Benjamin Dean was his senior officer.

Washington was built by Manuel, Jehu, and Benjamin Eyre in Kensington, Pennsylvania and was launched 7 August 1776, but she was never commissioned. She was scuttled off Whitehall, New Jersey on 2 November 1777 to prevent her capture by the British.

It may be that Brown and his Marines performed guard over the *Washington*. He may also have gone into the Jerseys with Captain Dean's company and remained with the artillery at Morristown until 1 April 1777. However, these are merely suppositions, for nothing is known of him after the date of his commission.

BUBIER [BEUBIER, BOOBYER, BOUBIER], WILLIAM. Baptized 30 March 1746 at Marblehead, Massachusetts, William Bubier was the son of Christopher and Margaret [Margarett] (LeVallier) Bubier [Boobyer]. With his brother-in-law, Captain Thomas Grant, Bubier was a goldsmith. He apparently also traded in land, for in 1775 he purchased about 200 acres from the town of Windham, Cumberland County, Massachusetts [now Maine].

Bubier married 19 June 1770, at Marblehead, Deborah, daughter of Joseph and Elizabeth Howard. She was born 10 December 1747 and died at Marblehead 18 September 1808, aged 62 years. They had two daughters. Elizabeth, who was baptized 18 November 1770, died a widow at Providence, Rhode Island, in June 1837. Her sister, Deborah, baptized 21 June 1772, died single at Salem, Massachusetts, 12 November 1857.

Bubier enlisted 24 April 1775 in Captain Thomas Grant's 3d Company, Colonel John Glover's 21st (Essex County) Regiment. The unit entered Continental service about 22 May 1775. He was reported commissioned a first lieutenant of Glover's regiment on 22 June 1775 and served in the Charles River, Cambridge. His boat was commanded by William Blackler, and, as they were firing at the enemy, a gun "busted," killing and wounding several of their own. Bubier was also a first lieutenant in the 14th Continental Infantry from 1 January to December 1776.

The date of Bubier's commission as a lieutenant of Marines is unknown, but he was the junior officer of frigate *Hancock*, Captain John Manley, when she sailed from Boston on 21 May 1777 with the frigate *Boston*. He participated in the capture of the British frigate *Fox* on 7 June 1777. However, he was captured in *Hancock* by the British 44-gun ship *Rainbow* on 8 July and imprisoned in Halifax.

On 9 November 1777, Lieutenant Bubier was among 50 American prisoners sent from Halifax to Boston in the cartel *Swift*. Unfortunately he was taken ill and died five days after reaching Marblehead.

His widow, Deborah, was appointed administratrix of his estate in September 1792.

BURKE, EDWARD. No information is available on either the origin or the later life of Edward Burke.

He was probably on board the ship *Columbus*, Captain Abraham Whipple, on the expedition to New Providence in early 1776.

On *Columbus'* return to New London, she lost most of her Marine detachment when all her Marine officers resigned. Lieutenants Cummings (or Cumming) and Trevett had to get an order from Esek Hopkins to obtain their clothes from the vessel. Lieutenant Matthew Parke, of *Alfred*, was promoted to captain and transferred to *Columbus*.

To replace one of the resigned lieutenants, sergeant of Marines Edward Burke was promoted to lieutenant of Marines.

CARR [KERR], JOHN. Details of the private life of Lieutenant of Marines John Carr are wanting. He should not be confused, however, with Captain John Carr, of Rhode Island, a Continental Army officer.

He is purported to have been appointed a lieutenant of Marines in the brig *Cabot* at Providence, Rhode Island on 30 April 1776, replacing Lieutenant of Marines James Hood Wilson, who was killed in the action with *Glasgow* on 6 April. Carr's senior Marine officer was Captain John Welsh.

Cabot made a short and uneventful cruise in May 1776. She went to sea again in early 1777 and in March, while cruising off the coast of Nova Scotia, was chased by the British frigate *Milford*. Her captain ran her ashore and escaped with his crew.

Lieutenant Carr's activities for some time after are unknown. It may have been that he was in some manner captured by the British and imprisoned in England. He may have escaped and made his way to France, or possibly he was exchanged.

He joined the brig *Lexington* at Bordeaux, France as a lieutenant of Marines on 8 May 1777, but his service was of short duration. He was discharged at Nantes, France on 22 May 1777, leaving only Lieutenant of Marines James Connelly and one enlisted Marine on board.

CATLIN, BENJAMIN. Benjamin Catlin was possibly from Connecticut.

He was listed as a lieutenant of Marines by Sweeny, Allen and Paullin, but none cited a source for this information.

One Benjamin Catlin, of Connecticut, was a corporal in the Lexington Alarm in April 1775 and was a sergeant in one Captain Chester's Company in the same year. Shortly after, he was a quartermaster in Arnold's expedition to Canada and was taken prisoner at Quebec, 31 December 1775. He was exchanged 10 January 1777.

If this Benjamin Catlin is one and the same, he was exchanged in time to serve in frigate *Trumbull*. Congress ordered two frigates to be built in Connecticut and, early in 1777, the Council of that state directed *Trumbull* to be built at Chatham, on the Connecticut River. She was commissioned in 1780.

Meanwhile officers and crew were being recruited by her first commander, Captain Dudley Saltonstall. Appended to a muster roll in the Connecticut State Archives, in the main body of which all naval rates were specified, was "A List of Names Entered on Board the *Trumbull*, by Benjamin Catlin." Eleven men were recruited by him between 5 March and 20 April 1777, but none were actually named as Marines.

If indeed Benjamin Catlin was one of the earlier lieutenants of Marines of *Trumbull*, further service of his is unknown. He was not on board her during the engagement of the frigate in the stand-off battle with HMS *Watt* on 1 June 1780.

CHAPIN, SETH. Seth Chapin was born 31 March 1746 in Mendon, Massachusetts, the son of Ensign Ebenezer and Abigail (Perry) (Wood) Chapin.

At the age of 30, Seth Chapin was a corporal in Captain John Albee's 1st Company, which marched from Mendon to the Lexington Alarm on 19 April 1775.

On 24 June 1776, he was appointed a second lieutenant of Marines by the recommendation of the Committee appointed to build two Continental frigates in Rhode Island. The Committee directed Lieutenant Chapin to go to Plymouth and elsewhere to "Interest as many Seamen and Marines as he can . . . ," and supplied him funds for that purpose.

He was subsequently assigned to the frigate *Providence*, built at Rhode Island, and on 14 September 1776 was instructed to enlist recruits for the ship.

According to a deposition in his pension records, one Stephen Earle was a sergeant on board *Providence* in July 1776. Earle asserted that Lieutenant Chapin was "broke" about December 1776 and that he [Earle] was advanced to his second lieutenancy. However, another source credits Second Lieutenant Chapin with having been a member of the court martial of Navy Lieutenant Richard Marvin on board frigate *Providence*, off Field's Point in the Providence River on 3 April 1777.

Nevertheless, Chapin did leave the Marine service. He was appointed a first lieutenant on 19 July 1777 in Captain James Webb's Company, Colonel Henry Sherburne's Battalion of Rhode Island troops. In July 1780, he was appointed captain in a three-month regiment raised by the Rhode Island Legislature.

Chapin, who possessed the title "Deacon" in his later years, married three times, first, on 10 September 1765, to Ruth Bullard, of Holliston, Massachusetts, who bore him no children and died shortly after. Second, on 27 October 1767, he married Elizabeth Rawson, daughter of Deacon Edward and Mary (Morse) Rawson, Mendon, Massachusetts, by whom he had four children. She died 17 November 1778. He married, third, Eunice Thompson, daughter of Moses Thompson, of Medway, Massachusetts, and she bore him an additional seven children. She died 27 June 1821.

Deacon Seth Chapin died in Mendon, 15 November 1833, aged 87.

CHILTON, JOHN. The origins of John Chilton are unknown.

He was commissioned a first lieutenant of Marines before 20 January 1776. On that day, he wrote to the Committee of Safety at Philadelphia, thanking them for the appointment:

> I apprehend the best return I can make for this Honour conferred on, and Trust reposed in me, is, to perform my Duty, which I hereby promise to do to the utmost of my abilities, and that no Earthly consideration shall make me swerve from it, ever keeping in view the Glorious cause of Liberty, that cause I now take up arms to defend.

Chilton's duties for the following three years are uncertain. On 31 July 1778, one John Chilton, a midshipman in frigate *Raleigh*, was reported to have attempted to recruit British soldiers from Prospect Hill at Cambridge, Massachusetts, for the ship. The matter was reported by Town Sergeant John Rice to Major General William Heath on the following day.

Lieutenant Chilton was apparently on board sloop *Providence*, Captain Hoysteed Hacker, when the vessel was ordered by the Massachusetts Council in April 1779 to cruise the bay for small enemy cruisers. Following this duty, *Providence* was sent to Delaware Bay to take on a supply of bread. Enroute, on 7 May, *Providence* engaged the British brig *Diligent*, 12 guns, off Sandy Hook. *Providence* had 12 wounded and two killed, among the latter Lieutenant of Marines John Chilton. Shortly after, he was replaced by Lieutenant Robert Davis.

CLARK, LEMUEL. Lemuel Clark was born about 1755, but his parents are not known. At the beginning of the American Revolution, he was living in Quincy, Norfolk County, Massachusetts.

His first service was in the troops of the Commonwealth of Massachusetts. He enlisted in May 1775 with a company of troops guarding the shores south of Boston. In May 1776, he commanded a unit that guarded stores at Watertown, Massachusetts. Between October 1776 and April 1777, he was first a sergeant and later adjutant in another state regiment. Following this tour, he served three months in the regiment of one Colonel Titcomb. And, finally, he was a lieutenant for a short period of time in the Massachusetts regiment of a Colonel Gill.

Clark's service as a lieutenant of Marines is qualified only in his pension application, in depositions dated 6 April 1818 and 19 June 1820. His written recollections in these two accounts vary by one year.

The second declaration conforms most closely to the movements of the Continental vessels concerned. He certified that he received his lieutenant's commission in January 1777, and was ordered to put himself under the command of Captain Seth Baxter, then recruiting for the frigate *Hancock*. It is interesting to note that Baxter himself was a resident of both Quincy and Braintree, Massachusetts. He may have been instrumental in the commissioning of Clark.

Before *Hancock* and *Boston* sailed on 21 May 1777, the deposition continues, he purportedly took command of the Marines on *Boston*, in the place of Captain Richard Palmes, who Clark reported "was left sick."

If so, he would have been senior to Marine Lieutenants John Harris and Robert McNeill. It is possible, although no muster roll has been found for *Boston* in that period, that he served in some other capacity at the time.

However, Captain Palmes did sail on the cruise. When the British frigate *Fox* was captured on 7 June 1777, Lieutenants Harris and McNeill were part of her prize crew, but were recaptured and carried prisoners into Halifax. At this time, Palmes or the captain of the ship may have elevated Clark to lieutenant of Marines for the remainder of the cruise.

Clark's personal knowledge of the events on board *Boston* that followed—cited 41 years later—leads one to believe that he was indeed aboard the vessel.

He further deposed that on the return of the frigate to Boston in November 1777, he was ordered to recruit men for the frigate *Deane*. However, because of a surplus of Marine officers for *Deane*, he was transferred to the sloop *Providence* in which he served until February 1779.

Reverting to his first deposition, he asserted that in March 1779 he became lieutenant of Marines of the Massachusetts brig *Tyrannicide*, which was subsequently destroyed at the fiasco of Penobscot.

Later, as a lieutenant of Marines of the privateer *Essex*, he was captured in June 1780 and imprisoned in England. He was released and returned to America shortly before the end of the war. The Secretary at War ordered him to settle his accounts with the Continental agent at Boston and to hold himself in readiness to go on board another vessel.

He never received further orders. Nor was he discharged. He was placed on the federal pension roll in 1821 at the sum of $240.00 per annum, retroactive to 6 April 1818. The amount of his pension was increased to $320.00 on 4 March 1831.

Clark wrote to the Honorable Lewis Cass in 1832 from his residence in Barre, Washington County, Vermont, asking for his back pay.

Lemuel Clark died 12 April 1834. Among his children was a son, Soper. He purportedly had another son, Lemuel, Jr., for which the Navy Department requested a commission as a lieutenant of Marines on 6 July 1798. He was to serve on board the USS *Constitution* during the Naval War with France. He was commissioned a captain of Marines on 3 August 1798 and commanded the Marines at Boston when he resigned his commission 30 November 1801.

One Lemuel Clark, probably the latter, aged 89, was the head of his own family at Southbridge, Worcester County, Massachusetts, on 1 June 1840.

COAKLEY [COKELY, COKELEY], JAMES. James Coakley's nativity and parentage are unknown. However, he enlisted as a sergeant of Marines in the company of Captain Robert Mullan in Philadelphia on 1 July 1776 for service on the frigate *Delaware*.

Early in December, however, Mullan's company boarded gondolas and proceeded up the Delaware River to Trenton to reinforce General Washington. In the middle of the month, Robert Morris, of Continental Congress, wrote Washington asking that the captain and seamen of *Delaware* be returned to Philadelphia. Major Samuel Nicholas assembled a 20-man detachment from the three Marine companies under Colonel Cadwalader. They were placed under the command of Lieutenants David Love and Daniel Henderson.

Sergeant Coakley was temporarily promoted to first lieutenant to replace Love, and apparently accompanied Washington in the battles of Trenton and Princeton and went into winter quarters at Morristown.

In February 1777, Mullan's company was relieved of artillery duty at Morristown and returned to the barracks in Philadelphia, ready to board *Delaware* on her completion. Lieutenant Coakley was purportedly reassigned to the frigate *Effingham*.

Effingham was laid down in Southwark in 1776 and was launched in the same year. She was never commissioned nor her armament mounted. In September 1777, Lieutenant Coakley went on board with his detachment and proceeded to a new berth at Burlington, New Jersey. She was scuttled off Whitehill on 2 November 1777.

Lieutenant Coakley and his Marines probably remained at the Middle District, Eastern Navy Board, Bordentown, New Jersey. He was a member of a court martial of prisoners of the ship *Repulse*, held 25 November 1777 on board the ship *Lion*. Perhaps both of these ships belonged to the Pennsylvania Navy or were privateers.

On or about 8 March 1778, Lieutenant Coakley was placed under the command of Captain John Barry, who officered a boat detachment of 28 men which made a spectacular capture of a British schooner off Port Penn, south of Philadelphia. The schooner was carried by boarding, 6 officers, 10 British soldiers, and 100 seamen and Marines were captured.

Nothing further is known of the career or life of Lieutenant Coakley although Doyle Sweeny in his 1794 Treasury report asserted that he resigned his commission.

CONNELLY [CONNOLLY], JAMES. James Connelly is reputed to have been a native of Ireland, but may have more recently been a resident of one of the colonies.

A muster roll of *Lexington*, dated March 1777, named James Connelly as a lieutenant of Marines. It must be surmised that he joined the vessel prior to this time, for the ship departed Baltimore on 27 February 1777 and arrived in France in April.

John Carr (or Kerr) was also a lieutenant of Marines in *Lexington*, joining 8 May, but departing 22 May 1777, leaving only Lieutenant Connelly and one enlisted Marine aboard.

Lexington sailed from St. Nazaire, France with *Reprisal* and *Dolphin* on 28 May 1777. Shortly after, they were chased by *Foudroyant*, 84 guns, but all escaped and cruised in the Irish Sea for prizes. They arrived back in France in July 1777.

British pressure on the French Government soon forced both *Lexington* and *Reprisal* to depart for America. *Reprisal* left St. Malo and *Lexington* sailed for Morlaix, hoping to meet at sea for the hazardous voyage homeward, but failed. *Reprisal* disappeared at sea, foundering with all hands, off the Banks of Newfoundland.

Lexington fell in with the British 10-gun cutter *Alert* on 19 September 1777. *Alert* chased the American ship for several hours, and after an engagement of only one hour *Lexington* struck her colors. Richard Dale, one of her Navy officers, attributed a lack of ammunition to her surrender.

Lexington lost 11 wounded and seven killed. One of the latter was Lieutenant of Marines James Connelly.

COOPER, WILLIAM. The birth and parentage of William Cooper, Continental Marine officer, have not been proved. He may have been the son of William Cooper, Town Clerk of Boston from 1761 to 1809, and of Katharine (Wendell) Cooper. They had a son who was born in February 1750 and who drowned in Passamaquoddy Bay, 7 February 1788.

There may be some substance in the assumption that William Cooper of Boston was an acquaintance of Captain of Marines Matthew Parke. If so, he may have been a brother of Judith Cooper, who married Captain Parke on 15 August 1781. Furthermore, Parke may have been influential in persuading Cooper to enter the Continental Marines.

Second Lieutenant William Cooper entered frigate *Boston*, Captain Samuel Tucker, on 28 March 1779. On 6 June of the same year, *Boston* sailed with *Confederacy*, and soon after captured the British frigate *Pole*, the schooner *Betsy*, and the sloop *William*. Lieutenant Cooper was ordered to the *Pole* as a member of her prize crew, and *Confederacy* escorted the three ships into port.

No further service is known to be credited to Lieutenant Cooper.

On 23 March 1780, one William Cooper married Rebecca Jenkins.

CRAIG, ISAAC. Born in Hillsborough, County Down, Ireland about August 1741, Isaac Craig is reported to have moved to America in 1765 or 1767. Although he was reported to have had only two brothers—Robert and Joseph Patterson—a James Craig was an occupant of the same house and had the same occupation as Isaac in 1775. On his arrival in Philadelphia, he carried on the trade in which he had served his apprenticeship in Ireland and became a master carpenter and cabinetmaker. He resided in the south part of the Dock Ward.

One source asserted that he had had a militia command early in the Revolutionary War, but this has not been verified.

At the organization of the Continental Marines, Craig was commissioned a first lieutenant on 29 November 1775, for service on board the converted brig *Andrew Doria*, Captain Nicholas Biddle. He entered her with his Marines in December and sailed 17 February 1776 for the Bahamas, where he participated in the first raid on New Providence in March.

On the return voyage, *Andrew Doria* fell in with HMS *Glasgow* and assisted in beating her off in a sharp engagement on 6 April 1776. Two days later, she reached port at New London where she was blockaded by a British squadron.

Craig remained on board *Andrew Doria* until at least 21 September 1776 when, as a first lieutenant, he sat on the court martial of Navy lieutenant Luke Matthewman of the brig *Lexington*.

He was commissioned a captain of Marines for the Continental galley *Champion* on 22 October 1776. On board her, he made only one short cruise, returning in November.

General Washington's desperate need for troops in his New Jersey campaign led to the sending of a small battalion of Marines under the command of Major Samuel Nicholas, with Captain Craig as adjutant. On 27 December the Marines joined Colonel John Cadwalader at Burlington, New Jersey. As a member of the battalion, Craig participated in the battles of Assunpink (second battle of Trenton) and of Princeton, 2 and 3 January 1777.

When Army artillery units departed from Washington's forces in February 1777, Captain Craig and his Marine company were ordered to serve the artillery pieces at Morristown.

On 3 March, Craig accepted a commission of captain in Colonel Thomas Proctor's Pennsylvania Regiment of Artillery (later 4th Continental Artillery). He asked to vacate his Marine commission, and his resignation was accepted.

With the Army, he participated in the battles of Brandywine on 11 September 1777 and Germantown on 4 October 1777. Shortly after, he was selected to be one of a small group of officers who were ordered to Carlisle, Pennsylvania for instruction in munitions laboratory techniques. He was on this assignment from 1 February to 1 August 1778.

He was back in Philadelphia on 18 March 1779, writing a letter to General Washington in support of his claim for promotion. In April he was located at Billingsport, in command of a fort on the Delaware River. He also served in Proctor's Artillery with General John Sullivan's army in the campaign against the Six Indian Nations in the Genesee River country of New York. That campaign lasted from July into September.

Craig was back with Washington's army at Morristown by January 1780. About 20 April he was ordered to Fort Pitt (now Pittsburgh) with artillery and military stores, arriving there 29 May. The next year, between July and November, he accompanied Geoge Rogers Clark down the Ohio River to the Falls of the Ohio (now Louisville, Kentucky). It was during this time that he received his promotion to major, dating from 7 October 1781. He returned to Fort Pitt on 26 December 1781. Between 13 November and 2 December 1782, he commanded a scouting party to Cayahoga and the Grand River. His military service in the Revolution concluded when he left the military on 17 June 1783.

Craig had cast his lot with the development of Pittsburgh. In 1782, while still in the Army, he had become the Worshipful Master of his Masonic Lodge, Military No. 19, which had moved to Pittsburgh. He bought the first land sold in the city on 22 January 1784, together with Stephen A. Bayard, who became his business partner.

A year later, in February 1785, he married Amelia Neville, daughter of John Neville and the only sister of Pressley Neville. His son, Neville B. Craig, was born 29 March 1787 in the Redoubt (the blockhouse built by Colonel Henry Bouquet), located near the Point in Pittsburgh.

Unexpected recognition came in May 1787 when he was elected a member of the prestigious American Philosophical Society. Born and reared a Presbyterian, it was also fitting that he became a trustee of the newly incorporated Presbyterian Congregation of Pittsburgh in September 1787.

In February 1791, Craig was offered the post of deputy quartermaster and military storekeeper at Pittsburgh by the infant American Army. A retaliation attack by a Captain Brady and some of his Virginia Rangers on a party of Indians near Fort Pitt received Craig's support in a letter to the Secretary of War on 16 March 1791.

He declined an appointment in November 1794 as Commissary General in the Army of Major General Anthony Wayne for a campaign against the Indians of the Northwest Territory.

Continuing his business ventures, in 1797 he joined with James O'Hara in establishing the first glass works erected west of the Allegheny Mountains.

Troubles with France in 1798 led to steps directed against Spain, ally of France. The Federal Government began preparations for control of the lower Mississippi. Issac Craig was appointed superintendent for the construction of two row galleys, the *President Adams* and *Senator Ross*, at Pittsburgh.

Although now quite elderly, Craig served during the War of 1812, preparing munitions for the Northwestern Army, a technique he had learned at Carlisle in 1778.

Isaac Craig died 14 May 1826 at his home on Montour's Island in the Ohio River, at the age of 85. First buried in the First Presbyterian Graveyard, Pittsburgh, his remains were moved to the Allegheny Cemetery on 23 October 1902.

He was an original member of the Society of the Cincinnati of Pennsylvania.

CUMMINGS [CUMMING], ROBERT. The parents, birth and birthplace of Robert Cummings are unknown.

According to the papers of Commodore Esek Hopkins, Robert Cummings was a lieutenant of Marines in the ship *Columbus* when she departed Philadelphia in February 1776 for the expedition to New Providence. His senior officer was Captain Joseph Shoemaker and John Trevett was *Columbus'* first lieutenant of Marines.

When *Columbus* returned to Providence, Rhode Island in April 1776, Lieutenants Cummings and Trevett appealed to Commodore Hopkins for permission to "get our cloathes, etc., from on board the ship Columbus." Hopkins addressed a short note to Captain Abraham Whipple directing him to allow the Marines to obtain their property.

A muster roll of *Columbus*, dated from her commissioning to 14 December 1776, asserted that Cummings had "deserted," but it is believed that he resigned his commission. Captain Shoemaker and Lieutenant Trevett were also listed as deserters, but Shoemaker also resigned, and Trevett transferred to the brig *Andrew Doria*.

DAVIS, ROBERT. The British brig, *Diligent*, 12 guns, was captured on 7 May 1779 by the sloop *Providence* off Sandy Hook. Lieutenant of Marines John Chilton was killed in the engagement. *Providence* put into New Bedford, Massachusetts for repairs, and Lieutenant Chilton's place was taken by Robert Davis.

One Robert Davis, of Massachusetts, had been a second lieutenant in the 15th Continental Infantry on 1 January 1776. He was promoted to first lieutenant of the 1st Massachusetts Infantry on 1 January 1777 and to captain on 4 November 1777. He resigned his commission 1 April 1779.

He may have been the same Robert Davis "of the continental army" whom Sergeant Thomas Philbrook of sloop *Providence* wrote of in 1832. Davis landed from *Providence* with 30 Marines on Banks Island in Penobscot Bay at sunset, 11 August 1779.

Davis' further career as a Continental Marine officer is uncertain. However, on 5 January 1780, one Robert Davis was identified as a "Volunteer Officer" of frigate *Providence*. With Midshipmen Thomas Bowen and Henry Young, and Samuel Hacker, another volunteer, he was granted parole from New Providence, British Bahama Islands, by Governor Montfort Browne.

It was asserted that they had been taken in the frigate *Providence*, but the frigate had lost no men in battle and was, in fact, not captured until the fall of Charleston, South Carolina on 12 May 1780.

She had sailed from Nantasket Roads on 23 November 1779, bound for Charleston. On 5 December she captured the privateer brig *Dolphin*, manned her with a prize crew, and sent her into Charleston.

One can only assume that Davis and his companions were members of the prize crew of *Dolphin*, which was perhaps recaptured by the British and carried into New Providence.

DAYTON, HENRY. Henry Dayton was probably from Newport, Rhode Island.

The date of his commission as a lieutenant of Marines is in doubt. Pehaps it is permissable to deduce that he may have joined John Trevett, also of Newport, when the latter "sailed from Providence in a sloop called the Catea [*Katy*] commanded by Abra[ha]m Whipple, Esq., of Providence, with a number of passengers, to sail with a fleet of armed vessels fixing at Philadelphia in the month of Nov. 1775."

Trevett, however, was commissioned first lieutenant of Marines in the ship *Columbus* for the New Providence expedition, and Dayton was first lieutenant of Marines in the sloop *Providence* [ex-*Catea* or *Katy*], Captain John Hazard.

Trevett, in describing his experiences ashore at New Providence, twice identified Dayton as "Captain," who was commanding one company of Marines. It is fairly obvious that his journal was written after the fact, for Captain Samuel Nicholas is also referred to therein as "Major." Trevett contended that he and "Captain" Dayton, acting on their own volition, took Governor Montfort Browne prisoner at Nassau.

Following the return of the fleet to Providence, Lieutenant Dayton served on two courts martial. The first was that of Captain Abraham Whipple on 6 May 1776, and the other, on 8 May, for the trial of Captain John Hazard.

Although the service of Lieutenant Dayton following the expedition is relatively unknown, Lieutenant Trevett is known to have been on board *Providence* from 24 October 1776 through February 1777.

On 7 January 1777, Commodore Esek Hapkins ordered the payment of 30 pounds and 13 shillings to Lieutenant Dayton as prize money.

Trevett's journal, dated January 1777, gives some detail of an expedition by the crew of the *Providence* to destroy the British frigate *Diamond* aground at Warwick Neck. It was a failure, but he wrote that at the time "Com. Hopkins sent over Capt. Henry Dayton to me, for to come to Providence immediately. . . ." Presumably Lieutenant Dayton

joined ship *Providence* at Providence, while Trevett led a party of two midshipmen on a short spying mission to Newport on board a cartel.

While there, in disguise and under a pretext, Trevett managed to get ashore in the vicinity of his home near the Long Wharf. There he sought the assistance of a Mrs. Battey, of Long Wharf, describing the conversation as follows:

> I goes to Mrs. Battey on the Long Wharfe finding She was A Lone and Making a fire I new her well I asked Mrs. Battey to Le[nd] Me a Funel She Anserd Very Short no as no Soul was [near] I Goes up to her and Told her She Shoul[d] Lend one S[he] New My Mode of Speking and Says for God Sake Whare are You from I informed her from Providence She Says How Did You Leve My Son Mening Capt. Henry Dayton I informed her well.

Trevett also recorded in his journal that several months before he had picked up Mrs. Battey's husband as a Tory, and he was at that time a pilot for the British at Newport.

No further mention is made of Dayton as a Continental Marine officer, but one Henry Dayton was a captain of the Rhode Island militia in 1779.

In 1774, a Henry Dayton of Newport, had a family of two females. In 1790, he was the head of a family of two males and five females.

Several persons of the surname "Battey" and "Batty" resided in Rhode Island, mostly in Newport and Providence Counties, both in 1774 and in 1790.

DE LA FALCONNIERE [FALCONIER], PANATIERE [PANATIER]. Panatiere de la Falconniere was a Frenchman from the West Indies who, for unknown reasons, was in Philadelphia in September 1776. One prominent source asserts that he "memorialized Congress and wheedled the Marine Committee into appointing him a second lieutenant in Captain [Samuel] Shaw's company." Shaw's Marines were then in barracks in the Northern Liberties, a suburb of Philadelphia.

Frigate *Randolph*, Captain Nicholas Biddle, was then being outfitted at the Kensington shipyard, and the Marines were there guarding her stores. It is probable that de la Falconniere went on board the frigate late in 1776, for she went to sea in February of the following year. Franklin Reid, reported to have been from Philadelphia, was *Randolph*'s other lieutenant of Marines.

No sooner had *Randolph* put to sea on her first cruise, seeking the British frigate *Milford*, than she sprung her foremast and put into Charleston, South Carolina for repairs.

Several occasions of past misconduct of Lieutenant de la Falconniere came to a head while repairs were being made. As a consequence, six officers of the ship signed a petition to Captain Biddle with several charges against the lieutenant. Biddle composed a letter to the Marine Committee and enclosed the charges of his other officers. On the following morning, 1 September 1777, he discharged de la Falconniere from the ship before again putting to sea. Whether de la Falconniere ever reported to Congress is unknown.

DE LA VALLETTE [DELAVOLLETTE], LOUIS [LEWIS]. He is presumed to have been a French citizen, but little else is known of him.

He was an officer of the sloop *Ariel*, Captain John Paul Jones, which was obtained from the French after Jones was abandoned in France by Captain Pierre Landais on 8 July 1780.

Ariel sailed from L'Orient for America on December 1780. A list of officers, men and passengers sent by Jones to Franklin on that date rates De la Vallette as the captain's clerk. *Ariel* arrived in Philadelphia with a load of military supplies on 18 February 1781.

On the 15th of August 1781, a report was taken under consideration by the Board of War regarding the pay of certain individuals on board *Ariel*. When the report passed on 25 August, Louis De la Vallette was named a "Lieut of Marines," and was paid $289.36 for his services to that date.

It is not known whether he saw further service in the Continental Marines.

DEAN [DEANE], BENJAMIN. Benjamin Dean is believed to have been a native of Philadelphia, but his vital statistics are unknown. Several families of the surname "Dean" and "Deane" were residents there during and after the Revolution, but their relationship, if any, is unknown. On 31 March 1775, Benjamin Dean, living in the Mulberry Ward, was assessed a provincial tax of 15 shillings.

On 25 June 1776, when the Continental Congress appointed new officers to command the Marine detachments of four frigates being built in Philadelphia, Benjamin Dean was designated captain of Marines on board the 32-gun *Washington*. Lieutenants Peregrine Brown and Abel Morgan were his subordinate officers.

Washington was built at Kensington, launched 7 August 1776, but was never commissioned. She was sunk by the Americans in November 1777 in the Delaware River off New Jersey to prevent her capture.

Captain Dean commanded one of the companies of Major Samuel Nicholas' battalion which accompanied General Washington into New Jersey in December 1776. He also accompanied General Washington to the winter encampment at Morristown. When the regular enlistment of Washington's artillery expired in January 1777, the Marine companies took over the duties. Dean's company served until 1 April 1777, but was not paid for the duty until August of that year.

Early in April the company returned to Philadelphia for duty on board the frigate *Washington*. She was still unprepared for sea, however, and the enlistment of the company expired in June. Dean apparently became discouraged and resigned his commission in July 1777.

Nothing further is known of the life of Dean, but one Benjamin Deane, of Philadelphia, departing for a voyage to Europe, made his will on 20 December 1780. It was probated 30 November 1781. Among the executors and administrators were his brothers, John and William. An inventory of his estate included "1 Steel mounted small sword . . . 1 Bayonet . . . [and] Simes Military Guide."

MacPherson's Directory of Philadelphia, 1785, listed two widows with the surname "Dean."

One Dean family, with similar given names, sold a house and lot on the west side of Fourth Street, between High and Chestnut Streets in 1811.

Final administration of the estate of Benjamin Dean did not take place until 1826, when he was identified as a hatter and also had an interest in a brewery in Springfield, Kentucky.

DEVOL [DEVALL, DEVOLL, DUVAL], SILAS. The committee appointed to build two Continental frigates in Rhode Island voted on 20 June 1776 that Silas Devol, of Tiverton, Rhode Island, be appointed captain of Marines of the frigate

Providence. It is apparent that appointments of this committee were not always approved or sanctioned by the Continental Congress.

Nevertheless, the committee also voted "That a Captain of Marines shall Inlist forty good men a Lieutenant thirty three, a second Lieutenant Twenty seven before they be intitled to their Commissions."

On 1 July 1776, the minutes of the Marine Committee in Philadelphia noted that "Silas Devall" had been recommended to Congress as captain of Marines for that vessel. William Jennison was recommended as first lieutenant of Marines.

On 3 April 1777, Captain Devol was a member of the court martial of Navy Lieutenant Richard Marvin on board *Providence*.

Stephen Earle, reputedly a first lieutenant of Marines in *Providence* at this time, asserted that Devol [or Devoll] was appointed a Navy second lieutenant about August or September 1777, and that he was appointed captain of Marines in his place.

Little else is known of Devol, with the exception that one "Silas Devel" of Tiverton, appears in the 1774 census of Rhode Island. His name does not appear in the 1790 census.

DICKENSON, JAMES. The vital statistics of James Dickenson are unknown.

On 24 January 1776, he was a lieutenant of Marines on board the ship *Columbus*, Captain Abraham Whipple. On that date, James Read, Paymaster of the Navy, paid Dickenson one month's advance wages.

There is no record, however, that Dickenson was still on board the ship when she departed for the New Providence expedition on 17 February 1776.

DIMSDELL [DIMSDALE], JOHN. John Dimsdell was named a lieutenant of Marines in Doyle Sweeny's Treasury Department listing of 1794, but no supporting evidence is available.

On what vessel Dimsdell may have served and under what circumstances he was captured by the British is unknown.

On 30 September 1778, however, one John "Dimsdale" was a passenger in the cartel *Swift* which transported American prisoners from Halifax to Boston.

DISNEY, JAMES. James Disney is believed to have been from Anne Arundel County, Maryland.

On Sunday, 20 July 1776, the Maryland Council of Safety ordered one Isaac Harris to deliver public arms in his possession to Captain James Disney of the Anne Arundel Militia for the use of his company during its stay in Annapolis. Further, on 5 August 1776, the Council ordered that Disney, an officer in Colonel Hall's battalion of militia for the Flying Camp, be commissioned a captain, dating from 13 July.

He was commissioned a captain of Marines on 22 October 1776, but was noted as "unemployed in Md." On 24 November 1776, Richard Henry Lee, writing from Philadelphia, notified Samuel Purviance, Jr., in Baltimore, that Disney had been assigned to the frigate *Virginia*, then being built in Baltimore. He replaced Captain John Stewart, whose commission on board *Virginia* had been confirmed on 25 June 1776.

Inasmuch as the frigate had been bottled up inside the Chesapeake Bay for nearly a year, discontent had risen in her crew. Disney apparently resigned his commission either late in 1777 or early 1778. Lieutenant Thomas Plunkett was promoted to his captaincy before *Virginia* sailed on her abortive mission on 30 March 1778.

The 1790 Census listed two adult James Disneys in Anne Arundel County. One had a family of seven and only one slave; the other a family of six but 20 slaves.

At the moment, little more is known of the life of James Disney, but documentary material concerning the Disney, Brewer, Baldwin, Maccubbin and other related families from 1690 to 1758, in the Maryland Historical Society, has not yet been searched.

EARL [EARLE], STEPHEN. Born 4 May 1754 in Dartmouth (now Westport), Massachusetts, Stephen Earl was one of eight children of William and Mary (Lawton) Earl.

According to his deposition for a pension, dated 21 April 1818, he enlisted at Providence, Rhode Island, in July 1776 as a sergeant of Marines in the frigate *Providence*. He served six months.

Earl asserted that following this service, Second Lieutenant of Marines Seth Chapin was "broke," and he was promoted to be second lieutenant of Marines.

After having served as a second lieutenant for about four months, Avery Parker, first lieutenant of Marines, was also "broke," and Earle advanced to that rank on board *Providence*.

Following an additional four months, Captain of Marines Silas "Devoll" was appointed Navy second lieutenant on the frigate, and Earl was appointed to his captaincy. He continued in this station until April 1778, when he was discharged at Providence.

Stephen Earl first married 5 May 1782, Mary Hicks, of Dartmouth, by whom he had one child. He married, second, Priscilla Hicks, sister of Mary, of Westport (formerly Dartmouth), in September 1789 and, by her, had eight children.

On 4 May 1779, Earl purchased 100 acres of land at Providence, Saratoga County, New York and moved there in 1800. He died in Milton, New York on 8 May 1834.

ELLIOT [Elliott], ROBERT. The origins of Robert Elliot are uncertain, but a prominent Virginia source asserts he was probably from Botetourt County, which was formed in 1770.

This Robert Elliot was a second lieutenant of the 12th Regiment, Continental Line, commissioned on 20 March 1777. He was apparently stationed at Fort Pitt late in 1777, under the command of General Edward Hand, of Pennsylvania.

There, he was appointed second lieutenant of Marines on the recommendation of General Hand for the armed boat *Rattletrap*. The latter was purchased for the expedition of Captain James Willing down the Ohio and Mississippi Rivers.

The expedition departed Fort Pitt on 10 January 1778. Though little is known of the activities of Lieutenant Elliot, he probably participated in Willing's raids against British settlements enroute to New Orleans.

On 23 February 1778, Marines under two other Marine lieutenants captured the British sloop *Rebecca* at Manchac, and carried her on to New Orleans. On their arrival there, Oliver Pollock, Continental Agent, appointed William Pickles captain of the vessel and began her conversion to a cruiser and renamed her *Morris*.

Robert Elliot was appointed the *Morris'* captain of Marines on 1 April 1778, but little mention is made of him in contemporary accounts for more than a year.

A hurricane on 18 August 1779 destroyed the incomplete *Morris*. Eleven men died. Captain Elliot survived and, on the ultimate failure of the expedition, made his way back to Kaskaskia and later to Philadelphia.

John W. Gwathmey, in *Virginians in the Revolution*, quotes an item in the Northwest Territory Expense Papers:

> May 23, 1781—Paid Captain Robert Elliot for his own and the expenses of three other men sent express to Col. Clark at the Kaskaskies from New Orleans by Oliver Pollock Esq. $275.

Elliot purportedly served also as a captain of militia.

Several claims by Captain Elliot to the Board of Admiralty in 1781 were referred to the Board of War, as not being within their purview. The Board of War, basing its decision on a "certificate of Capt. James Willing," issued Elliot a warrant "for eighty-nine dollars specie, that sum appearing to be due him from January 1, 1778 to the last of March" inclusive "for pay and retained rations." The warrant was issued 21 August 1781.

Elliot presumably returned to Virginia, for on 24 January 1806 he received a bounty land warrant of 2,666 acres of land for his service throughout the war. An additional 259 acres was awarded him on 12 February 1808 for service of seven months. In both instances, the land, in the Military District of Ohio, was granted an assignee, James Taylor.

Gwathmey asserts that Robert Elliot died on 4 January 1838.

ELLIOTT, JOHN. Nothing is known of the origin or personal history of John Elliott.

His first appearance as a lieutenant of Marines was on 30 April 1776 in the port of Philadelphia. There he witnessed the will of Captain of Marines Miles Pennington. Both were on board the brig *Reprisal*, Captain Lambert Wickes.

In early July 1776, *Reprisal* put to sea from Delaware Bay with several merchantmen bound for the West Indies, and had on board William Bingham, a Continental agent. Enroute, she took two prizes, but encountered the British sloop *Shark* at the entrance to Martinique. *Shark* fired several times, and *Reprisal* finally countered with a heavy broadside. A French battery forced the British ship to sea. *Reprisal* departed Martinique on 26 August and reached Philadelphia 13 September.

In November 1776, *Reprisal* put to sea again, carrying Benjamin Franklin, the newly-appointed American Commissioner, to France. She arrived at Nantes early in December, taking two small British brigs off the French coast. Franklin traveled overland to Paris, arriving there 22 December.

In February 1777, Captain Wickes made a cruise into the Bay of Biscay in which he captured five vessels. One was the British packet *Swallow* which she engaged for 45 minutes. Lieutenant Elliott was slightly wounded by a musket ball lodged in his wrist.

Frigate *Deane* was built at Nantes, France in 1777, as arranged by the American Commissioners. Samuel Nicholson was appointed her captain. On or about 15 August, Captain Wickes recommended Lieutenant Elliott as captain of Marines of *Deane* and Captain Nicholson accepted him.

Reprisal sailed from France to America on 14 September, never to be heard of again, but Captain Elliott had already made his transfer. *Deane* arrived at Portsmouth, New Hampshire in May 1778.

Captain Elliott made only one cruise in *Deane*. She departed Boston 14 January 1779 with frigate *Alliance*, but parted company three days later. One vessel, the letter of marque *Viper*, was captured during a four-month cruise in the West Indies and taken into Philadelphia.

Deane brought dispatches for Congress from Martinique. When she arrived off the mouth of the Delaware, weather prevented her passage up the river, therefore, a packet boat was sent up to the city with the dispatches. Captain Elliott was on board, asleep covered by a sail. By accident, he rolled overboard and was drowned.

ELTING, THOMAS. Thomas Elting was purportedly a lieutenant of Marines in the American Revolution, according to Doyle Sweeny in the list he compiled for the Treasury Department in 1794. Any supporting information must have been destroyed in the War Department fire of 1800.

ELWOOD [EHLENWOOD, ELLWOOD], THOMAS. Thomas Elwood was born 13 July 1754, the son of Richard and Rachel Elwood, of Fairfield, Connecticut.

In March 1776, Elwood signed on the Connecticut brig *Defence* as sergeant of Marines, serving first under Captain Seth Harding and later under Captain Samuel Smedley. Between 20 May 1776 and 20 April 1777, he participated in the capture of 10 British vessels, all laden with supplies for the British Army, which were quite welcome to the Continental forces. He apparently served in this capacity until the fall of 1777, when *Defence* was taken into wharf at Boston and altered from a brig to a ship.

Elwood was commissioned a first lieutenant of Marines 24 August 1778 in the frigate *Alliance*, which had been launched 28 April 1778. Her captain was Pierre Landais. Elwood's fellow Marine officers were Captain Matthew Parke and Second Lieutenant James Warren, Jr.

Carrying the Marquis de Lafayette, *Alliance* sailed for France on 14 January 1779, arriving at Brest on 6 February. Enroute a mutiny was discovered, in which Landais was to have been cast afloat, the Marine officers hanged and quartered, and Lafayette sent a prisoner to England. The mutineers were sent ashore in irons, later to be sent back to America and court martialed.

Alliance was then put under the orders of Captain John Paul Jones to cruise with his squadron around the British Isles. In *Bonhomme Richard's* engagement with *Serapis* on 23 September, *Alliance* appears to have been more of a detriment than an ally, allegedly firing into *Richard's* hull rather than at the enemy. It is recorded that several people on *Alliance* remonstrated with Landais and others refused to fire at all. There is no record of the actions of her Marine officers, although later Captain Parke refused to defend Landais.

On their return to the Continent, Jones took command of *Alliance* and brought charges against Captain Landais. She made one cruise off the coast of Spain and returned to L'Orient.

There on 12 June 1780, Jones departed for Paris, and Landais forcibly took command of *Alliance* and sailed for America on 8 July. Enroute, his unbalanced actions and behavior prompted the officers and crew to take command of the frigate. She arrived in Boston 16 August where she remained for the rest of the year. Landais was court martialed and dismissed from the Navy. Captain John Barry took command.

The three Marine officers were joined on 29 January 1781 by Lieutenant of Marines Samuel Pritchard, of Boston, formerly of frigate *Deane*.

Alliance sailed for France on 11 February, arriving at L'Orient 9 March. Loaded with supplies for the Continental forces, accompanied by a French letter of marque, she departed France on 29 March 1781. Unfortunately, another mutiny occurred soon after; three of the conspirators were confined in irons for the remainder of the voyage.

Enroute, *Alliance* had several engagements in the Atlantic, of which the most important was that with HMS *Atalanta* and *Trepassey* on 28 May 1781. Lieutenant Elwood apparently emerged unscathed, but one Marine officer was killed and another critically wounded.

Alliance departed America on 23 December 1781, again taking Lafayette to France, arriving there on 18 January 1782. Parke and Elwood were still on board, and Lieutenant of Marines William Morris joined the vessel on 1 December. Ten enlisted Marines had also been transferred to her from the frigate *Deane*.

While at L'Orient, Elwood was among those who expressed their appreciation to Benjamin Franklin in Paris for his efforts to secure their prize money for the 1779 cruise around the British Isles. *Alliance* returned to New London on 13 May 1782. Lieutenant Morris resigned 29 June, leaving Captain Parke and Lieutenant Elwood the only Marine officers on board.

Alliance's last voyage to France began on 14 August 1782 with a cruise in the Atlantic for prizes. After taking four, Captain Barry decided to make a French port rather than return immediately to America. While in port, Captain Parke and several of the other officers had a dispute with Captain Barry over the matter of prize money and past wages and, when Barry could not comply with their desires, they were left ashore. Lieutenant Elwood was the only Marine officer and Hezekiah Welch the only regular Navy officer on board on her return. Enroute, *Alliance* stopped at Martinique, West Indies. There she joined the ship *Duc de Lazun* with which she transported specie for the use of Congress. In March, she put into Providence where the majority of her crew was discharged.

Lieutenant Elwood spent the greater part of the summer of 1783 inventorying the stores of *Alliance*, and was discharged in September, being the last surviving Marine officer of the American Revolution on active duty.

In his affidavit for a pension in 1818, Elwood declared he was a saddler by trade, but that he had been unable to work for six years. He had been assisted by the town of Fairfield for two years and he had no family residing with him.

David Burr, also of Fairfield, deposed that he had been acquainted with Elwood for 50 years, having seen him in his uniform as a lieutenant of Marines on visits home. In fact, after the close of the war, Burr was a captain of militia and he had purchased Elwood's sword.

Gershom Burr testified that he had seen Elwood when he was enlisting men for the Continental Marines. Two other residents of Fairfield had served as Marines under Elwood.

The date and place of his death are unknown. On 26 April 1853, his only surviving child, Eunice Elwood Fairchild, of Stratford, Connecticut, applied to the U. S. Government for any prize money still due Lieutenant Elwood.

FARNHAM [FARNUM], ZEBEDIAH [possibly also VARNUM, ZEBULON]. The identity of Zebediah Farnham, Continental Marine officer, has not been absolutely established. A series of statistics of various individuals with the same given name and surname have been found, and only further research will reveal the facts.

One Zebediah Farnham, of Windham, Connecticut, is identified in records of the Connecticut Historical Society as having been a lieutenant of Marines on board the ship *Providence* in 1780. On 23 November 1779, frigate *Providence* sailed from Nantasket Roads, Massachusetts and, after capturing the privateer brig *Dolphin*, arrived in Charleston, South Carolina. Her lieutenant of Marines is reputed to have been Zebediah Farnham.

Many persons with the surname "Farnham" appear to have been associated with Windham, Connecticut, among them the following:

1. One Zebediah Farnham, born in 1721, served in the Connecticut militia from 1775 to 1776. He is reputed to also have been a Marine on *Providence* in 1780. He had five sons in the Revolution. He died in Windham in 1814.

2. One Zebediah Farnham was born in Windham in 1721 and married, on 27 July 1743, Mary Faber (or Fuller) and had three sons, Daniel, Abial, and Thomas, and possibly a fourth, Elisha. He died 8 August 1814.

3. One Zebediah Farnham, son of Zebediah and Mary Farnham, was born in Windham on either 10 January 1745 or 1746.

4. One Zebediah Farnham was married to Mary Hebbard in Windham on 9 November 1763.

5. One Zebediah Farnham was a corporal in Captain Caleb Trowbridge's company in the battle of Long Island, was captured, and died on a prison ship.

6. One Zebediah Farnham was in the 6th Company, 8th Connecticut Regiment of Militia in the Revolution. His son, Ebenezer, served in Rhode Island in 1778, was taken prisoner, and died in captivity on a prison ship.

7. One Zebediah Farnham was a first lieutenant in the 8th Connecticut Regiment from 6 July to 10 December 1775, and in the 17th Continental Infantry from 1 January to 31 December 1776. He subsequently served in the Navy and died in 1814.

8. One Zebediah Farnham was reported married to Betsey Crapon at Providence, Rhode Island on 14 January 1784.

9. One "Farnham, Capt. Zebediah" was reported married

to Betsey Crapon "both of this town [Providence, Rhode Island]" on 11 January 1784.

10. One "Farnham, Zebediah of Povidence, age 63 years, a soldier of the Revolution, living, July 4, 1827," was reported in "Deaths," *Vital Records of Rhode Island, 1636–1850*, volume 12, *Revolutionary Rolls and Newspapers, Providence Journal*, 1901.

FIELDING, WILLIAM. William Fielding's vital statistics are not known.

Frigate *Trumbull* lost her lieutenants of Marines Daniel Starr and Jabez Smith, Jr., in her fierce engement with the British letter of marque *Watt* on 1 June 1780.

Trumbull returned to Boston harbor for refitting, with Captain Gilbert Saltonstall the only Marine officer remaining, and he had been wounded several times. A recruiting rendezvous was also opened.

She sailed again on 11 August 1780, cruising the Atlantic with frigate *Deane*. In September they captured the British brig *Little William*.

A muster roll compiled at that time, mentioning the engagement, listed *Trumbull's* lieutenant of Marines as William Fielding.

FITZGERALD, THOMAS. Thomas Fitzgerald was named a lieutenant of Marines by Doyle Sweeny in his listing of 1794, and also by Allen and Paullin. None, however, provided supporting evidence.

FITZPATRICK, JOHN. Absolute origins of Continental Lieutenant of Marines John Fitzpatrick have not been established, but in view of later testimony by Samuel Nicholas, his commanding officer, he is believed to have been from Philadelphia, Pennsylvania.

Nicholas was captain of Marines and Matthew Parke the first lieutenant of Marines on board the ship *Alfred*, which was fitting out at Philadelphia in late 1775 and early 1776. In January 1776, the clerk of the Continental Congress wrote to Commodore Esek Hopkins requesting a list of officers and men on each of his Continental vessels. In addition to Nicholas and Parke, John Fitzpatrick was named a lieutenant of Marines on board *Alfred*.

He may have been the one John Fitzpatrick who appeared in the Provincial Tax Duplicate List for Philadelphia County in 1775, recorded as a laborer and who was taxed one shilling six pence. The Constable's Tax Return for the same year gave his profession as a tanner, situated in the South Part of the Dock Ward with four children, ages one to five.

It should be noted, in addition, that one John Fitzpatrick was married to Eleanor Pryor at Christ Church in Philadelphia on 8 February 1776, only a few days before the Hopkins' fleet departed on its first wartime cruise.

It is a fact that Fitzpatrick was a lieutenant of Marines on board *Alfred* in the New Providence expedition. On 10 April 1776, Captain Nicholas, in his report of the cruise, stated that during the engagement of *Alfred* and *Glasgow* on 6 April, the first broadside felled Fitzpatrick by his side "shot by a musketball through the head." Furthermore, he wrote, "I have lost a worthy officer, sincere friend and companion, that was beloved by all the ship's company."

Although little is known of Fitzpatrick, the following may be of value. In 1785, according to *McPherson's Directory*, Philadelphia, one Ann Fitzpatrick, Shopkeeper, kept her place of business on Water Street, between Catharine and Almond Streets, and her residence was at 394 Water Street. Also, one Elizabeth Fitzpatrick was a staymaker at the corner of Second and Front Streets.

GAMAGE, SAMUEL. Samuel Gamage may possibly have been the son of Joshua and Elinor (Forster) Gamage, of Bristol, Lincoln County, Maine, but the date of his birth is not known.

In his affidavit for a pension for services in the Revolutionary War, Gamage asserted that he was a volunteer at the age of 25 at the battle of Lexington and that he continued as such until he received a second lieutenant's commission in artillery on 11 May 1776. The same source credits him as having been a second lieutenant in Captain Winthrop Gray's 5th Company, Lieutenant Colonel Thomas Craft's Massachusetts Artillery Regiment. He was on a list of men to be commissioned on 9 October 1776 by order of the Massachusetts Council. This unit was raised for only one year.

Gamage was commissioned a second lieutenant of Marines on board frigate *Deane*, Captain Samuel Nicholson, on 26 August 1778, in Boston harbor. On the following 14 January 1779, *Deane* sailed from Boston in company with frigate *Alliance*. Soon after, *Deane* captured the ship *Viper* and sent her into port. She cruised thereafter in the vicinity of the West Indies, returning to Philadelphia on 17 April 1779.

During the following five months, Gamage still aboard, *Deane* sailed in company with *Boston* and *Confederacy*, ranging the Atlantic for prizes. Eight were taken after 29 July, including the ships *Sandwich* and *Thorn* and they returned to Boston on 6 September 1779.

Shortly afterward, *Deane* was ordered to sail for Charleston, South Carolina with *Boston* and *Queen of France*, but she could not be readied in time. This probably saved her from capture by the British in May 1780.

Meanwhile, Lieutenant Gamage resigned his commission on 2 November 1779, because of "infirmities contracted in the service. . . ."

His application for pension was made in 1818 from Hudson, Columbia County, New York. His death date is unknown.

GEORGE, ROBERT. The parents, and the date and place of birth of Robert George are unknown, but a Robert George served in the 5th Virginia Regiment, Continental Line, at an unknown date.

Perhaps it was this Robert George who was appointed a first lieutenant of Marines at Fort Pitt in late 1777 and, together with Second Lieutenant Robert Elliot, accompanied Captain James Willings' expedition to New Orleans in January 1778.

The main party was on board the galley-type armed boat *Rattletrap*, purchased especially for the voyage. Following the capture of Loyalist Anthony Hutchins above Natchez on 19 February 1778, Lieutenant George separated from the main body, scouting ahead in a canoe. All arrived in Natchez on 19 February, where property of the mixed population was confiscated.

On 23 February the British sloop *Rebecca* was captured and carried on to New Orleans, there to be refitted and renamed the *Morris*. Information concerning George's actions is scarce, but when wholesale desertions threatened the detachment in May 1778, he was instrumental in holding together the original company.

When it became evident in mid-1778 that Willing's presence in New Orleans was more of an embarrassment than a help to American Agent Oliver Pollock, arrangements were made for the company to return northward. Lieutenant George took an oath that he would proceed by the route defined by Spanish Governor Bernardo de Galvez and would not bother British subjects along the way.

He probably departed in August 1778, and by June 1779 he and the Marine company reached the camp of Colonel George Rogers Clark at Kaskaskia. The company was disbanded, and George became captain of an artillery company under Clark's command.

Captain George served in Colonel Clark's Illinois Regiment from 1779 to the end of the war, the majority of the time at Fort Jefferson, five miles below the mouth of the Ohio River. He and thirty men defended the fort for six days and nights in 1780 from an attack by Chickasaw and Choctaw Indians.

Following his service in the Revolution, George apparently took advantage of Virginia's bounty land provision, for on 29 April 1784 he received a warrant for 4,000 acres. The land, in the Illinois Country, had been set aside for veterans of Clark's Regiment.

Captain George died 31 March 1804, unmarried, at Woods Ferry, now the town of Utica, Clark County, Indiana, eight miles northeast of Louisville, Kentucky, on the Ohio River. Most of his heirs were the children and grandchildren of a sister.

It was not until 1836 that a commutation claim for the sum of $8,070.44 was pressed. On 19 February 1836, the United States Pension Office issued a half-pay certificate of $240.00 per annum to his heirs. It was recommended that it be paid from the end of the war—11 April 1783—to 15 March 1804.

GILLMORE [GILMORE], WILLIAM. William Gilmore, whose vital statistics are unknown, was appointed a second lieutenant of Marines on 25 June 1776.

Extant records do not reveal that he was assigned to any vessel. On 24 November 1776 he served on the court martial of Private Henry Hasson at the Philadelphia Barracks.

Doyle Sweeny, in his 1794 Treasury Department list, asserts that Gillmore "went into the land service," but no records have been found to qualify the assertion.

GIRTY, GEORGE. Simon Girty, Sr., an Irishman, came to America in his middle years and settled in Lancaster County, Pennsylvania. He was a pack horse driver engaged in Indian trade. He married Mary Newton about 1737. Their home was at Chamber's Mill on the east side of the Susquehanna River, five miles above Harrisburg.

George Girty was born in 1745, the youngest of four sons, the others being Thomas (1739), Simon (1741), and James (1743). When he was four years old, the family moved to Sherman's Creek in Perry County. His father was killed by a drunken Indian in 1751. Two years later, Mrs. Girty married one John Turner, by whom she had one son.

The family was captured by Indians at Fort Granville in August 1755, and at Kittanning, John Turner was tortured to death by the French and Indians. George was adopted by the Delawares but was released at Pittsburgh in 1759, where he was employed as a trader and interpreter during Lord Dunmore's War.

When Captain James Willing departed Fort Pitt in January 1778 on his expedition to New Orleans, his Marine company was under the command of Lieutenants Robert George and Robert Elliot. On 6 February 1778, George Girty joined the expedition and was appointed a second lieutenant of Marines. British sources maintain that he was captured by Willing, robbed and threatened with imprisonment, and joined only until he could make his escape.

Girty deserted the expedition on 4 May 1778 and, with several British soldiers whom he joined on the Illinois River, made his way to Detroit, arriving there 8 August 1779. Thereafter, with his brothers Simon and James, he served the British and was a disbursing agent to the Shawnee towns.

Girty apparently married a Delaware woman by whom he had several children. Just before the War of 1812, he died at a trading post on the Maumee River, about two miles below Fort Wayne, Indiana.

GRANNIS, JOHN. John Grannis was born in North Haven, Connecticut, 18 September 1739, the eldest son of John and Mary (Bradley) Grannis.

He was in the French and Indian War, enlisting in April 1761 as a sergeant in the 7th Company, Captain Amos Hitchcock, 2d Regiment. One John Grannis was also in the same company in August 1757 and again on 10 April 1758, but it is uncertain whether this John Grannis, Senior or Junior.

John Grannis, Jr. is believed to have been a fisherman and probably moved to Falmouth, Massachusetts. He is reputed to have been present at a County Congress at Barnstable on 16 November 1774, of which James Otis was the moderator.

He enlisted in the Revolution on 25 June 1775, serving in the Elizabeth Islands for six months and 22 days. He was commissioned a captain in the 3d Company, Seacoast Defense, for the state of Massachusetts, and served in the Elizabeth Islands and Martha's Vineyard for 7 months and 4 days, beginning 1 January 1776.

The committee appointed to build two Continental frigates in Rhode Island voted on 14 June 1776 to appoint John Grannis a captain of Marines in frigate *Warren*, which had been launched at Warren, Rhode Island on 15 May 1776. In the meantime, he recruited men for the *Warren*.

Grannis apparently acted in a dual capacity for a period of time for, on 18 June 1776, he was still serving in the Elizabeth Islands, having delivered there a prominent Tory. Furthermore, it was not until 1 July 1776 that the Marine Committee recommended him as captain of Marines for the *Warren*.

His service allegiance was not settled for several months. Exchanges of letters flew between the Marine Committee, Governor Cooke of Rhode Island, and the Council of the State of Massachusetts Bay, each loath to lose him from its service. Finally, the latter agreed to his relief from its duty, but they would not consent to Grannis enlisting his Marine complement from the ranks of Massachusetts troops. He relieved Lieutenant William Jennison, who had departed for the army, on board *Warren*.

Warren, however, had difficulty getting to sea, caused by a shortage of manpower and the occupation of Newport and the lower bay by the British. Commodore Esek Hopkins' pennant was on *Warren*, and Grannis was apparently instrumental in the charges brought against Hopkins which resulted in his subsequent dismissal.

Grannis, together with Lieutenants of Marines George Stillman and Barnabas Lothrop and other ships' officers, including the chaplain, wrote these charges on 11 February 1777 to Robert Treat Paine, Taunton, Massachusetts. Both the commodore and his son, Captain John B. Hopkins, were cast in the same undesirable mold.

Captain Grannis carried the officers' petition to the Marine Committee and remained to impart other information. Countercharges by Hopkins followed, but he was formally dismissed from the naval service on 2 January 1778.

Meanwhile, Grannis had been replaced as captain of Marines in *Warren* by Captain Richard Palmes, formerly of the frigate *Boston*, in November 1777.

It is possible that this long period of inactivity and dissension prompted Captain Grannis, together with his Marine lieutenants, to seek other more useful employment. Further information concerning them in the Continental Marines is elusive.

Perhaps John Grannis found a berth on a state or privateer vessel and was later captured by the British. On 3 October 1778, following an escape from prison in England, he received the sum of $240.00 from the U. S. Commissioner in France for his return to America.

His biographer says that he later went to Edenton, North Carolina, where in 1790 his household contained, besides himself, three males over 16 years of age, two females and two slaves.

GREEN, PETER. Vital statistics pertaining to Peter Green are unknown.

He was appointed a lieutenant of Marines on 25 September 1778 and was reported as serving at Boston. The 28-gun ship *Queen of France*, formerly *La Brune*, which had been purchased in France, was then in Boston harbor. It is possible that Peter Green was appointed her lieutenant of Marines at this time.

If so, he was probably on board when she departed for sea on 13 March 1779, joining the sloop *Ranger* and the frigate *Warren* on a cruise in which they captured seven of nine British ships on 6 and 7 April.

Shortly after, Captain John Peck Rathbun assumed command of *Queen of France*. She sailed again on 18 June and together with frigate *Providence* and sloop *Ranger*, captured an additional eight ships which netted well over a million dollars.

On 23 November 1779, she sailed with *Providence*, *Boston*, and *Ranger* enroute to Charleston, South Carolina, where they arrived 23 December.

In April 1780, *Queen of France* was sunk in the Cooper River in an attempt to retard the passage of the British fleet. Apparently, her captain of Marines, Edmund Arrowsmith, Lieutenant Green, and their Marines joined in the defense of Charleston with the land forces and were captured there when the city fell to the British on 12 May 1780.

Most of the officers were soon put on board a cartel for Chester, Pennsylvania, but the name of Lieutenant Peter Green is not among those of the officers paroled.

GUIGNACE, JOHN. John Guignace is listed simply as a lieutenant of Marines in Doyle Sweeny's Treasury List of 18 March 1794.

HADDOCK, ROGER. Roger Haddock was named a lieutenant of Marines by Sweeny, Allen, and Paullin, but none of them exhibited support of the allegation.

One Roger Haddock, however, was prizemaster of the brig *Hazard*, Captain John Foster Williams, from 1 March 1779 to 20 April 1779. He was also first lieutenant of the brig *Active*, Massachusetts State Navy, and took part in the Penobscot expedition.

HAMILTON, JAMES. Sweeny's list of 1794 records James Hamilton as a lieutenant of Marines. Allen and Paullin followed suit.

It may be that his signature on original records consulted by Sweeny was confused with that of one "Jonas" Hamilton, but research has not proved that either of this given name held that rank.

Perhaps both were mistaken for William Hamilton, whose biography follows.

HAMILTON, JONAS. Jonas Hamilton is recorded as a lieutenant of Marines in the 1794 list of Sweeny, and subsequently by Allen and Paullin. No information to support this allegation has been found in recent research.

His signature may have been confused with that of one "James" Hamilton. However, lieutenant of Marines William Hamilton is the only person of that surname for whom primary evidence of the rank has been discovered.

HAMILTON, WILLIAM. William Hamilton, like Captain of Marines John Welsh, is reputed to have been a native of Ireland, coming to America at the beginning of the Revolution.

Perhaps he came to Philadelphia, for on 14 December 1776, he entered ship *Alfred* as a sergeant of Marines. During the landing on New Providence in March 1776, he served under Marine officers Captain Samuel Nicholas, First Lieutenant Matthew Parke, and Second Lieutenant John Fitzpatrick.

Sergeant Hamilton was transferred to the ship *Columbus*, Captain Abraham Whipple, on 7 August 1776. Matthew Parke had been promoted to captain and was then the new commanding officer, Captain Joseph Shoemaker and Lieutenant Robert Cummings (or Cumming) having resigned following the expedition.

On 10 October 1776, Commodore Esek Hopkins "paid Sundry Men belonging to the Columbus on accot . . . Wages," William Hamilton receiving the sum of 21 dollars.

Apparently, Hamilton was commissioned a first lieutenant of Marines shortly after, for on 23 October 1776, he was a member of the court martial of Gunner James Bryant, brig *Hampden*, on board *Alfred* at Newport, Rhode Island.

In either September or October 1776, Hamilton went on board *Alfred*, where he joined Captain Edmund Arrowsmith and Second Lieutenant Alexander Neilson, who had also been promoted from sergeant. *Alfred* was in command of Captain John Paul Jones and, together with *Providence*, sailed on a cruise off Cape Breton.

On 11 and 13 November, the brig *Active* and the ship *Mellish* were captured, for which these officers were entitled to prize shares. *Alfred* returned to Boston on 14 December.

Lieutenant Hamilton apparently left *Alfred* shortly after on leave of absence or on other duty. Early in 1777, Captain Elisha Hinman in *Alfred* was trying to find officers and crew for a projected cruise to France with the frigate *Raleigh*, Captain Thomas Thompson.

Esek Hopkins, in Providence, wrote Hinman at Boston on 9 February . . . "have Inclosed you Mr. Richards [second lieutenant of Marines] Commission . . . the Lieutenant of

Marines Commission I have not Sent as I don't know at present where Lieutenant Hamilton is placed." Lieutenant Hamilton must have been found, for when *Alfred* sailed, he, Captain John Welsh, and Lieutenant Nathaniel Richards were on board.

Alfred and *Raleigh* attacked a 60-vessel merchant fleet on the way to France, hoping to cut some of them out, but the fleet was well-protected by four British men-of-war. *Raleigh* contented herself with badly mauling the 14-gun ship *Druid*, but *Alfred*, being a slow sailer, did not participate.

Returning from France to America on 9 March 1778, *Alfred* fell into a trap set by two smaller-gunned British ships and was captured. Captain Thompson, in *Raleigh*, lightened his lading and escaped. He was later cashiered for abandoning *Alfred*.

Captain Welsh and Lieutenant Hamilton were carried into Barbadoes, where they were transferred to the *Yarmouth*, and carried to Forton Prison at Portsmouth, England. Lieutenant Richards was paroled to take news of the capture to America, but had the misfortune to be captured again and imprisoned in Halifax.

Hamilton and Welsh arrived in England 18 July 1778, and after several months of captivity, escaped to France. Welsh received funds with which to return to America, but by what means Hamilton reached there is unknown.

In mid-May 1779, the frigate *Warren* was fitting out in Boston for the combined Continental and Commonwealth of Massachusetts effort to wrest control of Penobscot Bay from the British. Welsh became *Warren*'s captain of Marines, but the identity of his junior Marine officer was not known.

Approximately 30 Marines from the fleet captured Banks' Island from the British on 26 July 1779. Two days later, a larger force, which included 200 Marines landing on the strong right flank of Bagaduce Peninsula, met heavy resistance. Captain Welsh and eight of his Marines were killed outright.

On 2 September 1779, *The Continental Journal*, of Boston, printed the following salutary obituary, which informed the local populace of the participation of Lieutenant William Hamilton:

> We are informed that the brave Lieut. Hamilton, of the Continental marines, on board the ship Warren died of wounds he received in the attack of Majorbagaduce heights: This intrepid, placid young gentleman, left Ireland (which place he is a native of) in the beginning of these troubles, with an intent to enter into the service of these United States; in which he had continued, discharging his duty with fidelity and integrity till the ball deprived him of his life, and America of a valiant officer. In the assault at the Eastward, he discovered the greatest coolness and bravery.

Lieutenant Hamilton's parents and birthplace are unknown. Nor is it known whether he married and produced progeny. Considering his close association with Captain John Welsh in his Marine career, it is entirely possible that they were acquainted in Ireland and endeavored together to reach America.

HARDY, JOSEPH. Details of Joseph Hardy's early life are unknown, but he may have been from Philadelphia.

Hardy entered sloop *Providence*, Captain John Paul Jones, on 23 January 1776 as a midshipman and earned prize money as a result of captures made by the ship.

On 25 June 1776, he was appointed a captain of Marines by the Marine Committee sitting in Philadelphia. At the time he was serving on ship *Columbus* as the captain's clerk. The news of his appointment finally reached the ship, and Commodore Esek Hopkins wrote the Marine Committee on 14 February 1777 that Hardy had been given leave to report for orders.

His first assignment appears to have been to report for duty at Providence, Rhode Island, where two frigates were being built for the Continental Navy. Hardy served as a member of the court martial of Navy Third Lieutenant Richard Marvin of *Warren* on board *Providence*. Also on the board were Captain of Marines Silas Devoll and Second Lieutenant of Marines Seth Chapin.

On 27 October 1778, the Marine Committee directed the Navy Board of the Eastern Department to order Hardy on board *Confederacy*, Captain Seth Harding. The order was sent 3 November 1778.

Confederacy sailed from New London on 1 May 1779 and then from Delaware Bay on 26 October 1779 with John Jay and the French Minister, Alexandre Gerard, bound for Europe. Bad weather was encountered enroute, in which the frigate was dismasted, split its rudder, and was forced into the harbor at Martinique on 18 December 1779. After transferring her passengers to another ship, *Confederacy* made jury repairs and sailed from that port on 30 March 1780. She returned to Philadelphia where she was refitted between 27 April and 5 December 1780.

On 21 December 1780, *Confederacy* cleared Delaware Bay on her way to the Caribbean and the West Indies. Unfortunately, on her return to America, she was intercepted by the British frigates *Roebuck*, 44 guns, and *Orpheus*, 32 guns and was captured on 14 April 1778. Her officers were reported by one source as paroled and the men imprisoned on the prison ship *Jersey*.

Captain Hardy was shipped as a prisoner from New York to New England but arrived in Ireland by accident. He escaped to France and was at Nantes on 19 August 1782 when he wrote Benjamin Franklin in Paris for assistance in arranging his exchange.

Presumably he was able to arrange his exchange, for he was back in Philadelphia on 28 April 1783, where he signed a petition to the Continental Congress regarding pay due officers, seamen, and Marines.

Hardy kept a journal of his experiences on board *Confederacy*, which today is a primary source of information on the services and duties of a Marine officer. In it, several entries seem to infer that his home was in Philadelphia.

Early in 1790, he applied for membership in the New York State Society of the Cincinnati and was voted in on 4 July 1790. He served as secretary of this chapter from 1805 to 1810.

Hardy's son, Joseph L. C. Hardy, entered the U. S. Navy as a midshipman on 18 June 1812, serving in the Southern Squadron and in the battle of New Orleans during the second war with Great Britain. He left the Navy after receiving an injury to become a second lieutenant of Marines on 3 March 1823. He was promoted to first lieutenant 13 June 1830, to captain 3 March 1847, and died in Brooklyn, New York on 26 November 1853 after 42 years of combined service. The *Navy Register* gives New York as the state of his birth and South Carolina as the state of his appointment and citizenship.

The son's admission to the Society of the Cincinnati in 1820 indicated his father died between 1810 and 1820.

HARRIS, JOHN. The origins of John Harris have not been ascertained. In view of the fact that his future ship, the frigate *Boston*, was launched at Newburyport, Massachusetts in 1777, and that his Marine superiors, Captain Richard Palmes and First Lieutenant Robert McNeill both appear to have been from Boston, he may also have been from that area.

Neither is the date of his commission as a second lieutenant known. *Boston* came around to Boston for fitting out on 30 October 1776, but the date her Marine officers boarded is unknown. Captain Palmes was commissioned on 23 July 1776 and, in January 1777, was searching other vessels for deserters from *Boston*, Captain Hector McNeill.

The first intimation that Lieutenant Harris was on board *Boston* occurred on 24 April 1777. He, together with four naval officers and his fellow Marine officers, addressed a petition to their captain. Apparently, their rations had been inferior. They had not been paid, and they were not receiving the same benefits as officers in the land service. They prevailed on Captain McNeill to support their demands before sailing. It is not known whether he did.

Nevertheless, they were all on board when *Boston* sailed in company with frigate *Hancock*, Captain John Manley, on 21 May 1777, and on 30 May they eluded the British 64-gun ship *Somerset*. On 7 June, *Hancock* captured the 28-gun frigate *Fox*. Lieutenants McNeill and Harris were put on board the enemy frigate to control her crew. Sergeant William Jennison was promoted to acting lieutenant of Marines.

Enroute back to Boston, the two American frigates and their prize encountered three British ships with a combined armament of 94 guns. The former scattered, but on 8 July the British ship *Rainbow* captured *Hancock*. *Fox*, with her American crew on board, was engaged by the frigate *Flora*, 32 guns, but ran aground. Both Lieutenants McNeill and Harris were captured and sent prisoner into Halifax. *Boston* escaped into a cove in Maine and later made her way back to Boston.

On 8 November 1777, Lieutenant William Bubier, of the *Hancock*, was released from Halifax and sailed on a cartel to Boston. Accompanying him and others released was a letter from one Edward Brooks to James Bowdoin, notifying him that Harris and McNeill were still prisoners with him, but that Sir Gorge Collier had promised their exchange if British prisoners were sent for them.

Lieutenant McNeill was sent back to Boston on 29 January 1778, but when and if Lieutenant Harris was exchanged remains unknown, nor has documentation been found regarding further service of Harris in the Continental Marines.

HARRISON, RICHARD. Richard Harrison is believed to have been from Maryland, but no data of his origins have been found.

While the frigate *Virginia* was being built in Baltimore in 1776, Richard Harrison was recommended by Maryland delegates to the Continental Congress as a second lieutenant of Marines. Congress gave its approval on 25 June 1776, and he was commissioned and assigned to *Virginia*.

The frigate was confined to the Chesapeake Bay until early in 1778, and in the meantime Harrison appears to have been replaced as second lieutenant by Thomas Plunkett on 9 December 1776.

Only two persons named Richard Harrison appear in the 1790 Census of Maryland, one in Anne Arundel County and the other in Frederick County. Further research of either has not been undertaken to discover whether he was the Continental Marine officer.

HARRISON, RICHARD. Richard Harrison's parentage and birth are unknown. In light of later documentation of his life, he may well have been a resident of the huge territory that was pre-Revolutionary Virginia.

Captain James Willing's expedition from Fort Pitt to New Orleans reached Natchez, on the Mississippi, in February 1778. Natchez had been the home of Willing from 1774 to 1777, and he knew the inhabitants well. It was populated by a mixture of American, English, and French settlers who signed an oath of neutrality with Willing .

Willing ordered the single men of the community to join him, in consequence of which Richard Harrison, an American settler, was appointed a second lieutenant of Marines on 12 February 1778.

On the following 23d, Lieutenant Harrison, together with a Lieutenant McIntyre and 18 Marines, captured the British sloop *Rebecca*. She was carried into New Orleans for refitting and renamed *Morris*. Shortly after, Harrison participated in the captue of the British lumber brig *Neptune*, which was taken into New Orleans as a prize.

In April, Willing sent Lieutenant Harrison with a detachment of Marines to Natchez to see if the oath of neutrality was being observed. Despite having been forewarned that the settlers were arming themselves for his arrival, Harrison entered a trap. After losing five men killed and several captured, he returned to New Orleans with the remainder.

Shortly after, recruits began deserting, disappointed that booty did not materialize. Harrison, with Lieutenant Robert George, managed to hold together the original Marine company.

Faced with the failure of Willing's expedition, and with growing discontent of the Spanish authorities in New Orleans, he and Lieutenant George departed in August, making their way by river and overland northward, joining Colonel George Rogers Clark at Kaskaskia in June of the following year.

There Harrison became a captain-lieutenant in Captain Robert George's newly-formed company of artillery in the Virginia State Troops' Illinois Regiment on 4 June. He was stationed successively at Fort Jefferson and Fort Nelson, and was promoted to captain in January 1782.

His wife's name is unknown, but his son, Philip B. Harrison, was the administrator of his estate. The latter stated that his father was the eldest brother of James Harrison who, as a lieutenant, fell in the battle of Saratoga in October 1777. Both James and Richard were the recipients of 2,666 acres of Virginia bounty land, the first awarded on 26 March 1784 and the latter on 29 April 1784. Richard was his brother's legal representative.

Richard Harrison died 29 December 1799, but the place of residence of his later years has not been ascertained. In 1841, however, James A. Girauld, of Natchez, Mississippi, was appointed attorney for his heirs. They included Philip B., Richard, and Hay B. Harrison; James Dunbar and his wife, Mabella, formerly Mabella Harrison, and Philip B. Dougherty, the only child of Caroline Dougherty, formerly Caroline Harrison.

HEMPSTED [HEMPSTEAD, HEMSTEAD], SAMUEL. The vital statistics of Samuel Hempsted are unknown. Basic genealogical research leads one to believe that he was one of the numerous family of that surname from New London, Connecticut.

The date of his original commission as lieutenant of Marines is also unknown. He served on the court martial of three enlisted men of the frigate *Alliance* on 27 July 1782. They had been accused of "Mutiny and Sedition" while the frigate was at New London on 16 May 1782.

Alliance had just returned from a voyage to France. Lieutenant of Marines William Morris had been sent to Philadelphia with dispatches for Congress, and on his return he presented his resignation to Captain John Barry. He also had presented a copy of it to Robert Morris in Philadelphia, and the latter accepted the resignation on 30 July 1782.

On the same day, Matthew Parke, captain of Marines in *Alliance*, wrote Barry, notifying him "As the ship is now vacant of a Lieut. of Marines on account of Mr. Morris's absence; I leave to your obsion [option] whether another shall be appointed in his room. Mr. Hemstead of this place, is now present, that belongs to the service. If agreeable to you, and in your power to order him, shall be glad."

Hempsted was not approved by Captain Barry and his career thereafter is unknown.

HENDERSON, DANIEL. The birth, parentage, and place of the birth of Daniel Henderson are not known. Inasmuch as several of the Continental Marine officers appointed by Congress on 25 June 1776 came from the Philadelphia area, it may be plausible to assume that it might also have been his native city.

Frigate *Effingham* was built and launched in 1776 in Southwark, the southern district of Philadelphia, but was never commissioned. Andrew Porter, of Philadelphia, was appointed her captain of Marines, with Daniel Henderson and James McClure, his lieutenants of Marines. The frigate was scuttled to prevent her capture on 2 November 1777.

In December 1776, an additional detachment was formed for frigate *Delaware* at Philadelphia from three of the four companies of Marines recently formed. A 20-man detail for her was placed under the command of Lieutenants Henderson and David Love, James McClure remained with *Effingham*.

Three companies of Marines served with Washington in the New Jersey campaign from December 1776 into April 1777. Whether Henderson accompanied them is unknown.

Captain Robert Mullan's company returned in February 1777, awaiting completion of the *Delaware*. Captain Andrew Porter transferred to the Continental artillery, taking Lieutenant James McClure and nine Marines with him. Lieutenant David Love probably resigned about this time. *Delaware* was ready to sail on 5 April 1777, and Henderson was promoted to David Love's position, Alexander Neilson became her second lieutenant of Marines.

Later in the year, *Delaware* was stationed above the chevaux-de-frise on the Delaware River when Henderson accused Neilson of theft of liquor from their common sea chest. Neilson was acquitted in a court martial and replaced the liquor. The case was carried on by charges and countercharges in the local press, until finally dropped.

On 26 September 1777, after the British had occupied Philadelphia and set up batteries along the river, *Delaware* proceeded upriver, firing on the emplacements. British return fire soon forced the frigate aground in mid-river. Lieutenant Neilson and a few men escaped to the Jersey shore, but the remainder of the officers and crew were captured and imprisoned in Philadelphia. One must assume that Henderson was among them.

Nothing is known of a further Marine career of Daniel Henderson. One Daniel Hendeson was a lieutenant in a list of officers and crew of the Massachusetts brig *Griffin*, Captain Gideon Henfield, on 23 May 1780, and he was entitled to two and one-half shares in prizes she might capture.

Doyle Sweeny asserts Lieutenant Daniel Henderson was "lost at sea," but no documentation has been found to substantiate the declaration.

HOLT, SAMUEL. Samuel Holt was born at New London, Connecticut, 27 February 1754, the son of William, Jr. and Mary Holt.

He served as a drummer boy in the 5th company of Captain James Chapman's Continentals in 1775. The following year, he was in Captain [Hezekiah?] Brackett's company of Connecticut state troops.

From his entry on 13 May 1777 to his discharge on 22 September 1777, Holt was sergeant of Marines of the Connecticut ship *Oliver Cromwell*, Captain Seth Harding. In a cruise to the eastward, *Oliver Cromwell* captured six vessels, all but two of which reached American ports.

Holt apparently became a Continental lieutenant of Marines before December 1777, for one Andrew Palmes asserted in his deposition for a pension "About December, 1777 . . . he served as a Sergeant of Marines under Lieut Samuel Holt and served aboard the ship Warren lying at Boston till he was ordered on board the Confederacy lying at New London." When *Confederacy* was delayed in sailing, Palmes signed aboard a privateer.

It may be that Lieutenant Holt signed on board the *Con-*

federacy to be with his former captain, Seth Harding, who was given command of her in September. With him were Marine Captain Joseph Hardy, Lieutenant Gurdon Bill, and possibly Lieutenant Ephraim Bill.

Further proof of his having been on board *Confederacy* in 1779 appears in a muster roll of that year, in which he, together with Gurdon Bill, were listed under "Men's Names" but with no rank recorded.

His family biographer asserts that he was in *Confederacy* when she surrendered to the British frigates *Roebuck* and *Orpheus* on 14 April 1781, and that he was confined to a prison ship off the New Jersey coast. Another prominent source asserts that at least some of the officers—including Gurdon Bill—were sent from New York to prison in England.

Samuel Holt married Margaret Warnick, of Philadelphia, at the Swedes' Church in that city on 19 August 1780. Following the war, he lived successively in New London, Philadelphia, Bristol, and again in Philadelphia. Ten children were born of this union. He died in Philadelphia in March or April 1818.

HOLTON, WILLIAM. Lack of information hinders writing even a token biography of William Holton. The lone citation of Doyle Sweeny in his Treasury List of 1794 "broke 11th of Dec., 1778" identifies him as a captain of Marines.

On 28 March 1777, while in barracks at Philadelphia "Wm. Holton of th. Marines" was issued one-half cord of wood. Only three days before Captain Robert Mullan, with three officers and 50 privates, was issued one and one-half cords of wood and two pounds of candles.

HUDDLE, BENJAMIN. One Benjamin Huddle was purportedly a lieutenant of Marines in the Revolution, according to Sweeney, Allen, and Paullin, and perhaps to McClellan. Possibly he was confused with William Huddle, whose biography follows.

HUDDLE, WILLIAM. The origins of lieutenant of Marines William Huddle are unknown.

By order of the Continental Congress, the schooner *Wasp*, 8 guns, Captain William Hallock, was outfitting in Baltimore, Maryland, late in 1775. On or about 5 January 1776, Captain William Stone, commanding sloop *Hornet*, also at Baltimore, was ordered to take both vessels into Chesapeake Bay and await the orders of Commodore Esek Hopkins.

On 11 February, *Hornet* and *Wasp* joined the Continental fleet near Cape Henlopen for the New Providence expedition. Lieutenant William Huddle commanded the five Marines on board *Wasp*. The fleet got underway 17 February 1776, but shortly after, *Hornet* and *Fly* collided and remained behind.

Wasp continued on with the fleet and, during the landing of Marines southeast of Fort Montagu, she covered the landing. It is unknown whether Huddle and his Marines remained with the vessel or joined their fellow Marines ashore. *Wasp* remained at the Fort Montagu anchorage through 4 March, but departed Fort Nassau with the fleet on 16 March.

Enroute to America, the fleet encountered a severe gale on 22 March. *Wasp* was separated from the fleet and made her own way into Chesapeake Bay.

In October 1777, *Wasp* was one of the several Continental vessels guarding the chevaux-de-frise in the Delaware River below Philadelphia. On the forceful advance of the British upriver, she was one of the ships destroyed to prevent her capture. Whether Lieutenant William Huddle was still on board the schooner at that time is unknown.

HUNTER, ROBERT. One Robert Hunter was reported by Doyle Sweeny to have been a lieutenant of Marines in the American Revolution. His name also appeared in the lists of Allen and Paullin, however, this service cannot be ascertained.

One Robert Hunter was a sergeant of Marines in the brig *Andrew Doria* during the expedition to New Providence. On the return of the vessel to Rhode Island, he was reported sick in the hospital at Providence.

The same Robert Hunter was later a commissioned officer in the Continental and New York State land forces from 1777 to 1779.

JENNISON, WILLIAM. Born 4 August 1757 at Milford, Massachusetts, William Jennison was the son of William and Mary (Staples) Jennison. His father was a physician, but engaged in business in Mendon, a near-by community. The younger William graduated from Harvard in 1774. In the same year, his father was a member of the Provincial Congress from Mendon.

Jennison's first service in the Revolutionary War was as a sergeant in a company of minutemen commanded by his father who marched on the alarm of 19 April 1775 to Roxbury and Cambridge, Massachusetts. Shortly after, he enlisted in the Army, serving as quartermaster of Colonel Joseph Read's 20th Massachusetts Regiment, as confirmed by returns dated 18 May and 26 September 1775.

He was appointed a first lieutenant of Marines in April 1776 by the Navy Board, Eastern Department and ordered to the frigate *Warren*, then under construction at Providence, Rhode Island. He proceeded to enlist Marines for the vessel, but resigned his commission in June, and is reported to have re-entered the Army. He was recommended by the Marine Committee for an appointment as first lieutenant of Marines for frigate *Providence* but did not serve as he was then in the Army.

By 13 May 1777, he had become a sergeant of Marines, for on that date he signed a "List of Marines on board" frigate *Boston*, Captain Hector McNeill. *Boston* made a cruise with frigate *Hancock*, capturing the British ship *Fox* on 7 June 1777. Lieutenants of Marines John Harris and Robert McNeill went aboard as part of her prize crew. Jennison was then made acting lieutenant of Marines under Captain Richard Palmes. The ship returned to Boston 21 August 1777.

By mid-February 1778, he was a lieutenant of Marines, sailing in *Boston* to France with John Adams, the newly appointed American Commissioner, as a passenger. He returned to Boston on 23 October of the same year.

After the frigate's return, Jennison was appointed her purser on 14 November. While *Boston* was being refitted for sea, Jennison secured a leave of absence and sailed on a privateering cruise on boad the schooner *Resolution*. Unfortunately, the privateer was captured on 8 May 1779 and her crew imprisoned in Halifax.

After his exchange and return, Jennison rejoined *Boston* on 29 September and sailed 23 November as a midshipman. Captain Samuel Tucker now commanded her and several other ships as a squadron enroute to the relief of Charleston, South Carolina. On 16 January 1780, Captain Tucker suggested to Jennison that he act as "volunteer Marine officer."

At Charleston, Jennison served for a short time on board the South Carolina ship *Brecole* and ashore at the fort at Governor's Bridge, until captured 12 May 1780.

Jennison's confinement ended 10 June 1780, when he went on board a cartel ship, arriving at Chester, Pennsylvania on 23 June. After only a short time residing in a house on Spruce Street, Philadelphia, he returned to Boston on 16 August. He apparently saw no further service in the Continental Marines.

His intention to marry Mary Wibird was published in Boston 19 November 1783. Although he had studied law before the war, he became a teacher establishing academies at Boston, St. Johns, Nova Scotia, Vicksburg, Mississippi, and Baton Rouge, Louisiana. While at Vicksburg, he published a small pamphlet on political economy. On 15 September 1797, one Samuel Webber, of Cambridge, Massachusetts, completed his preparation for college in the private school kept by Jennison.

He was a federal pensioner, being placed on the rolls 10 July 1819 with his pension dating from 2 September 1818.

Jennison died in Boston 24 December 1843 at the age of 86 years. His widow, Mary, also died in Boston almost ten years later, on 11 April 1853 at the age of 90 years. They were reported to have had several children, of whom no information has been found.

JONES, WILLIAM. The fourth of five children of William and Elizabeth (Pearce) Jones, of Newport, Rhode Island, William Jones was born 8 October 1753. His father, who died in 1759, was a first lieutenant on the *Duke of Marlborough* in England's war with France.

In January 1776, at the age of 23, he received a lieutenant's commission in Colonel Henry Babcock's (later Colonel Christopher Lippitt's) Rhode Island State Regiment and was promoted to captain in September of that year.

He participated in the battle of Harlem Heights in October, and successively in the battle of White Plains, and in Washington's retreat into New Jersey. He spent the early part of the winter there, his regiment's term due to expire in December. In response to an appeal of Washington, officers and men of the regiment signed on for another month, confronting the British at the bridge of Assunpink and in the battle of Princeton.

In February 1777, he returned to Rhode Island, remaining with his family until early 1778. In March, Captain Abraham Whipple recommended Jones as captain of Marines in frigate *Providence*. John Deshon of Boston described him as follows:

> He is a man well acquainted with the service, having been in it most of the War, and has great interest in raising a comp[any] of Marines. His appearance and character is such I apprehend as will do honor to the service.

On 30 April 1778, *Providence* set sail for France with dispatches for the American Commissioners pertaining to the Treaty with France. Arriving at Paimboeuf on 30 May, Whipple sent Captain Jones to Paris with the dispatches. Jones waited there until 11 June, receiving in turn dispatches for Congress.

Providence sailed to Brest on 8 August, arriving there in six days. There she found *Boston* and *Ranger*, in whose company she sailed for America on 22 August. On 26 September they were off the Banks of Newfoundland, and arrived in Portsmouth with three prizes on 15 October 1778.

In June, *Providence*, together with *Queen of France* and *Ranger*, made a cruise eastward. In mid-July, they fell in with a large Jamaica fleet under armed convoy and, sailing in company with them, cut out eleven of the merchant vessels. Three were recaptured later, but the Americans made port at Boston with eight prizes of considerable value.

Frigate *Providence* sailed again on 23 November 1779, with *Boston*, *Queen of France*, and *Ranger*, arriving at Charleston, South Carolina 23 December. Their crews, with the exception of those of *Ranger*, were sent ashore to man the land batteries, but they were too feeble to stem the superior force of the British. Captain Jones was captured on 12 May 1780.

Although paroled immediately, he was never exchanged and spent the remainder of the war in a non-combatant status. In Providence he engaged with his brothers in the hardware mercantile business, which he continued to his death.

On 17 March 1787, he married Nancy [Anne?] Dunn, the daughter of Samuel Dunn, of Providence. He was for a time a Justice of the Peace, and in 1807 was elected one of four members of the General Assembly from Providence. From 1809 to 1811 he was Speaker of the Assembly and, in April 1811, was elected Governor of Rhode Island by the Federal Party. Successive terms continued until 1817.

At that time he retired from public life, but was a Fellow of Brown University, President of both the Peace Society and the Rhode Island Bible Society. He was also a charter member of the Society of the Cincinnati.

Jones died 9 April 1822, aged 67, leaving his widow and an only daughter, Harriet. It is said that on his tombstone that he was the first officer to have worn the American uniform in Paris, perhaps because he was the first to arrive after the ratification of the Treaty, and therefore the first to be recognized as such.

KIRKPATRICK, HUGH. Hugh Kirkpatrick's family statistics are not known, but he was probably from Philadelphia.

In November 1780, he was appointed lieutenant of Marines on board the sloop of war *Saratoga*, Captain John Young. He relieved Lieutenant Abraham Van Dyke, who probably retired because of his age.

On 15 December 1780, *Saratoga* departed Philadelphia with the frigate *Confederacy*, convoying merchant vessels to the West Indies. Enroute she captured the British privateer *Resolution* with only one broadside. Afterwards she engaged the letter of marque *Tonyn* for a full hour, the musket fire of Kirkpatrick's Marines on the quarterdeck contributing to the enemy's discomfort before her capture. She arrived at Cape François 27 January 1781.

Saratoga, accompanied by *Confederacy*, *Deane*, and the privateer *Fair American*, departed the West Indies on 15 March 1781, again with a convoy returning to the United States. *Saratoga* spied two small vessels on 18 March and, by noon, captured one of them. After putting a prize crew aboard, Captain Young stood forth to capture the second vessel.

Midshipman Nathaniel Penfield, on board the prize, described the *Saratoga* as disappearing on the horizon before his very eyes. Lieutenant Kirkpatrick, together with the remainder of *Saratoga's* crew, apparently was lost at sea.

Before the departure of *Saratoga* from Philadelphia, Kirkpatrick had issued a power of attorney to William Geddis of that city. On 15 January 1785, a bond of administration was issued on the estate of one Hugh Kirkpatrick. His administrators were James Kirkpatrick, John Wilson and John Nesbit, but no records survive of the disposition of the estate.

LEARY, DENNIS. Dennis Leary was probably from Pennsylvania, but details of his birth and parents are unknown.

Brig *Andrew Doria*, ex-*Defiance*, was puchased for the Continental Navy in Philadelphia in November 1775. She was converted at the Kensington shipyard and commissioned by 4 January 1776.

Leary entered *Andrew Doria* in January as a midshipman and participated in the expedition to New Providence. Following the return of Hopkins' fleet to America, Leary was one of many reported sick in New London on 8 April 1776.

She was cruising off the New England coast in June 1776 and captured two of several Scottish transports loaded with Highland troop replacements. Issac Craig and John Trevett were her lieutenants of Marines. On 3 August 1776, Leary was paid a midshipman's wages of four pounds and 10 shillings.

Sometime in September 1776 Issac Craig left the ship and Leary was appointed lieutenant of Marines in his place. In October she sailed for Dutch St. Eustatius for a cargo of military supplies, and there was saluted as a merchant vessel, although Dutch authorities later denied having done so. Returning to Philadelphia with her stores, the brig captured the British 12-gun sloop *Racehorse* after an engagement of two hours.

It has been surmised by one authority that *Andrew Doria* arrived back in Philadelphia just in time for her Marines to debark and to march with Major Samuel Nicholas' battalion into New Jersey with General Washington. No documentary proof can be found of this, however.

The British took Philadelphia in the fall of 1777, but the Americans were still in possession of the Delaware River. Stationary defenses consisted of forts on the New Jersey and Pennsylvania shores and chevaux-de-frise sunk in the channel. Floating defenses consisted of vessels of the Pennsylvania Navy and included the *Andrew Doria*. Most of her efforts to defeat the British afloat were to little avail, and she was destroyed to prevent her falling into British possession. Lieutenant Leary is reported to have been promoted to captain of Marines about this time.

Leary's activities for sometime thereafter are uncertain. However, it appears that Bordentown, New Jersey, the headquarters of the Middle District Navy Board, was the gathering place for unemployed Navy and Marine officers and their detachments, and he may have been awaiting orders there.

Employment was found for him in September 1780. After recruiting a detachment of Marines in Philadelphia, he proceeded to Reading, Pennsylvania to superintend the cutting of masts for Continental and French ships. Troubles with Indians necessitated relief by militia and no further touble ensued.

In 1785, a settlement of claims pointed out that "Capt. Dennis Leary . . . has been paid in full to Jul. 1, 1779, since when he has been but a short time in service and has asked no settlement. . . ."

One Dennis Leary was married to Ann Cassel in the Swedes' Church, Philadelphia 8 July 1780. Nothing further is known of his career or life.

LONGSTREET, DANIEL. Daniel Longstreet's origin and vital statistics are unknown.

In January 1778, Captain James Willing, who had been commissioned a captain in the Navy in mid-1777, departed Fort Pitt for his expedition down the Ohio and Mississippi Rivers to New Orleans.

On 23 February 1778, Willing captured the British sloop *Rebecca* at Manchac and, taking her to New Orleans, proceeded to convert her to a cruiser and renamed her *Morris*.

There, Robert Elliot, who had accompanied Willing the entire voyage, was named captain of her Marine company. Daniel Longstreet, who may originally have been from the south, was appointed her first lieutenant of Marines on 1 April 1778.

After laboring for nearly a year to convert *Morris*, her crew suffered the disappointment of seeing her sunk in a hurricane on 18 August 1779. Captain Elliot resigned and returned northward to Kaskaskia and then to Philadelphia. Lieutenant Longstreet chose to remain behind and take command of the surviving members of the Marine detachment.

Shortly after, Continental Agent Oliver Pollock obtained an armed schooner from the Spanish governor, Bernardo de

Galvez, and Longstreet and his Marines were transferred to the new vessel. In early September, they joined Galvez with a force of American, Spanish, and Indian troops to raid British forces at Manchac, Pensacola, and Mobile.

Enroute, on 10 September the schooner engaged the British armed sloop *West Florida* on Lake Pontchartrain and captured her after a short engagement. Captain William Pickles, who had been in command of *Morris* and the schooner, transferred a portion of his command to *West Florida*. Lieutenant Longstreet, in command of her Marines, participated in raiding several settlements on the north shore of the lake, placing them under American rule.

On 4 March, *West Florida* set sail with Spanish forces for Mobile, and on the 17th Fort Charlotte capitulated to the combined forces. *West Florida* also captured a small sloop, with which she then departed fo Philadelphia.

The date of arrival of *West Florida*, with Longstreet and his Marines, is unknown. On 20 November 1780, the vessel's master, Philip Lyon, presented a petition to Congress asking remuneration for the property they had lost in the sinking of *Morris* and for wages and prize money accrued. Lieutenant Longstreet, together with Sergeants of Marines John Clarkey and John Murry, and Lawrence Smith, Marine, was among the petitioners.

Congress referred the matter to the Board of Admiralty, whose members met on 5 December 1780, recommending that the crew be paid and Captain Pickles relieved. On 21 December, the matter was postponed and its final outcome is not known.

Further information concerning Lieutenant Daniel Longstreet is unknown.

LOTHROP [LATHROP], BARNABAS. Barnabas Lothrop was born on 25 February 1757, the son of Edward and Abigail (Howard) Lothrop, West Bridgewater, Massachusetts.

One Barnabas Lothrop was a sergeant in Captain Elisha Nye's Company from 25 January to 3 February 1776 and from 5 April to 1 June 1776. The company was stationed at Elizabeth Islands, Massachusetts, in the same area that Captain John Grannis' 3d Company was stationed for seacoast defense. Grannis was later Lothrop's commanding officer in frigate *Warren*.

Lothrop was commissioned a second lieutenant of Marines in *Warren*, then fitting out at Providence, Rhode Island, on 14 June 1776.

He joined Captain Grannis, First Lieutenant of Marines George Stillman and other officers of *Warren* in composing and signing a petition to the Marine Committee on 19 February 1777 making accusations against *Warren's* flag officer, Commodore Esek Hopkins.

In March 1777, during the hearing on the charges, Captain Grannis testified that both Lothrop and Stillman were from Bridgewater. The information is true for Lothrop, but the surname "Stillman" is not to be found in records of that locality.

Lothrop, together with Grannis and Stillman, appears to have sought other duty late in 1777 or early in 1778, but further Continental service for Lothrop cannot be found. Perhaps he resigned.

Lothrop married twice. First, on 27 November 1777, Sarah, the daughter of Jonathan Bozworth [or Bosworth], by whom he had seven children. One given name particularly attracts attention. His eldest son, Stillman, may have been named for his former brother officer on board *Warren*, First Lieutenant George Stillman. He later became a prominent clergyman in Cambridgeport, Massachusetts.

Sarah Lothrop died 18 February 1813, aged 51. He married, second, Rachel, the daughter of Samuel Bartlett, of Bridgewater. They had no children, and she died 13 March 1862. Barnabas Lothrop's death is unrecorded.

LOVE, DAVID. David Love's parents and his date of birth are unknown, but it is probable that he was a native of the Philadelphia area.

Research in Philadelphia revealed little of the surname "Love," however, on 20 October 1772, one David Love was indentured "by Capt. George Forten to serve Samuel Cooper of Newtown Township Gloucester County West Jersey & Assigns seven years from the day the time of his arrival" for a consideration of 20 pounds.

Congress commissioned David Love a first lieutenant of Marines on 25 June 1776. He was assigned to the frigate *Delaware*, which was launched on 12 July 1776. Captain Robert Mullan was his senior officer, and Hugh Montgomery was *Delaware's* second lieutenant of Marines.

When Mullan's company went with General Washington into New Jersey in December 1776, Lieutenant Love did not accompany it, but probably remained aboard *Delaware*, outfitting off Philadelphia. Nevertheless, he was on the payroll of Mullan's company through 1 April 1777.

Lieutenant Love appears to have resigned his commission at some time after 1 April 1777, for there is no further mention of him as a Marine officer.

Conclusive biographical data of a further career of this David Love is questionable. David Love, an iron founder of Philadelphia, sold three tracts of land between 1811 and 1814 in an area bounded by Second Street Schuylkill, Vine Street, Callowhill Street, and Front Street Schulykill, in what is now downtown Philadelphia.

Furthermore, the records of Philadelphia show a bond of administration of the estate of one David Love, merchant, of Lexington, Kentucky, dated 24 January 1820. One of the administrators was his brother, John Love, of Franklin, Howard County, Missouri, and the other two, William Hood and Robert Foland, were merchants in Philadelphia.

MacCARTHY [MACARTHY, McCARTY], EUGENE. Eugene MacCarthy was born at Ochtermony, County Kerry, Ireland, 17 August 1757, the son of Timothy and Betty (O'Connell) MacCarthy.

He entered the French service in 1773 as a cadet in the Berwick Infantry Regiment of the Irish Brigade. He became a sub-lieutenant in the Regiment of Walsh on 27 June 1776, and a second lieutenant 13 November 1779.

On 5 February 1779, his commanding officer wrote Captain John Paul Jones, recommending the services of MacCarthy, together with his brother officer, Edward Stack. MacCarthy himself wrote Jones on 3 March from Quimber, France, that he "wishes to make a Campaigne . . . under command of a Gentleman who has Distinguished himself by his Superior talents."

MacCarthy received a brevet commission of lieutenant of Marines probably dating from 4 February and, with Stack, reported on board *Bonhomme Richard*. Jones and his squadron sailed on the cruise around the British Isles 14 August 1779.

When Jones made his assignments in preparation for the

engagement with the frigate *Serapis* on 23 September 1779, he stationed Lieutenant MacCarthy in the powder magazine. Overhead, Jones had mounted six old 18-pounder cannon. The bursting of three of these at the outset of the battle wounded Lieutenant MacCarthy. Following the surrender of *Serapis*, *Bonhomme Richard* sank, and MacCarthy transferred with other survivors to the British ship, losing his personal possessions.

During his absence from his regiment, MacCarthy was promoted to captain in the French Army 12 December 1779. Discharged from United States service on 13 February 1780, he returned to the Irish Brigade.

He participated in several French-British naval actions in 1781 and 1782. The Regiment of Walsh promoted him to senior lieutenant 14 July 1785, to captain-lieutenant on 6 April 1788, and captain-commandant 10 July 1788. With the permission of the French government, he was in Dutch service in 1785 and 1786.

He was appointed a lieutenant colonel in the Berwick Regiment, Irish Brigade in 1791, but never served with the unit. In the following year he served on the Rhine River with the brigade.

The brigade was taken into the regular British establishment before 1794, and he was appointed lieutenant colonel of the 4th Regiment 1 October 1794, serving first in Ireland and later in Jamaica. He was placed on the half-pay list 25 December 1797 and was promoted to colonel 1 January 1801.

Apparently John Paul Jones was instrumental in MacCarthy being elected to the Society of Cincinnati in France.

MacCarthy died unmarried and childless in London, March 1801.

McCLURE, JAMES. James McClure's date of birth and parents are unknown, but he is believed to have been from Pennsylvania.

McClure was one of four persons commissioned a second lieutenant of Marines by Continental Congress on 25 June 1776. He was assigned to the Marine company of Captain Andrew Porter, ostensibly for the frigate *Effingham*, then being built at Southwark, Pennsylvania, however, it appears that he may have served for a short time on a sister frigate being built at Kensington, north of Philadelphia city.

On 4 August 1776, Christopher Marshall, Chairman of the Committee of Inspection and Observation, "wrote 2 passes . . . one to James M'Cluer Lieutenant Marines Ship Randolph, to the Camp." On the following 24 November, he served on the court martial of Private Henry Hasson at the Philadelphia Barracks.

Early in December 1776, Lieutenant McClure accompanied Captain Porter's company into New Jersey, campaigning with General Washington through the battles of Assunpink and Princeton and into winter camp at Morristown. In the meantime, he was promoted to first lieutenant, replacing Daniel Henderson who had been assigned to the frigate *Delaware*.

In early February 1777, Washington's army artillerymen went home, their terms of enlistment having expired. The three companies of Marines took over the duty at Morristown. Of the three, Porter's company served the longest period, to 28 April. About that time, both Porter and McClure resigned their Marine commissions and joined Colonel John Lamb's artillery regiment.

McClure's commission as captain-lieutenant in the artillery was apparently back-dated to 1 January 1777. From some later date until 26 March 1780, he was in Captain Andrew Porter's company, Major Thomas Proctor's 4th battalion, Lieutenant Colonel Thomas Forrest's 2d Continental Regiment of Artillery. He was probably stationed at Newtown and North Windsor, Pennsylvania.

By 5 April 1782 he was on duty at West Point, and on 19 April 1781 was promoted to captain, brought about by Proctor's resignation the day before. On 31 January 1783, a return of Brigadier General Edward Hand's command at James Island, South Carolina showed that following the surrender of Cornwallis, Captain McClure commanded his own company, but he was also reputed to have retired from the sevices on 1 January 1783.

McClure's life following the Revolution has not been determined. A survey of the 1790 census in Pennsylvania lists two persons of that name in both Chester and Cumberland Counties, one each in Franklin, Mifflin, Northumberland, and Westmoreland Counties, and one in the southern district of the city of Philadelphia.

McCLURE, RICHARD, JR. Richard McClure's origins are uncertain, but he may have been a resident of Boston or Essex County, Massachusetts.

On 5 November 1777, one Lieutenant McClure was in Captain Phillips' company of the regiment of artillery raised for the defense of the town of Boston. He was ordered to duty at Castle Island in Boston harbor.

The 16-gun armed merchantman brig *General Gates*, ex-*Industrious Bee*, was built in Liverpool, England in 1777. She was captured by the *Lee* on 29 August of that year, and commissioned in the Continental Navy by 24 May 1778.

About that time, Richard McClure, Jr. was commissioned lieutenant of Marines in *General Gates*, Captain John Skimmer. He was entered on her undated prize list and on her bounty roll of 1778. Another "Richard McClure" was her quartermaster.

The *General Gates* was sold in 1779, and further information of the career and life of Richard McClure is unknown.

McHARRON, CHARLES. Doyle Sweeny, Charles Oscar Paullin, and Gardner W. Allen all state that Charles McHarron was a lieutenant in the Continental Marines. Current research shows no record of him as such.

McNEILL [McNEIL], ROBERT. Robert McNeill's date of birth is not known, but it is possible that he was the son of one William McNeill, who was chosen surveyor of the town of Boston 11 March 1732. Also, one Robert McNeill was in a "Catalogue of Scholars at the South Writing School," Boston, 1768.

The date of his commission as a lieutenant of Marines is likewise unknown, but that of his superior Marine officer, Captain Richard Palmes, frigate *Boston*, was 3 July 1776.

Boston was built at Newburyport, Massachusetts in 1776 and was brought around to Boston harbor for fitting out by her captain, Hector McNeill. At that time, her Marines were said to have been a militia company.

Captain Hector McNeill, originally a Bostonian, was in Quebec when the Revolution began, but returned to his native city and received a commission in the Navy. Some evidence points to a relationship with William McNeill. It may be that Hector McNeill was an uncle of Robert McNeill, and he may have been instrumental in the commissioning of Robert as a Marine first lieutenant.

There was some discontent among the officers of *Boston* prior to her departure for sea. Robert McNeill, with Palmes, Second Lieutenant of Marines John Harris, and others, petitioned their captain on 27 April 1777 for their back wages, one month's advance wages, and equal footing with officers of the land forces. They noted that they were allowed "neither Chocolate Coffee or Surgar, the Stinking New England Rum which has been furnished the Ship is little better than Grogg, the Mollasses so sour that even the Swine refuse it, Salt provisions for near a month together while laying in a plentifull Country, is what we did not expect to be reduced to."

Whether their complaints were reconciled is unknown, but *Boston* departed Nantasket Road on 21 May 1777, sailing with frigate *Hancock*, Captain John Manley, together with nine privateers which soon parted company. On 30 May they fell in with the 64-gun ship *Somerset*, but managed to elude her during the night.

On 7 June *Hancock*, with but very little assistance from *Boston*, subdued the 28-gun frigate *Fox*. McNeill's attempt to put his first lieutenant aboard as prize master was frustrated by Manley, but among the prize crew were Marine Lieutenants McNeill and Harris.

Shortly after, Manley ordered the three ships to proceed for Massachusetts. On 16 July, they ran afoul of *Rainbow*, 44 guns, *Victor*, 18 guns, and, on the following morning, *Flora*, 32 guns. *Hancock*, pressed by *Rainbow*, was forced to surrender on 8 July, again receiving little help from McNeill in *Boston*. *Flora* re-took *Fox*, and her prize crew was imprisoned in Halifax. *Boston* escaped, putting into Wiscasset, Maine, and Hector McNeill sent his prisoners from *Fox* overland to Boston.

Writing to his wife in Boston on 23 July, he gave his first clue to his relationship with William McNeill:

> I have wrote the Council that those Gentlemen [British prisoners] may be detained until my Arivall That with them I may redeem my Own officers, who I fear are taken and carried to Halifax. You must tell My Friend [?] William, That he must make a point of obtaining one of them to redeem his Son Robert who with Mr. Gross, Mr. Harris, Mr. Knowles and Mr. Mullen together with 53 men I put on board that Ship to help man her.

Back in Boston in October 1777, Captain McNeill wrote to one John Butler, Halifax, relating that "Fortune of war has drawn a young kinsman of mine to Halifax." He enclosed a letter to Lieutenant Robert McNeill, informing him that Butler would give him 20 pounds sterling "to help your fellow prisoners (your ship mates in the BOSTON), and I hope if any of them stand in Need you will not see them suffer, . . . You may Acquaint Mr. Gross and Mr. Harris That I have remitted Cash to both their familys." Further, "I would have sent you more credit but I have hopes of your being soon relieved."

On 22 December 1777, Daniel McNeill, a sea captain and brother of Robert, informed the Massachusetts Council that a British prisoner, Donald McBean, was available for the exchange of Robert McNeill. The latter arrived in Boston aboard the cartel ship *Favorite* on 29 January 1778.

Robert McNeill apparently served no further in the Continental Marines, for Doyle Sweeny, in his 1794 Treasury List, asserts that he resigned his commission on 5 April 1778. One Robert McNeill, however, was capain of Marines of the privateer ship *General Mifflin* sometime after 29 January 1778.

A Robert McNeill and Sally Cunningham were married in Boston on 7 November 1779 by the Reverend John Eliot. From 1783 to 1790, one Robert McNeill—the only person of that name in contemporary Boston records—was successively clerk of the market of Boston and, as a merchant, was requested as a bondsman on three separate occasions.

In the 1790 Census of Massachusetts, one Robert McNeill was one of two free white males of that name 16 years and upwards who had one free white male under 16, and six free white females.

MANIFOLD, PETER. Vital statistics of Peter Manifold are not known. Doyle Sweeny and three other authorities of the American Revolution assert that he was a Continental Marine officer, but this cannot be ascertained.

One Peter Manifold, from Pennsylvania, was a cornet in the 4th Continental Dragoons on 14 April 1778, a lieutenant on 1 May 1778, and a captain on 14 April 1779. He resigned the last commission on 30 October 1780.

A "List of Officers proposed for exchange . . . Americans belong to CONFEDERACY," dated 25 September 1781, includes one Peter Manifold "Passenger." Frigate *Confederacy* was captured 15 April 1781 by HMS *Roebuck* and *Orpheus* off the Virginia Capes.

The Journals of Congress state that Peter Manifold was a volunteer Marine officer of frigate *Trumbull* in 1781. On 9 August 1781, *Trumbull* was captured by HMS *Iris* and *General Monk* off the Delaware Capes and Manifold was reputed to have been held a prisoner on Long Island.

MATTHEWMAN, WILLIAM. William Matthewman is assumed to have been a resident of New York.

One William Matthewman was a second lieutenant in the 4th New York Regiment from 28 June to November 1775.

Frigate *Montgomery* was one of the two ships authorized to be built by Lancaster Burling at Poughkeepsie, New York in 1776. Frigate *Congress* was the other. Lieutenant Victor Bicker, Jr. was Marine officer of the *Congress* and was commissioned 4 December 1776.

The date of William Matthewman's commission as captain of Marines of the *Montgomery* is not known. He advertised in five issues of the *New York Journal and General Advertiser* from 8 September to 6 October 1777 for a deserter named David Hardy of the *Montgomery*. Matthewman's physical description of the fugitive included the doubtful attribute of his having "sore shins, [and being] very talkative when in liquor."

Purportedly, the search for Hardy was discontinued on 6 October 1777, for both the frigates were burned on that day to prevent capture by the British. Nothing further is known of Captain Matthewman.

One Luke Matthewman, however, was a Navy lieutenant on board the brig *Lexington* and was court martialed on *Andrew Doria* at Philadelphia on 21 September 1776. He later saw service in both Continental and state vessels and was a prisoner in England for some time. He wrote a narrative of his Revolutionary experiences in the *New York Packet* in 1787.

MEADS [MEAD, MEADE, MEEDS], STEPHEN. Vital statistics of Stephen Meads remain uncertain. His native state is believed to have been New York, but it is possible that he was from New Hampshire.

His only known service was on board the frigate *Raleigh*, which he joined at Portsmouth, New Hampshire on 19 August 1776 at a salary of $18.00 per month. He was probably commissioned a first lieutenant of Marines on either 24 June or after 22 July 1776. On the former date, John Langdon, Continental Agent at Portsmouth, wrote William Whipple that Meads, together with Captain George Jerry Osborne and Second Lieutenant Nathaniel Thwing had accepted the Marine commissions for the frigate. All were recommended to Congress on 1 July, and on the 22d the Marine Committee asked that the commissions be granted.

Apparently, some of the officers of *Raleigh* received few provisions during their first few months on board. On 1 April 1777, Thomas Thompson, commanding the frigate, requested Langdon to pay "Mr. Meeds" 225 pounds of bread, 337 pounds of beef, 56 pounds of rice or flour, 14 1/16 gallons of rum, and 32 gills of vinegar.

Raleigh sailed with *Alfred* on 12 August 1777 for France. Enroute, they infiltrated a convoy of British merchantmen, but, unable to cut any of them out, *Raleigh* contented herself with badly damaging one of their escorts, the *Druid*.

Loaded with military stores, the frigate departed France for America in February 1778. *Alfred* was captured by three British ships on 9 March, *Raleigh* taking no part in the action. Captain Thompson took his supplies on into Boston, where he was relieved of his command and subsequently cashiered.

Second Lieutenant Nathaniel Thwing had resigned by the time *Raleigh* sailed again late in September 1778 under the command of Captain John Barry, but Osborne and Meads were joined by Lieutenant Jabez Smith, Jr. Shortly after leaving Boston harbor, *Raleigh* was intercepted by two large British ships, *Experiment* and *Unicorn*, the latter pursuing in a running battle to the northward. *Raleigh* ran ashore on Wooden Ball Island, off Maine, two days later.

During the ensuing night, the crew was evacuated, but fifteen men were missing. Barry and Osborne took fifty men to the mainland, but Lieutenant Smith was captured with the remainder. The fate of Lieutenant Meads, whether killed in action or escaped to the Maine shoreline, is not known. *Raleigh* was taken into the British Navy.

If indeed, Meads did escape, he may have been the Stephen "Mead," who, in 1790 was a resident of North Salem, Westchester County, New York, on Long Island Sound. Another New Yorker of the same year was Stephen "Mead" of Frederickstown, Dutchess County.

One Stephen "Meades" lived in New Durham, Strafford County, New Hampshire in 1790, and Stephen "Meeds" resided in Washington, Cheshire County, New Hampshire.

MIX, JONATHAN. Jonathan Mix was purportedly a Continental lieutenant of Marines, according to Doyle Sweeny. Allen and Paullin echo Sweeny's original deposition, but no record can be found of him.

MONTGOMERY, HUGH. The positive identity and vital statistics of Hugh Montgomery, lieutenant of Marines in the American Revolution, will be difficult to ascertain.

He should not be confused with Captain Hugh Montgomery of the Pennsylvania Navy, although they may have been related in some manner.

For instance, one Hugh Montgomery, a mariner living in the north part of the Dock Ward, Philadelphia, paid fourteen shillings constable's tax on 30 October 1775. He had four children, ranging from 12 to 27 years of age. One may have been the future lieutenant of Marines.

Also, one Hugh Montgomery married Amelia Irwine at the First Presbyterian Church of Philadelphia on 30 August 1769. Another married Hannah Hanzeley at the Swedes' Church on 31 May 1775.

Hugh Montgomery was appointed a second lieutenant of Marines on 25 June 1776 and was the junior officer in the company of Captain Robert Mullan. He was scheduled to serve in the frigate *Delaware*, launched in the following month and then outfitting at the docks of Southwark, Pennsylvania.

Lieutenant Montgomery is reputed to have recruited Marines for Mullan's company in Lancaster County, Pennsylvania, and in Kent County, Maryland. On 24 November 1776, he was one of the four members of the court martial of Henry Hassan, a private at the Philadelphia Barracks, for desertion and for quitting his guard.

Early in December 1776, Lieutenant Montgomery apparently departed Philadelphia with Mullan's company up the Delaware River to join the forces of Colonel John Cadwalader in support of General Washington's crossing of the river. He probably participated in the battles of Trenton (Assunpink) and Princeton, and accompanied the army into winter quarters at Morristown. There, he was assigned the duty of quartermaster for the Marines.

He was carried on the muster roll and pay roll of Mullan's company through 1 April 1777, although the latter returned to Philadelphia in February 1777. It is recorded elsewhere that on 14 March, he became a first lieutenant in the 4th Continental Artillery, Colonel Thomas Proctor. The exact date of his resignation from the Continental Marines is uncertain.

A prominent source asserts that Lieutenant Montgomery died on 15 May 1777, but details of his death are not provided.

Two Philadelphia directories for the year 1785 list several widows surnamed "Montgomery," but neither sheds any light on their deceased husbands.

On 9 October 1829, a bond of administration was issued on the estate of Amelia Montgomery, and her sons, Robert and William, both storekeepers in Philadelphia, were two of the administrators. Her deceased husband was not named.

MORGAN, ABEL. The birth, parentage and residence of Abel Morgan are unknown.

In the reorganization of the affairs of the Marines in 1776, Abel Morgan was one of four men appointed second lieutenant of Marines on 25 June.

Morgan is reputed to have commanded the Marines of the brig *Lexington* on 7 April 1776 in her engagement with the sloop *Edward*, tender of frigate *Liverpool*, in Delaware Bay.

He was on a regimental court martial ordered by Major Samuel Nicholas when it sat at the Philadelphia Barracks 24 November 1776 in the trial of Private Henry Hasson.

He was also a lieutenant of Marines in frigate *Washington*, built at Kensington, Pennsylvania and launched 7 August 1776, but she was never commissioned.

Doyle Sweeny, in his 1794 Treasury list, relates that Morgan resigned from the service, but gives no date.

MORRIS, WILLIAM. The date and place of birth of William Morris are unknown, and his parentage is uncertain. One prominent source asserted that he was a nephew of the artist, Benjamin West.

One William Morris was a third lieutenant of the Maryland ship *Defence* on 19 September 1776, but whether this individual was later the Continental lieutenant of Marines is unknown.

Following the engagement of sloop *Ranger* with *Drake* on 24 April 1778 and the death of lieutenant of Marines Samuel Wallingford in the battle, sergeant of Marines John Ricker took temporary command of *Ranger's* Marines. Both *Ranger* and her prize arrived at Brest, France on 8 May.

Jonathan Williams, Continental Navy Agent at Nantes, France wrote Captain John Paul Jones on 26 May 1778, enclosing a letter from William Morris. The latter expressed his appreciation that Jones considered him "worthy of being Lieut. of the Merrines [on *Ranger*]; [that he] had no other view than to serve his country; [and] will proceed to Breast as sone as posoble." It is perhaps possible that Morris had been a prisoner of the British in England, had escaped or been exchanged, and was thus available at this opportune time.

Lieutenant Morris joined *Ranger* in Brest in July 1778 and served in her until the capture of Charleston, South Carolina on 12 May 1780. Apparently, he was among the Navy and Marine officers who were paroled to Pennsylvania and was exchanged at an unknown time.

Frigate *Alliance* had had an engagement in the Atlantic on 28 May 1781 with HMS *Atalanta* and *Trepassey*, in which lieutenant of Marines Samuel Pritchard was mortally wounded, and lieutenant of Marines James Warren, Jr. was severely wounded. This important frigate was left with only Captain Matthew Parke and Lieutenant Thomas Elwood commanding her Marines. *Alliance* put into Boston for refitting on 6 June 1781.

Morris was appointed *Alliance's* second lieutenant of Marines on 1 December 1781, and the vessel departed America 23 December, carrying Lafayette to France and arrived there 18 January 1782. She returned to New London, Connecticut on 13 May 1782.

The same month, Lieutenant Morris carried the dispatches of American Commissioner Benjamin Franklin, in Paris, from New London to the Continental Congress in Philadelphia. On his return trip he carried money from Joseph Pennell to Captain John Barry of the *Alliance*.

Citing personal and business reasons, Lieutenant Morris submitted his resignation from the Continental Marines at New London on 29 June 1782, and a copy went to Robert Morris, in Philadelphia. The latter's secretary, John Brown, accepted the resignation, with regrets, on 30 July 1782.

MOULTON [MOLTEN, MOLTON], MICHAEL. Michael Moulton was born at Newport, Rhode Island 17 March 1757, the son of Michael and Hannah (Pierce) Moulton. His father, a sea captain, died at Jamaica, West Indies on 30 January 1763.

Moulton apparently had a checkered career in the Revolution if the following entries of him are correct. He is not to be confused with the Moulton family of Massachusetts, most of whom followed the trade of silversmiths.

One Michael Moulton was a private in Captain Aaron Hayn's Company, Colonel Asa Williams' Regiment, at Fort Ticonderoga from 1 January to 1 December 1776. On the latter date he was reported "sick on board the gondola."

During this same period, one Michael "Molten," of "Swanzey," [Swansea, then in Rhode Island, but later in Massachusetts] was listed in a pay abstract made up to 1 April 1776, as an ensign in Captain Joshua Read's Company, Colonel Varnum's Regiment. The Michael Molten [Molton] in question was also married 7 November 1776 in Newport.

From 29 September to 30 October 1777, Moulton was a sergeant in Captain Caleb Peck's Company, Colonel George Williams' Regiment at Tiverton, Rhode Island, with the roll sworn to at Swansea.

In an affidavit dated 6 October 1818, "Michael Molton Senr" deposed that he had entered sloop *Providence*, Captain John Peck Rathbun, at New Bedford, Massachusetts sometime in November 1777 as a lieutenant of Marines.

In December 1777, *Providence* put into Georgetown, South Carolina after the capture of a British privateer. In a coffee house in Charleston, Captain Rathbun gathered intelligence concerning the defenses of Nassau, New Providence Island, British Bahamas. Returning to *Providence* at Georgetown, he outlined a plan for a second raid on Nassau to his officers, who included Captain of Marines John Trevett and Lieutenant of Marines Michael "Moulten."

After proceeding to sea, *Providence* paused at the Island of Abaco for water and supplies. At midnight, 27 January 1778, 26 Marines filled their pockets with extra cartridges and prepared to go ashore at Nassau.

One source states, "The barge would not hold all the landing party at one time, so Lieutenant Moulten went in first with 15 men, and the barge then returned to the sloop for Trevett and the other 11 Marines."

From the 27th through the 29th of January, the Marine officers held Fort Nassau by means of a series of subterfuges against both the military and civilian populace, until Captain Rathbun could take as many of the vessels in the harbor as possible.

Captain Rathbun released about 30 American prisoners, spiked the cannon of the fort, stripped the magazine of all its powder, and, with several South Carolina merchantmen, departed the island on 29 January.

Following this foray, Moulton apparently returned to the Rhode Island forces where, on 1 May 1778, he was a lieutenant in Captain Jacob Fuller's Company, Colonel John Jacobs' Regiment, for service of eight months. He was also recorded on 23 September 1778 as being a first lieutenant in the same company and regiment, for the defense of the New England states. For the short period of 2 to 9 August 1780, he was a sergeant in Captain Caleb Peck's Company, Colonel Thomas Carpenter's Regiment.

As noted above, Moulton was married on 7 November 1776, in Newport, to Dorothy Brown, daughter of Ezekiel and Rachel (Cole) Brown. She was born in 1759 and died 20 August 1837. They were the parents of nine children.

Moulton's biographer related that he "was a soldier in the Revolution, serving from May, 1775, until February, 1779, as Sergeant, Ensign, Lieutenant, and Lieutenant at sea." He was at the siege of Boston, and in the battles of Harlem, White Plains, Trenton, Princeton, the Island of Rhode Island, and at sea.

Moulton was entered on the pension roll of the United States in 1818 for his service as lieutenant in the Rhode Island line. He died at Newport, 18 December 1820, at the age of 64.

MULLAN [MULLEN, MULLIN], ROBERT. The birth and parentage of Robert Mullan are uncertain, but there is no doubt that he was a resident of Philadelphia.

One Thomas "Mullin," who married Ann Bowles at Christ Church on 21 October 1758, may perhaps have been a son of Thomas "Mullen" the elder, and a brother of Robert.

Several published accounts, nearly all undocumented, credit "The Tun" [Tavern] [originally Carpenter's Tavern], on the corner of Tun Alley and South Water Street, as the principal rendezvous of the Continental Marines. The name Thomas "Mullen" appeared in the *Pennsylvania Gazette* on 6 September 1770 as its proprietor, and other accounts attribute its ownership to Margaret or "Peg" Mullen.

One author asserts its owner was Robert Mullan. A record book found at Milbank, Upper Darby, Pennsylvania in 1891 had in it accounts of both the Free Masons and those of Tun Tavern. It credits Robert Mullan, a member of the Masons and the Continental Marines, with ownership of the tavern. On the other hand, it was also the "day book" of one Nathan Sellers up to May 1833, and he is not known to have been a member of either. Perhaps a shortage of paper in the two centuries compelled each organization and the latter individual to use whatever accounting document or stationery that could be obtained.

If indeed Thomas and Peg Mullan (or Mullen) were husband and wife and were the parents of Robert Mullan, "The Tun" may well have been the Marine rendezvous.

One Thomas "Mullan" was a tavernkeeper in 1767 in the Lower Delaware Ward of Philadelphia, and he paid a tithe of five shillings to the "Overseers of the Poor" in that year.

Furthermore, one Robert "Mullan," the eldest son of Thomas "Mullen," with Robert Morris, merchant, and Robert Bridges, sailmaker, posted a large bond with the county Register of Wills on 25 February 1775 "to make an Inventory of the property of Thomas 'Mullen,' deceased, Innkeper." No such inventory has been found in present-day Philadelphia records.

When the Continental Marines were reorganized on 25 June 1776, Robert "Mullin" was commissioned a captain of Marines on the recommendation of the Marine Committee of Congress. He then proceeded to enlist his company of Marines in Philadelphia and its environs and assembled them in barracks.

It is likely that the first barracks of the Marines were the original barracks that were constructed for Highlanders brought to the Northern Liberties after the defeat of Braddock's army. They were located north of the city of Philadelphia, across Race Street and the swamps of Pegg's Run over which a causeway was erected connecting the barracks to the city. They extended from Second to Third Street and from St. Tammany Street to Green Street. The artillery park along Duke Street later became known as "Artillery lane." The whole area was known as Campington.

On 24 November 1776, Captain Mullan was president of the court martial of Private Henry Hassan at the barracks.

In December three companies of Major Samuel Nicholas' battalion of Marines, including Mullan's company from the frigate *Delaware*, were ordered to join Washington's army for the New Jersey campaign. Captain Mullan took part in the battle of Assunpink Bridge on 2 January and of Princeton on the following day.

In February 1777, Mullan and his company were in Morristown serving as artillery, and on the 20th were ordered to escort 25 British and Hessian prisoners to jail at Philadelphia. Arriving there on the 27th, they were reported to have again gone into barracks, for the *Delaware* was still uncompleted.

The Marine barracks were no longer in Campington. Instead it appears that the house of Abraham Wilt, on the east side of Second Street, between Mulberry (now Arch) and Sassafras (now Race) Streets, had been leased by the government and were utilized as quarters for the Marines. On 21 March 1777, Deputy Barrack Master Peter Ozeas filed a report regarding the unsatisfactory sanitary conditions of the quarters.

On 1 April 1777, Captain Mullan had accrued a total of four months service and his pay amounted to 45 pounds. According to his pay roll, his "Amount of Cloathes at Mr. Tods" came to 15 shillings and 10 pence, and the amount of clothing he drew at Morristown totaled eight shillings and nine pence.

The Journals of the Continental Congress, dated 8 August 1777, entered the notation that Major Nicholas' three companies of Marines on artillery duty, including Mullan's company from 1 February to 1 March 1777, were due extra compensation for that duty.

In early April, Mullan and his Marines went on board the frigate *Delaware*, which was ready to sail at last, and the vessel dropped down to Fort Island. She remained above the chevaux-de-frise at Billingsport until August, sailing through it occasionally in pursuit of British ships.

In September the British occupied Philadelphia from the west and erected batteries along the Delaware River to prevent the American fleet from "annoying the city." On the morning of 27 September, the *Delaware* tried to pass the batteries, firing as she sailed northward, but on receiving shot in return, she was set on fire and grounded in midchannel, where she struck to the enemy.

Her Marines and sailors hurriedly attempted to escape to the New Jersey shore. Marine Lieutenant Alexander Neilson and a few others succeeded, but apparently Captain Mullan and the remainder of the crew were captured.

They were imprisoned first in the Philadelphia city jail and later in the State House. One Marine, Stephen Archer, later recounted that "all us that were not killed, made prisoners and put into the new jail, where out of 109 only 18 or 19 escaped starving."

When Philadelphia was abandoned by the British in June 1778, the Americans reoccupied the city. The Marine Barracks, presumably those in the Wilt house on North Second Street, were reestablished. Mullan was instructed to recruit a company of 50 Marines, which he accomplished between 9 August 1779 into early 1780. Desertion was still a problem for Mullan advertised for deserters "To be delivered at my Quarters, the Corner of Pine-Street in Front-Street."

At some time thereafter, Captain Mullan was apparently captured and imprisoned by the British, but the circumstances under which it occurred are not known. On 16 February 1781, Mullan wrote from the *Jersey* prison ship at New York to Thomas Bradford, Commissary of Naval Prisoners in Philadelphia:

> . . . Mr. David Sproat, Commissary of Naval Prisoners here, has promised me that if you will order a person for me, he will set me at Liberty, if you have a Marine [or] officers in the Privateering Day [duty], I beg you to send him immediately for me or let me know the reason I am forgot by you. If you have not a Marine officer, send Some Boddy, Else by his work [word ?] he will let me go.

A Mr. John Ebert, in the custody of Commissary Abraham Skinner, Elizabethtown, New Jersey, was being considered for the exchange in March 1781. The date of the exchange is unknown.

Captain Mullan was one of the signers of a petition to the Continental Congress on 28 April 1783 concerning the provisions for back pay of the Navy and Marines.

In 1785, Robert "Mullen" was the proprietor of a "beefstake and oyster house" on Walnut Street, between Front and Second Streets. Another directory of the same year listed Robert "Muller" at 312 Walnut Street.

He was voted into Masonic Lodge No. 2, A. Y. M. (Moderns) on 29 March 1762. There is also record of a Robert Mullan being admitted to Masonic Lodge No. 2 on 8 December 1778 and chosen steward, 14 December 1779. They may have been father and son.

Robert Mullan's death is unrecorded in Philadelphia records, but Doyle Sweeny asserted that he died before 1794.

The surname "Mullen" and its derivatives were not too common in Revolutionary and post-war Philadelphia, and a list of those which have come to attention may be beneficial for further research:

1. J. Mullen was a justice of the peace, when on 23 October 1780, he wrote to Commissary of Prisoners Thomas Bradford.

2. A bond of administration was issued on the estate of John Mullen, late of the Pennsylvania regiment of artillery on 24 October 1783. Andrew Porter was one of the administrators.

3. James Mullan was a shopkeeper on Lombard Street, between Front and Second Streets, in 1785.

4. Patrick Mullen was listed at 481 Catharine Street in 1785.

5. William Mullen lived at 545 Third Street in 1785.

6. A bond of administration was issued on the estate of Arthur Mullan on 5 March 1789. Barney Mullan was an administrator.

7. One Robert Mullen died 19 September 1793, aged 27, and was buried in Christ Church graveyard.

8. Thomas Mullen, a house carpenter, bought a lot on the north side of Locust Street, west of Tenth Street, on 31 March 1807.

9. Thomas Mullin, formerly a private in the "Marine Corps art." began receiving a federal pension on 9 March 1817, as listed in the Pension Roll of 1835.

10. A bond of administration was issued on the estate of Margaret Mullen, a widow, in Southwark, on 12 May 1836.

NEILSON [NELSO, NELSON, NIELSON, NILSON], ALEXANDER. The birthdate and parents of Alexander Neilson are unknown. He is believed to have been a resident of Philadelphia, but may have been from Maryland.

In 1775, one Alexander "Nelson," of the North Ward, Philadelphia, was assessed a county tax of about two shillings.

On 8 December 1775, Alexander "Nilson" entered the ship *Alfred* under Captain of Marines Samuel Nicholas as a sergeant of Marines. He served at the capture of Forts Montagu and Nassau at New Providence in March 1776.

Before the departure of the fleet for the island, he may have been married to Mary Clouser at Philadelphia's Second Presbyterian Church on 12 January 1776.

On what date Neilson became a second lieutenant of Marines is uncertain, but on 14 October 1776, Commodore Esek Hopkins paid "Alexr Nelso Lt Marines Alfred" an order on Mr. Manley for the sum of 20 dollars. Furthermore, "Alexand.r Nelson . . . 2.d Lieut" was on a "List of Officers & Seamen belonging on board the Ship Alfred who are entitled to a Share in the Ship Mellish & Brig Active" in November 1776.

When Captain John Paul Jones left *Alfred* late in 1776,

Lieutenant Neilson may have accompanied him. His day-to-day movements are unknown, but on or about 1 April 1777, First Lieutenant of Marines David Love is believed to have resigned from Captain Robert Mullan's company of Marines in Philadelphia. Second Lieutenant Daniel Henderson was then promoted to Love's first lieutenancy, and Neilson was assigned on frigate *Delaware* as second lieutenant of Marines.

In the same month, after outfitting for sea, *Delaware* dropped from Philadelphia southward to Fort Island, above the chevaux-de-frise in the Delaware River. On 25 June, Neilson apparently had guests aboard and, to serve them refreshments, borrowed some from the sea chest of his fellow officer, Henderson. Henderson subsequently charged Neilson with theft, but a court martial found him not guilty, provided he replace the liquors.

Late in September 1777, *Delaware* ran upriver past Philadelphia, bombarding British gun emplacements on the western bank. Counterfire from the enemy set her afire and she grounded in mid-channel. Just before British troops boarded the frigate, Lieutenant Neilson escaped to the Jersey shore with some of the crew, but the remainder were captured and imprisoned in Philadelphia.

Alexander Neilson resigned his commission on 5 April 1778. If so, he may have been the person of the same name, a resident of Maryland, who was commissioned a second lieutenant in the 1st Continental Artillery, but who resigned the latter commission on 23 May 1779.

Even though the career of Alexander Neilson is obscure, it is possible that he was one Alexander "Nelson" who wrote Francis Gurney of Philadelphia on 21 July 1781, from "On board the Pris[on] [ship ?] at N. York" . . . "no parole to go ashore in the city. Use your influence to procure an exchange or parole."

NICHOLAS, SAMUEL. The only son of Anthony, a blacksmith, and Mary (Shute) Nicholas, he was born in Philadelphia about 1744. His father died in 1751. From 1 January 1752 to 17 December 1759, he received his education at the Academy of Philadelphia, forerunner to the University of Pennsylvania. After his graduation, he was apparently admitted into the society of the city, becoming member No. 102 of the State in Schuylkill (the Schuylkill Fishing Company of the State in Schuylkill) on 1 May 1760, and 1766 he became a member of the Gloucester Fox Hunting Club. One source asserts that he sailed as supercargo on the colonial mechant vessels of Robert Morris.

On 10 November 1775, Continental Congress authorized two battalions of Marines for the Continental service. The first officer commissioned for these Marines was Captain Samuel Nicholas on 28 November 1775, thus, he was the first officer commissioned in the naval service. He proceeded to enlist his company in Philadelphia between 6 December 1775 and 14 January 1776, signing up 62 men in 40 days for service on board the ship *Alfred*.

At New Providence Island, the Marines of the squadron landed on 3 March 1776 under the command of Captain Nicholas, the senior Marine officer, capturing Forts Montagu and Nassau.

Leaving New Providence for New England on 17 March 1776, the squadron was off the coast of Rhode Island when it was intercepted by HMS *Glasgow*, and *Alfred*, among others, engaged her in combat. Two days later, on 8 April, *Alfred* reached port at New London, Connecticut. Nicholas' account of the expedition added much to the annals of the Marines' first amphibious landing.

Nicholas was sent by Commodore Esek Hopkins with dispatches from New London to Congress in Philadelphia. On reporting there, he was detached from *Alfred* and promoted to major of Marines on 25 June 1776. His orders were to remain in Philadelphia and to recruit and discipline four new companies of Marines for frigates then on the stocks. On 13 August and again on 16 September, the Pennsylvania Council of Safety asked to substitute Nicholas' Marines as guards for the frigates, in relief of units of the Philadelphia Associators. In December 1776, three of the four companies of Marines were ordered to join General Washington for his projected campaign in New Jersey.

Major Nicholas and his small battalion of 80 men joined Washington and participated in the Second Battle of Trenton (Assunpink Creek) and in the Battle of Princeton, 2–3 January 1777.

Nicholas returned to Philadelphia from service with the Army about 1 March 1777. In September 1777, he was on special duty at Lebanon, Connecticut, obtaining arms and lead for the Army. Between July 1777 and the year 1780, he executed orders for the Marine Committee and the Board of Admiralty as major of Marines and as muster master.

He married Mary Jenkins, of Jenkintown, Pennsylvania, believed to have been the daughter of Dr. Charles Jenkins, at Friends Meeting in 1778. They had three sons: Samuel, Charles Jenkins, and William, and two daughters, Sarah and Mary. He was recorded in 1778 as a member of the "Patriotic Association of Philadelphia."

Following the British evacuation of Philadelphia in 1778, the Americans reoccupied the city. The Marine Barracks were reestablished, and recruiting was renewed by Major Nicholas and Captain Robert Mullan. In March 1777, the "Marine Barracks" was a house on the east side of Second Street, between Mulberry (now Arch) and Sassafras (now Race) Streets, owned by one Abraham Wilt. Although extant Philadelphia records show neither leases or rentals, deeds therein indicate this site as its location. Perhaps in 1778 the "Marine Barracks" was again located there.

An additional duty was given Major Nicholas on 8 August 1778 when he was elected an authorized signer of Continental currency, on the recommendation of the Board of Treasury. About this time, it also appears that he was sent to Boston to transport specie just arrived from France, carefully avoiding British-held New York, to Philadelphia. The mission was accomplished. He was again sent to Boston in September 1778 to convoy clothing from France for the use of the army. He was recommended for an additional assignment on 29 May 1779 as a signer of Continental bills of credit.

His services as a Marine officer were revived on 7 August 1779 when the Navy Board, Middle District, directed him to raise a company of 50 Marines at Philadelphia "as expeditiously as possible" although the reason therefor is not apparent. Captain Robert Mullan, assigned as company commander, recruited these men from 9 August 1779 into 1780.

Nicholas made an effort to return to sea on 10 August 1781 when he requested assignment to the new frigate *America*. He also asked Congress for pay and prize money, pointing out this his promotion to major had deprived him of all opportunity to obtain the latter in the land service.

After some consideration Congress resolved that his accounts be settled up to 25 August 1781, at which time he would be considered as retiring from the service. Although technically retired, he was appointed on 30 November 1781 to a court to inquire into the loss of the frigate *Trumbull*. On 28 April 1783, he joined in a memorial to the Continental Congress concerning pay, describing himself as a "late Major of Marines."

In the fall of 1782, Nicholas stood for election as Sheriff of the City and County of Philadelphia, but results of the election are unknown. It may be worthy of note here to relate that his sister-in-law, Sarah Jenkins, was married to General Josiah Harmar, U. S. Army, at Swedes' Church, Philadelphia, on 10 October 1784.

As a citizen, both before and after the Revolution, he was a tavernkeeper at the "Sign of the Conestogoe Waggon" located at No. 78 Market (or High) Street, between Fourth and Fifth Streets, owned by his mother-in-law. One source states that it was the meeting place of the Society of the Cincinnati. His residence appears to have been adjacent or at least in the same block. General Harmar purchased the tavern on 19 September 1798.

Nicholas also speculated in considerable tracts of land in the area which is now adjacent to South Broad Street in the heart of midtown, modern Philadelphia.

Between 1785 and 1788, he served on the Standing Committee of the State Society of the Cincinnati of Pennsylvania. There are indications that the revived Schuylkill Fishing Company held its social meetings at the "Connostogoe" as late as March 1789. Nicholas was also a member of Lodge No. 13, A.Y.M. (Masonic) according to its 1783 records.

Samuel Nicholas died in Philadelphia on 27 August 1790 at the early age of 46 years. He was buried in the Friends Burial Ground, Fifth and Arch Streets, Philadelphia, in an unmarked grave, as is the custom of the Society of Friends. His burial is recorded in their records.

Papers were granted on 8 September 1790, for the administration of will to his widow, Mary Nicholas, for the settlement of his estate. No inventory of his belongings accompanies the Court records, but some of his holdings are accounted for in the will of his widow, dated 28 September 1818.

NICHOLSON, WILLIAM. Nothing of the origins of William Nicholson is known. In fact, it has not been absolutely determined that he was a captain of Marines.

Sweeny, in 1794, was the first to list him as such, perhaps from the documentation that does not exist today. He was followed in order by Paullin, Allen, and McClellan.

Captain Nicholson is reputed to have been on board the frigate *Warren* shortly after 18 April 1779. On that date, *Warren*, *Queen of France*, and *Ranger* sailed from Boston on a cruise. They subsequently captured seven of nine vessels, the *Jason*, *Maria*, *Hibernia*, *Patriot*, *Prince Frederick*, *Bachelor John*, and *Chance*.

There is reason to doubt that William Nicholson was indeed on board at this time. On 17 February 1779, Richard Palmes, in his capacity as captain of Marines in *Warren*, advertised in a Boston newspaper for "Marines, Seamen or Landsmen." It is not likely that a frigate would have two captains of Marines on board at one time. It was not until May 1779 that Palmes left the ship, and he was replaced at that time by Captain John Welsh, just before the Penobscot expedition.

NICOLSON [NICHOLSON], WILLIAM. William Nicolson is assumed to have been a native of France. One Murray de Nicolson, "lord of the chateau de Nicolson," Sennonville, France, wrote Captain John Paul Jones on 14 May 1780, asking the pleasure of a day with him at his estate.

The relationship of the two is unknown, but the Countess de Nicholson has been reputed to have been one of Jones' mistresses while in France. William was purportedly either her brother or brother-in-law.

On 2 June 1780, Jones received a letter from another acquaintance in Paris that he understood Jones was thinking of accepting the services of William Nicolson. Further, an artist in Paris informed Jones that he would like him to sit for a portrait that the Countess had commissioned of him.

Frigate *Alliance*, after being seized by Captain Pierre Landais, put to sea for America on 8 July 1780, leaving behind Jones and many of his men from the sunken *Bonhomme Richard*. To take his passage home, Jones obtained the sloop-of-war *Ariel*, a captured British ship, from the French, and proceeded to outfit her and load with military supplies for the Continental cause.

Many of his former officers and enlisted men joined him. The remainder of the crew was made up of French citizens. To command the French Marines, he appointed William Nicolson "Captaine de Marine," and W. [or M.] Jorris, the lieutenant of Marines.

Ariel put to sea on 7 October 1780, accompanied by two brigs also loaded with military stores. They were caught the following day in a violent storm which nearly wrecked the vessel. They put back into L'Orient for repairs.

During November, Captain Nicolson was in Paris, using the influence of the nobility to obtain another ship for Jones. His efforts were evidently unsuccessful, for *Ariel* sailed again on 18 December 1780.

Enroute she had a short skirmish with a British 20-gun privateer which appeared to have surrendered and then fled. *Ariel* reached Philadelphia on 18 February 1781.

Officers of the *Ariel* had considerable difficulty obtaining their pay, Jones being tardy in submitting the pay roll in May and June 1781. However, William "Nicholson" received his pay of $392.36 to the 25th of August 1781, the day that the pay report passed the Board of War.

In the meantime, the Board of Admiralty had taken into consideration on 6 July 1781 a report on the appointment of William "Nicholson" as a captain of Marines. Apparently, a blank commission signed by John Hancock, President of Congress, had been issued Nicolson [or Nicholson] in France, and confirmation was now being requested.

On 16 July 1781, the Board ruled that commissions granted by United States Ministers abroad be considered temporary brevets and that Nicolson [or Nicholson] be appointed a lieutenant of Marines. The appointment was approved the following day.

Further information of the career of this Marine officer is unknown.

O'KELLY, JAMES GERALD. James Gerald O'Kelly's exact provenance is unknown, but he was of Irish origin. The circumstances under which he became affiliated with the French military are not known.

On 14 June 1779, he was a sub-lieutenant of grenadiers in the Irish regiment of marine artillery, French Army, at Paimboeuf. On that date, Colonel Francois Jacques Walsh-Serrant wrote Captain John Paul Jones, informing him that O'Kelly wished to join his two comrades, Edward Stack and Eugene MacCarthy in *Bonhomme Richard*.

James Moylan, United States commercial agent at L'Orient

also wrote Jones on 18 June, stating that O'Kelly had resigned his commission in Walsh's regiment. O'Kelly was subsequently given a brevet lieutenant's commission signed by John Hancock.

In the engagement of *Bonhomme Richard* and *Serapis* on 23 September 1779, Lieutenant O'Kelly was stationed on the poop deck with Lieutenant Colonel Paul de Chamillard, a French volunteer and 20 Marines.

Lieutenant O'Kelly was killed in the battle, but ironically, not by the enemy. Captain Pierre Landais, in frigate *Alliance*, contributed little assistance to Jones. Nor did he assist other ships of the squadron against the *Countess of Scarborough*, *Serapis*'s consort. Rather, he poured three broadsides into *Bonhomme Richard*, one of which killed O'Kelly.

O'CONNELL, MAURICE. The origins of Maurice O'Connell are in doubt, but he may have been of the Clan O'Connell, of County Kerry, Ireland. He may have been related to Eugene MacCarthy, also of that county, whose mother was an O'Connell, and may have been acquainted with Edward Stack, whose forebears were natives of Kerry. All were in the French service at the beginning of the American Revolution.

The French frigate *Pallas*, of twenty-six 9-pounders and six 4-pounders, Captain Denis-Nicolas Cottineau Kerloguen, was assigned to John Paul Jones' squadron in 1779. His Marine officer was Captain Maurice O'Connell. The ship's captain was given a United States commission, but it is not known whether O'Connell received a brevet Marine commission.

From 28 July through 2 August 1779, Captain O'Connell was a member of the court martial of Quartermaster Robert Towers, of the *Bonhomme Richard*, for mutiny.

On 23 September 1779, *Pallas* engaged and captured the 22-gun sloop *Countess of Scarborough*, at the same time *Bonhomme Richard* was defeating *Serapis*.

O'Connell wrote Captain Jones a letter from Rotterdam, Holland on 17 December 1779. In it, he expressed the belief that he was too old to be under the orders of Lieutenant Colonel Paul de Chamillard, a French volunteer with Jones, but had he been under Jones himself, he would not have left his service. He also informed Jones that he had been mistaken for him several times and felt complimented. He further asserted that he would join Jones again if he so desired.

OSBORNE, GEORGE JERRY. The date of Osborne's birth and the names of his parents are presently unknown. Evidence from several primary sources established his residence as Portsmouth, New Hampshire. One source asserts it was Exeter, only a short distance away. He married Olive, daughter of Captain Thomas and Dorothy Pickering, of Portsmouth on 13 May 1756. He was a schoolteacher and resided there for several years before the Revolution. New Hampshire vital statistics for the years 1765 and 1770 include the name on the tax list, with the prefix of "Captain."

In a return of officers and men of New Hampshire military units dated 5 November 1775, Osborne was captain of a company of artillery at Portsmouth.

Frigate *Raleigh*'s keel was laid at Portsmouth, 21 March 1776, and she was launched two months later. Captain Thomas Thompson was given her command about 10 June, and he set about recruiting his crew. New Hampshire Naval Agent John Langdon sent a list of selected officers to the Marine Committee at Philadelphia, noting that all had accepted the offers. Osborne was her captain of Marines. First Lieutenant Stephen Meads [or Meeds] of New York and Second Lieutenant Nathaniel Thwing of Boston were the other two Marine officers.

Captain Osborne, described as an "Englishman" and of a brown complexion, reported on board *Raleigh* 12 August 1776 at wages of twenty-six and two-thirds dollars per month. Between 13 August 1776 and 26 June 1777, five enlisted men of Osborne's original artillery company signed on board *Raleigh* as Marines.

Undoubtedly Osborne attempted to recruit many others from New Hampshire's land forces, for considerable correspondence exchanged hands among the State Committee of Safety, Captain Thompson, and Captain Osborne regarding the latter's right to recruit Marines and seamen.

Raleigh got to sea on 12 August 1777, sailing with *Alfred* to France for military stores. On 2 September they took a small ship from the British West Indies fleet and, on the following day, fell in with the fleet itself, using the enemy's code signals. *Raleigh* singled out a man-of-war, the *Druid*, and engaged her for about 20 minutes, completely disabling her. *Raleigh* was forced to break off the engagement when other escort vessels attacked her. Private John McCoy, *Raleigh*, was wounded in the battle.

Raleigh and *Alfred* departed France in February 1778 with an additional 17 French Marines on board the former. On 9 March, HMS *Ariadne*, 20 guns, and *Ceres*, 14 guns, gave battle to *Alfred*, a slow sailer, and captured her after a few broadsides. When he determined that he could not retake *Alfred*, Captain Thompson sailed on for America, feeling that the military stores were of more importance.

Command of *Raleigh* was given to Captain John Barry on her return to Boston. Second Lieutenant of Marines Nathaniel Thwing apparently resigned and his place was taken by Lieutenant Jabez Smith, Jr., of Groton, Connecticut.

Captain Barry sailed from Boston, 25 September 1778, with a brig and a sloop under convoy. At noon she sighted two heavily-armed British ships, *Unicorn* and *Experiment*. For the following 48 hours, Barry tried to elude them, but being badly damaged by cannonading, *Raleigh* was run ashore on Wooden Ball Island, off Maine.

Barry and Captain Osborne transported a large part of the crew in boats to the mainland, leaving Lieutenant Smith in charge of those still on the island and with instructions to destroy the frigate. A subordinate officer surrendered the ship and Smith, with those under his charge, was captured.

Meanwhile, the keel of the largest vessel thus far to be built in this country—the *America*—had been laid at Langdon's Island, Kittery, Maine. She was to mount 74 guns.

Captain Osborne was appointed on 20 November 1779 to command her Marines. Captain John Paul Jones arrived at Portsmouth in June 1781 to superintend her construction and eventually command her. It is likely that Osborne's Marines mounted the guard over her at night, for the British tried to destroy her several times. She was not launched until November 1782, but was immediately presented to the French government.

Little more definitive information is known of Captain Osborne, but it is evident that he and his wife had at least one son. George Jerry, Jr., a printer, was a partner in the *New Hampshire Gazette* at Portsmouth and later worked in Newburyport, Massachusetts. The son died there in June 1800, leaving a son, Nathaniel.

Taxes were levied on Captain George Jerry Osborne in 1807 and 1808 at Portsmouth, but in following years the levy was made on his heirs for the buildings of his estate.

PALMES, RICHARD. Existing genealogies of the Palmes families place the majority of that surname in the vicinity of Boston, Massachusetts. In 1765, one Richard Palmes and Mehitable, his wife, deeded to a third party a house and tract of land on the south side of Elm Street in that city. One Richard Palmes also held town office there from 1770 to 1779 and testified at the trial of the defendants of the Boston Massacre on 5 March 1770. A positive origin of Richard Palmes, the Continental Marine officer, however, is elusive.

Following the recommendation of the Marine Committee, Continental Congress appointed Richard Palmes a captain of Marines on 23 July 1776. He apparently recruited other Marines, for, on the following day, Congress commissioned Benjamin Thompson a lieutenant of Marines in Palmes' company.

It is likely that Palmes was assigned to the frigate *Boston*, Captain Hector McNeill, late in 1776. In January 1777, he was ordered to search the privateer *Rising States* at Marblehead, Massachusetts, for deserters, but he found none.

Boston sailed from Boston in May 1777 in company with frigate *Hancock*, Captain John Manley. They captured a small brig on 29 May and the British frigate *Fox* on 7 June. However, on 7 July, both the prize *Fox* and the *Hancock* were lost to a British squadron of three ships. *Boston* sought refuge in Maine's Sheepscot River, where she was blockaded until 10 August. On that date, she slipped out and made the port of Boston on 21 August.

Unfortunately, all efforts in the meantime were not in harmony to extricate *Boston* from her predicament. On the day she left the Sheepscot, Captain McNeill placed Palmes under arrest for "unofficer-like behaviour and repeated breach of my Orders . . ." This was followed by several other petulant letters from McNeill until on 9 September 1777, McNeill charged Palmes with "Misaplication of the Ships Stores, Neglect of duty, disobedience of orders, and attempts to Excite Murmuring and Mutiny among the Ship's Company," and requested his court martial.

Palmes, however, was sent on or about 12 November 1777 by the Marine Committee to the Navy Board, Eastern Department, on an errand. There he apparently presented his own case rather succinctly and was not court martialed. Instead, he was reassigned as the Marine officer of frigate *Warren*, exchanging posts with Captain of Marines John Grannis. Captain McNeill even tried to withhold prize money due Palmes in October 1777, but Paymaster General James Warren ruled against the move.

Shortly after, Captain Samuel Tucker took command of *Boston*, and when she sailed 15 February 1778, with John Adams a passenger for France, Palmes was again her captain of Marines. She captured the British vessel *Martha* on 11 March and arrived at Bordeaux on 1 April. Three days later, Palmes accompanied Adams to Paris and for a time acted as an officer-courier between *Boston* and the French capital. *Boston* arrived back in the harbor of Boston on 23 October 1778, after taking a number of prizes enroute.

By 17 February 1779, Palmes was back on board *Warren*, Captain John B. Hopkins, for on that date he advertised in a Boston newspaper for Marines, seamen, and landmen for a cruise to take place six days later. It was a short cruise, after which Palmes delivered dispatches from Captain Hopkins to the Eastern Navy Board at Boston on 17 April. He informed the board that eight vessels had been captured, and that the ship was then in Nantasket Roads.

Captain Palmes' seeming penchant for trouble asserted itself again in mid-1779 when "he left his duty at a critical time without the leave and against the judgment of Mr. [William] Vernon." On 7 June, the Marine Committee wrote the Eastern Navy Board suggesting that Palmes be court martialed if they felt he had "transgressed in any manner" as they "will never countenance any indecency of behaviour in their Officers to their Superiors." Palmes was replaced on board *Warren* on 18 May 1779 by Captain John Welsh.

Again, there is no evidence that Captain Palmes was court martialed, and he sailed on *Boston* from Nantasket Roads for Charleston, South Carolina on 23 November 1779.

This was not a successful voyage, and it proved the last for frigate *Boston* in the American service, for she was captured at Charleston on 12 May 1780. Action was taken on a petition submitted by Palmes on 5 July 1780, when he was paroled and soon after exchanged for captain of Marines Hector McNeill of HMS *Somerset*.

His next assignment appears to have been in frigate *Deane*, he being the fourth person to enter her on 1 May 1781, serving in her until 31 May 1782. Two other men with the surname of "Palmes" were also aboard *Deane*, one being Midshipman Samuel Palmes and the other Francis Palmes, a boy, but their relationship, if any, is unknown.

Captain Palmes continued serving on this vessel, renamed *Hague*, Captain John Manley, from 10 May to 29 August 1783.

No trace has been found of Richard Palmes after the American Revolution. However, one Richard Palmer, which misspelling was often seen in Revolutionary records, was one of two free white males of that name 16 years of age and upwards and had two free white females in Salem, Essex County, Massachusetts in 1790.

PARKE, MATTHEW. Matthew Parke was born near Ipswich, England in 1746. His grandfather, a British Army colonel and former aide to the Duke of Marlborough, was later Governor of the Windward Islands. He went to Virginia for a time, taking his grandson with him. He later returned to England and left Matthew in the colonies.

Parke probably moved to Philadelphia where, on 28 November 1775, he was commissioned a first lieutenant of Marines in Captain Samuel Nicholas' company on board the ship *Alfred*.

His service on *Alfred* took him to the Bahamas with Commodore Esek Hopkins' squadron where he participated in the landing at New Providence. On 6 April 1776, *Alfred* and the brig *Cabot* engaged the British ship *Glasgow* in Block Island Sound. Parke was uninjured, but his fellow lieutenant, John Fitzpatrick, was killed.

After his return to Providence, Rhode Island, Parke was a member of the court martials of Captain Abraham Whipple, ship *Columbus*, and Captain John Hazard, sloop *Providence*, for their supposed inactivity in the engagement with *Glasgow*.

On 26 May 1776, Parke was promoted to captain of Marines by Congress and assigned to *Columbus*, replacing Captain Joseph Shoemaker, who had resigned his commission. However, Joseph Hardy had already been appointed to that vessel by the Marine Committee, and Commodore Hopkins released Parke for reassignment on 14 February 1777.

He was next assigned to sloop *Ranger*, Captain John Paul Jones. On 27 July orders were sent to Parke at Providence to go to Boston to enlist seamen and Marines. On 5 September he proceeded to Portsmouth to join *Ranger*.

When *Ranger* sailed from Portsmouth on 1 November 1777 for France, she carried a large contingent of Marines with two officers, Captain Parke and Lieutenant Samuel Wallingford, of Somersworth, New Hampshire. On their arrival in France, their fellow officers complained that the 20-gun ship did not rate a captain of Marines, and that Parke's large share of prize money reduced that of others. Consequently, Parke received his discharge from Jones and returned to America in the frigate *Deane* in February 1778.

Parke was then assigned to the frigate *Alliance*, Captain Pierre Landais, and sailed for France on 11 January 1779 with Lafayette as a passenger. A mutiny almost succeeded enroute, but it was frustrated and *Alliance* anchored at Brest on 6 February 1779.

In a letter to his personal friend, Colonel Paul Revere, dated 18 February, Parke wrote that the attempted mutiny took place on 3 February and that advance information of only one hour enabled the officers to subdue the uprising and place 38 men in irons.

About June 1779, *Alliance* was ordered to L'Orient to join the squadron of Captain John Paul Jones, then commanding *Bonhomme Richard*. She sailed shortly in company with several French ships on a cruise around the British Isles.

On 23 September 1779, *Bonhomme Richard* defeated the frigate *Serapis*, and the French ship *Pallas* took the sloop *Countess of Scarborough* off Flamborough Head. Captain Landais in *Alliance* contributed little to either victory and even damaged Jones' ship with several broadsides. When the matter came up in Landais' court martial in October 1780, however, Captain Parke defended his captain in his testimony.

Bonhomme Richard sank after the battle, and Jones sailed *Serapis* into the Texel, Holland. There Jones took command of *Alliance* for a cruise off the African Coast and back to France. During Jones' absence ashore, Landais again took control of the frigate and set sail for America on 8 July 1780.

Enroute, Landais became irrational in his actions and, on 11 August the crew, aided by the officers and passengers, took over the ship and placed Landais under arrest. When *Alliance* arrived at Boston, Captain John Barry relieved Landais. Captain Parke and his Marines had to forcefully remove the ex-captain from his cabin to shore.

When *Alliance* put to sea again on 11 February 1781, she was carrying diplomatic passengers to France. Captain Parke, assisted by Lieutenants Thomas Elwood, James Warren, Jr. and Samuel Pritchard, commanded the Marines. Another mutiny was quelled enroute to America.

Several prizes were taken by the frigate before she encountered two strong British ships, the *Atalanta* and *Trepassey* on 27 May 1781. On the following day, a three-hour battle ensued. Parke again escaped injury, but Pritchard was mortally wounded, and Lieutenant Warren sustained a wound that later necessitated amputating his leg. The frigate reached Boston harbor on 6 June.

On 28 June 1781, Parke's intention to marry was published in Boston. The bride was Judith Cooper, the only daughter of William Cooper, the Town Clerk of Boston from 1771 to 1809, and of Katherine Wendell Cooper. The marriage took place on 15 August 1781.

Alliance departed Boston for France again on 25 December 1781 with Lafayette aboard. Parke and Elwood had been joined by Lieutenant William Morris, who had been captured at Charleston in May 1780. After arriving back in New London, another mutiny occurred on 16 May 1782, and Captain Parke was a witness at the court martial held on 27 July.

Parke's last voyage in *Alliance* began on 4 August 1782. She took four prizes by 23 September, and in October she put into L'Orient. On 17 November the officers asked for their prize money and back wages, for some had been paid very little for several years. When Captain Barry was unable to provide it, six of the officers moved to quarters ashore. Only Navy lieutenant Hezekiah Welch and Lieutenant Elwood remained on board. Captain Barry ordered them back by late afternoon of the 24th. When they refused, Barry left them ashore and put to sea on 9 December.

Captain Parke was tried by court martial on board the *George Washington* in Philadelphia on 16 May 1783. He was charged with disobedience of orders, to which he pled guilty. Second, he was charged with "detaining the Ship in Port [and] causing an additional Expence," to which he pled not guilty. He was found guilty of only the first charge, for which he was to "forfeit his Commision Provided however that this sentence shall not affect any wages or Monies due . . . on or before the twenty fourth day of November" 1782. Nevertheless the court "beg leave from the long & faithful Services of Captain Matthew Park to recommend him to the Honble the Congress of the United States."

William Cooper Parke, the only son of Matthew and Judith Parke had been born in Boston on 7 August 1782. Judith died on 14 September 1782. The son later became a shipping merchant of Boston and died there 11 November 1857. His only male heir and descendants later lived in Honolulu.

Following the war, Matthew Parke was a grocer and seedsman, with his place of business on the north side of the Town Dock. Among other things, he was approved by the Selectmen of Boston as a retailer of rum, and in 1786 was "appointed to weigh bunches of Onions."

On 8 December 1796 he was married a second time, to one Jane Inglish (or English). In the following year, he bought a one-third interest in a house on Fleet Street, formerly the property of one Alexander English. In 1798, he occupied, but did not own, a brick and wooden dwelling on Fleet Street, valued at $1,500.00.

Matthew Parke died in Boston on 28 December 1813.

PARKER, AVERY. Little is known, including the origins, of Avery Parker.

On 6 July 1776, the Committee appointed to build two Continental frigates in Rhode Island voted that Avery Parker be appointed a first lieutenant of Marines in the frigate *Providence* "on his inlisting thirty three good Men in Twenty Days from this Date."

Parker apparently made good on the qualifying requirement, for it appears that he was on board for nearly a year.

One Stephen Earle asserted in his pension affidavit, dated 21 April 1818, that he had served on board *Providence* as a sergeant and second lieutenant of Marines.

He further declared that "Avery Parker, the first Lieutenant of Marines was broke [circa April or May 1777] and this deponent was then advanced to the first Lieutenancy of Marines . . ."

PENNINGTON, MILES. Miles Pennington was from the District of Southwark, Philadelphia, Pennsylvania. His parentage and birth date are unrecorded.

Sometime during the period from 1766 to 1776, Pennington apparently purchased property, perhaps his residence, from one Richard Boardman. Unfortunately, the record of the deed has been lost. In 1775, he paid an assessment of two shillings for County Tax and four shillings six pence Provincial Tax on the Southwark property.

Southwark was then, as now, just south of the Dock Ward, and both were adjacent to the Delaware River. Considerable shipbuilding took place here, and seafaring people made their homes in close proximity. It is quite likely that Pennington was acquainted with others of the neighborhood who became Continental Marines.

The exact date of Pennington's commission in the Continental Marines is not recorded. On 30 April 1776 "Miles Pennington of the City of Philadelphia, Captain of Marines on Board the good Ship Reprisal now lying in the Port of Philadelphia," made his will.

To his wife, Catharine, he left all his worldly belongings, including "all my Wages and Prize Mony which shall be due me from said Ship or any other Vessel." Witnesses were William Woodhouse, John Ord, Jr., and John Elliott.

The last witness was lieutenant of Marines of the brig *Reprisal*. He was not present at the proving of the will, however, for both he and John Ord, Jr., were "absent at Sea."

Reprisal, Captain Lambert Wickes, was of 16 guns and operated much of the time as a packet ship. She patrolled the Atlantic Coast during the absence of Commodore Hopkins' fleet in the Bahamas, and carried William Bingham, America's commercial and naval agent, to the West Indies.

On 21 September 1776, Captain Pennington was a member of the court martial of Navy Lieutenant Luke Matthewman, brig *Lexington*, on board the brig *Andrew Doria*, in Philadelphia harbor.

Late in 1776, *Reprisal* was dispatched to France, carrying Benjamin Franklin, one of the American Commissioners, arriving there in December. Two British brigs captured enroute and sent into Nantes were the first American prizes sent into a French port.

Shortly after her arrival, she cruised in the Bay of Biscay, and Lieutenant Elliott was wounded in the wrist while engaging one of the two vessels captured. Sailing with *Lexington* and *Dolphin* from February to July 1777, *Reprisal* sailed around Ireland, taking several prizes. Pressure from England on France precipitated the departure of these vessels from France and their return to America. Shortly before leaving, Lieutenant Elliott was transferred to the frigate *Deane*.

Reprisal departed in September, but foundered with all hands off the coast of Newfoundland about 1 October 1777.

Sums of money, presumably from the wages of Captain Pennington, were paid to Catharine Pennington from 18 June 1776 to 23 July 1777, and an inventory of his estate was made on 14 September 1778. Although many of the items of household goods and furniture were above average for the era, there was nothing to indicate his seafaring or military service. Only one entry, "One, Large Bible, Eleven Books of History," would today be of historical significance to the Marine Corps.

Catharine Pennington made two appeals to Benjamin Franklin for relief, one dated 15 December 1785 and the other 13 January 1786, but no record exists that she received the requested assistance.

PLUNKETT, THOMAS. Thomas Plunkett's origins are unknown. He may have been from Maryland. Nearly all of the frigate *Virginia*'s Marine officers—appointed, resigned or dismissed, and remaining—were from that state.

Plunkett was commissioned a second lieutenant of Marines for the *Virginia*, being built at Baltimore, on 9 December 1776. He replaced Lieutenant Richard Harrison. Captain James Disney, his senior officer, had been appointed on 22 October 1776, but probably resigned in late 1777 or early 1778. Plunkett was then promoted to the captaincy by the 4 March 1778 order of the Marine Committee. Lieutenant William Barney, another Marylander, came from Bordentown, New Jersey to fill the position of second lieutenant.

Virginia was blockaded in the Chesapeake Bay from the time of her launching on 12 August 1776. On 30 March 1778 she tried to escape to open sea from Annapolis, but ran aground in the darkness on the 31st.

Lieutenants of Marines Thomas Pownal and William Barney were captured by HMS *Emerald* that morning, but Plunkett escaped this indignity. In his haste to depart, the captain of the vessel, James Nicholson, had left the senior Marine officer behind with the tender.

Regretfully, nothing further is known of Plunkett, and Maryland records consulted show no trace of him.

PORTER, ANDREW. Andrew Porter was born 24 September 1743, one of the nine sons and five daughters of Robert Porter, on a farm four miles outside Norristown, Pennsylvania. His father had emigrated from the Isle of Bert, a short distance from Londonderry, Ireland, in 1720.

Disdaining the professions of carpenter or farmer, Andrew displayed a proclivity for the study of mathematics and science, and opened a successful school in Philadelphia in 1767.

He was married to Elizabeth McDowell on 10 March 1767, by whom in the following six years he had five children. She died in the childbirth of twins on 9 April 1773. Only the month before, Porter had purchased a lot from Thomas

Shoemaker, a house carpenter, on the south side of Spruce Street, between Third and Fourth Streets, in the Dock Ward.

Apparently, he built a house there, for in 1775 he paid constable's taxes on the property. In addition to his five children, he had a bound servant 16 years of age and a roomer. He married second, 20 May 1777, Elizabeth Parker, by whom he had four sons and three daughters, the last born in 1793.

Following the expedition to New Providence, Captain Samuel Nicholas was promoted to major, dispatched from the *Alfred*, and ordered by Congress to organize four companies of Marines to man new frigates being built in Philadelphia and its environs. *Effingham*, built at Southwark, just south of Porter's residence in Dock Ward, was launched 7 November 1776. Andrew Porter was commissioned her captain of Marines on 25 June 1776.

Effingham never got to sea. British troops advancing through New Jersey toward Philadelphia in November 1776 threw the city into a panic. The Hessians halted in Trenton, and General Washington determined to make a daring raid to bolster American spirits.

On 2 December 1776, Porter's Marines with two other companies of Marines, were ordered to sail up the Delaware River in gondolas and to join Colonel John Cadwalader's brigade. Washington envisioned Cadwalader's command being one of the prongs across the river on Christmas Eve to attack the German mercenaries.

Ice prevented Cadwalader's so doing, and Washington's other forces had the pleasure of the rout. The general returned to the Pennsylvania side shortly after, but Cadwalader, with the Marines, finally made his crossing on 27 December. They remained at Bordentown until Washington returned to New Jersey.

Subsequently, Captain Porter's Marines took part in the repulse of the British at the bridge of the Assunpink Creek on 2 January 1777, and on the following day at Princeton. On 6 January, they reached Washington's winter encampment at Morristown, New Jersey. There they were at first garrisoned apart from the main army. In February, however, when army artillery companies departed for home, the Marines were ordered into the artillery service at the main encampment.

Porter left the Continental Marines in the spring of 1777 to become a captain of the 2d Continental Artillery of Brigadier General Henry Knox, but the exact date of his commission is uncertain. Issac Craig left about the same time, entering Colonel Thomas Proctor's Pennsylvania artillery regiment, his commission as a captain being dated 14 March 1777. This unit later became the 4th Continental Artillery.

Captain Porter transferred to the 4th on 1 January 1781, and on 19 April was promoted to major. Captain Craig apparently contested the promotion, for in the following two years, he wrote a series of letters to General Washington protesting that he was ranked by a subordinate. More seriously, he claimed that Porter had drawn salaries simultaneously for both his Marine and army commissions. On 2 March 1782, Major Nicholas refuted his assertion. Based on the recommendation of General Knox, Washington settled the issue in favor of Major Porter.

Porter took part in the battles of Brandywine and Germantown and, in April 1779, joined Brigadier General James Clinton in operations against the northern Indians. He was promoted to lieutenant colonel on 1 January 1782. Prior to the siege of Yorktown, he was ordered to superintend the production of munitions at Philadelphia.

Following the Revolution, Porter retired to his farm at Norristown, shunning several lucrative offers of employment. In 1784, the state of Pennsylvania persuaded him to assist in determining the boundary line between his home state and Virginia. While serving on the commission, he probably maintained a city residence at 5 North Street, Philadelphia. In 1785 and 1786, he spent approximately six months in the actual physical survey of the western boundary between Pennsylvania and Ohio, northward to Lake Erie.

In 1800, he was appointed brigadier general of a Pennsylvania militia brigade and later major general of its parent division. In 1809, he became surveyor-general of Pennsylvania, holding the office until his death.

Because of his advanced age, Porter declined appointments by President James Madison to both brigadier general in the Army and Secretary of War in 1812 and 1813. He died on 16 November 1813 at the age of 70.

POWARS, SAMUEL. One Samuel Powars was listed by Sweeny in his 1794 Treasury Department list as a lieutenant of Marines. Allen, Paullin, and McClellan also listed him.

No record can be found of him in extant documents.

POWNAL [POWNALL], THOMAS. Thomas Pownal's parentage and vital statistics are not known, but he is believed to have been a resident of Maryland at the beginning of the Revolution.

On 25 June 1776, a member of the Marine Committee in Philadelphia notified Samuel Purviance, Jr., who was supervising construction of the frigate *Virginia* in Baltimore that Maryland delegates had recommended three of their citizens as Marine officers. One of them, First Lieutenant Thomas Pownal, was the only original appointee to serve on board *Virginia* throughout her short existence.

As of 9 August, Pownal had not actually received his Marine commission for *Virginia*, which was launched three days later. In a letter to the Maryland Council of Safety on that date, Pownal stated that he was a first lieutenant in a militia company of the 31st Maryland Battalion, but "now beg leave to resign the same (being inclosed) having a daily expectation of a marine commission from the Honourable Continental Congress."

On 30 March 1778, Captain James Nicholson attempted to run the British blockade of the Chesapeake Bay, but early the following morning ran the *Virginia* aground. At daybreak, the *Virginia* found herself under the guns of two British frigates and surrendered without firing a shot.

Lieutenant Pownal, with naval Lieutenant Joshua Barney, was sent on board HMS *Emerald*, but on 12 April was transferred to HMS *St. Albans*. It is reported that he was later sent northward to the enemy flagship *Ardent*. On 23 June he was supplied sundries amounting to the sum of 25 pounds.

On 12 August 1778, the Marine Committee and Board of Admiralty wrote a letter to one Colonel John Beatty (presumably a commissary of prisoners) notifying him that Lieutenant Pownal had been sent to Philadelphia from New York. They instructed Colonel Beatty to send the lieutenant of Marines of HMS *Mermaid*, captured on 28 July 1778, to New York in exchange for Pownal.

Lieutenant Pownal apparently had no further Marine duty during the Revolution. On 8 May 1787, writing from New York, he informed the United States Congress of his service on board *Virginia* and that his pay should have amounted to more than eight hundred dollars. Under the terms of depreciation of Continental currency, he had received only sixty-two dollars. In view of the hardship he claimed to have suffered, he asked remuneration for the additional sums authorized before 11 July 1780. His memorial was acted upon 1 August 1787, but the sum he finally received is not recorded.

Search through available secondary records of both Maryland and New York reveals nothing further of Thomas Pownal or Pownall; there are several references to similar surnames of Powal, Powall, Powel and Powell in both states.

PRITCHARD [PRICHARD], SAMUEL. Samuel Pritchard was from Boston, Massachusetts, but his vital statistics have not been ascertained.

Pritchard was a lieutenant of Marines on board frigate *Deane* from 20 September 1777 to 20 October 1780. He was also a member of the court martial of Captain Pierre Landais on board *Alliance* on 20 November 1780.

He transferred to the frigate *Alliance*, Captain John Barry, at Boston on 28 January 1781. His fellow Marine officers were Captain Matthew Parke, First Lieutenant Thomas Elwood, and Second Lieutenant James Warren, Jr.

Alliance departed Boston for France on 11 February 1781, arriving at L'Orient 9 March. There she loaded with supplies for Continental forces, and, with the French letter of marque *Marquis de Lafayette*, sailed for America on 29 March.

Alliance engaged and captured two privateer brigs on 2 April, one of which she turned over to the *Lafayette*. Both the latter and her prize parted company in a storm shortly after and were later captured. *Alliance* also took a brig and a snow on 2 May.

On 28 May 1781, she fell in with the British ship *Atalanta*, 16 guns, and brig *Trepassey*, 14 guns, and engaged them both at the same time. They were badly shattered when they struck to the American frigate. *Trepassey* was sent to Halifax as a cartel ship, but *Atalanta* was outfitted with a prize crew and proceeded with *Alliance*, but she was recaptured by the British before reaching port.

Alliance had five men killed and 22 wounded in the engagement, three of them mortally. Among the latter was lieutenant of Marines Pritchard, who probably died on 30 May. Lieutenant of Marines James Warren was also wounded and had his right leg amputated later.

Doyle Sweeny, of the Treasury Department in Philadelphia, made an audit of the accounts of Pritchard on 24 April 1793, in which it was determined that the sum of $821.32 was due his estate.

In 1818, Pritchard's widow, Martha, applied for a pension from Boston, but the outcome of her deposition is unknown, nor has further documentation about her been discovered.

REED [REID], FRANKLIN. No vital statistics of Franklin Reed are currently known, although one prominent source asserts that he was a Philadelphian.

He was appointed a first lietenant of Marines by the Continental Congress on 25 June 1776 in the company of Captain Samuel Shaw. Shaw's Marines were in barracks in the Northern Liberties, north of Philadelphia for a time, and were probably guarding both the stores and the frigate *Randolph*, being built at the Kensington shipyard nearby.

On Wednesday, 14 August 1776, Lieutenant Reid [Reed] signed an advertisement in the *Pennsylvania Gazateer* for a deserter from Shaw's company on board the *Randolph*.

Nothing further is known of the career of Franklin Reed, although he may have still been on board when *Randolph* sailed with Captain Shaw and Second Lieutenant Panatiere de la Falconniere as her Marine officers. He did not sign the petition of 31 August 1777 at Charleston, South Carolina which accused de la Falconniere of misconduct. Nor is it known whether he perished when *Randolph* blew up in her engagement with the British 74-gun ship *Yarmouth* on 7 March 1778.

REED, JEREMIAH. Jeremiah Reed is believed to have been born on 3 October 1757 at Marblehead, Massachusetts, the son of Richard and Elizabeth (Tucker) Reed.

He was a sergeant in Colonel John Glover's Massachusetts Regiment from May to December 1775. He was promoted 1 January 1776 to ensign in Glover's 14th Regiment, originally the 21st, or "Marine" regiment. It is interesting to note that William Bubier, later a Continental Marine officer, was a second lieutenant in the 2d Company of this same regiment. He may have been influential in determining Reed's later decision to join the Continental Marines.

Following the first battle of Trenton, Major William R. Lee, of Glover's Regiment, was promoted to colonel and returned to Massachusetts to recruit his own Continental regiment. He commissioned Jeremiah Reed a lieutenant sometime after 1 January 1777, but Reed resigned on 26 September 1778.

Reed was appointed a first lieutenant of Marines on 3 December 1778 in frigate *Boston*, Captain Samuel Tucker, at which date the roll was made up for advance pay of one month Seth Baxter was his captain and William Cooper second lieutenant of Marines.

In the following March, *Boston* sailed south and, at the end of May, accompanied frigates *Deane* and *Confederacy* on a cruise of the Atlantic Coast. On 9 June 1779, *Boston* took the British privateer ship *Pole* and Lieutenant Cooper was placed in her prize crew.

On 29 July, *Boston* and *Deane* convoyed merchantmen out of the Chesapeake Bay and a following cruise netted eight prizes. They arrived back in Boston on 6 September.

Boston, *Deane*, and *Queen of France* began fitting out for service at Charleston, South Carolina in September 1779 and sailed from Nantasket Roads 24 November. Enroute they took a privateer brig and arrived at their destination 23 December.

At Charleston, *Boston's* crew, presumably including her Marines, destroyed the lighthouse and Fort Johnston on the south side of the harbor. Crews of some of the larger vessels were sent ashore to reinforce the batteries on the Cooper River. When Charleston fell on 12 May 1780, Lieutenant Reed was apparently among those captured. It is surmised that he was also among those placed in a cartel for Philadelphia and exchanged soon after.

Reed signed on frigate *Deane*, Captain Samuel Nicholson, as first lieutenant of Marines on 15 November 1781. He was the 43d person to enter for the cruise. Captain of Marines Richard Palmes entered on 1 May, and Second Lieutenant of Marines William Waterman signed on 20 October.

Between 1 May 1781 and 1 March 1782, 76 enlisted Marines signed on board, although some of them were transferred to the frigate *Alliance* for her voyage to France with Lafayette. Others were lost through desertion and death.

Deane sailed from Boston in March 1782 on a short cruise to the West Indies and captured four vessels. She returned to Boston in May, and the crew was paid off. Lieutenant Reed received upwards of 36 pounds, after slops were deducted.

In September 1782 *Deane's* name was changed to *Hague*, and Captain John Manley assumed command of her. The exact date that Jeremiah Reed was appointed her first lieutenant of Marines is unknown, but on 3 July 1783 he signed a receipt for wages to 10 May 1783. On 5 December 1783, he signed another receipt for wages from 10 May to 29 August 1783.

Hague, under Manley, made a West Indies cruise and, in January 1783, was run aground and cannonaded by four British ships. She survived and returned to Boston shortly after and was put out of commission.

No definitive details are available of the later life of Jeremiah Reed. The 1790 Census of Massachusetts listed one Jeremiah Reed of Windsor, Berkshire County, the head of a family of three females.

RICE, ALPHEUS. The birth and parentage of Alpheus Rice are unknown, but he may have been from either Rhode Island or Great Barrington, Massachusetts.

One Alpheus Rice, Great Barrington, Massachusetts, enlisted as a private in Captain William King's 1st Company, Colonel John Fellows' 8th Massachusetts Regiment on 20 May 1775, and served two months and 16 days. He was also on the company return, dated Dorchester, 7 October 1775, and received either a bounty coat or the equivalent in money on 17 November 1775.

On 21 May 1776, one Alpheus Rice petitioned Commodore Esek Hopkins from Providence, Rhode Island, that he had "Quit'd the land Service with a design to get into the Fleet under Your Comd & as your Petir is well versed in the Small Arm exercise" . . . and . . . "as the COLUMBUS is Now a fitting out I should be very much oblidged to your Honour [if] you would give Me a birth aboard of [h]ir as I want to be in Business. . . ."

Some evidence exists that Rice was a lieutenant of Marines on board *Andrew Doria* before 10 June 1776, but it is certain that he was appointed lieutenant of Marines in the sloop *Providence* on 12 June 1776.

He apparently had some difficulty with either Captain John B. Hopkins or John Paul Jones, the latter then commanding *Providence* for, on 9 August 1776, he wrote from Chester, Pennsylvania, while under confinement, that "I have given Mr. Hopkins Sattisfaction, & there now subsists, a good understanding between Us.—I am now Conscious that I was Culpable.—but have this to Say in excuse—that I was in a passion.—am sorry for it.——& pray that you would overlook it.— . . ."

The Marine Committee recommended that Rice be appointed first lieutenant of Marines aboard the brig *Hampden*, Captain Hoysted Hacker, on 29 August 1776, and it was apparently approved. In October 1776, *Hampden* sailed with Jones, then in command of *Alfred*, for Cape Breton, but she ran on a ledge and her crew transferred to *Providence*. Shortly thereafter *Providence* returned to Rhode Island.

Nothing is known of the further career or life of Alpheus Rice.

RICHARDS, NATHANIEL. Nathaniel Richards was a native of New London, Connecticut, born there on 25 May 1756, the son of Guy and Elizabeth Richards.

Charles Buckley, a boyhood friend of Richards in New London, had been a midshipman on board the ship *Alfred* since 1776. Nathaniel's brother, Peter, was a navy lieutenant on board the vessel. Either may have recommended Nathaniel Richards as a second lieutenant of Marines for the ship. Commodore Esek Hopkins wrote her captain, Elisha Hinman, in Boston, on 9 February 1777 that he was enclosing "Mr. Richards Commission."

Entering *Alfred* in the spring, he joined captain of Marines John Welsh and First Lieutenant William Hamilton. Shortly after they sailed, they met frigate *Raleigh* outside Portsmouth harbor. Later, they witnessed the action of *Raleigh* and the British 14-gun ship *Druid*, before continuing on to Europe. Most of their winter was spent in France.

On 9 March 1778 both ships were returning to America when *Alfred* was overtaken by the British ships *Ariadne* and *Ceres*, while *Raleigh* escaped and made her way to Boston harbor.

All the prisoners of *Alfred* were taken into Barbadoes, and most were transported in the *Yarmouth* to prison in England. Lieutenant Richards was recognized by Captain Thompson of *Yarmouth*, who had known Nathaniel as a child and was intimate with his father's family.

By Thompson's intercession, Richards was paroled and sailed in a cartel for Martinique, in order that he might report the news of *Alfred's* capture to the Navy Board and to inform its members of the desertion by *Raleigh*. From there he took passage for America in the brig *Charming Sally*. Unfortunately, the latter was captured by HMS *Ambuscade*, and Lieutenant Richards was imprisoned in Halifax. On 7 July he was released and sailed in a cartel to New London, where he arrived on 28 July 1778. He was probably exchanged about the first of October 1778.

Shortly after, he received an appointment as purser of the frigate *Confederacy*, Captain Seth Harding, outfitting at New London. From February through March 1779, he carried on considerable correspondence with Major Joshua Huntington, of Norwich, Connecticut, soliciting supplies for *Confederacy*.

The frigate sailed on 1 May 1779 and, with the frigate *Boston*, captured the British frigate *Pole*, on 6 June. Unfor-

tunately, Richards' personal narrative in his application for a pension on 23 March 1818, contains no detail of *Confederacy's* three cruises, nor does it reveal, after the *Confederacy's* final voyage and capture by two British frigates on 14 April 1781, whether he was again committed to prison or paroled in New York with the remainder of her crew.

One Nathaniel Richards and Elizabeth Coit, both of New London, were married there on 22 January 1784. Vital Records of New London record only one Richards of this given name.

The 1790 Census of Connecticut disclosed that he was one of two free white males of that name 16 years of age and upwards, that he had one free white male under 16 years of age, four free white females, and two slaves. His pension deposition of 1818 suggests that he had been in the commercial mercantile business, but that he had been reduced from "affluence to poverty."

SALTONSTALL, GILBERT. Born in New London, Connecticut 27 February 1751/2, he was the son of Gurdon and Rebeckah (Wanton) Saltonstall. His bothers, Winthrop and Dudley, also served in the American Revolution. Gilbert graduated from Harvard College in 1770.

Gardner W. Allen asserted that Saltonstall received his commission as captain of Marines in 1776. However, one must refer to Gilbert's own recollection of events, written to Silas Deane in London, 16 August 1783, for the complete story:

> Give me leave to sketch out the Ghost of my departed Hours from that Time [early in the war] to the present. My country will be amenable I hope for the Murther of my Time, since it was in her Service the Crime was committed; the Emolument if any accrued, She certainly has, for I have rec[d] nothing during Seven Years Service. In January 1777, I was appointed Cap[t] Marines in the American Navy, and was Order'd on board the Trumbull where I remained 'till '79.

On 19 July 1779, the ill-fated fleet of New England privateers, merchant vessels and three Continental ships—the frigate *Warren*, brig *Diligent*, and sloop *Providence*—departed Boston to take Majabagaduce from the British in the Penobcot River.

Saltonstall wrote further to Silas Deane:

> As there was no prospect of her [*Trumbull*] getting to Sea, went as a Volunteer in the Penobscot Business, on my return took my old Station again.

Gilbert Saltonstall's brother, Dudley, was commodore of the expedition, with his flag on *Warren*. John Welsh was her captain of Marines, but Gilbert was also on board as secretary to Dudley. Councils of War on board *Warren* dated 27 and 29 July and 14 August 1779 were signed by him. He apparently did not serve in the landings on Banks Island or Majabagaduce. Captain Welsh was killed in the assault of 28 July, and Lieutenant of Marines William Hamilton, also of *Warren*, was wounded and later died of his wounds.

Trumbull was finally gotten across the bar of the Connecticut River in 1779, fitted for sea at New London, was commissioned and got to sea in 1780. On 1 June 1780, *Trumbull* fell in with the British letter of marque *Watt* and, fought one of the bitterest engagements of the war. *Trumbull's* captain, James Nicholson, made an excellent report of the battle. Two letters from Captain Gilbert Saltonstall to his father dated 14 and 19 June 1780, give details of the officers and crew that were not contained in Nicholson's report.

He related that David Bill, a "volunteer," was killed at the beginning of the engagement. Daniel Starr, who probably was a lieutenant of Marines, was wounded with a grape shot and died on 5 June. Both were from New London and probably were well known to Saltonstall. Lieutenant of Marines Jabez Smith, of Groton, Connecticut, was also killed. Of the numerous enlisted personnel killed and wounded, it is not known which were Marines.

Captain Saltonstall also suffered "eleven different wounds, from my shoulder to my hip; some with buck-shot, others with the splinters of the after quarter-deck gun. I had one shot through the brim of my hat; but was not so disabled as to quit the quarter-deck till after the engagement; and am now as well as ever.—Have one buck-shot in my hip yet."

He also asserted that "the marines fired pistols during the engagement exclusive of which they fired near 1200 rounds."

Notwithstanding, Saltonstall reported to his father that the entire crew was exuberant about going back to sea. *Trumbull* was refitting, a recruiting rendezvous had been opened, and she expected to sail with the frigate *Deane*.

Saltonstall was still *Trumbull's* captain of Marines when she sailed on 11 August 1780. On the 29th of that month, she began a cruise of the Atlantic Coast with *Deane*. In September, they captured the British brig *Little William*. A muster roll compiled at the time listed William Fielding as lieutenant of Marines of *Trumbull*.

His account to Silas Deane is the only surviving memoir of his further service:

> . . . Order'd to Philad[a] where remaind till July 81. left the Capes of Delaware on July 9th in four Hours were taken, after an Action of three Glasses, by the Iris formerly the Hancock, and the Monk of 20 Guns. were carried into New York. I was Parold from the Prison Ship to New London just time eno' [enough] to see the destruction of the Town, a Scene too distressing to dwell upon. May 1st [1783] an Order from the Warr Office liberated all Prisoners from Parole.

Gilbert Saltonstall and one D. Mumford, Jr. engaged in mercantile business in New York following the war, and he wrote Deane for contacts on the Continent.

Saltonstall was married, the time and place unknown, to Harriet Babcock, by whom he had two sons. One, Gurdon, was later a professor of mathematics at the University of Alabama, and died there, leaving no issue. The second son, Gilbert, was born in 1791, married in 1814, and died at Tuscaloosa, Alabama on 6 February 1833. His progeny had a line of descent into the early 1900's.

The date and place of death of Captain Gilbert Saltonstall is unknown.

SCULL, ———. One "Mr. Scull" has not been determined to have been a Continental Marine officer. Only Major Edwin N. McClellan, USMC, so designates him.

However, when Captain Joshua Barney commanded the Pennsylvania ship *Hyder Ally* in her action with the British ship *General Monk* off Cape May about 8 April 1782, Mr. Scull was the marine officer in charge of Bucks County riflemen acting as Barney's Marines.

Scull may have been acting under false pretenses, but in June 1780, he "Enlisted men in to the service of the United States on the Continental Establishment as Marines" at Reading, Pennsylvania.

John Fox and Jacob Hlees, identified only as from Burks,

County, and Michael Spatz, of Reading, were recruited by Scull. From there, they marched to Wilmington, where they entered on the *South Carolina*, of that state's navy. They were captured outside the Virginia Capes and confined in the Jersey prison ship for approximately eleven months.

SHAW, SAMUEL. The surname "Shaw" was a prominent one in Philadelphia during and after the American Revolution, but the majority of them appear to have been engaged in rather prominent mercantile pursuits.

In 1775, one Samuel Shaw was an occupant of the household of Jacob Good, a laborer, in the east part of the Mulberry Ward in Philadelphia. Living in this vicinity, he may have been acquainted with any of the other four Philadelphians who were appointed captains of Marines on 25 June 1776, Andrew Porter, Joseph Hardy, Benjamin Dean, and Robert Mullin (Mullan).

Frigate *Randolph*, built, launched, and commissioned at Kensington, a northern suburb of Philadelphia, was the first duty assignment of Captain Shaw.

As early as 14 August 1776, First Lieutenant of Marines Franklin Reid (Reed) advertised in a local newspaper for a deserter from Shaw's company on board *Randolph*. In November of the same year, Shaw himself advertised for deserters from his company "lying in barracks at Philadelphia . . ."

Randolph, Captain Nicholas Biddle, put to sea in February 1777, only to be dismasted and forced to put into Charleston, South Carolina for repairs. She was out again in midsummer, taking four vessels before returning to Charleston in September.

When *Randolph* reached Charleston Second Lieutenant Panatiere de la Falconniere, Shaw's junior officer, was put under arrest for misconduct in the last cruise, the result of a petition of six officers of the ship. Captain Biddle sent him off to Congress for determination of his case.

Randolph sailed for France in the fall of 1777, arriving at L'Orient in December. She remained only a short time, for the British agents there objected strenuously to France harboring the warships of an enemy. Biddle returned directly to Charleston.

There, Captain Biddle organized a squadron of *Randolph* and four ships of the South Carolina Navy. South Carolina troops composed the Marines of the squadron.

Sailing off the West Indies and east of Barbadoes, they met with the British 74-gun ship *Yarmouth* on 7 March 1778. Only *Randolph* and the *General Moultrie* engaged her. *Randolph* appeared to be inflicting most of the damage on her opponent, when the American frigate suddenly blew up. All but four of the crew perished. The survivors were found floating on wreckage on 12 March 1778.

Search of the surname "Shaw" in the various records of Philadelphia reveal no direct relationship of the families there to Captain of Marines Samuel Shaw.

SHOEMAKER, JOSEPH. The surname "Shoemaker" abounded in Revolutionary Philadelphia and its environs, especially in the area of the Northern Liberties and Germantown. The prevalence of several with the given name "Joseph" following the war nevertheless makes the identity of the Continental Marine officer difficult to ascertain.

There is little doubt that Joseph Shoemaker, the senior Marine officer in ship *Columbus*, Captain Abraham Whipple, in January 1776, was a native of the Philadelphia area.

Captain Shoemaker, with Lieutenant of Marines Robert Cummings [or Cumming] commanded 60 Marines of the ship in the landing at New Providence in March 1776. Their individual actions are unknown.

Columbus returned to New London, Connecticut in April 1776, and almost immediately Captain Shoemaker tendered his resignation to Commodore Esek Hopkins. Whether he saw further service in the Revolution in any capacity is unknown.

Shoemaker tried to obtain additional pay later in the Revolution. Major Samuel Nicholas, in a sworn affidavit dated 12 September 1782, affirmed to the Chief Justice of Pennsylvania that he had examined the records of the Marine Committee of Congress and that he had found the following entry:

> It is the opinion of this Committee that Captain Shoemaker should not receive pay longer than till his arrival in Philadelphia—nor be allowed for Rations after he left the Ship, and his pay was accordingly settled and pay'd up to the Second Day of May 1776.

Nicholas also swore in his affidavit that he was a candidate for the office of Sheriff in the City and County of Philadelphia at that time. He informed the judge that he therefore would be unable to attend a hearing of Joseph Shoemaker in Richmond [Northern Liberties] Court on 8 October 1782.

In post-Revolutionary Philadelphia several persons named Joseph Shoemaker were represented in various public and legal notices. One was a shopkeeper—and apparently a tanner as well—with his shop on Second Street, between Vine and Race Streets. Between 1798 and 1808, he and his wife, Abigail, sold several pieces of property in the city and in its northern and southern suburbs.

One Joseph Shoemaker married Mary Priest at St. Paul's Church on 28 September 1790. He was a silversmith. They bought and sold property in Philadelphia in the early 1800's. He apparently died before 1829, for on 30 September of that year a bond of administration was issued on his estate, valued at only $24.94. One Joseph Shoemaker a druggist, was one of the administrators.

Joseph Shoemaker, a hatter, made his will on 25 November 1793, leaving all his unlisted estate to his wife, Anna.

In 1819, a farmer in Germantown named Joseph Shoemaker made his will. His wife was also named Abigail.

A bond of administration was issued on the estate of one Joseph Shoemaker on 10 April 1830, he being the son of Abraham Shoemaker who died in 1798. His profession was not given.

In none of the foregoing documents was there any substantial evidence as to the identity of Captain of Marines Joseph Shoemaker.

SMITH, JABEZ, JR., Jabez Smith, born 31 August 1751, was a native of Groton, Connecticut. His parents are currently unknown.

He is first reputed to have been on board the Connecticut ship *Oliver Cromwell*, but on 21 August 1777, he transferred as lieutenant of Marines to the Continental brig *Resistance*, Captain Samuel Chew. On 26 December 1777, *Resistance* captured a ship bound from Scotland to the West Indies and sent her into Boston. Her cargo was valued at about 7,000 pounds.

Frigate *Raleigh* escaped from two British ships, the *Ariadne* and *Ceres*, on 9 March 1778 and made her way into Boston, but in so doing sacrificed the *Alfred* to them. Her captain, Thomas Thompson, was court martialed and dismissed, and Captain John Barry took command. *Raleigh's* second lieutenant of Marines, Nathaniel Thwing, resigned and his place was taken by Jabez Smith. His superiors were Captain George Jerry Osborne and First Lieutenant Stephen Meads.

Raleigh sailed again on 25 September. At noon, she was pursued by the *Experiment*, 50 guns, and *Unicorn*, 22 guns. After two days Captain Barry thought he had eluded them, but on the morning of the 27th *Unicorn* again took up the chase, each vessel trading broadsides. *Raleigh* was considerably damaged in her tops, losing control of her movements, and Barry determined to run her ashore to prevent capture.

She was run aground on Wooden Ball Island, south of the mouth of Penobscot Bay, on the night of the 27th. Barry decided to get his crew ashore and to destroy the frigate. He, and probably Captain Osborne and Lieutenant Meads, and 85 men got to the mainland, leaving Lieutenant Smith with the remainder of the crew on the island. Ignoring orders to destroy the ship, a midshipman struck her colors, and Lieutenant Smith was forced to surrender himself and 133 men to the British.

Smith was apparently soon paroled, but was not exchanged immediately. On 13 January 1779, William Vernon and John Deshon, of the Navy Board at Boston, wrote the Massachusetts Council urging that Smith and others from *Raleigh* and *Resistance* be exchanged "as their services are much wanted in the Navy."

When frigate *Trumbull* sailed from New London in May 1780, Smith was her second lieutenant of Marines. Captain Gilbert Saltonstall was his superior Marine officer. On 1 June Smith was a participant in the desperate engagement with the British letter of marque *Watt*, in which the latter had 40 killed and wounded.

Captain Saltonstall, writing to his father on 14 June 1780, noted simply "Jabez Smith, Jr., of Groton, lieutenant of marines, killed."

If not buried at sea, perhaps Smith's body was returned to Massachusetts. It is possible that relatives or friends may have acquired a plot for him in Boston's Granary Burying Ground in Tremont Street. A stone there identifies him as a Marine officer, but records his death as having occurred on 28 June 1780, aged 29 years.

SPENCE, ———. One Spence, with no given name, was listed by Sweeny, Allen, Paullin, and McClellan as a captain of Marines in the American Revolution.

Neither primary or secondary records available today support the allegation.

STACK, EDWARD [EDMOND, EDMUND]. Edward Stack was born 28 April 1756 at Kealand, County Kerry, Ireland, the son of "Stack of Crotts." His early education is unknown, nor is it known under what circumstances he entered the military service of the French government. However, he was commissioned a second lieutenant in the Regiment of Walsh, Irish Brigade, French Army on 23 March 1777.

In appreciation of the exploits of Captain John Paul Jones in *Ranger*, King Louis XVI gave him the command of *Duc de Duras* in 1779 and granted permission to rename her the *Bonhomme Richard*. Moreover, he promised assistance in arming and manning her.

Following Louis XVI's example, Captain Fitz-Maurice of Walsh's regiment recommended to Jones the services of Edward Stack on 5 February 1779. Stack apparently went on board immediately. His brevet lieutenant's commission, signed in blank by John Hancock and issued in France, was dated 4 February.

Lieutenant Stack performed an active part in the engagement of *Bonhomme Richard* and *Serapis* off Flamborough Head on 23 September. He commanded the main top with 15 Marines and four seamen. Their continuous fire of muskets, swivel guns, and grenades on *Serapis'* upper deck and hatches greatly contributed to the surrender of the British ship.

Following the battle, *Bonhomme Richard* sank and Jones transferred his remaining crew to *Serapis* and sailed to the Texel, Holland on 3 October. The wounded British were sent ashore, but the remainder were held captive in *Serapis* under trying conditions. When a party of them attempted escape on 14 October, Lieutenant Stack and a force of armed Marines pursued them. All but four were killed or drowned.

On 21 October 1779, Jones received a letter from Stack's father, a captain in the Regiment of Walsh. The latter expressed the fear that his son was dead, and confided that if he had served like a "gentleman and a Soldier, I Shant half regret his death," but "his loss will lie heavy 'pon me the rest of my days."

Shortly after 25 November 1779, Stack was released from his duties with Jones, for his regiment was shipping for the Antilles. Jones presented him with a testimonial of his good conduct and he was awarded a stipend for the personal belongings he had lost in *Bonhomme Richard*.

While absent from his regiment, Stack had been promoted to first lieutenant and, upon his return, was promoted to brevet captain and received a lifetime annuity from the king for his service to the American cause.

Stack came to the United States with his regiment, and probably served in the campaigns of the Carolinas. Successive promotions carried him to lieutenant colonel of the 87th Regiment, a sister unit of Walsh's, on 25 July 1791. He subsequently served on the Rhine, at Coblentz, and in the Irish

Grave Marker of Marine Lieutenant Jabez Smith, Jr.

Brigade of the House of Bourbon. From thence, he went into the British Army, serving in Ireland and Jamaica.

He was promoted to colonel 1 January 1801, serving part of the time on half-pay, and was appointed brigadier general 1 September 1803. Between then and 1814, he was again in France, where twice he was detained by Napoleon during hostilities.

He was promoted to major general in 1808, lieutenant general in 1813, and general 22 July 1830.

He became a member of the Society of Cincinnati in France upon the recommendation of John Paul Jones, dated 13 April 1785.

Stack died in Calais, France in December 1833.

STARR, DANIEL. Born 26 December 1741, Daniel Starr was the son of Daniel and Elizabeth (Hempstead) Starr, of New London, Connecticut.

One Daniel Starr was an enlisted Marine on board the Connecticut ship *Oliver Cromwell*, entering her on 6 December 1777, but he was reputed to have deserted on 9 June 1778.

The Daniel Starr who was listed by Doyle Sweeny in 1794 as a lieutenant of Marines, probably signed on board frigate *Trumbull* while she was fitting for sea at New London in 1779. She sailed in May 1780.

Trumbull engaged the British letter of marque *Watt* on the morning of 1 June 1780 in a drawn battle that rivalled the engagement of *Bonhomme Richard* and *Serapis* as one of the fiercest in the American Revolution.

Captain of Marines Gilbert Saltonstall wrote of the battle on 14 June 1780:

> We have suffered greatly from New London. Mr. Starr was wounded in the latter part of the engagement with a grape shot, which went in just above the right hip bone, and was cut out behind. He lived till 4 o'clock Monday morning following when he died without a struggle. I was with him most of the time after he was wounded till he died. The day after he was wounded, he was out of his head, and so continued until his decease. I suppose his bowels mortified, as he was insensible of pain.

Captain Saltonstall made an accounting of Lieutenant Starr's effects in Boston harbor on 19 June and asked his father for directions from Starr's friends respecting them.

Daniel Starr is believed to have been married before the Revolution, on 5 January 1764, to Lucy Douglass, the daughter of William Douglass, of New London. They had one son, Daniel, born 25 February 1765.

STEWART, JOHN. The state of Maryland appears to have been the home of John Stewart, but his vital statistics are not known.

Maryland delegates to the Continental Congress recommended Stewart to the Marine Committee as a captain of Marines. On 25 June 1776, approval was given and he, together with Lieutenants Thomas Pownal and Richard Harrison, were appointed to the frigate *Virginia*, then being built at Baltimore.

On 7 August following, the Maryland Council of Safety at Annapolis wrote to its delegates in Congress that Stewart already held a lieutenant's commission in a Maryland company and wished confirmation of which commission he was to fill.

Perhaps he continued in the Maryland capacity, for he is reputed to have been relieved and James Disney was appointed in his place on 22 October 1776.

The surname "Stewart" abounded in Maryland following the Revolution. In fact, in the 1790 Census of Maryland, two adult John Stewarts lived in Anne Arundel County, one in Baltimore County, one in Harford County, and two, including a doctor, in Prince Georges County.

STILLMAN, GEORGE. The birthdate of George Stillman is believed to have been 24 February 1752. His parents are unknown and his birthplace is in doubt. The only clue to the latter is in the examination of Captain John Grannis in March 1777. In his testimony against Commodore Esek Hopkins and Captain John B. Hopkins, he stated that both Second Lieutenant of Marines Barnabas Lothrop and First Lieutenant of Marines George Stillman were from Bridgewater [Massachusetts]. Proof of the former's residence there has been found, but available records do not support the case for Stillman.

Nevertheless, George Stillman was appointed a first lieutenant of Marines on 14 June 1776, by the committee appointed to build two continental frigates in Rhode Island. He was assigned to frigate *Warren*, with Grannis as his captain and Lothrop the second lieutenant of Marines.

Warren was launched at Warren, Rhode Island on 15 May 1776, but was not commissioned until 18 April 1778. The idleness of her crew and Marines and discontent with the brothers Hopkins probably contributed to these Marine officers looking elsewhere for useful employment.

When Stillman left *Warren* is not known, but the petition against her Commodore and Captain was signed on 11 February 1777. In August 1776, however, one Major George Stillman—believed to have been the same person—commanded a company of militia troops against the depredations of the fleet of Sir George Collier on the coast and town of Machias, Maine [then Massachusetts].

On 1 November 1782, he was a lieutenant colonel in the 6th Regiment of Militia, Colonel Alexander Campbell, in Lincoln County, of which Machias was a part.

In 1784, Major George Stillman was described as "one of the principal inhabitants of the town of Machias, and . . . was ordered by the Legislature to oversee the incorporation of the town. He was later the Town Treasurer."

He was appointed colonel of the 1st Regiment, 2d Brigrade, 8th Division, of the Massachusetts Militia, on 27 September 1791, with his responsibility lying in Lincoln County. On 20 August 1795, he was a brigadier general, 2d Brigade, 10th Division of the militia, but resigned his commission on 7 May 1803.

Two biographical sources differ on the marriage of George Stillman. One states that he married Rebecca Crocker at Hartford, Connecticut, in 1770. The other declared he married the same woman at Hartford on 7 November 1781. One is inclined to accept the former, for the biographer also confirms his removal to Machias, Maine. Nearly all his six children were apparently born in Machias. Furthermore, the former date would justify his having had, according to the 1790 Census, one son over the age of 16 years.

Further proof of his Machias residence is the fact that, in his capacity as a town official, he united several couples in marriage.

He should not be confused with the Reverend Samuel Stillman, of Boston, although they may have been related.

His death date is unrecorded.

STROBAGH [STROBACH, STROGBOCH], JOHN MARTIN. John Martin Strobagh was probably from Philadelphia, but his family background is unknown.

He was lieutenant of Marines of the 10-gun sloop *Hornet*, Captain William Stone, which departed Delaware Bay on 17 February 1776 on the New Providence expedition. Shortly after the departure of the fleet, *Hornet* had a collision with *Fly*, and had to put back into the bay for repairs.

Finding that the sea service disagreed with him, Strobagh applied for a discharge on 14 May 1776, feeling that he would be more useful in the land service.

In fact, he had already been appointed a lieutenant in Proctor's Pennsylvania Artillery Battalion on 13 May. He was promoted to captain on 5 October 1776. He retained the latter rank when the unit become the 4th Continental Artillery, and was promoted to lieutenant colonel on 3 March 1777. He died 2 December 1778.

A bond of administration of his estate was issued in Philadelphia on 3 September 1784, but no inventory of the estate accompanied the papers filed.

THOMPSON, BENJAMIN. Nothing is known of the origins of Benjamin Thompson.

After the frigate *Boston* was launched at Newburyport, Massachusetts on 3 June 1776, Richard Palmes was commissioned her captain of Marines on 23 July. On the following day, the Marine Committee recommended Benjamin Thompson to be a lieutenant of Marines "in Captain Palmes's company."

Thompson does not appear to have served long in *Boston* for when she sailed on her maiden cruise 21 May 1777 Robert McNeill and John Harris were Palmes' first and second lieutenants, respectively.

Major Edwin N. McClellan, in his *History of the United States Marine Corps*, notes that Second Lieutenant Benjamin Thompson also served in the ship *Ranger* in 1778 and 1779, but that his source, an early Maine Corps historian, was "extremely inaccurate regarding Marine personnel of the Revolution."

THWING, NATHANIEL, The pedigree of lieutenant of Marines Nathaniel Thwing is difficult to ascertain. The surname itself abounds in the Colonial and Revolutionary eras in eastern Massachusetts, and to a lesser degree in southern Maine. The given name "Nathaniel" was also quite popular.

For instance, one Nathaniel Thwing was born 30 January 1758 in Boston, the son of Nathaniel and Abigail (Greenough) Thwing. He was later a tailor in Augusta, Maine and moved to Woolwich, Maine in 1783. He married Mary Eastman on 29 March 1792. Perhaps he was the Justice of the Peace at Thwing's Point, Woolwich, who performed a considerable number of marriages until 1816.

Another Nathaniel Thwing, the son of James and Martha (Clapp) Thwing, Boston, was born 8 December 1757. He became a merchant and married his second wife, Mary Greenleaf, of Boston, on 14 August 1794.

Still another Nathaniel Thwing married Margaret Dunn (or Dun) 8 April 1770, in Boston.

Considered unlikely are Thwings with the given name of Nathaniel who served as enlisted personnel in Massachusetts land forces, both during and after the known service of the lieutenant of Marines.

One Nathaniel Thwing, of Boston, was recommended before 24 June 1776 by New Hampshire naval agent John Langdon to be second lieutenant of Marines in frigate *Raleigh*, then outfitting at Portsmouth. His name was submitted to the Marine Committee in Philadelphia on 1 July, 1776, and his commission was approved 22 July of that year.

On 21 August 1776, Lieutenant Thwing entered *Raleigh*, with a monthly salary of $18.00. He probably made only one cruise in the frigate, departing Portsmouth 12 August 1777 with ship *Alfred*, bound for France to obtain military stores. Enroute, *Raleigh* engaged the 20-gun ship *Druid*, nearly destroying her before being forced to turn away from her other armed escorts.

On the return trip, the following February, *Raleigh* is said to have abandoned *Alfred* during the latter's capitulation to HMS *Ariadne* and Ceres for the sake of the military equipment she was carrying. She reached Boston in April 1778.

Thwing apparently resigned his commission at the end of this cruise, for he was replaced on *Raleigh* in August by Second Lieutenant Jabez Smith, Jr.

Thwing's further life is only a matter of conjecture. He may have been the Nathaniel Thwing who was appointed and confirmed as the naval officer of the Middlesex County seacoast town of Townsend on 28 April 1778, by the Massachusetts House of Representatives and Council.

TREVETT, JOHN. John Trevett, the son of Eleazar and Mary (Church) Trevett, was born at Newport, Rhode Island in 1747. He was reputed to have been in the merchant service before the Revolutionary War.

In November 1775, Trevett sailed from Providence, Rhode Island for Philadelphia on board the sloop *Catea* [or *Katy*], Abraham Whipple commanding. On their arrival in Philadelphia, *Catea's* name was changed to *Providence*. Trevett was commissioned a midshipman in the ship *Columbus*.

During the cruise of Hopkins' fleet to New Providence, Trevett was commissioned a first lieutenant of Marines and was a participant in the taking of Forts Montagu and Nassau. With Lieutenant Henry Dayton and another lieutenant of Marines from Philadelphia, he took Governor Monfort Browne captive.

The fleet returned to New London, Connecticut in April 1776 and Trevett was transferred to the brig *Andrew Doria*, Captain Nicholas Biddle. She departed Providence with brig *Cabot* on a short cruise on 19 May. On the 29th, *Andrew Doria* captured two British transports, the *Oxford* and *Crawford*, each containing a company of Scottish troops. Lieutenant Trevett went on board *Oxford* as prize-master's mate, second in command to Navy Lieutenant John McDougall. They proceeded to Providence in company with *Andrew Doria* and the other prize.

Enroute, pursuit by five enemy ships forced them to separate, *Oxford* sailing westward. On 11 June the prisoners rose and retook the ship and sailed her into Virginia's James River. There, aided by the Virginia Navy, *Oxford* was retaken and the enemy troops were marched off to prison.

Lieutenants Trevett and McDougall were given funds for travel by Virginia authorities and proceeded overland to Philadelphia. There they heard Thomas Mifflin read the Declaration of Independence in the square before the State House on 6 July 1776. Leaving Philadelphia on 12 July, they reached Newport in August and reported on board *Andrew Doria*.

The brig sailed almost immediately on another cruise, in company with *Columbus*. She captured four ships, of which one, the merchant brig *Lawrence* with an assorted cargo, was taken into Providence on 7 September, with Lieutenant Trevett as master.

Trevett then joined the brig *Hampden* as her Marine officer. She sailed from Newport on 27 September with *Alfred* and

Providence, but struck a sunken ledge. Trevett and his Marines were transferred to *Providence*, Captain Hoysted Hacker. On 12 and 13 November 1776, they took the brig *Active* and ship *Mellish* off Cape Breton. The latter was loaded with clothing for Burgoyne's army, which was sent to General Washington, then retreating across New Jersey.

Providence put back into Rhode Island the last of November, and Trevett apparently took a month's leave of absence. During early December, the British took Newport, and Trevett's family was forced to leave its home. He reported back on board ship and burned a large quantity of stacked hay on Gould Island to prevent its use by the enemy. *Providence* then stood up the bay for Warwick Neck.

The sloop attempted to sink the British frigate *Diamond* which was aground there but failed. Shortly after, he and two midshipmen disguised themselves and went aboard a cartel ship for Newport. Trevett landed on the Long Wharf, where he knew most of the residents, and conducted a short spying mission mingling with the British ashore.

In February 1777, the sloop ran the blockade. Shortly after, she took a transport brig loaded with British soldiers, and Lieutenant Trevett took her into New Bedford, Massachusetts.

Captain John Peck Rathbun took command of *Providence* in May 1777 and began fitting her for sea. At this time Trevett heard of the capture of his brother, Constant Church Trevett, in a merchant vessel bound from the West Indies to the Carolinas. He was sent to the Jersey prison ship in New York. John Trevett immediately procured a British captain for an exchange, but it was too late, for his brother had died from ill treatment.

Trevett was apparently promoted to captain of Marines in mid-June 1777. In November, *Providence* sailed again, bound for South Carolina. Michael Moulton, also of Newport, had since been commissioned her lieutenant of Marines. Outside Charleston harbor, she took a British privateer and carried her into Georgetown, South Carolina.

There, Captain Rathbun conceived the idea of raiding New Providence a second time. Captain Nicholas Biddle tried to talk him out of it, and implored Trevett to come on board frigate *Randolph*, both to no avail.

Early on the morning of 28 January 1778, Trevett, Moulton, and 26 Marines landed outside Fort Nassau and took possession of the fort from only two sentries. They held it for three days by deceiving the citizens as to their true strength. Five vessels were taken in the harbor, including the 16-gun ship *Mary*. American prisoners were released, and the magazine was stripped of all its powder. *Providence* and *Mary*, the latter under the command of Captain Trevett, departed the Bahamas, and arrived at New Bedford early in March.

In July 1779, *Providence* was fitting out at Boston for the Penobscot expedition. Captain Trevett foresaw that the British would be ready for the invasion and stated that he had no inclination to participate. He remained behind to transact some personal business.

In February 1780, he joined the privateer brig *Rattlesnake*, 14 guns, at Newport, for a cruise. Only four days out, two British ships overtook her, and the *Galatea* ran *Rattlesnake* ashore off Shrewsbury, New Jersey. Most of the crew made it to the beach, but the brig was burned by the enemy. Trevett reached Newport again only two weeks after sailing.

Trevett sailed from London early in May 1780 as a volunteer in the frigate *Trumbull*, Captain James Nicholson. Off Bermuda, she captured the British privateer *Queen Charlotte* and sent her into Philadelphia.

On 1 June, *Trumbull* engaged the letter of marque *Watt*, in which David Bill, another volunteer, and Lieutenants Jabez Smith, Jr. and Daniel Starr were killed, and Captain of Marines Gilbert Saltonstall was wounded. Trevett lost his right eye and was wounded in the right foot. He received only his prize money from the privateer, and had to hire a horse to carry him to his temporary residence at East Greenwich, Rhode Island, arriving there 23 June.

His wounds still healing, he signed on board the frigate *Deane*, Captain Elisha Hinman on 2 July 1780, but his station is unknown. He left his father's family 30,000 Continental dollars and some Rhode Island state securities to tide them over. Off the Banks of Newfoundland, the brig *Elizabeth* was taken. A Navy officer and Trevett were assigned to take her into Boston, but they were retaken shortly after by another armed British ship.

Both officers were carried into St. Johns, Newfoundland and sent on board the prison ship *Porteas*. There, Trevett met Winslow Warren, a brother of lieutenant of Marines James Warren, Jr., who was a "prisoner at large," having been left there as a hostage for one "Capt. MacNeail."

It appears that Trevett was given considerable freedom, both on board the prison ship and ashore. He continually resisted British attempts to put him in the crews of British merchant vessels sailing from there. Through the efforts of a friendly British Navy captain, he managed to get on board a vessel going to the West Indies. The ship was subsequently taken by the French frigate *Amazon* and Trevett was taken into L'Orient, France.

Captain John Paul Jones endeavored to sign him on board *Ariel* for America, but Trevett sensed animosity in his old commander and declined. He made his way to Brest, where he went aboard a prize ship commanded by an Irishman named Luke Rion [Ryan?]. In her, he made two cruises in the English Channel, returning each time to France.

With another American, he traveled overland to Amsterdam, arriving there in March 1781. They went on board the frigate *South Carolina*, Commodore Alexander Gillon, and sailed in September. A roundabout cruise took them to Spain, the Canary Islands, and the British Bahamas where, with a Spanish fleet, New Providence was captured on 11 May 1782. *South Carolina* arrived at Philadelphia 28 May 1782 and Trevett was paid off at the rate of four pounds sterling per month.

Trevett returned to Newport, where he was a mariner and later a ship's joiner, until incapacitated by illness.

In 1786, he figured in a landmark legal case testing the validity of the issue and circulation of paper money. A butcher named Weeden refused to take Trevett's script at par value in payment of a debt. Both apparently had able and prominent attorneys, and William Channing was the Attorney General. The court ruled in favor of Trevett, to the great joy of witnesses.

Trevett was pensioned by the United States in 1818, and in 1820 his family consisted of his wife, Elizabeth, a daughter 30 years of age, and a grandchild of five. His wife died 22 January 1823, aged 73, and Trevett, who had been totally blind the past four years, died 3 November 1823, at the age of 76.

TROWBRIDGE [TROBRIDGE], ELIHU. Elihu Trowbridge was born in Southbury, Connecticut in 1748, the son of Joseph and Trial (Morehouse) Trowbridge.

His earliest Revolutionary service was in the land forces. He enlisted 20 April 1775 at Hatfield, Massachusetts in Captain Israel Chapin's Company, Colonel John Fellows' Regiment and was discharged 8 October 1775.

He reenlisted 1 January 1776 at Cambridge, Massachusetts in Captain Samuel Hunt's Company, Colonel Ward's Regiment of the Massachusetts Line.

Trowbridge was chosen in February 1776 as one of General George Washington's Life Guards, as a sergeant under the command of one Captain Gibbs and served about nine months.

On 1 January 1777, he was commissioned a second lieutenant in Captain Samuel Granger's Company, Colonel Charles Webb's 2d Connecticut Regiment, but resigned on 4 December 1777.

In January 1778, Trowbridge was commissioned a captain of Marines on board the frigate *Warren*, Captain John B. Hopkins. He asserted in a deposition for his pension some years later that he recruited a complement of Marines. The cruise following was unsuccessful, and the ship returned to port where his Marine detachment was discharged.

He reported having recruited another guard and on a second cruise several prizes were taken, but all were recaptured by the British before they reached port. On 8 February 1779, he petitioned the Continental Congress in Philadelphia for funds to aid in recruiting a third complement.

It appears that Captain Hopkins had previously bestowed upon Trowbridge the sum of $390.00 and sent him to Springfield, Massachusetts to obtain the required recruits. In correspondence dated 13 March 1779, the Eastern Navy Board at Boston admitted to the Navy Board at Philadelphia that it had given Trowbridge $200.00 for the same purpose.

On the basis of testimony of a sergeant of Marines engaged in recruiting with Trowbridge, the Navy Board at Boston suspected that Captain Trowbridge had gambled away the money and deserted the service. On 22 March 1779, he was dismissed from the service.

A Trowbridge family biographer cited a different sequence of the latter events. He asserted that Trowbridge served on board frigate *Warren* until she was destroyed at Penobscot. No proof of this allegation exists in contemporary records. The author further asserted that the captain was on leave of absence from the Marine Committee until peace was declared.

The biographer related that Trowbridge followed the trade of pump-making following the end of the Revolution, living in Earl, Lancaster County, Pennsylvania, and at Derby Landing and Washington, Connecticut, before returning to Southbury.

Trowbridge was also reputed to have received a Federal pension for a few years prior to his death, unmarried, in Southbury, on 23 March 1826.

TRUMBULL, GEORGE. The origins of George Trumbull are unknown. From one of the many sources at hand in 1794, Doyle Sweeny asserted that he was a lieutenant of Marines in the Revolution. Allen, Paullin, and McClellan followed suit.

No record can be found that Trumbull served in Continental Marines. However, on 3 September 1776, one George Trumbull was ordered by the Maryland Council of Safety to receive the sum of 514 pounds from the Western Shore Treasurer to pay the crew of the Maryland ship *Defence*.

Sweeny asserts that Lieutenant George Trumbull was discharged 4 March 1778.

TURNER, THOMAS. Thomas Turner, whom Doyle Sweeny asserted to have been a lieutenant of Marines, was possibly from Massachusetts.

It cannot be ascertained that he was a Continental Marine officer. One Thomas Turner was a captain of Marines on the Massachusetts ship *Mars*, commanded by Simeon Samson in 1780. He was discharged on 12 March 1781.

VAN DYKE, ABRAHAM. Born in 1719 at Mapleton, Middlesex County, New Jersey on the Raritan River, Abraham Van Dyke was the son of Jan and Anna Verkerk [another source says "Anna Verkerk Van Buren"] Van Dyke.

In earlier years he had been a lieutenant of Marines aboard a privateer in the French and Indian War. At the outbreak of the Revolution, he was living in New York City, and there organized a militia company called the "Granadiers." The unit was attached to Colonel John Lasher's 1st New York Battalion. Van Dyke and some of his men were captured by the British there in September 1776.

In 1778, the American Commissary General of Prisoners visited New York under a flag of truce and discovered that Van Dyke had been imprisoned and treated with "uncommon vigor" for the past 18 months. The British had accused him of having set fire to the city when the Americans evacuated it, but the Commissary's investigation revealed that Van Dyke had been closely confined for four days before the fire occurred.

Van Dyke was exchanged in May 1778. Armed with a letter from General Washington, he appeared before the Board of Admiralty in Philadelphia in June. Washington had recommended him for a captaincy of Marines and he found a champion in Francis Lewis, Chairman of the Board.

Lewis presented a petition for a captain of Marines, on board sloop *Saratoga*, to Congress. The latter sent it back unsigned, for *Saratoga*, then outfitting in Kensington, a northern suburb of Philadelphia, rated only a lieutenant of Marines.

Lewis persisted, recommending the captain's commission, but allowed that Van Dyke would receive only the pay of a lieutenant. Again, Congress declined to act on it.

A third proposal, submitted on 24 July 1778, asked for only the commission of a lieutenant of Marines. Congress concurred, and magnanimously back-dated the commission to 7 July 1778.

Van Dyke apparently was already on board *Saratoga*, preparing for a cruise. She dropped down the Delaware on 13 August, escorting the packet *Mercury*, with Henry Laurens on board, beyond the danger of British frigates off the shore, before returning to the Capes.

Two days later, *Saratoga* met frigates *Deane* and *Trumbull*, with which she had been ordered to cruise, off Fort Penn. The frigates continued on to Philadelphia for they were short of supplies. When they did not return, *Saratoga* proceeded to sea with *Mercury* on 18 August.

After the departure of *Mercury*, Captain John Young in *Saratoga* proceeded to cruise, on orders from Laurens. During the night of 9 September, she fought an inconclusive engagement with the British brig *Keppel*. Enroute back to the Delaware Capes, she captured a merchantman on 12 September and arrived in Chester the following day.

She was in port only five days, departing on 18 September, endeavoring to join *Deane* and *Trumbull* for a cruise between Sandy Hook and Charleston. She found neither. Between 25 September and 11 October 1780, *Saratoga* captured six vessels, sending them into the Chesapeake or the Delaware River. As fortune would have it only two of the ships arrived in Philadelphia. The others were recaptured and their prize crews imprisoned.

Saratoga returned to Chester in mid-October, terminating

what was probably the last cruise of Lieutenant Abraham Van Dyke. He was then 61 years of age and perhaps the duty on board *Saratoga* was too strenuous. He resigned his commission in November 1780 and was replaced by Lieutenant of Marines Hugh Kirkpatrick, of Philadelphia.

Nothing definite is known of the further life of Van Dyke. An inventory of the estate of one "Abraham Van Dike" was filed in Hunterdon County, New Jersey in 1804.

WADSWORTH, ———. Little of consequence is known of Lieutenant Wadsworth who, for a short period of time, was with the Continental Marines. He was listed in Doyle Sweeny's 1794 Treasury Department report in that capacity.

When the brig *Andrew Doria* returned from the New Providence expedition in April 1776, she put in New London, Connecticut, where many of her crew were transferred sick to the hospital. Commodore Esek Hopkins acquired temporary replacements from Washington's army.

On 10 May 1776, one Lieutenant Wadsworth was on a list of "New Marines" on board the *Andrew Doria*, under the command of Captain of Marines Isaac Craig. On 12 May, *Andrew Doria* and *Cabot* were ordered to sea to annoy the enemy.

General Washington did not allow his soldiers to remain long in the fleet, for on 12 May John Paul Jones in *Providence* set sail from Newport, Rhode Island to New York with the troops. Whether Lieutenant Wadsworth was among them is a matter of conjecture.

WARREN, JAMES, JR. James Warren, Jr. was born at Plymouth, Massachusetts, 18 October 1757, the eldest of five sons of James and Mercy (Otis) Warren. James, Sr. was a member of the lower house of the Massachusetts General Court and Provincial Congress from 1766 to 1778. He was also Paymaster General of the Continental Army in 1776, and from that year until 1781 he served on the Navy Board, Eastern Department, at Boston.

James, Jr. graduated from Harvard College in 1777, and apparently was engaged in the mercantile business for a short time. He was commissioned a first lieutenant of Marines on 2 October 1778, his commission signed by President of Congress Henry Laurens.

It appears that he signed on board frigate *Alliance*, Captain Pierre Landais, as a second lieutenant of Marines, his superiors being Captain Matthew Parke and Lieutenant Thomas Elwood.

He served on *Alliance* successively under Landais, John Paul Jones, and John Barry during two voyages to France and in Jones' cruise around the British Isles in 1779. One source states that he was severely wounded in the engagement of *Bonhomme Richard* and *Alliance* with *Serapis*, but no substantiation can be found in contemporary records.

Lieutenant Warren was also in the engagement of *Alliance* with *Atalanta* and *Trepassey* off the Atlantic Coast on 28 May 1781. He was severely wounded in the right knee, and the leg was subsequently amputated. *Alliance* reached Boston on 6 June 1781.

On 30 September 1781, Charles Warren, one of the brothers of James, wrote another brother, Winslow, then traveling in Holland and France:

> . . . we had reasons to expect a summer of felicity, when on sudden those gilded hopes were tarnished by an adverse stroke of fortune towards our eldest brother . . . A severe wound in the right knee had cost him many hours of excrutiating torture before it was thought necessary to be removed by amputation, to which the unhappy youth at length consented, & bore the stroke with magnanimity & perseverance that did him honour both as a soldier & a man. I am happy to inform you that he behaved in the engagement with all that calm resolution & that determined bravery of which the Veteran would be proved . . . he seems to be dispirited & dull, but perhaps time will smooth his brow & rouse him from his dispondence; methinks he has a claim to a more than ordinary share of attention as well from his country as his relations for what can alleviate distress like the kind aid of friendship. every time we behold him, it ought to remind us of the frailty of human nature, every time he is beheld by his country, it ought to remind her of the obligations she is under toward him.

On 1 November 1781, the State of Massachusetts placed James Warren on a muster roll of pensioners of the Navy of the United States at the rate of three dollars per month, and at the end of 1787, he had been paid a total of 216 dollars.

His father also made efforts to assist him. On 2 March 1782, he interceded with Major General Benjamin Lincoln, Secretary at War, for money due his son either from wages or from prize money. In 1783, he wrote two letters to the judge of the Massachusetts Supreme Court, seeking the position of clerk of the Court for his son.

James Warren, Jr., postmaster of Plymouth, Massachusetts, died unmarried on 5 August 1821, aged 64.

WALLINGFORD, SAMUEL. Born 4 February 1755 at Somersworth, New Hampshire, Samuel Wallingford was the son of Judge Thomas Wallingford and his third wife, Elizabeth (Swatt), the widow of Dr. Mark Prime, of York, Maine.

Wallingford spent his youth at his father's residence on the old road from Dover to Salmon Falls. It was the last house on the left, near the Somersworth meeting house, by the Salmon Falls River, which separates New Hampshire and Maine.

He married, 22 July 1775, at the age of 20, Lydia, the daughter of Colonel Otis and Tamsen (Chesley) Baker, of

Dover. Of this union, only one child, George Washington Wallingford, was born on 19 February 1776.

On 5 November 1775, Wallingford was a first lieutenant in Captain Moses Yeaton's Company, Colonel Joshua Wingate's Regiment, on Pierce Island, guarding Portsmouth harbor. In the following month, he was a second lieutenant in the 25th company of New Hampshire troops, which was raised to replace defecting Connecticut troops at Winter Hill, Charlestown, Massachusetts.

In July and August 1776, he was second lieutenant of Captain Joseph Badger's Company, raised for Canada, but sent instead to reinforce Major General John Sullivan's Northern Army in New York.

Finally, he was captain of his own company in the second regiment of militia, New Hampshire, reinforcing the Continental Army in New York on 17 December 1776. The company was at Exeter, New Hampshire on 4 January 1777.

Circumstances which induced Wallingford to quit the New Hampshire service are unknown. The Continental ship *Ranger*, built at Badger's Island, in Portsmouth harbor, was launched 10 May 1777. Captain John Paul Jones was appointed to command her. Samuel Wallingford was chosen by Colonel John Langdon, Navy agent at Portsmouth, as *Ranger's* lieutenant of Marines. Most of his fellow officers were also from the vicinity of Portsmouth. Wallingford's recruiting was apparently conducted among his friends and neighbors in Strafford and Rockingham Counties and in nearby York County, Maine. His superior officer, the only "outsider" besides Jones, was Captain Matthew Parke, who did his recruiting in Providence, Rhode Island.

Ranger sailed from Portsmouth for France on 1 November 1777, carrying the news of Burgoyne's surrender. She arrived at Brest on 13 February 1778, where a salute to the American flag was given by the French on the following day.

There was resentment on board among the naval officers concerning the effect of Captain Parke's rank on the distribution of possible prize money. Therefore, Parke left the ship and returned to America in the frigate *Deane*. Wallingford was the only Marine officer remaining.

On 10 April 1778 *Ranger* sailed from France for the Irish Channel. When she arrived off Whitehaven, England, Captain Jones called for men to make a landing to burn the shipping in the harbor. Wallingford was one of 31 volunteers. He was directed to fire one side of the harbor, but failed to do so. Meeting his Marine lieutenant back on the pier, Jones inquired the reason for the failure. Wallingford replied that his torch had gone out, and that he could not see that anything would have been accomplished by burning poor people's property.

After withdrawing from Whitehaven, *Ranger* touched next at St. Mary's Isle, Galloway, Scotland. Lieutenant Wallingford, Master David Cullam, and about a dozen of the crew landed at Kirkcudbright Bay, seeking to take the Earl of Selkirk hostage in exchange for American prisoners held by the British. In the Earl's absence the raiders took only Lady Selkirk's silver service.

Lady Selkirk left a pen-picture of Wallingford in a letter to her husband. She wrote that "of the two officers, one was a civil young man [Wallingford] in a green uniform, an anchor on his buttons which were white. He came to the house in a blue great-coat." Master Cullam "had a vile blackguard look, still kept civil as well he might."

Lord Selkirk, in a later letter to Jones, remarked: "We were all sorry to hear afterwards that the younger officer in green uniform was killed in your engagement with the *Drake*, for he in particular showed so much civility, & so apparent a dislike at the business he was then on, that it is surprising how he should have been one of the composers of it."

Just one day later, 24 April 1778, *Ranger* fell in with HMS *Drake* off Belfast Lough, near Carrickfergus. A well vouched-for story is that when *Ranger* was clearing for action, Wallingford, looking very pale, came on deck dressed conspicuously in his "regimentals." Another officer asked why he came on deck so dressed, as he would be a mark to shoot at, and suggested that he change into sailor's clothing. Wallingford did so, but it failed to save his life. During the battle, he was killed by a musket ball in the head. His body was committed to the deep with due honors on the evening of 25 April 1778.

Ranger returned to Piscataqua harbor at Portsmouth on 16 October 1778. There, according to the log of the officer of the deck, "sent Lieut. Wallingfords trunks, etc., ashore."

Lydia Wallingford was remarried on 20 November 1785, to Colonel Amos Cogswell, a lawyer, of Dover. The New Hampshire Legislature voted on 17 January 1787 ". . . that Lydia Wallingford, alias Cogswell. be allowed half pay, as the widow of officers who d.[ied] in the land service, provided it can be ascertained that Congress will admit the same as a charge against the United States." The petition was submitted to Congress on 9 June 1788, but there is no record that it was honored. She died at Sandwich, New Hampshire 14 February 1828.

George Washington Wallingford graduated from Harvard in 1795 and was thereafter a lawyer in Kennebunk, Maine. He was married three times and built a large dwelling there which survives today. He died there 19 January 1824.

Samuel Wallingford's progeny continued only in the female line, through a granddaughter, Lucretia Wallingford Sabine. Her grandson, Roger Wallingford Wood, died a few years ago. His widow, Mrs. Roswell P. Averill, of Old Town, Maine, donated the officer's green waistcoat of Lieutenant Samuel Wallingford to the Maine State Museum Commission.

This very rare and highly valued article of clothing, discovered in an inventory of the personal estate of the "Late Captain Samuel Wallingford," dated 27 November 1787, in the Probate Court records at Dover, is now on extended loan to the United States Marine Corps Museum.

WATERMAN, WILLIAM. William Waterman was from Cranston, Rhode Island and was born about 1759. His parents may have been any of five families or one widow by that surname from Cranston in the 1774 Census of Rhode Island.

Waterman was commissioned a first lieutenant of Marines on 29 April 1778 at the age of 19. He was assigned to the frigate *Providence*, Captain Abraham Whipple, his senior officer was Captain William Jones, of Providence, and Jonathan Woodworth, probably from New London, was second lieutenant of Marines.

On 30 April or 1 May 1779, *Providence* sailed for France with dispatches for the American Commissioners. While Captain Jones delivered them, Waterman and Woodworth stayed on board the frigate at Paimboeuf.

Providence joined *Boston* and *Ranger* at Brest, France in August and sailed with them to America on the 22d. Enroute they took three prizes off the Banks of Newfoundland and took them into Portsmouth.

In June 1779, *Providence* made a cruise with *Ranger* and *Queen of France* in which they cut eleven merchant vessels out of the Jamaica fleet. Only eight were gotten into port; the others were captured.

Providence sailed on her last voyage for the United States 23 November 1779, to the relief of Charleston, South Carolina, with *Boston, Queen of France*, and *Ranger*.

Waterman, with Captain Jones and Lieutenant Woodworth, was captured there by the British on 12 May 1780. Jones was paroled immediately, but was never exchanged. The fate of Woodworth is unknown. Waterman was kept a prisoner for about 20 months, at the end of which he was paroled and exchanged.

Frigate *Deane* arrived at Boston 17 April 1781 with a cargo of military stores. Her crew was paid off and recruiting begun anew. She was also laid up for some time undergoing repairs.

Captain of Marines Richard Palmes signed on 1 May 1781. William Waterman was summoned from Providence as her first lieutenant of Marines, being the twenty-eighth person to enter, on 29 October. Second Lieutenant Jeremiah Reed signed on 15 November. This completed her Marine officer complement. She did not receive her full quota of enlisted personnel until 1 March 1782, and some of these were lost to *Alliance*.

Deane left harbor the same month and, before returning to Boston on 17 May, captured five ships. Lieutenant Waterman's whole wages for the cruise amounted to more than 42 pounds, of which he collected about 39 pounds.

In September 1782, the name of the frigate was changed to *Hague*, and Captain John Manley took command. Lieutenant Waterman remained on board and participated in the subsequent cruise to the West Indies. In January 1783, the ship ran aground off Guadaloupe where she was bombarded by five British ships. After sustaining considerable damage to her hull and rigging, she got off and put into a French port for repairs.

In May, *Hague* put into Boston again, and on 5 December 1783, Lieutenant Waterman was paid his final wages.

In the meantime, Waterman apparently returned to Cranston where, on 13 July 1783, he married Esther, the daughter of Captain William Field, also of Cranston. According to the 1790 Census of Rhode Island, William and Esther Waterman, living close to her father's family, appear to have had two sons under 16 years of age and one daughter.

William Waterman died in Cranston at the age of 62, on 18 June 1821. On 17 October 1851, the two sons, William G. and Nathan, were the only survivors of the family.

WELSH, JOHN. John Welsh, born in Ireland, is reported to have come to America specifically to join the American service during the Revolution. Nothing is known of his Irish ancestry.

He was one of the first three captains of Marines to be commissioned, Samuel Nicholas and Joseph Shoemaker being the other two. With Philadelphia as the colonial capital, and Nicholas and Shoemaker known natives of the area, it is quite possible that Welsh had also settled there.

Captain Welsh and a company of about 40 Marines entered the brig *Cabot*, Captain John B. Hopkins, at Cuthbert's wharf, Philadelphia, 4 January 1776. First Lieutenant John Hood Wilson was the junior Marine officer. On 9 January, Welsh advertised in the *Pennsylvania Evening Post* for a deserter.

Cabot sailed on 17 February 1776 as part of Commodore Esek Hopkins' fleet on the New Providence expedition. On completion of the capture of Forts Montagu and Nassau, and the loading of their armament and military stores, the fleet departed for New England on 17 March.

Enroute, during the night of 6 April, *Cabot* was the leading vessel of one of Hopkins' columns in Block Island Sound. She was the first to engage the 20-gun frigate, HMS *Glasgow*. Her first broadside was answered at once, the reply mortally wounding Lieutenant Wilson and two Marines. Captain Welsh is not reported to have been injured. *Cabot* bore away for repairs and left the engagement to the other Continental ships. *Cabot* put into New London, Connecticut on 7 April.

On 6 May, Captain Welsh was a member of the court martial of Captain Abraham Whipple, *Columbus*, and on 8 May, of that of Captain John Hazard, *Providence*. Both were tried aboard *Alfred* in Providence harbor, for their supposed inaction in the engagement with *Glasgow*.

When *Cabot* sailed again on 19 May, she was accompanied by the brig *Andrew Doria*. Captain Welsh still commanded the Marines of the former, but Lieutenant John Kerr (or Carr) had replaced the deceased John Hood Wilson. *Cabot* parted company soon and was not seen again until her arrival in New London on 26 June. No record is available of her engagements or prizes.

Sometime in March 1777, *Cabot* put to sea from Boston under Captain Joseph Olney. Several days later, off Nova Scotia, she was chased by the British frigate *Milford* and was run ashore 14 days later. Her crew barely had time to escape before the brig was captured. It is only a matter of conjecture whether Captain Welsh and Lieutenant Kerr (or Carr) were still her Marine officers; there is no record of their movements since June 1776.

Again, a supposition must be made that Welsh entered *Alfred* as her captain of Marines before she sailed for France in 1777. Lieutenant of Marines Nathaniel Richards boarded her in Boston in the spring of 1777, and Lieutenant William Hamilton was on board when she and *Raleigh* weighed anchor.

Enroute, using captured signals, they infiltrated a British fleet of 60 vessels, conveyed by four armed ships. *Alfred* was not swift enough to take part in the ensuing action, but *Raleigh*, having no such disadvantage, engaged the 14-gun *Druid*, damaging her heavily. Unable to cut out any merchant vessels, they soon continued on to France.

On 29 December 1777, *Alfred* and *Raleigh* departed France for America and, on 9 March 1778, encountered *Ariadne*, 20 guns, and *Ceres*, 16 guns. Underestimating the enemy's firepower, Captain Thompson of *Raleigh* ordered Captain Hinman in *Alfred* to pursue one of the vessels. Both were fast, and jointly captured *Alfred* while Captain Thompson pondered what to do. Then, pursued herself, *Raleigh* threw weight overboard and escaped.

Captain Welsh and Lieutenant Hamilton were taken to Barbadoes, where they were transferred to the *Yarmouth* for transport to England. Lieutenant Richards was paroled. On 18 July, Welsh and Hamilton reached Forton Prison at Portsmouth, but escaped several months later. On 29 September 1778, Welsh received the sum of $480.00 from the U. S. Commissioners in France for his return to America. He sailed from Bordeaux, but was captured off the Capes of Virginia and taken into Boston. By 30 April 1779, he had been released and sent to New London. On 18 May 1779, he was ordered to the frigate *Warren* to replace Captain Richard Palmes.

Warren was under the command of Captain Dudley Saltonstall, and sailed shortly on the abortive expedition to Penobscot Bay. Although personal accounts of the expedition are sketchy, it is now evident that Lieutenant William Hamilton was also on board *Warren*. Captain of Marines Gilbert Saltonstall, of frigate *Trumbull*, was a "volunteer" aboard, apparently acting in the capacity of secretary to his brother, Dudley, but taking no part in the fighting.

Marines of the Continental ships and those of the Commonwealth of Massachusetts, landed on Banks' Island on 26 July 1779. The enemy was driven off, four pieces of artillery and a small quantity of ammunition were captured, and the Americans suffered no losses.

The next landing on Bagaduce Peninsula on 28 July was made against strong resistance. About 200 Marines, including those from *Warren* under Welsh and Hamilton, landed on the right wing where the enemy's greatest strength lay. Militia and artillery landed on the left. The Americans lost more than one hundred officers and men killed and wounded. Captain Welsh and eight of his Marines lost their lives immediately, and Lieutenant Hamilton died later of his wounds.

Anna Welsh, widow of the Captain, of whom nothing is known, applied for a pension in 1797, only to discover on 7 February of that year "that the resolution of Congress promising half-pay to widows of officers who fell in service, *did not extend to officers of the Navy*." [Italics added.]

WHITE, JACOB. The background of Jacob White is uncertain. In light of his appointments of officers for the frigate *Trumbull*, he may well have been a native of Connecticut.

Trumbull was launched at Portland, Connecticut on 5 September 1776. Dudley Saltonstall was her first captain. His brother, Gilbert, captain of Marines, was appointed about August 1776. Jacob White was selected as *Trumbull*'s lieutenant of Marines on 16 October.

Due to her deep draft, *Trumbull* was unable to pass over the bar at the mouth of the Connecticut River and did not reach the sea until August 1779. In the meantime, Captain Gilbert Saltonstall served as a volunteer on board *Warren* at Penobscot. Jabez Smith, of Groton, Connecticut, who had been captured with a portion of the crew of *Raleigh*, paroled and exchanged, became *Trumbull*'s lieutenant of Marines.

It is possible that Jacob White forsook his Marine commission sometime in 1777, for a muster roll of the vessel for that period listed one Jacob White as naval third lieutenant. Also, when *Trumbull* captured the brig *Little William* in September 1780, Jacob White was a navy lieutenant on board the frigate, and William Fielding was junior Marine officer to Gilbert Saltonstall.

The date of Jacob White's departure from the service is not known. Nor is his life following the Revolution. Six persons of that given name lived in Connecticut in 1790, one in Litchfield County, two in Windham County, and three in Fairfield County. One other Continental Marine officer, Lieutenant Thomas Elwood, who served until late 1783, was also from Fairfield County.

WILSON, JAMES HOOD. No vital statistics of James Hood Wilson are available. Philadelphia and its environs would be a likely area in which to search his background.

The brig *Cabot*, ex-*Sally*, was purchased in Philadelphia in November 1775 for the Continental fleet. She was converted and fitted out by John Wharton and Joshua Humphreys at their Kensington shipyard. It is probable that her officers and crew were recruited from the surrounding neighborhood and the adjoining Northern Liberties.

On 4 January 1775, *Cabot* was alongside Cuthbert's Wharf on the Delaware River. Aboard her were 43 Marines under the command of Captain John Welsh and Lieutenant John Hood Wilson. They soon dropped downriver to Reedy Island and then on to Cape Henlopen. On 17 February the fleet cast off for the March landings at New Providence Island.

Returning to New England in April, *Cabot* was foremost in one column of Hopkins' squadron, and early in the morning of the 6th was the first to fire on the British 20-gun frigate *Glasgow* off Block Island.

Glasgow returned a more effective broadside immediately, and Lieutenant Wilson, with two of his Marines, was mortally wounded.

WOODWORTH, JONATHAN. Jonathan Woodworth was probably from New London, Connecticut.

Frigate *Providence*, commissioned at Providence, Rhode Island in December 1776, had lain there until 1778. All her original Marine officers had departed for other pursuits. Early in 1778, her captain, Abraham Whipple, was faced with discovering new officers.

First, he found William Jones, formerly of the Army, whom he made captain. On 1 April 1778, John Deshon, of the Eastern Navy Board, sent Jonathan Woodworth to Whipple with 20 men he had enlisted as Marines, and for which he was paid a bounty of $1,200.00. Woodworth was subsequently commissioned a lieutenant of Marines.

Woodworth apparently sailed in *Providence* to France either on 30 April or 1 May 1778; the vessel arrived on 30 May. He probably returned with her reaching Portsmouth, New Hampshire on 15 October 1778. Nothing further is known of his career.

The state of Connecticut abounded with 30 families surnamed Woodworth in 1790, mostly in New London and surrounding counties. Jonathan Woodworth is not recorded in the Census.

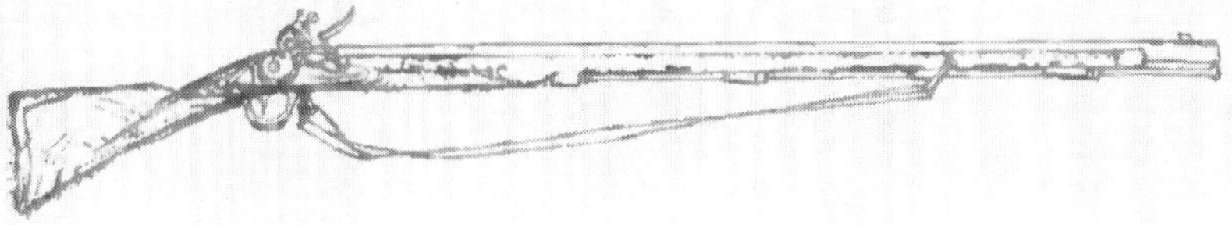

INDEX

C

D

E

F

G

H

I

J

K

L

M

N

O

P

T

X

Y

☆U.S. GOVERNMENT PRINTING OFFICE: 1975 O—555-928

Made in United States
North Haven, CT
10 November 2026

22097469R00278